T0134987

Communications in Computer and Information Science 1712

More information about this series at https://link.springer.com/bookseries/7899

Wenhui Fan · Lin Zhang · Ni Li ·
Xiao Song (Eds.)

Methods and Applications for Modeling and Simulation of Complex Systems

21st Asia Simulation Conference, AsiaSim 2022
Changsha, China, December 9–11, 2022
Proceedings, Part I

Editors
Wenhui Fan
Tsinghua University
Beijing, China

Lin Zhang
Beihang University
Beijing, China

Ni Li
Beihang University
Beijing, China

Xiao Song
Beihang University
Beijing, China

ISSN 1865-0929 ISSN 1865-0937 (electronic)
Communications in Computer and Information Science
ISBN 978-981-19-9197-4 ISBN 978-981-19-9198-1 (eBook)
https://doi.org/10.1007/978-981-19-9198-1

This Springer imprint is published by the registered company Springer Nature Singapore Pte Ltd.
The registered company address is: 152 Beach Road, #21-01/04 Gateway East, Singapore 189721, Singapore

Preface

These two volumes contain the papers from the 21st Asia Simulation Conference (AsiaSim 2022), which is an annual simulation conference organized by the ASIASIM societies: CSF (China Simulation Federation), JSST (Japan Society for Simulation Technology), KSS (Korea Society for Simulation), SSAGsg (Society for Simulation and Gaming of Singapore), and MSS (Malaysian Simulation Society). The conference started in the 1980s and is held each year in a different Asian country. This conference provides a forum for scientists, academicians, and professionals from around the world. The purpose of the AsiaSim conference is to provide a forum in Asia for the regional and national simulation societies to promote modelling and simulation in industry, research, and development.

This year AsiaSim was held in Changsha China, together with the 34th China Simulation Conference. Research results on various topics, from modeling and simulation theory to manufacturing, defense, transportation, and general engineering fields, which combine simulation with computer graphics simulations, were shared at AsiaSim 2022. Three reviewers evaluated each contribution. A total of over 200 submissions were received and only 96 papers were accepted and presented in the online and offline oral sessions. The selected papers were finally accepted for this CCIS volume.

We thank the keynote speakers for giving great insights to the attendees. Furthermore, we wish to thank the external reviewers for their time, effort, and timely responses. Also, we thank the Program Committee and Organizing Committee members who made the conference successful. Finally, we thank the participants who participated remotely despite the difficult circumstances.

Due to the Covid pandemic AsiaSim 2022 has been postponed to January 2023

October 2022

Wenhui Fan
Lin Zhang
Ni Li
Xiao Song

Organization

Honorary Charis

Bo Hu Li	Beihang University, China
Axel Lehmann	Universität der Bundeswehr, München, Germany

General Chair

Jianguo Cao	China Simulation Federation, China

General Co-chairs

Wenhui Fan	Qinghua University, China
Lin Zhang	Beihang University, China
Satoshi Tanaka	University of Tokyo, Japan
Yahaya Md Sam	UTM, Malaysia
Gary Tan	National University of Singapore, Singapore

Organizing Committee Chair

Ni Li	Beihang University, China

Publication Committee Chair

Xiao Song	Beihang University, China

International Program Committee

Lin Zhang (Chair)	Beihang University, China
Kyung-Min Seo	Korea University of Technology Education, South Korea
Jangwon Bae	Korea University of Technology Education, South Korea
Kyoungchan Won	Center for Army Analysis & Simulation, South Korea
Gyu M. Lee	Pusan National University, South Korea
Bohu Li	Beijing University of Aeronautics and Astronautics, China
Liang Li	Ritsumeikan University, Japan

Satoshi Tanaka	Ritsumeikan University, Japan
Lin Zhang	Beihang University, China
Terence Hung	Rolls Royce, Singapore
Dong Jin	Illinois Institute of Technology, USA
Farzad Kamrani	KTH Royal Institute of Technology, Sweden
Helen Karatza	Aristotle University of Thessaloniki, Greece
Sye Loong Keoh	University of Glasgow, UK, and Singapore Campus, Singapore
Yun Bae Kim	Sungkyunkwan University, South Korea
Ge Li	National University of Defence Technology, China
Zengxiang Li	Institute of High Performance Computing, A*STAR, Singapore
Malcolm Low	Singapore Institute of Technology, Singapore
Linbo Luo	Xidian University, China
Imran Mahmood	National University of Science & Technology, Pakistan
Yahaya Md Sam	Universiti Teknologi Malaysia, Malaysia
Zaharuddin Mohamed	Universiti Teknologi Malaysia, Malaysia
Navonil Mustafee	University of Exeter, UK
Bhakti Stephan Onggo	University of Southampton, UK
Ravi Seshadri	Singapore-MIT Alliance for Research and Technology, Singapore
Xiao Song	Beihang University, China
Yuen Jien Soo	National University of Singapore, Singapore
Claudia Szabo	University of Adelaide, Australia
Sun Teck Tan	National University of Singapore, Singapore
Wenjie Tang	National University of Defense Technology, China
Yifa Tang	Chinese Academy of Sciences, China
Simon Taylor	Brunel University, UK
Yong Meng Teo	National University of Singapore, Singapore
Georgios Theodoropoulos	Southern University of Science and Technology, China
Stephen John Turner	Vidyasirimedhi Institute of Science and Technology, Thailand
Bimlesh Wadhwa	National University of Singapore, Singapore
Yiping Yao	National University of Defense Technology, China
Allan N. Zhang	Singapore Institute of Manufacturing Technology, Singapore
Jinghui Zhong	South China University of Technology, China

Contents – Part I

Modeling Theory and Methodology

Continuous System/Discrete Event System/Hybrid System/Intelligent System Modeling and Simulation

Complex Systems and Open, Complex and Giant Systems Modeling and Simulation

Integrated Natural Environment and Virtual Reality Environment Modeling and Simulation

Big Data Challenges and Requirements for Simulation and Knowledge Services of Big Data Ecosystem

Artificial Intelligence for Simulation

Contents – Part II

Application of Modeling/Simulation in Energy Saving/Emission Reduction, Public Safety, Disaster Prevention/Mitigation

Modeling/Simulation Applications in the Military Field

Modeling Theory and Methodology

Research on Reuse and Reconstruction of Multi-resolution Simulation Model

Xiaokai Xia[1,2(✉)], Fangyue Chen[1], Gang Xiao[2], and Zhiqiang Fan[3]

[1] North China Institute of Computing Technology, Beijing 100083, China
xiaxiaokai1986@126.com
[2] Beijing Institute of System Engineering, Beijing 100010, China
[3] Artificial Intelligence Institute of China Electronics Technology Group Corporation, Beijing 100041, China

Abstract. With the rapid development of military simulation technology, simulation systems for different levels of strategy, campaign, tactics, and technology have emerged. This manuscript intends to use model mapping transformation and machine learning technology to automatically reconstruct the low-resolution simulation model from the high-resolution simulation model data. The premise of learning is to have rich and effective learning data. Therefore, this project uses the existing data cultivation technology/learning data generation technology based on a large sample simulation experiment to generate the simulation data for machine learning through the continuous iterative process of scene design, high-resolution simulation scenario construction, simulation experiment design, and sample generation, efficient parallel simulation, data analysis, and application.

Keywords: Cross-platform multi-resolution · Deep learning · Artificial neural network

1 Introduction

With the rapid development of military simulation technology, simulation systems for different levels of strategy, campaign, tactics, and technology have emerged. The simulation models of these simulation systems have different resolutions and can meet different simulation experiment requirements. This project studies the reuse and reconstruction technology of cross-platform multi-resolution simulation model, realizes the automatic model reconstruction of a low-resolution simulation platform based on a high-resolution simulation platform, to realize the reuse of simulation model between different resolution simulation platforms, which can not only improve the efficiency of low-resolution simulation model construction but also improve the consistency of simulation model construction Credibility and credibility.

To research the reuse and reconstruction technology of cross-platform multi-resolution simulation model is a realistic demand in the process of large-scale application of simulation technology and continuous development of new technology:

W. Fan et al. (Eds.): AsiaSim 2022, CCIS 1712, pp. 3–17, 2022.
https://doi.org/10.1007/978-981-19-9198-1_1

(1) The complexity of military and war problems requires using simulation systems with different resolutions to solve different application problems. Other simulation systems for researching war problems include strategic, campaign, and tactical levels. These systems often have a large number of simulation models. Whether the models of the same object in these different platforms are consistent and reliable is an important issue.
(2) It is an effective way to improve the simulation model construction and experiment efficiency to realize automatic model reconstruction in a different resolution simulation platform. Simulation research units often deploy and use more than one simulation system, and each simulation system has different types of models. Suppose the models of a high-resolution simulation platform can be automatically transformed into models that can be used in a low-resolution simulation platform. In that case, the efficiency of model construction of the low-resolution simulation platform will be significantly improved, and the low-resolution simulation platform will be better promoted and applied. In addition, it can effectively ensure the model consistency of the two resolution platforms and lay a good foundation for the simulation of different resolution platforms.
(3) The development of machine learning and artificial intelligence technology provides new ideas and means for reusing cross-platform simulation models. Cloud simulation and other technologies support the generation of large-scale simulation experimental data. In contrast, machine learning and other uncertain model-building methods provide the possibility of automatically learning a low-resolution simulation model based on the operation data of a high-resolution simulation model. With large sample data, the learning-generated model can have high accuracy.

The rest of the paper is organized as follows: Sect. 2 describes our approach methodology of architecture analysis and optimization based on data farming. Section 3 presents the case study of the combat system based on the proposed approach. And in Sect. 4, we conclude the paper and propose future works.

2 Related Works

With the in-depth exploration of the winning mechanism of information war, the urgent need for combat experiments or training simulation is increasingly prominent. Using experiments to pretest equipment, combat plan, and combat method, the pre-practice of combat action is realized to provide data support for combat power generation. Many principled and experimental simulation application systems emerge as the times require, which have achieved good results in training, experiment, and research but also exposed apparent shortcomings. The first mock exam system lacks a systematic development concept, leading to the severe development of different applications and models. Especially when sharing basic class models, almost every application in the development stage needs to be redeveloped and share the basic model, resulting in severe problems of low development efficiency and high development cost. In addition, different simulation systems have different granularity requirements for simulation models. The urgent needs of complex system simulation are how to fully use the model resources to achieve multi-domain and different granularity model reuse and ensure the high efficiency, low cost, and high reliability of simulation application development.

The research of this topic involves multi-resolution modeling technology, simulation model reuse technology, simulation meta model technology, etc. the following is an analysis of their research status and the existing problems in supporting the research of this topic.

3 Methodology

This topic intends to solve the problem of reuse and reconstruction of cross-platform simulation models through the extensive use of significant sample simulation experiment technology, machine learning technology, model transformation technology, data mining technology, etc. the overall technical scheme is shown in the figure below.

Based on the investigation and analysis of cross-platform military requirements and related technology research status, this paper focuses on the two different resolution simulation platforms of campaign level and strategic level deployed in the laboratory, analyzes the specific problems in cross-platform model reconstruction, and drives the research of related key technologies by combining with reconstruction cases (Fig. 1).

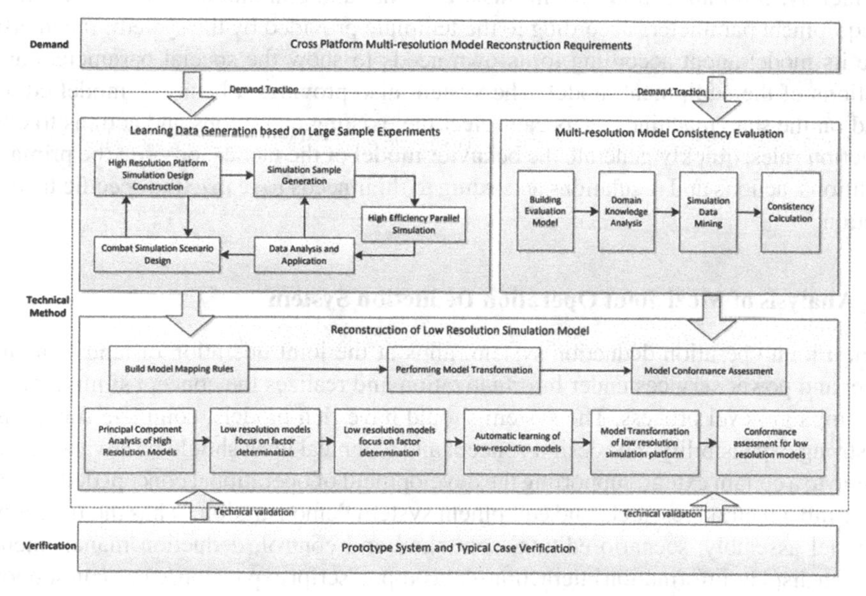

Fig. 1. Architecture of technical approach

This paper intends to use model mapping transformation and machine learning technology to automatically reconstruct the low-resolution simulation model from the data of the high-resolution simulation model. The premise of learning is to have rich and effective learning data. Therefore, this project uses the existing data cultivation technology/learning data generation technology based on a large sample simulation experiment to generate the simulation data for machine learning through the continuous iterative process of scene design, high-resolution simulation scenario construction, simulation experiment design, and sample generation, efficient parallel simulation, data analysis, and application.

3.1 Maxsim Simulation System and Model Analysis

Maxsim is a simulation system that adopts GBB (general blackboard) + DMAS (multi-agent) Architecture, rich development interfaces, powerful and flexible expansion, and secondary development capabilities. It supports rich combat platform design and visual tactics design covering live, virtual, and structural simulation. It can be widely used in campaign tactics and equipment simulation of various services and arms. It is an ideal simulation application to support campaign tactics simulation training, combat research, and weapon equipment demonstration And development system. Maxsim, Xsim, and VBS are widely used in our army's combat experiment and simulation training.

The system simulation model of Maxsim is divided into three categories: equipment model, behavior model (including war method model and atomic action model), and environment model. It has the function of fine equipment modeling. The equipment model consists of seven main categories: motion platform, sensor, weapon system, reaction facility, communication system, mast antenna, and ammunition. Users can fill in the equipment parameters according to the template provided by the system. It can also write its model agent according to its own needs to show the special parameters and functions of the equipment model. The system also provides a behavior model editor based on the state machine. Users can select the existing conditions and actions to edit the action rules, quickly generate the behavior model of the tactics, or write the primary conditions, actions and regulations according to their needs to realize the specific tactics function.

3.2 Analysis of Mozi Joint Operation Deduction System

Mozi, a joint operation deduction system, aims at the joint operation of land, sea, air, space, and power services under informatization and realizes the concept simulation of the campaign-level process. The system should have rich models, complete functions, and strong expansibility. The design concept and technical level should be advanced and flexible to a certain extent, supporting the development of operational concept deduction, new combat method research, and equipment system demonstration. It has the functions of model assembly, scenario editing, command and control, deduction management, situation display, information interaction, data output, script expansion, etc. it can support multiple combat styles such as sea attack, land attack, anti-submarine operation, air defense, and anti-missile, amphibious projection, etc., and has various resource data such as equipment model database, weapon performance database, and typical operational scenario database.

(1) Equipment entity model

Mozi system uses a web equipment entity editing tool to edit and model entity types and physical characteristics. It stores them in the MySQL database for scenario creation and deduction.

(2) Entity behavior model

The model system configures the entity behavior through the dialog box related to the rules, as shown in the figure below. It can define each entity's event, action, condition, trigger, special action, etc., to define the specific behavior of the entity. The entity's behavior model can be edited by script editing Lua language. The built entity's rules can be stored to facilitate subsequent reuse.

3.3 Analysis of Mo Subsystem Model Based on Maxsim Model

From the introduction of the simulation models of Maxsim and Mozi, we can see that there are not only similarities but also many differences between them, which is a common problem between different resolution simulation platforms. The details are as follows:

(1) The simulation entity modeling of two platforms adopts the idea of parametric and component modeling. Still, the number of model parameters for the same model type (such as an attack helicopter) of the two platforms is different, and the granularity is inconsistent.
(2) There is a big difference in the basic physical parameters, such as flight speed, of the same model type in the two platforms' model libraries, such as Zhi-8.
(3) There are significant differences between the two platforms in modeling entity behavior. Maxsim describes entity behavior based on a mechanism similar to the state machine. The ink subsystem is set through a series of dialog boxes or scripts.
(4) There are differences in the granularity of model description between the two platforms. On the whole, Maxsim pays more attention to the parameters, input, and output when modeling.

The above problems also need to be solved in the follow-up research on this topic. The follow-up research on this topic will also be closely combined with these two platforms so that the related technologies can be more effectively implemented and applied.

4 Effect Model Reconstruction Based on Machine Learning

4.1 Overall Idea of Model Reconstruction Based on Machine Learning

There is complex calculation logic in high-resolution platforms that aim at the atomic action models of simulation entities, such as maneuver and damage. For example, the maneuver model needs to accurately calculate the maneuver characteristics of simulation entities under different conditions, such as terrain, weather, atmosphere, temperature, etc. In a low-resolution simulation platform, external factors have a relatively weak influence on maneuver, generally through influence factors or shadow factors. The response meter is adjusted to the reference speed to obtain the corresponding maneuver model. To solve this kind of problem, this project plans to realize the automatic generation of a cross-platform low-resolution simulation model through the comprehensive operation of machine learning, data farming, and other technologies. Its basic idea is to approach the input-output behavior of the simulation model in a black box, to achieve the purpose of coalescing or reducing the dimension of the high-resolution simulation model and reconstructing the low-resolution simulation model.

The experimental data of the high-resolution simulation model is the key input of low-resolution simulation model reconstruction, and its quality directly determines the efficiency and accuracy of subsequent model construction. There may be invalid data or ambiguous data in the simulation experiment data. Therefore, before the follow-up work, it is necessary to preprocess the simulation data, including data cleaning, data integration, data transformation, and other methods in data analysis technology, to form a standardized data set that can be directly applied to machine learning and principal component analysis.

To reconstruct the low-resolution simulation model based on machine learning, it is necessary to determine the input, output, and model parameters of the low-resolution simulation model, which will be used to design the basic framework of the machine learning model. Therefore, principal component analysis and other methods can be used to analyze the input-output relationship of the preprocessed simulation data, extract the sensitive or core input variables, model parameters, and output variables, and finally determine the input variables, output variables, and model parameters that need to be retained in the low-resolution simulation model according to the requirements of the low-resolution simulation model.

The model learning framework can be determined based on the quasi-reconstructed low-resolution simulation model's input, output, and model parameter information. Taking the neural network model as an example, the input and model parameters of the quasi-reconstructed low-resolution simulation model can be regarded as the input of the neural network model, and the output of the quasi-reconstructed low-resolution simulation model can be regarded as the output of the neural network model. Due to the simple structure and strong adaptability of the BP neural network and RBF neural network, they can approximate the continuous non-linear function of the task. Therefore, it is expected that BP neural network and RBF neural network are effective machine learning methods to solve the problem of low-resolution model reconstruction. However, how to determine the number of hidden neurons in the neural network, how to avoid excessive fitting and insufficient fitting, and how to solve the problem such as How to improve the accuracy and calculation speed of the neural network by enhancing the learning method is a problem to be solved in the process of research. In this paper, we also introduce other suitable machine learning methods, compare the accuracy and efficiency of the model construction, and analyze which machine learning method is more suitable for different types of simulation model reconstruction.

4.2 Research on Model Reconstruction Method Based on Neural Network

The neural network is a large-scale nonlinear dynamic system consisting of many parallel distributed neurons interconnected to a particular structure to complete different intelligent information processing tasks. Synaptic weights represent the interaction between different neurons. The neural network learning process is constantly adjusting the weights to make the output of the neural network approach the desired result. Each neuron is a multi-input, single-output nonlinear element. It is the simplification and Simulation of biological neurons and the primary processing unit of a neural network. Its nonlinear characteristics can be approximated by closed value, piecewise, and linear functions.

Unlike the traditional simulation mathematical model fitting method, neural network as a simulation meta-modeling method has many obvious advantages. First of all, the requirements of neural networks for data are low. A neural network adjusts the system parameters through adaptive network training and learning, so it has strong adaptability to data. The neural network can be used whether linear or nonlinear, discrete or continuous. Secondly, the neural network has a solid ability to process information and fit a large number of data; thirdly, the neural network has broad applicability and can select different types of neural networks to fit according to the characteristics of simulation data; finally, it has a high degree of parallel processing ability, which makes it possible to use in a unified way In the framework of simulation, the simulation model can be replaced by the more complex meta-model established by the neural network, which makes the system simulation accurate and efficient. Because BP neural network and RBP neural network have simple structure and strong applicability, they can approximate any continuous nonlinear function. Therefore, BP neural network and RBP neural network will become a vital choice to solve the problem of building a data simulation model for information service. The following takes BP neural network as an example to illustrate how to carry out the fitting learning of the mathematical simulation model.

In mathematics, BP neural network can be regarded as a highly nonlinear mapping from R ^ n input space to R ^ n output space. It can fit the input-output relationship of the system by adjusting the internal weights and has good generalization ability and self-organization. Therefore, BP neural network can obtain much information by learning a group of experimental data and then establish a model to approximate the input-output relationship and apply this model to predict. In addition, the network's unlimited number of input and output endpoints is very suitable for modeling statistical data with multiple independent and dependent variables. The following figure shows the construction diagram of the simulation model based on neural network training (Fig. 2).

In applying the neural network training simulation mathematical model, we should focus on determining the number of neurons in the hidden layer of the neural network, how to learn the algorithm, and how to solve the problem of insufficient fitting and excess fitting.

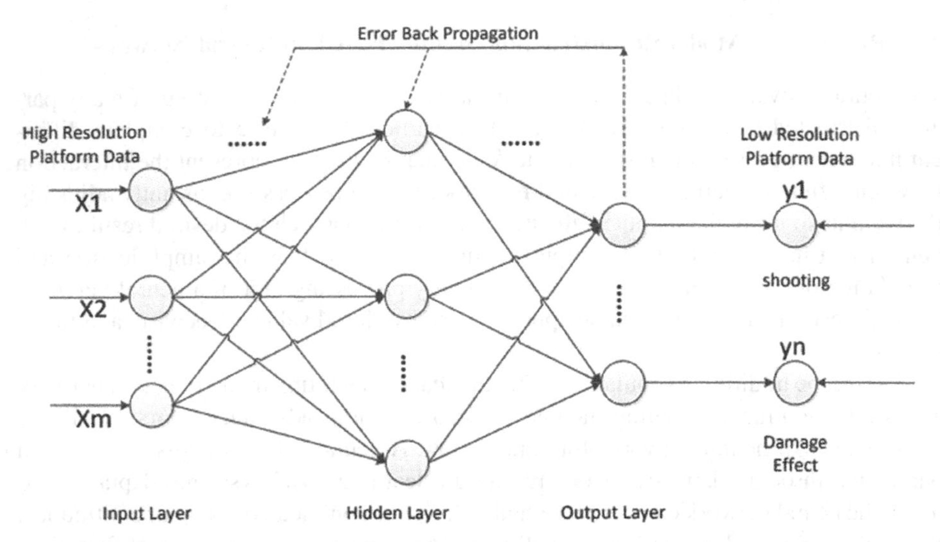

Fig. 2. Simulation model construction based on neural network training

1) Network hierarchy selection

The number of hidden layers in BP neural networks also dramatically influences the network's performance. If the number of hidden layers is too large, not only does the network's learning speed become slow, but it is also over-adapted, which makes its generalization ability worse. That is to say, it will produce good prediction results for the learned samples but poor prediction results for the unlearned data. A single hidden layer network with an appropriate number of hidden layer nodes sometimes has better results than a network with two hidden layers. Therefore, the best choice in the specific simulation experiment is three layers. If three layers are selected, four can be chosen if the accuracy is insufficient.

2) Determination of the number of neurons in the hidden layer

The selection of the number of hidden layer elements is a very complex problem, which is often determined according to the design experience and experiments, so there is no accurate analytical expression to express it. The number of hidden layer units is directly related to the requirements of the problem and the number of input and output units. Too many hidden layer units lead to too long a learning time, and the fitting accuracy is not necessarily the best, which quickly leads to excessive network fitting; too few hidden layer units lead to poor network fitting, poor fault tolerance, and poor data generalization ability. So far, there are many non-theoretical methods to select hidden layer neurons, among which there are general methods, that is, methods for specific problems, which are generally divided into two categories: deletion method and trial and error method.

Firstly, a neural network with a larger structure than needed is trained, and redundant connections and neurons are removed in the training process. In this method, some clear the least affected connection and neuron according to the error equation of the network, and some add a penalty term to the objective function to make the unnecessary connection weights tend to zero.

The trial-and-error method determines the number of hidden layer neurons. First, a small initial number of hidden layer neurons is given and connected to a BP neural network with a small structure for training. If the training times are too many or the convergence conditions are not met within the specified training times, stop the training and gradually increase the number of hidden layer neurons to form a new network for retraining. Currently, most neural network models use trial and error to determine the number of hidden layer neurons.

3) Insufficient fitting and excess fitting

Poor and excessive network fitting are two common cases of poor performance of the generated simulation meta-model. Inadequate fitting means that the network does not thoroughly learn the complexity of the forced function, and the fitting accuracy of the training data is not high, which leads to the poor generalization ability of the network. Overfitting means that the learning speed of the network is breakneck, and the fitting accuracy of the training data is very high. Still, the generalization ability of the network is inferior. These two situations mainly include the number of hidden layer units, the number of samples, the number of training times, and so on.

The number of hidden layer neurons is too small, the number of connection weights is limited, and the network cannot fully remember the information of training samples, so no matter how many times of training, no matter how many groups of training samples, it cannot meet the accuracy requirements of network training. Too many hidden layer neurons will lead to many weights that need to be adjusted. However, the complexity of the function to be forced is fixed, and only a part of hidden layer neurons and their connection weights need to be adjusted according to the training sample values. In this case, random factors exist in the weight, which is not easy to detect in the training process. Therefore, for the training samples, the learning accuracy of the network is excellent, but the network's generalization ability is inferior due to random factors.

Secondly, the number of training samples and training times also impact the network performance. When the number of samples is too small and the number of training times is too small, it is easy to lead to insufficient fitting; when the number of samples is too large and the number of training times is too large, it is easy to lead to excessive fitting. A group of correction samples is used in training to avoid the influence of training times and training samples. Each pair of training samples is trained, and the current connection weight is used to detect the error of the correction samples. In the experiment, the error of the correction samples always has a process of gradually decreasing and then increasing, So we can decide whether the network training is finished or not according to the error of the correction samples.

4.3 Introduction of Model Reconstruction Based on Neural Network

In a particular project, to support the development of air defense and anti-missile experiments, a more detailed simulation model of air search radar is realized in the Maxsim simulation platform, especially how to find and track missile targets, which has a more complex mechanism. In the Mozi simulation system, a low-resolution simulation model will be reconstructed using the provided method based on machine learning to realize the same type of air search radar. The focus is to build two behavior models of the average radar detection range and the average radar tracking range of the air search radar.

To realize the two simplified simulation models of radar average discovery range and radar average tracking range in the Mozi system, firstly, the Maxsim platform and its developed high-precision air search radar model and typical air defense scenarios anti-missile system are used. RCS, relative reference target distance, high altitude cruise altitude, and high-altitude cruise speed of a specific type of high-altitude missile are taken as experimental factors. The preliminary design of each factor is shown in the table below (Table 1).

Table 1. Experimental factors

Experimental factor	Factor value
Relative target distance	From 100 km to 180 km, take a value every 10 km
High altitude cruise altitude	From 6 km to 18 km, take a value every 1 km
High altitude cruise speed	From Mach 3.0 to Mach 3.8, take a value every Mach 0.1
RCS	0.1 to 0.8, take one value every 0.1

Experimental samples are formed according to the experimental factors, such as uniform or orthogonal experimental design methods. These experimental samples are executed to obtain the experimental data. Due to randomness in the simulation process, each sample is run 20 times, and the average radar discovery distance and radar average tracking distance of the obtained index data is taken as the average value.

Because the number of nodes in the input layer and output layer of the BP neural network is determined according to the needs of research problems, the selected network should have three input nodes and two output nodes. This project first uses a three-layer BP neural network to reconstruct the simulation model. If the effect is not good, then consider a four-layer network. In fact, due to the small number of experimental factors in the simulation model of air search radar and the weak correlation between them, the low-resolution simulation model based on a three-layer BP neural network can meet the accuracy requirements. In general, the fewer the number of output nodes of the network structure, the better the calculation results, and the faster the training speed of the network. Therefore, multiple single and multiple neural networks are generally used to build different output models rather than one neural network solution. So, for the searchability of air-to-air radar, two three-layer BP neural networks with three input units and one output unit are established.

This project directly uses the mature neural network model in MATLAB to carry out the training of this project. The setting of the hidden layer is the focus of neural network learning. In the experiment of selecting the number of hidden layer units, the maximum number of training batches of the network is set to 1000 so that the network can be fully trained until the maximum relative error of the network is less than the predetermined accuracy value after the verification of the corrected data, Network termination training. According to the general principle of the neural network's hidden layer setting, the range of the number of hidden layer neurons is 2 to 9. To illustrate the problem, we can compare the training process of single hidden layer neurons from 2 to 9 in a circular way, treat each network with the same number of hidden layer neurons as a kind of network, and train each type of network several times. To confirm the effectiveness of the neural network, some data (such as 2%) in the experimental samples can be used as test data to verify the fitting degree of the neural network.

The training process of single hidden layer neuron numbers from 2 to 9 is compared circularly. The network with the same number of neurons in each hidden layer is regarded as a kind of network. K test data are constructed, and each network is trained for m times. In this way, there will be K * m relative error values. If n% of the relative error values are less than ε, then the experimental effectiveness of this kind of network is n%. The verification results of radar average discovery distance are shown in the following table. The higher the validity, the better the performance of the trained neural network. Otherwise, the performance of the network is poor (Table 2).

Table 2. Verification data of hidden layer unit

Number of hidden layer units	Maximum relative error	Experimental validity
2	0.030	39%
3	0.287	86%
4	0.023	91%
5	0.018	96%
6	0.020	95%
7	0.014	89%
8	0.013	82%
9	0.009	70%

5 Prototype System and Typical Application Case Verification

5.1 Background of Case

The research of this project will be carried out by combining Maxsim and Mozi, two simulation platforms. Maxsim is the simulation platform with higher resolution, while Mozi is the simulation platform with lower resolution. Some models in Maxsim can be used in Mozi simulation tools through model mapping transformation and machine learning.

The rotating wing aircraft in the platform model, the radar model in the sensor model, and the missile model in the weapon model are selected as the research objects, respectively. Not only the mapping transformation of the equipment entity model is studied, but also the reconstruction method of some key behavior models based on the proposed neural network and polynomial regression methods is studied.

5.2 Validation Process of Case

The basic process of reconstructing the Maxsim simulation platform and Mozi simulation tool is as follows.

(1) Firstly, according to the method described in Sect. 5.3, the mapping and transforming of the equipment entity parameters of the three models are carried out. The prototype system reads the TKB file (XML file) of the equipment technology knowledge base of the Maxsim platform, extracts the parameter values involved in the mapping rules in Sect. 5.3, and forms the corresponding equipment entity attribute values of the Mo subsystem according to the mapping formula, and then saves the data Save to the MySQL database file of Mozi simulation platform.
(2) Aiming at the reconstruction of the action/effect model, firstly, in the Maxsim simulation platform, the scenarios of the rotary wing aircraft model, radar model, and missile model are designed, and different simulation samples are formed according to the experimental factors. For example, in the flight action model of the rotary wing aircraft model, the flight scenario from point a to point B is set first. Then the corresponding parameter settings in the scenario are adjusted according to the experimental factors such as atmosphere, altitude, wind, rain, and snow, such as wind force level 1 to level 12, and each level is taken as a factor value point to form a series of executable simulation samples.
(3) The virtual machine template of the Maxsim simulation platform is made, and the prototype system is applied to automatically set up ten experimental nodes in the cloud environment. Then the simulation samples formed in step 2 are distributed to the experimental nodes, and the experimental nodes are controlled to execute 20 times for each sample. The uploaded experimental data is saved in real-time. When the executing nodes are idle, new unexecuted nodes are distributed Samples until all samples are executed.
(4) The prototype system's data processing and analysis function module is applied to preprocess a large number of simulation sample data generated in step 3 to form data structure types that the learning module can directly use.
(5) The neural network, support vector machine, and polynomial regression method of MATLAB integrated into the prototype system are used to process the learning data and form the reconstructed mathematical model.
(6) The reconstructed model's prediction effect is verified and analyzed through the reserved test data.
(7) Through code writing and code reconstruction in the Mozi simulation system, the reconstructed mathematical model formed by the machine learning method is realized, and the running test is carried out.

5.3 Analysis of Case Validation Results

The previous section describes the equipment entity models of platform, sensor, and weapon models in detail. The mapping transformation, the Mozi simulation platform, forms the low-resolution simulation platform. The rules, processes, and results of equipment entity models are not detailed here.

For the method of reconstructing action/effect model by using machine learning methods such as neural network, support vector machine, and polynomial regression, the prediction/use accuracy of these models is generally evaluated by the quantitative indexes such as mean square error MSE, mean absolute error AAE and maximum fundamental error. Generally, the smaller MSE and AAE, the higher the accuracy of the reconstructed model. As a supplement of MSE and AAE, the more petite Mae reflects, the better the fitting effect of the reconstructed model.

$$\mathrm{MSE} = \frac{\sum_{i=1}^{N}(y_y - \hat{y}_i)^2}{N}$$

$$\mathrm{AAE} = \frac{\sum_{i=1}^{N}|y_y - \hat{y}_i|}{N}$$

$$\mathrm{MAE} = \max\{|y_y - \hat{y}_i|\,|i = 1, 2, \ldots, N\}$$

The following table shows the model accuracy verification results after the flight action model of the rotary wing aircraft model, the detection range model of the radar model, and the strike effect model of the missile model are reconstructed respectively by using neural network, support vector machine, and polynomial regression. It mainly reserves part of the data as the test verification data during the model learning (Table 3).

Table 3. Reconstruction model accuracy

Name		Neural network	Support vector machine	Polynomial regression
MSE	T1	64.74	632.26	90.08
	T2	26.33	514.57	63.31
	T3	218.83	333.14	201.02
AAE	T1	4.65	22.71	6.77
	T2	3.91	20.07	6.20
	T3	10.46	14.10	11.63
MAE	T1	25.06	45.87	27.53
	T2	14.53	34.61	18.02
	T3	34.51	32.60	27.48

6 Conclusion and Future Works

This paper focuses on the transformation and reconstruction of simulation models (entity model, action/effect model) from campaign tactical simulation platform to strategic simulation platform based on model mapping transformation, machine learning and other methods, so as to improve the efficiency, consistency and accuracy of building simulation models in low-resolution simulation platform The technique of model mapping transformation is used to reuse the equipment entity model data of the campaign tactical simulation platform into the strategic low-resolution simulation platform; (2) support extensive sample experiments and machine learning of part of the action/effect models in the campaign tactical simulation platform to form the required models for the low-resolution strategic simulation platform; (3) through two different precision simulation models Part of the model reuse and reconstruction in the simulation platform is taken as an experimental case to verify the effectiveness of the method and technology proposed in this project.

The research results of this project are expected to be applied to internal and external units carrying out simulation experiment research to a certain extent. When multiple sets of simulation platforms with different resolutions coexist, the low-resolution simulation model of the target platform can be reconstructed quickly based on the high-resolution simulation model to a certain extent, which can not only improve the efficiency and consistency of simulation model construction but also provide a reference for future development It also lays a good foundation for the simulation experiments of different resolution simulation models.

References

1. Peng, G., Mao, H., Zhang, H.: BMRSS: BOM-based multi-resolution simulation system using components. In: Tan, G., Yeo, G.K., Turner, S.J., Teo, Y.M. (eds.) AsiaSim 2013. CCIS, vol. 402, pp. 485–496. Springer, Heidelberg (2013). https://doi.org/10.1007/978-3-642-45037-2_53
2. Zhang, L., Zhang, X.S., Song, X., et al.: Model Engineering for complex system simulation. J. Syst. Simul. **25**, 2719–2735 (2013)
3. Robinson, S., Nance, R.E., Paul, R.J., et al.: Simulation model reuse: definition, benefits and obstacles. Simul. Model. Pract. Theory. **12**, 479–494 (2014)
4. Tolk, A.: Engineering Principles of Combat Modeling and Distributed Simulation. Wiley, Hoboken (2011)
5. Sarjoughian, H.S., Cellier, F.E.: Discrete Event Modeling and Simulation Technologies: A Tapestry of Systems and AI-Based Theories and Methodologies, pp. 61–100. Springer, New York (2001). https://doi.org/10.1007/978-1-4757-3554-3
6. Balcl, O., Arthur, J.D., Ormsby, W.F.: Achieving reusability and composability, with a simulation conceptual model. J. Simul. **5**, 157–165 (2011)
7. Horne, G., Akesson, B., Meyer, T., et al.: MSG-088 Data Farming in Support of NATO[R]. NATO Science and Technology Office, March 2014
8. NATO Science and Technology Office: MSG-124 Developing Actionable Data Farming Decision Support for NATO[R], NATO Science and Technology Office, July 2018
9. Xia, X.K., Qu, K., Shi, J., Fan, Z.Q., Xu, L.: The construction of effectiveness evaluation model based on system architecture. In: 2017 International Symposium on System Engineering, pp. 1–4 (2017)
10. Tran, H.T., Domercant, J.C., Mavris, D.: Trade-offs between command and control architectures and force capabilities using battlespace awareness. In: International Command and Control Research and Technology Symposium, pp. 1–21 (2014)

11. Domercant, J.C., Mavris, D.: Understanding and evaluating command & control effectiveness by measuring battlespace awareness. In: 19th International Command and Control Research and Technology Symposium, pp. 1–21 (2014)
12. Repast Simphony. https://repast.github.io/repast_simphony.html

Simulation Experiment Factor Screening Method Based on Combinatorial Optimization

Peng Zhang[1], Wei Li[1], Qingao Chen[1], and Jiahui Tong[2](✉)

[1] Control and Simulation Center, Harbin Institute of Technology, Harbin, China
[2] Science and Technology on Complex System Control and Intelligent Agent Cooperation Laboratory, Beijing, China
tjh80825@126.com

Abstract. When conducting complex simulation experiments, experimental factor screening is of great significance to improve the effectiveness and efficiency of the simulation. Aiming at the problems of unsatisfactory accuracy and efficiency of existing experimental factor screening methods, a simulation experiment factor screening method based on combinatorial optimization (EFS-CO) was designed. EFS-CO abstracts the experimental factor screening problem into a combinatorial optimization problem, designs the experimental factor evaluation criteria based on the support vector machine model, uses it as the fitness function of the genetic algorithm, and finally obtains the significant experimental factor. At the same time, to eliminate the influence of the randomness of the genetic algorithm on the screening results, the screening results of the experimental factors were determined based on the results of multiple experiments. Finally, the effectiveness of the method is verified by example analysis and comparing the results with the traditional experimental factor screening method.

Keywords: Experimental factor screening · Combinatorial optimization · Genetic algorithm · Support vector machine

1 Introduction

The experimental factor screening method refers to the method of identifying a few significant factors from the full set of experimental factors. The pareto theorem shows that only a few factors are important in the whole complex experiment. At present, experimental factor screening methods are mainly divided into three directions, namely the method based on sequential bifurcation (SB), the frequency domain experimental design method (FDE), and the supersaturated experimental design method (SED).

The SB method [1] is a sequential process method. The first step of the method aggregates all experimental factors into a group and tests whether there is an experimental factor that has a significant effect on the experimental output in this group. If so, the second step divides the group into two subgroups and tests for the presence of significant factors in both groups. The next steps are carried out similarly, that is, subgroups with significant factors are divided into smaller subgroups until all significant factors are

W. Fan et al. (Eds.): AsiaSim 2022, CCIS 1712, pp. 18–25, 2022.
https://doi.org/10.1007/978-981-19-9198-1_2

screened. Later, Hong Wan [2] improved the SB method and proposed an improved SB method (CSB), which improved the efficiency of experimental factor screening. Hua Shen [3] proposed a Controlled Sequential Factorial Design (CSFD) method for discrete event simulation experiments to eliminate the effect of heterogeneous variance in experimental factor screening. Sanchez [4] proposed a mixed two-stage screening method (FFCSB), which first uses the fractional factorial method to group the main effect directions of the experimental output by experimental factors, and then uses the CSB method to sieve the two groups of experimental factors respectively.

The FDE method was proposed by Schruben [5] in 1987 and applied to second-order polynomial models with an arbitrary number of experimental factors. The FDE method takes the value of each experimental factor level according to different frequency changes and then analyzes the oscillation frequency of the experimental output to find out which experimental factors will affect the frequency of the experimental output, to screen out the significant factors.

In recent years, the construction and analysis of SED is one of the current research hotspots [6]. To measure the efficiency of supersaturated design, scholars have proposed several evaluation criteria (r-rank criteria, B-optimal, $\chi^2(D_n)$-optimal, etc.) [7–9], constructed two-level/multi-level supersaturated designs based on the above criteria, and analyzed and estimated the experimental results., to identify significant factors.

However, there are many problems with the existing experimental factor screening methods. The screening experiments based on the SB method are only suitable for system models below the second-order; the screening efficiency of the FDE method is unstable and is affected by the number of experimental factors and the structure of the simulation system; the method is suitable for cases where the number of samples is less than the number of experimental factors, which is not widely applicable.

This paper first introduces the mathematical description of the experimental factor screening method, and gives the corresponding mathematical description of the combinatorial optimization; then compares three optimization algorithms, and determines the optimization algorithm used in this method; then introduces the support vector machine model, which is applied to evaluate experimental factor subsets; finally, the method of EFS-CO is given, and the effectiveness of the method is verified by an example.

2 Experimental Factor Screening Problem Description and Optimization Method Selection

2.1 Experimental Factor Screening Mathematical Description

The complex experimental model can be defined as $T = (O, F, C)$, where $O = \{o_1, o_2, ..., o_n\}$ is the set of experimental factor data, $F = \{f_1, f_2, ..., f_m\}$ and $C = \{c_1, c_2, ..., c_k\}$ are the set of experimental factors and a collection of experimental outputs. In the above model, the experimental data o_j can be represented by the Cartesian product $f_1 \times f_2 \times ... \times f_m$ of the experimental factors, that is a feature vector $o_j = <\alpha_1, \alpha_2, ..., \alpha_m>$, where α_i is the specific value of o_j on f_i.

Experimental factor screening means that the simulation analyst selects a subset of the experimental factors' data $o = \{o_i, o_j, ..., o_k\}$ from the full set of experimental factors' data. By selecting the appropriate experimental factor subset evaluation function

$f(o)$, the above factor screening problem can be transformed into a combinatorial optimization problem, and the best experimental factor subset for evaluation can be selected through the optimization algorithm, and then the experimental factors that have a significant impact on the experimental output can be obtained. Combinatorial optimization can be represented by two-tuple (U, f), where U represents the feasible solution area of combinatorial optimization, that is, $U = \{o|o \subseteq O\}$; f represents the evaluation function of experimental factors, and all experimental factors in the feasible solution μ^* satisfying $f(\mu^*) = \max\{f(\mu)|\mu \in U\}$ are the experimental factors that have a significant impact on the experimental output.

2.2 Optimization Algorithm Selection

Use the traveling salesman problem (TSP) problem to test the performance of the genetic algorithm, ant colony algorithm, and particle swarm algorithm, and conduct a comparative analysis. Thirty city coordinates are randomly generated, as shown in Table 1.

Table 1. City coordinate

City index	X coordinate/km	Y coordinate/km
1	0.064	0.48
2	0.54	0.93
3	0.42	0.078
⋮	⋮	⋮
28	0.23	0.77
29	0.85	0.45
30	0.0055	0.34

The optimization performance of the three algorithms on this problem is shown in Fig. 1, Fig. 2, Fig. 3, and Table 2.

From the above comparison, it can be seen that the number of iterations required by the ant colony algorithm to achieve the optimal solution is less, but the time spent on each iteration is much longer than in the other two optimization algorithms; in the genetic algorithm and the particle swarm optimization algorithm, the genetic algorithm achieves the optimal solution. The solution has fewer iterations. Therefore, this paper chooses the genetic algorithm as the optimization algorithm of the factor screening method.

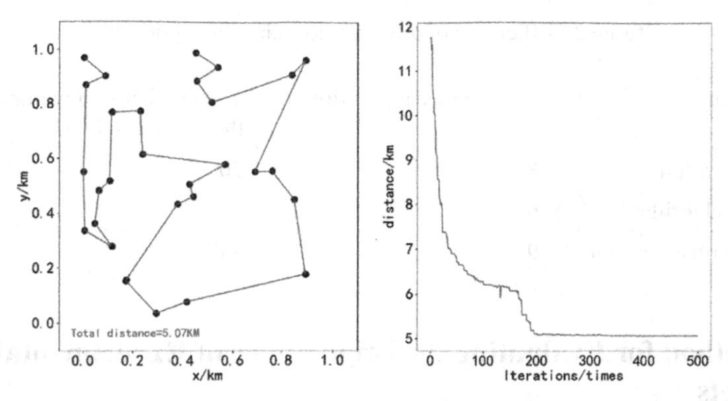

Fig. 1. Genetic algorithm

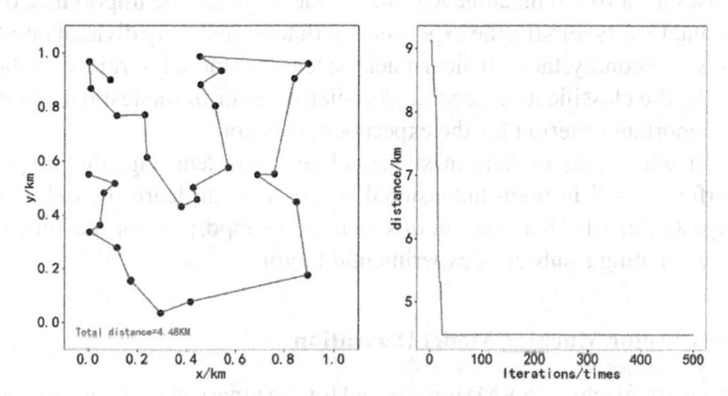

Fig. 2. Ant colony algorithm

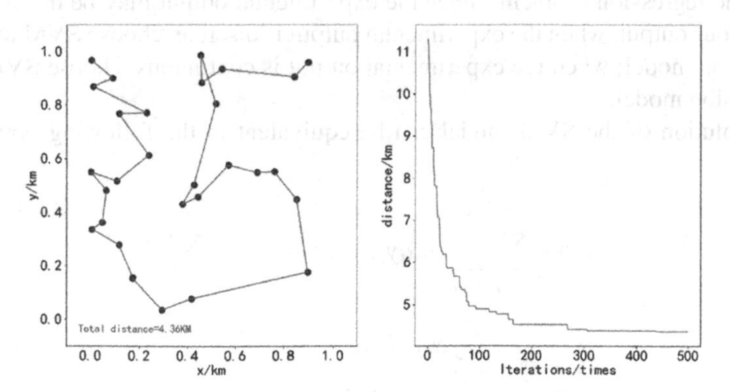

Fig. 3. Particle swarm algorithm

Table 2. Efficiency of different optimization algorithms

Optimization	Average time per iteration/ms	The number of iterations to reach the optimal solution
Genetic algorithm	18	200
Ant colony algorithm	876	30
Particle swarm algorithm	9	440

3 A Method for Evaluating the Importance of Experimental Factor Subsets

This paper uses a statistical machine learning model to judge the importance of a subset of experimental factors. Firstly, the experimental data is randomly divided into a training set and a test set, secondly, the statistical machine learning model is trained on the training set, and finally, the classification accuracy/prediction error of the test data on the model is the most important criterion for the experimental factor.

Compared with other models in statistical machine learning, the support vector machine performs well in high-dimensional space, and can learn the only hyperplane with the largest interval. Therefore, in this lecture, a support vector machine is used as a model for evaluating a subset of experimental factors.

3.1 Support Vector Machine Model Derivation

The support vector machine (SVM) is proposed for the binary classification problem, and support vector regression (SVR) is an important branch of the support vector machine to solve the regression problem. Since the experimental output may be discrete output or continuous output, when the experimental output is discrete, choose SVM to train the classification model; when the experimental output is continuous, choose SVR to train the regression model.

The solution of the SVM model can be equivalent to the following optimization problem:

$$\begin{aligned} &\min \frac{1}{2}\sum_{i=1}^{N}\sum_{j=1}^{N}\alpha_i\alpha_j y_i y_j \varphi(x_i^T)\varphi(x_j) - \sum_{i=1}^{N}\alpha_i \\ &s.t. \quad \sum_{i=1}^{N}\alpha_i y_i = 0 \\ &\quad 0 \le \alpha_i \le C,\ i = 1, 2, ..., N \end{aligned} \tag{1}$$

The solution of the SVR model can be equivalent to the following optimization problem:

$$f(x) = \sum_{i=1}^{m}(\overline{\alpha}_i - \alpha_i)\varphi(x_i^T)\varphi(x_j) + \mathrm{b}$$

$$s.t. \sum_{i=1}^{m} (\overline{\alpha}_i - \alpha_i) = 0,$$

$$0 \le \alpha_i,\ \overline{\alpha} \le C,\ i = 1, 2, ..., N \tag{2}$$

Among them, $\{\alpha_1, \alpha_2, ..., \alpha_N\}$ is a set of parameters that make the above formula true, $\overline{\alpha}$ is the average value of this set of parameters, x, y is the model input data and output data, respectively, $\varphi(\bullet)$ is the mapping from low-dimensional space to high-dimensional space.

3.2 Support Vector Machine Model Solving

The sequence minimization algorithm is a solution algorithm that uses the characteristics of the support vector machine, the basic idea is coordinate descent, that is, by cyclically using different coordinate directions, fixing other elements each time, and optimizing only along one coordinate direction to achieve the local minimum of the objective function.

Algorithm 1 SMO

Input: Optimize the target f

Output: μ, minimize $f(\mu)$

1: while does not converge do

2: for $i \longleftarrow 1$ to n do

3: $\mu_i \longleftarrow \arg\min_{u_i} f(\mu)$

4: end for

5: end while

6: return μ

4 Factor Screening Method and Validation

4.1 Factor Screening Method

In this paper, the genetic algorithm is used as the optimization algorithm of the experimental factor screening method, and the support vector model is used as the evaluation model for evaluating the subset of experimental factors. The method is shown in Fig. 4.

Among them, the genetic algorithm coding method selects binary coding, and each bit on the chromosome represents whether the experimental factor of that bit is selected; the crossover method selects the single-point crossover method, and by selecting two chromosomes, segmentation is performed at a randomly selected position point. And exchange the right part to obtain different sub-chromosomes; the mutation method selects the binary single-point mutation method. First, all individuals in the population are judged whether to mutate with a preset mutation probability and then randomly select the mutation bit for the individuals undergoing mutation. The mutation is carried out so that the genetic algorithm can obtain local random searchability.

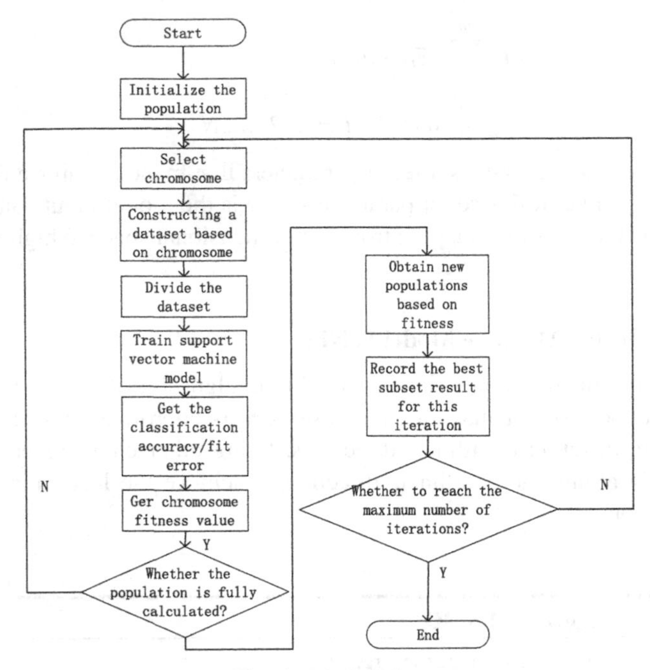

Fig. 4. Method flow

4.2 Case Study

To illustrate the effectiveness of the method, that is, the proposed method is suitable for solving the factor screening problem of high-dimensional complex models, this section uses a second-order polynomial function for verification. The function expression is as follows:

$$\eta(x_1, x_2, \ldots, x_{20}) = \beta_0 + \sum_{i=1}^{20} \beta_i x_i + \sum_{i=1}^{5} \sum_{j=i+1}^{5} \beta_{i,j} x_i x_j \tag{3}$$

This function is a deterministic second-order interaction model. First, use the quasi-Monte Carlo method to sample in the input space, set the sampling point to 1000, and then use the SB-based experimental factor screening method, the Sobol' method, and the method in this section respectively. After factor screening, the obtained significant factors or factor importance rankings are shown in Table 3.

For the above second-order polynomial functions, the literature [10] adopts the SB method and CSB method respectively, and the significant factors obtained are $x_3, x_6, x_{13}, x_{14}, x_{15}$. Compared with Sobol' method, the traditional factor screening method fails to correctly screen two significant factors x_1 and x_2, while the significant factors obtained by this section are consistent with the top ten important factors obtained by Sobol' method.

Table 3. Screening results of experimental factors by different methods

The experimental factor screening method	Significant factors/factor ordering
Factor screening method based on SB	$x_3, x_6, x_{13}, x_{14}, x_{15}$
Sobol' method	$x_3 > x_1 > x_{13} > x_2 > x_{15} > x_{14} > x_6 > x_{10} > x_4 > x_5 > x_{18} > x_{17} > x_{11} > x_{12} > x_8 > x_{16} > x_9 > x_{20} > x_{19} > x_7$
EFS-CO	$x_1, x_2, x_3, x_4, x_5, x_6, x_{10}, x_{13}, x_{14}, x_{15}$

5 Conclusion

Aiming at the problem that the accuracy and efficiency of the existing experimental factor screening methods are not ideal, an experimental factor screening method based on combinatorial optimization is designed. This method is based on experimental data rather than designing screening experiments to screen experimental factors. Compared with the traditional SB-based experimental factor screening method, this method's accuracy is higher. At the same time, compared with Sobol' sensitivity analysis method, EFS-CO only needs simulation experiments under the same sample size N, while Sobol' method needs to run simulation experiments at least $(k+2)N$ times, k is the number of experimental factors. That is, under the condition of reducing the sample size, the same experimental screening effect as Sobol' method is achieved.

References

1. Bettonvil, B., Kleijnen, J.P.C.: Searching for important factors in simulation models with many factors: Sequential bifurcation. Eur. J. Oper. Res. **96**(1), 180–194 (1997)
2. Wan, H., Ankenman, B.E., Nelson, B.L.: Controlled sequential bifurcation: a new factor-screening method for discrete-event simulation. Oper. Res. **54**(4), 743–755 (2006)
3. Shen, H., Wan, H.: Controlled sequential factorial design for simulation factor screening. Eur. J. Oper. Res. **198**(2), 511–519 (2009)
4. Sanchez, S.M., Wan, H., Lucas, T.W.: Two-phase screening procedure for simulation experiments. ACM Trans. Model. Comput. Simul. (TOMACS) **19**(2), 1–24 (2009)
5. Schruben, L.W., Cogliano, V.J.: An experimental procedure for simulation response surface model identification. Commun. ACM **30**(8), 716–730 (1987)
6. Georgiou, S.D.: Supersaturated designs: a review of their construction and analysis. J. Stat. Plann. Inference **144**, 92–109 (2014)
7. Qin, H., Chatterjee, K., Ghosh, S.: Extended mixed-level supersaturated designs. J. Stat. Plann. Inference **157**, 100–107 (2015)
8. Weese, M.L., Edwards, D.J., Smucker, B.J.: A criterion for constructing powerful supersaturated designs when effect directions are known. J. Qual. Technol. **49**(3), 265–277 (2017)
9. Cheng, C.S., Das, A., Singh, R., et al.: E (s2)-and UE (s2)-optimal supersaturated designs. J. Stat. Plann. Inference **196**, 105–114 (2018)
10. Wang, Q.: Research on factor screening methods and tools in simulation experiments. Harbin Institute of Technology, Harbin (2015)

Development of a Generic Mesh Converter for Numerical Simulations

Minjuan Liu, Guyu Deng, Jifeng Li, Haifeng Li, and Yonghao Xiao(✉)

Institute of Computer Application, China Academy of Engineering Physics, Mianyang 621900, Sichuan, China
18729226863@163.com

Abstract. Mesh models from various numerical simulation programs have great diversity in the data formats and structures, leading to difficulties in sharing and reusing of mesh data for collaborative simulations between commonly used commercial programs and in-house codes. To break the barrier against the free exchange of mesh models, in this work a unified mesh model with full coverage of simulation-related data is deliberately designed. Based on such mesh model and the open source mesh data management framework MOAB (Mesh-Oriented database), a generic mesh model converter is developed, enabling the high-efficient conversions of mesh models between various simulation programs. The converter could be easily extended for conversion between new pair of mesh models by integration of the corresponding IO interfaces. The correctness and scalability of the converter is validated through a series of examples.

Keywords: Mesh models · Mesh format transformation · IO interfaces

1 Introduction

In academic and engineering fields a rich variety of numerical simulation programs are available. Among them, some are commercial, such as MSC.Nastran, ANSYS, ABAQUS, FLUENT, LS-DYNA, and some are open-sourced such as Saturne and Open-Foam. Depending on the features and specialties, different programs may be adopted for different application scenarios or domains [1–3]. For most simulation programs, mesh plays a core role throughout the modeling-computing-analysis processes in a numerical simulation. However, due to the lack of a unified mesh format standard in the community, the description and storage of mesh data differ notably from one simulation program to another, resulting in difficulties in the sharing and reusing of mesh data between different programs [4, 5]. Although some commercial programs such as HyperMesh provide tools for mesh conversions between different formats, their functionalities, scalabilities and the customizabilities are usually limited, especially for dealing with mesh formats used for in-house simulations codes. Therefore, it is of great significance to develop a powerful and easy-to-use tool to enable convenient mesh format transformations and mesh data exchanges between various simulation codes for efficiency of collaborative simulations.

W. Fan et al. (Eds.): AsiaSim 2022, CCIS 1712, pp. 26–39, 2022.
https://doi.org/10.1007/978-981-19-9198-1_3

To implement the mesh transformation and exchange between commercial or in-house simulation programs [6], in this work a simulation-oriented mesh data model is deliberately designed. In this model, the structure and storage form of mesh-related data and other data for descriptions of boundary conditions, initial conditions, constraints, loads, material properties are clearly defined. By implementation of transformation interfaces between this mesh data model and the ones for commonly used simulation programs, an efficient and extensible mesh data transformation and exchange tool is developed, enabling the mutual transformation and exchange of mesh data between a wide range of programs.

2 Diversity in Commonly Used Mesh Models

2.1 Formats of Mesh Data

A mesh model typically contains mesh-related data such as nodes, elements, and mesh sets, and in general sense data on material properties, geometric information, boundary conditions and some other information for description of the physical problem is also included. In what follows the term mesh model refers to its general meaning unless specified otherwise. For mesh models from various simulation programs, huge diversity exists in mesh file formats and data structures, as illustrated in Table 1.

Table 1. Comparison on mesh models in different simulation programs.

Mesh type	File format	Data structure	Element type
CGNS	Binary	Tree-like structure	Structured/unstructured
ABAQUS	Text	Assembly structure	Rich engineering elements
NASTRAN	Text	Data card	
ANSYS	Text	APDL script/data card	
FLUENT	Text & binary	Lisp statement	Not explicitly defined
PLOT3D	Text & binary	Mesh block	Structured
LS-DYNA	Text	Data card	Unstructured, degradable

It should be mentioned that in Table 1, the mesh data in CGNS is organized in a hierarchical way as ZONE-BLOCK-CELL and stored in a tree-like structure. As the file format is binary, a library needs to be called to access mesh data. In ABAQUS the mesh data for a complex structure is treated as the assembly of the mesh instances of the consisting parts by translation and rotation operations. In FLUENT the mesh model is described in a LISP-like script language with strict semantics and grammar.

2.2 Representation Capability of Mesh Models

Mesh models from different programs show dramatic differences in the representation capability and corresponding data structures due to the diverse rules they follow for descriptions of element connectivity, element degradation, etc., as will be briefly introduced below.

Element Connectivity. Different mesh models may follow different rules for node numbering order or element connectivity for elements with similar geometric shapes, as illustrated in Fig. 1. Conversion of mesh data in these mesh models needs to map the element connectivity accordingly.

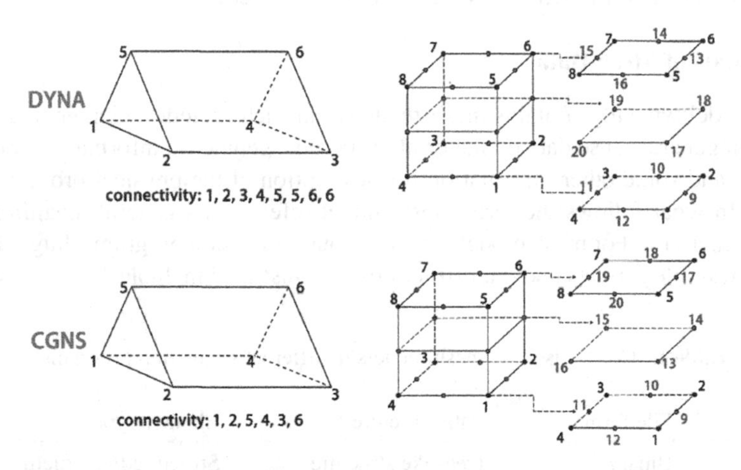

Fig. 1. Illustration on element connectivity in DYNA and CGNS.

Element Degradation. In most CAE programs, the element name and type are strictly matched with no ambiguity. While in other programs such as ABAQUS and LS-DYDA, some element types such as the quadrilateral and hexahedron elements may be degradable. For instance, a tetrahedral element can be regarded as the degradation of a hexahedral element [7], as shown in Fig. 2. When conversion to mesh model that does not support element degradation, the element type and the connectivity need to be re-determined according to the node numbering order and the degradation rules.

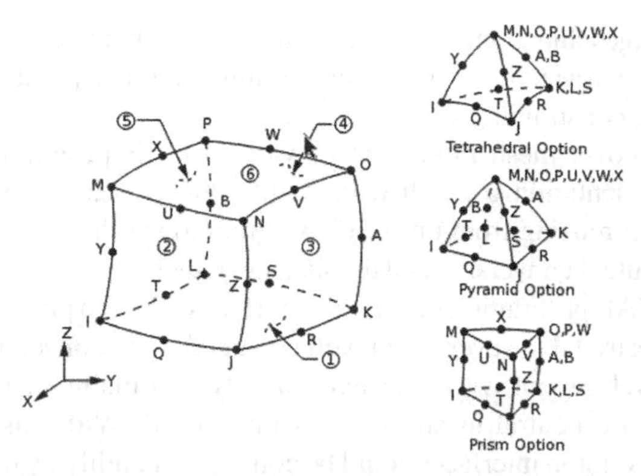

Fig. 2. Hexahedral element and degraded elements.

Element Representation. In most CAE programs, an element is represented by its associated nodes and their connections, while in Fluent there is no explicit element definition. A 3D element in Fluent is expressed by directional bounding faces. When conversion to element in other mesh models, the associated nodes and the connections need to be reconstructed (Fig. 3).

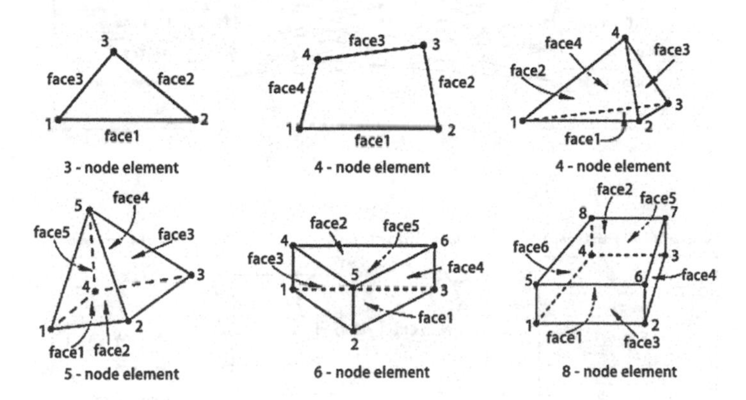

Fig. 3. Elements and the bounding faces in Fluent.

Representation of Boundary Surface. Some mesh models do not have explicit definitions of boundary surfaces, such as NASTRAN and ANSYS. In these models, the boundary conditions are usually discretized and applied on the boundary nodes, which is inconvenient for direct updates or modifications. In LS-DYNA, a boundary surface is described with a segment set consisting of directional quadrilateral element faces defined by boundary nodes. While in ABAQUS, a boundary surface is defined by an element set and the face number indicating the location and direction of the boundary surface.

Element Type. A couple of specialized types of elements for specific purposes may be found in different programs, including dynamic and kinematic constraint elements such as springs, hinges and engineering elements such as beams and shells. These elements often contain extra auxiliary nodes for defining reference points, centerlines, local coordinate systems, constraints, etc.

As discussed above, mesh models from various CAE programs may have huge differences in representation capability, data structure or data format [8], making it impossible for direct reusing from one CAE program to another.

To enable the mutual conversion and the sharing of these heterogeneous mesh models between different CAE programs, conversion interfaces for each pair of source and target mesh models are desired. However, if not well designed, such conversion interfaces may easily form a cobweb layout with poor extensibility and maintainability. To this end, in this work a generic neutral mesh model is introduced. With this proposed model, development of conversion interfaces could be conducted handily by the following steps: First, implementation of an importing interface for conversion from the source mesh model to the neutral model, and then an exporting interface for conversion from the neutral model to the target one, resulting in sharp reduction of the complexity of the conversion tasks from O (m * n) to O (m + n) (Fig. 4).

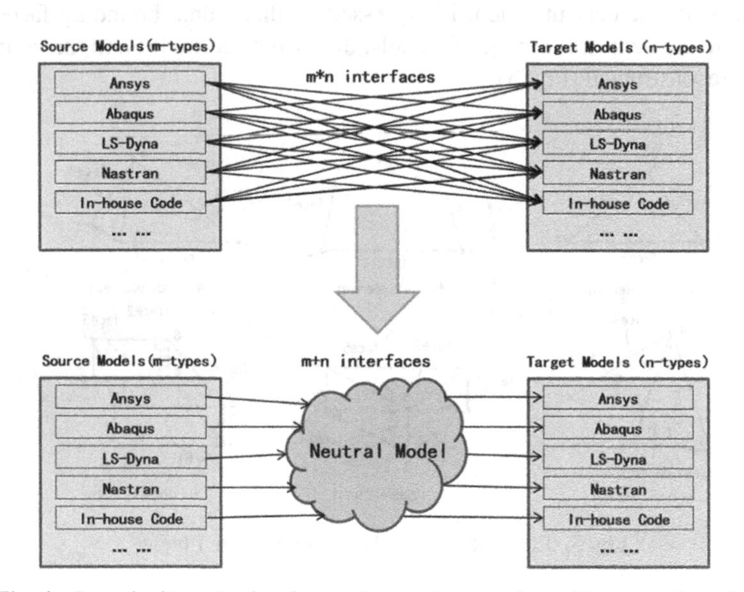

Fig. 4. Complexity reduction for mesh model conversion with a neutral model.

3 Design of the Neutral Mesh Model

The neutral mesh model presented in this work contains two types of data: the mesh data and the attribute data. The former mainly refers to information on mesh entities such as nodes, elements and mesh sets, and the latter includes data for description of various physical attributes.

3.1 Representation of Mesh Entities

Node

A node entity contains data on its ID, coordinates, as well as the coordinate system it is defined if necessary. It is described with a list as:

<node ID, coordinates, ...>

Element

An element entity mainly consists of ID, geometric and topology information, element order, as well as extra information such as the associated material. The geometric information mainly refers to the geometric shape and dimension, usually including: point, line segment, quadrilateral, pyramid, hexahedron, etc. Topological information, also known as connectivity, describes the hierarchical relationship of associated sub entities such as nodes, edges, and faces of the element (Fig. 5).

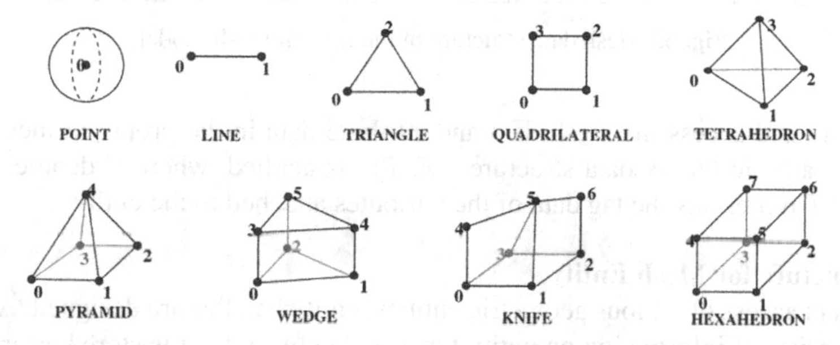

Fig. 5. Node numbering for various types of elements.

In summary, an element entity could be described using the following n-components list as:

<element ID, geometric type, connectivity, ...>

Mesh Set

A mesh set entity, including node set and element set, mainly consists of set ID, set name, set type, and the sequence of the associated geometric entities, which could be described as

<set ID, set name, set type, geometric entities...>

The set type can be material set, node set, boundary surface set, element set and extended as desired.

3.2 Mesh Data Structure

The mesh data structure needs not only to be powerful to fully represent the mesh entities described above, but also be efficient and flexible so that the data could be stored compactly, accessed efficiently and modified dynamically. To this end, a mesh data structure is designed as shown in Fig. 6.

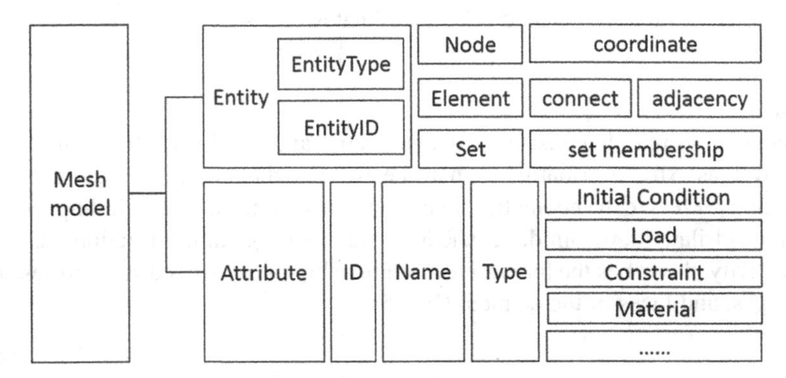

Fig. 6. Mesh data structure of the neutral mesh model.

To store and access the mesh data and attribute data in the proposed mesh model efficiently, a basic binary data structure <*E*, *T*> is applied, where *E* denotes a mesh entity and *T* represents the tag data of the attributes attached to the entity.

Data Structure for Mesh Entity

For efficient access of various geometric entities, entity handles are designed. An entity handle consists of information on entity types and entity IDs. It is stored as an integer with the upper four bits for entity types (including nodes, quadrilaterals, hexahedrons, node sets, etc.) and the other lower bits for entity IDs, as demonstrated below in Fig. 7.

Fig. 7. Design of mesh entity handle.

Mesh entities can be classified into three types, namely node entities, element entities, and mesh set entities, depending on the data type and storage structure. For mesh set entities, the data of entities in the set is compactly stored in a sequence with a continuous memory for compactness. For other mesh entities, a double-linked list is used to store piecewise consecutive data, enabling efficient and flexible access, insertion and deletion, as illustrated in Fig. 8.

Data Structure for Attribute Data

Attribute data is attached to mesh entities in the form of tags. This decoupled design can reduce the costs of attaching or detaching attribute data to/from mesh entities for

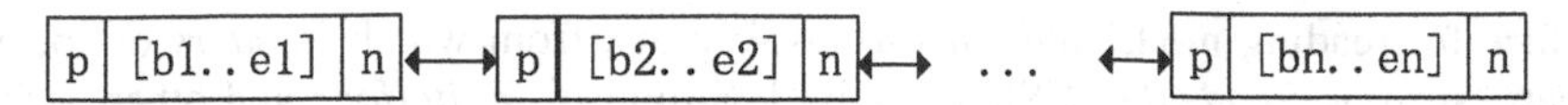

Fig. 8. Storage structure of mesh entity handle.

multiple times, and improve the flexibility of mesh data management. Attribute data could be attached to a mesh entity, an entity set or the entire mesh with tag data. The composition of a piece of tag data is shown in Table 2.

Table 2. Composition of tag data

Item name	Type	Meaning
Name	String	Attribute name
TagHandle	Integer	Attribute handle
DataType	Enum	Integer/double/string/struct
DataSize	Integer	Data size of the entity
DefaultValue	Void*	Default data value
DefaultSize	Integer	Default data size in bytes

To further improve the storage efficiency, tag data is classified into three types, namely dense, sparse and bit field. The dense type is used for storage of each entity in an entity set, the sparse type is used for storage of entities with auxiliary data, and the bit-field type for storage of attributes of an entity set in the form of bits for compactness.

Attribute data in a mesh model is typically used for definition of loads, constraints, boundary conditions, material properties, section constants for engineering elements such as beam and shells.

4 Implementation of the Mesh Model and Conversion Interfaces

Following the design scheme of the proposed mesh model, a generic mesh data conversion tool is developed for numerical simulations. In this converter the mesh data and other simulation-relevant attribute data constituting a mesh model is managed in a centralized manner. A couple of interfaces are implemented for mutual conversion between mesh models of commonly used CAE programs and the mesh model implemented here, enabling an efficient transfer of mesh models between various commercial, open sourced or in-house CAE programs.

The tool is implemented using the open source mesh data management framework MOAB [9]. The class diagram for the functionalities is shown in Fig. 9, where the class *Interface* is served as the main API interface of the mesh model, and the class *Core* is responsible for the implementation of *Interface*. As is introduced in the design scheme of the mesh model, EntityHandle is used to represent mesh entities such as nodes, elements, and sets, and Tag is used to attach attribute data to mesh entities. *ReadIface* is the main interface for reading model files in various formats, from which *Readers* of specific formats (such as *ReadABAQUS*) are derived. Similarly, *WriteIface* and other derived *Writers* are responsible for writing model files in corresponding formats.

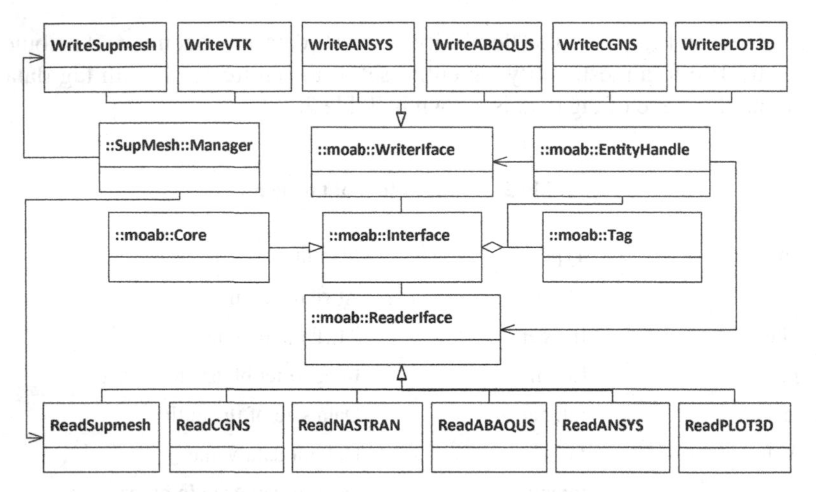

Fig. 9. Class diagram for the conversion tool

All *Readers* and *Writers* are registered in the *MeshIO* class so that reading or writing mesh files with different formats could be performed through a unified interface, which are internally done by calling corresponding *Readers* or *Writers* depending on the mesh file type passed in by the caller.

By using the centralized reading and writing interfaces, the mesh conversion tool we developed gains notable flexibility and extensibility. Mesh data conversions between a series of file formats are supported, as listed in Table 3, where SuperMesh is an in-house numerical simulation code for large scale complex engineering problems, which has customized mesh data format. Conversions between more pairs of file formats could be easily integrated in the tool by simple implementations of the specific reading and writing interfaces.

Table 3. File formats supported by the mesh conversion tool

Format	Read	Write	File name extension
NASTRAN format	Yes	Yes	.bdf
ABAQUS format	Yes	Yes	.inp
ANSYS format	Yes	Yes	.cdb, .dat
Kitware VTK	Yes	Yes	.vtk

(continued)

Table 3. (*continued*)

Format	Read	Write	File name extension
PLOT3D	Yes	Yes	.xyz, .g
DYNA format	Yes	Yes	.k
FLUENT format	Yes	Yes	.cas, .msh
CGNS	Yes	Yes	.cgns
EXODUS II	Yes	Yes	.exo, .exoll, exo2
HDF5	Yes	Yes	.h5m, .mhdf
SuperMesh	Yes	Yes	.J

5 Validation

5.1 Conversion Correctness

The correctness of the mesh conversion tool is roughly demonstrated by observations of the generated mesh files with various formats. Given a mesh file generated in PLOT3D with 922816 nodes, 846660 elements and 11 material sets, it is converted to mesh files with CGNS, DYNA and ANSYS formats, respectively, using the conversion tool. Such mesh files with four different formats are then imported to FLUENT, TECPLOT, LS-PREPOST and ANSYS respectively. All these models could be interpreted correctly in corresponding programs and look similar to each other as shown in Fig. 10, which to some extent implies that the developed mesh converter could correctly convert mesh model in PLOT3D format to the ones in CGNS, DYNA and ANSYS formats.

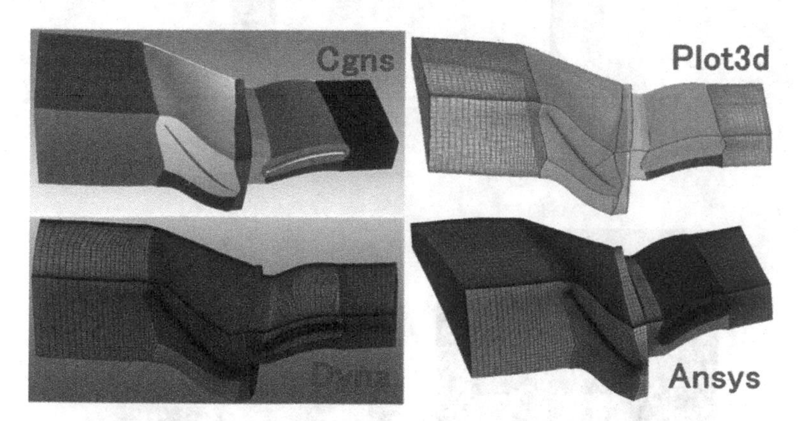

Fig. 10. Comparison of the mesh models generated by the conversion tool

To further validate the capability of the conversion tool, a series of comparative numerical simulations are conducted using various formats of mesh models generated by the tool. Considering the deformation of a simple structure subjected to a uniformly-distributed pressure of 20 MPa on the top face, as depicted in Fig. 11. The four edges in the base are fixed. The material of the structure is assumed to be steel, with a Young's modulus of 210000 MPa and Poisson's ratio of 0.3.

Fig. 11. Geometry model.

A mesh model with data on nodes, elements, material properties, as well as constraints and loads is firstly setup in ANSYS, and then exported in the cdb format. With the conversion tool, this mesh data is further converted to inp and bdf formats used by ABAQUS and NASTRAN, respectively. Numerical simulations are conducted using these three mesh data files in corresponding simulation programs and a comparative contour plot of the deformation of the structure in the Z direction are shown in Fig. 12.

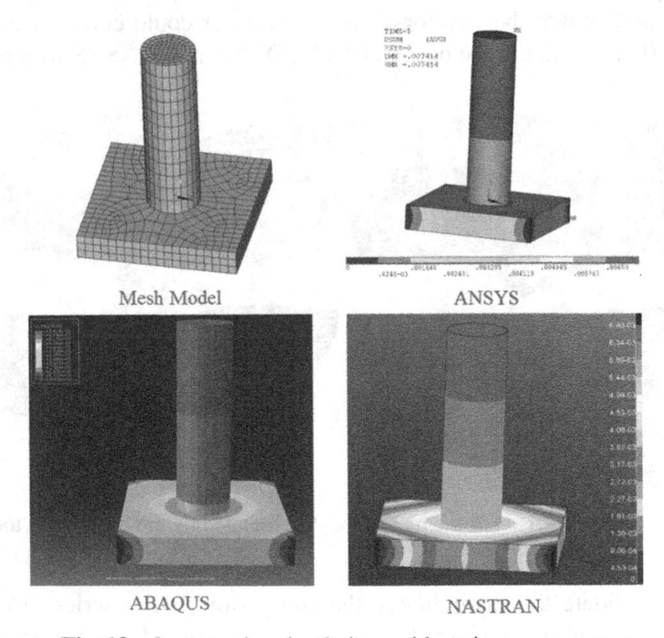

Fig. 12. Comparative simulations with various programs.

The results of maximum deformation obtained from the three programs are listed in Table 4. It could be seen that the results from ANSYS and ABAQUS are identical, and the results from ANSYS and NASTRAN are also in good agreement, which suggests that mesh files converted from the developed tool are correct.

Table 4. Comparison of calculation results of three software.

Program	Result (Max UZ in mm)	Relative error
ANSYS (baseline)	0.0007414	–
ABAQUS	0.0007414	0.0%
NASTRAN	0.0006808	8.1%

The mesh model is further converted to mesh models in msh format used for a heat transfer analysis with Fluent, and k format for an impact analysis with DYNA, both give reasonable results as depicted in Fig. 13, which implies the correctness of the converted mesh models, and the applicability of the developed tool to conversion between mesh models used in ANSYS to those in Fluent and DYNA.

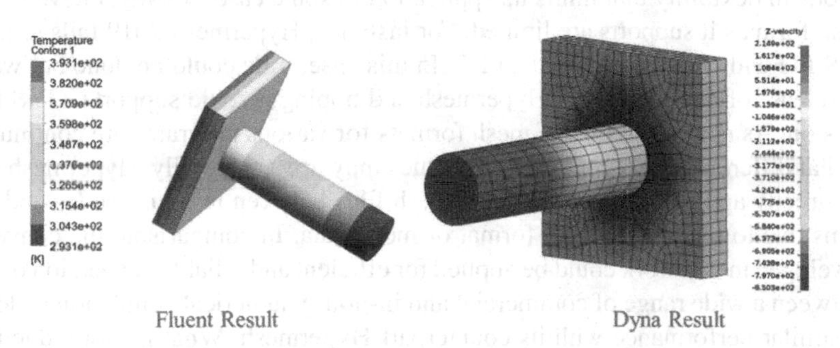

Fig. 13. Simulation results of Fluent and DYNA.

Based on the above descriptions, the conversion tool in this paper can realize the mutual conversion of various formats of mesh files, and the conversion results can be used in common commercial software for corresponding numerical simulation analysis. In summary, the conversion tool developed in this work could be successfully applied to the conversion of mesh files in various formats used in prevailing numerical simulation programs. The correctness and the applicability of the tool has been extensively validated.

5.2 Conversion Scalability

To validate the scalability of the mesh conversion tool for large-scale mesh data, a CGNS format mesh file with roughly twenty million elements is converted to mesh file in format of DYNA. The time costs for reading the CGNS file and converting it to DYNA format are 79s and 61s with Intel I7–7700 processor, respectively, which is acceptable. The generated mesh data in DYNA format is plotted in Fig. 14.

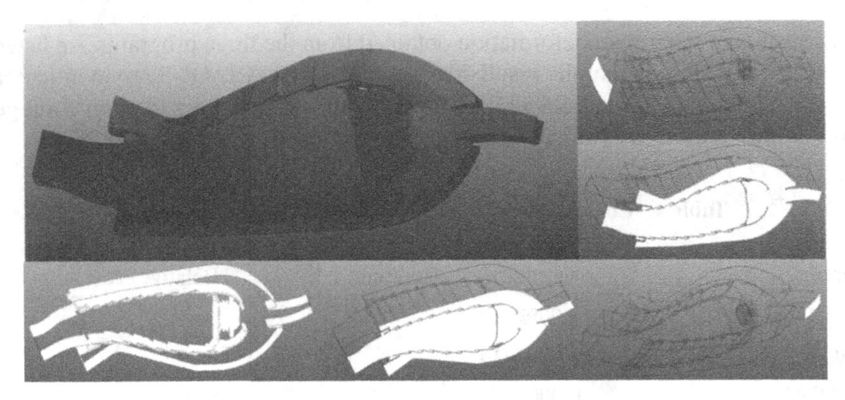

Fig. 14. Application of the conversion tool at a scale of 20 million elements.

5.3 Conversion Comparison with Hypermesh

As a dominating commercial mesh data converter, Hypermesh provides a rich set of conversion interfaces and is widely used for converting mesh files between commonly used CAE programs such as NASTRAN, ABAQUS, DYNA, etc. However, it also has limitations in flexibility that limits its application in some cases. Firstly, the versions of the mesh formats it supports are limited. For instance, Hypermesh 2019 fails to import a CGNS file with version higher than 2.5. In this case, little could be done but waiting for the release of new version of Hypermesh and hoping it could support CGNS file in that version. As newer versions of mesh formats for various programs are continuously released at different paces, compatibility issues may arise. Secondly, Hypermesh could not be directly applied to conversions of mesh files between in-house codes and other programs due to the user-defined format of mesh data. In comparison, the conversion tool developed in this work could be applied for efficient and reliable conversion of mesh data between a wide range of commercial and in-house numerical simulation codes and shows similar performance with its counterpart Hypermesh. What is more, due to the well-designed structures, this tool could be straightforwardly extended for dealing with mesh files with higher versions or in-house codes with user-defined formats, in which cases Hypermesh may be hardly functioned.

6 Concluding Remarks

Mesh serves as the key role in a typical numerical simulation. While due to the lack of a unified standard, mesh models in prevailing simulation programs are diverse notably in formats and structure of mesh data, prohibiting the free exchange of mesh models between various programs. In this work, a unified mesh data model with full representation of mesh data for general-purposed simulations is designed. The application of such a neutral model in conversion process could avoid the typical cobweb-structured layout of conversion work and gain great extensibility and maintainability. Based on this mesh model and the open source mesh management framework MOAB, a generic mesh converter is developed, enabling the efficient conversion of mesh models between a couple of commonly used simulation programs. This tool has preliminarily bridged the isolation and prompted the collaboration between various programs. In the future work, further investigation will be made on AMG (algebraic multi grid) and mesh stitching methods to enrich the functionalities of the developed mesh conversion tool.

References

1. Chen, Z.: The finite element analysis software of great projects and its application. Electro-Mech. Eng. **93**(5), 25–28 (2001)
2. Lu, S., Zheng, Q., et al.: A comparative study on computational accuracy and efficiency of welding residual stress and deformation in a Q390 steel thick plate T joint among three kinds of different FEM software. J. Mech. Eng. **55**(6), 11–22 (2019)
3. Wang, C.: Numerical simulation and experimental study on the synthesis of large-size and high-quality diamond single crystal at high temperature and high pressure. Jilin University (2021)
4. Li, X., Zhou, D., Feng, L.: Mesh transform and transaction between UG and ProCAST. Foundry Technol. **36**(03), 722–731 (2015)
5. Iványi, P.: Finite element mesh conversion based on regular expressions. Adv. Eng. Softw. **51**(6), 20–39 (2012)
6. SuperMesh [EB/OL], 06 June 2022. http://www.caep-scns.ac.cn/SuperMesh
7. Xue, F.: Study of degeneration models of isoparametric element in finite element method. Eng. J. Wuhan Univ. **39**(1), 58–62 (2006)
8. Li, A., Shen, W., Zhang, Y.: Research of the patched-grid technology based on CFX-Pre. Sci. Technol. Inf. **22**(082), 82–86 (2018)
9. Tautges, T.J., Meyers, R., Merkley, K., Stimpson, C., Ernst, C.: MOAB: a mesh-oriented database. Sandia National Laboratories (2004)

Design and Development of a Simulation Model Validation Tool

Hongjia Su[1], Zhifeng Lu[2], Fei Liu[1](✉), Rufei Li[2], and Feng Ye[3](✉)

[1] South China University of Technology,
Guangzhou 510006, Guangdong, People's Republic of China
feiliu@scut.edu.cn
[2] Shanghai Electro-Mechanical Engineering
Institute, Shanghai 201109, People's Republic of China
[3] National Key Laboratory for Complex Systems Simulation,
Beijing 100000, People's Republic of China
13701384998@139.com

Abstract. Model validation is one of the core issues in modeling and simulation, which aims to check whether the model output correctly reflects the output of the corresponding real system. Model validation is also a complicated process, which requires powerful software to achieve its purpose. In this paper, we describe a simulation model validation tool (shortly SMVT) that we developed recently for the validation purpose of complex simulation models. We first talk about the functions and architectural design of SMVT and then employ a missile simulation model to demonstrate the application of SMVT. This tool has so far been widely used in many scenarios.

Keywords: Simulation model validation · Modeling and simulation · Validation tool

1 Introduction

With the technological development of aerospace, there is an increasing demand for research and development on missiles, launching vehicles and other flight vehicles. The traditional development process of these vehicles is highly dependent on physical objects, which takes a lot of time and money and suffers from high risks. In contrast, modeling and simulation can substantially reduce the cost by analyzing the constructed models of these real entities on computers. Now, simulation has been widely used in different areas such as aerospace, military and manufacturing. Furthermore, verification, validation, and accreditation (shortly VV&A) have to be done for any simulation models before they are used to assure their credibility. In fact, VV&A has become one of the essential issues of modeling and simulation, of which model validation is the core problem [1, 2].

Model validation aims to check whether the model output correctly reflects the output of the corresponding real system. Due to the variety of model outputs or performance parameters and the availability of real data (also called reference data), so far

W. Fan et al. (Eds.): AsiaSim 2022, CCIS 1712, pp. 40–50, 2022.
https://doi.org/10.1007/978-981-19-9198-1_4

many validation methods have been proposed, which can be basically classified into two categories: subjective methods such as Turing test and face validation, and objective methods such as hypothesis test and time series analysis methods. Moreover, these validation methods can also be grouped according to the types of performance parameters, namely static (time free) and dynamic (time-dependent). For the detailed description of these validation methods, please refer to [3–7].

To support model validation, different validation tools have been developed, and some typical ones can be summarized as follows. Abroad, Charlow et al. gave a tool to help to standardize the documentation of VV&A [8]. Balci et al. developed a collaborative evaluation environment for assessing simulation models including model validation [9]. Holmes et al. [10] developed Random Tool Box (RTB), which contains several simulation validation methods like Turing test, Theil's inequality coefficient test, and spectrum analysis. At home, Liu et al. [11] developed a simulation validation tool developed for complex simulation systems. Jiao et al. [12] designed a simulation credibility evaluation aid tool. Ma et al. [13] developed a simulation model validation tool and gave the internal implementation of this tool. In [2], Liu et al. presented a model VV&A platform that comprises a suite of tools. In contrast, there is still a gap between China and western countries in terms of the VV&A tool development due to many reasons.

This paper is structured as follows. It first gives the functions and architectural design of our simulation model validation tool. After that, a case study is used to demonstrate the usage of the tool, followed by the conclusion.

2 Simulation Model Validation Tool

Simulation model validation tool (shortly SMVT) aims to support sufficient validation of different types of complex models with a number of popular validation algorithms and deals with both static performance parameters like kill probability and dynamic performance parameters like traces of missile models. SMVT is used as a standalone software tool that can access different data sources from popular databases such as Oracle, MySql, SQLite, and Sqlserver. Therefore, it can be easily integrated into many simulation platforms to achieve validation work. In the following, we briefly describe the functions and architectural design of SMVT.

2.1 Main Functions

The main functions implemented in the current SMVT are listed as follows:

(1) Project management. This includes the creation, editing, and deletion of both projects and validation tasks. SMVT manages the validation work at two levels: project and task. A user has to first create a project, in which one or more tasks can be created further depending on the purpose.
(2) Data management. This includes the management of both simulation data and reference data, such as importing, editing, and lookup functions. Many different sources of data, such as Oracle, MySql, SQLite, Sqlserver databases, csv, and txt, can be easily imported.

(3) Model validation index construction. This allows users to freely create a validation index system for a specific model. Besides, a number of validation index templates are offered to reuse the existing experience of model validation. The creation of an index system is performed by dragging an icon to the canvas and connecting these icons by edges.
(4) Pre-processing of simulation and reference data. We offer a couple of pre-processing methods to handle raw simulation and reference data to assure that the data can be correctly used with appropriate statistical algorithms. Besides, these pre-processing methods can be usually classified as two groups in terms of the types of specific validation performance parameters, namely static and dynamic. The former category includes normal test, abnormal data test and elimination, etc. and the latter includes missing value check and data interpolation.
(5) Validation of static performance parameters. Usually, hypothesis testing can be used to validate this type of index and typical methods include T-test for checking the mean of a parameter and an F-test for checking the variance of a parameter. If the reference data cannot be obtained, face validation has to be used.
(6) Validation of dynamic performance parameters. Here such methods as TIC can be employed, but strict pre-processing should be performed before applying such methods. Again, if the reference data cannot be obtained, we also have to resort to other time series analysis methods and domain experts to make a reasonable reasoning.

2.2 Architectural Design

In order to support efficient validation of complex simulation models with a large volume of data, we deliberately design the architecture (see Fig. 1) of our tool, which includes five main layers: GUI, business, service, storage and external interface layer.

(1) GUI layer. SMVT can run on Windows or Linux due to its cross-platform design. SMVT can also run on different Windows platforms, such as Win 7 and Win 10.
(2) Business layer. SMVT can perform almost all validation tasks including validation index construction and computation, validation result display and report generation, and general project and data management.
(3) Server layer. SMVT offers popular pre-processing algorithms for both sample data and time series data, different types of hypothesis test algorithms, and different types of time series validation methods.
(4) Storage layer. SMVT employs the SQLite database to store the internal data as it does not need to be installed separately. This facilitates the installment of the software tool.
(5) External interface layer. SMVT employs python to support complicated computational capabilities and different database interfaces to accomplish the import of different database types. This allows users to flexibly design their own computational models for specific issues.

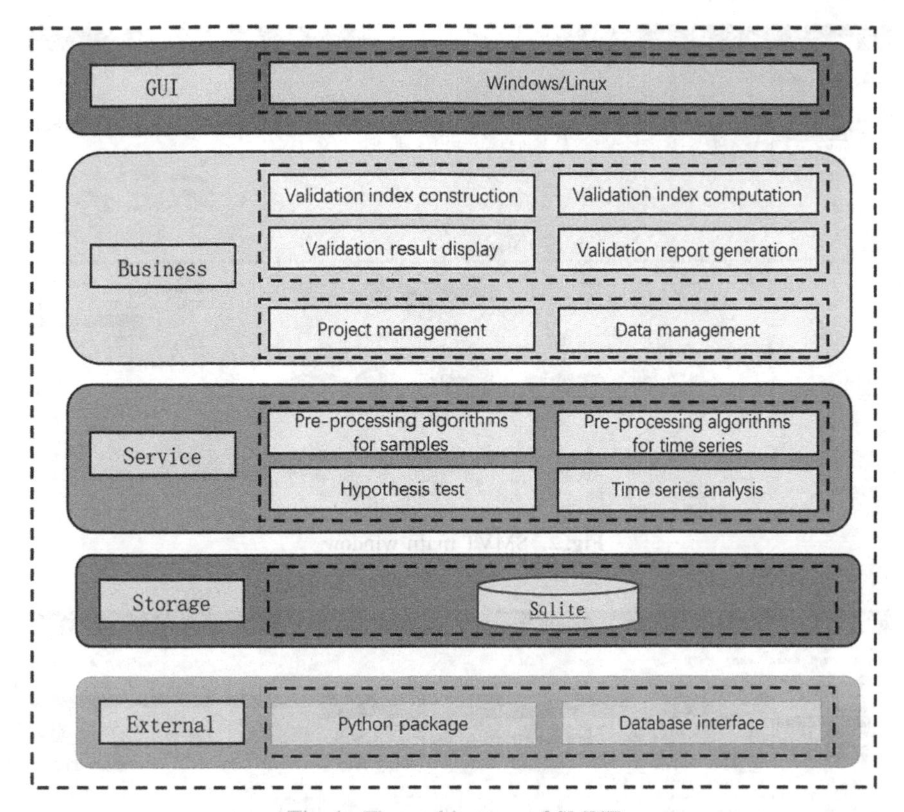

Fig. 1. The architecture of SMVT.

2.3 Main Interfaces

We employ qt, C++ and python to implement our software, which offers friendly interfaces for users to perform validation of models. Figure 2 gives the main window of our tool, which is also the validation index creation window. In this window, the user can perform different operations to create an index system.

After the construction of the index system of a model, we can enter the computation window (see Fig. 3) to perform validation for each performance parameter given in the index system.

From the computation window, we can see many preprocessing methods for samples and time series are offered and the execution of these methods is also friendly by clicking the corresponding buttons with clear notes. Moreover, an intelligent recommendation algorithm of these preprocessing methods is also offered to assist in the selection of these methods.

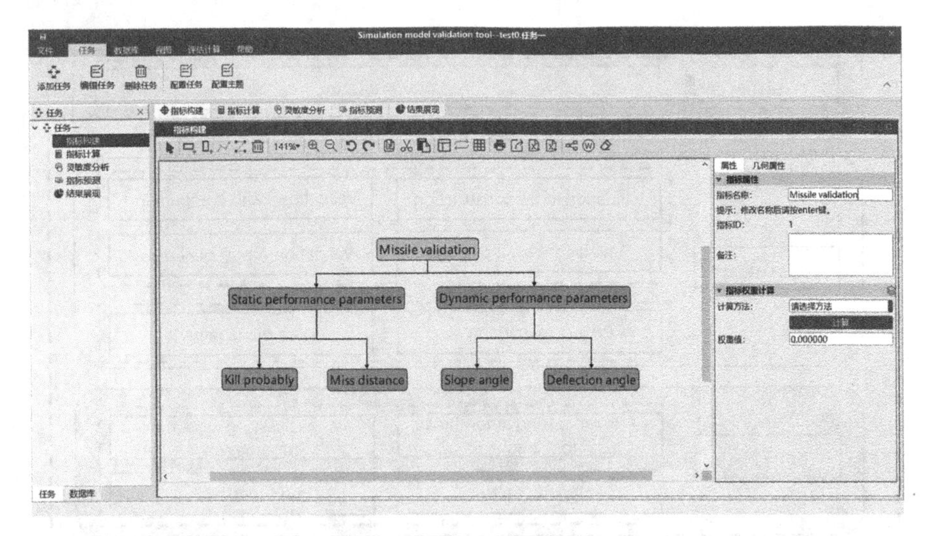

Fig. 2. SMVT main window.

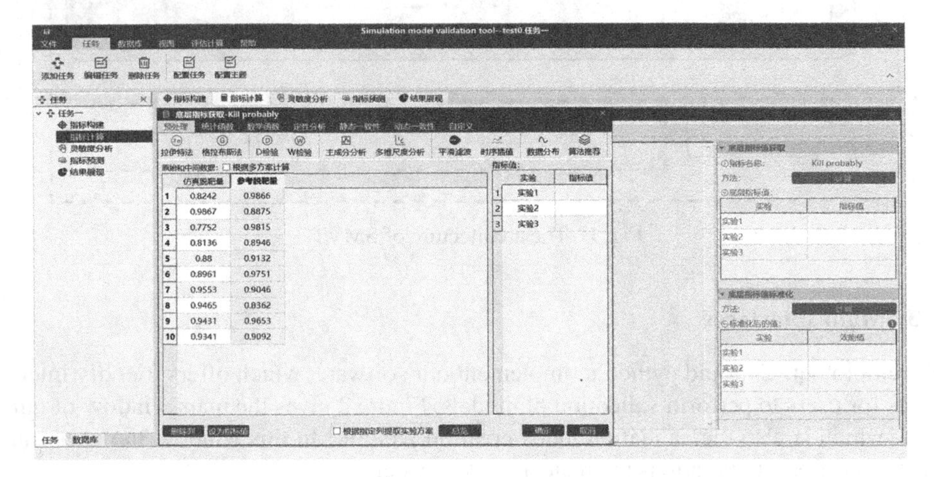

Fig. 3. SMVT computation window.

2.4 Workflow

The workflow of the tool is usually given as follows:

(1) Create a project and at least a task. All the validation work is organized by a task. A task usually corresponds to a model validation job.
(2) Construct an index system for a validation issue. In terms of the validation purpose, such an index system can be constructed from scratch or with the existing index system templates.
(3) Perform validation in terms of the category of a validation parameter. Each of the two classes of parameter follows a distinct procedure. For a static parameter, this

usually includes pre-processing of data, normal test of data, variance consistence test and mean consistency test. For a dynamic parameter, this includes pre-processing of time series and consistency analysis of time series.

(4) Display validation results. When the validation is finished, all the validation results can be collected and displayed. The user can also choose to output a validation report.

3 Case Study

3.1 Model and Data

We build a 3-DOF missile model for simulating the interception of objects in order to validate our software tool. We assume that the missile is initially positioned at (0 m, 1000 m, 0 m), and flies at a fixed speed 700 m/s and the speed slope angle and deflection angle is 45° and 10°, respectively. To simulate the errors due to the environmental effect, we associate 5% random Gaussian disturbance for the flight speed, initial speed slope angle and deflection angle, respectively. Besides, we also assume the object flies at a fixed speed of 500 m/s and the initial speed slope angle and deflection angle is 60° and 0°, respectively. Besides, we set the simulation step to 0.001 s.

We perform 20 simulation runs and obtain 20 simulation samples of the miss distance, which is then used as the static performance parameter validation (Table 1).

Table 1. Simulation samples of the miss distance for the missile model.

Sample number	1	2	3	4	5
Missile miss distance	0.8242	0.9867	0.7752	0.8136	0.88
Sample number	6	7	8	9	10
Missile miss distance	0.8961	0.9553	0.9465	0.9431	0.9341
Sample number	11	12	13	14	15
Missile miss distance	0.9866	0.8875	0.9815	0.8946	0.9132
Sample number	16	17	18	19	20
Missile miss distance	0.9751	0.9046	0.8362	0.9653	0.9092

We aim to demonstrate our software tool with this model, so we divide the above-mentioned simulation samples into two groups, each having 10 samples, with the first group being the simulation data and the second group being considered as the reference data.

Besides, we choose the speed slope angle to demonstrate the validation of time series data. To do this, we run two simulations, one with random noise (see Fig. 4a) and one without noise (see Fig. 4b). The former is considered as simulation data and the latter is considered as reference data (Table 2).

Table 2. Simulation data and reference data of the speed slope angle.

Time t/s	**0**	**1**	**2**	**3**	**4**
Simulation data	41.5332	48.4936	53.2320	56.3416	58.2697
Reference data	45.0070	50.8008	54.4860	56.6583	57.8404
Time t/s	**5**	**6**	**7**	**8**	
Simulation data	59.3905	60.0167	60.3902	60.6723	
Reference data	58.4524	58.8025	\	\	

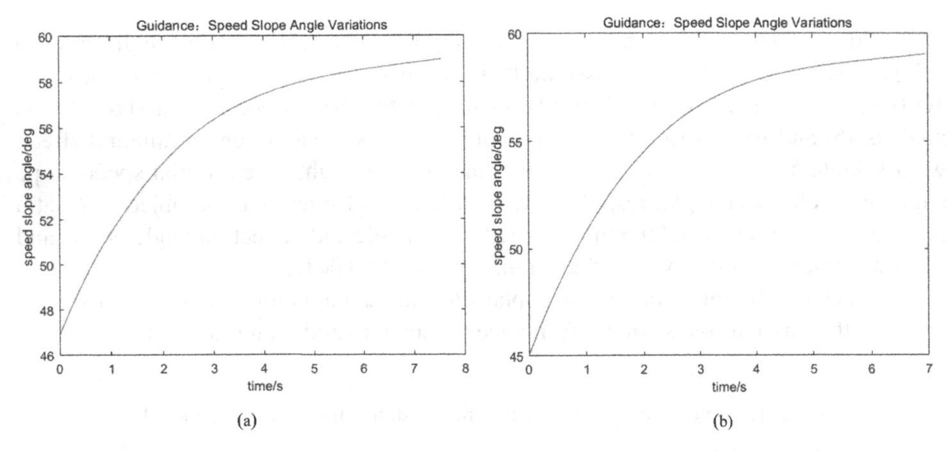

Fig. 4. Simulation data (a) and reference data (b) of the speed slope angle.

3.2 Validation of Static Performance Parameters

The validation of static performance parameters usually follows a strict procedure, namely, pre-processing, normal test and hypothesis test, which is now fully supported by our tool. We will demonstrate this procedure as follows.

(1) Pre-processing of data. As the sample number of both simulation and reference data is 10, we have to choose Grubbs method to check if there are abnormal values. With the given significance level $\alpha = 0.05$, we find no abnormal values in both groups of data.
(2) Normal test of data. Similarly, the sample number of both simulation and reference data is 10, so we have to choose W test and obtain the p values are 0.39873 and 0.293195, respectively, for the simulation data and reference data. Compared with the confidence level $\alpha = 0.05$, both group of data can be considered to follow the normal test.
(3) Variance consistency test. We continue to test the variance consistency of both simulation and reference data by considering the significance level $\alpha = 0.05$, and obtain the test result given in Fig. 5. From this figure, we can see the significance

of this test is 0.270301, obviously is higher than the given 0.05, so we can conclude that the variance of both group of data is consistent.

(4) Mean consistency test. From the above-mentioned test results, we can see both groups of data follow the normal test and satisfy the homogeneity in terms of their variance. Therefore, we can use the setting given in Fig. 6 to continue to test the mean consistency of both groups of data, and the result is also shown in Fig. 6. That is the statistics t is −1.09672, and the significance level p is 0.29, higher than the given level of 0.05. Therefore, the means of both groups of data are consistent.

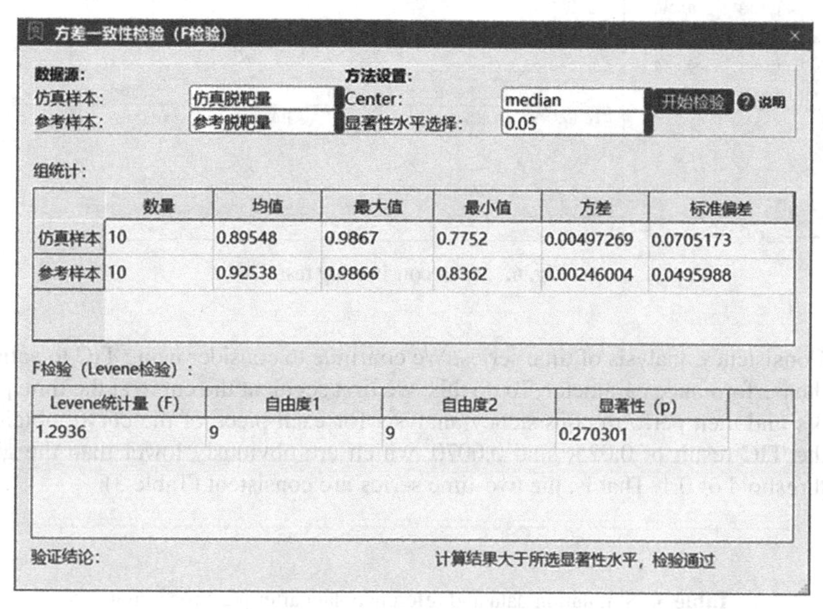

Fig. 5. The variance consistency test result.

3.3 Validation of Dynamic Performance Parameters

Validation of dynamic performance parameters usually follows two big steps: pre-processing and consistency analysis. In the following, we take the TIC method as an example to illustrate how to use our tool to achieve such validation of time series. For this purpose, we obtained two groups of data on the trajectory slope angle with two noisy simulation runs, which are shown in Fig. 7 (the top right corner).

(1) Pre-processing of time series. We first pre-process the data and find that there are no missing values in both groups of data and the time points are consistent for both data. To obtain time series with identical length, we discard two data points that time = 7 s and time = 8 s from simulation data.

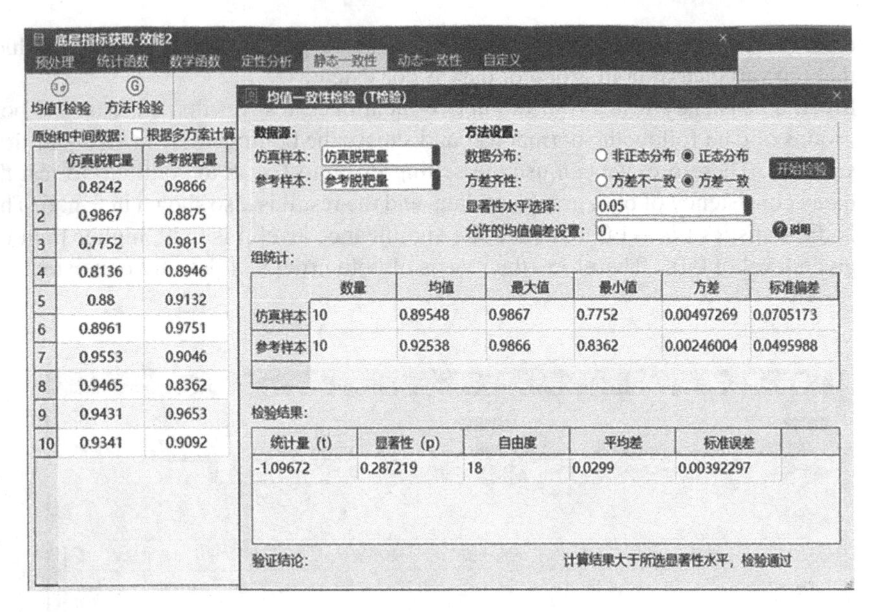

Fig. 6. Mean consistency test.

(2) Consistency analysis of time series. We continue to consider using TIC to validate the performance parameter. To do this, we first segment the curve at the time point 3 s and then perform consistency analysis for each piece of the curve, obtaining the TIC result of 0.0256 and 0.0070, which are obviously lower than the given threshold of 0.1. That is, the two-time series are consistent (Table 3).

Table 3. Simulation data and reference data after pre-processing.

Time t/s	**0**	**1**	**2**	**3**	**4**
Simulation data	41.5332	48.4936	53.2320	56.3416	58.2697
Reference data	45.0070	50.8008	54.4860	56.6583	57.8404
Time t/s	**5**	**6**			
Simulation data	59.3905	60.0167			
Reference data	58.4524	58.8025			

From the analysis above, we can see that the tool offers a friendly interface for users to validate both static and dynamic performance parameters. Moreover, this tool supports a complete validation workflow from preprocessing to displaying the result. The tool also can be applied to many other kinds of models.

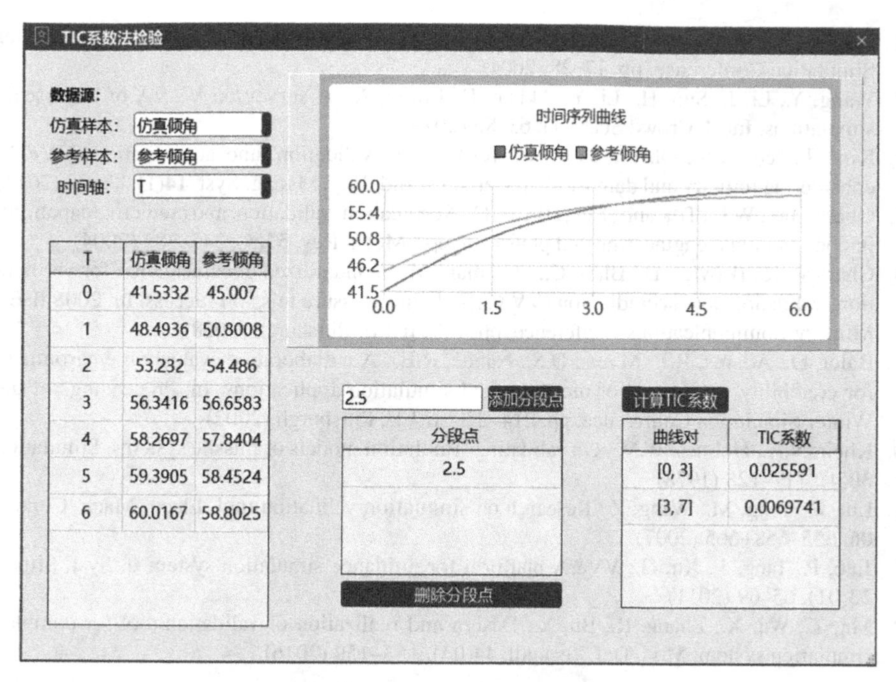

Fig. 7. The TIC analysis result.

4 Conclusions

This paper describes a new simulation model validation tool, designed by considering many practical validation requirements of some simulation systems. This tool offers functions ranging from general project management to specific validation index system construction, index computation and result display. This tool also provides many methods, including pre-processing of sample data and time series, and validation methods of both static and dynamic performance parameters. We also use a missile model to demonstrate the use of our tool. All these show that our tool is powerful and easy to use.

In the next step, we will continue to improve the tool by adding new functions and new model validation algorithms. Besides, introducing artificial intelligence into the tool to support the selection of validation methods and to merge different sources of assessment results is also an option to improve our tool.

References

1. Pace, D.K.: Modeling and simulation verification and validation challenges. J. Hopkins APL Tech. Dig. **25**(2), 163–172 (2018)
2. Liu, F., Yang, M.: The management of simulation validation. In: Beisbart, C., Saam, N.J. (eds.) Computer Simulation Validation. Simulation Foundations, Methods and Applications, Springer, Cham (2019). https://doi.org/10.1007/978-3-319-70766-2_24
3. Sargent, R.: Verification and validation of simulation models. J. Simul. **7**, 12–24 (2013)

4. Sargent, R.: Verification and validation of simulation model. In: Proceeding of 2004 Winter Simulation Conference, pp. 17–28 (2004)
5. Wang, Y., Li, J., Sun, H., Li, Y., Akhtar, F., Imran, A.: A survey on VV&A of large-scale simulations. Int. J. Crowd Sci. **3**(1), 63–86 (2019)
6. Kim, J., Jeong, S., Oh, S., Jang, Y.: Verification, validation, and accreditation (VV&A) considering military and defense characteristics. Ind. Eng. Manag. Syst. **14**(1), 88–93 (2015)
7. Oberkampf, W.L., Trucano, T.G., Hirsch, C.: Verification, validation, and predictive capability in computational engineering and physics. Appl. Mech. Rev. **57**(5), 345–384 (2004)
8. Charlow, K., Broyles, D., Blais, C., Stutzman, M.: Standardized documentation for verification, validation, and accreditation (VV&A)—helping assure mission success. In: 2008 IEEE Military Communications Conference, pp. 1–7. IEEE, Pittsburgh (2008)
9. Balci, O., Adams, R.J., Myers, D.S., Nance, R.E.: A collaborative evaluation environment for credibility assessment of modeling and simulation applications. In: Proceedings of the Winter Simulation Conference, pp. 214–220. IEEE, Pittsburgh (2002)
10. Kheir, N.A., Holmes, W.M.: On validating simulation models of missile systems. Simulation **30**(4), 117–128 (1978)
11. Liu, F., Yang, M., Wang, Z.: Research on simulation validation tool. Electr. Mach. Control **06**, 655–658+665 (2007)
12. Jiao, P., Tang, J., Xu, G.: VV&A platform for guidance simulation system. J. Syst. Simul. **23**(01), 65–69 (2011)
13. Ma, Z., Wu, X., Zhang, R., Bu, X.: Design and realization of validation tool for complex simulation system. Mod. Def. Technol. **44**(04), 153–159 (2016)

Research on Graphical Modeling and Simulation Method of Control System Based on Python

Yongxuan Xie[1,3], Xiao Song[2,3](✉), Yuchun Tu[2,3], Yong Cui[1,3], Junhua Zhou[1,3], and Yanjun Zhai[1,3]

[1] School of Automation Science and Electrical Engineering, Beihang University, Beijing, China
[2] School of Cyber Science and Technology, Beihang University, Beijing, China
songxiao@buaa.edu.cn
[3] Beijing Simulation Center, Beijing, China

Abstract. Compared with traditional control system modeling, directed graph, as one of the graphical model representation methods, can be introduced into the control system modeling process to effectively improve the modeling efficiency. Based on this, a graphical modeling and simulation method of control system based on directed graph is proposed. Specifically, a control system modeling method based on hierarchical model elements and directed graph description is proposed. Secondly, a control system parsing method oriented to graphical description is designed and implemented, and an integrated environment of graphical modeling and simulation execution is realized. On this basis, control systems based on directed acyclic graphs and directed cyclic graph descriptions are constructed, and the effectiveness of the proposed method in control system modeling and simulation is demonstrated by cases of first-order system, second-order system, follower systems and hydraulic circuits.

Keywords: Graphical modeling · Control system · Directed graph · Python · Simulation · Web

1 Introduction

Control system [1] plays a very important role in modern industry [2, 3], agriculture [4], national defense [5], etc. With the rapid development of computer-related science and technology, the design of control system has also been transferred from paper to computer [6]. Graphical modeling and simulation of control system based on computer graphics technology has gradually become the main way for researchers to study and design control systems.

Along with the development of low-code [7, 8] and visual development-related technologies, graphical-based control system modeling and simulation platforms have become one of the research hotspots in the modeling and simulation community. This is not only because the graphical-based control system modeling and simulation platform can improve work efficiency, but also because it can further reduce the threshold for research and design of control systems. Since the 1980s, the research on graphical

W. Fan et al. (Eds.): AsiaSim 2022, CCIS 1712, pp. 51–73, 2022.
https://doi.org/10.1007/978-981-19-9198-1_5

modeling and simulation has gradually begun at home and abroad. MMS [9] (Modular Modeling System), developed by the Electric Power Research Institute (EPRI), is the earliest modular modeling software for thermal systems. It pioneered the idea of modular modeling and has been cited and developed to date. It uses topological analysis method to solve the fluid network, determines the connection mode and parameter transfer relationship between each device, and classifies each device module so as to standardize the connection mode of the device module, thus ensuring the stability and reliability of the connection of each device module. APROS [10] (Advanced PROcess Simulator), a high-precision dynamic simulation software for energy power systems, was jointly developed by VTT Technical Research Centre of Finland and Fortum Engineering and Technology Finland Ltd. It provides a flexible user interface and extensive software libraries, taking into account simulation engineering development and real-time running environment, with visualization, online interaction, and graphical modeling functions. It can not only be used to create a simulation training simulator for thermal power plant operators, but also can be used for design verification and operation control optimization of other general thermal power systems and their automation systems, and is widely used in dynamic simulation of energy power systems. Simulink, a visual simulation tool in MATLAB [11] launched by Mathworks, is a block diagram environment for multi-domain simulation and model-based design. Simulink provides graphical editors, customizable block libraries and solvers, capable of modeling and simulating dynamic systems, and is widely used in various fields such as automotive, aerospace, and industrial automation. Xcos in the software SCILAB [12] developed by the French National Institute of Computing and Automation (INRIA) is dedicated to the modeling and simulation of continuous and discrete hybrid dynamic systems. It includes a graphical editor for designing models of hybrid dynamic systems, which can be easily represented as graphical block diagrams by connecting modules to each other. In addition, there are also many related studies in China. GNET, an integrated graphical modeling environment for thermal power plant simulators developed by Tsinghua University, fully adopts a graphical user interface, enabling modelers to complete the entire process from modeling to debugging through a convenient and friendly graphical interface. The integrated and visual simulation support system STAR-90 independently developed by Huafang Technology Co., Ltd. provides users with a convenient modular and graphical modeling environment. It supports online model modification, multi-model simultaneous running environment and automatic splicing of models, etc. It is widely used in aerospace, electric power, petroleum, chemical and other industries. Nowadays, the mainstream graphical modeling and simulation platforms mainly include MATLAB Simulink from MathWorks, and SCILAB Xcos developed by INRIA.

Although MATLAB, Simulink, etc. can well realize the modeling and simulation of the control system, these softwares all have risks of being disabled. Considering that Python is an open-source language and has scientific computing capabilities similar to those of mainstream software such as MATLAB and SCILAB, this paper proposes and implements a method for graphical modeling and simulation of control systems based on Python.

The graph is a structure used to represent a certain relationship between objects, and its representation is in the form of nodes and the relationship between nodes that is by connecting lines. Directed graphs are graphs with directional connections between nodes, which are used in many fields such as physics, chemistry, computers, and control systems. In this paper, the model of the control system and its subsystems is also established by means of a directed graph.

Python [13] is an object-oriented, open source general-purpose programming language that provides a large number of scientific computing libraries and packages [14], such as Numpy (numerical computing), SciPy (mathematical, scientific and engineering computing), Matplotlib (a plotting tool similar to plot in MATLAB), and more. Python has a stricter and clearer syntax, which can fulfill high-level requirements such as user interface, files, and encapsulation. It has better portability and is also very suitable for AI, networking and other related applications. Using third-party libraries of Python such as Numpy, Scipy, Matplotlib, etc., can replace the functions of MATLAB and SCILAB to realize the modeling and simulation of control systems such as linear time-invariant systems, linear time-varying systems, and nonlinear time-varying systems [15, 16].

This paper adopts the design and development technology that separates the front and back ends. The front end is based on HTML, CSS and JavaScript [17] to design and realize the construction environment of the graphical control system modeling and simulation on the Web side. The back end is based on Python to design and implement the method of parsing and running the graphical control system model. Finally, based on this graphical control system modeling and simulation tool, the control system modeling and simulation verification experiments are carried out, while the results are verified by comparison experiments with MATLAB simulink.

2 Hierarchical Model System Design of Control System

To build complex and diverse control system models, it is necessary to design an effective model system, so that engineers can quickly build corresponding models and systems and improve work efficiency.

The hierarchical model system includes basic model, specialized model and extended model. The model system has high model reusability and expansibility. In addition to satisfying the basic modeling requirements, the subsystem structure can be used to realize the encapsulation of sub-models and participate in the construction process of other models in the form of composite models, so as to realize the reuse of models. Besides, the model system can also improve the scalability of the entire model system by using the external model interface, thereby ensuring the diversification of model sources.

2.1 Basic Models

The basic model is the basis of graphical modeling and is the meta-component for building a graphical system. The modeling simulation tool implemented in this paper is designed with basic data type models and their related computational model. The main basic data type models include number model, string model, bool model, array model, object model, and matrix model. In order to satisfy the modeling of possible complex systems, this paper also designs a model of subsystem encapsulation to solve problems

Table 1. Basic models

Name	Symbol	Name	Symbol
Number	number [123]	String	string $[abc]$
bool	boolean $bool$	Array	array [...]
object	object {...}	Matrix	matrix $[mat]$
Pow	pow x y x^y out	Ln	ln x $\ln(x)$ out
abs	abs x $\lvert x \rvert$ out	Sin	sin x $\sin(x)$ out
arctan	arctan x $\arctan(x)$ out	radians	radians deg rad rad
And	and a1 a2 and out	xor	xor a1 a2 xor out
transpose	transpose mat $[M]^T$ res	inv	inv mat $[M]^{-1}$ res
if	IF value pass true false	subsystem	子系统 in in2 out
In port	in	Out port	out
step signal	Step	S transfer function	TransferFunc $\frac{1}{s+1}$
Z transfer function	ZTransferFunc $\frac{1}{1+z^{-1}}$	State space	StateSpace $\dot{x} = Ax + Bu$ $y = Cx + Du$
saturation	saturation x Sa y	Exp(t)	exp(t) e^t

such as model nesting and model reuse. Based on the above-mentioned types of models, the modeling and simulation tool can build various types of basic mechanism models by means of graphical programming.

In order to satisfy the modeling and simulation of control systems, this paper designs and provides various types of time-function class models as models for system input signals or time-domain system modeling, linear control system models such as S-domain transfer function models, Z-domain transfer function models, state space models and difference equation models, and nonlinear control system models such as common nonlinear characteristic models and general differential equation models (Table 1).

2.2 Professional Models

In addition to basic models similar to basic mathematical operations, the graphical modeling and simulation system also provides some professional models, allowing engineers to easily and simply construct corresponding simulation models without having to deeply understand the corresponding professional knowledge.

For example, the nonlinear time-varying models of hydraulics are difficult to describe with transfer functions or state space equations, and these models are relatively complex. If you want to implement these models independently, you need to have a deep understanding of the relevant disciplines. The professional hydraulic model provided by the system allows engineers to build the corresponding hydraulic system model more simply and conveniently.

2.2.1 Hydraulic Models

See Table 2.

The oil temperature calculation model of the hydraulic oil tank is shown in formula (1),

$$\frac{dT}{dt} = \frac{1}{c_p m}[\sum_i \dot{m}_i \overline{c}_p (T_i - T_o) - \dot{Q}_w] \tag{1}$$

in the formula, m, T and c_p are the mass, temperature and specific heat capacity of the oil in the hydraulic tank, respectively; $\dot{m}_i$ is the mass of the oil flowing into the hydraulic oil tank, $\overline{c}_p$ is the average specific heat capacity of the oil flowing into and out of the hydraulic oil tank; T_i and T_o are the temperature of the oil at the inlet and outlet of the hydraulic tank, respectively; $\dot{Q}_w$ is the convective heat exchange between the oil and the tank.

The oil temperature calculation model of the hydraulic pipeline is shown in formula (2),

$$C_p m_p \frac{dT}{dt} = \sum \rho C_p q_{in} T_{in} - \sum \rho C_p q_{out} T_{out} - h_p A_p (\frac{T_{in} + T_{out}}{2} - T_e) + \Delta p_l Q_l \tag{2}$$

in the formula, C_p, m_p, h_p and A_p are the oil specific heat capacity, mass, heat transfer coefficient, and heat transfer area of the hydraulic pipeline, respectively; q and T are

Table 2. Hydraulic models

Name	Symbol	Description
Hydraulic tank	Tank Tin Tout	Oil tank for hydraulic system
Hydraulic pipeline	Pipe Tin Tout	Pipeline for hydraulic system
Hydraulic pump	pump Tin Tout Tbl	Pump for hydraulic system
Heat exchanger	HeatExchange Tin Tout	Heat exchanger for hydraulic system
Pipeline branch	Branch	Pipeline branch
Pipeline confluence	merge	Pipeline confluence

the pressure and temperature of the oil respectively, where the subscript *in* corresponds to the input oil, and the subscript *out* corresponds to the output oil; T_e is the ambient temperature; Δp_l and Q_l are flow losses through the pipe (differential pressure between inlet and outlet) and pipe flow, respectively.

The formula for calculating the oil temperature of the hydraulic pump is shown in formula (3).

$$\begin{cases} c_V m_{cvp}\frac{dT_{out}}{dt} = \rho c_p(q_{in}T_{in} - q_{out}T_{out}) + N_{pl} - (1-\eta_v)q_t(p_{out} - p_{bl}) \\ -(1-\eta_v)\rho q_t c_p T_{out} - h_{pb}A_{pb}(T_{out} - T_{bl}) \\ c_V m_{cvp}\frac{dT_{bl}}{dt} = h_{pb}A_{pb}(T_{out} - T_{bl}) - h_{be}A_{be}(T_{bl} - T_{be}) \\ +\rho(1-\eta_v)q_{in}[c_p T_{out} + \frac{p_{out}-p_{bl}}{\rho}] \end{cases} \quad (3)$$

In the above two formulas, the subscript bl represents the shell oil return port. h_{pb} and A_{pb} are the heat transfer coefficient and heat transfer area between the pump working chamber components and the hydraulic oil in the working chamber, respectively. h_{be} and A_{be} are the heat transfer coefficient between the hydraulic oil of the pump casing and the environment and the heat transfer area between the pump casing and the environment, respectively. $q_t = q_{in}$ is the theoretical output flow of the hydraulic pump. p_{in}, p_{out} and p_{bl} are the pump inlet, outlet and case return port pressures, respectively.

The formula for calculating the oil temperature of the hydraulic heat exchanger is shown in formula (4).

$$\begin{cases} c_{vh} m_h \frac{dT_h}{dt} = \dot{m}_h c_{vh}(T_{hi} - T_{ho}) - h_e A_e(\frac{T_{hi}+T_{ho}}{2} - T_e) \\ - h_{fh} A_{fh}(\frac{T_{hi}+T_{ho}}{2} - \frac{T_{fi}+T_{fo}}{2}) \\ c_{vf} m_f \frac{dT_f}{dt} = \dot{m}_f c_{vf}(T_{fi} - T_{fo}) + h_{fh} A_{fh}(\frac{T_{hi}+T_{ho}}{2} - \frac{T_{fi}+T_{fo}}{2}) \end{cases} \quad (4)$$

in the formula, the subscript h corresponds to the hot side of the heat exchanger, the subscript f corresponds to the cold side of the heat exchanger, and the subscript e corresponds to the external ambient temperature of the heat exchanger. c, m, h, A and T correspond to specific heat capacity, mass, heat transfer coefficient, heat transfer area and temperature, respectively.

The oil temperature calculation formulas of the hydraulic branch and confluence nodes are shown in formula (5) and formula (6) respectively.

$$q_i = \sum_{k=1}^{n} q_{ok}, T_i = T_{ok}, p_i = p_{ok} \quad (5)$$

$$q_o = \sum_{k=1}^{n} q_{ik}, T_o = \frac{\sum_{k=1}^{n} q_{ik} T_{ik}}{\sum_{k=1}^{n} q_{ik}}, p_{ik} = p_o \quad (6)$$

In the formula, q, T, and p are the oil flow, temperature and pressure, respectively.

2.3 Models for Extension

Simply relying on the built-in model system of the system often cannot meet various needs. Therefore, it is particularly important to provide extended models to expand the functionality and diversity of the entire graphical modeling system.

The built-in model system mainly provides two types of extension models: extension through code and extension through external API (Tables 3 and 4).

Table 3. Code extension models

Name	Symbol	Description
JavaScriopt	JavaScript JavaScript	Extending through JavaScript code
Python	Python	Extending through Python code

Table 4. Interface extension models

Name	Symbol	Description
API	API Node API	Extending through external APIs.

3 Description and Parsing of Directed Graph Models of Control Systems

For the modeling of the control system, if it is a deterministic continuous system, the common mathematical models are differential equations, transfer functions, and state space expressions and if it is a discrete-time system, its common mathematical models include difference equation, z transfer function, and discrete state space expression. Whether it is a deterministic continuous system or a discrete-time system, their mathematical models can be converted to each other. Therefore, the essence of simulation execution for deterministic continuous systems is the solution of differential equations, while the essence of simulation execution for discrete-time systems is the solution of difference equations.

For the difference equations, which are themselves mathematical models executed for discrete computer simulations, iterative calculations can be performed directly from the initial values of the system as well as from the difference equations. The simulation calculation of differential equations is mainly solved by various numerical integration algorithms, such as Euler method, Runge-Kutta method, Adams method and so on. Because the accuracy and stability zone of various algorithms are different, the reliability of the simulation results of the control system is strongly related to the numerical integration algorithm used. The modeling and simulation tools implemented in this paper mainly use the real-time fourth-order Runge-Kutta method to solve the transfer function and state space of the continuous system. In addition, the modeling and simulation tool also provides the variable-step Runge Kutaferberger method RKF45 as an alternative to improve the simulation speed.

In addition to mathematical models such as difference equations or differential equations, the control system can be represented by a more intuitive system structure block diagram. In order to realize the graphical description of the control system model, this paper proposes a method to describe the block diagram of the control system through a directed graph. That is, a directed graph model of the control system similar to the block diagram of the control system is constructed through the Web graphical control system modeling environment, and then the directed graph is parsed to simulate the control system (Fig. 1).

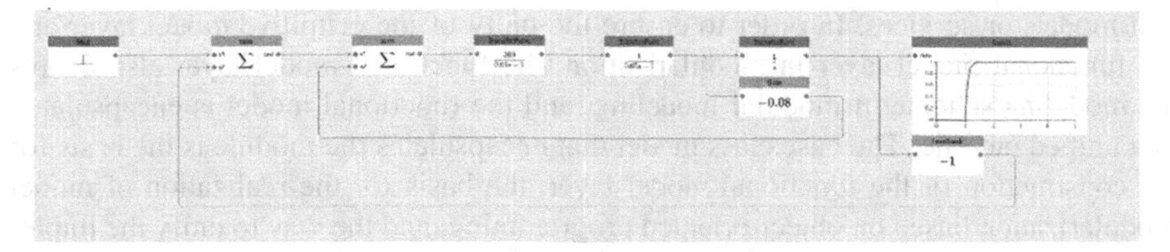

Fig. 1. Example of a control system based on directed graph description

3.1 Parsing of Directed Graphs of Control Systems

The Web graphical modeling and simulation tool researched and implemented in this paper adopts the idea of layered architecture, and divides the entire modeling and simulation system into three parts, corresponding to the primitive model layer, the functional model layer and the parsing runtime layer respectively (Fig. 2).

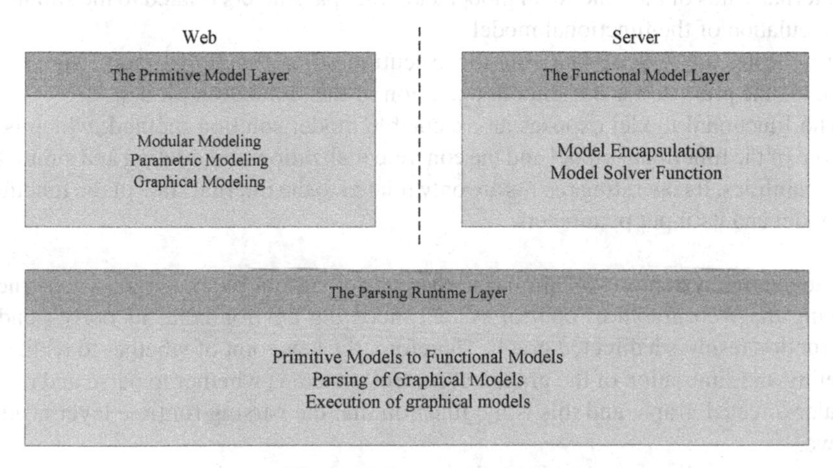

Fig. 2. System architecture

The primitive model layer corresponds to the user-oriented web-side draggable modeling and simulation primitive resources. It provides users with draggable primitive nodes, and provides parameterized configuration capabilities for primitive nodes while ensuring the user's graphical editing experience. The primitive model layer combines the modeling ideas of modular modeling, parametric modeling and graphical modeling, processes each graphical modeling and simulation primitive as independent functional modules (One-to-one correspondence with the module model in the functional model layer), and provides corresponding parameter configuration capabilities, to realize the functionalization and diversification of modules.

The model of the primitive model layer is a visual display of modeling and simulation capabilities on the Web side, and is an unexecutable virtual model. The models of the functional model layer are the concrete realization of the modeling and simulation capability on the server side. They have complete logic realization and are executable real models or services. In order to ensure the unity of the primitive model layer and the functional model layer in modularization, the functional model layer also adopts the modeling scheme of modular modeling, and the functional model is encapsulated in a unified module. The base class model that encapsulates the module is the basis for the construction of the functional model layer, the basis for the realization of model modularization based on object-oriented programming, and the way to unify the implementation logic of the functional model and reduce the iterative expansion cost of the system.

A functional model is a collection of input parameters (InParams), output parameters (OutParams), internal states (InnerStates), connection relationships (Edges), and model solving methods (ParseFunc). The base class abstract encapsulation of the functional model mainly implements the following function.

1) The functional model uses the primitive node data of the primitive model to initialize the functional model, mainly initializing the input parameters, output parameters and internal states of the functional model and other parameters related to the simulation calculation of the functional model.
2) It indicates the flow of data after the execution of the functional model simulation, and is the premise for the smooth operation of the simulation model.
3) The functional model exposes an executable model solution method, which is the core of the functional model and the concrete realization of modeling and simulation capabilities. Its operating results are only related to the internal state of the functional model and its input parameters.

The control system model similar to the control system block diagram constructed by using the Web graphical control system modeling environment, its corresponding construction result is a directed graph. Therefore, the key point of whether to realize the modeling and simulation of the graphical control system is whether to parse and run the modular directed graph, and this is the function that the parsing runtime layer needs to achieve.

A graph is a collection of nodes and edges that represent various relationships, where nodes are vertices corresponding to objects, edges are connections between objects, and directed graphs are graphs with edges with directions (Fig. 3).

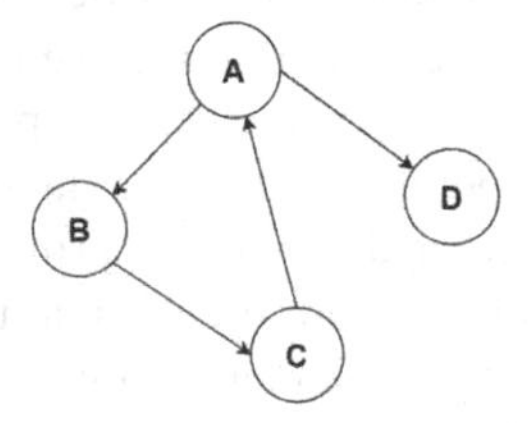

Fig. 3. Directed graph

The control system model similar to the control system block diagram constructed by using the Web graphical control system modeling environment, its construction result is either a simple directed acyclic graph or a complex directed cyclic graph. Therefore, in order to realize the complete parsing of the directed graph of the control system, it is necessary to be able to parse and simulate the control system described by the directed acyclic graph and the control system described by the directed cyclic graph.

Whether it is a directed acyclic graph or a directed cyclic graph, they are essentially directed graphs. A directed acyclic graph can also be regarded as a complex "directed acyclic graph". Therefore, as long as a complete parsing logic for the directed acyclic graph can be realized, the parsing logic of the directed cyclic graph can also be realized.

Before introducing the parsing algorithm, first introduce several concepts related to the parsing algorithm.

1) Node: The primitive model in the graphical simulation model is also the functional model in the real simulation model. It can be regarded as a black box model or black box service. As long as the running parameters (input parameters) required by the model are given, the simulation of the model can be performed and the corresponding running results (output parameters) can be given.
2) Directed edge: The directed connection between the primitive model and the primitive model in the graphical simulation model is also the input-output relationship between the functional model and the functional model in the real simulation model. It describes the flow direction of the running results after the functional model runs, and it is the hub for the combination of functional models to form a complex simulation model.
3) Input port: The input port of the primitive model in the graphical simulation model is also the external input parameter of the functional model in the real simulation model. The running result after the previous node (model) runs will flow into the current node through such ports along the directed edge.
4) Output port: The output port of the primitive model in the graphical simulation model is also the external output parameter of the functional model in the real simulation model. After the node runs, the operation result will flow out from this type of port, and then act on the next nodes along the directed edge.
5) Activate: "Activate" is a concept for input ports. When non-empty data flows into the input port and acts on the external input parameters of the functional model, it means that the input port is activated.
6) Deactivate: "Deactivate" is a concept for input ports. When there is empty data (None) flowing into the input port and then acting on the external input parameters of the functional model, or when the external input parameters of the functional model are set to null, it means that the input port is deactivated.
7) Executable node: The node whose operating parameters (input parameters) required by the node are all satisfied, that is, the node whose input ports are all active.
8) Self-executing node: In the initial state, a node with no input ports or all input ports are active, that is, a node that can be executed in the initial state.
9) Control port: The control port is a special type of input port. When the node is running, the control port will be deactivated, that is, the input port will be set to an empty data (None). Only when the control port is reactivated can the node become an executable node again, so it can be used to truncate the directed ring to ensure that the entry node in the directed ring will not be repeatedly executed or enter an

infinite loop. The design of the control port is the key to process the directed cyclic graph into a directed acyclic graph.
10) Executable queue: A first-in, first-out data structure that stores executable nodes to ensure orderly execution of nodes.

The core processing logic of the directed graph parsing algorithm of the control system mainly includes.

1) The simulation execution of the graphical simulation model takes the self-executing node as the entry of the program.
2) After the node runs, the result data will flow out from the output port, and then act on the next-level node along the directed edge to activate the corresponding input port of the next-level node.
3) When all input ports of a node are activated, it becomes an executable node and is then added to the executable queue to be executed.

Based on the above concepts and processing logic, the pseudo-code for implementing the directed graph parsing algorithm of the control system is as follows (Table 5).

Table 5. The directed graph parsing algorithm of control system

The directed graph parsing algorithm of control system
Input: The **graph_data** corresponding to the graphical simulation model, including the primitive model (**node_data**) and the directed edge model (**edge_data**)
1. Initialize a hash table (**id2node**) of node's ID to functional models, initialize the executable queue (**exec_queue**)
2. for node_data in graph_data:
a. Instantiate the corresponding functional model according to node_data
b. Store the functional model in the hash table id2node
3. for edge_data in graph_data:
a. According to edge_data, bind the input and output relationship of the functional model
4. Traverse the functional model in id2node and add all self-executing nodes to the executable queue (exec_queue)
5. while exec_queue is not empty:
a. Shift the first node from the executable queue (exec_queue)
b. Execute the corresponding node, and then deactivate the node's control ports
c. Apply the result data of node execution to the next-level node along the input-output relationship of the directed edge, and activate the input port corresponding to the next-level node
d. if next-level node is executable:
Add the corresponding next-level node to the executable queue (exec_queue)
Output: Returns the **response results**, **error messages** and **function outputs** generated during the execution of the graphical simulation model

Based on the above-mentioned directed graph analysis algorithm of the control system, a general programming mathematical model can be parsed and executed. Usually, the simulation process of the system model is time-dependent, so the algorithm needs to consider the simulation time. According to the foregoing algorithm description, the cyclic execution of the simulated model can be realized through a directed ring. Therefore, adding a directed ring on the basis of the existing graphical simulation model can also realize the loop iterative execution of the simulation model controlled by the simulation time. However, the implementation based on directed rings complicates the construction of time-dependent simulation models. In order to ensure that the graphical modeling and simulation tool can effectively improve the work efficiency of engineers and technicians, the control of simulation time is added to the above-mentioned directed graph analysis algorithm. The improved algorithm realizes that the time-related simulation parameters are injected into the simulation execution process of the graphical simulation model as public parameters. The overall process is as follows (Table 6):

Table 6. Improved directed graph parsing algorithm of control system

The directed graph parsing algorithm of control system
Input: The **graph_data** corresponding to the graphical simulation model, including the primitive model (**node_data**) and the directed edge model (**edge_data**)
1. Initialize the current time (**cur_time**), the step size of the simulation (**cur_step**), the total duration of the simulation (**total_time**)
2. while cur_time <= total_time:
a. simulation execution
b. update cur_step = new_step
c. update cur_time += cur_step
Output: Returns the **response results**, **error messages** and **function outputs** generated during the execution of the graphical simulation model

3.2 Control System Based on Directed Acyclic Graph Model Description

Based on the directed graph parsing algorithm of the control system in Sect. 2.1, the following uses the web graphical control system modeling environment to build a control system based on the directed acyclic graph description to verify the validity and accuracy of the algorithm.

A system described by a first-order differential equation is called a first-order system. Some control components and simple systems, such as RC networks, generators, air heaters, liquid level control systems, etc., are first-order systems.

The differential equation of the first order system is:

$$T\frac{dc(t)}{dt} + c(t) = r(t) \tag{7}$$

the corresponding system transfer function is:

$$\Phi(s) = \frac{C(s)}{R(s)} = \frac{1}{Ts + 1} \tag{8}$$

in the formula, T is called the time constant.

The main performance indicator of the step response of a first-order system is the adjustment time t_s. The theoretical adjustment time for $\pm5\%$ error band is $3T$, and the theoretical adjustment time for $\pm2\%$ error band is $4T$.

The effectiveness of the graphical modeling simulation tool and its algorithm is verified by the unit step response experiment of the first-order system below.

For a unit step input, there is

$$R(s) = \frac{1}{s} \tag{9}$$

Then, from Eq. (8), we get

$$C(s) = \Phi(s)R(s) = \frac{1}{Ts+1} \cdot \frac{1}{s} \tag{10}$$

And the unit step response of the first-order system can be obtained by doing the inverse pull transformation of $C(s)$.

$$h(t) = L^{-1}[\frac{1}{Ts+1} \cdot \frac{1}{s}] = 1 - e^{-\frac{1}{T}t}, t \geq 0 \tag{11}$$

For the convenience of calculation, the time constant T is taken as 0.5 for the experiment. Therefore, the system transfer function $\Phi(s)$ is:

$$\Phi(s) = \frac{1}{0.5s+1} \tag{12}$$

The time response of the system $h(t)$ is:

$$h(t) = 1 - e^{-2t}, t \geq 0 \tag{13}$$

The above control system is discretized, and the bilinear transformation method is used to process $\Phi(s)$ into a discrete Z-domain transfer function $\Phi(z)$. Based on the bilinear transformation method, we get

$$s = \frac{2}{T}\frac{z-1}{z+1} \tag{14}$$

in the formula, T is the sampling period and also the simulation step size, taking $T = 0.01$, we get

$$\Phi(z) = \frac{0.01 + 0.01z^{-1}}{1.01 - 0.99z^{-1}} \tag{15}$$

Based on the above derivation, the models of the continuous system described by the above S-domain transfer function (Eq. 12), the time-domain unit step response of the system (Eq. 13) and the discrete system described by the Z-domain transfer function (Eq. 15) are constructed in the Web graphical modeling and simulation tool. The construction result is shown in Fig. 4.

In addition, the system described by the corresponding S-domain transfer function was constructed using MATLAB Simulink for comparative experiments, as shown in Fig. 5.

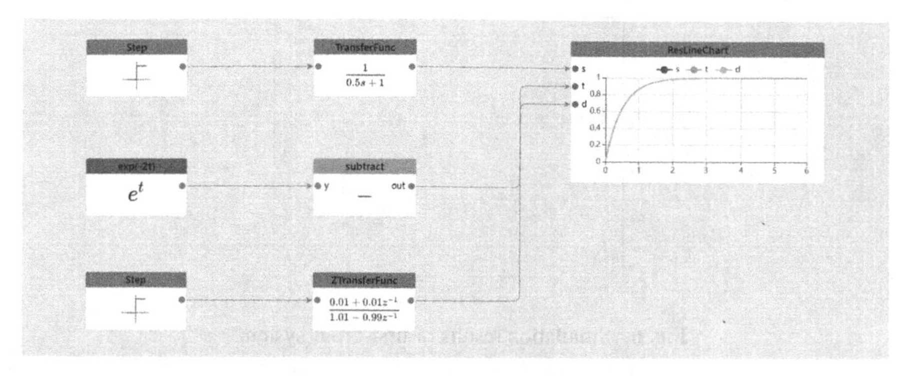

Fig. 4. System block diagram of web environment

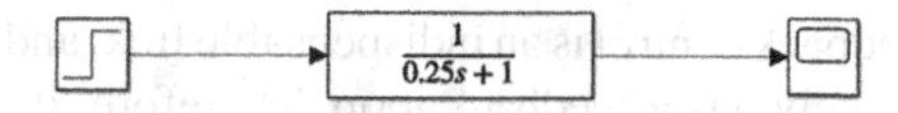

Fig. 5. System block diagram of MATLAB Simulink

The result data of the experiment is shown in the table below, in which the result of the adjustment time is calculated according to the 5% error band (Table 7).

Table 7. First-order system step response

	Adjustment time	Error
Theoretical	1.5 s	0
S-domain	1.5 s	0
Time-domain	1.5 s	0
Z-domain	1.51 s	0.67%
Simulink	1.498 s	0.13%

The result curve of the experiment is shown in Fig. 6.

By using the system model and simulation results of the first-order control system built in the Web graphical modeling environment in different domains, and by comparing the simulation results with MATLAB simulink, it can be found that the algorithm has excellent performance in both precision and accuracy in the simulation execution of the control system described by the directed acyclic graph. The directed graph parsing algorithm of control system can well analyze and simulate the control system described based on the directed acyclic graph.

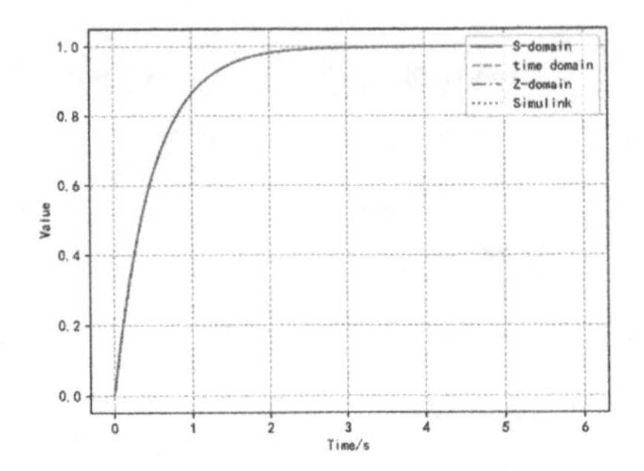

Fig. 6. Simulation results of first-order system

3.3 Control System Based on Directed Cyclic Graph Description

In the control system, feedback control is an indispensable link, and the structure diagram of the control system usually has a feedback loop. Therefore, it is very important that the algorithm can parse the control system described by the directed cyclic graph.

In this section, an experiment on a second-order system with negative feedback will be used to verify the effectiveness of the algorithm in the control system described by the directed cyclic graph.

Given an arbitrary second-order system, its closed-loop transfer function is

$$\Phi(s) = \frac{C(s)}{R(s)} = \frac{1000}{s^2 + 20s + 1000} \tag{16}$$

and the corresponding open-loop transfer function is

$$G(s) = \frac{1000}{s(s + 20)} \tag{17}$$

According to the transfer function of the above second-order system, it is not difficult to obtain that the undamped natural oscillation frequency of the system is $\omega_n = 10\sqrt{10}$ and the damping ratio is $\zeta = 1/\sqrt{10}$. Therefore, the theoretical performance indicators of the unit step response corresponding to the system mainly include:

1) Peak time: $t_p = \frac{\pi}{\omega_n\sqrt{1-\zeta^2}} \approx 0.105\,\text{s}$;
2) Overshoot: $\sigma\% = e^{-\frac{\pi\zeta}{\sqrt{1-\zeta^2}}} \times 100\% \approx 35.092\%$.

Use the Web graphical modeling environment and MATLAB Simulink to build the corresponding system model, as shown in Fig. 7 and Fig. 8 respectively.

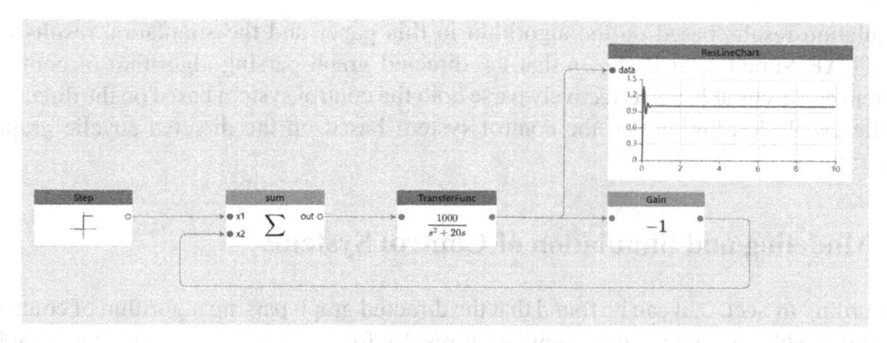

Fig. 7. System block diagram of web environment

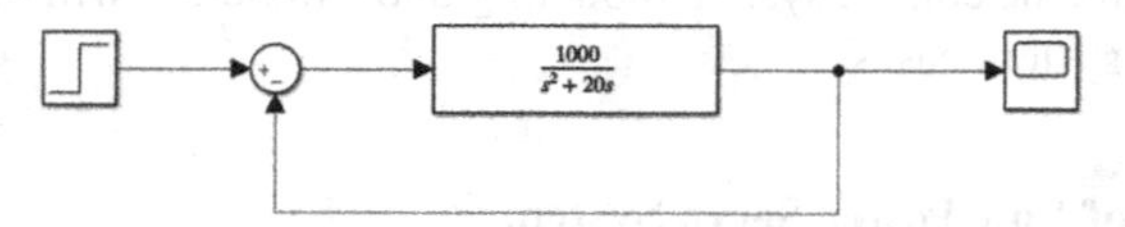

Fig. 8. System block diagram of MATLAB Simulink

The results of the experiment are shown in the table below (Table 8).

Table 8. Simulation result data

	Theoretical value	Exp value	Exp error	Simulink value
t_p	0.105 s	0.104 s	0.95%	0.105 s
$\sigma\%$	35.092%	35.097%	0.01%	32.091%

The experimental result curve is shown in Fig. 9.

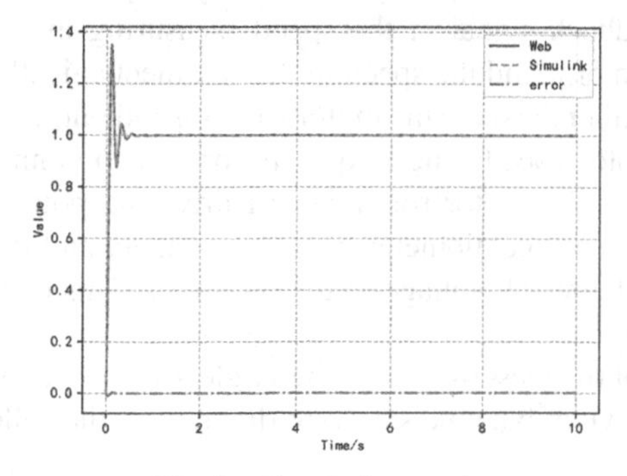

Fig. 9. Simulation result

According to the experimental results in Sects. 2.2 and 2.3, the comparison between the experimental results and the theoretical results, and the comparison between the simulation results based on the algorithm in this paper and the simulation results of MATLAB Simulink, it is shown that the directed graph parsing algorithm of control system can accurately and effectively parse both the control system based on the directed cyclic graph description and the control system based on the directed acyclic graph description.

4 Modeling and Simulation of Control Systems

According to Sect. 2, it can be found that the directed graph parsing algorithm of control system is able to achieve good simulation results for various forms of directed graph control systems.

In the following, the control system modeling and simulation will be verified by two more real control system cases.

4.1 Simulation of Low Power Servo System

The servo system is an automatic control system that reproduces the change of the input value with a certain precision. The servo system occupies a prominent position in the control of the production process and moving objects, as well as in positioning, aiming, tracking, signal transmission and reception and other devices, and has now become an integral part of various automatic adjustment systems. This section will build a low-power servo system by using the Web graphical control system modeling simulation environment to demonstrate the practicability and effectiveness of the directed graph parsing algorithm of control system in the actual control system.

The following figure shows an angle servo system that reproduces the rotation angle, which consists of a potentiometer, an operational amplifier (op-amp), a power amplifier, a torque motor and a tachometer motor.

In the figure, the input signal is the voltage signal of the given potentiometer. After comparing and amplifying with the voltage signal of the feedback potentiometer in the op-amp I, it is input to the op-amp II for secondary amplification, and the op-amp II also amplifies the feedback signal of the speed measuring motor for comprehensive operation. The torque motor and the speed motor are mechanically connected together, and the speed motor will understand the motor rotation and output in the form of voltage proportional to the motor speed. The torque motor is also connected with a position feedback potentiometer, the motor rotation will drive the potentiometer at the same time, and the slider of the potentiometer will also change, and the output voltage will be used as the system feedback voltage to the op-amp I. This constitutes an automatic tracking loop (Fig. 10).

According to the above description of the angle servo system, the servo system is modeled, and the following dynamic structure diagram of the follower system is given (Fig. 11).

The dynamic parameters are $T_1 = 0.01$, $T_2 = 0.05$, $K_0 = 1$, $K_1 = 300$, $K_2 = 1$ and $K_c = 0.08$.

Similarly, we use the Web graphical control system modeling simulation environment and MATLAB Simulink to conduct comparative experiments.

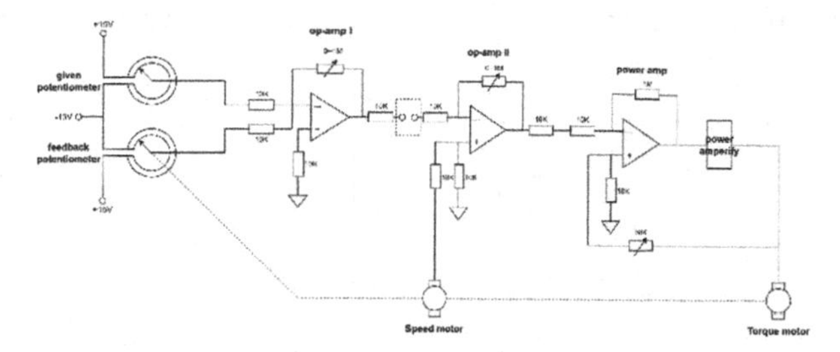

Fig. 10. Servo system

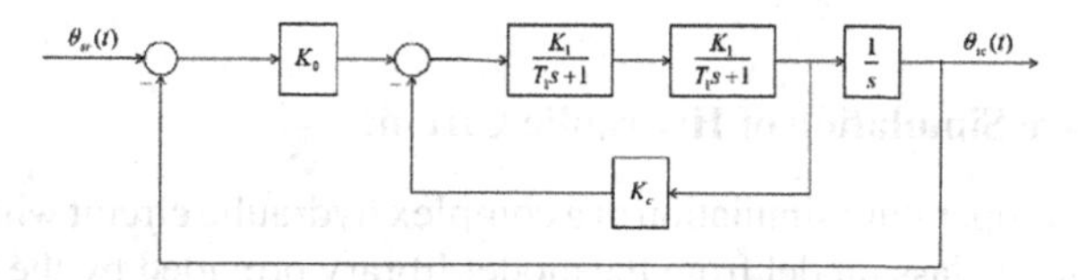

Fig. 11. System structure diagram of servo system

According to the system dynamic diagram, the corresponding system model is constructed as shown in Fig. 12 and Fig. 13.

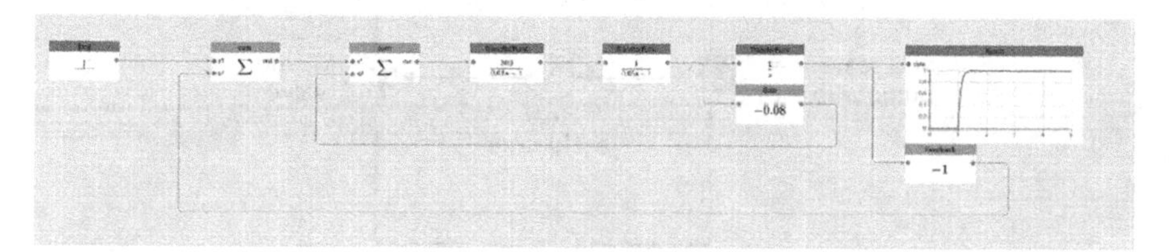

Fig. 12. System block diagram of web environment

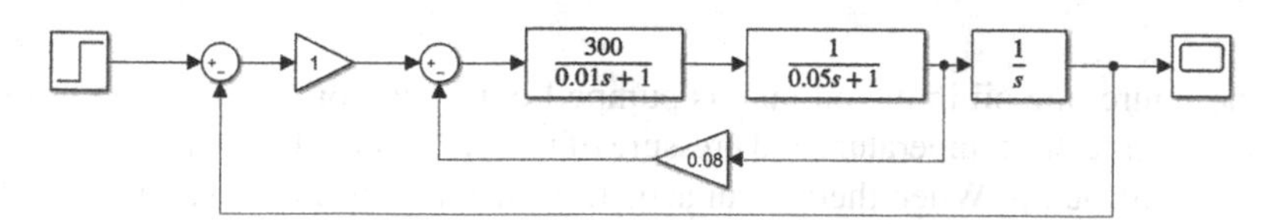

Fig. 13. System block diagram of MATLAB Simulink

In order to test the following effect of the angle servo system, the system input is a step signal of $\theta_{sr}(t) = 1(t-1)$ and a ramp signal of $\theta_{sr}(t) = t$ is used for experiments. Through the experiment, the result curve shown in Fig. 14 is obtained.

The experimental data show that the simulation results of the Web graphical control system modeling simulation environment can quickly and accurately follow the input signal. It is shown that The directed graph parsing algorithm of control system also has excellent performance for the complex directed graph control system with multiple loops like the servo system.

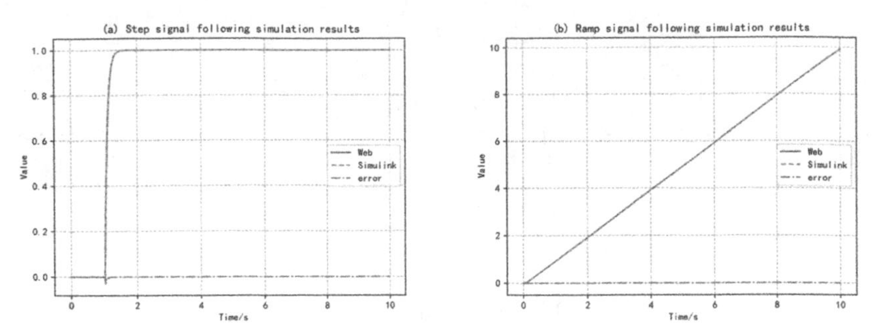

Fig. 14. Simulation results of servo system

4.2 Temperature Simulation of Hydraulic Circuits

In the following, temperature simulation of a complex hydraulic circuit will be performed by using a hydraulic class model from the model library provided by the web graphical modeling environment.

The schematic diagram of the hydraulic circuit is shown below (Fig. 15).

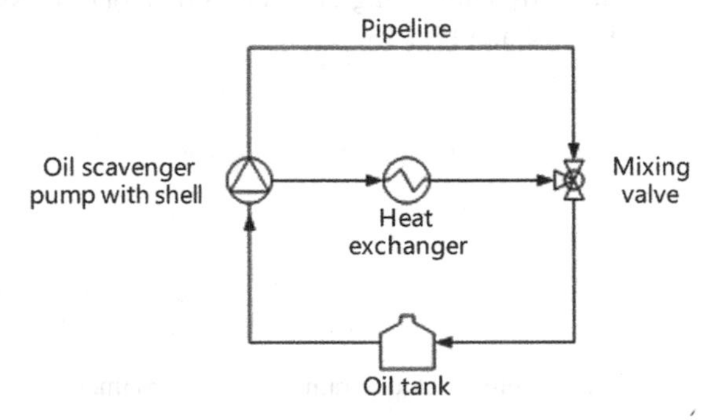

Fig. 15. Schematic diagram of hydraulic circuit

In the figure, the oil in the oil tank is pumped out by the oil scavenger pump with shell. At this time, the temperature and pressure of the hydraulic oil increase after passing through the oil pump. When there is an actuating mechanism in the pipeline, such as a steering gear, etc., the hydraulic oil will push the actuating mechanism such as the steering gear to perform work. For the oil scavenger pump with shell, the hydraulic oil cavity in the pump body has a working cavity and a return oil cavity. Most of the hydraulic fluid will flow out of the working volume and through the pipelines. There is also a small amount of high-temperature and high-pressure hydraulic oil flowing out from the oil scavenger pump with shell, and then cooled by the heat exchanger and merged with the hydraulic oil in the pipeline at the mixing valve, and finally returned to the oil tank. Then repeat the above process.

In the hydraulic circuit, after the hydraulic oil comes out of the oil tank, after passing through the hydraulic pump, the temperature of the hydraulic oil will rise, and the temperature of the hydraulic oil in the return oil chamber of the casing is higher than that of the working chamber. Then, the hydraulic oil in the oil return chamber of the shell is cooled down to a temperature close to the working chamber through the heat exchanger, and finally returns to the oil tank through the mixing valve. Of course, when the system is stable, the temperature of each hydraulic component in the hydraulic system will tend to a certain stable value.

By using the model library provided by the Web graphical modeling environment, the corresponding graphical model is constructed according to the above schematic diagram, as shown in Fig. 16.

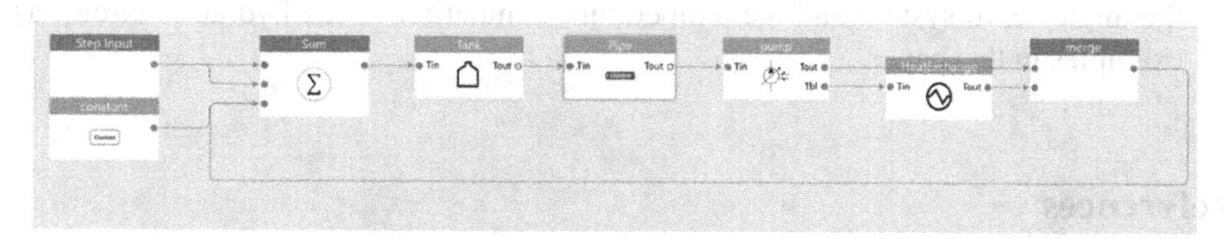

Fig. 16. System block diagram of web environment

After execution, the simulation result diagram shown in Fig. 17 is obtained.

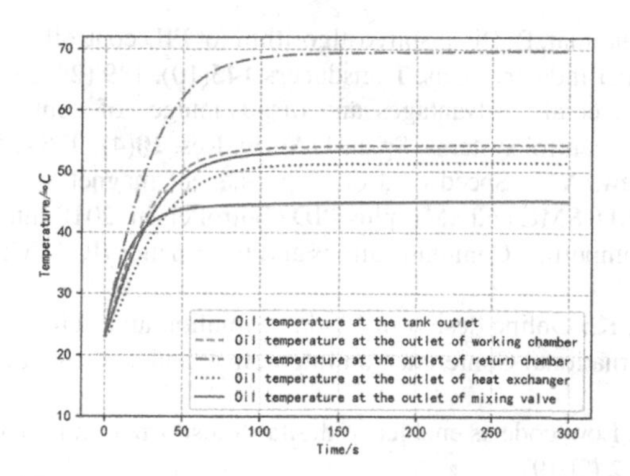

Fig. 17. Simulation results of hydraulic circuit temperature

The simulation results in the above figure are consistent with the theoretical analysis. This shows that the directed graph parsing algorithm of control system can also achieve accurate simulation results under complex professional models.

5 Conclusion

1) Using the development technology that separates the front and back ends, the modeling and simulation construction environment of the graphical control system is implemented based on Web design, and the back-end.
2) Various basic models that are convenient for users to construct and use are designed and implemented in the graphical control system modeling and simulation environment.

3) The directed graph parsing algorithm of control system shows the same excellent effect as MATLAB Simulink in both the control system based on the directed acyclic graph and the control system based on the directed cyclic graph. For example, the error under the analytical simulation of the control system described based on the directed acyclic graph is not greater than 0.2%, and the error under the analytical simulation of the control system described by the directed cyclic graph is not greater than 1%.
4) The directed graph analysis algorithm of the control system can also achieve the expected effect in the application of the actual system, such as the simulation of the angle servo system and the temperature simulation of the hydraulic circuit as examples in this paper.

References

1. Ang, K.H., Chong, G., Li, Y.: PID control system analysis, design, and technology. IEEE Trans. Control Syst. Technol. **13**(4), 559–576 (2005)
2. Ko, J.-S., Huh, J.-H., Kim, J.-C.: Improvement of temperature control performance of thermoelectric dehumidifier used Industry 4.0 by the SF-PI controller. Processes **7**(2), 98 (2019)
3. Nagaraj, B., Vijayakumar, P.: Bio inspired algorithm for PID controller tuning and application to the pulp and paper industry. Sens. Transducers **145**(10), 149 (2012)
4. Duarte-Galvan, C., et al.: Advantages and disadvantages of control theories applied in greenhouse climate control systems. Span. J. Agric. Res. **10**(4), 926–938 (2012)
5. Suman, K., Mathew, A.T.: Speed control of permanent magnet synchronous motor drive system using PI, PID, SMC and SMC plus PID controller. In: 2018 International Conference on Advances in Computing, Communications and Informatics (ICACCI), pp. 543–549. IEEE (2018)
6. Janik, Z., Žáková, K.: Online design of Matlab/Simulink and SciLab/Xcos block schemes. In: 2011 14th International Conference on Interactive Collaborative Learning, pp. 241–247. IEEE (2011)
7. Sanchis, R., et al.: Low-code as enabler of digital transformation in manufacturing industry. Appl. Sci. **10**(1), 12 (2019)
8. Sahay, A., et al.: Supporting the understanding and comparison of low-code development platforms. In: 2020 46th Euromicro Conference on Software Engineering and Advanced Applications (SEAA), pp. 171–178. IEEE (2020)
9. Smith, L.P., Dixon, R.R., Shor, S.W.W.: Modular Modeling System (MMS): a code for the dynamic simulation of fossil and nuclear power plants: overview and general theory. Final Report (1983)
10. Silvennoinen, E., et al.: The APROS software for process simulation and model development. VTT Technical Research Centre of Finland (1989)
11. Chaturvedi, D.K.: Modeling and Simulation of Systems Using MATLAB® and Simulink®. CRC Press, Boca Raton (2017)
12. Nagar, S.: Introduction to Scilab. In: Introduction to Scilab, pp. 1–14. Apress, Berkeley (2017)
13. Srinath, K.R.: Python–the fastest growing programming language. Int. Res. J. Eng. Technol. **4**(12), 354–357 (2017)
14. Nagpal, A., Gabrani, G.: Python for data analytics, scientific and technical applications. In: 2019 Amity International Conference on Artificial Intelligence (AICAI), pp. 140–145. IEEE (2019)

15. Ranjani, J., Sheela, A., Pandi Meena, K.: Combination of NumPy, SciPy and Matplotlib/Pylab-a good alternative methodology to MATLAB-A Comparative analysis. In: 2019 1st International Conference on Innovations in Information and Communication Technology (ICIICT), pp. 1–5. IEEE (2019)
16. Kumar, R.: Future for scientific computing using Python. Int. J. Eng. Technol. Manag. Res. **2**(1), 30–41 (2015)
17. Robbins, J.N.: Learning Web Design: A beginner's Guide to HTML, CSS, JavaScript, and Web Graphics. O'Reilly Media, Inc., New York (2012)

Continuous System/Discrete Event System/Hybrid System/Intelligent System Modeling and Simulation

One-Dimensional Photonic Crystal Filter with Multiple Defect Layers Based on Particle Swarm Optimization

Kaizi Hao[1(✉)], Jian Du[1], Jing Ma[1], Ying Zhang[1], Yiyuan Ma[1], and Chen Wan[2]

[1] Science and Technology on Special System Simulation Laboratory, Beijing Simulation Center, Beijing 100854, China
862562660@qq.com

[2] North University of China, Taiyuan 030051, China

Abstract. In this paper, the defect mode of one-dimensional photonic crystal was studied. For photonic crystal with one defect layer, the effects of refractive index and thickness of defect layer, photon period number and incident angle on defect mode were discussed. Photonic crystal structure with multiple defect layers was constructed, and multiple defect modes were obtained in the forbidden band. Based on the photonic crystal structure with two defect layers, a dual-channel narrow-band filter operating at 480 nm and 532 nm was designed. Particle swarm optimization algorithm was introduced to optimize the position and transmittance of defect modes, which mainly depended on the thickness of defect layer. The calculation results showed that the transmittance of the two defect modes reached 97.52% and 97.55%, respectively, while the full width at half maximum was only 1.0 nm and 1.8 nm. The simulation of the electric field distribution further proved the correctness of the design results. The research results had important value in the field of underwater laser communication.

Keywords: Photonic crystal · Defect mode · Particle swarm optimization · Narrow-band filter

1 Introduction

Underwater acoustic technology was the most commonly used underwater communication method, but it had the shortcomings of long transmission delay, serious signal attenuation and narrow bandwidth, which limited its application ADDIN EN.CITE [1, 2]. Underwater laser communication had been widely concerned for its advantages of low transmission delay, high carrier frequency and wide bandwidth [3, 4]. The attenuation of blue-green light in seawater was very small, so the laser of 450–570 nm bands was used in underwater communication [5]. In order to avoid the interference of stray light and improve the signal-to-noise ratio, a dual-band filter with high transmittance and narrow bandwidth needed to be designed. Photonic crystal (PhC) provided a promising method. It was an artificial material whose refractive index changed periodically.

W. Fan et al. (Eds.): AsiaSim 2022, CCIS 1712, pp. 77–89, 2022.
https://doi.org/10.1007/978-981-19-9198-1_6

Similar to the band structure of semiconductors, electromagnetic wave in the forbidden band could not pass through PhC [6–8]. Compared with two-dimensional PhC and three-dimensional PhC, one-dimensional (1D) PhC was most commonly used because of its simple structure [8–12].

In PhC, if the original periodicity was destroyed by a defect layer, a defect mode with a narrow bandwidth appeared in the forbidden band. The electromagnetic waves at the defect frequency could continue to be transmitted through the PhC, while other frequencies in the forbidden band were cut off [13–16]. The parameters of defect layer affected the performance of defect mode, so a high transmittance and narrow bandwidth could be obtained by adjusting them [14, 17, 18]. The refractive index of defect layer varied with temperature and pressure. By introducing temperature and pressure fields into the dispersion formula, their effects on the defect mode were analyzed [19, 20]. PhC with alternating positive and negative refractive index materials had multiple defect modes, but the use of negative refractive index materials limited its application. Moreover, the number and position of defect mode were difficult to be regulated [21]. In order to ensure that the PhC had good transmission performance, the parameters of defect layer needed to be optimized. A variety of evolutionary algorithm had been applied to PhC structure design and optimization of arbitrarily spaced dual-band selective filters [22]. In addition, genetic algorithm (GA) was used to realize a photonic heterostructure with wide omnidirectional reflection bandwidth [23, 24]. Compared with GA, particle swarm optimization (PSO) was an evolutionary algorithm with fast search speed and high search accuracy, which had been widely used in electromagnetic structure optimization since it was proposed [25–29].

In this paper, 1D PhC with defect mode was studied for underwater laser communication. Three PhC structures with multiple defect layers were constructed to obtain different number of defect modes. The expression of PhC could be expressed as Air$(HL)^m$M$(HL)^n$Substrate, Air$(HL)^q$M1$(HL)^p$M2$(HL)^q$Substrate and Air$(HL)^s$M01$(HL)^t$M02$(HL)^t$M03$(HL)^s$Substrate respectively, corresponding to one, two and three defect modes. Based on the analysis of PhC structure, a dual-band narrow-band filter was designed. PSO was used to optimize the transmission performance of two defect modes. The transmittance at 480 nm and 532 nm reached 97.52% and 97.55% respectively, and the bandwidth was only 1.0 nm and 1.8 nm.

2 Theoretical Formula

The electromagnetic wave passed through the 1D PhC composed of multilayer dielectric materials, its transmission performance could be calculated by transmission matrix method [30, 31]. Each layer medium of PhC could be equivalent to a transmission matrix, which could be written as

$$M_i = \begin{bmatrix} \cos\delta_i & \frac{j}{\eta_i}\sin\delta_i \\ j\eta_i\sin\delta_i & \cos\delta_i \end{bmatrix} \tag{1}$$

In Eq. (1), η_i was the wave impedance of i-th layer medium. For different polarization (TM mode and TE mode), it could be written as $\eta_{TM} = n/\cos\theta_i$ and $\eta_{TE} = n\cos\theta_i$, respectively. θ_i was the incident angle of electromagnetic wave at the interface of i-th layer, and it can be obtained from Snell theorem.

$$n_0 \sin\theta_0 = n_1 \sin\theta_1 = \cdots = n_i \sin\theta_i = \cdots \tag{2}$$

In Eq. (1), δ_i was the phase shift of electromagnetic wave in PhC and could be expressed as

$$\delta_i = \frac{2\pi}{\lambda} n_i d_i \cos\theta_i \tag{3}$$

For the PhC composed of N-layer medium, it could be regarded as a cascade of N transmission matrices. The transmission equation M_{PhC} could be expressed as

$$M_{\text{PhC}} = \prod_{i=1}^{N} M_i = \prod_{i=1}^{N} \begin{bmatrix} m_{11i} & m_{12i} \\ m_{21i} & m_{22i} \end{bmatrix} \tag{4}$$

When the substrate were considered, the transmission matrix could be expressed as

$$M_{\text{total}} = M_{\text{air}} M_{\text{PhC}} M_{\text{s}} = \begin{bmatrix} 1 & 0 \\ 0 & 1 \end{bmatrix} \prod_{i=1}^{N} \begin{bmatrix} m_{11i} & m_{12i} \\ m_{21i} & m_{22i} \end{bmatrix} \begin{bmatrix} 1 \\ \eta_{\text{s}} \end{bmatrix} = \begin{bmatrix} p_1 \\ p_2 \end{bmatrix} \tag{5}$$

In Eq. (5), η_{s} was the impedance of the substrate. The refractive index of the substrate was n_{s}. Assuming $p = p_2/p_1$, the reflectance of PhC to electromagnetic wave could be obtained from

$$R = \frac{\eta_0 - p}{\eta_0 + p} \times conj\left(\frac{\eta_0 - p}{\eta_0 + p}\right) \tag{6}$$

The transmittance could be obtained from

$$T = \frac{4\eta_0 \eta_{\text{s}}}{(\eta_0 p_1 + p_2) conj(\eta_0 p_1 + p_2)} \tag{7}$$

In Eqs. (6) and (7), η_0 was the air impedance. For TM and TE modes, it could be expressed as $1/\cos\theta_i$ and $\cos\theta_i$, respectively. The absorptance could be expressed as

$$A = 1 - R - T \tag{8}$$

Based on the $\lambda/4$ criterion, the dispersion relation of 1D PhC composed of ***H*** and ***L*** could be expressed as [8].

$$\cos k(w)d = \cos\delta_{\text{H}} \cos\delta_{\text{L}} - \frac{1}{2}\left(\frac{n_{\text{H}} \cos\theta_i}{n_{\text{L}} \cos\theta_{i+1}} + \frac{n_{\text{L}} \cos\theta_{i+1}}{n_{\text{H}} \cos\theta_i}\right) \sin\delta_{\text{H}} \sin\delta_{\text{L}} \tag{9}$$

In Eq. (9), w was the angular frequency. δ_{H} and δ_{L} were the phase shifts of light wave in PhC. n_{H} and n_{L} were the refractive index of ***H*** and ***L***. When $\cos\boldsymbol{k}(w)d < 1$, it corresponded to the pass band. When $\cos\boldsymbol{k}(w)d > 1$, it corresponded the forbidden band. The boundary of the forbidden band and the pass band could be calculated by solving the equation $\cos\boldsymbol{k}(w)d = 1$. When the optical thickness of the two materials satisfied the requirements of $n_H d_H = n_L d_L = \lambda_0/4$ (λ_0 was the reference wavelength), the forbidden band was the widest. When the incident angle θ_i was 0°, Eq. (9) could be simplified as

$$\cos^2\delta - \frac{1}{2}\left(\frac{n_H}{n_L} + \frac{n_L}{n_H}\right)\sin^2\delta = 1 \tag{10}$$

In Eq. (10), δ satisfied the condition $\delta = \frac{2\pi}{\lambda} n_H n_H = \frac{2\pi}{\lambda} n_L n_L = \frac{2\pi}{\lambda}\frac{\lambda_0}{4}$. By solving Eq. (10), the forbidden band was obtained from

$$\Delta\lambda = \frac{\lambda_0}{1 - \Delta g} - \frac{\lambda_0}{1 + \Delta g} \tag{11}$$

The forbidden band ranged from $\lambda_0/(1 + \Delta g)$ to $\lambda_0/(1 - \Delta g)$. Δg could be calculated from

$$\Delta g = \frac{2}{\pi}\sin^{-1}\left(\frac{n_H - n_L}{n_H + n_L}\right) \tag{12}$$

When the refractive index of ***H*** and ***L*** were 2.30 and 1.45, the band structure of 1D PhC was simulated as shown in Fig. 1. We could see that the two forbidden bands appeared in the frequency (a/λ, a was lattice constant) range of 0.24–0.32 and 0.80–0.88, respectively. Here, the first forbidden band was mainly considered. According to $\lambda/4$ criterion, the thickness of high and low refractive index materials was $d_H = \lambda_0/4n_H$ and $d_L = \lambda_0/4n_L$ respectively. The forbidden band calculated from Eqs. (11) and (12) was $0.87\lambda_0$–$1.17\lambda_0$. When the reference wavelength λ_0 was 0.48 μm, the forbidden band was in the range of 0.42–0.56 μm.

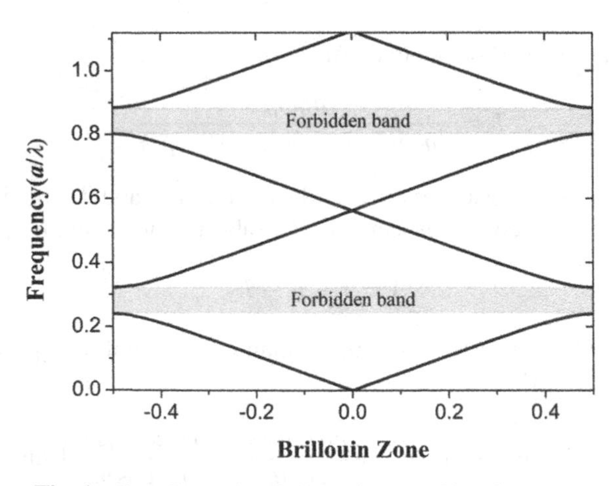

Fig. 1. One-dimensional photonic crystal band structure

3 Discussion on Defect Layer of Photonic Crystal

3.1 Photonic Crystal Structure with One Defect Layer

Figure 2 showed the 1D PhC structure with one defect layer. Its expression could be written as Air(HL)^{m}M(HL)^{n}Substrate. ***H*** and ***L*** were high and low refractive index materials, and ***M*** was defect layer. *m* and *n* were the left and right photon period number of the defect layer, respectively. In the calculation, the refractive index of the substrate was taken as 1.52.

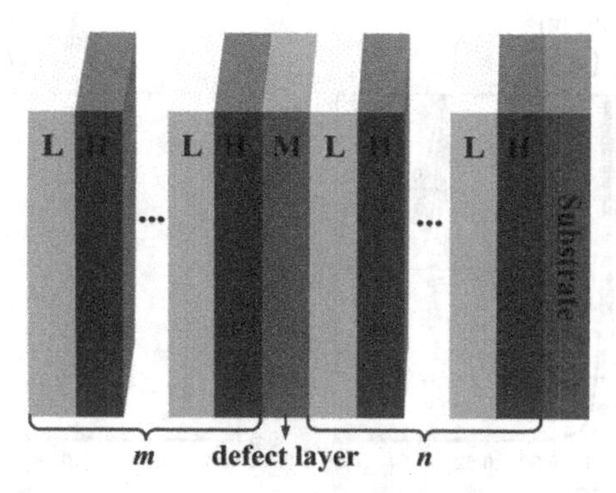

Fig. 2. The structure of one-dimensional photonic crystal with one defect layer

To analyze the properties of defect mode, the number of photon period *m*, *n* was set to 6 and the refractive index of defect layer was 1.38. As shown in Fig. 3a, when the thickness of the defect layer was 50.7 nm, 86.9 nm, 123.2 nm, the defect mode appeared at 444 nm, 480 nm and 516 nm, respectively. With the increase of the thickness of the defect layer, the position of the defect mode moved to the long wavelength direction.

Figure 3b showed the defect mode with different refractive index (n_M) of defect layer. The thickness of the defect layer satisfied condition $n_M d_M = \lambda_0/4$. When n_M was 1.38, the transmittance reached 98.74%, while when n_M was 4.00, the transmittance decreased to 31.69%. With the increase of n_M, the transmittance of defect mode decreased.

Figure 3c showed the defect mode with symmetry layout when the number of photon period was 4, 5, 6 and 7. When the number of photon period (*m* and *n*) increased from 4 to 7, the full width at half maximum of defect mode decreased from 7.2 nm to 0.4 nm. The bandwidth of defect mode became narrower with the increase of photon period number. So the bandwidth of defect mode could be adjusted by changing the number of photon period.

Figure 3d showed the defect modes of TE polarization at incident angles of 0°, 15°, 30° and 45°. With the increase of incident angle, the defect mode moved to the short wavelength direction.

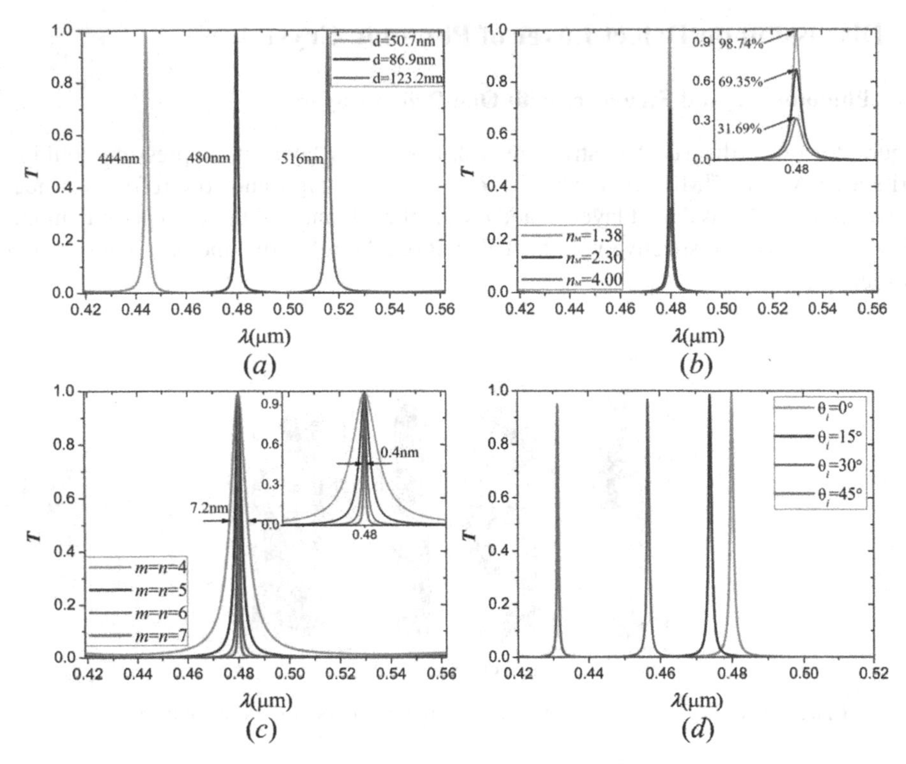

Fig. 3. Defect mode with different thickness, refractive index, photon period number and incident angle

3.2 Photonic Crystal Structure with Two Defect Layers

From the above analysis, we could draw a conclusion that the position, transmittance and bandwidth of single defect mode could be regulated by adjusting the parameters of PhC. However, the PhC structure with one defect layer had only one defect mode. In order to obtain multiple defect modes, a new PhC structure needed to be studied. Figure 4 showed a 1D PhC structure with two defect layers $\boldsymbol{M_1}$ and $\boldsymbol{M_2}$. The number of photon period between the two defect layers was *p*, and the other two sides were *q*. The expression could be written as $\text{Air(HL)}^q\text{M}_1\text{(HL)}^p\text{M}_2\text{(HL)}^q\text{Substrate}$.

In the PhC structure shown in Fig. 4, the thickness of ***H*** and ***L*** satisfied the condition of $n_\text{H}d_\text{H} = \lambda_0/4$ and $n_\text{L}d_\text{L} = \lambda_0/4$, respectively. To analyze the properties of the two defect modes, the thickness of the defect layer was set to $d_\text{M} = \lambda_1/4\ nM$, λ_1 was reference wave of defect layer. Its refractive index was 1.38. Figure 5a showed the defect mode with different photon period number *p* (1, 2 and 4). There were two defect modes in the forbidden band, and they gradually approached with the increase of *p*. Therefore, the distance between two defect modes could be regulated by adjusting *p*. Figure 5b showed defect mode distribution with different reference wavelength λ_1. Different λ_1 corresponded to different thickness of defect layer. We could see that the two defect modes moved to the long wavelength direction with the increase of the thickness of the defect layer. Therefore, the position of the two defect modes could be regulated by adjusting λ_1.

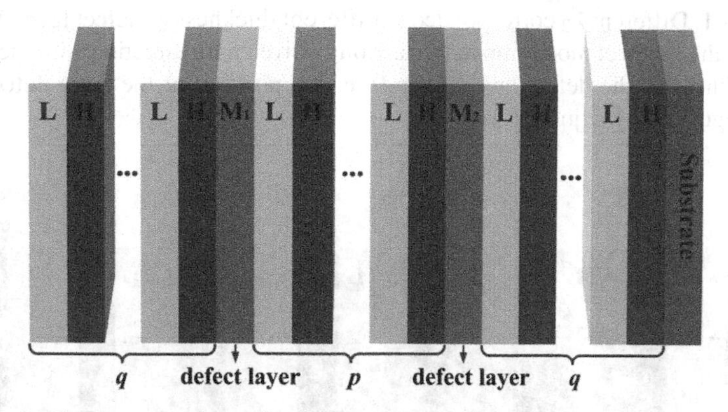

Fig. 4. The structure of one-dimensional photonic crystal with two defect layers

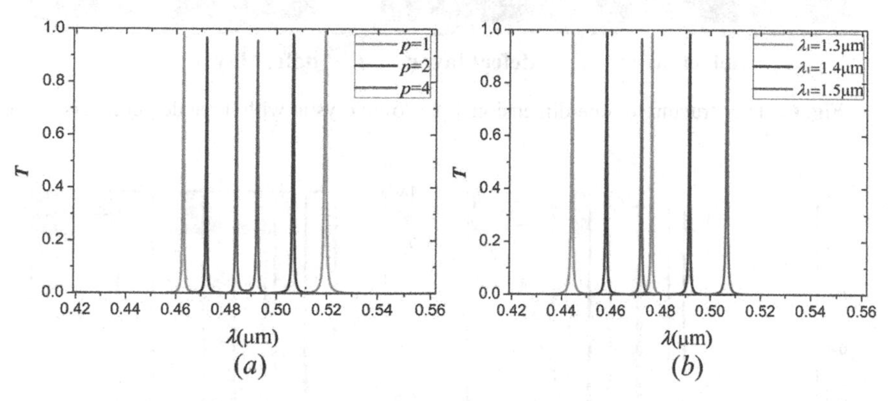

Fig. 5. Defect mode with different photon period number p and reference wavelength λ_1

3.3 Photonic Crystal Structure with Three Defect Layers

Similarly, in order to obtain three defect modes, a PhC structure with three defect layers needed to be designed. Figure 6 showed the PhC structure with three defect layers $\boldsymbol{M_{01}}$, $\boldsymbol{M_{02}}$ and $\boldsymbol{M_{03}}$. The number of photon period on the inner side was t and the other side was s. The expression could be written as Air$(HL)^s M_{01}(HL)^t M_{02}(HL)^t M_{03}(HL)^s$Substrate. The thickness of $\boldsymbol{H}$ and $\boldsymbol{L}$ satisfied the condition of $n_H d_H = \lambda_0/4$ and $n_L d_L = \lambda_0/4$, respectively. The thickness of the defect layer was set to $d_M = \lambda_2/4\ nM$, λ_2 was reference wave of defect layer. Its refractive index was 1.38.

Figure 7a showed the defect mode with different photon period number t (1, 2 and 3). In the calculation, the reference wavelength λ_2 of the defect layer was 2.5 μm. Three defect modes were obtained in the forbidden band, and the outer defect modes gradually approached the middle defect mode with the increase of t. The position of the middle defect mode was almost unchanged. Figure 7b showed defect mode distribution with different reference wavelength λ_2. In the calculation, the number of photon period s was 5, and t was 1. Different λ_2 corresponded to different thickness of defect layer. We could see that the three defect modes moved to the long wavelength direction with the increase of the thickness of the defect layer. Therefore, the position of the three defect modes could be regulated by adjusting λ_2.

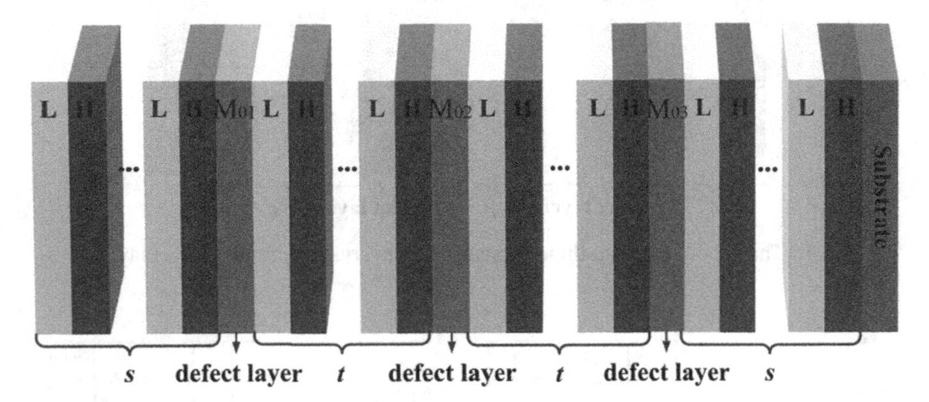

Fig. 6. The structure of one-dimensional photonic crystal with three defect layers

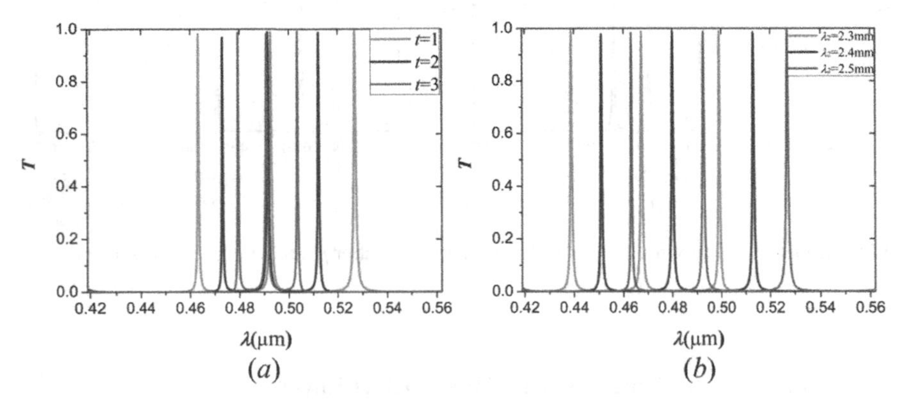

Fig. 7. Defect mode with different photon period number t and reference wavelength λ_2

4 Photonic Crystal Filter Design

4.1 Particle Swarm Optimization Algorithm

In order to change the position and transmittance of the defect mode, it was necessary to select the appropriate thickness of the defect layer. Therefore, it could be regarded as an objective optimization problem. PSO algorithm was introduced to solve this problem.

The PSO randomly initialized a population in the search space, and searched for the optimal value by controlling the flight position and speed of population particles. The position of each particle was represented by a vector $x_i = (x_{i1}, x_{i2}, \cdots, x_{iD})$, $i = 1, 2, \cdots, N_p$, where D was the dimensionality of the search space and N_p was the number of particles in the population. Particle had a flight velocity, represented by $v_i = (v_{i1}, v_{i2}, \cdots, v_{iD})$, $i = 1, 2, \cdots, N_p$. Each particle corresponded to the fitness value determined by the objective function [26, 27]. The personal and global best positions of the particle were recorded as p_{id} and p_{gd}, respectively. The particle updated its velocity and position according to the following formula Eq. (13) and Eq. (14).

$$v_{id} = w_0 v_{id} + c_1 r_1 (p_{id} - x_{id}) + c_2 r_2 (p_{gd} - x_{id}) \tag{13}$$

$$x_{id} = x_{id} + v_{id} \tag{14}$$

In Eq. (13), w_0 represented inertia weight. c_1, c_2 were acceleration constants. r_1, r_2 were random numbers uniformly distributed in [0, 1]. The flight velocity v_{id} of particle should be limited to [−1, 1]. x_{id} was the position of particle in D-dimensional search space. The inertia weight was set to be varied with iteration.

$$w_0 = w_{start} - (w_{start} - w_{end}) \cdot (iter/ger) \tag{15}$$

where w_{start} and w_{end} were the initial and final inertia weights, respectively. ger represented the maximum numbers of iteration. $iter$ was a counter for recording the iterative number.

Based on the suggested values in previous works and our many attempts [26, 32], we had chosen the parameters used in the PSO algorithm as follows: the population size N_p was 50; the maximum iteration number ger was 200; the inertia weight w_0 decreased linearly from 1.2 to 0.3 during the iteration; the acceleration constant of c_1 and c_2 was 2. The number of variables (the thickness of defect layers) in the optimization problem was equal to the dimension D of search space.

In order to obtain good transmission performance at all defect modes, the objective function could be set to

$$f = \max\left(\frac{T_1 + T_2 + \cdots + T_D}{D}\right) \tag{16}$$

In Eq. (16), T_D was the transmittance at the defect mode, respectively. The objective function represented the average transmittance of all defect modes.

4.2 Dual-Channel Narrow-Band Filter Design

High performance filter could be designed by using the characteristic of 1D PhC with defect layer. In this section, a dual-channel narrow-band filter with high transmittance at 480 nm and 532 nm was designed based on PSO algorithm. TiO_2 and SiO_2 were selected as high and low refractive index materials, respectively. They had refractive index of 2.30 and 1.45 in the visible band. The defect layer material used MgF_2 and had refractive index of 1.38. The number of photon period q and p were 5 and 1, respectively. The expression of

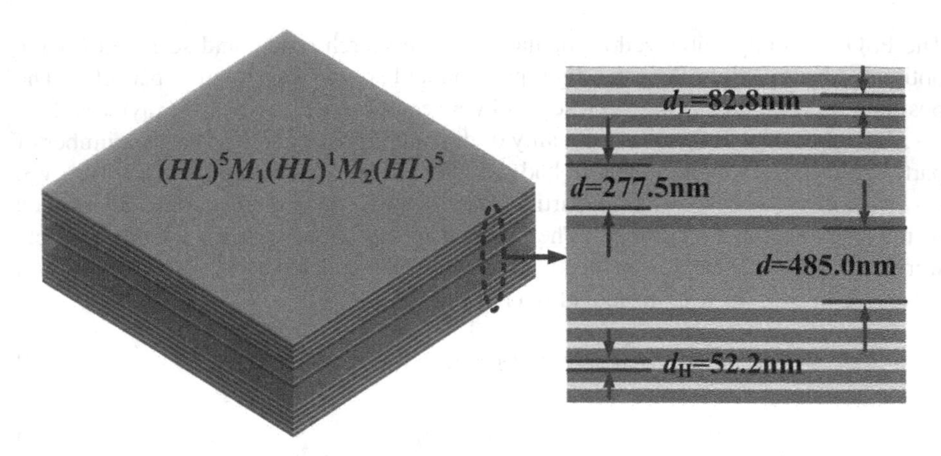

Fig. 8. The structure of dual-channel narrow-band filter

PhC could be written as Air$(TiO_2/SiO_2)^5MgF_2(TiO_2/SiO_2)^1MgF_2(TiO_2/SiO_2)^5$Glass. As shown in Fig. 8, the thickness of TiO_2 was 52.2 nm and the thickness of SiO_2 was 82.8 nm.

The thickness of defect layer $\boldsymbol{M_1}$ and $\boldsymbol{M_2}$ satisfied $n_{M1}d_{M1} = \lambda_{01}/4$ and $n_{M2}d_{M2'} = \lambda_{02}/4$, where λ_{01} and λ_{02} were regard as variable to be optimized.

$$\begin{cases} T_1 = f(\lambda_{01}, \lambda_{02}) \\ T_2 = f(\lambda_{01}, \lambda_{02}) \end{cases} \tag{17}$$

The objective function could be written as

$$f = \max(\frac{T_1 + T_2}{2}) \tag{18}$$

The iteration process of PSO was recorded as shown in the Fig. 9a. The optimal variable λ_{01} and λ_{02} of 1.532 μm and 2.677 μm were obtained at the end of the iteration. The objective function value was 0.975. The thickness of two defect layers could be calculated, which were 277.5 nm and 485.0 nm respectively. Figure 9b showed the transmittance of the designed dual-channel narrow-band filter. Two defect modes in the forbidden band of 0.42–0.56 μm were obtained. The transmittance at 480 nm and 532 nm reached 97.52% and 97.55%, respectively. The full width at half maximum of the two defect modes were only 1.0 nm and 1.8 nm, respectively.

Figure 10 showed the band structure of dual-channel narrow-band filter. Two defect modes appeared in the range of forbidden band. The defect mode shape was close to straight line, which indicated that the filter had a narrow bandwidth. In order to further verify the performance of the dual-channel narrow-band filter, the electric field distribution of PhC at 480 nm and 532 nm was simulated. As showed in Fig. 11, the electric field could pass through the PhC, which further proved the correctness of the designed high performance filter.

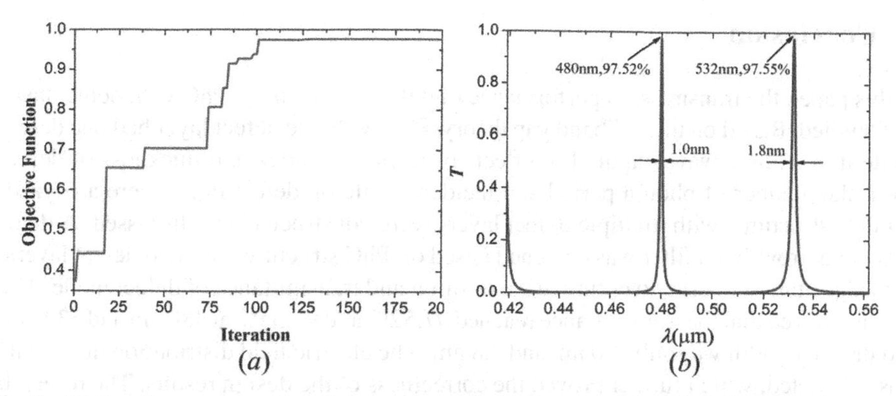

Fig. 9. (a) Iterative process of particle swarm optimization algorithm. (b) The transmittance of the designed dual-channel narrow-band filter

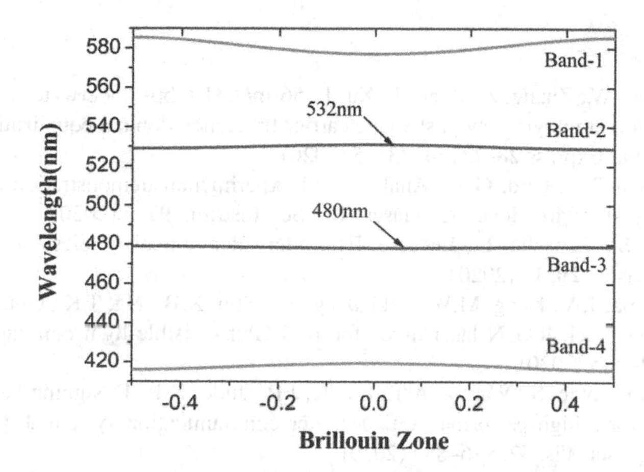

Fig. 10. Band structure of dual-channel narrow-band filter with two defect modes

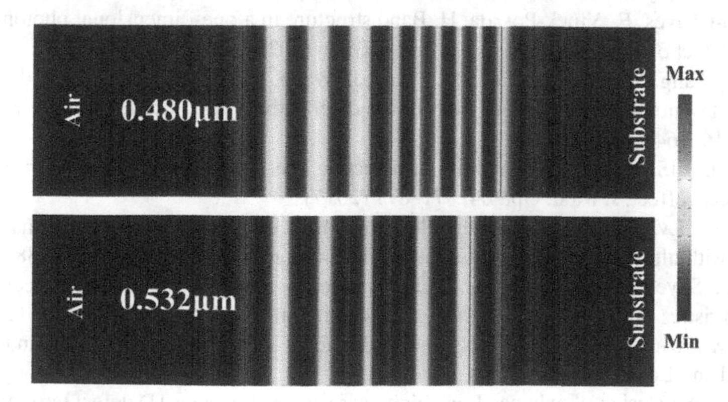

Fig. 11. Electric field distribution of incident wave at 480 nm and 532 nm

5 Conclusion

In this paper, the transmission performance of defect mode in 1D PhC with defect layer was studied. Based on the λ/4 band gap theory, PhC with one defect layer had one defect mode in reference wavelength. The effects of refractive index and thickness of defect layer, the number of photon period and incident angle on defect mode were analyzed. 1D PhC structures with multiple defect layers were constructed and discussed. A dual-channel narrow-band filter was designed based on PhC structure with two defect layers. PSO algorithm was used to optimize the position and transmittance of defect mode. The results showed that the transmittance reached 97.52% and 97.55% at 480 nm and 532 nm, and the bandwidth was only 1.0 nm and 1.8 nm. The electric field distribution in the PhC was calculated, which further proved the correctness of the design results. The research results had important significance in the field of underwater laser communication.

References

1. Chen, X., Lyu, W., Zhang, Z., Zhao, J., Xu, J.: 56-m/3.31-Gbps underwater wireless optical communication employing Nyquist single carrier frequency domain equalization with noise prediction. Opt. Express **28**, 23784–23795 (2020)
2. Hou, D., Chen, J.Y., Guo, G.K.: Analysis and experimental demonstration of underwater frequency transfer with diode green laser. Rev. Sci. Instrum. **91**, 5 (2020)
3. Spagnolo, G.S., Cozzella, L., Leccese, F.: Underwater optical wireless communications: overview. Sensors **20**, 14 (2020)
4. Holguin-Lerma, J.A., Kong, M.W., Alkhazragi, O., Sun, X.B., Ng, T.K., Ooi, B.S.: 480-nm distributed-feedback InGaN laser diode for 10.5-Gbit/s visible-light communication. Opt. Lett. **45**, 742–745 (2020)
5. Avramov-Zamurovic, S., Watnik, A.T., Lindle, J.R., Judd, K.P.: Designing laser beams carrying OAM for a high-performance underwater communication system. J. Opt. Soc. Am. A-Opt. Image Sci. Vis. **37**, 876–887 (2020)
6. Lee, K.J., Wu, J.W., Kim, K.: Defect modes in a one-dimensional photonic crystal with a chiral defect layer. Opt. Mater. Express **4**, 2542–2550 (2014)
7. Segovia-Chaves, F., Vinck-Posada, H.:Band structure in a one-dimensional photonic crystal with a defect of Ga1−xAlxAs. Optik **205**, 163996 (2020)
8. Hao, K., Wang, X., Zhou, L., Yang, S., Zhang, J., Wang, Y., Li, Z.: Design of one-dimensional composite photonic crystal with high infrared reflectivity and low microwave reflectivity. Optik **216**, 164794 (2020)
9. Aly, A.H., ElSayed, H.A.: Tunability of defective one-dimensional photonic crystals based on Faraday effect. J. Mod. Opt. **64**, 871–877 (2017)
10. Zhang, W., Lv, D.: "Preparation and characterization of Si/SiO_2 one-dimensional photonic crystal with ultra-low infrared emissivity in the 3–5 μm band. Optik **202**, 163738 (2020)
11. Aly, A., Sayed, F., Elsayed, H.: Defect mode tunability based on the electro-optical characteristics of the onedimensional graphene photonic crystals. Appl. Opt. **59** (2020)
12. Wang, X., Lou, S., Xing, Z.: Loss characteristic of hollow core photonic bandgap fiber. Infrared and Laser Eng. **48** (2019)
13. Banerjee, A.: Design of enhanced sensitivity gas sensors by using 1D defect ternary photonic band gap structures. Indian J. Phys. **94**, 535–539 (2020)
14. Segovia-Chaves, F., Vinck-Posada, H.:One-dimensional photonic crystal with coupled InSb defects. Optik **203**, 164018 (2020)

15. Singh, P., Thapa, K.B., Singh, S.K., Gupta, A.K.: Study of design tunable optical sensor and monochromatic filter of the one-dimensional periodic structure of TiO2/MgF2 with defect layer of Liquid Crystal (LC) sandwiched with two silver layers. Plasmonics (2020)
16. Segovia-Chaves, F., Vinck-Posada, H., Gómez, E.A.:Superconducting one-dimensional photonic crystal with coupled semiconductor defects. Optik **209**, 164572 (2020)
17. Xiang, Y.J., Dai, X., Wen, S., Fan, D.: Properties of omnidirectional gap and defect mode of one-dimensional photonic crystal containing indefinite metamaterials with a hyperbolic dispersion. J. Appl. Phys. **102**, 093107–093107 (2007)
18. Singh, P., Nautiyal, V.K., Janma, R., Thep, K.B.: Theoretical investigation of enhanced sensing property in 1D TiO2/SiO2 periodic layers containing a defect layer of the nanocomposite with different radii of silver nanoparticles in the host liquid crystal. Phys. Scr. **95**, 11 (2020)
19. Segovia-Chaves, F., Vinck-Posada, H.: Dependence of the transmittance spectrum on temperature and thickness of superconducting defects coupled in dielectric one-dimensional photonic crystals. Optik **170**, 384–390 (2018)
20. Segovia-Chaves, F., Vinck-Posada, H.: Effects of pressure and thickness on the transmittance spectrum in a PS/PMMA photonic crystal. Optik **183**, 918–923 (2019)
21. Zhang, W., Fang, Q., Cheng, Y., Liu, J., Xia, G.: Narrow band interleaver based on one-dimensional photonic crystal with positive-negative index alternant multilayer. Acta Optica Sinica **27**, 1695–1699 (2007)
22. Chaker, H., Badaoui, H., Abri, M., Benadla, I.: Efficient synthesis of dual-band selective filters using evolutionary methods in a 1D photonic crystal slab for near-infrared applications. J. Comput. Electron. **19**, 353–358 (2020)
23. Jiang, L., Zheng, G., Shi, L., Yuan, J., Li, X.: Broad omnidirectional reflectors design using genetic algorithm. Opt. Commun. **281**, 4882–4888 (2008)
24. Qiang, H., Jiang, L., Li, X.: Design of broad omnidirectional total reflectors based on one-dimensional dielectric and magnetic photonic crystals. Opt. Laser Technol. **42**, 105–109 (2010)
25. Yang, Y., Zhang, T., Yi, W., Kong, L., Li, X., Wang, B., Yang, X.: Deployment of multistatic radar system using multi-objective particle swarm optimisation. IET Radar, Sonar Navig. **12**, 485–493 (2018)
26. Hao, K., Wang, X., Yang, S., Zhang, J., Li, Z.: Complex permittivity inversion algorithm with adaptive learning strategy and parameter balancing mechanism. Optik **223**, 165402 (2020)
27. Hao, K., Li, Z., Wang, X., Yang, S., Wang, Y., Xu, C., Zhou, L., Gao, Y.: Design of dual-band wide-angle RF/IR beam combiner based on impedance matching. IET Microwaves, Antennas & Propagation **14**, 7–14 (2020)
28. He, S., Liu, Q., Sa, T., Wang, Z.:Design of broadband reflector at the visible wavelengths using particle swarm optimization. AIP Adv. **9**, 075301 (2019)
29. Safdari, M.J., Mirjalili, S.M., Bianucci, P., Zhang, X.: Multi-objective optimization framework for designing photonic crystal sensors. Appl. Opt. **57**, 1950–1957 (2018)
30. Abadla, M., Tabaza, N., Tabaza, W., Ramanujam, N., Wilson, K.S., Dhasarathan, V., Taya, S.: Properties of ternary photonic crystal consisting of dielectric/plasma/dielectric as a lattice period. Optik **185**, 784–793 (2019)
31. Rahmani, Z., Rezaee, N.: The reflection and absorption characteristics of one-dimensional ternary plasma photonic crystals irradiated by TE and TM waves. Optik **184**, 134–141 (2019)
32. Xu, W., Duan, B.Y., Li, P., Hu, N., Qiu, Y.: Multiobjective particle swarm optimization of boresight error and transmission loss for airborne radomes. IEEE Trans. Antennas Propag. **62**, 5880–5885 (2014)

Linear Constant Discrete System Based Evaluation of Equipment System-of-Systems

Chen Dong(✉), Shu He, Zhi-feng Lu, Jun-nan Du, and Peng Lai

Shanghai Institute of Electromechanical Engineering, Shanghai 201109, China
dongchenhit@163.com

Abstract. To meet the demand that analyzing the impact of architecture and equipment performance on capability and cost of equipment system-of-systems, an evaluation method is investigated based on linear constant discrete system. Model of equipment system-of-systems is established using state space equation of linear time invariant discrete system, which describes equipment composition, connection relation, operation process and equipment performance of the equipment system-of-systems. Operation characteristic of the equipment system-of-systems is analyzed using state motion equation of linear time invariant discrete system. Then, evaluation models for adaptation capability, survivability capability, task capability and response capability are established, as well as a cost evaluation model for equipment system-of-systems are given. The method proposed in this paper is demonstrated with a hypothetical air defense equipment system-of-systems. Evaluation results of the above capabilities and cost of the air defense equipment system-of-systems are derived. These results can support optimization of the equipment system-of-systems architecture and equipment performance.

Keywords: Equipment system-of-systems · Capability evaluation · Architecture · Equipment performance · Linear constant discrete system

1 Introduction

With the development of military science and technology, weapon equipment system-of-systems presents characteristics of networking and informatization. The relation between equipments is becoming more and more complex. Traditional evaluation methods for equipment system-of-systems are based on tree structure index systems. These methods do not fully consider the impact of the relation among equipments on the capability of equipment system-of-systems. These methods are lack of effectiveness [1].

In view of the above problems, modeling and evaluation methods for equipment system-of-systems based on complex network are widely investigated and applied in various domains [2–4]. Main idea of these methods is to abstract the equipment in equipment system-of-systems as node, and the relation among equipments as edge. Then, the network model of equipment system-of-systems is established. Base on this model, complex network analysis method [5–7] can be used to analyze and evaluate the equipment system-of-systems. The analysis methods include safety analysis methods [8,

W. Fan et al. (Eds.): AsiaSim 2022, CCIS 1712, pp. 90–101, 2022.
https://doi.org/10.1007/978-981-19-9198-1_7

9], adaptability analysis methods [10], robustness and invulnerability analysis methods [11–15], task capability analysis methods [1, 16], etc.

At present, modeling and evaluation methods based on complex network focus more on topological properties of equipment system-of-systems, but pay less attention to operation process of an equipment system-of-systems. Capability of the equipment system-of-systems is determined by the relation among equipments and the performance of the equipment [17]. So, it is necessary to consider both the equipment performance and the topological properties of the equipment system-of-systems when carrying out evaluation of equipment system-of-systems.

In this paper, an evaluation method for equipment system-of-systems is investigated based on linear constant discrete system. A model of equipment system-of-systems is established using a state space equation of linear time invariant discrete system, which describes equipment composition, connection relation, operation process and equipment performance of the equipment system-of-systems. And operation characteristic of the equipment system-of-systems is analyzed using a state motion equation of linear time invariant discrete system. Then, evaluation models for adaptation capability, survivability capability, task capability and response capability are established, as well as a cost evaluation model for equipment system-of-systems is given. A hypothetical air defense equipment system-of-systems is taken as an example to demonstrated the method proposed in this paper. Evaluation results of the above capabilities and cost of the air defense equipment system-of-systems are derived. These results verify the effectiveness of the method proposed in this paper, which can support optimization of the equipment system-of-systems architecture and equipment performance.

2 Modeling and Analysis of Equipment System-of-Systems

2.1 Fundamental

Definition of OODA chain. Referring to the "observe, orient, decide, act" (OODA) activity model [18], equipments of equipment system-of-systems perform observe, orient, decide, act and other activities in a specific order to jointly complete specific tasks. Such a chain composed of several equipments and operating orderly is called an OODA chain.

Equipment system-of-systems may contain multiple OODA chains. The collection of all the OODA chains present equipment composition, connection relation and operation process of the equipment system-of-systems. Equipment performance affects operation effect of an OODA chain, such as operating time and probability, and ultimately affects capabilities of the equipment system-of-systems. It should be noted that activities and equipments are not one-to-one correspondence. One equipment may perform multiple activities. Similarly, one activity may be jointly performed by several equipments in some order.

2.2 Modeling of Equipment System-of-Systems

Equipment system-of-systems are modeled based on a linear constant discrete system. Equipments in the equipment system-of-systems are abstracted as system states, and relation among equipments is abstracted as a state matrix of the linear constant discrete system. There is

$$\begin{aligned} \boldsymbol{x}(k+1) &= \boldsymbol{A}\boldsymbol{x}(k) + \boldsymbol{B}u(k) \\ y(k) &= \boldsymbol{C}\boldsymbol{x}(k) \\ k &= 0, 1, 2, \cdots \end{aligned} \tag{1}$$

In Eq. 1, k is a discrete time variable. $\boldsymbol{x}$ is a n-dimensional state vector. u is an input variable. y is an output variable. $\boldsymbol{A}$ is a $n \times n$-dimensional system matrix. $\boldsymbol{B}$ is a $n \times 1$-dimensional input matrix. $\boldsymbol{C}$ is a $1 \times n$-dimensional output matrix. By default, the first $n - 1$ states of $\boldsymbol{x}$ correspond to all the equipments constituting the equipment system-of-systems, and the nth state is a termination state.

The matrix $\boldsymbol{A}$ can describe the OODA chain contained in the equipment system-of-systems. Mark the element in the ith row and the jth column of $\boldsymbol{A}$ as a_{ij}, when $a_{ij} \neq 0$, it represents that the jth equipment is in an OODA chain and has a connection relation with the ith equipment. The operation process is from the jth equipment to the ith equipment. Otherwise $a_{ij} = 0$. When the value of non-zero a_{ij} is taken as 1, it only presents the structure of the OODA chain. The non-zero a_{ij} can also take the value related to the performance of the jth equipment to present the impact of equipment performances on operation effect of the OODA chain. An OODA chain is driven by u, which acts on the state corresponding to the equipment at the beginning of the OODA chain. The state corresponding to the equipment at the end of an OODA chain points to the termination state, which is presented by y.

2.3 Analysis of Equipment System-of-Systems

Based on Eq. 1, there is

$$\boldsymbol{x}(k) = \boldsymbol{A}^k \boldsymbol{x}(0) + \sum_{i=0}^{k-1} \boldsymbol{A}^{k-i-1} \boldsymbol{B}u(i) \tag{2}$$

$$y(k) = \boldsymbol{C}[\boldsymbol{A}^k \boldsymbol{x}(0) + \sum_{i=0}^{k-1} \boldsymbol{A}^{k-i-1} \boldsymbol{B}u(i)] \tag{3}$$

Let the initial state vector $\boldsymbol{x}(0)$ be the zero vector. And let the input $u(k)$ be the unit pulse signal, which is

$$u(k) = \begin{cases} 1, & k = 0 \\ 0, & k > 0 \end{cases} \tag{4}$$

Bring Eq. 4 into Eq. 3, there is

$$y(k) = \boldsymbol{C}\boldsymbol{A}^{k-1}\boldsymbol{B},\ k > 0 \tag{5}$$

The k and y satisfying $y(k) > 0$ can be solved based on Eq. 5. When the non-zero element of $\boldsymbol{A}$ is marked as a_{ij}, the sum of all $y(k) > 0$ is

$$\sum y(k) = \sum_{i=1}^{n_c} (\prod a_{ij}) \tag{6}$$

In Eq. 6, n_c is the number of OODA chains included in the equipment system-of-systems. Each $\prod a_{ij}$ represents the equipments involved in an OODA chain, as well as the connection relation and operation process of the equipments. If a_{ij} represents a certain type of performance parameters related to the cooperation between the jth equipment and the ith equipment, $\prod a_{ij}$ is the product of such performance parameters.

In particular, when the non-zero element a_{ij} has the following form,

$$a_{ij} = \sigma^{b_{ij}} \tag{7}$$

$\sigma > 0$ is a constant, the sum of all $y(k) > 0$ is

$$\sum y(k) = \sum_{i=1}^{n_c} \sigma^{\sum b_{ij}} \tag{8}$$

If b_{ij} represents a certain type of performance parameters related to the cooperation between the jth equipment and the ith equipment, $\sum b_{ij}$ is the sum of such performance parameters.

In particular, when the non-zero elements $a_{ij} = 1$, there is

$$n_c = \sum y(k) \tag{9}$$

Each $y(k) > 0$ represents that there are y OODA chains operating at the end with the discrete-time variable k. And the number of equipments involved in each OODA chain is $n_e = k - 1$.

3 Evaluation Method and Models

Based on the analysis of operation characteristics of an equipment system-of-systems, evaluation methods and models of adaptation capability, survivability capability, task capability, response capability and cost of an equipment system-of-systems are investigated in this section.

3.1 Capability Evaluation

Adaptation Capability. Adaptation capability is the capability of an equipment system-of-systems to adapt to changes. When the equipment system-of-systems contains more OODA chains, the equipment system-of-systems has more ways to perform a task, its adaptation capability is stronger. Adaptation capability index is I_a. When a_{ij} takes 0 or 1, there is $I_a = n_c$ according to Eq. 9.

Survivability Capability. Survivability capability is the capability of an equipment system-of-systems to withstand damage. Under the condition of ensuring the existence of an OODA chain, the greater minimum number of equipment loss or information link interruption that the equipment system-of-systems can withstand, the stronger the survivability capability of the equipment system-of-systems. Survivability capability index is I_r, which is divided into two cases:

Case 1 (Equipment Loss). Search j ($1 \le j < n$) for a_{ij} that minimizes I_a when all $a_{ij} = 0$ ($1 \le i \le n$). If the $I_a = 0$, it means that the loss of one equipment may lead to the loss of all OODA chains of the equipment system-of-systems, then $I_r = 0$. If the $I_a > 0$, based on the previous step, search the remaining j for a_{ij} that minimizes I_a when all $a_{ij} = 0$, until $I_a = 0$. Set the number of iteration as n_s, then $I_r = n_s - 1$. The pseudo code is as follows:

```
Begin
ns=1
While
   Search j (1 ≤ j < n) for aij that minimizes Ia when all aij = 0 (1 ≤ i ≤ n)
   If Ia=0 Then
      Ir=ns-1
      Break
   Else
      Take all aij=0
   EndIf
EndWhile
End
```

Case 2 (Information Link Interruption). Search i ($1 \le i \le n$) and j ($1 \le j < n$) for a_{ij} that minimizes I_a when $a_{ij} = 0$. If the $I_a = 0$, it means that the interruption of one information link may lead to the loss of all OODA chains of the equipment system-of-systems, then $I_r = 0$. If the $I_a > 0$, based on the previous step, search the remaining i and j for a_{ij} that minimizes I_a when $a_{ij} = 0$, until $I_a = 0$. Set the number of iteration as n_s, then $I_r = n_s - 1$. The pseudo code is as follows:

```
Begin
ns=1
While
   Search i (1 ≤ i ≤ n) and j (1 ≤ j < n) for aij that minimizes Ia when aij = 0
   If Ia=0 Then
      Ir=ns-1
      Break
   Else
      Take all aij=0
   EndIf
EndWhile
End
```

Task Capability. Task capability is the probability of the equipment system-of-systems completing its tasks. The higher the average probability of all OODA chains operating completion, the stronger the task capability. Task capability index is I_q. Set $n \times n$-dimensional matrix $\boldsymbol{Q}$, whose element in the ith row and the jth column is q_{ij}. Q_{ij} is the probability that the jth ($j < n$) equipment successfully executes the corresponding activity and turns to the ith ($i < n$) equipment to start executing the next activity or to the termination state ($i = n$). Set $n \times n$-dimensional matrix $\boldsymbol{A}_q = \boldsymbol{A} * \boldsymbol{Q}$. The operator "*" represents the multiplication of the same row and column elements of two matrices with the same dimension. According to Eq. 5 and Eq. 6, there are

$$y_q(k) = \boldsymbol{C}\boldsymbol{A}_q^{k-1}\boldsymbol{B} \tag{10}$$

$$\sum y_q(k) = \sum_{m=1}^{n_c} (\prod q_{ij}) = \sum_{m=1}^{n_c} q_m \tag{11}$$

In Eq. 11, q_m is the probability of successful operating of the mth OODA chain. There is

$$I_q = \frac{1}{n_c} \sum y_q(k) \tag{12}$$

Response Capability. Response capability is the timeliness of the equipment system-of-systems to complete its tasks. The shorter the average time of running all the OODA chain, the stronger the response capability. Response capability index is I_t. Set $n \times n$-dimensional matrix $\boldsymbol{P}$, whose element in the ith row and the jth column is p_{ij}. There is

$$p_{ij} = \begin{cases} \sigma^{t_{ij}}, 1 \le i \le n, 1 \le j < n \\ 0, \quad j = n \end{cases} \tag{13}$$

In Eq. 13, $\sigma > 0$ is a constant. T_{ij} is the time-consuming for the jth equipment to perform corresponding activities. Set $n \times n$-dimensional matrix $\boldsymbol{A}_p = \boldsymbol{A} * \boldsymbol{P}$. According to Eq. 5 and Eq. 8, there are,

$$y_p(k) = \boldsymbol{C}\boldsymbol{A}_p^{k-1}\boldsymbol{B} \tag{14}$$

$$\sum y_p(k) = \sum_{m=1}^{n_c} \sigma^{\sum t_{ij}} = \sum_{m=1}^{n_c} \sigma^{t_m} \tag{15}$$

In Eq. 15, t_m is the operating time of the mth OODA chain. Set n_c different σ, and form the following equations

$$\begin{cases} \sum y_p(k, \sigma_1) = \sum\limits_{m=1}^{n_c} \sigma_1^{t_m} \\ \sum y_p(k, \sigma_2) = \sum\limits_{k=1}^{n_c} \sigma_2^{t_m} \\ \vdots \\ \sum y_p(k, \sigma_{n_c}) = \sum\limits_{m=1}^{n_c} \sigma_{n_c}^{t_m} \end{cases} \tag{16}$$

t_m ($1 \le m \le n_c$) can be solved through Eq. 16. There is

$$I_t = \frac{1}{n_c} \sum_{m=1}^{n_c} t_m \tag{17}$$

3.2 Cost Evaluation

Adding equipment with different performance in the equipment system-of-systems, as well as changing connection relation among equipments, will have an impact on the overall cost of the equipment system-of-systems. A cost evaluation model for the equipment system-of-systems is established. Cost index of the equipment system-of-system is I_c, there is

$$I_c = \boldsymbol{I}_n^{\mathrm{T}}(\boldsymbol{F} * \boldsymbol{A})\boldsymbol{I}_n \tag{18}$$

In Eq. 18, $\boldsymbol{I}_n$ is a n-dimensional unit vector. $\boldsymbol{F}$ is a $n \times n$-dimensional equipment cost matrix, whose element in the ith row and the jth column is f_{ij}. F_{ij} is the cost of adding the jth equipment to the equipment system-of-systems and establishing a connection relation with the ith equipment.

4 Evaluation Examples

Taking a hypothetical air defense equipment system-of-systems as an example, evaluation models for adaptation capability, survivability capability, task capability, response capability and cost of the air defense equipment system-of-systems are established. Based on the models, evaluation results are derived in this section.

4.1 Model of Air Defense Equipment System-of-Systems

The air defense equipment system-of-systems consists of four types of equipment: search equipment, command equipment, guidance equipment and interception equipment. The search equipment includes S1 and S2. The command equipment includes C1 and C2. The guidance equipment includes G1 and G2. And the interception equipment includes F1 and F2.

A general form of an OODA chain is "search equipment → command equipment → guidance equipment → interception equipment". The search equipment is the starting equipment of the OODA chain, and the interception equipment is the terminal equipment of the OODA chain. The equipment composition, connection relation and operation process of the air defense equipment system-of-systems in centralized, independent and network-centric operation modes are as shown in Fig. 1.

In Fig. 1, a solid line circle represents an equipment, a dotted line circle represents a termination state, and an arrow represents the connection relation and operation process between two equipments or between an equipment and the termination state.

Using the evaluation methods proposed in this paper, the models of the air defense equipment system-of-systems in centralized, independent and network-centric operation modes are established respectively. In these models (see Eq. 1), $\boldsymbol{x}$ is an 9-dimensional state vector. It's states corresponds to S1, S2, C1, C2, G1, G2, F1, F2 and T. $u(k)$ is the unit pulse signal. $y(k)$ takes the termination state.

In models of the air defense equipment system-of-systems, $\boldsymbol{B}$ and $\boldsymbol{C}$ are

$$\boldsymbol{B} = \begin{bmatrix} 1 & 1 & 0 & 0 & 0 & 0 & 0 & 0 & 0 \end{bmatrix}^{\mathrm{T}} \tag{19}$$

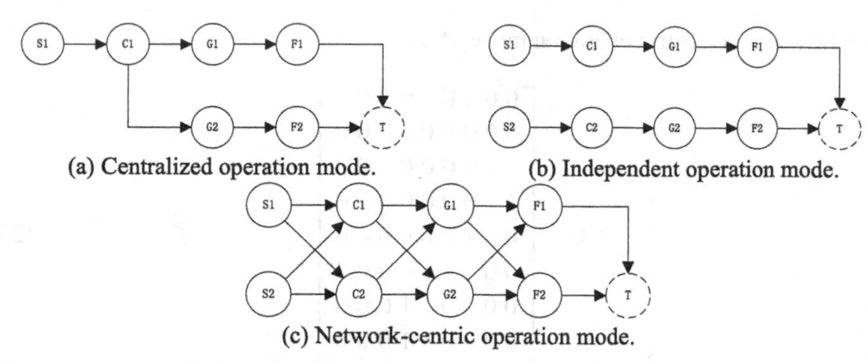

(a) Centralized operation mode. (b) Independent operation mode.

(c) Network-centric operation mode.

Fig. 1. Equipment composition, connection relation and operation process of the air defense equipment systems in three different operation modes.

$$\boldsymbol{C} = \begin{bmatrix} 0 & 0 & 0 & 0 & 0 & 0 & 0 & 0 & 1 \end{bmatrix} \tag{20}$$

In the centralized operation mode, $\boldsymbol{A}$ is

$$\boldsymbol{A} = \begin{bmatrix} 0&0&0&0&0&0&0&0&0 \\ 0&0&0&0&0&0&0&0&0 \\ 1&0&0&0&0&0&0&0&0 \\ 0&0&0&0&0&0&0&0&0 \\ 0&0&1&0&0&0&0&0&0 \\ 0&0&1&0&0&0&0&0&0 \\ 0&0&0&0&1&0&0&0&0 \\ 0&0&0&0&0&1&0&0&0 \\ 0&0&0&0&0&0&1&1&0 \end{bmatrix} \tag{21}$$

In the independent operation mode, $\boldsymbol{A}$ is

$$\boldsymbol{A} = \begin{bmatrix} 0&0&0&0&0&0&0&0&0 \\ 0&0&0&0&0&0&0&0&0 \\ 1&0&0&0&0&0&0&0&0 \\ 0&1&0&0&0&0&0&0&0 \\ 0&0&1&0&0&0&0&0&0 \\ 0&0&0&1&0&0&0&0&0 \\ 0&0&0&0&1&0&0&0&0 \\ 0&0&0&0&0&1&0&0&0 \\ 0&0&0&0&0&0&1&1&0 \end{bmatrix} \tag{22}$$

In the network-centric operation mode, $\boldsymbol{A}$ is

$$\boldsymbol{A} = \begin{bmatrix} 0\,0\,0\,0\,0\,0\,0\,0\,0 \\ 0\,0\,0\,0\,0\,0\,0\,0\,0 \\ 1\,1\,0\,0\,0\,0\,0\,0\,0 \\ 1\,1\,0\,0\,0\,0\,0\,0\,0 \\ 0\,0\,1\,1\,0\,0\,0\,0\,0 \\ 0\,0\,1\,1\,0\,0\,0\,0\,0 \\ 0\,0\,0\,0\,1\,1\,0\,0\,0 \\ 0\,0\,0\,0\,1\,1\,0\,0\,0 \\ 0\,0\,0\,0\,0\,0\,1\,1\,0 \end{bmatrix} \tag{23}$$

The matrixes describing equipment performance and cost are

$$\boldsymbol{Q} = \begin{bmatrix} 0 & 0 & 0 & 0 & 0 & 0 & 0 & 0 & 0 \\ 0 & 0 & 0 & 0 & 0 & 0 & 0 & 0 & 0 \\ 0.5 & 0.6 & 0 & 0 & 0 & 0 & 0 & 0 & 0 \\ 0.8 & 0.7 & 0 & 0 & 0 & 0 & 0 & 0 & 0 \\ 0 & 0 & 0.8 & 0.8 & 0 & 0 & 0 & 0 & 0 \\ 0 & 0 & 0.7 & 0.9 & 0 & 0 & 0 & 0 & 0 \\ 0 & 0 & 0 & 0 & 0.8 & 0.7 & 0 & 0 & 0 \\ 0 & 0 & 0 & 0 & 0.5 & 0.6 & 0 & 0 & 0 \\ 0 & 0 & 0 & 0 & 0 & 0 & 0.8 & 0.7 & 0 \end{bmatrix} \tag{24}$$

$$\boldsymbol{P} = \begin{bmatrix} 0 & 0 & 0 & 0 & 0 & 0 & 0 & 0 & 0 \\ 0 & 0 & 0 & 0 & 0 & 0 & 0 & 0 & 0 \\ \sigma^5 & \sigma^3 & 0 & 0 & 0 & 0 & 0 & 0 & 0 \\ \sigma^3 & \sigma^4 & 0 & 0 & 0 & 0 & 0 & 0 & 0 \\ 0 & 0 & \sigma^8 & \sigma^7 & 0 & 0 & 0 & 0 & 0 \\ 0 & 0 & \sigma^5 & \sigma^4 & 0 & 0 & 0 & 0 & 0 \\ 0 & 0 & 0 & 0 & \sigma^4 & \sigma^5 & 0 & 0 & 0 \\ 0 & 0 & 0 & 0 & \sigma^6 & \sigma^8 & 0 & 0 & 0 \\ 0 & 0 & 0 & 0 & 0 & 0 & \sigma^3 & \sigma^5 & 0 \end{bmatrix} \tag{25}$$

$$\boldsymbol{F} = \begin{bmatrix} 0 & 0 & 0 & 0 & 0 & 0 & 0 & 0 & 0 \\ 0 & 0 & 0 & 0 & 0 & 0 & 0 & 0 & 0 \\ 0.37 & 0.51 & 0 & 0 & 0 & 0 & 0 & 0 & 0 \\ 0.68 & 0.56 & 0 & 0 & 0 & 0 & 0 & 0 & 0 \\ 0 & 0 & 0.48 & 0.52 & 0 & 0 & 0 & 0 & 0 \\ 0 & 0 & 0.52 & 0.63 & 0 & 0 & 0 & 0 & 0 \\ 0 & 0 & 0 & 0 & 0.64 & 0.52 & 0 & 0 & 0 \\ 0 & 0 & 0 & 0 & 0.35 & 0.36 & 0 & 0 & 0 \\ 0 & 0 & 0 & 0 & 0 & 0 & 0.68 & 0.52 & 0 \end{bmatrix} \tag{26}$$

4.2 Evaluation Results of Air Defense Equipment System-of-Systems

The following Table 1 gives evaluation results of the air defense equipment system-of-systems. The evaluation of survivability capability considers the case of equipment loss, and the constant σ in Eq. 16 are

$$\sigma_1 = 0.5$$

$$\sigma_i = \sigma_{i-1} + \frac{1}{2^i},\ i > 1 \tag{27}$$

t_m ($1 \leq m \leq n_c$) in Eq. 16 are solved by an improved genetic algorithm.

Table 1. Evaluation results of the air defense equipment systems.

Operation mode	Adaptation capability	Survivability capability	Task capability	Response capability	Cost
Centralized mode	2	0	0.2105	21.50	3.57
Independent mode	2	1	0.2603	20.50	4.24
Network-centric mode	16	1	0.2586	19.44	7.34

In terms of adaptation capability, the air defense equipment system-of-systems includes 2 OODA chains in both centralized and independent operation modes. In the network-centric operation mode, due to the significant increase in the complexity of the correlation between various equipment, the number of OODA chains has increased to 16, so that the equipment system-of-systems has more ways to complete tasks.

In terms of survivability capability, the number of OODA chains of the equipment system-of-systems in centralized operation mode will be reduced to 0, as long as any equipment in S1 and C1 is destroyed. So, the number of equipment losses that can be tolerated is 0, and the survivability in centralized operation mode is 0. In independent operation mode and network-centric operation mode, each type of equipment includes 2 equipments. It is necessary to destroy at least 2 equipments of the same type to reduce the number of OODA chains to 0. The number of equipment losses that can be tolerated is 1, and the survivability in independent mode and network-centric operation mode is both 0.

In terms of task capability and response capability, the average probability and average time of OODA chain operation are calculated according to equipment performance in the three operation modes. In independent operation mode and network-centric operation modes, the average probability of OODA chain operation is higher than that in the centralized operation mode, and the average time is shorter than that in the centralized operation mode, due to adding new equipment connection relation and changing operation process.

In terms of cost, from centralized operation mode to independent operation mode, and then to network-centric operation mode, the number of equipment and the connection between equipment have been gradually increased. So, the cost of the equipment system-of-systems is gradually rising.

5 Conclusions

A linear constant discrete system based evaluation method for equipment system-of-systems is investigated in this paper. Based on state space equation of linear constant discrete system, model of equipment system-of-systems is established to describe the equipment composition, connection relation, operation process and equipment performance. Operation characteristic of equipment system-of-systems is analyzed using state motion equation of linear constant discrete system. Then, evaluation models of adaptation capability, survivability capability, task capability and response capability, as well as cost of the equipment system-of-systems are proposed. Taking a hypothetical air defense equipment system-of-systems as an example, the evaluation method of equipment system-of-systems are demonstrated. Evaluation results are obtained. These results can provide decision support for optimization of the equipment system-of-systems architecture and equipment performance.

References

1. Liang, J.L., Xiong, W.: Capabilities assessment of the weapon system based on combat ring. Syst. Eng. Electron. **41**(8), 1810–1819 (2019)
2. Li, K., Wu, W.: Research status of weapon equipment system-of-systems based on complex network. J. Acad. Armored Force Eng. **30**(4), 7–13 (2016)
3. Wang, W., Shan, L.P., Wu, K.: Analysis of ability of combating with formation aircrafts of aerial defense system based on theory of complex networks. Command. Control. Simul. **34**(6), 4–8 (2012)
4. Huang, S.C., Zhou, Y.Y., Wei, G.: Operational effectiveness analysis of anti-TBM system with space-based information support. Syst. Eng. Electron. **31**(10), 2414–2417 (2009)
5. Watts, D.J., Strogatz, S.H.: Collective dynamics of 'small-world' networks. Nature **393**(6684), 440–442 (1998)
6. Barabasi, A.L., Albert, R.: Emergence of scaling in random networks. Science **286**(5439), 509–512 (1999)
7. Latora, V., Marchiori, M.: Efficient behavior of small-world networks. Phys. Rev. Lett. **87**(19), 1–4 (2001)
8. Jiang H Q, Gao J M, Gao Z Y, et al. Safety analysis of process industry system based on complex networks theory. In: Proceedings of 2007 IEEE International Conference on Mechatronics and Automation, pp. 480–484. IEEE, Harbin, China (2007)
9. Zhang, W.X.: A Weapon System of Systems Safety Analysis Method Based on Complex Interaction Networks. National University of Defense Technology, Changsha, China (2015)
10. Xiao, B., Liu, F.Z., Qin, Y.S.: Research on the structure adaptability of early warning intelligence system-of systems based on complex network. Mod. Def. Technol. **50**(2), 1–10 (2022)
11. Crucitti, P., Latora, V., Marchiori, M., et al.: Efficiency of scale-free networks: error and attack tolerance. Phys. A: Stat. Mech. Its Appl. **320**, 622–642 (2003)
12. Herrmann, H.J., Schneider, C.M., Moreira, A.A., et al.: Onion-like network topology enhances robustness against malicious attacks. J. Stat. Mech: Theory Exp. **1**, 1–9 (2011)
13. He, S., Yang, K.W., Liang, J.: Research on contribution of single equipment to weapon system-of-systems based on network invulnerability. Fire Control. Command. Control. **42**(8), 87–96 (2017)

14. Wang, H.Y., Wu, W., Wei, Y.Y.: Weapon system-of-systems invulnerability analysis based on super network model. Syst. Eng. Electron. **39**(8), 1782–1787 (2017)
15. Wang, Z., Li, J.H., Kang, D.: Robustness of two-layer heterogeneous interdependent network model for networked information system of system. Syst. Eng. Electron. **43**(4), 961–969 (2021)
16. Zhao, Q.S., Shang, H.L., Zhang, X.K., et al.: Capability analysis for weapon system of systems based on network. Fire Control. Command. Control. **42**(6), 17–21 (2017)
17. Li, M.H.: Research on Capability Generation Oriented Weapon Network Portfolio Optimization. National University of Defense Technology, Changsha, China (2018)
18. Révay, M., Líška, M.: OODA loop in command & control systems. In: Proceedings of 2017 Communication and Information Technologies (KIT), pp. 1–4. IEEE, Vysoke Tatry, Slovakia (2017)

Online Identification of Gaussian-Process State-Space Model with Missing Observations

Xiaonan Li, Ping Ma, Tao Chao(✉), and Ming Yang

Control and Simulation Center, Harbin Institute of Technology, Harbin 150080, China
chaotao2000@163.com

Abstract. When the state-space model is black-box, it is difficult to identify the system based on the input and observation. In predictive control and other fields, the state and model need to be updated in real time, and online identification becomes very important. However, compared with offline learning, online learning of black-box model is more difficult. This paper proposes an online Bayesian inference and learning method for state-space models with missing observations. When the state-space model is black-box, we expressed it as basis function expansions. Through the connection to the Gaussian processes (GPs), the state and basis function coefficients are updated online. The problems of missing observations caused by the sensor failure are often encountered in practical engineering and are taken into consideration in this paper. In order to keep the online algorithm from being interrupted by missing observations, we update the states and unknown parameters according to whether the observation is missing at the current time. This conservative strategy makes the online learning continuous when the observation is missing, and makes full use of the available statistics in the past. Numerical examples show that the proposed method is robust to missing data and can make full use of the available observations.

Keywords: Online Bayesian learning · State-space model · System identification · Basis function · Missing observations

1 Introduction

The purpose of system identification is to learn mathematical models from data [1, 2]. Markov Chain Monte Carlo (MCMC) method is often used to infer the unknown parameters of state-space model. In addition, Expectation Maximization (EM) algorithm is also used to system identification [3, 4]. However, when likelihood is intractable, Variational Inference [5–7] are commonly used to infer the unknown parameters. On the other hand, learning becomes time-consuming when the model is partially unknown. Kullberg uses an extended Kalman filter method to jointly estimate the state and learn the partially unknown models [8], via the basis function expansion. However, when the model is black-box, learning and inference will be difficult. Tobar presents an offline learning method based on basis function expansion and learns the coefficients of basis function through the MCMC method [9].

W. Fan et al. (Eds.): AsiaSim 2022, CCIS 1712, pp. 102–114, 2022.
https://doi.org/10.1007/978-981-19-9198-1_8

In order to avoid the problem of overfitting, Svensson proposes a flexible state-space model to learn the black-box model and estimate the state [10]. The unknown black-box state-space space model is modeled as a basis function expansion. The prior of the coefficients is inspired by GPs [11]. In other words, the unknown model is modeled as a nonlinear state-space model with regularized coefficients of basis function. The learning methods for the black-box model are often offline. In practical engineering, such as state filtering, predictive control and other fields, it is necessary to update the state and state-space model in time. However, online learning and state filtering of black-box model are difficult. Berntrop improves the offline learning method of the state-space model proposed by Svensson [12]. The online Bayesian inference and learning method of state-space model is presented based on basis function expansion. Online inference of unknown parameters is based on Bayes theorem and inspiration from GPs. The statistical information is updated online and constitutes the key parameters of the distribution function of unknown parameters. The states are updated by particle filter.

However, the observations will fail occasionally in practical engineering when sensor fails. The missing observation will interrupt the online learning algorithm, which makes it difficult for the learning method to use all the available information. Gopaluni presents a system identification method with missing observations [13, 14]. This method is off-line and aims to solve the problem of system identification when there are uncertain parameters and missing observations in the state-space model. When the state-space model black-box and the observation is missing, system identification and state filtering are more difficult. In order to learn the black-box model online and make the learning method robust to missing observations, we improve the methods proposed by Svensson [10] and Berntrop [12] and propose an online learning method for state-space model with missing observations. The proposed method makes the online learning continue even if the observation is missing. In addition, the method makes full use of the available observation information. Numerical examples show that the proposed method is robust to missing observations.

2 Problem Formulation

2.1 State-Space Model

When the process noise and measurement noise are Gaussian distribution, the state-space model is usually expressed as

$$\begin{aligned} x_t &= f(x_{t-1}, u_{t-1}) + v_{t-1} \\ y_t &= h(x_t, u_t) + e_t \end{aligned} \tag{1}$$

where, $x_t \in \mathbb{R}^{n_x}$ is the latent state at time step t, $y_t \in \mathbb{R}^{n_y}$ is the measurement, u_t is the input, $f : \mathbb{R}^{n_x} \rightarrow \mathbb{R}^{n_x}$ is the state transition function. $h : \mathbb{R}^{n_x} \rightarrow \mathbb{R}^{n_y}$ is the measurement function. v_t is the process noise, e_t is the measurement noise, where $v_t \sim N(0, Q)$ and $e_t \sim N(0, R)$. Considering that the state space model is a black box model, the state transition function is unknown, and the process noise is also unknown. The measurement function and measurement noise are known. When a parameterized model is used, the model may be too rigid to describe the behavior of the model. However, when the data is limited, using a flexible model may cause the problem of overfitting.

2.2 Mathematical Description

The problem of missing observations is often encountered in practical engineering caused by sensor failure, which is non-ignorable. Therefore, it is taken into consideration in this work. The observations are written as $\boldsymbol{y}_{0:t}^{o_1:o_\alpha}$, where t is the current step, $o_1 : o_\alpha$ is the available observation sequence. Our goal is to learn the state-space model with missing observations online. This is a challenging job. Based on the derivation of Svensson [10], the state-space model is expressed by the basis function

$$\hat{f}_i(\boldsymbol{x}) = \sum_{j=1}^{M} \gamma_{ij}\phi_j(\boldsymbol{x}) \tag{2}$$

where, the coefficients are inspired by the GPs. There are some choices of basis functions. The alternatives include polynomials [13], Gaussian kernels [14]. We choose the eigenfunctions with associated eigenvalues, that allow the prior assumptions on $f(\cdot)$ inspired by the GPs. Taking the one-dimensional ($n_x = 1$) model for example, the basis function defined on a closed interval $[-L, L] \in \mathbb{R}$ is written as

$$\phi_j(\boldsymbol{x}) = \frac{1}{\sqrt{L}} \sin\left(\frac{\pi j(x+L)}{2L}\right) \tag{3}$$

The eigenvalues are written as

$$\lambda_j = \left(\frac{\pi j}{2L}\right)^2 \tag{4}$$

Similarly, in the multi-dimensional case, the eigenfunctions will be defined in a hypercube space $[-L_1, L_1] \times \cdots \times [-L_n, L_n] \in \mathbb{R}^{n_x}$. On the other hand, the state transition function is modeled as GPs. The square exponential covariance is chosen as kernel function.

$$\kappa(r) = \sigma^2 \exp\left(-\frac{r^2}{2\ell^2}\right) \tag{5}$$

The connection between the basis function expansion and the GPs can be written as:

$$f(\boldsymbol{x}) \sim GP(0, \kappa(\boldsymbol{x}, \boldsymbol{x}\prime)) \Leftrightarrow f(\boldsymbol{x}) \approx \sum_{j=1}^{M} \boldsymbol{\gamma}_j \phi_j(\boldsymbol{x}) \tag{6}$$

Therefore, the coefficients of the basis function can be inspired by GPs. The state transition function is written as

$$x_{k+1} = \underbrace{\begin{bmatrix} \gamma_{11} & \cdots & \gamma_{1m} \\ \vdots & & \vdots \\ \gamma_{n_x1} & \cdots & \gamma_{n_xm} \end{bmatrix}}_{\boldsymbol{A}} \underbrace{\begin{bmatrix} \phi_1(\boldsymbol{x}_k) \\ \vdots \\ \phi_M(\boldsymbol{x}_k) \end{bmatrix}}_{\boldsymbol{\varphi}(\boldsymbol{x}_k)} + \omega_k \tag{7}$$

Coefficient satisfies normal distribution

$$\gamma_{ij} \sim \mathrm{N}(\mathbf{0}, S(\lambda_j)) \tag{8}$$

where

$$S(\omega) = \sigma^2 \sqrt{2\pi \ell^2} \exp\left(-\frac{\pi^2 \ell^2 \omega^2}{2} \right) \tag{9}$$

Coefficient matrix $\boldsymbol{A}$ satisfies matrix normal distribution $\boldsymbol{A} \sim MN(\boldsymbol{M}, \boldsymbol{U}, \boldsymbol{V})$. The right covariance $\boldsymbol{U}$ is set as the variance of the process noise $\boldsymbol{Q}$, and the left covariance $\boldsymbol{V}$ is the diagonal matrix with $S(\lambda_j)$ as the diagonal element. The prior of process noise $\boldsymbol{Q}$ is IW distribution. We obtain the *MNIW* distribution, which is the combination of the *MN* distribution and the *IW* distribution

$$MNIW(\boldsymbol{A}, \boldsymbol{Q}|\boldsymbol{M}, \boldsymbol{V}, \boldsymbol{\Lambda}, \nu) = MN(\boldsymbol{A}|\boldsymbol{M}, \boldsymbol{Q}, \boldsymbol{V}) IW(\boldsymbol{Q}|\nu, \boldsymbol{\Lambda}) \tag{10}$$

3 Online Bayesian Inference and Learning with Missing Observations

Our goal is to estimate the state and learn the state-space model with missing observations. In other words, we want to obtain posterior distribution $p\big(\boldsymbol{x}_{0:t+1}, \theta_t | \boldsymbol{y}_{0:t}^{o_1:o_\alpha}\big)$. According to Bayes theorem, the posterior can be written as

$$p\big(\boldsymbol{x}_{0:t+1}, \theta_t | \boldsymbol{y}_{0:t}^{o_1:o_\alpha}\big) = p\big(\theta_t | \boldsymbol{x}_{0:t+1}, \boldsymbol{y}_{0:t}^{o_1:o_\alpha}\big) p(\boldsymbol{x}_{0:t+1} | \boldsymbol{y}_{0:t}^{o_1:o_\alpha}) \tag{11}$$

The online Bayesian inference and learning is divided into two steps

(1) Use the particle filter to infer the state $p(\boldsymbol{x}_{0:t+1} | \boldsymbol{y}_{0:t}^{o_1:o_\alpha})$
(2) Estimate the posterior of the unknown parameters $p(\theta_t | \boldsymbol{x}_{0:t+1}, \boldsymbol{y}_{0:t}^{o_1:o_\alpha})$

3.1 SMC for State Inference

Importance sampling methods are often nested in Bayesian inference of state-space models. SMC is a numerical approximation method for state estimation. SMC approximate state probability density using particle weight. Similarly, Lindsten proposes the Particle Gibbs with ancestor sampling (PGSA) method to learn the state [15]. Svensson uses PGSA algorithm to approximate state sequence [10]. Berntorp improved the online learning method to update the states online [12]. We improve the method proposed by Berntorp and propose an online Bayesian learning method with missing observations. Based on SMC method, the posterior density can be approximately written as

$$p\big(\boldsymbol{x}_{0:t} | \boldsymbol{y}_{0:t}^{o_1:o_\alpha}\big) \approx \sum_{i=1}^{N} w_t^i \delta_{\boldsymbol{x}_{0:t}^i}(\boldsymbol{x}_{0:t}) \tag{12}$$

When $\boldsymbol{y}_t$ is available, the weights are updated according to

$$w_t^i \propto w_{t-1}^i p\left(\boldsymbol{y}_t | \boldsymbol{x}_t^i\right) \tag{13}$$

When $\boldsymbol{y}_t$ is missing, the weights stay the same

$$w_t^i = w_{t-1}^i \tag{14}$$

To infer the state $p\left(\boldsymbol{x}_t | \boldsymbol{x}_{0:t-1}, \boldsymbol{y}_{0:t-1}^{o_1:o_\alpha}\right)$, we marginalize out the unknown parameters according to

$$p\left(\boldsymbol{x}_t | \boldsymbol{x}_{0:t-1}, \boldsymbol{y}_{0:t}^{o_1:o_\alpha}\right) = \int p(\boldsymbol{x}_t | \boldsymbol{\theta}_{t-1}, \boldsymbol{x}_{t-1}) p\left(\boldsymbol{\theta}_{t-1} | \boldsymbol{x}_{0:t-1}, \boldsymbol{y}_{0:t}^{o_1:o_\alpha}\right) \mathrm{d}\boldsymbol{\theta}_{t-1} \tag{15}$$

Suppose the second part of the integral satisfies the *MNIW* distribution

$$\boldsymbol{A}_{t-1}, \boldsymbol{Q}_{t-1} \sim \mathcal{MNIW}\left(\boldsymbol{M}^*, \sum\nolimits^*, \boldsymbol{\Lambda}^*, v_{t|t-1}\right) \tag{16}$$

where $\boldsymbol{M}^*$, $\boldsymbol{\Sigma}^*$, $\boldsymbol{\Lambda}^*$, $v_{t|t-1}$ are the statistics

$$\begin{aligned} \boldsymbol{M}^* &= \boldsymbol{\Psi}_{t|t-1}\left(\boldsymbol{\Sigma}_{t|t-1} + \boldsymbol{V}^{-1}\right)^{-1} \\ \boldsymbol{\Sigma}^* &= \left(\boldsymbol{\Sigma}_{t|t-1} + \boldsymbol{V}^{-1}\right)^{-1} \\ \boldsymbol{\Lambda}^* &= \boldsymbol{\Lambda}_0 + \boldsymbol{\Phi}_{t|t-1} - \boldsymbol{\Psi}_{t|t-1}\left(\boldsymbol{\Sigma}^*\right)^{-1}\boldsymbol{\Psi}_{t|t-1}^{\top}. \end{aligned} \tag{17}$$

where, $\boldsymbol{\Psi}$, $\boldsymbol{\Sigma}$, $\boldsymbol{\Phi}$ are the statistics related to states

$$\begin{aligned} \boldsymbol{\Phi}_{t|t} &= \Phi_{t|t-1} + \boldsymbol{x}_{t+1}\boldsymbol{x}_{t+1}^{\top} \\ \boldsymbol{\Psi}_{t|t} &= \boldsymbol{\Psi}_{t|t-1} + \boldsymbol{x}_{t+1}\varphi(\boldsymbol{x}_t)^{\top} \\ \boldsymbol{\Sigma}_{t|t} &= \boldsymbol{\Sigma}_{t|t-1} + \boldsymbol{\varphi}(\boldsymbol{x}_t)\boldsymbol{\varphi}(\boldsymbol{x}_t)^{\top} \\ v_{t|t} &= v_{t|t-1} + 1 \end{aligned} \tag{18}$$

Berntorp uses the principle of exponential forgetting to describe the parameters that are slowly time-varying [12].

$$\begin{aligned} \boldsymbol{\Phi}_{t|t-1} &= \lambda\Phi_{t-1|t-1}, \\ \boldsymbol{\Psi}_{t|t-1} &= \lambda\boldsymbol{\Psi}_{t-1|t-1}, \\ \boldsymbol{\Sigma}_{t|t-1} &= \lambda\boldsymbol{\Sigma}_{t-1|t-1}, \\ v_{t|t-1} &= \lambda v_{t-1|t-1} \end{aligned} \tag{19}$$

where, λ is the forgetting factor, which aims to pay more attention to new data and forget older data. Forgetting factor is important for online updating model. Therefore, $p(\boldsymbol{x}_t | \boldsymbol{\theta}_{t-1}, \boldsymbol{x}_{t-1})$ can be written as

$$\boldsymbol{x}_t | \boldsymbol{x}_{t-1}, \boldsymbol{A}_{t-1}, \boldsymbol{Q}_{t-1} \sim \mathcal{N}\left(\boldsymbol{A}_{t-1}\boldsymbol{\varphi}(\boldsymbol{x}_{t-1}), \boldsymbol{Q}_{t-1}\right) \tag{20}$$

3.2 Inference the Unknown Parameters

After obtaining the state estimation, the posterior of unknown parameters can be written as

$$\begin{aligned} p\left(\boldsymbol{\theta}_t|\boldsymbol{x}_{0:t+1}, \boldsymbol{y}_{0:t}^{o_1:o_\alpha}\right) &= p(\boldsymbol{\theta}_t|\boldsymbol{x}_{0:t+1}) \\ &\propto p(\boldsymbol{x}_{t+1}|\boldsymbol{\theta}_t, \boldsymbol{x}_{0:t})p(\boldsymbol{\theta}_t|\boldsymbol{x}_{0:t}) \end{aligned} \tag{21}$$

where

$$p(\boldsymbol{x}_{t+1}|\boldsymbol{\theta}_t, \boldsymbol{x}_t) = \mathcal{N}\left(\boldsymbol{x}_{t+1}|\boldsymbol{A}_t\boldsymbol{\varphi}(\boldsymbol{x}_t), \boldsymbol{Q}_t\right) \tag{22}$$

When observation $\boldsymbol{y}_t$ is available at the current time t

$$\begin{aligned} p\left(\boldsymbol{\theta}_t|\boldsymbol{x}_{0:t+1}, \boldsymbol{y}_{0:t-1}^{o_1:o_\alpha}, \boldsymbol{y}_t\right) = \mathcal{MNIW}\Big(&\boldsymbol{\theta}_t|\Psi_{t|t}\boldsymbol{\Xi}^{-1}, \\ &\boldsymbol{\Xi}^{-1}, \boldsymbol{\Lambda}_0 + \boldsymbol{\Phi}_{t|t} - \boldsymbol{\Psi}_{t|t}\boldsymbol{\Xi}^{-1}\boldsymbol{\Psi}_{t|t}^{\top}, \nu_{t|t}\Big) \end{aligned} \tag{23}$$

The posterior distribution of unknown parameters can be obtained by marginalizing out the state trajectory according to

$$p\left(\boldsymbol{\theta}_t|\boldsymbol{y}_{0:t}^{o_1:o_\alpha}\right) = \int p\left(\boldsymbol{\theta}_t|\boldsymbol{x}_{0:t+1}, \boldsymbol{y}_{0:t}^{o_1:o_\alpha}\right)p\left(\boldsymbol{x}_{0:t+1}|\boldsymbol{y}_{0:t}^{o_1:o_\alpha}\right)\mathrm{d}\boldsymbol{x}_{0:t+1} \tag{24}$$

The integral is difficult to obtain. We use the particles and the corresponding weights to approximate the integral

$$p\left(\boldsymbol{\theta}_t|\boldsymbol{y}_{0:t}^{o_1:o_\alpha}\right) \approx \sum_{i=1}^{N} q_k^i p\left(\boldsymbol{\theta}_k|\boldsymbol{x}_{0:t+1}^i, \boldsymbol{y}_{0:t}^{o_1:o_\alpha}\right) \tag{25}$$

When $\boldsymbol{y}_t$ is missing at the current time t, we adopt a conservative strategy not to update the unknown parameters

$$\boldsymbol{\theta}_t = \boldsymbol{\theta}_{t-1} \tag{26}$$

Our strategy avoids the interruption of online learning when observation is missing. On the other hand, the algorithm makes use of all available observations.

3.3 Algorithm Introduction

In this section, we summarize the algorithm proposed in this paper and introduce the improved algorithm in the form of pseudo code.

Algorithm 1 Online Bayesian inference and learning with missing observations

Input: Sample N particles $\{x_0^i\}_{i=1}^N$ from $p_0(x_0)$, set $\{q_{-1}^i\}_{i=1}^N = 1/N$, $\{\Phi_0^i, \Psi_0^i, \Sigma_0^i, \Lambda_0^i, v_0^i\}_{i=1}^N = \{0, 0, 0, \Lambda_0, v_0\}_{i=1}^N$

1:**for** t=0,1,... **do**
2: **if** $y(t)$ is missing **then**
3: **for** $i \in \{1,...,N\}$
4: update weight $\bar{w}_t^i$ using (14)
5: **end for**
6: **else**
7: **for** $i \in \{1,...,N\}$
8: update weight $\bar{w}_t^i$ using (13)
9: **end for**
10: **end if**
11: Compute the normalized weights $w_t^i = \bar{w}_t^i / (\sum_{i=1}^N \bar{w}_t^i)$
12: Compute $Neff = 1/(\sum_{i=1}^N (w_t^i)^2)$
13: **if** $Neff \le N_{threshold}$ **then**
14: Resample and copy the corresponding statistics. Set $w_t^i = 1/N$
15: **end if**
16: **for** $i \in \{1,...,N\}$
17: Sample x_{t+1}^i from (20)
18: Compute statistics $\Phi_{t|t}^i, \Psi_{t|t}^i, \Sigma_{t|t}^i, \Lambda_{t|t}^i, v_{t|t}^i$ using (18)
19: **end for**
20: **if** $y(t)$ is missing **then**
21: Compute θ_t using (26)
22: **else**
23: Approximate the posterior of θ_t using (25)
24: **end if**
25: **end for**

4 Numerical Example

We consider the following example to illustrate the effectiveness of the proposed Algorithm. What is more important, we consider systems with missing output observations.

$$\begin{aligned} x_{t+1} &= \tanh(2x_t) + v_t, \ v_t \sim N(0, 0.1) \\ y_t &= x_t + e_t, \qquad e_t \sim N(0, 0.1) \end{aligned} \tag{27}$$

The measurement function $g(\cdot)$ and the noise variances are known. Our goal is to learn $f(\cdot)$ and the process noise. However, y is randomly missing due to sensor failure. The observations missing percentages are set as 5%, 10%. The number of basis function is set as $M = 16$. We find that M does not need to be strictly set. We set $N = 50$ particles. All the learning results are shown in Figs. 1 and 2. The black solid line is the true function. The blue solid line is the mean function of the learning results. The gray region is the 3σ confidence of the learning results. The red dots are the true state samples.

In order to fully demonstrate the robustness of our method to missing observations, we compare the learning results with those of Berntorp. It is worth noting that Berntorp's method is based on the full observations [12].

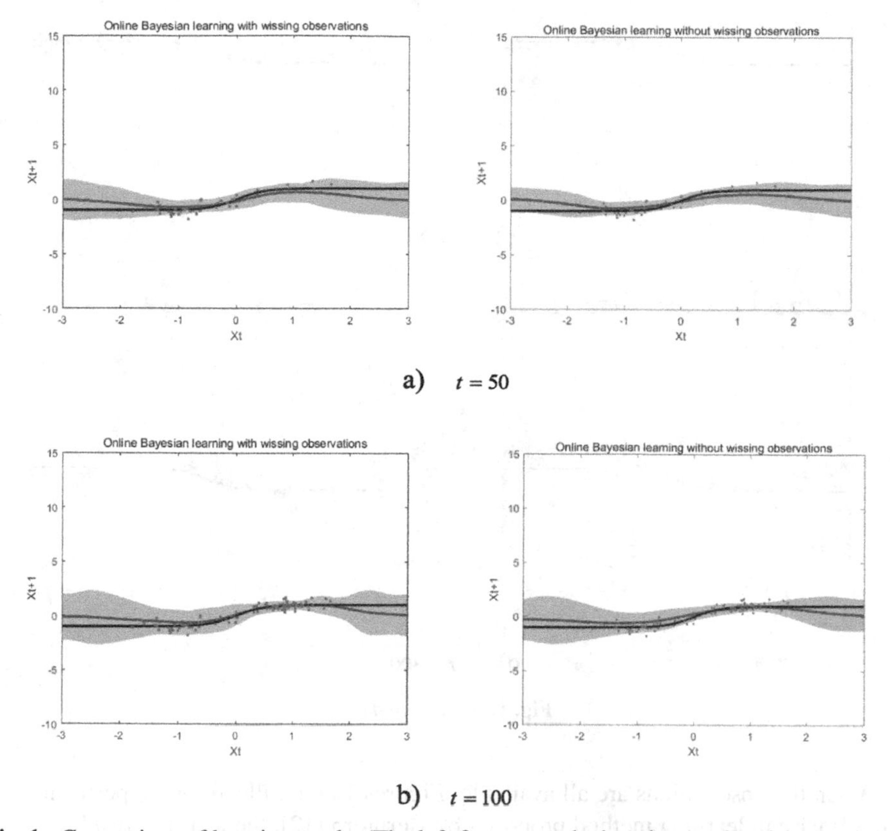

Fig. 1. Comparison of learning results (The left figures are the learning results of the black-box model with 5% missing observations obtained by our method, the right figures are the learning results without missing observation obtained by Berntorp's method [12])

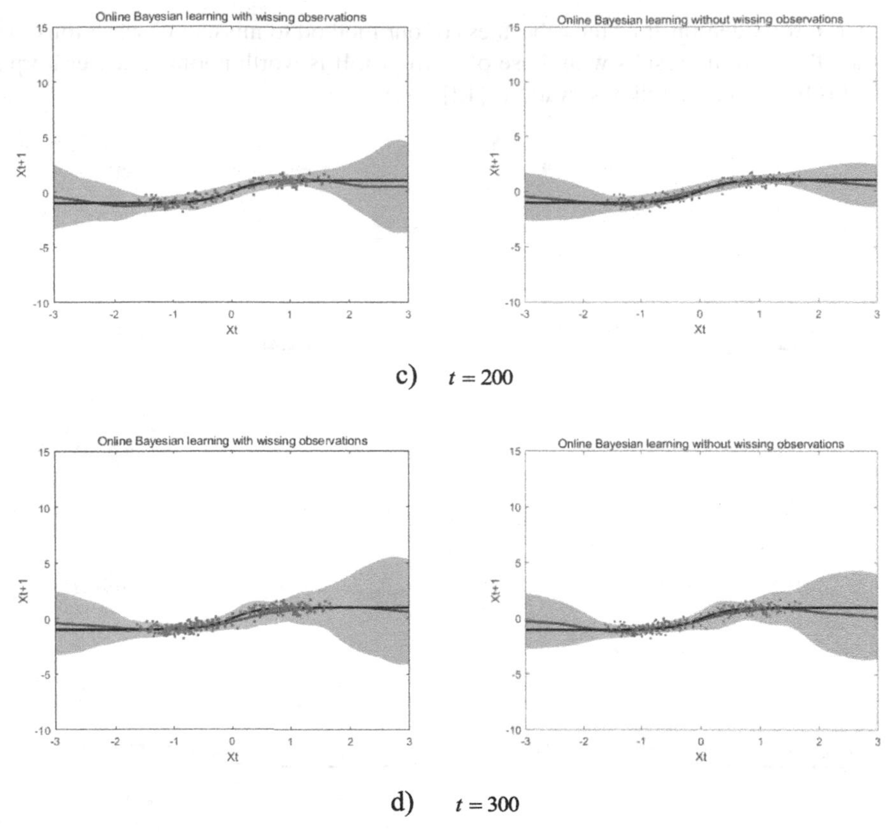

Fig. 1. (*continued*)

When the observations are all available. Figures 1 and 2 illustrate the performance of the Bayesian learning method proposed by Berntorp [12], the learning result is very close to the true function when there are enough observations. This shows that the method proposed by Berntorp is a data-driven method. When 5% and 10% observations are randomly missing. Our method is used to identify $f(\cdot)$ online. Figure 1 and Fig. 2 illustrate the performance of the proposed method. We find that our learning result is also close to the true function. In addition, the comparison with the results obtained by the method proposed by Berntorp shows that the proposed method is robust to incomplete observations.

State filtering is another important goal. When all observations are available, we use Berntorp's method and estimate the state based on particle filter. When there are missing observations, we use the proposed method for state filtering. The comparison between estimations and true states are shown in Fig. 3, 4 and 5. When there is no missing observation, the state filter of Berntorp's method can obtain state estimations close to true states. When there are missing observations (5% and 10% are considered), our method also obtains satisfactory state estimation. These results show that our method is robust to missing observations when estimating states.

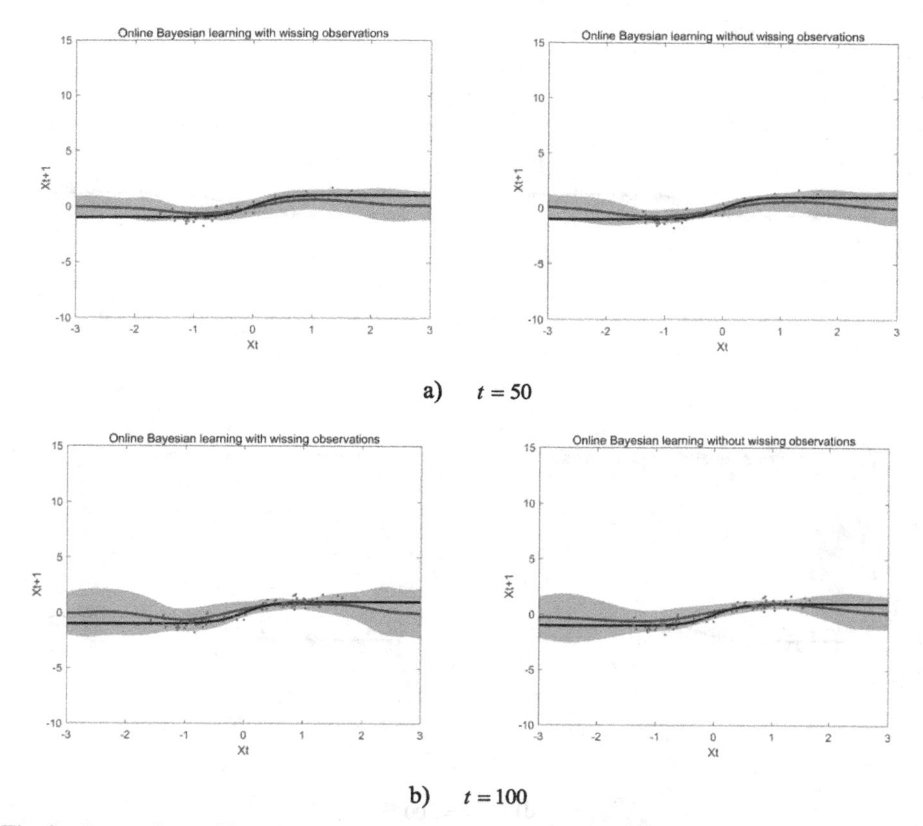

Fig. 2. Comparison of learning results (The left figures are the learning results of the black-box model with 10% missing observations obtained by our method, the right figures are the learning results without missing observation obtained by Berntorp's method [12])

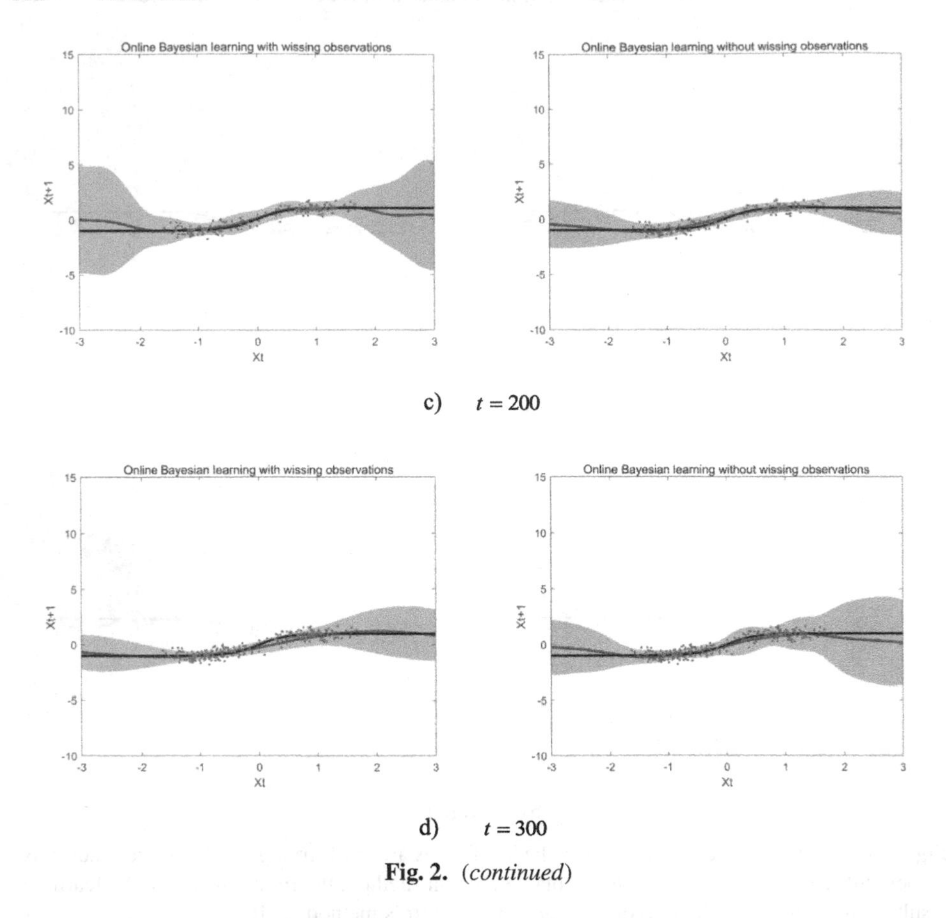

c) $t = 200$

d) $t = 300$

Fig. 2. (*continued*)

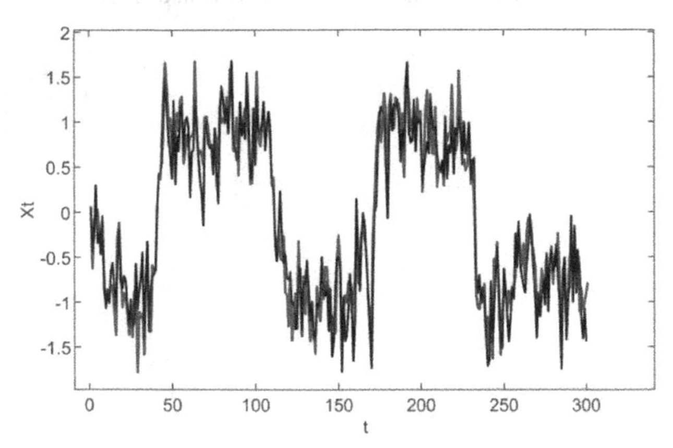

Fig. 3. State filtering without missing observation based on Berntorp's method

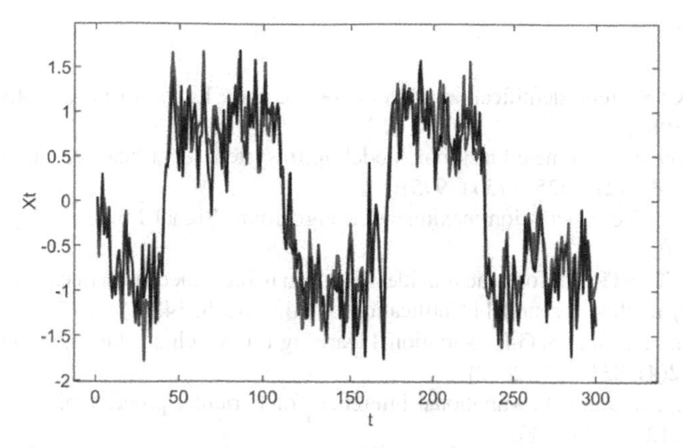

Fig. 4. State filtering with 5% missing observation based on our method

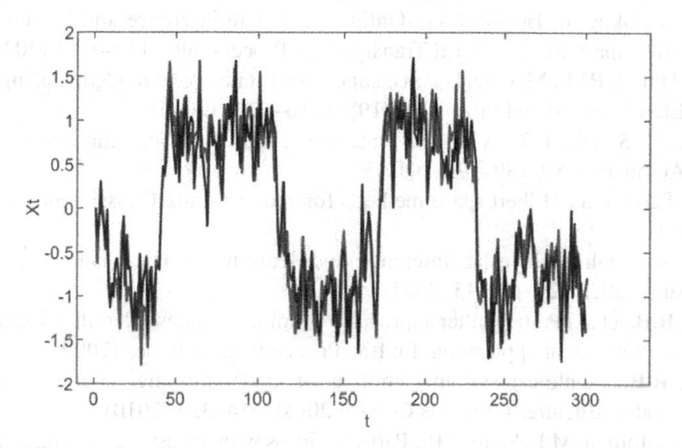

Fig. 5. State filtering with 10% missing observation based on our method

5 Conclusion

The online identification method of GP state-space Model with missing observations is proposed in this paper. When the SSM is a black box and the observations are missing. Learning the model and estimating the states are difficult. We improved the method proposed by Berntorp, and propose two different learning strategies according to whether the observation is missing or not. We demonstrate the effectiveness of the proposed method through a numerical example. The learning result is similar to the true function. On the other hand, our learning result is also close to that of Berntorp, which demonstrates that the proposed method is robust to incomplete observations. In addition, the results of state filtering show that our method is also robust to missing observations when estimating states.

References

1. Sinha, N.K.: System identification—theory for the user: Lennart Ljung. Automatica **25**(3), 475–476 (1989)
2. Sjöberg J., et al.: Nonlinear black-box modeling in system identification: a unified overview. Automatica **31**(12), 1725–1750 (1995)
3. Moon, T.K.: The expectation-maximization algorithm. Signal Process. Mag. IEEE **13**(6), 47–60 (1996)
4. Matarazzo, T.: STRIDE for structural identification using expectation maximization: iterative output-only method for modal identification. J. Eng. Mech. **142**(4) (2015)
5. Ghahramani, Z., Hinton, G.E.: Variational learning for switching state-space models. Neural Comput. **12**(4), 831–864 (2000)
6. Blei, D.M., Jordan, M.I.: Variational inference for Dirichlet process mixtures. J. Bayesian Anal. **1**(1), 121–143 (2006)
7. Jacobs, W.R., et al.: Sparse Bayesian nonlinear system identification using variational inference. IEEE Trans. Autom. Control. **63**, 4172 (2018)
8. Kullberg, A., Skog, I., Hendeby, G.: Online joint state inference and learning of partially unknown state-space models. IEEE Trans. Signal Process. **69**, 4149–4161 (2021)
9. Tobar, F., Djuric, P.M., Mandic, D.P.: Unsupervised state-space modeling using reproducing kernels. IEEE Trans. Signal Process. **63**(19), 5210–5221 (2015)
10. Svensson, A., Schön, T.B.: A flexible state space model for learning nonlinear dynamical systems. Automatica **80**, 189–199 (2017)
11. Solin, A., Särkkä, S.: Hilbert space methods for reduced-rank Gaussian process regression. arXiv (2014)
12. Berntorp, K.: Online Bayesian inference and learning of Gaussian-process state–space models. Automatica **129**, 109613 (2021)
13. Gopaluni, R.B., et al.: Particle filter approach to nonlinear system identification under missing observations with a real application. In: Ifac Proceedings Volumes (2009)
14. Gopaluni, R.B.: Nonlinear system identification under missing observations: the case of unknown model structure. J. Process Control **20**(3), 314–324 (2010)
15. Lindsten, F., Jordan, M.I., Schn, T.B.: Particle Gibbs with ancestor sampling. J. Mach. Learn. Res. **15**(15), 2145–2184 (2014)

Complex Systems and Open, Complex and Giant Systems Modeling and Simulation

System Identification of Nonlinear Dynamical System with Missing Observations

Xiaonan Li, Ping Ma, Tao Chao(✉), and Ming Yang

Control and Simulation Center, Harbin Institute of Technology, Harbin 150080, China
chaotao2000@163.com

Abstract. We consider a nonlinear state-space model with unknown state transition function and process noise. The state transition function is modeled as a Gaussian process, besides, it is also expressed as basis function expansion. Using the connection to the Gaussian process, the prior of the coefficients of basis function can be obtained. The posterior of the state and unknown coefficients can be obtained through Bayesian inference. Sequential Monte Carlo (Particle Gibbs with ancestor sampling) is used to estimate the states. The coefficients are modeled as random variables related to state statistics. The Markov Chain Monte Carlo (MCMC) method is used to repeated iteratively sample from parameter posterior and state posterior. The problems of missing observations due to sensor failure are often encountered in practical engineering and are taken into consideration in this paper. We propose a learning method for nonlinear dynamical system with missing observations. According to whether the observation data is missing or not, the state and parameters are updated respectively. The proposed nonlinear system identification method is robust to incomplete dataset. A numerical example is used to demonstrate the effectiveness of the proposed method.

Keywords: State-space model · System identification · Basis function expansion · Markov chain Monte Carlo · Bayesian learning

1 Introduction

Nonlinear system identification aims to learn mathematical models from data [1, 2]. State-space model is a common model in the control system. When there are unknown parameters in the state-space model, MCMC method is often used for Bayesian inference to obtain the posterior of unknown parameters [3]. MCMC method is usually expensive. Therefore, Variational Inference (VI) has been proposed recently [4–6]. In addition, Expectation-Maximization (EM) Algorithm is also used to identify parameters [7]. Under the assumption on the nonlinear function, the learning problem the of nonlinear state-space model will use Bayesian framework. The GP process is a popular machine learning method [8, 9]. GPs are also used to encode prior to prevent overfitting [10].

The state-space model is usually expressed as

$$x_k = f(x_{k-1}, u_{k-1}) + w_{k-1}$$

W. Fan et al. (Eds.): AsiaSim 2022, CCIS 1712, pp. 117–128, 2022.
https://doi.org/10.1007/978-981-19-9198-1_9

$$y_k = h(x_k, u_k) + e_k \tag{1}$$

where, $x_k \in \mathbb{R}^{n_x}$ is the latent state at time step k, u_k is the input, $y_k \in \mathbb{R}^{n_y}$ is the measurement. $f : \mathbb{R}^{n_x} \to \mathbb{R}^{n_x}$ is the state transition function. $h : \mathbb{R}^{n_x} \to \mathbb{R}^{n_y}$ is the measurement function. w_k and e_k are the process noise and the measurement noise, where $w_k \sim N(0, Q)$ and $e_k \sim N(0, R)$.

When a parameterized model is used, the model may be too rigid to describe the behavior of the model. However, when the data is limited, using a flexible model may cause the problem of overfitting. GP is a probability model, which can provide prediction uncertainty. GP has been widely used in the field of system identification. In order to avoid overfitting problem of flexible model, Gaussian Processes is used to encode prior assumption of state transition function and observation function [10].

Svensson [11] expands the state transition function into the form of basis function expansion to parameterize the function

$$\hat{f}_i(\boldsymbol{x}) = \sum_{j=1}^{M} \gamma_{ij}\phi_j(\boldsymbol{x}) \tag{2}$$

When the observation function is unknown, a similar method is used.$\{\phi_j(\cdot)\}_{j=0}^{M}$ are the basis functions. $\{\gamma_{ij}(\cdot)\}_{j=0}^{M}$ are the coefficients, whose prior $p(\gamma_{ij})$ is connection to GPs[10]. The introduction of a priori makes it possible to use Bayesian framework for learning. M is the number of basis function. Setting M small will limit the expressiveness of the surrogate model. However, when M is set too large, there will be too many parameters to learn, and the calculation will be expensive.

The problem of missing observations caused by sensor failure is non-ignorable. We improve the method proposed by Svensson [11], and propose a Bayesian learning method with missing observations. Similarly, we use the Particle Gibbs with ancestor sampling (PGSA) to learn the state [12]. When using PGSA algorithm for particle filter, the weight and ancestor sequence will be updated respectively according to whether the observation is missing. The observations missing percentages are selected as 5%, 10%, 15%, 20% to illustrate the effectiveness and robustness of our method.

2 Modeling

In order to learn the model from the data and avoid the problem of overfitting, the unknown functions in the state-space model are expressed as basis functions expansions, and GP-inspired prior is applied to Bayesian inference. We focus on $f(\cdot)$ and Q to detail our approach. Assuming that the observation function and measurement noise are known.

2.1 Basis Function Expansion of State-Space Model

We use basis function expansion to model the function

$$x_{k+1} = \underbrace{\begin{bmatrix} \gamma_{11} & \cdots & \gamma_{1m} \\ \vdots & & \vdots \\ \gamma_{n_x 1} & \cdots & \gamma_{n_x m} \end{bmatrix}}_{A} \underbrace{\begin{bmatrix} \phi_1(\boldsymbol{x}_k) \\ \vdots \\ \phi_M(\boldsymbol{x}_k) \end{bmatrix}}_{\varphi(\boldsymbol{x}_k)} + \omega_k \tag{3}$$

where, the coefficient matrix A is inspired by the GP. There are some choices of basis functions that allow the prior assumptions on $f(\cdot)$ inspired by the GP. The alternatives include polynomials [13], Gaussian kernels [14]. We choose the eigenfunction with associated eigenvalues, when $n_x = 1$, the basis function defined on a closed interval $[-L, L] \in \mathbb{R}$ equal

$$\phi_j(\boldsymbol{x}) = \frac{1}{\sqrt{L}} \sin\left(\frac{\pi j(x+L)}{2L}\right) \tag{4}$$

$$\lambda_j = \left(\frac{\pi j}{2L}\right)^2 \tag{5}$$

When $n_x > 1$, the eigenfunction is defined on the closed interval $[-L_1, L_1] \times \cdots \times [-L_{n_x}, L_{n_x}] \in \mathbb{R}^{n_x}$,

$$\phi_{j_1,\ldots,j_{nx}} = \prod_{n=1}^{n_x} \frac{1}{\sqrt{L_n}} \sin\left(\frac{\pi j_n(x_n+L_n)}{2L_n}\right) \tag{6}$$

The corresponding eigenvalues

$$\lambda_{j_1,\ldots,j_{nx}} = \sum_{n=1}^{n_x} \left(\frac{\pi j_n}{2L_n}\right)^2 \tag{7}$$

From (2), it is clear that the number of weights to be inferred increase exponentially with n_x. The computational complexity can be reduced by two assumptions. A direct assumption is that some dimensions are independent. The other one is to choose another set of basis function such as radial basis function expansion [11].

2.2 Connection to the GP

Gaussian process is a non-parametric approach for regression. We use GPs to encode prior assumptions. Modeling the state transition functions as GPs, we have

$$f(\boldsymbol{x}) \sim GP(m(\boldsymbol{x}), \kappa(\boldsymbol{x},\boldsymbol{x}')) \tag{8}$$

where, $m(\boldsymbol{x})$ is the mean function,$\kappa(\boldsymbol{x},\boldsymbol{x}')$ is the covariance function. We choose the squared exponential covariance function

$$\kappa(r) = \sigma^2 \exp\left(-\frac{r^2}{2\ell^2}\right) \tag{9}$$

For isotropic covariance function $\kappa(\cdot)$, There is a connection between basis function expansions and GPs.

$$f(\boldsymbol{x}) \sim GP(0, \kappa(\boldsymbol{x},\boldsymbol{x}')) \Leftrightarrow f(\boldsymbol{x}) \approx \sum_{j=1}^{M} \gamma_j \phi_j(\boldsymbol{x}) \tag{10}$$

In particular, when we choose the eigenfunction, the prior of the coefficient can be expressed by the spectral density S

$$\gamma_{ij} \sim N(\mathbf{0}, S(\lambda_j)) \tag{11}$$

The spectral density of the squared exponential covariance function is expressed as

$$S(\omega) = \sigma^2 \sqrt{2\pi \ell^2} \exp\left(-\frac{\pi^2 \ell^2 \omega^2}{2} \right) \tag{12}$$

where, σ and l are hyperparameters. Obviously, $S(\omega)$ tends to 0 when $\omega \to \infty$,which means that prior for large j is close to 0. From (10) and (11), the prior of the coefficient of the basis function can be obtained. For convenience, we express the prior of the coefficients Matrix as a Matrix normal (MN) distribution [15]. The MN distribution is parameterized by a mean matrix, a right and a left covariance. Firstly, we define the coefficient matrix $\boldsymbol{A}$ as

$$vec(\boldsymbol{A}) \sim N(vec(M), U \otimes V) \tag{13}$$

where, $\otimes$ is the Kronecker product, $U \in \mathbb{R}^{n_x \times n_x}$ is the right covariance, and the $V \in \mathbb{R}^{M \times M}$ is the left covariance. Then $\boldsymbol{A} \sim MN(\boldsymbol{M}, \boldsymbol{U}, \boldsymbol{V})$. We set $\boldsymbol{M} = \mathbf{0}$ and $\boldsymbol{V}$ be a diagonal matrix with diagonal elements $S(\lambda_j)$,which connects to (11). We set the right covariance $\boldsymbol{U} = \boldsymbol{Q}$. The elements in $\boldsymbol{A}$ are scaled by $\boldsymbol{V}$ and $\boldsymbol{Q}$.

Process noise covariance matrix is often unknown. Inverse Wishart distribution is always used as the prior distribution of the covariance matrix of multivariate normal distribution [15]. We formulate prior over the process noise $\boldsymbol{Q}$ as an IW distribution. We obtain the compound distribution

$$MNIW(\boldsymbol{A}, \boldsymbol{Q}|\boldsymbol{M}, \boldsymbol{V}, \boldsymbol{\Lambda}, \nu) = MN(\boldsymbol{A}|\boldsymbol{M}, \boldsymbol{Q}, \boldsymbol{V}) IW(\boldsymbol{Q}|\nu, \boldsymbol{\Lambda}) \tag{14}$$

Equation (14) provides the joint prior of $\boldsymbol{A}$ and $\boldsymbol{Q}$.

2.3 Problem Formulation

In this paper, we assume $u_{1:T}$ are measurable, and they are available all the time. We omit u_t and the observation function to avoid notational clutter, the state transition function can be written as

$$\boldsymbol{x}_{t+1} = \boldsymbol{A}\boldsymbol{\varphi}(\boldsymbol{x}_k) + \boldsymbol{w}_t \tag{15}$$

$$\omega_t \sim N(\mathbf{0}, \boldsymbol{Q}) \tag{16}$$

The priors of the unknown parameters are

$$[\boldsymbol{A}, \boldsymbol{Q}] \sim MNIW(\mathbf{0}, \boldsymbol{V}, \boldsymbol{\Lambda}, \nu) \tag{17}$$

The system observations are randomly missing. We divide the observations into two groups, one is the available output data $Y_{obs} = \{y_{o_1:o_\alpha}\}$, the other is the missing output data $Y_{obs} = \{y_{m_1:m_\beta}\}$. Therefore, our goal in this paper is to learn the parameters $\Theta = \{\boldsymbol{A}, \boldsymbol{Q}, \boldsymbol{x}_{1:T}\}$ based on the available output dataset $Y_{obs} = \{y_{o_1:o_\alpha}\}$.

3 Particle Filter with Missing Output Observations

Sequential Monte Carlo (SMC) is a useful method for learning the unknown parameters in state-space models. SMC provides a numerical method to the state filtering problem. SMC approximates the density of state according to the weights of particles

$$\hat{p}(x_t|y_{1:t}) = \sum\nolimits_{i=1}^{N} q_t^i \delta_{x_t^i}(x_t) \tag{18}$$

where, q_t^i is the weight of $i-th$ particle at time step t. Sequential Monte Carlo renew the weights of particles with respect to the measurement. Then resample particles according to the weights. The particle at the next time is obtained through state transition function. We use the PGSA Markov kernel to reduce the affection of the particle degeneration. The key idea of PGSA is to set a fixed state-space trajectory.

When the system observations are randomly missing, the numerically approximation of the posterior density function (PDF) of state is expressed as

$$\hat{p}(x_t|y_{o1:o\alpha}) = \sum\nolimits_{i=1}^{N} q_t^i \delta_{x_t^i}(x_t) \tag{19}$$

The weights for the new particles are derived as

$$q_t^i \propto q_{t-1}^i p(y_t|x_t) \tag{20}$$

When the observations are missing, the new particles are generated from the PDF of state transition

$$p(x_t|x_{t-1}) \tag{21}$$

The weights of the new particles stay the same

$$q_t^i = q_{t-1}^i \tag{22}$$

The PGSA algorithm with missing observations is outlined by Algorithm 1.

Algorithm 1 PGAS Markov kernel with missing observations

Input: $x_{1:T}[k]$, number of particles N, known model (f,g,Q,R)
Output: state trajectory $x_{1:T}[k+1]$
1:Sample $x_1^i \sim p(x_1)$ for $i=1,...,N-1$.
2:Set $x_1^N = x_1(k)$
3:**for** $t=1$ to T **do**
4: **if** the observation y_t is available **then**
5: set $q_t^i = N(y_t \mid g(x_t^i),R)$
6: Sample a_t^i with $P(a_t^i = \text{j}) \propto q_t^i$ for $i=1,...,N-1$
7: Sample $x_{t+1}^i \sim N(f(x_t^{a_t^i}),Q)$ for $i=1,...,N-1$
8: set $x_{t+1}^N = x_{t+1}(k)$
9: Sample a_t^N with $P(a_t^N = j) \propto q_t^i N(x_{t+1}^N \mid f(x_t^j),Q)$
10: **else**
11: set $q_t^i = q_{t-1}^i$ for $i=1,...,N$
12: set $a_t^i = i$ for $i=1,...,N$
13: Sample $x_{t+1}^i \sim N(f(x_t^{a_t^i}),Q)$ for $i=1,...,N-1$
14: set $x_{t+1}^N = x_{t+1}(k)$
15: **end if**
16:**end for**
17:Sample J with $P(J=i) \propto q_T^i$ and set $x_{1:T}[k+1]=x_{1:T}^J$

4 Posterior of the Unknown Parameters

We infer the unknown parameters and the unknown states repeatedly and alternately. We use the MCMC approach for exploring the full posterior $p(\theta|y_{o_1:o_\alpha})$. The PGAS Markov kernel is used to sample from $p(x_{1:T}|y_{o_1:o_\alpha}, \theta)$. In this section, we will focus on inferring the posterior of the unknown parameters. The posterior is expressed as

$$p(\theta|x_{1:T}, y_{o_1:o_\alpha}) \sim MNIW(\boldsymbol{A}, \boldsymbol{Q}|\Psi(\Sigma + V^{-1})^{-1}, (\Sigma + V^{-1})^{-1}, \boldsymbol{\Lambda} + \Phi - \Psi(\Sigma + V^{-1})^{-1}, l + Tn_x) \tag{23}$$

where, Φ, Ψ, Σ are the statistics

$$\Phi = \sum_{t=1}^{T} \boldsymbol{x}_{t+1}\boldsymbol{x}_{t+1}^T$$

$$\Psi = \sum_{t=1}^{T} \boldsymbol{x}_{t+1}\boldsymbol{\varphi}(\boldsymbol{x}_t, \boldsymbol{u}_t)^T$$

$$\Sigma = \sum_{t=1}^{T} \varphi(x_t, u_t)\varphi(x_t, u_t)^T \tag{24}$$

Algorithm 1 is used to obtain the states $x_{1:T}$ with missing output observations. We use a particle Gibbs sampler as shown in Algorithm 2.

Algorithm 2 Bayesian inference and learning

Input: $y_{o_1:o_\alpha}$, priors on A, Q

Output: K MCMC-samples of $x_{1:T}$

1:Initialize $A[1], Q[1]$

2:**for** $k=1$ to $K+1$ **do**

3: Sample $x_{1:T}[k+1] \mid A[k], Q[k]$ based on Algorithm 1

4: Sample $Q[k+1] \mid x_{1:T}[k+1]$ based on (23)

5: Sample $A[k+1] \mid x_{1:T}[k+1], Q[k+1]$ (23)

6: **end for**

Algorithm 2 is a combination of a Gibbs sampler and Metropolis-within-Gibbs. If there are other parameters are unknown, it can be included in Algorithm 2.

5 Numerical Example

An example is given to illustrate the effectiveness of the proposed Algorithm. Consider systems with missing output observations.

$$x_{t+1} = 10 \sin c(\frac{x_t}{7}) + w_k \quad w_k \sim N(0, 0.1) \tag{25}$$

$$y_t = x_t + e_k \quad e_k \sim N(0, 0.1) \tag{26}$$

$g(\cdot)$ and the noise variances are known. Our goal is to learn $f(\cdot)$. However, y is obviously disturbed by noise. On the other hand, some measurements are randomly missing due to sensor failure. Now we have $T = 40$ observations. The observations missing percentages are selected as 5%, 10%, 15% and 20%. Setting $M = 40$ to obtain excellent expression ability of basis function expansion. We find that M does not need to be strictly set. All the learning results are presented in Fig. 1, 2, 3, 4 and 5. The black solid line is the true function. The black solid line is the true function. The blue dash line is the mean function of the learning results. The gray region is the 3σ confidence of Bayesian learning. The red dots are the state samples underlying data.

When the observations are all available. Figure 1 illustrates the performance of the Bayesian learning method proposed by Svensoon [11]. When the state belongs to $[-10, 10]$, the learning result is very close to the true function. The samples in this interval are dense, so the potential information of the model can be learned. This shows that the method proposed by Svensoon [11] is a data-driven method.

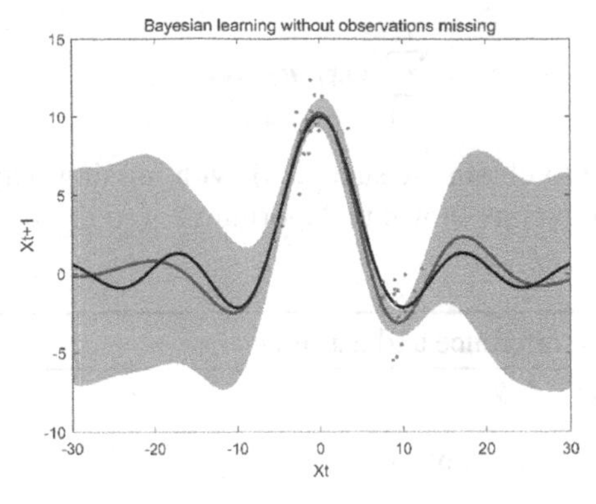

Fig. 1. Bayesian learning without outputs missing based on Svensoon's method [11].

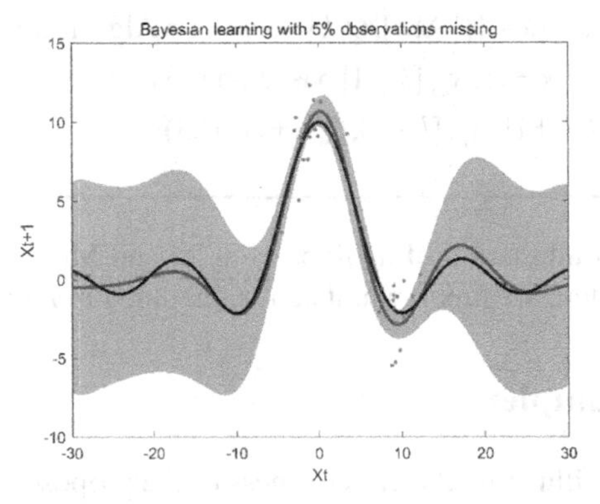

Fig. 2. Bayesian learning with 5% outputs missing based on our method.

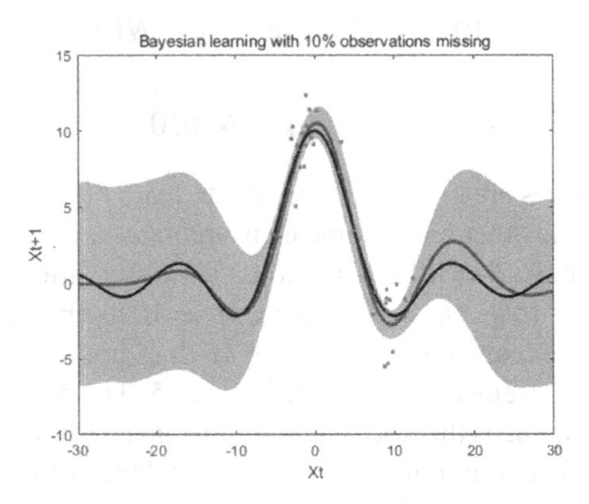

Fig. 3. Bayesian learning with 10% outputs missing based on our method.

When 5% and 10% observations are randomly missing. Our method (Algorithm 2) is used to learn $f(\cdot)$. Figures 2 and 3 illustrate the performance of our method. We find that the learning result is also close to the true function, which illustrates that our method

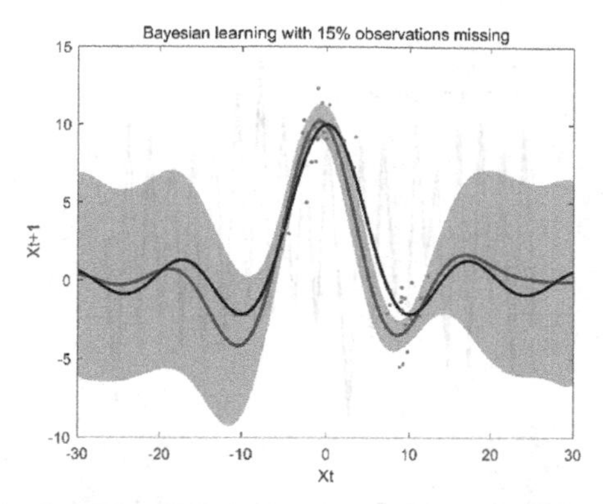

Fig. 4. Bayesian learning with 15% outputs missing based on our method.

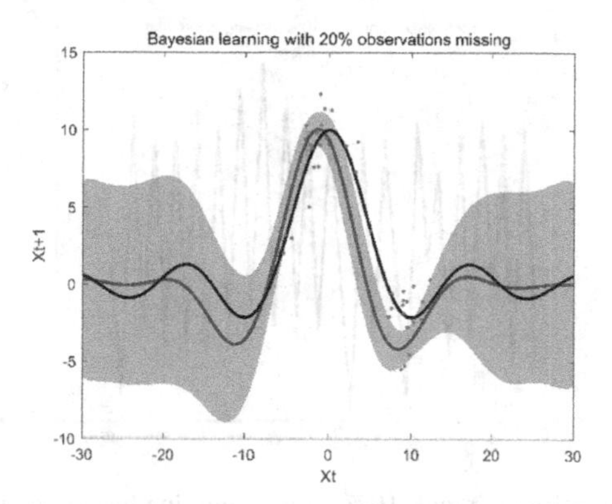

Fig. 5. Bayesian learning with 20% outputs missing based on our method.

is robust to incomplete observations. When 15% and 20% observations are randomly missing. Figures 4 and 5 illustrate that the learning results are similar to the true function, but the deviation increases significantly compared with that of 5% missing observations. When the state belongs to $[-10, 10]$, the learning result is very close to the true function. The samples in this interval are dense, so the potential information of the model can be learned. This shows that the method proposed by Svensoon [11] is a data-driven method.

Another important goal of the proposed method is to estimate the state. The state estimation will be interrupted when the observation is missing, which makes it difficult to use all the statistical information. Our method adopts a conservative strategy to use all the statistical information as much as possible. The results are shown as follows.

As shown in Fig. 6, the blue line is the true state, the black line is an estimate of the state. Our method is used to estimate states when observations are missing. The results show that the state estimations obtained by our method are close to the true states. On the other hand, the proposed method is robust to missing observations (Figs. 7, 8 and 9).

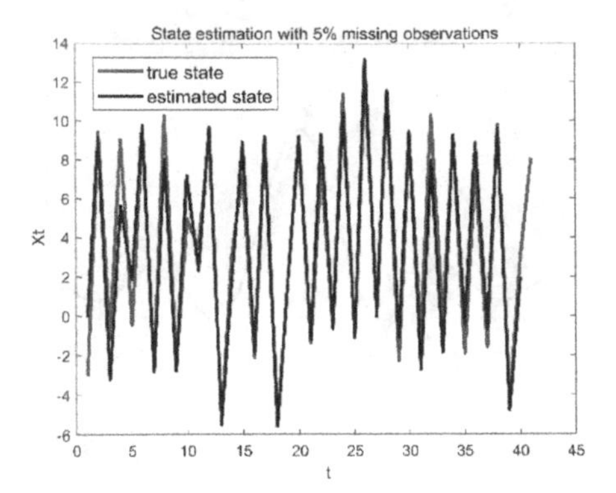

Fig. 6. State estimation with 5% outputs missing based on our method.

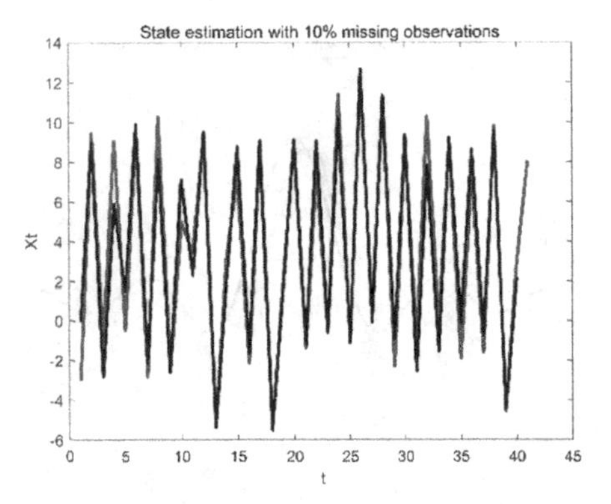

Fig. 7. State estimation with 10% outputs missing based on our method.

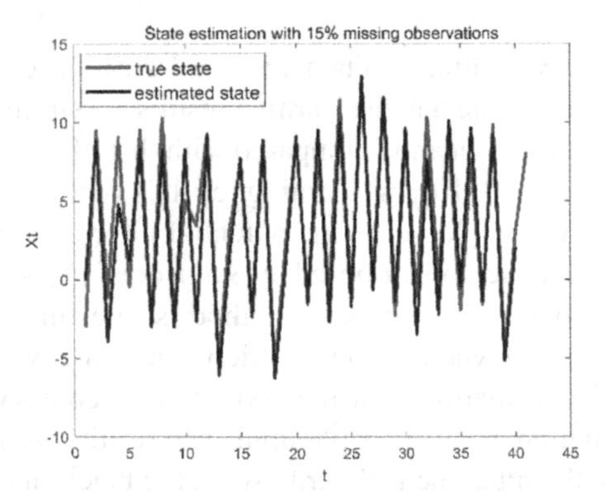

Fig. 8. State estimation with 15% outputs missing based on our method.

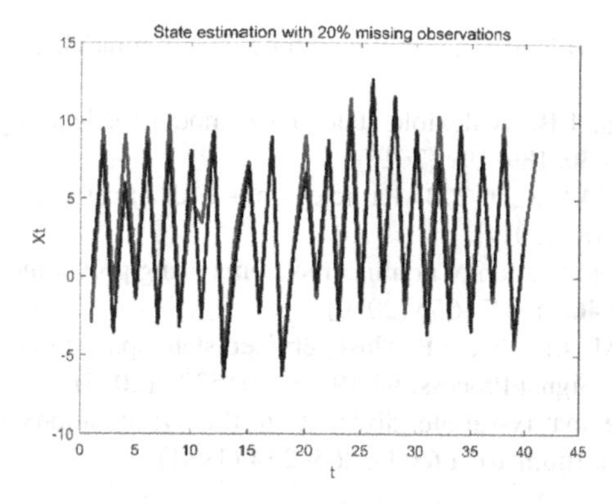

Fig. 9. State estimation with 20% outputs missing based on our method.

6 Conclusion

The learning methods of nonlinear state-space models have been greatly developed. However, when the SSM is a black box and the observations are missing, learning will become very difficult. We improve the method proposed by Svensoon [11], and propose a learning algorithm to learn a state-space model with missing observations. The coefficients of the basis function expansions are learned from missing observations. The effectiveness of the proposed method is demonstrated through a numerical example. The learning result is close to the true function when the observations missing percentages are less than 15%. Our learning result is also close to that of Svensoon [11], which shows that our method is robust to incomplete data. On the other hand, the deviation of learning results increases significantly when the observations missing percentages are larger than 15%.

References

1. Sinha, N.K.: System identification—theory for the user: Lennart Ljung. Automatica **25**(3), 475–476 (1989)
2. Sjöberg, J., et al.: Nonlinear black-box modeling in system identification: a unified overview. Automatica **31**(12), 1725–1750 (1995)
3. Godsill, S.J., Clapp, T.C.: Sequential Monte Carlo methods in practice (2001)
4. Ghahramani, Z., Hinton, G.E.: Variational learning for switching state-space models. Neural Comput. **12**(4), 831–864 (2000)
5. Blei, D.M., Jordan, M.I.: Variational inference for Dirichlet process mixtures. J. Bayesian Anal. **1**(1), 121–143 (2006)
6. Jacobs, W.R., et al.: Sparse Bayesian nonlinear system identification using variational inference. IEEE Trans. Automatic Control (2018)
7. Lei, C., et al.: Multiple model approach to nonlinear system identification with uncertain scheduling variables using EM algorithm. J. Process Control **23**(10), 1480–1496 (2013)
8. Rasmussen, C.E., Williams, C.: Gaussian Processes for Machine Learning (Adaptive Computation and Machine Learning). The MIT Press (2005)
9. Rasmussen, C.E., Nickisch, H.: Gaussian Processes for Machine Learning (GPML) toolbox. J. Mach. Learn. Res. **11**(6), 3011–3015 (2010)

10. Solin, A., Särkkä, S.: Hilbert space methods for reduced-rank Gaussian process regression. arXiv (2014)
11. Svensson, A., Schön, T.B.: A flexible state space model for learning nonlinear dynamical systems. Automatica **80**, 189–199 (2017)
12. Lindsten, F., Jordan, M.I., Schn, T.B.: Particle Gibbs with Ancestor sampling. J. Mach. Learn. Res. **15**(15), 2145–2184 (2014)
13. Paduart, J., et al.: Identification of nonlinear systems using polynomial nonlinear state space models. Automatica **46**(4), 647–656 (2010)
14. Tobar, F., Djuric, P.M., Mandic, D.P.: Unsupervised state-space modeling using reproducing kernels. IEEE Trans. Signal Process. **63**(19), 5210–5221 (2015)
15. Dawid, A.P.: Some matrix-variate distribution theory: notational considerations and a Bayesian application. Biometrika **68**(1), 265–274 (1981)

Parameter Identification of Nonlinear Systems Model Based on Improved Differential Evolution Algorithm

Liu Qian[1,2](✉), Lv Jianhong[2], Zhang Qiusheng[1], and Zhuo Hua[1]

[1] CHN Energy New Energy Technology Research Institute Co., Ltd., Beijing 102206, China
liuqian0614@126.com

[2] School of Energy and Environment, Southeast University, Nanjing 210000, China

Abstract. Aiming at the difficulty of optimizing the parameter estimation of nonlinear models, a new method for parameter identification of nonlinear system models based on improved differential evolution algorithm based on diversity evaluation index is proposed. By establishing a population reconstruction mechanism, based on the population diversity index, the concept of population similarity is proposed to guide the selection of evolutionary strategies and the adaptive adjustment of process parameters, thereby balancing the global and local search functions at different stages. The population size decreasing strategy effectively reduces the amount of computation and improves the convergence speed and algorithm efficiency. In order to verify the performance of the algorithm, several types of standard functions with typical complex mathematical characteristics are simulated and applied to the identification of a type of thermal system model parameters. The results show that the improved algorithm has high model parameter identification accuracy and faster convergence speed, which effectively improves the accuracy and efficiency of model establishment, and provides a feasible way to solve the model parameter identification problem in practical systems.

Keywords: Differential evolution algorithm · Nonlinear system · Parameter identification · Thermal processes

1 Introduction

System models and model parameter estimation are the basis of all control problems, but practical industrial processes are almost always nonlinear. At present, the identification of linear systems is very mature, and there are a large number of well-defined and very effective identification models and algorithms. Linearization, however, is a mathematical ideal, only an approximation to the actual system that may be effective in certain situations. Therefore, nonlinear system identification has become an important research topic [1]. Nonlinear models are generally complex, and it is not easy to obtain estimates of their parameters. Common methods for parameter estimation of nonlinear system models include least squares method and maximum likelihood method [2].

W. Fan et al. (Eds.): AsiaSim 2022, CCIS 1712, pp. 129–142, 2022.
https://doi.org/10.1007/978-981-19-9198-1_10

Evolutionary computation (EC), as another branch of artificial intelligence, adopts the computational model of the evolutionary process as the problem solving system, and provides a random search method to find the optimal solution to the target problem. The differential evolution (DE) algorithm was proposed by Storn et al. It belongs to the branch of evolutionary computing. It has few control parameters and is simple to implement. Which is an evolutionary algorithm based on population parallel random search, has shown good search ability in various test problems. But like other evolutionary algorithms, the standard DE algorithm will also fall into the local optimum point of the problem, which is easy to cause the algorithm to converge prematurely. By establishing the population reconstruction mechanism, this paper proposes an improved differential evolution algorithm based on the diversity evaluation index, performs numerical simulation on several typical standard functions, and performs nonlinear identification in one type of thermal process.

2 Problem Description

Parameter identification is to obtain parameter estimates by optimizing the objective function according to the system output and actual sampling data under a certain system input. Its essence is to transform the model parameter estimation problem into a nonlinear function optimization problem. The traditional optimization method is difficult to solve. Therefore, this paper uses the improved differential evolution algorithm to estimate the parameters of the nonlinear model. The general form of description for a nonlinear model is:

$$y(t) = f(x, \theta) + e, e \sim N(0, \delta^2) \tag{1}$$

where $y(t)$ is the output of the system; x is the input of the system; $\theta = (\theta_1, \theta_2, \cdots, \theta_k)$ is the parameter to be identified; e is the white noise with a mean of 0 and a variance of σ^2; f is a nonlinear function, which can be a transfer function, a state space, or an ARMA model. The principle of model parameter identification is shown in the Fig. 1.

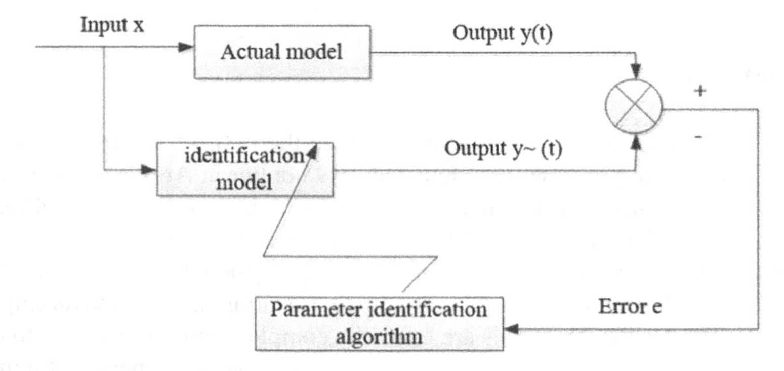

Fig. 1. Model parameter identification principle

3 Differential Evolution Algorithm

3.1 DE Algorithm Principle

The basic differential evolution algorithm is a heuristic search optimization algorithm based on the genetic algorithm to simulate the natural genetic and evolutionary mechanism [3]. The DE algorithm is simple and practical, and has the ability to memorize the individual optimal solution [4, 5]. The basic principle is: starting from a randomly generated initial group, a new individual is generated by weighting the vector difference of any two individuals in the population and then summing it with the third individual according to certain rules, and then the new individual is compared with the contemporary one. Through continuous iterative calculation, excellent individuals are retained, and inferior individuals are eliminated, and the search process is guided to approach the optimal solution until the accuracy requirements or stopping conditions are met [6, 7]. The differential evolution algorithm uses real number coding, which is easier to implement without decoding operation. It generates a mutation vector through the parent differential vector, crosses with the parent individual vector to generate the child vector, and directly selects with the parent individual vector, thus ensuring that the The algorithm has better convergence [8]. The basic steps of the DE algorithm are as follows:

a) Initialize the population. The initial population is randomly generated according to formula (2) in the feasible solution space.

$$X_{i,0} = X_l + rand(0, 1) \times (X_h - X_l)\ i = 1, 2, \cdots, NP \tag{2}$$

Among them, $X_{i,0}$ is the ith individual in the initial population; X_h and X_l are the upper and lower boundaries of the individual, NP is the population size.

b) Mutation operation. DE realizes individual mutation through the difference strategy, that is, randomly selects two different individuals in the population, scales their vector differences, and performs vector synthesis with the individual to be mutated [9]. Perform mutation operation on according to formula (3):

$$V_{i,t} = X_{r3,t} + F \cdot (X_{r1,t} - X_{r2,t}) \tag{3}$$

Among them, $X_{r1,t}, X_{r2,t}, X_{r3,t}$ are random individuals in the current population; F is the variation scaling factor, and $X_{r1,t} - X_{r2,t}$ is the difference vector.

c) Cross operation. Crossover is an operation performed on the generated mutant individuals, and it is a mechanism to ensure the exchange of information between individuals. Through gene exchange, new experimental individuals can inherit the excellent characteristics of the parent and mutant individuals, thereby further increasing the diversity of individual vectors [10]. Commonly used crossover operations are as follows:

$$U_{i,t}(j) = \begin{cases} V_{i,t}(j) \text{ if rand}(0, 1) \leq CR \, or \, j = j_{rand} \\ X_{i,t}(j) \text{ otherwise} \end{cases} \tag{4}$$

Among them, $j_{rand} \in [1, 2, \cdots, D]$ is a random number, CR is the crossover probability, and the value of CR represents the probability that the generated experimental individual inherits the gene of the variant individual.

d) Choose an action. The selection operation refers to selecting excellent individuals with a certain probability to form the next generation population, and its function is to ensure the reproduction and renewal of the population [11]. The selection operation is as follows:

$$X_{i,t+1} = \begin{cases} U_{i,t} \text{ if } f(U_{i,t}) \leq f(X_{i,t}) \\ X_{i,t} \text{ otherwise} \end{cases} \tag{5}$$

where f is the fitness function. $f(U_{i,t})$ is the fitness value corresponding to the experimental individual.

3.2 Improved Differential Evolution Algorithm

Population Similarity. The degree of individual difference in the population will gradually decrease with the increase of evolutionary algebra, which will lead to the premature convergence of the algorithm to the local extreme point and the phenomenon of premature maturity [12]. This paper takes population diversity as an important indicator, and proposes the concept of population similarity to guide the selection of evolutionary strategies and the adaptive change of parameters. The standardized Euclidean distance is introduced to calculate the similarity between individuals [13]. The similarity between individuals X_1 and X_2 in the population is defined as follows:

$$S_{X_1,X_2} = \frac{1}{1 + d_{X_1,X_2}} \tag{6}$$

where d_{X_1,X_2} is the standardized Euclidean distance between the individual X_1 and X_2. The smaller the distance between the two bodies, the higher the similarity, and if the similarity is 1, it means that the information of the two bodies is exactly the same. According to the individual similarity, the population similarity of S_z the following form can be defined:

$$\begin{aligned} S_{ib} &= \frac{\sum_{i=1}^{Np} S_{X_i,X_{best}}}{NP} \\ S_{iw} &= \frac{\sum_{i=1}^{Np} S_{X_i,X_{worst}}}{NP} \\ S_z &= \frac{\min(S_{ib}, S_{iw})}{\max(S_{ib}, S_{iw})} \end{aligned} \tag{7}$$

Among them, X_{best} is the best individual of the current population; X_{worst} is the worst individual of the current population; S_{ib}, S_{iw} are the average similarity between other individuals in the population and the best and worst individuals; S_z is the similarity of the current population.

Parameter Adaptation. The main process parameters of the differential evolution algorithm include scaling factor F, crossover probability CR and population size NP. Under normal circumstances, directly selecting fixed parameter values based on experience to complete the entire calculation process has a high convergence rate. It is necessary to adjust online parameters according to the information feedback in the calculation process, and combine the parameter adaptive mechanism with the population diversity index, so as to meet the different needs of global search and local development capabilities at different stages.

In this paper, the adaptive adjustment method of scaling factor F and crossover probability CR is selected as follows:

$$F = F_0 \cdot 2^{\exp(1-1/S_z(t))} \tag{8}$$

$$CR = CR_0 \cdot e^{-\lambda/S_z(t)} \tag{9}$$

Among them, F_0 and CR_0 are respectively the scaling factor and the initial value of the crossover probability; $S_z(t)$ is the t-th generation population similarity; λ is the coefficient.

The strategy that the population size NP dynamically decreases with the increase of the number of iterations can improve the local optimization ability in the later stage of the algorithm, and can effectively reduce the amount of online calculation and speed up the convergence speed. Therefore, this paper adopts the strategy of decreasing population size as follows:

$$NP_{t+1} = NP_0 - round(\frac{NP_0 - NP_{\min}}{e^{\mu \cdot S_z(t)}}) \tag{10}$$

Among them, NP_0 is the initial population size; $NP_{\min}$ is the set minimum population size; μ is the coefficient.

Evolution Strategy. Different mutation strategies of DE algorithm have their advantages. Therefore, in order to improve the performance of the DE algorithm, it is necessary to balance the search ability and development ability of the mutation strategy to meet the performance requirements of the algorithm at different stages. This paper proposes a mixed mutation strategy based on population diversity:

$$V_{i,t} = \begin{cases} X_{r1,t} + F \cdot (X_{r2,t} - X_{r3,t}) \text{ if } rand(0,1) > (1 - S_z(t))^{1.2} \\ X_{i,t} + F \cdot (X_{tbest} - X_{i,t}) + F \cdot (X_{r1,t} - X_{r2,t}) \text{ otherwise} \end{cases} \tag{11}$$

Among them, $X_{r1,t}$, $X_{r2,t}$, $X_{r3,t}$ are random individuals in the current population, and $r1 \neq r2 \neq r3$.

The hybrid mutation strategy can maximize the use of the known information in the calculation process, adapt to the performance requirements of the algorithm at different stages, balance the global convergence ability and the local optimal search ability on the whole, and improve the performance of the DE algorithm. In addition to the improvement of mutation operator, this paper also adopts the operation of population reconstruction to further improve the adaptability of evolution strategy in different stages.

A threshold σ is set as the judgment condition for population reconstruction. When the similarity of the population is higher than the threshold, that is $\sigma < S_z(t)$, the following population reconstruction operations are performed:

$$X_{i,t+1}(j) = X_{tbest}(j) + rand(-1,1) \cdot (X_h(j) - X_l(j)) \tag{12}$$

IDE Algorithm Flow. The improved DE algorithm flow is shown in Fig. 2, and the specific steps are as follows:

(a) Set the algorithm parameters, including the maximum number of iterations $T_{\max}$, the precision ε, the initial value of the population size NP_0, the initial value of the scaling factor F_0, the initial value of the crossover probability CR_0, the population similarity threshold σ and the correlation coefficient, etc.
(b) Set the corresponding fitness function $f(X)$ according to the specific optimization problem
(c) According to formula (2), the initial population is generated and the similarity of the initial population is calculated.
(d) Determine whether the current population meets the accuracy requirements or reaches the maximum number of iterations. If the conditions are met, the iteration is terminated and the optimal solution is output; otherwise, continue to step (5).
(e) The mutation operation is performed according to formula (11), wherein the adaptive scaling factor F value is calculated according to formula (8).
(f) The crossover operation is performed according to formula (4), wherein the value of the adaptive scaling factor CR is calculated according to formula (9).
(g) Select operation according to formula (5).
(h) Calculate the next-generation population size NP_{t+1} according to formula (10), and select NP_{t+1} individuals according to the fitness value to obtain a new population.

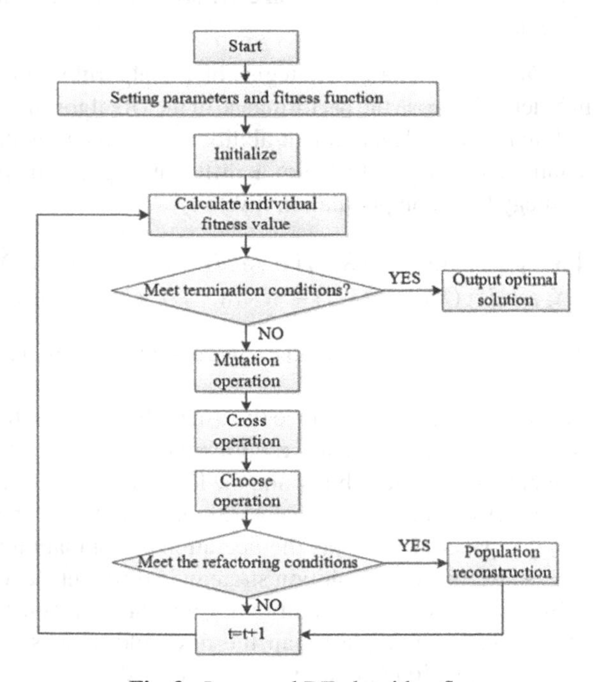

Fig. 2. Improved DE algorithm flow

(i) Calculate the similarity value of the current population according to formula (7), and judge whether the reconstruction conditions are met. If it matches, perform population reconstruction operation according to formula (12) to obtain the next-generation population; otherwise, directly use the current population as the next-generation population, and go to step (4).

3.3 IDE Algorithm Performance Test

In order to verify the effectiveness of the improved algorithm, this paper selects four standard functions with more complex mathematical characteristics to test the performance of the optimization algorithm. The selected test functions are all multi-extremum nonlinear functions, and the specific information is shown in Table 1.

Table 1. Test function

Function	Boundary range	Dimension	Form	Global extrema
Camel	[−2, 2]	2	$f_2(x, y) = (4 - 2.1x^2 + \frac{x^4}{3})x^2 + xy + (-4 + 4y^2)y^2$	−1.0316
Griewank	[−500, 500]	20	$f_3(x) = \sum_{i=1}^{D} \frac{x_i^2}{4000} - \prod_{i=1}^{D} \cos(\frac{x_i}{\sqrt{i}}) + 1$	0
Rosenbrock	[−10, 10]	20	$f_6(x) = \sum_{i=1}^{D} \left[100\left(x_i^2 - x_{i+1}\right)^2 + (1 - x_i)^2 \right]$	0
Sphere	[−10, 10]	20	$f_8(x) = \sum_{i=1}^{D} x_i^2$	0

In order to better verify the effectiveness of the improved DE algorithm, this paper chooses the basic differential evolution algorithm, basic genetic algorithm (GA), particle swarm algorithm (PSO) to compare with the improved DE algorithm, and jointly optimize the above selected standard functions. Performance Testing. In order to ensure the fairness of the experiment, Matlab 2018a software was used for performance testing on the same computer. The computer configuration is: memory 8G, processor AMD Ryzen5 2500U@2.00 GHz, 64-bit WIN10 operating system. All algorithm termination conditions are set to the maximum number of iterations, and the iteration precision. In addition, the genetic algorithm uses the single-point crossover operator, the basic mutation operator and the roulette selection operator in the test. The population size of the three basic algorithms is set to $NP = 100$, the improved DE algorithm is set to the initial population size $NP_0 = 100$, and the other parameters are set see Table 2.

In order to test the performance of the algorithm in terms of convergence, stability and computational efficiency, based on the above conditions and parameter settings, formula (2) is used as the initialization method, and 20 times of optimization calculations

Table 2. Algorithm control parameter settings

Optimization	Parameters	Value
PSO	Inertia weight ω	0.759
	Learning factor c_1	1.49445
	Learning factor c_2	1.49445
GA	Crossover probability P_c	0.9
	Mutation probability P_m	0.01
DE	Scaling factor F	0.5
	Crossover probability CR	0.3
IDE	Initial scaling factor F_0	0.5
	Initial Crossover Probability CR_0	0.3
	Lower limit of population size NP_{min}	50
	Population similarity threshold σ	0.84

are performed on the four standard test functions selected above, and the records are recorded. The best value, worst value, average value and total optimization time of each result. The specific data are shown in Table 3, and the convergence of the fitness values of the four standard functions during the test is shown in Figs. 3, 4, 5, and 6.

Table 3. Standard function test result record table

Test function	Fitness value	PSO	GA	DE	IDE
Camel	Optimal	−1.03E+00	−1.03E+00	−1.03E+00	−1.03E+00
	Average	−9.91E−01	−1.03E+00	−1.03E+00	−1.03E+00
	Worst	−2.15E−01	−1.03E+00	−1.03E+00	−1.03E+00
	Time/s	6.52	30.03	22.47	18.95
Griewank	Optimal	1.48E−02	2.46E−05	0	0
	Average	8.66E−02	3.08E−03	0	0
	Worst	1.82E−01	3.01E−02	0	0
	Time/s	8.20	31.99	22.01	17.63
Rosenbrock	Optimal	0	1.03E−02	2.75E−02	3.50E−03
	Average	1.97E−01	2.54E−01	2.86E−01	1.76E−01
	Worst	3.93E+00	7.75E−01	8.69E−01	3.73E−01
	Time/s	4.19	17.86	15.49	13.38

(continued)

Table 3. (*continued*)

Test function	Fitness value	PSO	GA	DE	IDE
Sphere	Optimal	6.92E−07	5.28E−19	3.85E−22	9.02E−41
	Average	6.18E−05	1.35E−17	1.22E−20	5.88E−29
	Worst	3.17E−04	3.01E−16	2.60E−16	2.88E−27
	Time/s	3.81	17.16	14.84	11.57

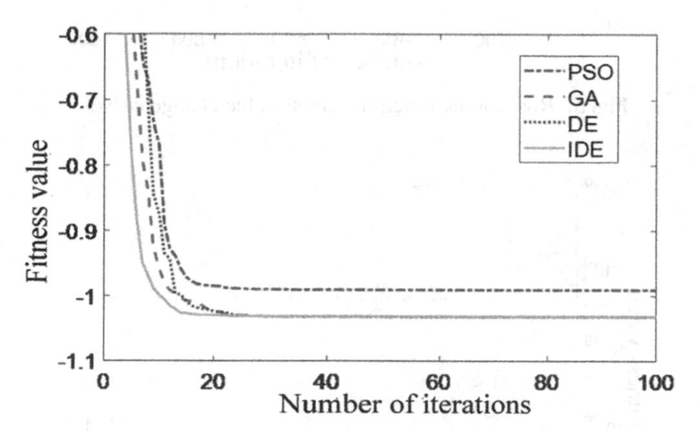

Fig. 3. Camel function fitness value change curve

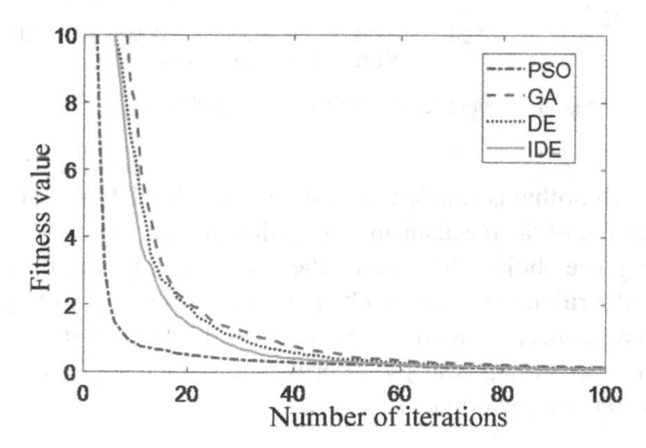

Fig. 4. Griewank function fitness value change curve

It can be seen from the data in the table that the optimal values of the IDE algorithm for the test results of the four standard functions are all the smallest, which are closer to the theoretical optimal solution, so it can be verified that the improved algorithm can effectively improve the convergence accuracy. Comparing the best, worst and average values of each group of test results, it can be seen that the difference between the

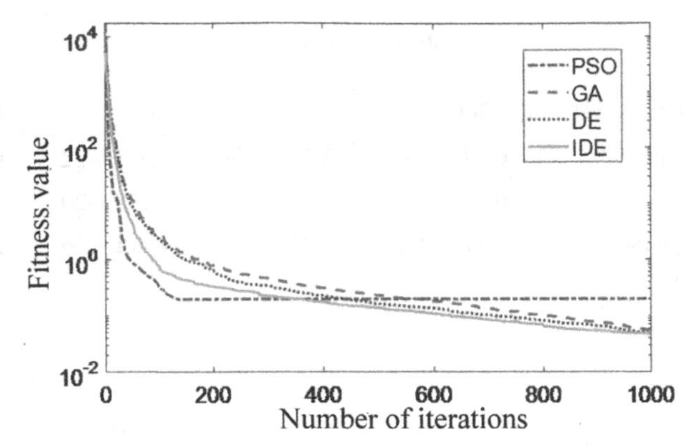

Fig. 5. Rosenbrock function fitness value change curve

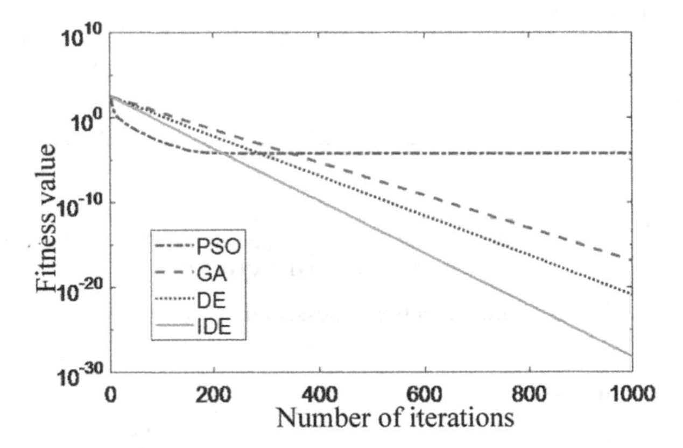

Fig. 6. Sphere function fitness value change curve

results of the IDE algorithm is smaller. This shows that the IDE algorithm has stronger stability due to the use of the mechanism of population reconstruction, which improves the global convergence ability and reduces the risk of falling into the local optimum. Overall, the IDE algorithm has obvious advantages in the accuracy of optimization and the rapidity of convergence, and it effectively reduces the amount of calculation through the population size decreasing strategy, improves the efficiency of the algorithm, and has better comprehensive performance.

4 Application of IDE Algorithm in Thermal System Model Parameter Identification

Parameter identification in thermal systems usually refers to the use of optimization algorithms to determine unknown parameters based on collected industrial field data under the premise of assuming or known object model structure. In this paper, the outlet temperature of the final superheater of a 1000 MW unit is selected as the object for model parameter identification. Taking the opening of the desuperheater valve as the control quantity, the outlet temperature of the final superheater is the controlled quantity. According to the characteristics of the steam temperature object, the opening of the desuperheating water valve U versus the amount of desuperheating water D, the amount of desuperheating water D versus the superheater inlet temperature T_i (leading area) of the final superheater, and the final superheater inlet temperature T_i versus the superheater are constructed respectively. The transfer function model structure of the three links of the outlet temperature T_o (inert zone) is as follows:

$$W_{U-D} = G_1(s) = \frac{K_1}{(T_1 s + 1)^{n_1}} \tag{13}$$

$$W_{D-T_i} = G_2(s) = \frac{K_2}{(T_2 s + 1)^{n_2}} \tag{14}$$

$$W_{T_i-T_o} = G_3(s) = \frac{K_3}{(T_3 s + 1)^{n_3}} e^{-\tau s} \tag{15}$$

Among them, n_1, n_2, n_3 is a positive integer, representing the model order; T_1, T_2, T_3 is the inertia time; K_1, K_2, K_3 is the static gain; τ is the delay time.

When using the field experimental data to identify the model parameters, it is necessary to carry out the dynamic characteristics test of the temperature of the final superheater. The command remains unchanged, and the opening of the desuperheating water valve is increased by 10% in steps. After the system is stable, the opening of the desuperheating water valve is reduced by 10% stepwise, and the experiment is terminated when the system resumes stable operation. The PSO algorithm, GA algorithm, DE algorithm and IDE algorithm are used to identify the parameters of the transfer function of the superheated steam temperature object of the unit. The model structure is shown in formulas (13–15), and the parameters to be identified are $(K_1, K_2, K_3, T_1, T_2, T_3, n_1, n_2, n_3, \tau)$. In order to ensure the accuracy of parameter identification, the goal is to minimize the deviation between the model output and the actual operating data of the unit, and the fitness function is set as follows:

$$f_{\min} = \sum_t \frac{y^r_{T_o}(t) - y_{T_o}(t)}{|T_{o0}|} \tag{16}$$

Among them, $y^r_{T_o}$ is the actual superheated steam temperature value of the unit; y_{T_o} is the model output superheated steam temperature value; T_{o0} is the superheated steam temperature value under rated operating conditions.

The range of values to be determined is as follows:

$$K_1 \in (0, 5), K_2 \in (-1, 1), K_3 \in (-10, 10)$$
$$T_1 \in (1, 10), T_2 \in (1, 50), T_3 \in (1, 100)$$
$$n_1 \in [1, 2], n_2 \in [1, 3], n_3 \in [1, 5], \tau \in (5, 80)$$

When performing parameter identification, set the maximum number of iterations $T_{\max} = 100$ of the algorithm and the iteration accuracy $\varepsilon = 10^{-1}$. The results of parameter optimization are shown in Table 4 and Table 5:

Table 4. Parameter identification results of four algorithms

Optimization	Load/MW	K_1	K_2	K_3	T_1	T_2	T_3
PSO	980	0.197	−0.518	6.243	3.55	23.12	31.38
	750	0.432	0.512	−3.052	2.07	7.694	65.11
	540	0.852	−0.522	2.683	7.62	42.92	47.73
GA	980	0.131	0.231	−6.443	2.21	11.68	52.22
	750	0.152	0.641	−6.684	6.37	18.69	73.12
	540	0.725	−0.691	4.365	3.34	13.38	83.79
DE	980	0.352	−0.358	3.944	1.67	6.69	67.11
	750	0.231	−0.658	6.244	1.37	5.64	48.73
	540	0.241	0.472	7.874	3.2	18.55	90.32
IDE	980	0.939	−0.425	1.125	1.89	13.01	55.67
	750	1.17	−0.859	0.87	3.92	16.74	61.46
	540	2.12	−0.874	3.12	2.74	21.85	87.52

Table 5. Parameter identification results of four algorithms

Optimization	Load/MW	n_1	n_2	n_3	τ	$f(X)$
PSO	980	1	2	5	7.02	0.035
	750	1	2	4	37.05	0.027
	540	1	2	4	32.47	0.031
GA	980	1	2	4	19.83	0.028
	750	1	2	4	16.97	0.03
	540	1	3	4	4.96	0.022
DE	980	1	2	4	26.59	0.013
	750	1	3	5	7.04	0.018
	540	1	2	4	8.39	0.015
IDE	980	1	2	4	23.75	0.009
	750	1	3	4	25.88	0.012
	540	1	2	4	40.47	0.01

The model output curve when the desuperheating water valve opening degree changes by 15% is compared with the actual temperature response curve on site, and the results are shown in Figs. 7, 8, and 9.

Although the model output is affected by factors such as on-site noise and disturbance, there is a certain deviation from the actual curve, but under different working conditions, the parameter models identified by the four optimization algorithms can

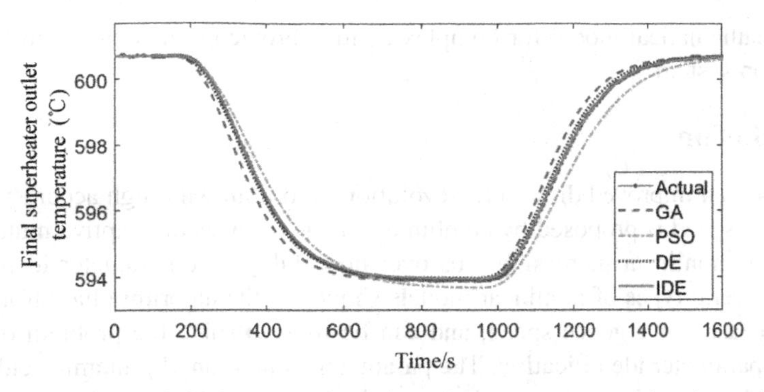

Fig. 7. Comparison of 980 MW superheated steam temperature model and actual output

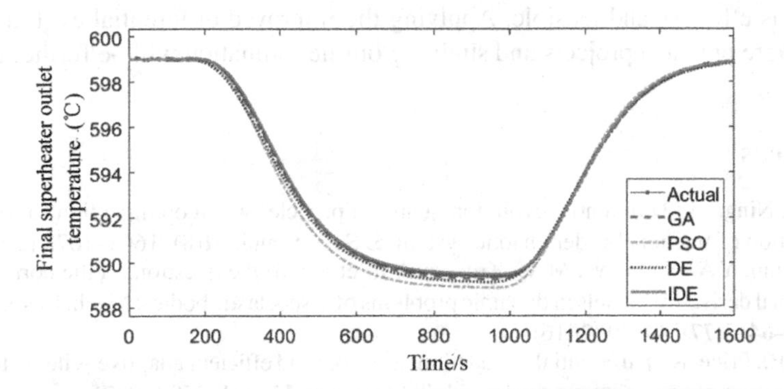

Fig. 8. Comparison of 750 MW superheated steam temperature model and actual output

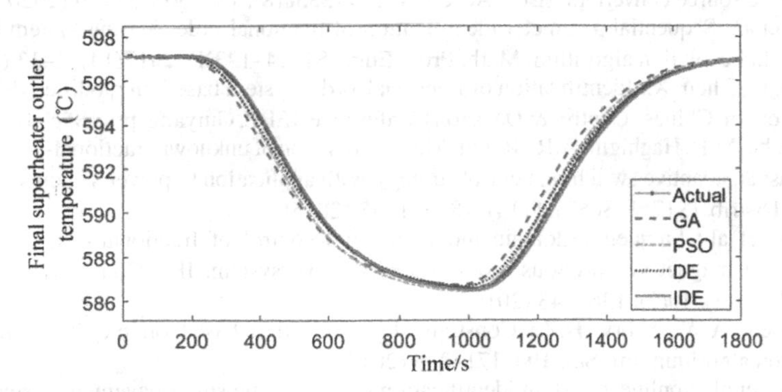

Fig. 9. Comparison of 540 MW superheated steam temperature model and actual output

more accurately describe the superheated steam temperature with the opening of the desuperheating water valve. Change trend, and ensure that the deviation is within the allowable range. It can be seen from the above comparison curves that, compared with the other three algorithms, the dynamic and steady-state deviations of the models identified by the IDE algorithm are smaller and have higher accuracy. Therefore, using IDE algorithm for model parameter identification of thermal system can provide a more accurate mathematical model for complex control problems such as machine-furnace coordination system.

5 Conclusion

In this paper, an improved differential evolution algorithm with high accuracy and fast convergence speed is proposed by combining the advantages of adaptive mutation rate and dynamic nonlinear increasing crossover probability. The parameter identification results of several types of nonlinear models show that the algorithm has strong search ability and fast convergence speed, and can be used to solve the problem of nonlinear model parameter identification. The parameter estimation of glutamic acid bacteria growth model is used as an example to verify the application, which proves that the nonlinear model parameter identification based on the improved differential evolution algorithm is effective and feasible. Applying the improved differential evolution algorithm to more practical projects and studying online estimation will be further research goals.

References

1. Ze, D., Ning, M.: Differential evolution quantum particle swarm optimization for parameter estimation of fractional-order chaotic system. J. Syst. Simul. **31**(8), 1664–1673 (2019)
2. Rossikhin, Y.A., Shitikova, M.V., Krusser, A.I., et al.: To the question on the correctness of fractional derivative models in dynamic problems of viscoelastic bodies. Mech. Res. Commun. **S0093–6413**(77), 44–49 (2016)
3. Storn, R., Price, K.: Differential evoduation: a simple and efficient adaptive scheme for global optimization over continuous spaces. Global Optimiz. **11**, 341–359 (1997)
4. Sharma, M., Rajpurohit, B.S., Agnihotri, S., et al.: Development of fractional order modelling of voltage source converters. IEEE Access **S2169–3536**(8), 131750–131759 (2020)
5. Li, L., et al.: Sequential parameter identification of fractional-order duffing system based on differential evolution algorithm. Math. Prob. Eng. (S1024–123X). **2017**(14), 1–13 (2017)
6. Huang, L., Zhou, X.: Identification of fractional-order system based on modified differential evolution. In: Chinese Control & Decision Conference. IEEE, Guiyang, pp. 1856–1861 (2013)
7. Aghababa, M.P., Haghighi, A.R., Roohi, M.: Stabilisation of unknown fractional-order chaotic systems: an adaptive switching control strategy with application to power systems. IET Gen. Trans. Distrib. (S1751–8687). **9**(14), 1883–1893 (2015)
8. Yu, W., et al.: Frequency domain modelling and control of fractional-order system for permanent magnet synchronous motor velocity servo system. IET Control Theory Appl. (S1751–8644). **10**(2), 136–143 (2016)
9. Mohamed, A.W., Sabry, H.Z.: Constrained optimization based on modified differential evolution algorithm. Inf. Sci. **194**, 171–208 (2012)
10. Kim, Y., et al. Nonlinear system identification of large-scale smart pavement systems. Exp. Syst. Appl. **40**(9), 3551–3560 (2013)
11. Wang, L., Li, L.P.: Fixed-structure H∞ controller synthesis based on differential evolution with level comparison. IEEE Trans. Evolut. Comput. (S1089–778X). **15**(1), 120–129 (2011)
12. Lionel, L.S., Godpromesse, K., Andrew, M.F.: A new static synchronous series compensator control strategy based on RBF neuro-sliding mode technique for power flow control and DC voltage regulation. Electr. Pow. Comp. Syst. **46**(4), 456–471 (2018)
13. Jiang, T., Li, J., Huang, K.: Longitudinal parameter identification of a small unmanned aerial vehicle based on modified particle swarm optimization. Chin. J. Aeronaut. **28**(3), 865–873 (2015)

Research and Implementation of Model Engineering Environment Integration Based on OpenMBEE

Junjie Xue[1,2], Junhua Zhou[1,2,3], Guoqiang Shi[1,2,3], Chaoqun Feng[1,2], Lin Xu[4], Penghua Liu[5], and Hongyan Quan[5](✉)

[1] Engineering Research Center, Beijing Simulation Center, Beijing 100854, China
[2] State Key Laboratory of Intelligent Manufacturing System Technology, Beijing Institute of Electronic System Engineering, Beijing 100854, China
[3] Science and Technology on Space System Simulation Laboratory, Beijing Simulation Center, Beijing 100854, China
[4] Yunnan Chensheng Tendering Consulting Co., Ltd., Kuming 650102, China
[5] School of Computer Science and Technology, East China Normal University, Shanghai 200062, China
hyquan@cs.ecnu.edu.cn

Abstract. Aiming at the problem of integrating open tools in a user development environment in complex product design, an integration strategy of model engineering environment based on OpenMBEE is studied. It is based on the technologies of user identity unified authentication and user interaction data collaboration. Specifically, in the solution of unified identity authentication on cross-platform, the extended controller method is provided based on the message queue information flow mechanism. Then, the separation of business logic, data information, and interactive view of user unified authentication are realized, on the basis, that the function of unified identity authentication is realized by using the Software Development Kit (SDK) of OpenMBEE. Besides, in the study of user interaction data collaboration technology, three-layer collaboration mechanisms are proposed, they are the presentation layer collaboration, the business layer collaboration, and the persistence layer collaboration. When the jQuery programming framework is integrated into the bootstrap front-end framework, combing with the OpenMBEE back-end server interface, the collaboration of user interaction data is realized. Through experiments and project practice analysis, the user unified authentication strategy and user interactive collaboration strategy are tested respectively for proving the effectiveness and feasibility of the proposed method.

Keywords: OpenMBEE · Model · Identity authentication · Coordination

1 Introduction

Complex products have the characteristics of complex customer demand, complex design, and complex manufacturing processes. In recent years, with the rapid development of computer and artificial intelligence technology, complex products as the

W. Fan et al. (Eds.): AsiaSim 2022, CCIS 1712, pp. 143–157, 2022.
https://doi.org/10.1007/978-981-19-9198-1_11

national strategy and the pillar of the national economy, its production, and its research and development technology are undergoing tremendous changes. Among the existing technologies, since the concept of the digital twin was put forward by Professor Michael Grieves in 2005 [1], digital twin technology has been widely used and become a new technology attracting much attention in complex product design and manufacturing.

As complex product engineering involves multi-disciplinary and multi-domain knowledge, the project of digital twin technology is often organized by Model-based Systems Engineering (MBSE) [2] to solve the problems of high integration and the difficulty in coupling multi-domain tool modeling. Complex product design is realized through the evolution and iteration of model design, so as to realize the integration and collaboration from different tool models. In order to facilitate the engineering development of complex products, NASA developed OpenMBEE [2], an open-source design environment for complex products based on MBSE. It integrates various modeling tools and provides components conducive to team collaboration, including the components of a front-end View Editor (VE) for product design, Model Management System (MMS), and Model Development Kit (MDK).

Recently, people have carried out some research works on the basis of the OpenMBEE environment, such as the agile modeling technology based on MBSE [3], executable model strategy in the open environment [4], and document flow technology research based on model engineering [5], etc. This paper explores the integration method based on unified user identity authentication and user interaction data collaboration. In order to make full use of the OpenMBEE environment to realize user task collaboration in complex product design, although there have been some related studies, such as heterogeneous data collaboration technology [6, 7], heterogeneous model data fusion method for complex product collaborative design [8], the integration of OpenMBEE with user development environment has not been discussed deeply. If the OpenMBEE platform tools can combine with the user design environment seamlessly, cross-platform software development, integration verification, and other bi-directional interaction can be realized.

In this work, the model-based engineering environment integration study is divided into two key technical problems: unified user identity authentication and user interaction data collaboration. On the basis of solving the problem of unified authentication, it further explores the technical problem of interaction collaboration. We take the user development platform "Skylark" as an example of a design environment to study.

In the existing related works, although there are some unified user authentication technologies on the internet environment [9–12], compared with them, the integration problem of OpenMBEE is more complex and it needs to solve the communication problem between different platforms and the communication of different components, so the research problem is more complicated. Based on Model View Controller (MVC) architecture, OpenMBEE integration of complex design background is different from existing data synchronization and collaboration problems [13, 14], the integration of the complex system also needs to increase the communication function between the application development platform and OpenMBEE to realize the dynamic collaboration and sharing of cross-platform model data.

The innovation of this research work is as follows. Firstly, a model engineering environment integration method based on unified user identity authentication technology and user interaction data collaboration technology is proposed. Secondly, a solution of cross-platform unified identity authentication based on a message queue flow mechanism is proposed, and an extended controller mechanism is proposed to effectively solve the problem of unified authentication cross-platform. At last, it studies the cooperation strategy of user interaction, namely, realizing the dynamic cooperation of the presentation layer, the business layer, and the persistence layer respectively, realizing the synchronization of user interaction in the three-layer cooperation function.

2 Unified Authentication Method Based on Extended Services

In order to realize unified authentication based on the complex product design, this paper analyzes the existing architecture of OpenMBEE and then studies the mechanism of the unified authentication based on the controller layer extension service.

2.1 Three-Layer Service Framework

In OpenMBEE environment integration, we design functional structure of a unified user authentication system based on the inherent system structure of OpenMBEE, which is on MVC architecture (shown in the left part of Fig. 1) and has good scalability providing the necessary conditions for extending the controller layer.

In the study, we take full use of the controllability to extend the controller functions. The unified authentication system architecture is set up based on RabbitMq [15], shown in the right part of Fig. 1. We design the authentication interface as a controller service layer in the authentication system within the system framework. In addition, the view layer that corresponds to the front-end service of MVC architecture can provide the basic conditions of data presentation and information interaction for the unified authentication interface service.

We take full use of model-multi-view reusability and independence of modules in MVC three-tier services [16, 17], that is, changing some functions of a module will not affect the functions and logic of content at other levels. Besides, the loosely coupled components are established and extended through the service of the control layer, and strong decoupling is used for extension without affecting the functions of other model layers. In the design of the authentication service interface, the authentication interface class SyncUserContr is defined, as shown in Fig. 2. This enables the system can respond to the unified user authentication service of the Skylark platform.

In the design of the extension service, the SDK of OpenMBEE is employed to realize the function of the class SyncUserContr. In addition, other classes are also defined including RabbitMQUtils, SyncUserAndOrg and ConvertMsg, as shown in Fig. 3.

These classes are used in establishing the connections of message queue components in RabbitMq [15], synchronizing data such as user groups and users, converting data formats, and cleaning data, and so on. When we design the unified authentication function, the organization and the user identity user information are considered. To achieve the data format compatibility, the RabbitMq [15] message in the queue data needs to be

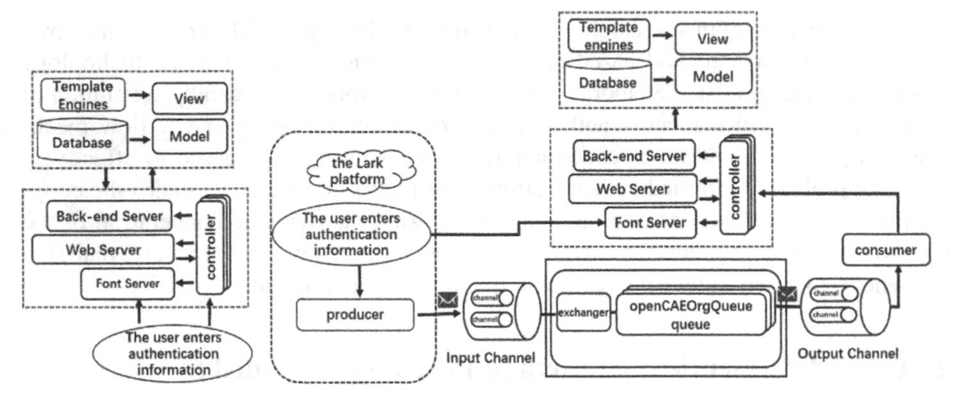

Fig. 1. The architecture of the OpenMBEE system and unified authentication system. The left one is the inherent architecture of the OpenMBEE platform and the right one is the architecture of our proposed unified authentication system.

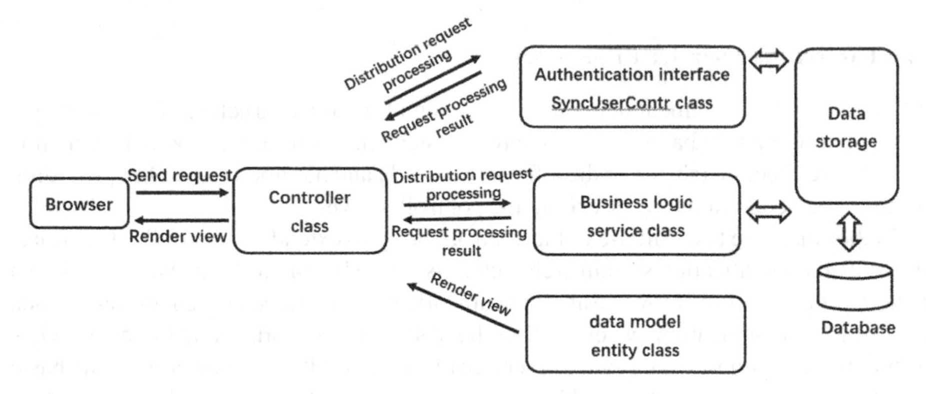

Fig. 2. The designed structure with controller function expansion.

converted to the OpenMBEE data format (called the standard format for simplification). So, we define the class ConvertMsg to convert data formats, where the dynamic function converUserPo converts user identity information into a standard format, then the dynamic function converOrgPo converts user group information into a standard format.

In addition, the unified authentication interface class SyncUserContr defines some dynamic members used for logical processing and data synchronization of standard user organization information and user identification information. These members are described in Table 1.

The user authentication process has some differences after the expansion of functional services. Before the expansion, the business process of user login function in the OpenMBEE inherent architecture is described as follows.

Step 1: The user inputs authentication information and sends an authentication request.
Step 2: The authentication system sends the user request to the controller to obtain the authentication information.

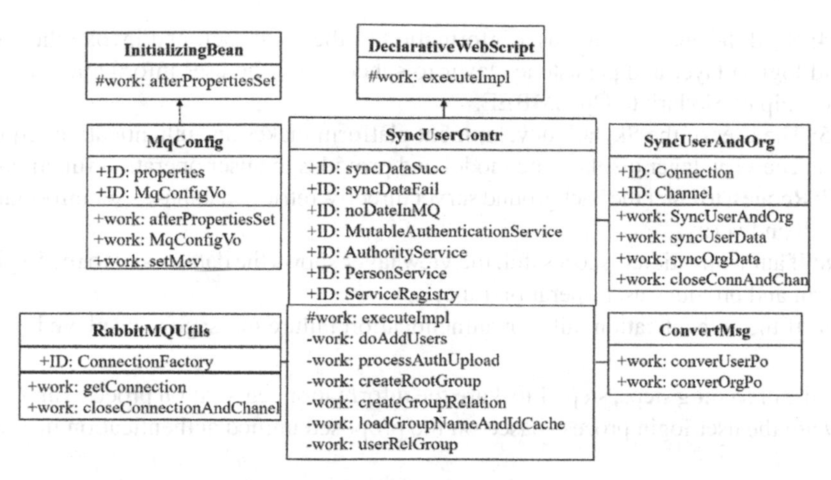

Fig. 3. The structure of the unified authentication interface class.

Table 1. The main dynamic member of the SyncUserContr class.

Dynamic member name	Function
doAddUsers	Add a new user
processAuthUpload	Upload the authentication information to server
createRootGroup	Create the group information
createGroupRelation	Create the relationship between user groups
loadGroupName-AndIdCache	Load the identity of the user group to serve
userRelGroup	Associate an identity with a user group

Step 3: The system accesses the database and retrieves the user information.
Step 4: Compare authentication data to verify the validity of users.
Step 5: If authentication is successful, the view layer presents the data model and provides user operation functions.
Step 6: If the authentication fails, an authentication failure message is displayed.

Whereas in our extended structure, the user login function needs to be decoupled and packaged through messages, and the user authentication process is described as follows:

Step 1: The Skylark user asks for the OpenMBEE synchronization authentication.
Step 2: The view layer displays the data model and provides the administrator with a synchronization entry.
Step 3: The RabbitMq message queue receives user information and decouples and encapsulates the information.

Step 4: Send the encapsulated user information to the controller, and invoke the background logical layer and persistence layer to synchronize the user information and the relationship of Skylark to OpenMBEE.
Step 5: The user of the Skylark development platform makes an authentication request.
Step 6: The view layer displays the model and provides the user operation functions.
Step 7: Request to enter the background server directly, retrieve database user information in MMS and verify.
Step 8: If authentication is successful, the view layer shows the data model to the Skylark platform and provides user operation functions.
Step 9: If the authentication fails, an authentication failure message is displayed.

In the preceding steps, steps 1 to 4 are the information registration process, and steps 5 to 9 are the user login process based on the proposed unified authentication method.

2.2 Authentication Interface Design Based on RabbitMQ

In this study, we take the software package RabbitMq [15] to designed. Based on the framework of RabbitMq, decoupling function of communication between processes. The design is based on the RabbitMq authentication interface and uses RabbitMq messaging middleware to receive, store and forward messages. We design the RabbitMq's authentication interface as the following components: producer program, input channel, switch, message queue, output channel, and consumer program, as shown in Fig. 1. After OpenMBEE integration, the service process of the authentication interface is designed as follows:

1) *Initialize the configuration information.* It establishes the connection between the RabbitMq authentication interface and OpenMBEE, and the connection between the RabbitMq authentication interface and Skylark platform respectively. That establishes the bidirectional connections respectively, and it is necessary to build the message propagation channel between them, including configuring RabbitMq, setting the VM name, as well as providing the connection factory hostname, the port, virtual host, user name, and password. In our study, the virtual machine is set to /MMS, switch identifier is set as webAdminUserAndOrgExchange. Besides, two queues are established, named openCAEOrgQueue and openCAEUserQueue. These bidirectional connections are established by class RabbitMQUtils based on TCP protocol.
2) *Authentication information processing and conversion in the switch.* To be specific, as shown in Fig. 2, the producer program pushes the authentication information of the "Skylark" user to the input channel, and then, it classifies and extracts the information of user group and user identity, and stores the information in the openCAEOrgQueue and openCAEUserQueue to be collected by the consumer.
3) *Information extraction and cleaning.* Specifically, the consumer program pulls the user uniform authentication information (including user group and user identity information) from the output pipeline and further checks if the organization data for each new user exists in the system. it is treated as dirty data and discarded.

4) *Encapsulate the data in OpenMBEE standard format*. In the last step, if it is successful, the cleaned authentication information, and related data are further encapsulated and processed into OpenMBEE standard data format. Then, user identity and organization information set up the association between them, and then they are inserted into the MMS database.

The following details are also noted in the design of the unified user authentication. For example, considering that the user and organization information is sensitive information, only the access control with system administrator permission is adopted in the security mechanism design, that is, only the user with administrator permission can trigger the synchronization operation of the user and group data.

In addition, in the design of synchronous authentication information, to ensure the reliability of the system, the system needs to feedback on the synchronization status and error information to the system administrator in time, once dynamic phenomena such as abnormal synchronization status and receiving message abnormally during the process of receiving messages.

3 Integration and Collaboration of User Interaction Data

In this study, based on the MVC architecture of OpenMBEE, three layers of collaboration strategies are explored in user interaction data collaboration technology, namely the presentation layer, the business layer, and the persistence layer. The user-customized model data of the "Skylark" platform is used as an example to design the model data, and the collaboration between the user and the integrated OpenMBEE system is realized through the dynamic customization of the model data. Figure 4 shows the scheme structure of collaboration between user interaction data and the integration platform designed.

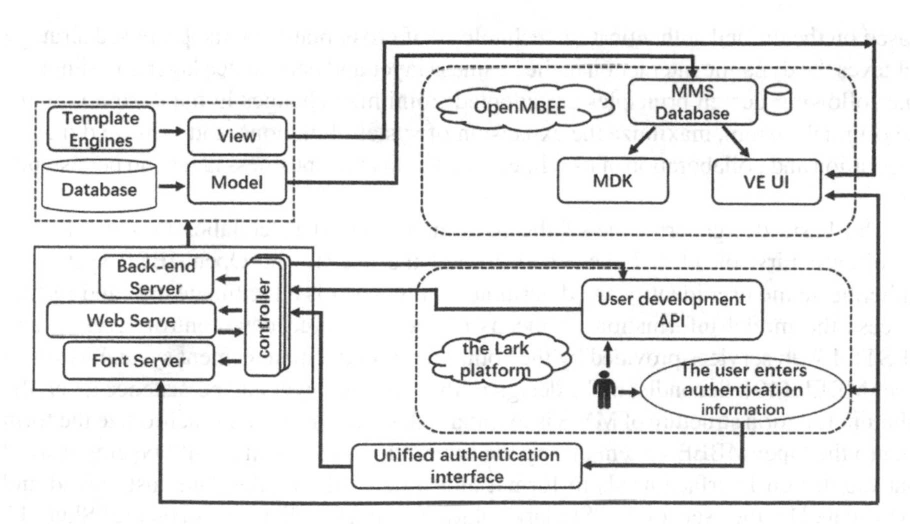

Fig. 4. The scheme diagram of user interaction data collaboration.

3.1 Collaborative Strategy of Presentation Layer

In the collaborative mechanism of the dynamic interaction. we consider how to extract dynamic customization information from the user customization data model of "Skylark" and how to communicate between the user dynamic interaction information and the server in coordination.

In order to obtain dynamic customization information from the user customization process of "Skylark", the research is designed based on the view layer of the "Skylark" platform, using CSS, HTML, JavaScript technology, and the front-end framework. In addition, the static HTML content of the page in development includes the header and the body part. The header contains the description information of user dynamic customization information and the main part contains the specific description information of the customized content, which is conducive to the extraction of the data information of the user-customized model. In addition, in the dynamic design, the jQuery programming framework of the front-end page is employed, and the event processing mechanism and encapsulation mechanism are used to control the DOM elements and page events of the page, so as to achieve the purpose of obtaining the customized information and web elements from various interactive behaviors of users. Based on the BootStrap static architecture [18] and dynamic rendering of jQuery pages, the purpose of dynamic collaboration in the presentation layer is realized in the Skylark platform.

In order to synchronize the user-customizing information to the OpenMBEE integrated environment, the MMS RESTful Web service framework in OpenMBEE is used for design [19]. For maintaining the representational state transition design style, the messages of data objects and attributes are transmitted in JSON format, and the user-defined HTTP methods are used to process the customized information.

3.2 Layer and Persistence Layer Design in Collaboration Strategy

Based on the unified authentication technology of cross-platform, the proposed strategy achieves the dynamic interaction in the business layer and persistence layer coordinately. The following design principles are adopted: minimize changes in the design logic of the original system, maximize the expansion of system functions, and achieve dynamic integration and collaboration of user interaction data in the business layer and persistence layer.

The basic design principle of the user interaction data collaboration function is as follows: First of all, it is necessary to add the user to the OpenMBEE system to authenticate the user identity and determine that the user is a legitimate user and further process the model information by means of the structured data control services and RESTful Web services provided by the component model management system MMS of OpenMBEE [19]. Secondly, in the design of the business layer and persistence layer, the inherent function structure of MMS is maintained, so that users can synchronize the form data to the OpenMBEE system during the interaction process. It should be emphasized that the design interface needs to format and process the model data customized and coordinated by the users of the "Skylark" platform, that is, edit the form on the "Skylark" platform, while maintaining the logic of the business layer and persistence layer of the data created by MMS.

3.3 User Interaction Process Based on Three-Layer Collaboration

In this study, the user from the "Skylark" platform customizing project model is taken as an example to realize the sharing of the OpenMBEE model. The dynamic process of user dynamic customization model data is designed, as shown in Fig. 5.

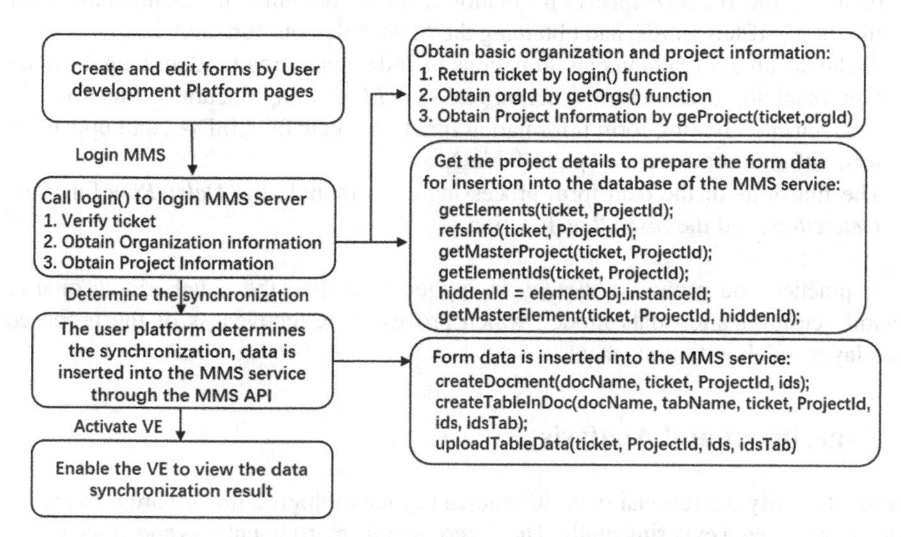

Fig. 5. User interaction flow based on collaboration mechanism.

The user interaction process based on the three-layer collaboration mechanism is described as follows:

1) Under the function of cross-platform unified authentication, users of the "Skylark" platform need to log in MMS system to obtain the user organization information and user project information, specifically, obtaining the effective user organization information to ensure that the login information of the authenticated user is valid and obtaining detailed information of the project to prepare for data collaboration.
2) User creates and edits data forms on the Skylark platform page.
3) The "lark" user sends a request for synchronous processing on the page, and the business layer cooperative persistence layer updates the user's synchronous data to MMS database system.
4) We start the VE and can view the data synchronization result.

In order to realize the collaborative processing function, the dynamic jQuery programming framework is employed to embed the static BootStrap framework [20] to design a function package for synchronous processing, completing the request interaction of user-customized data, and achieving the purpose of collaborative data processing. The designed function includes the following categories:

1) *Authentication information processing functions* include the user information login verification function, the *getOrgs* function is designed to obtain organization identifiers, and the getProject function is designed to obtain project information.
2) *The element obtaining class functions* include obtaining the element information function *getElements*, obtaining the stored *Refs* information function *refsInfo*, obtaining the *MasterProjects* information function, obtaining the element identifier function getElementIds, and obtaining the MasterElement function.
3) 3)The document fetching class functions include generating random identity function, obtaining user identity function *getDefineId*, creating document function createDocment, creating form information function createTableInDoc, and uploading form information function uploadTableData.
4) The functions of the data form processing class include the *DeleteRow* function, *DeleteRow*, and the *InsertRow* function.

In practice, the proposed strategy is applied to realize the collaboration of user dynamic customization data model, which proves the effectiveness of the proposed three-layer collaboration mechanism.

4 Experiment and Analysis

In order to verify the rationality of the studied key technologies, the research strategy is tested and analyzed experimentally. The experimental environment was constructed, and Centos7 was used as the operating system. Cameo19.0 SP, MMS3.4.2 and VE4.2.2 were installed. The cross-platform unified identity authentication strategy and user interaction data integration collaborative strategy are verified by experiments respectively.

4.1 Verifying the Unified Identity Authentication Policies

In order to verify the effectiveness of the cross-platform unified identity authentication strategy, we set up a unified identity authentication environment, the "Skylark" platform is selected as the user application design platform in this experiment. The following three groups of experiments were conducted.

(1) Tests of synchronizing user group and user ID information

First, the experiment simulates user registration information, including user identity information and organization information, further transfers the simulated user data to RabbitMq, and the visual tools provided by RabbitMq show that the received personnel information is correct, as shown in Fig. 6.

We use our studied approach to build a unified authentication system that receives user information from RabbitMq and synchronously updates it to OpenMBEE's MMS database. In our experiment, the iFresco tool [20] is can display the result of user information synchronization successfully, as shown in Fig. 7.

Overview				Messages		
Virtual host	**Name**	Features	**State**	**Ready**	**Unacked**	**Total**
/mms	**openCAEOrgQueue**	D	idle	1	0	1
/mms	**openCAEUserQueue**	D	idle	1	0	1

Fig. 6. The result of received messages from RabbitMq

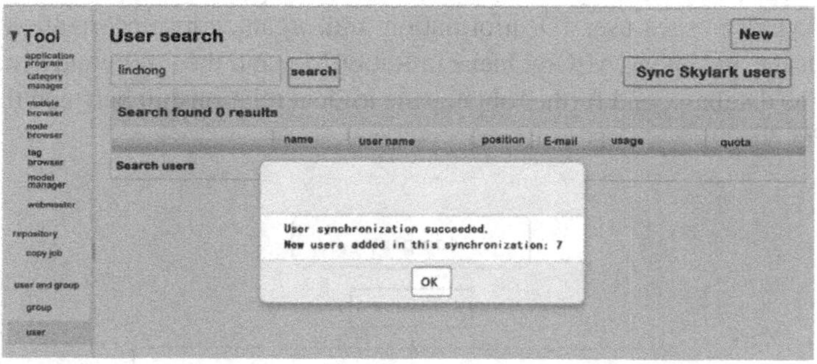

Fig. 7. The successful synchronization result of user information.

In addition, we synchronize and correlate user identity information and organizational information. Experiments show that user identity information and group information can be effectively synchronized and associated. Figure 8. shows the successful association results. From these results, it can be seen that the unified authentication provided in this paper can effectively obtain the basic information and organization information pushed by the Lark platform from the RabbitMq message queue, and can update the obtained synchronized authentication information to the MMS database. As the data pushed by the external system meets the format requirements specified by the system, the basic user information and organizational data can be successfully synchronized to the OpenMBEE system.

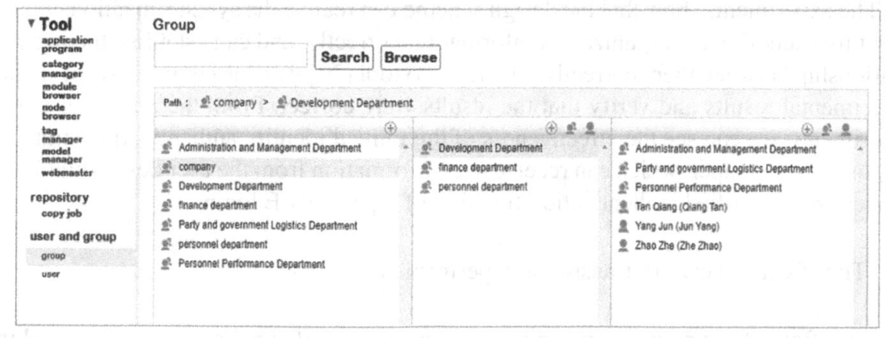

Fig. 8. The result of the user identification and group information being associated successfully

(2) Tests of correlating the organization and user identification

In this test, we extract the organization and user relationship and test the logical relationship of synchronization. In particular, the establishment of a test case, where the "corporation" contains "personnel department", "finance department" and "development department", and so on. This information will be used as valid information for user authentication of the "Skylark" platform. In the experiment, before data synchronization, Skylark pushes user ID information and organization information to openCAEUserQueue and openCAEOrgQueue in RabbitMq, and then pulls and synchronizes the data to the database, and further obtains the logical relationship between the data, as shown in Fig. 9.

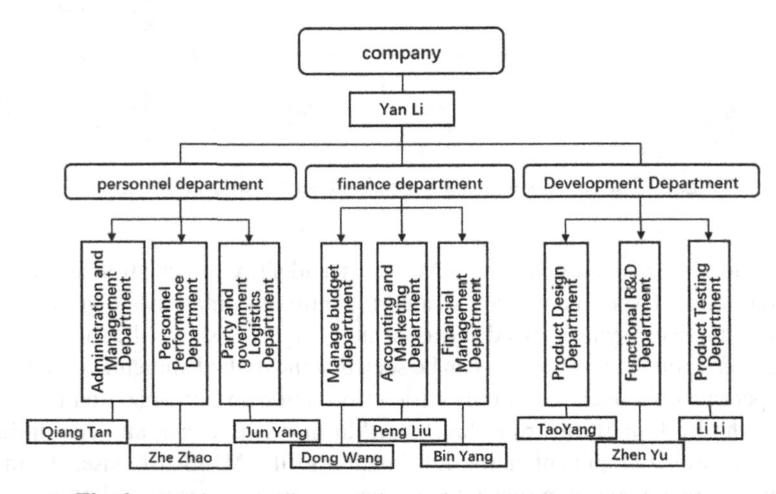

Fig. 9. Architecture diagram of organization and user relationship

The experiments show that the design scheme can realize the synchronization updating of user identity and organization information correctly, and can establish the logical relationship between them correctly. The iFresco tool [20] was further used to check the experimental results and verify that the results were correct. From the results of these experiments, we can see the effectiveness of the unified identity authentication strategy proposed in this paper, which can receive user information from the user design platform and realize the unified authentication function of OpenMBEE login.

(3) The effect of network transmission performance

The effect of the network transmission performance of synchronization is tested in the experiment. We have tested the design scheme under different conditions of stable and unstable network signal transmission. The test results are shown in Table 2, which shows three groups of experimental results: The first group of experiments is the results of stable network signal transmission, and the second and third groups of experiments are the results of unstable transmission.

In the column *Queue* of Table 2, 1 indicates the openCAEOrgQueue user organization queue, and 2 indicates the openCAEUserQueue user identity information queue.

Table 2. The synchronization results of RabbitMq reception information.

Group No.	Virtual machine	The queue	Synchronous state	The pushed Number	The retransmission number	Total number
1	/mms	1	Success	50	0	50
	/mms	2	Success	13	0	13
2	/mms	1	Success	10	50	200
	/mms	2	Success	50	0	200
3	/mms	1	Success	50	0	200
	/mms	2	Success	100	50	200

The pushed Number in the table represents the number of data pushed to the message queue by the "Skylark" platform but not synchronized, that is, the number of data produced but not consumed, the *The pushed Number* in the table denote the total number of synchronized messages. Because the mechanism of data retransmission is taken into account when designing the key technology algorithm, the column data of *The retransmission number* in the table represents the number of information automatically retransmitted by the system when the network signal is unstable. From the experimental results, we can see that although the environment is unstable and other factors, but the designed unified identity authentication strategy has good robustness, can still achieve successful synchronization of authentication data information.

4.2 Verification of User Interaction Data Collaboration

In order to verify the effectiveness of the user interaction data collaboration scheme, in the experiment, we establish the user form on the user design platform, edit the data of the form, and then use the research strategy to verify the cooperation function of user interaction data, and check the data synchronization results through OpenMBEE's component VE.

Specifically, on the Web page where the user edits the form in the "Skylark" platform, customize the row and column structure of the form, and edit the data of the form, as shown in Fig. 10. During user interaction, the user login identity information is verified first. After the authentication is passed, the request created form is submitted by using the "Upload form" button on the page.

Furthermore, using OpenMBEE backend server side, it can be observed that the synchronization data has been successfully inserted into the database component MMS, and at the same time, the user interaction page can display the synchronization result, as shown in Fig. 11, which verifies that the algorithm synchronization result is correct.

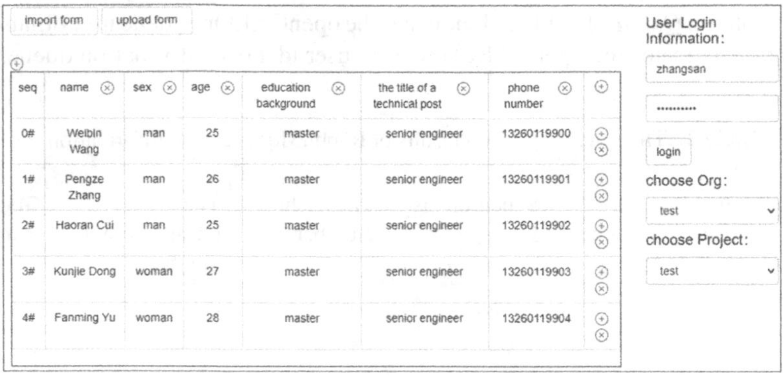

Fig. 10. The user customized form in the Web page

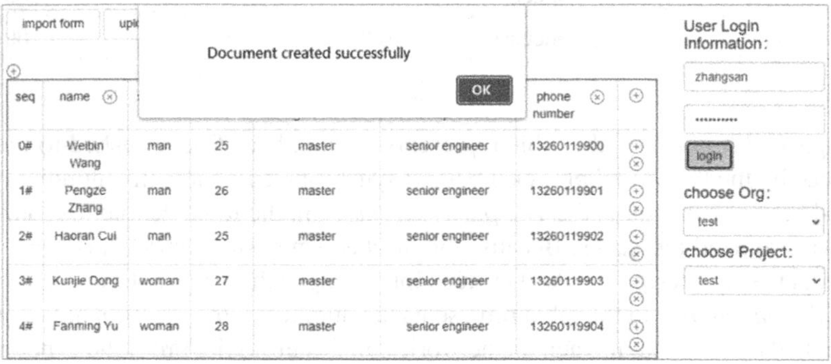

Fig. 11. The tips of successful collaboration in user interaction

5 Conclusion

In the complex product design, the integration method of model engineering environment based on OpenMBEE is studied. In the research of cross-platform unified authentication technology, the function of sharing user authentication information between the user-customized development platform is realized through the function expansion of the controller layer. In addition, a three-layer collaboration mechanism is proposed, namely, the performance layer collaboration, the business layer collaboration, and the persistence layer collaboration strategy. A data collaboration interface based on HTTP protocol is designed to realize the collaboration function between web user data and OpenMBEE back-end MMS database.

In further research, the collaboration problem of design tools based on OpenMBEE integration and the performance problem of customization and communication of multi-user model data based on information integration in complex product design can be considered. This further in-depth research will be beneficial to improving the quality of complex product design.

References

1. Grieves, M.W.: Product life cycle management: the new paradigm for enterprises. Int. J. Prod. Dev. **2**(1–2), 71–84 (2005)
2. OpenMBEE Homepage. http://www.openmbee.org/. Last Accessed 7 August 2022
3. Daniel, S.: Bridging the gap between OpenMBEE and Git. In: 2021 ACM/IEEE International Conference on Model Driven Engineering Languages and Systems Companion (MODELS-C), pp. 465–466, IEEE, Fukuoka, Japan (2021)
4. Karban, R.: Creating system engineering products with executable models in a model-based engineering environment. In: Modeling, Systems Engineering, and Project Management for Astronomy, vol. VII, p. 99110B. SPIE, Edinburgh (2016).
5. Thomas, B.C.: Assisted authoring of model-based systems engineering documents. In: MODELS '20: Proceedings of the 23rd ACM/IEEE International Conference on Model Driven Engineering Languages and Systems: Companion Proceedings, pp. 1–7. Association for Computing Machinery, New York (2020).
6. Xue, J.: A model based heterogeneous data collaboration method. In: 2020 IEEE 3rd International Conference on Electronics and Communication Engineering (ICECE), pp. 145–148. IEEE, Xi'An, China (2020)
7. Xue, J.: Key technique of constructing collaborating environment based on OPENMBEE. In: 2020 IEEE 6th International Conference on Computer and Communications (ICCC), pp. 797–801. IEEE, Chengdu, China (2020)
8. Xue, J.: Effective strategy of heterogeneous model data fusion in product collaborative design. Journal. **48**(6), 995–1003 (2022)
9. Deng, Y.: A distributed identity authentication scheme for differential fault attack. In: 2021 IEEE 21st International Conference on Communication Technology (ICCT), pp. 731–735. IEEE, Tianjin, China (2021)
10. Ohmori, M.: A case study of captive-portal detection for web authentication on wired LAN in a campus network. In: 2019 20th Asia-Pacific Network Operations and Management Symposium (APNOMS), pp. 1–4. IEEE, Matsue, Japan (2019)
11. Zheng, L.:A new mutual authentication protocol in mobile RFID for smart campus. In: IEEE Access, pp. 60996–61005. IEEE, IEEE Access (2018)
12. Zhang, G.: Research and Implementation of Unified Identity Authentication and Authorization System in Digital Campus. Yunnan University (2019)
13. Shang, F.: Application research of collaborative design method of knowledge-based oil testing scheme. Journal. **31**(01), 192–197 (2021)
14. Shi, H.: Multi-discipline forward collaborative design technology based on BIM interaction and data-driven. Journal. **56**(1), 176–181(2021)
15. RabbitMQ Homepage. https://www.RabbitMq.com/. Last Accessed 7 August 2022
16. Ozkaya, M.: MVCLang: a software modeling language for the model-view-controller design pattern. In: Proceedings of the 15th International Conference on Software Technologies, pp. 75–83. SciTePress, conference was held as a web-based event due to the COVID-19 pandemic (2020)
17. Paolone, G.: Empirical assessment of the quality of MVC web applications returned by xGenerator. Journal. **10**(2), 20–38 (2020)
18. BootStrap Homepage. https://www.bootcss.com/. Last Accessed 7 August 2022
19. Hu, B.Y.: A RESTful API enabled database for composite polymer name standardization. Journal. **13**(1), 22–35 (2021)
20. Ifresco Homepage. https://ifresco.com/. Last Accessed 7 August 2022

Research on the Construction Method of Digital Twins of Complex Products

Kai Xia[1(✉)], Wenjin Zhang[1], Liang Gao[2], Caihua Fang[1], Xuan Jiang[1], and Chi Hu[1]

[1] Digital Engineering Innovation Research Center, Wuhan Second Ship Design Institute, Wuhan 430200, China
kai_xia@126.com

[2] State Key Laboratory of Digital Manufacturing Equipment and Technology, Huazhong University of Science and Technology, Wuhan 430074, China

Abstract. The digital twins of the whole life cycle of complex products have been widely used. The abstract concept of digital twins provides principles for the modeling of digital twins, but it cannot effectively guide the modeling and application of digital twins. In this paper, a method to construct the digital twins of complex products is proposed by studying the components, relationships, construction processes and carrier forms of the digital twins of complex products.

Keywords: Complex products · Digital twins · Modeling and simulation · Virtual parallel system

1 Introduction

The system complexity of complex products is reflected in the number of system's elements, the number of relationships between system's elements, and the number of relationships between system's elements and the external environment. With the continuous improvement of science and technology, the products from the traditional mechanical system products, gradually developed and extended to the mechanical and electrical system products. The system complexity of the products increased exponentially. At present, the system complexity evolution of many large-scale and complex products has reached an unprecedented level, which has brought great challenges to the design, operation and guarantee of complex products. Development, production, operation and support of complex products is gradually upgrading to digital transformation, through strengthening the construction and application of digital models, taking the advantage of the low cost, high efficiency and fast iteration of engineering activities in the digital space, to promote the improvement of development, production and operation support. On this basis, the concept of digital twins gradually formed, and has been widely studied and applied.

In 2009, the U.S. Air Force Research Laboratory applied digital twins technology to solve the problem of airframe structure life management [1]. They built an airframe digital twin (ADT) with high fidelity airframe, integrated with damage tolerance design rules, and continuously updated and improved the digital model through actual flight data

W. Fan et al. (Eds.): AsiaSim 2022, CCIS 1712, pp. 158–164, 2022.
https://doi.org/10.1007/978-981-19-9198-1_12

to achieve real-time and accurate assessment of airframe structure life and reliability, reduce aircraft maintenance costs, and extend aircraft service time. In response to the real-time management of complex materials, structures and systems of aircraft under complex missions, NASA proposed to build a diversified, accurate, physical and stable digital twin model that can be integrated with each other and aircraft sensor equipment. By integrating with aircraft health management system data, the aircraft's health status, remaining lifespan, and mission success are continuously predicted while performing a mission. Through the comparative analysis of the actual response of external incentives and the predicted response, the response prediction of the incentives of unknown events in advance is realized. Damage is mitigated through self-healing mechanisms or command changes to extend service life or increase mission success [2]. NASA expects that the application of the digital twins will halve the cost of aircraft maintenance support and extend the overall service life by 10 times by 2035 [3].

The concept of digital twins continues to evolve, and various authoritative definitions emerge in an endless stream, including the three-dimensional model definition of Professor Grieves [4], the definition of integrated multidisciplinary, multi-physics, multi-scale, and probability model of NASA [2], the five-dimensional model definition of Professor Tao [5], the definition of virtual representation of physical assets of the American Institute of Aeronautics and Astronautics (AIAA) [6], etc. The concept of digital twins is still evolving. However, the research on modeling technology of digital twins (digital models built for specific application scenarios) is still very lacking. The abstract concept of digital twins is not enough to guide digital model construction and scenario-oriented transformation applications. In this paper, the components and relationships of digital twins of complex products are studied, and a construction method is proposed.

2 Components of Digital Twins of Complex Products

Digital twins of complex products are sets of multidimensional digital models built in virtual space that map the structure, environment, and behavior of a physical product and can be dynamically updated using physical product data. They are accurate mappings of the whole life cycle of physical products. They are conceived in the stage of product demand development, and their genes come from the user's needs. In the development and production stage of growth, their skeletons come from geometric design, their fleshes and blood come from multi-dimensional functions and performance design, their nervous systems come from measuring design and control design, their nutrients come from physical test data and physical product operation and use data. They become adults in the verification stage, and become mature in the operation and use stage. The process of digital twins modeling is accompanied by the entire product life cycle, relying on the application of Model-Based Systems Engineering (MBSE) technology in the early stages to carry out requirements analysis, functional, performance, geometric modeling, integrated design and continuous verification. In the later stage, relying on the application of cyber-physical system (CPS) technology, using actual data such as production construction, test verification, operation and use to continuously revise the model, in order to mature the twin components. The complex product digital twin consists of a behavioral model, a performance model, a geometry model, an algorithmic model, and a measurement control system.

Behavior models are the genes of digital twins. Complex products use MBSE technology to build demand models, conceptual models and functional logic models from top to bottom. After running conceptual models and functional logic models, functional logic integrated digital models built from bottom to top can accurately map the system's dynamic response to external and internal environmental factors. Behavior models are often described by state diagrams. System behavior is represented by describing the state of the system and events that cause the state transition of the system.

Geometric model is the skeleton of digital twins. Geometric model is a digital model used to describe the geometric characteristics of product space layout, assembly relationship, component shape, size, tolerance, etc. Geometric models gradually shape with the three-dimensional design of products, and further approximate the geometric characteristics of formal products as products are manufactured. When the physical product is shaped and delivered, the three-dimensional design model of the product is rendered with high fidelity and processed lightly according to the real physical material, which makes the digital twin geometry model more realistic and useful.

Performance model is the blood of digital twins. Performance model is a digital model used to characterize the performance of a product. Product performance model is a set of multiphysical, multiscale, probability simulation models, generally including one-dimensional simulation model (system simulation model) and three-dimensional simulation model. The multiphysical nature of the performance model is reflected in the description of a variety of physical properties of physical products, including structural dynamics, thermodynamics, stress analysis, fatigue damage and material properties such as stiffness, strength, hardness, fatigue strength, etc. [7]. The multiscale nature of the performance model is reflected in not only describing the macro-characteristics of each system level of the product, but also the micro-characteristics such as the material structure and roughness of the product. The probability of performance model is reflected in the introduction of probability statistics ethics and methods in the process of calculation and simulation. The performance model depends heavily on the industrial software used for modeling and simulation, which seriously restricts the application of multi-dimensional performance model integration. In engineering practice, technologies such as unified modeling based on Modelica, standardized model encapsulation based on FMI standard, distributed communication, model degradation are widely used. Performance models are integrated based on behavioral models. Interface relationships, data transfer, external incentives of different system performance models are defined by related behavioral models.

Algorithmic model is the brain of the digital twins. It encapsulates standards, specifications and requirements into algorithms for different scenarios of complex products. It develops artificial intelligence algorithms, machine learning algorithms, large data analysis algorithms, intelligent optimization algorithms, fault diagnosis and prediction algorithms, health management algorithms, reliability evaluation algorithms, etc. It encapsulates these components and integrates them with other models through on-demand combination and rapid reconstruction, to achieve prediction, analysis, optimization, assistant decision-making and other purposes.

Measurement and control system is the nervous system of digital twins. The measurement and control system of digital twins is hyper-realistic. It can not only accurately reflect the status of the measurement and control system of the entity product, but also exceed the data collection and control range and precision of the measurement and control system of the entity product, because the data space of the measurement and control amount of the digital space is much larger than that of the entity measurement point and controller. It lays the foundation for digital twins to reflect entity product status more completely and control entity product more optimally.

3 Construction Process of Digital Twins of Complex Products

The construction of digital twins of complex products is based on the operation and use of physical products and physical test data, and is ultimately published as software to the internal or external software platforms. The construction steps mainly include actual data accumulation, digital model accumulation, parallel operation system construction, application scenario construction, softwareization and deployment. The main construction process is shown in Fig. 1.

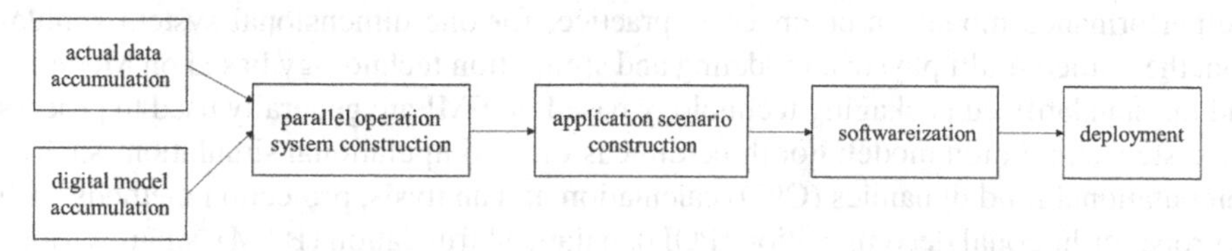

Fig. 1. Construction process of digital twins of complex products

The actual data accumulation and digital model accumulation are carried out synchronously, which is the basis of building a virtual parallel system. The actual data includes the operation and use of physical products and physical test data. The operation and use of physical products are collected through the product fault diagnosis and health management system, operation guarantee information system, equipment operation and use information, fault information, status information, etc. the physical test data is collected through the physical test data management system, including system, equipment function, performance verification data, etc., and the collected data are de-noising and standardized. In order to support the interoperability and mutual understanding of design, simulation and test data, it is necessary to build a shared data model, standardize the data format, name, dimension, etc. of all measurable data, and normalize the collected non standardized data through the shared data model. Digital model accumulation is the accumulation of behavior model, performance model, geometric model and algorithm model.

Virtual parallel system uses the accumulated actual data and digital models to build a virtual environment that runs parallel to physical products. In the virtual parallel system, the digital model is constantly corrected and improved through the actual data, and the operation of physical products is mapped and compared through the simulation operation of the digital model. Virtual parallel system is a development environment for building digital twins for application scenarios.

Digital twins' application scenario construction is aimed at selecting the required digital model in the virtual parallel system to carry out integrated application according to the specific application scenario requirements. Application scenarios generally include actual operation data duplication, full-dimensional display of operation and use process, equipment or system fault prediction, health management, fault isolation, fault recovery auxiliary decision-making, control optimization auxiliary decision-making, etc. The integrated application of digital models is realized by model bus technology, which is a distributed communication technology across systems, platforms and languages. It realizes the data sharing and integration of multi-dimensional digital models based on shared data models and various communication protocols.

The softwareizatization of digital twins is to process and package the integrated simulation digital model driven by application scenarios, so that the digital model can run independently in the form of executable programs after being separated from the relevant modeling and simulation platform. Behavior model, geometric model and algorithm model are relatively easy to realize independent operation. Behavior model can be transformed into finite state machine program that can run independently, and geometric model can be transformed and reused through general format. It is difficult to realize the performance model. In engineering practice, for one-dimensional system simulation, the unified multi physical modeling and simulation technology based on Modelica and the standardized packaging technology based on FMI are generally used to process the system simulation model; For three-dimensional computational simulations such as computational fluid dynamics (CFD) calculation and analysis, projection methods such as proper orthogonal decomposition (POD), balanced truncation (BTM), singular value decomposition (SVD) and fitting methods such as response surface (RSM), Kriging, long and short term memory network (LSTM), convolution neural network (CNN) are used to create reduced order models, so as to reduce the consumption of computational resources and avoid dependence on commercial software solvers.

The deployment and application of digital twins is to release the digital twins in the form of software to the internal or external software platform of the product for operation. When the internal software platform of the product is running, it is connected to the physical product measurement system in real time. Through dynamic data-driven simulation operation, it can realize the full-dimensional display of the operation process, equipment or system fault prediction, health management, fault isolation, fault recovery auxiliary decision-making, control optimization auxiliary decision-making and other scenario applications. When the external software platform of the product is running, the historical operation data of the physical product is accessed offline. Through the simulation operation driven by historical data, the actual operation data can be copied, the auxiliary decision-making of accurate guarantee, the optimization of digital model and other scenario applications can be realized.

4 Carrier Form of Digital Twins of Complex Products

The construction of digital twins of complex products is a continuous process, which continuously constructs, modifies and integrates various models for physical domain objects in the digital space, and synchronously uses actual data to promote its maturity. To integrate the construction process and application scenarios of digital twins, the following three aspects must be considered. First, digital twins cover different model languages such as behavior, performance, geometry, algorithm, measurement control and data. It is necessary to adopt open architecture technology to build a unified architecture, realize the standardization support of different digital models, and achieve the plug and play of digital models. Second, various models need to undergo repeated iterative modifications from construction to finalization. In order to save the time of deployment and upgrading, model data should be serviced and packaged into container services to realize continuous updating and rapid deployment of models; Third, digital twins need to be designed and implemented for different application scenarios. In order to avoid repeated work, further realize the reuse of capabilities on the basis of model data containerization, and quickly combine models to generate applications through the mechanism of service discovery and task scheduling. The bearing form of digital twins of complex products is shown in Fig. 2.

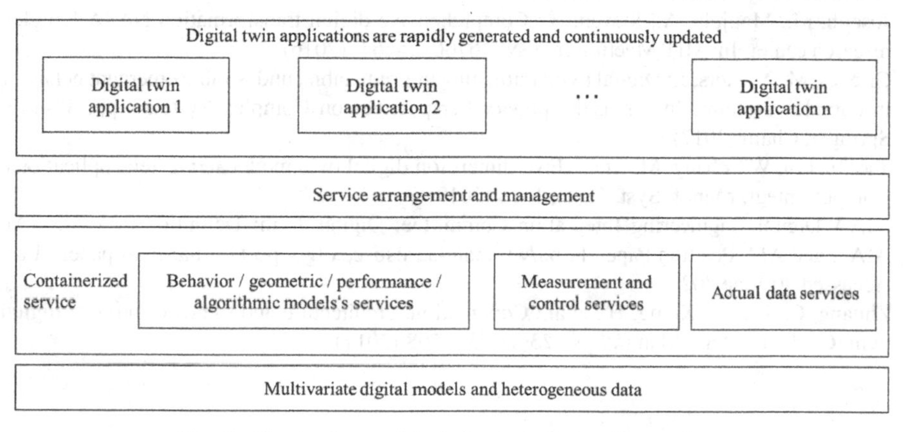

Fig. 2. Carrier form of digital twins of complex products

The container technology is applied to encapsulate the executable program formed by the transformation of various constituent elements of the digital twins to realize the shielding of environmental differences, that is, the behavior of the program is consistent no matter what environment it runs in. Container orchestration technology is applied to the composition and orchestration of container services, and digital twin applications are quickly generated for different scenarios. The automatic deployment, expansion and management technology of containers is applied to realize the deployment of new digital twin functions in real time during the operation and use of complex equipment, so as to improve the adaptability and adaptability of digital twin applications.

Relying on the lightweight characteristics of containers, thousands of container services can be installed and deployed on complex products. According to the needs of operation and use, computing resources can be deployed in real time to complete the rapid start, stop and switch of digital twin applications in different application scenarios in a short time, so as to better adapt to and support the efficient and agile operation and use of complex products.

5 Conclusion

This paper studies the construction method of complex product digital twins, and puts forward the constituent elements and their relationships, modeling process and bearing form of complex product digital twins, which plays a guiding role in the construction of complex product digital twins in engineering practice. Later, the construction method of complex product digital twins will be further refined and verified in engineering practice.

References

1. Tuegel, E.J., et al.: Reengineering aircraft structural life prediction using a digital twin. Int. J. Aero. Eng. **2011**, 14p (2011)
2. Glaessgen, E.H., Stargel, D.S.: The digital twin paradigm for future NASA and U. S. air force vehicles. In: 53rd AIAA/ASME/ASCE/AHS/ASC Structures, Structural Dynamics and Materials Conference, p. 1818 (2012)
3. Joseph, M., Manjula, A., Steven, B.: Comprehensive digital transformation NASA Langley research center. In: MIT Meeting (No. NF1676L-25664) (2016)
4. Grieves, M., Vickers, J.: Digital twin: mitigating unpredictable, undesirable emergent behavior in complex systems. In: Transdisciplinary Perspectives on Complex Systems, pp. 85–113. Springer, Cham (2017)
5. Tao, F., Liu, W., Zhang, M., et al.: Five-dimension digital twin model and its ten applications. Comput. Integr. Manuf. Syst. **25**(1), 1–18 (2019)
6. AIAA Digital Engineering Integration Committee: Digital Twin: Definition & Value—An AIAA and AIA Position Paper. http://www.aia-aerospace.org/report/digital-twin-paper/. Last Accessed 30 June 2021
7. Zhuang, C., Liu, J., Xiong, H., et al.: Connotation, architecture and trends of product digital twin. Comput. Integr. Manuf. Syst. **23**(4), 753–768 (2017)

A Portable Radar Jamming Simulation System Used for Flight Mounted Target

Xudong Pang[1(✉)], Beibei Cao[1], Liping Wu[1], and Xiaolong Zhang[2]

[1] Shanghai Publishing and Printing College, Shanghai, China
pxd210@163.com, bbc@sppc.edu.cn

[2] Shanghai Spaceflight Precision Machinery Institute, Shanghai, China

Abstract. Based on the Digital Radio-Frequency Memory (DRFM) technology, a design scheme of portable radar jamming simulation system is proposed in this paper. Here we mainly discuss the general design framework with detailed index, general compositions, and the jamming signal generation method, of which the DRFM module is analyzed in detail, and a digital simulation is made for some typical jamming signal forms. This portable scheme can satisfy the design requirements of the flight mounted target missile, and can make a fine developing example for such flight mounted applications of radar simulators.

Keywords: DRFM · Flight mounted target · Radar jamming simulator

1 Introduction

In modern warfare electromagnetic suppression is a key factor, where some missiles themselves have jamming release devices to improve the penetration capability [1]. At present, many tests have been carried out for drones to mount electronic countermeasure equipments [2], but drones are difficult to satisfy the requirements of future training in high-speed target simulation. In order to simulate the high-speed and complex practical electromagnetic interference environment as much as possible, the target missile should include the load design of portable jamming simulators [3]. The low-cost target missile with portable jamming simulators can also be used as a high-speed unmanned aircraft alone and thus become an expendable vanguard or a camouflage decoy for group cooperative penetration actions to achieve more valuable military purposes.

This paper mainly describes a compact radar jamming simulation system which can be mounted on flight targets. Based on the DRFM technology [4, 5], a design scheme of portable radar jamming simulation system is proposed, which can satisfy the actual demand of flight mounted target load and can be used as an example for the application of this kind of portable radar simulators [6].

2 Overview of the Jamming Simulation System

Considering about both the microwave anechoic chamber test and the installation mounted on the target missile, the composition of the jamming simulator uses a balanced scheme which can be used in both the In-field test or the out-field test of air

W. Fan et al. (Eds.): AsiaSim 2022, CCIS 1712, pp. 165–170, 2022.
https://doi.org/10.1007/978-981-19-9198-1_13

defense guided weapons. The simulator needs to synchronously receive both the signal from the semi-active ground radar and the signal from the active PD radar seeker and then identify the two types of signals. The simulator then uses the identification results as the guidance parameters to generate jamming signal forms for the semi-active radar and the active PD radar seeker respectively [6]. According to the test requirements, the simulator needs to realize the generation of dual-band radar jamming signals of both the K-band radar seeker and the X-band semi-active radar seeker, and the generated radar jamming signal must be consistent with the working mode of the tested jamming object.

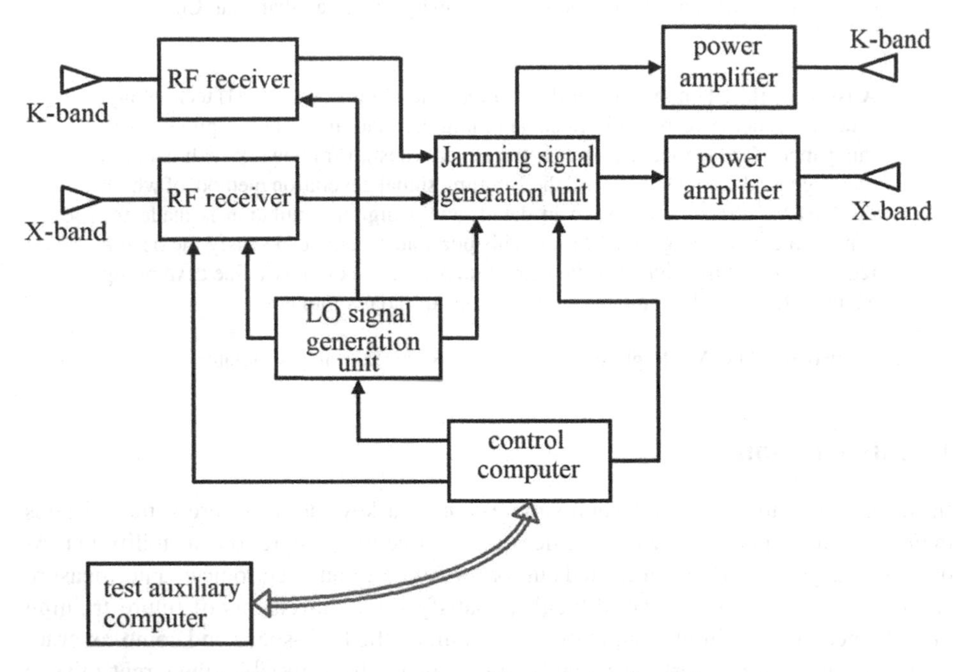

Fig. 1. The basic composition of the jamming simulation system

The composition of the simulation system is shown in Fig. 1, where the simulation system is mainly composed of RF signal receiver, jamming signal generation unit, LO signal generation unit, real-time control computer, antennas, power amplifier component and test auxiliary computer [5].

In Fig. 1 the dual-antenna time-division transceiver scheme is adopted in this simulator, which does not transmit jamming signals when receiving radar signals and does not receive transmitting signals from the radar under test when transmitting jamming signals. Here the polarization isolation and time division scheme are used to realize the isolation of signal transmission and reception. In the RF receiving unit, the signal is directly converted to a wideband signal of 50 ~ 450 MHz through the frequency conversion local oscillator (LO), and the signal is directly input into the jamming signal generating unit for various modulations, where the jamming signal is to be generated. The input of the jamming signal generating unit is the intermediate frequency signal output by the RF receiving unit, which is processed with distance modulation and digital frequency modulation of various jamming patterns by the DRFM module. In the jamming signal generating unit, DRFM components are used to generate responsive jamming signals such as frequency spot noise, distance pull-off, speed pull-off, multiple decoy targets and as well as possible combinations of the above signals. In this scheme, a solid-state power amplifier is used.

In order to adapt the installation mounted on the target missile, the basic size of the simulator body is limited within 280 mm * 240 mm * 160 mm, which can be installed inside the target missile, thus makes the jamming simulation system a viability of portable applications. The power amplifier components are also installed inside the missile body and the antennas are installed inside the radome of the missile, which also benefits the portable application. Through the quality control of components and materials, the simulator can meet the environment requirements of −55°C ~ +60°C. The power supply uses DC of 27 ± 10% V and the output power is greater than 1 W.

3 Jamming Signal Generation Method and Its Digital Simulations

In this system, in order to reduce the volume and weight of the simulator, the DRFM module is used to build DDS components and the arbitrary waveform generator in FPGA [4]. Using this digital signal processing method, various kinds of interference signals including frequency spot noise, comb spectrum and Doppler pull-off jamming are generated. The spectrums of these jamming signals which are generated by the real-time control computer are loaded into the DRFM component, where the received radar signal is processed by range delay, noise signal generation, and digital frequency domain modulation of the noise signal and Doppler frequency. Among which the doppler pull-off jamming range is ± 2 MHz while the doppler pull-off jamming accuracy is smaller than 10 Hz, and the frequency storage accuracy is better than 1 kHz while the minimum repeat delay is smaller than 0.5 μs.

As shown in Fig. 2, in the DRFM module we use high-speed devices such as ultra-high-speed A/D & D/A converters, FPGA, and dual-port SDRAMs. By using FPGA, the system has the ability to change the function sets on site in accordance with the test needs, which greatly improves the flexibility of the system's configurations. Various control signals of DRFM come from the main controller, which communicates directly with the main control computer with the ability to exchange data in duplex mode, where the DRFM module receives various control commands and transmits the working status data of the system synchronously. The main controller receives external control signals, and outputs three control signals of enable signal, read control signal and write control signal for delay control. The working status data input from the outside is converted into delayed data through the internal computing unit. Using the FPGA in the DRFM module and the digital noise method, a pseudo-random digital noise signal with a certain bandwidth is generated. The digital noise data is downloaded from the computer to the memory of the DRFM module before the test or directly loaded into the dual-port memory by using the solidified digital noise data. When the experiment starts, the digital noise data is sequentially read out from the digital noise memory, and then repeated in a loop and finally output to the digital up-converter and the DA converter to achieve the IF analog signals through the control of the logic circuit.

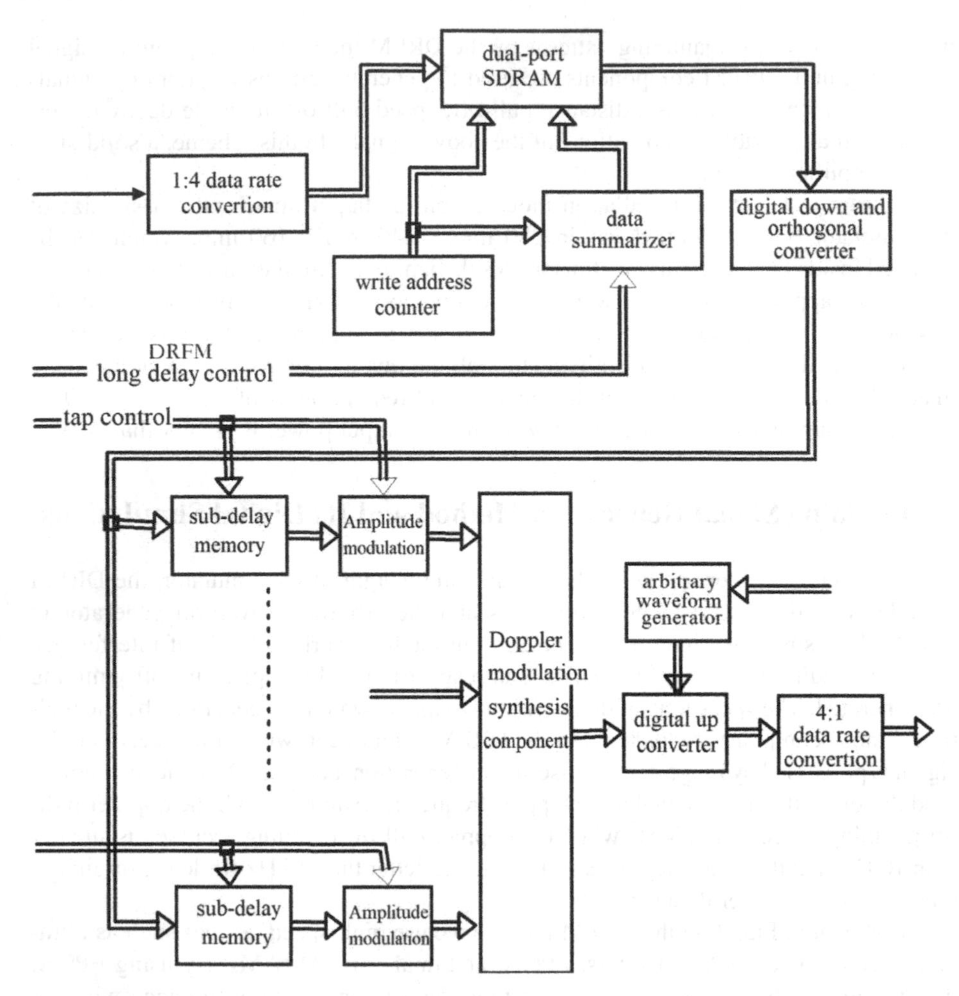

Fig. 2. Logic composition of DRFM jamming signal generation components

Digital signal processing methods can be used to generate various types of digital jamming signals. Figure 3 shows the digital simulation results of the distribution characteristics of the narrowband frequency spot noise signal in the time and frequency domains respectively. The phase randomization method is used to obtain an uniformly distributed noise signal spectrum in the bandwidth.

Figure 4 shows the digital simulation results of the characteristics of a typical comb spectrum jamming signal pattern in the time and frequency domains respectively. Comb spectrum jamming signal can be classified into two styles, which includes suppressed comb spectrum and deception comb spectrum. Suppressed comb spectrum which is actively generated by suppressing jamming signal sources does not need to receive radar signals. The jamming signal generator controls and suppresses the multiple FM jamming signals according to the start/stop frequency set, sub-band frequency center, sub-bandwidth and the dwell time, where up to 100 FM jamming signals can be generated. The deception comb spectrum however needs to receive, store, and replicate the radar signal by the DRFM module to generate multiple frequency points evenly distributed within a certain frequency range which is maximum between ±1MHz, where these frequency points can appear synchronously or asynchronously according to the control settings and the maximum number of frequency points is 8.

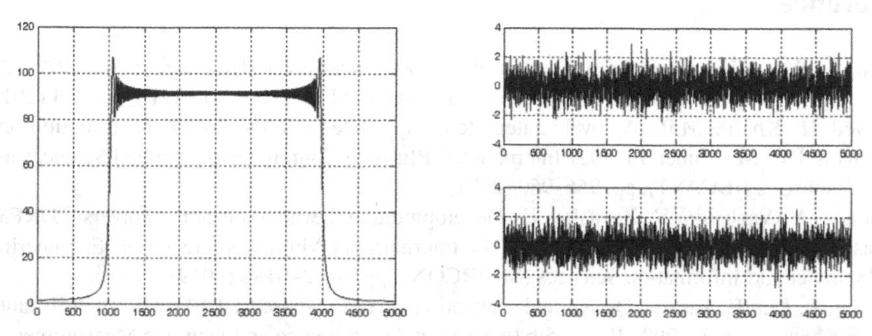

Fig. 3. Time domain (left) and frequency domain (right) distribution characteristics of narrowband frequency spot noise signal

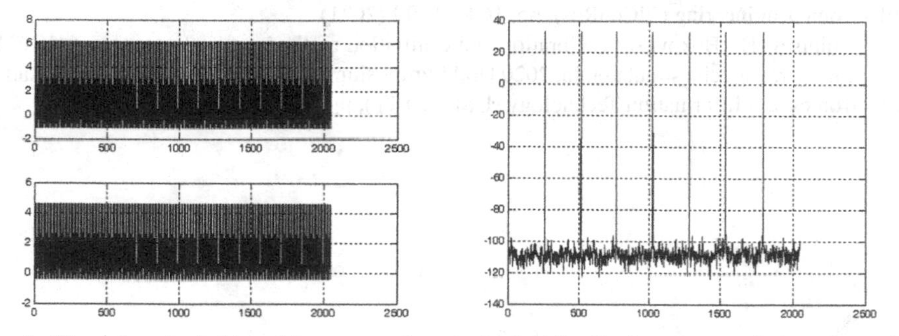

Fig. 4. Time domain (left) and frequency domain (right) distribution characteristics of a typical comb spectrum jamming signal

4 Conclusion

In this paper, the design scheme of a portable radar jamming simulation system is discussed, and some typical jamming pattern is simulated digitally. Compared with the traditional radar jamming simulator, the size of the system is reduced by the application of DRFM technology, which makes the jamming simulation system a viability of portable applications. The reconfigurable system design is more suitable for distributed module replacement, which can meet the flight mounting design requirements of the target missile. In the future, the processing capability and signal transformation capability of DRFM components can be upgraded in hardware and software to achieve higher

RF signal processing index, modulation loading capability of complex interference signals and simulation capability of multi-point false targets, so as to meet the constantly updated requirements of electronic countermeasures. The architecture of portable jamming simulator based on DRFM technology proposed in this paper can be extended to the development and design of various types of target-mounted signal simulators.

References

1. Xu, J., Wang, H.: Jamming simulation of synthetic aperture radar imaging. In: 2018 12th International Symposium on Antennas, Propagation and EM Theory (ISAPE), pp. 1–4 (2018)
2. Javed, H., Khalid, M.R.: A novel strategy to compensate the effects of platform motion on a moving DRFM jammer. In: 2021 International Bhurban Conference on Applied Sciences and Technologies (IBCAST), pp. 956–960 (2021)
3. Bokov, A., Vazhenin, V., Zeynalov, E.: Development and evaluation of the universal DRFM-based simulator of radar targets. In: 2019 International Multi-Conference on Engineering, Computer and Information Sciences (SIBIRCON), pp. 0182–0186 (2019)
4. Jiang, J., Liu, F., Hu, C.: Design and realization of FPGA-based DRFM with high instantaneous bandwidth. In: 2021 IEEE 15th International Conference on Electronic Measurement & Instruments (ICEMI), pp. 233–239 (2021)
5. Smolyakov, A.V., Podstrigaev, A.S.: Design of DRFM-based several radar targets simulator using FPGA. In: 2021 IEEE Conference of Russian Young Researchers in Electrical and Electronic Engineering (ElConRus), pp. 1694–1699 (2021)
6. Shishmintsev, D., Bokov, A.: Calibration and evaluation methods of performance of DRFM-technology radar echo simulator. In: 2020 Ural Symposium on Biomedical Engineering, Radio Electronics and Information Technology (USBEREIT), pp. 0372–0375 (2020)

A Decoupling Design Method of Compensated Active Disturbance Rejection Control for Multivariable System

You Wang[1], Chengbo Dai[1], Yali Xue[1,2](✉), and Donghai Li[1]

[1] State Key Laboratory of Power Systems, Department of Energy and Power Engineering, Tsinghua University, Beijing 100084, China
xueyali@tsinghua.edu.cn

[2] Shanxi Research Institute for Clean Energy, Tsinghua University, Taiyuan 030032, China

Abstract. To solve the coupling problem of multivariable systems, this paper proposed a decoupling design method of compensated active disturbance rejection control that can detect the disturbance among coupling loops with the extended state observer and design a control law to eliminate it. Besides, a compensation element is added to the traditional active disturbance rejection controller so that the observer can estimate the state of the system and the influence of disturbance accurately. The comparative simulation results show that the decoupling effect of the proposed method is good. The Monte Carlo experiment validates its robustness and good application potential.

Keywords: Active disturbance rejection control · Extended state observer · Multi-variable system · Decoupling design

1 Introduction

Multivariable systems exist widely in various industrial process systems. For example, binary rectification tower process in chemical process [1], combustion control system in thermal power generation process [2], and cold, heat and electricity load control in distributed energy system [3]. Compared with the single variable system, the multivariable system is characterized by coupling between different loops. The change of the input of one loop will cause the disturbance of the output of other loops to different degrees. If the control system cannot eliminate the disturbances caused by coupling promptly and effectively, the safety and economic operation of the industrial production process will face adverse effects.

The decentralized PID control method is the most used method in multivariable system control engineering. It has a simple structure and is easy to be deployed and adjusted. However, when the multivariable system is tightly coupled, the control effect of the decentralized PID controller is not satisfactory. Besides, some model-based control strategies proposed for multivariable systems need the precise mathematical models. Unfortunately, the precise mathematical model is too hard or too expensive to build because of the system complexity and wide working condition variation.

W. Fan et al. (Eds.): AsiaSim 2022, CCIS 1712, pp. 171–180, 2022.
https://doi.org/10.1007/978-981-19-9198-1_14

As an increasingly popular control algorithm, active disturbance rejection control (ADRC) has been validated in many simulations [4], experiments [5], and large industrial processes [6]. Therefore, it is of great significance to further study whether ADRC can meet the challenges of industrial multivariable systems. Based on a compensated ADRC [7], this paper proposed a decoupling design method of compensated ADRC for multivariable systems, and compared with decentralized PI and compensated ADRC by simulation. The results show that the proposed method has great decoupling ability, a simple design process, and good engineering application potential.

The rest of this paper is organized as follows. The compensated ADRC is introduced briefly in Sect. 2. Section 3 describes the proposed decoupling design method for the multivariable system. In Sect. 4, the simulation comparison is performed, and conclusions are given in Sect. 5.

2 Compensated Active Disturbance Rejection Control

2.1 Design of Regular ADRC

The basic principle of ADRC is to use the extended state observer (ESO) to estimate the internal and external disturbances of the system in real-time, and then design the state feedback control law (SFCL) to eliminate them. The structure of regular ADRC is shown in Fig. 1, where r, u, d, y, and z are the setpoint, the control signal, the external disturbance, the system output, and the system states, respectively.

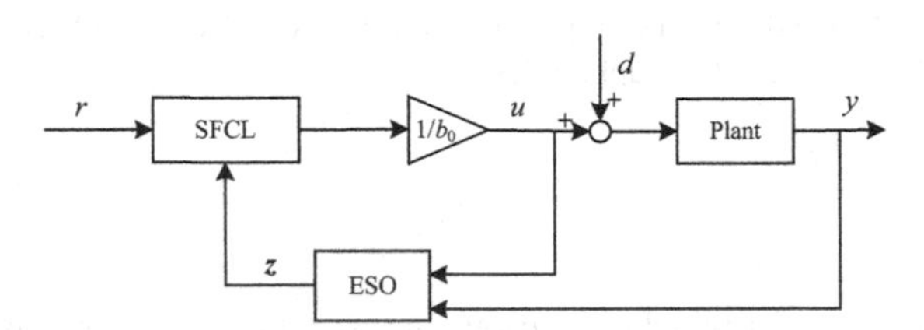

Fig. 1. The structure of regular ADRC.

For n-order ADRC, ESO is designed as:

$$\begin{cases} \dot{z} = Az + Bu + L(y - z_1) \\ y = Cz \end{cases}, \tag{1}$$

where

$$A = \begin{bmatrix} 0 & 1 & & \\ \vdots & & \ddots & \\ 0 & & & 1 \\ 0 & \cdots & \cdots & 0 \end{bmatrix}_{(n+1)\times(n+1)}, B = \begin{bmatrix} 0 \\ \vdots \\ 0 \\ b_0 \\ 0 \end{bmatrix}_{(n+1)\times 1}, C = \begin{bmatrix} 1 \\ 0 \\ \vdots \\ 0 \\ 0 \end{bmatrix}_{(n+1)\times 1}, L = \begin{bmatrix} \beta_1 \\ \beta_2 \\ \vdots \\ \beta_n \\ \beta_{n+1} \end{bmatrix}_{(n+1)\times 1} \tag{2}$$

The state feedback control law is:

$$u = \frac{k_1(r - z_1) + k_2(\dot{r} - z_2) + \cdots + k_n(r^{(n-1)} - z_n) - z_{n+1}}{b_0} =: \frac{K(\overline{r} - z)}{b_0}, \tag{3}$$

where $K = [k_1, k_2, \cdots, k_n, 1]_{1\times(n+1)}$, $\overline{r} = \left[r, \dot{r}, \cdots, r^{(n-1)}, 0\right]^T_{1\times(n+1)}$, and K is the feedback gain matrix of the system. The bandwidth-parameterization method [8] is usually used to simplify parameter tuning, and the controller parameters are obtained by tuning the controller bandwidth w_c and the observer bandwidth w_o. The simplified parameter tuning formulae is Eq. (4).

$$\begin{aligned} k_i &= \frac{n!}{(i-1)!(n+1-i)!} w_c^{n+1-i}, i = 1, 2, \cdots, n \\ \beta_j &= \frac{(n+1)!}{j!(n+1-j)!} w_o^j, j = 1, 2, \cdots, n+1 \end{aligned} \tag{4}$$

2.2 Design of Compensated ADRC

When the controller order does not match the order of the process to be controlled, a compensation element can be designed to make the order of the equivalent plant estimated by ESO match the controller order, which could improve the observation accuracy. Then the expected dynamics of the closed-loop system with the state feedback control law design can meet the design expectations. The compensated ADRC was introduced in Ref. [9] firstly and is omitted here.

Considering a linear time-invariant system, its transfer function can be expressed in the form of Eq. (5).

$$G(s) = \frac{K(1 + p_1 s)(1 + p_2 s)\cdots(1 + p_k s)}{(1 + q_1 s)(1 + q_2 s)\cdots(1 + q_l s)} e^{-\tau s}, \tag{5}$$

where K, τ, p_i, and q_i are the system gain, the delay time of the system, the pole and zero of the system, respectively. q_i is positive number so the system is stable.

When designing m-order compensated ADRC for Eq. (5), it is expected that the equivalent system observed by ESO is a series of m inertial elements with real poles. If m is even, the equivalent process should have conjugate poles and an even number of real poles. If m is odd, the equivalent process should have conjugate poles and an odd number of real poles. The designed compensation element should be able to satisfy the above conditions.

If the transfer function of the compensation element $G_{cp}(s)$ is designed as Eq. (6), the structure of compensated ADRC is shown in Fig. 2.

$$G_{cp}(s) = \frac{(1 + p_1 s)(1 + p_2 s)\cdots(1 + p_k s)}{(1 + q_{i_{m+1}} s)(1 + q_{i_{m+2}} s)\cdots(1 + q_{i_l} s)} e^{-\tau s} \tag{6}$$

After adding the compensation element, the system observed by the ESO is compensated to Eq. (7) whose order is equal to the order of ADRC.

$$G'(s) = \frac{Y(s)}{U_f(s)} = \frac{K}{(1 + q_{i_1} s)(1 + q_{i_2} s)\cdots(1 + q_{i_m} s)} \tag{7}$$

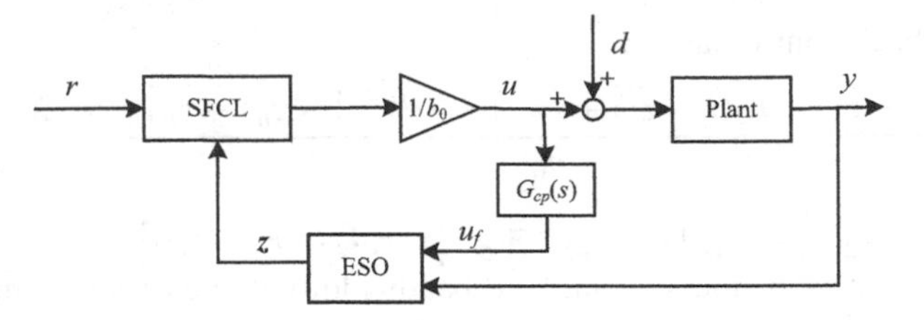

Fig. 2. The structure of compensated ADRC.

With zero initial conditions, Eq. (7) can be converted into a time-domain expression shown as Eq. (8).

$$Ku_f(t) = y^{(n)} \prod_{k=1}^{m} q_{i_k} + f \tag{8}$$

Combining Eq. (8) with the expression of total disturbance of m-order ADRC, we can see that:

$$b_0 = \frac{K}{\prod_{k=1}^{m} q_{i_k}} \tag{9}$$

If the bandwidth-parameterization method is used, the controller bandwidth w_c is set as the mean value of the poles of the process observed by the ESO to reflect the distribution of the poles of the system, shown as Eq. (10).

$$w_c = \frac{m}{\sum_{k=1}^{m} q_{i_k}} \tag{10}$$

If the bandwidth-parameterization method is not used, based on the characteristics of the observed system, the controller bandwidth w_c is tuned by Eq. (11) to make the expected dynamics of the system consistent with its open-loop response dynamics.

$$k_1 = \frac{1}{\prod_{k=1}^{m} q_{i_k}}, k_2 = \frac{k_1}{\sum_{l=1}^{m} 1/q_{i_l}}, \cdots \tag{11}$$

To ensure the observation speed of ESO, the observer bandwidth w_o is designed to be at least 10 times the average of the poles of the system. With the bandwidth parameterization method, we have

$$w_o = 10w_c = \frac{10m}{\sum_{k=1}^{m} q_{i_k}} \tag{12}$$

3 Decoupling of Compensated ADRC for Multivariable System

An n input n output system can be expressed in the form of the transfer function matrix of Eq. (13) and assumed that the matrix is diagonally dominant.

$$\begin{bmatrix} y_1 \\ y_2 \\ \vdots \\ y_n \end{bmatrix} = \begin{bmatrix} G_{11} & G_{12} & \cdots & G_{1n} \\ G_{21} & G_{22} & \cdots & G_{2n} \\ \vdots & \vdots & \ddots & \vdots \\ G_{n1} & G_{n2} & \cdots & G_{nn} \end{bmatrix} \begin{bmatrix} u_1 \\ u_2 \\ \vdots \\ u_n \end{bmatrix} \tag{13}$$

where $Y = [y_1, y_2, \cdots, y_n]^T$ is the output matrix of the system, and $U = [u_1, u_2, \cdots, u_n]^T$ is the input matrix of the system, and G_{ij} is the transfer function between the input u_j and the output y_i.

Since the decentralized ADRC has inverse decoupling ability for multivariable systems [10], the decentralized compensated ADRC can be designed directly for multivariable systems. The design details are omitted here due to page limit.

To fully utilize the coupling information between the loops of the multivariable system, the multivariable system design method in Ref. [11] can be combined with the compensated ADRC to design a multivariable compensated ADRC. For a 2 × 2 system, the control system structure is shown in Fig. 3.

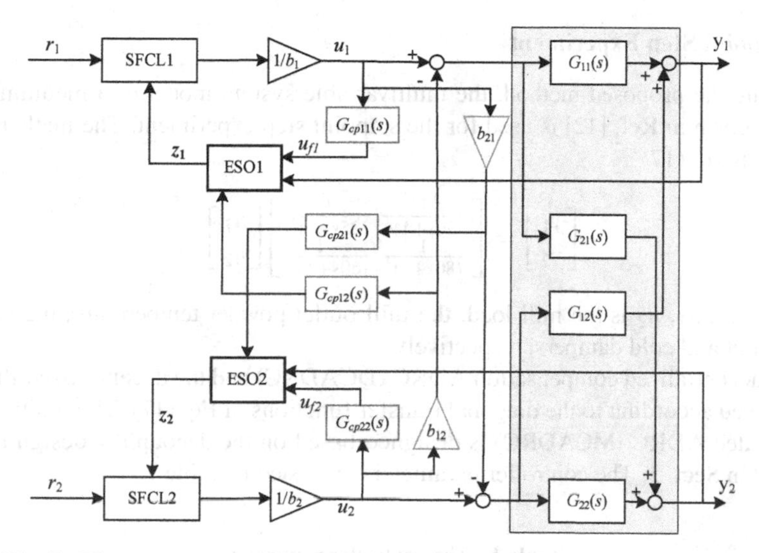

Fig. 3. The structure of multivariable compensated ADRC for a 2 × 2 system.

The one-order compensated ADRC is adopted in this paper. Taking the loop u_1-y_1 as an example, the design principle and process of multivariable compensated ADRC is given below, while the other loops can be designed in the same way.

From Eq. (13), we can see that

$$y_1 = G_{11}u_1 + G_{12}u_2 + \cdots + G_{1n}u_n \tag{14}$$

Since the control signal of each loop u_i ($i = 1, \ldots, n$) has an effect on y_1, the control signal of each loop should be observed. ESO1 can be designed as

$$\begin{cases} \dot{z}_{11} = z_{12} + \sum_{i=1}^{n} b_{1i}u'_{1i} + \beta_{11}(y_1 - z_{11}) \\ \dot{z}_{12} = \beta_{12}(y_1 - z_{11}) \end{cases}, \tag{15}$$

where u'_{1i} is the output of signal of u_{1i} after passing through the compensation element, and b_{1i} is the estimated value of the derivative gain of u_i to y_1.

To ensure that the system is free of steady-state errors and the disturbances from the other loops are completely cancelled, the control law of the first loop is designed as Eq. (16).

$$u_1 = \frac{k_{11}(r_1 - z_{11}) - \sum_{i=2}^{n} b_{1i}u_i - z_{12}}{b_{01}} \tag{16}$$

The compensation element for each control channel is designed to make the equivalent system observed by the ESO a one-order system.

4 Simulation Experiment

4.1 Setpoint Step Experiment

To validate the proposed method, the multivariable system model of a medium-speed coal mill given in Ref. [12] is used for the setpoint step experiment. The mathematical model is in Eq. (17).

$$\begin{bmatrix} y_1 \\ y_2 \end{bmatrix} = \begin{bmatrix} \frac{1}{(20s+1)^3} & \frac{1}{(25s+1)^3} \\ \frac{1}{(80s+1)^3} & \frac{-1}{(60s+1)^3} \end{bmatrix} \begin{bmatrix} u_1 \\ u_2 \end{bmatrix} \tag{17}$$

where y_1, y_2, u_1, u_2 is the mill load, the mill outlet powder temperature, the opening value of hot and cold dampers, respectively.

The decentralized compensation ADRC (DCADRC) and the decentralized PI (DPI) are designed according to the diagonal transfer functions of Eq. (17). The multivariable compensated ADRC (MCADRC) is designed based on the decoupling design method proposed in Sect. 3. The controller parameters are listed in Table 1.

Table 1. The controller parameters

Controllers	Parameters
DCADRC	Loop 1: $G_{cp11} = 1/(20s+1)^2, b_1 = 0.05, w_{c1} = 0.05, w_{o1} = 10w_{c1}$
	Loop 2: $G_{cp22} = 1/(60s+1)^2, b_2 = -1/60, w_{c2} = 1/60, w_{o2} = 10w_{c2}$
MCADRC	Loop 1: $G_{cp11} = 1/(20s+1)^2, b_1 = 0.05, w_{c1} = 0.05, w_{o1} = 10w_{c1}$
	$G_{cp12} = 1/(25s+1)^2, b_{12} = 0.04$
	Loop 2: $G_{cp22} = 1/(60s+1)^2, b_2 = -1/60, w_{c2} = 1/60, w_{o2} = 10w_{c2}$
	$G_{cp21} = 1/(80s+1)^2, b_{21} = 0.0125$
DPI	Loop 1: $k_p = 0.3785, k_i = 7.347 \times 10^{-3}$
	Loop 2: $k_p = -0.6595, k_i = -2.874 \times 10^{-3}$

Note that the setpoint r_1 and r_2 vary in unit step at 20 s respectively. When one setpoint changes, the other is set as zero. The simulation results are shown in Figs. 4, 5, 6 and 7.

The control performance indices are listed in Table 2. From Fig. 4, Fig. 6 and Table 2, it is obvious that compared with DPI, DCADRC improves the control performance of each loop and eliminates the coupling between loops slightly. MCADRC eliminates the coupling between loops almost completely and achieves a desirable control effect. Therefore, the superiority of MCADRC for multivariable systems is confirmed

Table 2. The statistical control performance indices.

	r_1 step change				r_2 step change			
	T_s/s	$\sigma/\%$	T_d/s	A_d	T_s/s	$\sigma/\%$	T_d/s	A_d
DCADRC	516.06	12.43	584.44	0.29	472.76	10.86	407.28	0.36
MCADRC	165.65	0.05	430.01	0.03	469.07	0.09	63.87	0.03
DPI	682.02	14.36	902.30	0.41	937.59	7.39	647.75	0.49

The indexes $\{T_s, \sigma, T_d, A_d\}$ denote the settling time and overshoot of tracking setpoints, the settling time and output variations amplitude caused by the coupling from other loops, respectively.

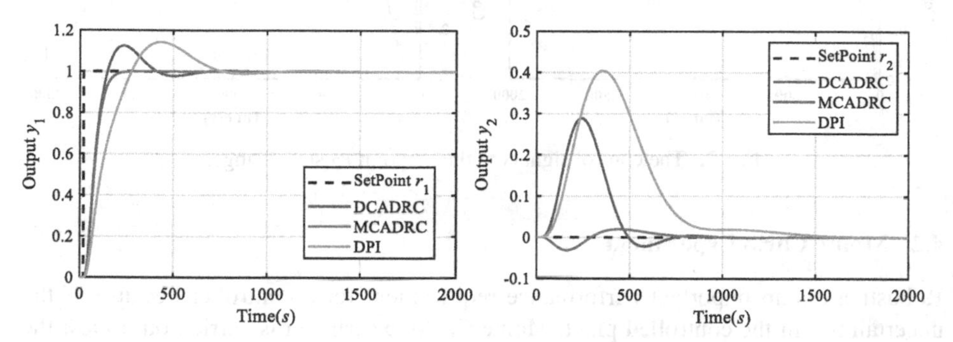

Fig. 4. The system outputs of the setpoint r_1 step change.

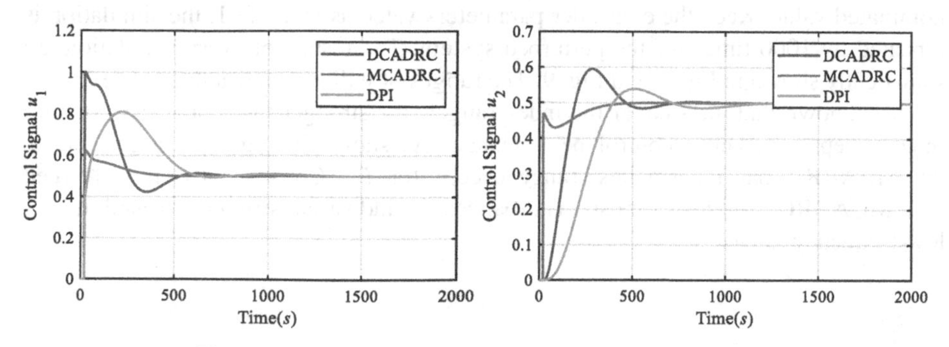

Fig. 5. The control signals of the setpoint r_1 step change.

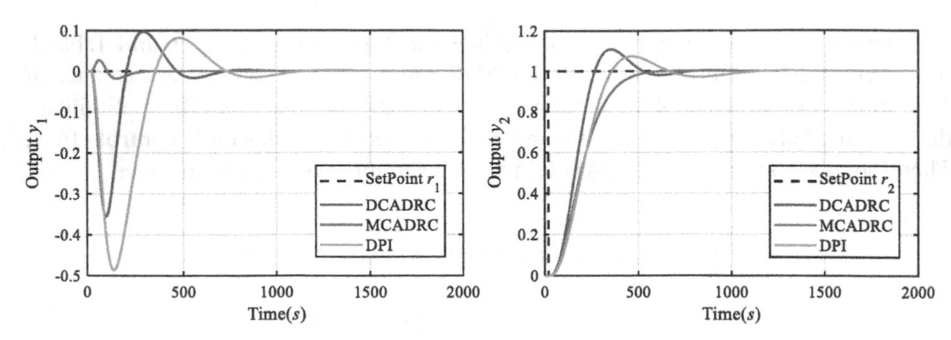

Fig. 6. The system outputs of the setpoint r_2 step change.

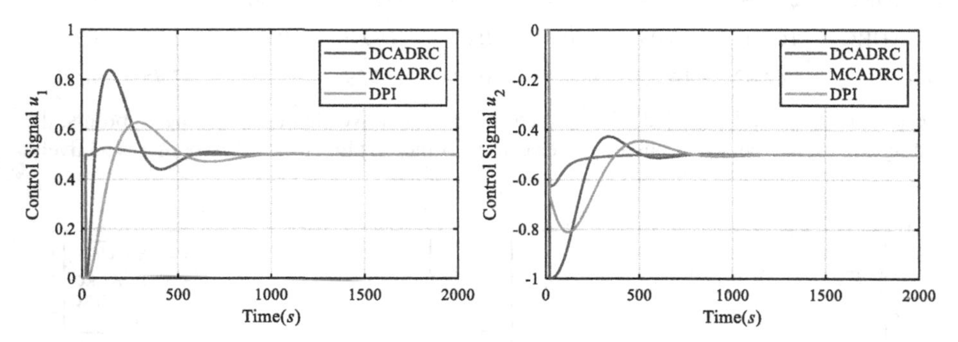

Fig. 7. The control signals of the setpoint r_2 step change.

4.2 Monte Carlo Experiment

Robustness is an important performance requirement for a controller because of the uncertainties in the controlled plant. Monte Carlo experiment is carried out to test the robustness of the above control methods by randomly perturbing the steady-state gain k and the time constant T of every transfer function in Eq. (17) with a range of $\pm 10\%$ nominated value. Keep the controller parameters value as in Table 1, the simulation is repeated by 1000 times for the perturbed system. The results of every simulation are statistically shown in Fig. 8 and Fig. 9. The ranges of indices are listed in Table 3.

It is known that the smaller the index ranges, the stronger the robustness. Smaller indexes represent better control performance. As shown in Fig. 8, 9 and Table 3, the MCADRC has the smallest range except for T_d. Consequently, it is learned that MCADRC can resist most disturbances in industrial systems located in the low-frequency range.

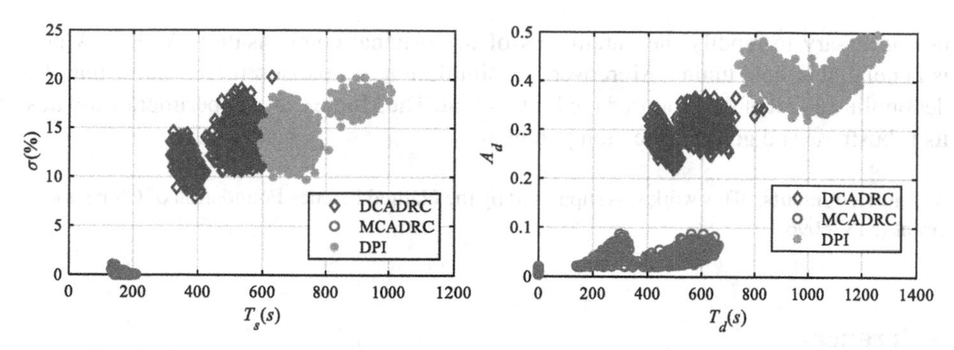

Fig. 8. The index records of the setpoint r_1 step change for the perturbed system.

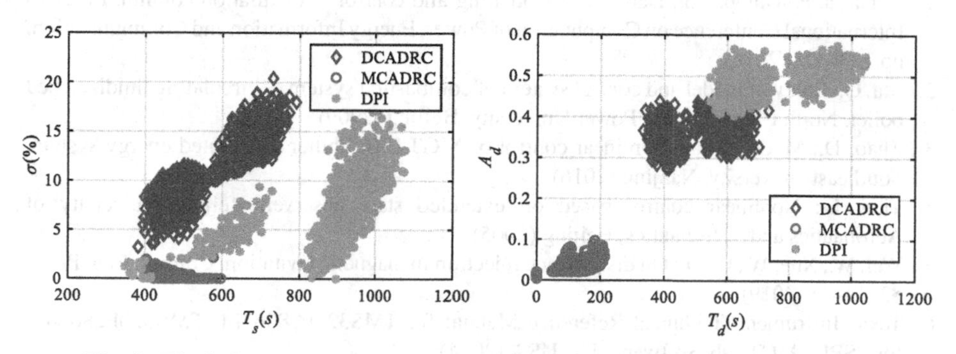

Fig. 9. The index records of the setpoint r_2 step change for the perturbed system.

Table 3. The index ranges of the perturbed system.

	r_1 step change				r_2 step change			
	T_s/s	$\sigma/\%$	T_d/s	A_d	T_s/s	$\sigma/\%$	T_d/s	A_d
DCADRC	[647.2, 314.5]	[20.20, 7.925]	[830.0, 398.4]	[0.369, 0.220]	[852.0, 382.7]	[20.40, 3.212]	[733.8, 333.0]	[0.455, 0.272]
MCADRC	[207.8, 133.3]	[1.050, 0]	[663.4, 0]	[0.087, 0.004]	[596.4, 390.1]	[2.145, 0]	[214.3, 0]	[0.098, 0.004]
DPI	[1008, 594.4]	[20.17, 9.734]	[1278, 735.3]	[0.496, 0.321]	[1080, 348.8]	[19.57, 0]	[1045, 544.6]	[0.575, 0.403]

5 Conclusions

To solve the coupling problem of multivariable systems, this paper proposed a decoupling design method of compensated active disturbance rejection control. The essence is to detect the disturbance among coupling loops with the extended state observer and design a control law to eliminate it. Besides, a compensation element is designed to increase the observation accuracy of observer. The decoupling design of multivariable compensation ADRC does not affect the equivalent control system of the original controller, so it is not necessary to modify the parameters of the original compensation ADRC, which is beneficial to field tuning. Moreover, the simulation comparison results show that the decoupling effect of the proposed method is great. The Monte Carlo experiment validates its robustness and good application potential.

Acknowledgments. This work was supported by the Natural Science Foundation of China under Grant (51876096).

References

1. Acharya, P., Dumpa, G., Dan, T.K.: Modelling and control of distillation column. In: 2016 International Conference on Computation of Power, Energy Information and Communication, pp.123–128 (2016)
2. Ma, Q.: Study of model and control system of combustion system of circulating fluidized bed boiler. North China Electric Power University, Beijing (2009)
3. Zhao, D.: Modeling and nonlinear control of MGT-LiBr chiller distributed energy system. Southeast University, Nanjing (2016)
4. Tian, L.: Nonlinear control based on extended state observer. Beijing University of Aeronautics and Astronautics, Beijing (2005)
5. Wei, W., Xue, W., Li, D.: On disturbance rejection in magnetic levitation. Control. Eng. Pract. **82**, 24–35 (2019)
6. Texas Instrument: Technical Reference Manual for TMS320F28069M, TMS320F28068M InstaSPIN-MOTION Software, TX, USA (2014)
7. Wang, Y., Wu, Z., Xue, Y., et al.: Design of linear active disturbance rejection controller for high order large inertia system. Control Decis. https://doi.org/10.13195/j.kzyjc.2021.1576
8. Gao, Z.: Scaling and bandwidth-parameterization based controller tuning. In: Proceedings of the American Control Conference, pp. 4989–4996 (2006)
9. Wu, Z., Li, D., Xue, Y., Sun, L., He, T., Zheng, S.: Modified active disturbance rejection control for fluidized bed combustor. ISA Trans. **102**, 135–153 (2020)
10. He, T.: Active disturbance rejection control design and application in thermal energy system. Tsinghua University, Beijing (2019)
11. Wang, S.: Active disturbance rejection control for multivariable systems. North China Electric Power University, Beijing (2021)
12. Cui, T., Liu, X., Shen, J.: Multivariable decoupling control of medium speed pulverizer based on PID control. Ind. Control Comput. **31**(04), 21–23 (2018)

Integrated Natural Environment and Virtual Reality Environment Modeling and Simulation

Performance Degradation of Multi-level Heterogeneous Middleware Communication in Virtual-Reality Simulation

Ziquan Mao[1], Jialong Gao[1], Jianxing Gong[1](✉), and Miao Zhang[2]

[1] College of Intelligence Science and Technology, National University of Defense Technology, 109 Deya Road, Kaifu District, Changsha, Hunan, China
fj_gjx@nudt.edu.cn

[2] College of Systems Engineering, National University of Defense Technology, 109 Deya Road, Kaifu District, Changsha, Hunan, China

Abstract. This paper proposes a layered framework for combining virtual and real simulation. The framework combines a physical system based on ROS with a simulation system based on a simulation engine, builds a multi-level agent middleware between virtual and real heterogeneous systems, and analyzes the latency and packet loss rate metrics of multi-level agent communication through experiments. When sending messages at 50 Hz as frequency, reducing the number of agent levels leads to a significant reduction in the average value of latency and jitter. More stable communication performance and increased bandwidth reduce the packet loss rate. The variation of communication performance with frequency for the most complex three-level agent is analyzed with emphasis. Finally, the conclusions of the experiments are given, and future work has prospected.

Keywords: Middleware · Virtual-reality simulation · Communication · ROS

1 Introduction

With the development of society and the advancement of technology, Unmanned Aerial Vehicle (UAV) is deeply integrated into human life. It shines in military, rescue, communication, detection, photography, and other fields. The idea of UAV clustering comes from the flocks of fish, birds, and bees in nature. Through clustering behavior, multiple UAVs can achieve functional complementarity, producing a cluster effect of $1 + 1 > 2$.

The current research on UAV swarm simulation methods primarily uses computer simulation techniques [1, 2]. The advantages of computer simulation are low cost, low risk, easy editing, easier development, and control of the simulation speed by manually increasing or decreasing the simulation step. Matlab, Gazebo, and simulation software are used more often. However, computer simulation is too ideal due to its simulation environment, object, and model, resulting in the algorithm trained in the virtual simulation environment not being directly applied to the real UAV cluster. Longer cycles of algorithm debugging and physical testing are also required, from computer simulation

W. Fan et al. (Eds.): AsiaSim 2022, CCIS 1712, pp. 183–196, 2022.
https://doi.org/10.1007/978-981-19-9198-1_15

to the actual deployment of algorithms. Cyber-Physical System (CPS) can integrate the network and physical entities [3] to realize the network of physical entities, and virtual reality simulation is a simulation method developed based on CPS.

Virtual-reality simulation refers to a simulation system in which there are both virtual and physical resources. The interaction behavior based on information interaction can be carried out between the two—typical application scenarios such as LVC simulation, parallel systems, etc. The virtual-reality simulation in this paper means that the virtual entity and the physical entity can interact with information. In other words, the physical entity can perceive the virtual environment through the virtual entity so that the UAV swarm algorithm can be directly arranged on the physical entity, and the physical entity can be trained in the virtual environment. Through this simulation method, the advantages of virtual simulation and physical stimulation can be effectively integrated, with both the environment's and airborne sensors' easy editability. At the same time, the addition of physical entities also makes the algorithm training results more credible and effectively reduces the distortion of algorithm training.

A suitable simulation system communication method is necessary and fundamental for designing a simulation system, especially in cluster simulation; the good or bad communication effect often determines whether a simulation system's performance is superior. High-Level Architecture (HLA) and Data Distribution Service (DDS) are widely used simulation system communication architectures. HLA is widely used in the construction of complex simulation systems, but HLA does not perform well when there is physical involvement. Simulation systems using DDS communication architecture can transmit data quickly and accurately in heterogeneous distributed environments and achieve high performance and low data transmission latency through Quality of Service (QoS) [4]. AL-Madani et al. [5] developed robust control of UAV clusters based on DDS, enabling control algorithms with good flexibility, reliability, and portability. The literature [6] implemented real-time communication between an autonomous vehicle model in virtual space and a hardware-in-the-loop simulation (HILS) module through DDS to achieve distributed communication for hardware-in-the-loop simulation. Based on these studies, this paper proposes a DDS-based communication method for HILS and measures the metrics of this communication method to verify the applicability for HILS of UAV clusters, which has particular guiding significance for the development of future HILS systems.

2 The Simulation Framework

The simulation resources in the virtual-real simulation system are distributed in virtual and physical space. The virtual space is editable, the environment is easy to change, and collisions between UAVs and the environment or between UAVs do not damage the equipment. The operation in the physical space is necessary for the UAV algorithm to move from simulation to deployment, serving as a test of the algorithm's effectiveness. The distribution of simulation resources in virtual space and physical space is shown in Fig. 1.

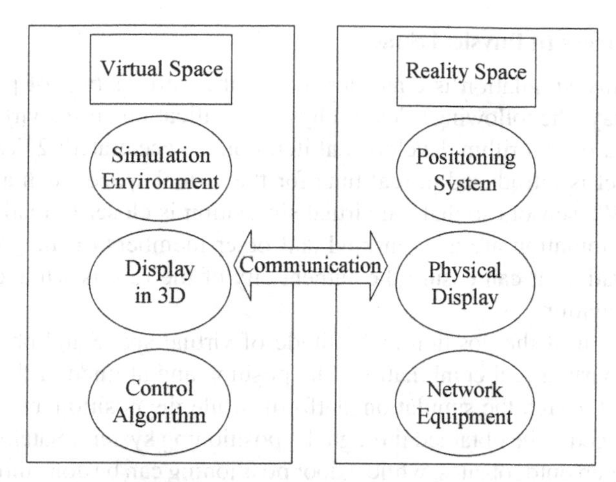

Fig. 1. Resources in virtual and reality space

2.1 The Resources in Virtual Space

In the virtual-real UAV swarm simulation system, the virtual space mainly provides the services of creating a simulation environment, 3D real-time display, and control algorithm simulation.

The simulation environment is where the UAV performs its mission, including objective factors such as terrain, landforms, and features. In the case of military operations, it also includes enemy armed forces, such as anti-aircraft fire, UAV attack targets, and air-to-air reconnaissance radar. These elements are often difficult to obtain in the physical world, such as weapons and equipment, or difficult to change in the physical environment, such as terrain and terrain features. However, they can be easily obtained and edited in the virtual space.

The 3D display is to realize the real-time display of the position and attitude of the virtual entity in the virtual space with the help of a perfect simulation engine. The current UAV cluster simulation results are primarily displayed in two-dimensional planes utilizing point and line diagrams [7, 8], which have the advantage of good quantification and can clearly show the performance of the control algorithm. However, the disadvantage is that it is not intuitive enough and difficult to observe when the cluster size is large. A three-dimensional display can intuitively show the motion of the UAV swarm. It can also output two-dimensional diagrams of user interest as needed during the simulation process, which has a better display effect. However, development is challenging, and the simulation should not advance too fast. Otherwise, it is not easy to observe.

The core algorithm is the content of the simulation, and the current research on UAV clusters mainly includes path planning, formation control, collision avoidance, mission planning, etc. These algorithms are simulated in virtual space to verify whether the results and performance meet theoretical expectations.

2.2 The Resources in Physical Space

The virtual reality simulation is characterized by the introduction of physical space, which mainly plays the following roles: 1. Physical verification during virtual simulation so that the speed of algorithm development iteration is accelerated; 2. The state of the simulation object is introduced in real-time for the virtual space, such as the working state of the UAV's sensors so that the virtual simulation is closer to reality; 3. The key objects in the simulation are used in-kind and other members of the group. Through computer generation, it can ensure the authenticity of the results while expanding the scale of the simulation.

The unification of the position and attitude of virtual space and physical space is the basis of the virtual-real combination. The position and attitude of the virtual entity can be obtained through the simulation platform, while the position and attitude of the physical entity need to be obtained through the positioning system. Satellite positioning can be used in open outdoor sites, while indoor positioning can be done through a motion capture system or other means.

The physical display is the second dimension of the virtual-real result display and is also the verification means of whether the algorithm can be deployed to the physical object. The synchronous or synergistic relationship between the physical entity and the virtual entity means that corresponding control algorithms need to be designed for the physical entity as well so that it can achieve the same state as the virtual entity.

Network equipment is the basis of communication. Indoor communication can be done with routers, WIFI modules, etc., while outdoor communication in a wide range requires larger radio modules. Network equipment performance is decisive for the quality of communication and is also the background and condition for the design of the virtual-real simulation communication method.

2.3 Layered Model of the Virtual-Reality Simulation System

Based on the resource composition of the virtual-real simulation system, this paper proposes a layered model of the virtual-real simulation system; the model consists of a hardware layer, connection layer, and application layer; the specific structure is shown in Fig. 2.

The hardware layer contains the various instruments and equipment that make up the simulation system and is responsible for environment construction, physical networking, and result display. The connection layer is the communication architecture for connecting the virtual space with the physical space or interconnecting within the space, such as HLA or DDS architecture; the communication protocol for communication between entities; the topology of the internal members of the network and the external interface of the simulation system, etc. The application layer is responsible for realizing the functions of the virtual-real simulation system in providing simulation services, simulation object modeling, simulation environment editing, simulation effect evaluation, and other application function development.

The communication method studied in this paper is located in the connection layer, explores the topology and communication protocols of the virtual-real simulation system under the DDS framework, and gives the test and evaluation of the communication performance.

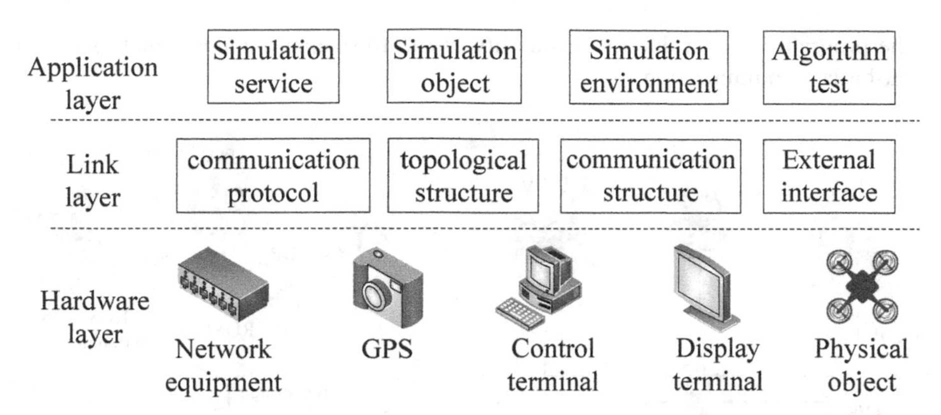

Fig. 2. Layered model of the virtual-reality simulation system

3 Communication Mode and Index of the Simulation System

In the physical test, it often faces the problems of insufficient equipment, limited space, and complicated and time-consuming operation due to the large number of physical UAVs. The objective of the virtual-reality simulation system for UAVs is to supplement the number of physical UAVs with virtual UAVs and then complete the simulation work that originally requires a large number of physical UAVs with a small number of physical UAVs and a certain number of virtual UAVs. Therefore, the number of physical drones that can be accommodated by the communication network of a simulation system is a key factor in the simulation scale of the virtual-real simulation system.

Centralized control has the advantages of strong stability, centralized information, and simple communication. In the UAV simulation system, the surrounding environment is stable and controllable, the space for UAV movement is small, the ground station centralized control method can be used, and the computer and the UAV communicate by means of a wireless network, and the centralized architecture is shown in Fig. 3. In the wireless network, the factors affecting the communication quality mainly include the frequency of the wireless network, the surrounding network environment, the number of terminals in the communication network, and the frequency of information communication.

In the virtual reality simulation system, the physical UAV and the ground control computing method use the UDP protocol to communicate under the WIFI connection. The computer has two ways to process the received Mavlink message. One is to parse the frame directly after receiving, get the information, and hand it over to the control algorithm for processing and feedback control information, which is called direct connection communication, as shown in Fig. 3.

The second is a bridge approach within the framework of the Robot Operating System (ROS), where data is forwarded from ROS1 to ROS2 via ros1bridge and then to the software of the display and control terminal via rosbridge. The disadvantage of this approach is that forwarding generates time delays and packet loss, which degrades data transmission performance; the advantage is that communication is always carried out within ROS, which is less difficult to develop and easier to understand, and the portability of the program is better due to the stability of the ROS environment, which is called agent-based communication.

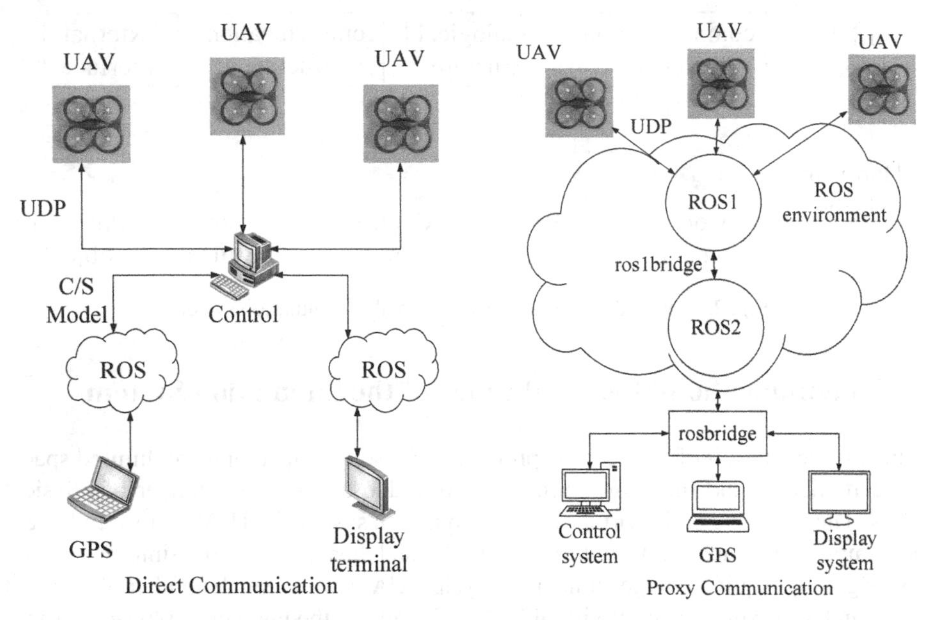

Fig. 3. The comparison of direct communication and proxy communication

In this paper, we built a UAV simulation system with the above two communication architectures, tested the UAV communication under the same conditions, and explained the advantages and disadvantages of the two communication architectures through the comparison of communication parameters.

3.1 Delay

Delay is the total time required for data to travel from one end of a network or link to the other and consists of transmission delay t_s, propagation delay t_m, processing delay t_c, and queuing delay t_a. It can be calculated by the formula (1).

$$Delay = t_s + t_m + t_c + t_a \tag{1}$$

When the geographical size of the simulation system is small, the propagation delay t_m is negligible, the processing delay t_c is of the order of microseconds (μ_s)or less, and the total delay is often of the order of milliseconds (m_s), then the processing delay is also negligible in the calculation. Therefore, the final total delay depends on t_s and t_a. *Delay* can be calculated by the formula (2).

$$Delay = t_s + t_a \tag{2}$$

3.2 Packet Loss Rate

The packet loss rate is a measure of the integrity of the information propagated in the network. The calculation formula is formula (3).

$$P_L = \frac{N_s - N_r}{N_s} \times 100\% \tag{3}$$

P_L represents the packet loss rate, N_s is the number of packets sent, N_r is the number of packets received. Unlike bandwidth, the correlation between packet loss rate and latency is weak [9]. The packet loss rate is related to the network condition as well as the hardware device status, and under the premise of good hardware operation, this paper considers that the packet loss rate is only related to the network condition.

3.3 Jitter

Jitter is the time difference between the maximum delay and the minimum delay, which is used to measure the stability of the network.. The calculation formula is formula (4).

$$T_\Delta = Delay_{\max} - Delay_{\min} \tag{4}$$

In this formula, T_Δ is jitter, $Delay_{\max}$ is the maximum delay and $Delay_{\min}$ is the minimum delay. If the network is congested, the queuing delay will affect the end-to-end delay, resulting in a delay jitter.

Jitter is a performance parameter that has a significant impact on voice transmission and image transmission. When the UAV is equipped with a camera module or voice recording module in the virtual reality simulation, ensuring that the time delay jitter is maintained at a level that does not affect the service is an important condition for the design of the communication network of the simulation system.

4 Evaluation

In order to study the association of the communication network performance ros agent of the virtual reality simulation system, comparative experiments were designed by setting up communication in the ROS1 environment only, communication in the combined ROS1 and ROS2 environment through ros1bridge forwarding, and communication with the simulation platform built by Unreal Engine on another computer through ros1bridge and rosbridge communication in three groups. The relationship between the communication performance and the ros agent can be obtained by the variation of the indicators.

4.1 Experimental Environment and Configuration

The experiments were conducted on a Windows OS computer running the Ubuntu system on a virtual machine. ROS1 internal communication, ROS1 and ROS2 combined environment communication were conducted on the virtual machine, and the simulation platform built by Unreal Engine was located inside Windows, and the ROSIntegration inside Unreal Engine was connected through rosbridge so that Unreal Engine receives messages from ROS1 in the virtual machine. The configuration of the experimental computer and the virtual machine is shown (Table 1).

Table 1. Computer configuration

Computer configuration		Virtual machine configuration	
CPU	Intel(R) Core(TM) i7-10700K CPU @ 3.80 GHz 3.79 GHz	CPU	2
RAM	16.0 GB (15.8 GB available)	RAM	4 GB
OS	Windows 10 Enterprise Edition	OS	Ubuntu 18.04.1

4.2 Test Methods for Network Performance Parameters

The Talker in ros sends the moment and the serial number of that moment, which is the time in milliseconds (ms) since the ros system was started. When the talker sends its own information to the control station at a certain frequency, the delay and jitter can be obtained by the delay calculation method with the following algorithm.

$$Delay = \Delta_{time\_stamp} - \frac{1}{\nu}$$

Δ_{time_stamp} is the difference between timestamp values of two consecutive packets.

By designing the talker_seq value bound to the sending time at the talker sending end, the seq value will be +1 for each packet sent. And the listener at the receiving end will +1 the internal listener_seq value for each packet received, calculate the difference between listener_seq and talker_seq, and get the number of packet losses. Then, the packet loss rate can be obtained by the packet loss rate calculation formula.

4.3 Experimental Results and Analysis

Experiment on ROS1 and ROS1-ROS2 Communication

The experiment was conducted in two sessions. The first test was only to publish and receive messages inside ROS1. The second test was done in the ros1bridge bridge between the ROS1 environment and the ROS2 environment, and the sending and receiving were done in the ROS2 and ROS1 environments respectively, transmission frequency $\nu = 50$ Hz, the number of sending and receiving is shown in Table 2.

Table 2. Communication data

ROS1 internal communication				ROS2-ROS1 communication			
No	Send times	Receiving times	Packet loss	Send times	Receiving times	Packet loss	Packet loss rate
1	529	529	0	534	521	13	2.43%
2	550	550	0	522	520	2	0.38%
3	531	531	0	560	523	37	6.61%
4	535	535	0	556	522	34	6.12%
5	535	535	0	532	520	12	2.26%
6	537	537	0	594	547	47	7.91%
7	574	574	0	535	518	17	3.18%
8	547	547	0	518	516	2	0.39%
9	539	539	0	535	516	19	3.55%
10	564	564	0	554	517	37	6.68%

By observing the data in Table 2, the two-level proxy has more unstable communication performance due to packet loss during transmission than the one-level proxy.

The experimental results of the latency measurements are shown in Table 3. The second column is the delay of ROS1 internal communication, and the third is the delay after bridging through ros1bridge. It can be found that the message delay has been significantly improved after being forwarded by ros1bridge.

Table 3. Average delay measurement results

No	ROS1 (ms)	ROS1-ROS2 (ms)
1	1.252	1.610
2	1.229	1.609
3	1.199	1.580
4	1.198	1.636
5	1.229	1.656
6	1.229	1.586
7	1.237	1.608
8	1.221	1.634
9	1.251	1.622
10	1.227	1.630

Table 4. Jitter and standard deviation of two experiments

No	ROS1		ROS1-ROS2	
	Jitter	Standard deviation	Jitter	Standard deviation
1	2.957	0.302	2.752	0.234
2	2.683	0.291	3.194	0.346
3	2.377	0.244	1.149	0.190
4	2.377	0.244	1.862	0.229
5	2.580	0.247	5.003	0.420
6	2.580	0.246	1.846	0.205
7	2.516	0.290	1.575	0.204
8	2.455	0.262	2.379	0.227
9	2.893	0.303	2.248	0.226
10	4.593	0.313	1.719	0.191

By analyzing the time delay data, the standard deviation and jitter of ten sets of data are obtained, as shown in Table 4.

The curves of jitter and standard deviation can be obtained from the data in Table 4.

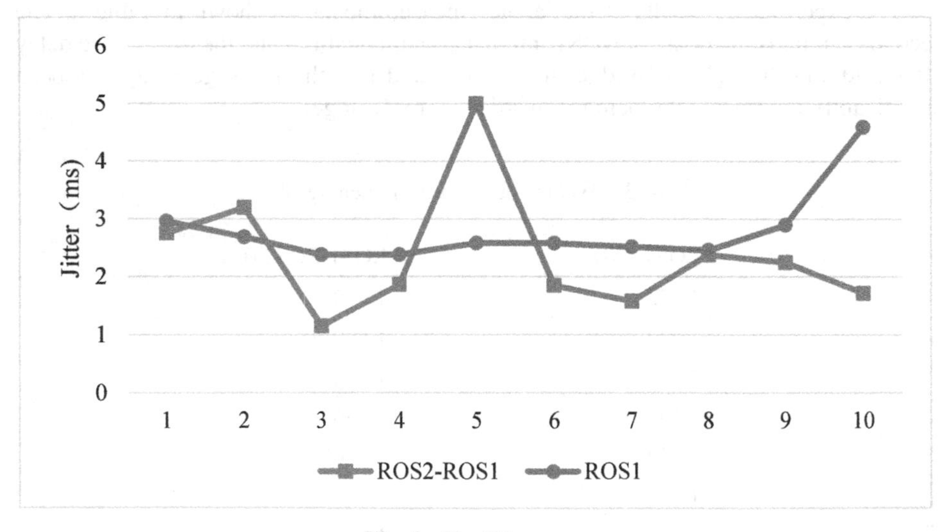

Fig. 4. The jitter curve

From Table 4, Fig. 4, and Fig. 5, it can be seen that the jitter and standard deviation of the delay are not significantly different in the case of the first- and second-level proxy

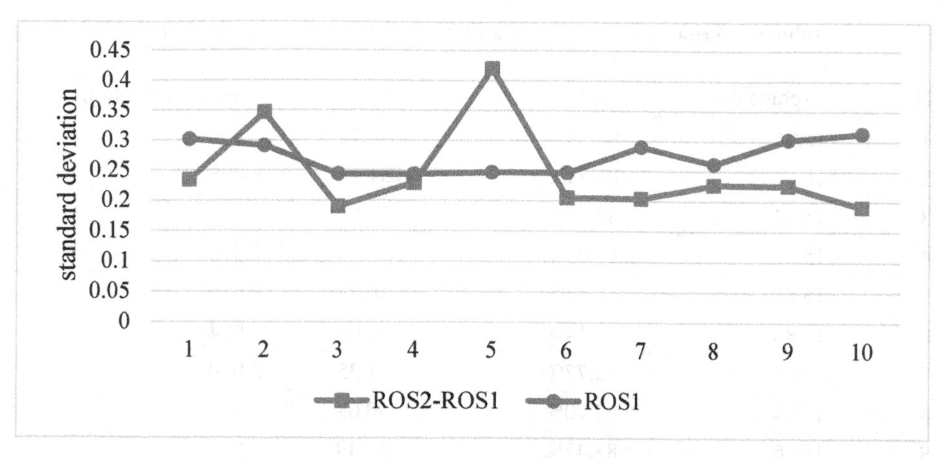

Fig. 5. The curve of standard deviation

communication. The main impact of the two levels of communication on the performance is the size of the average delay and the packet loss rate.

Experiment on ROS1-ROS2-Simulation Engine Three Level Agent Communication

The experiment uses ros1bridge and rosbridge to connect ros1, ros2 to the simulation engine so that the data is sent from the ros1 environment to the simulation engine through rosbridge in ros2, realizing the data from the virtual to the real. After receiving the message, the unreal engine sends it back to ros1 through the same path in reverse, realizing the data is virtual to real. From this, the communication system's Round-Trip Time (RTT) can be obtained, and thus the one-way delay.

Under the condition of control variables, ten experiments were conducted for three-level proxy communication with 50 Hz as the sending frequency and the data are shown in Table 6, and the standard deviation of average delay and jitter are in milliseconds (ms).

As seen from Table 5, at 50 Hz, all performance indicators of the three-level agent are significantly higher than those of the one-level and two-level agents. This phenomenon indicates that the communication performance of the three-level agent is considerably lower than that of the one-level and two-level agents at high transmitting frequencies. Although the three-level agent increases the system's scalability and enables the simulation system's data to flow across systems and platforms, the communication performance is sacrificed.

In the case of the three-level agent, the communication frequency is minimal. So the sending frequency is adjusted in later experiments, and in one experiment, there are nine groups of experiments with sending frequencies of 1, 2, 3, 4, 5, 10, 15, 20, and 30 Hz. Each group of experiments takes 500 received data to obtain the delay of each sending and receiving and the total packet loss rate. The three-level agent's experimental results are more complicated than the first and second levels. The nine sets of experimental data are taken in order of frequency from small to large, with the delay as the vertical coordinate and the sequence of packets as the horizontal coordinate. Figure 6 shows one

Table 5. Performance metrics data for three-level proxy communication

No	Average delay	Packet loss rate	Jitter	Standard deviation
1	15.96	83.39%	30.72	7.862
2	16.02	82.71%	31.25	8.095
3	15.74	83.50%	30.81	7.821
4	15.96	83.03%	31.54	7.976
5	15.87	82.97%	30.87	7.919
6	15.88	83.14%	30.05	7.924
7	16.04	82.77%	31.35	8.020
8	15.94	82.70%	30.08	7.879
9	16.06	83.42%	31.49	7.991
10	15.73	83.06%	31.67	8.006

results obtained for each group of 500 data, the horizontal axis is divided into 9 parts, every part has 500 data, which represents different frequencies.

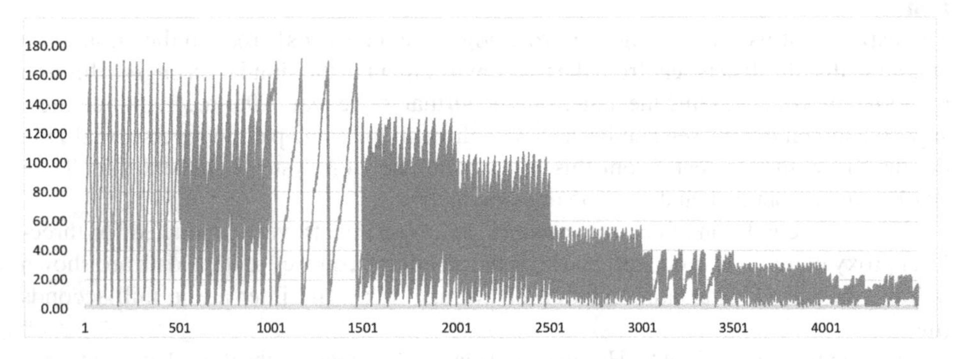

Fig. 6. Nine groups of data in different frequency

In the two experiments, each data group has a more obvious periodicity, and the period varies with the sending frequency. Table 6 shows each data group's maximum value, minimum value, and packet loss rate.

Data from Table 6 produce the line graph of maximum value, minimum value, jitter and packet loss rate, they can be seen in.

The limited number of bytes in the first experiment generated a truncation error when the talker sent the data at the millisecond level. This truncation error was eliminated by adjusting the data format in the second experiment. Therefore, in the minimum performance, there is a significant jitter in the time delay minimum of the first experiment due to the truncation error generated by the retention of valid digits.

As can be seen from Fig. 7, the jitter and the maximum value of the delay are inversely proportional to the packet loss rate, i.e., as the packet loss rate increases, the delay of the

Table 6. Communication metrics for two experiments

Frequency	Max delay (ms)		Min delay (ms)		Jitter (ms)		Packet loss rate	
	First	Second	First	Second	First	Second	First	Second
1	171.92	171.05	0.88	1.87	170.05	169.18	0.00%	0.00%
2	175.22	161.48	2.75	2.03	173.18	159.45	0.00%	0.20%
3	173.73	170.99	0.58	1.85	171.88	169.15	0.00%	0.60%
4	133.25	130.94	2.89	1.78	131.47	129.17	18.10%	18.74%
5	109.56	107.27	0.25	1.36	108.20	105.91	35.56%	34.25%
10	56.05	56.71	0.26	1.37	54.68	55.34	64.03%	64.55%
15	41.25	40.15	0.08	1.47	39.78	38.68	74.64%	73.72%
20	33.70	31.64	3.33	1.32	32.38	30.32	78.62%	78.08%
30	25.71	23.82	0.59	1.66	24.05	22.16	85.72%	84.19%

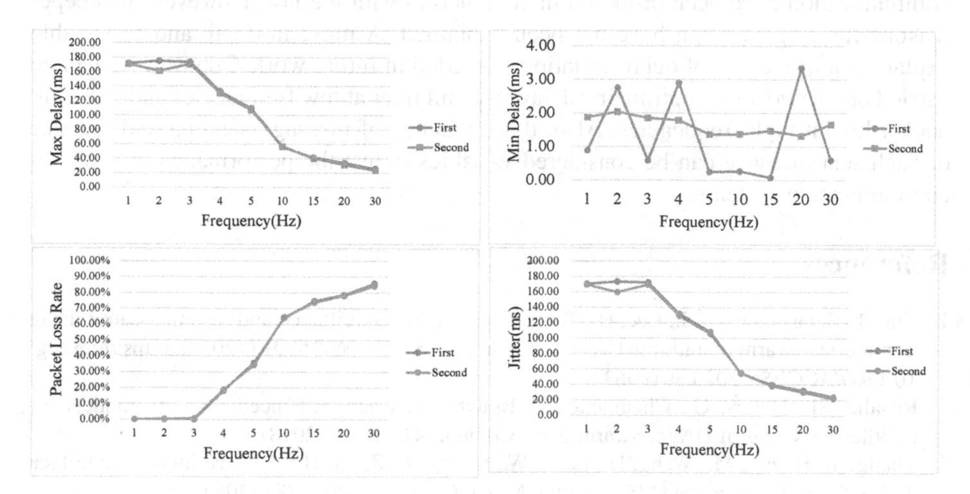

Fig. 7. Communication performance index chart

communication system decreases instead, and there is a significant reduction in the jitter of the delay. This property can reduce the delay by sacrificing the packet loss rate and increasing the transmission frequency in the simulation system that requires low delay.

5 Conclusion and Future Work

The virtual-real simulation system is one of the development directions of the future simulation system. When designing the simulation system, the communication structure of the simulation system can be designed according to the different resources required by the physical space and the virtual space of the system. In the direct-connected communication structure, the effective bandwidth can be significantly increased, and the time delay is greatly reduced because there is no bridge limitation, but the resources accessible to the virtual space are less, and often only simple processing and display of data can be performed. With the adoption of agent-based communication architecture, the resources accessible to the virtual space also increase with the increase of the agent level, and with it, the communication performance decreases.

In this paper, we study the communication structure of multi-level agents of the virtual-reality simulation system and test the communication performance of three agent methods based on some existing network parameters design procedures. As the level of agent increases, although the communication system is easier to design and the scalability is better than the lower-level agent, the communication performance represented by delay and packet loss rate also decreases. Therefore, when designing a simulation system, a suitable communication structure should be selected in conjunction with the design requirements to harmonize communication performance and simulation system functions.

The work in this paper has certain limitations, as only the law of delay variation has been observed so far, and the application scenarios of different levels of proxy communication have been proposed in conjunction with the law. However, the deeper reasons for its generation have not been explained. A more in-depth and reasonable explanation of the law of delay variation is needed in future work. Future work can be carried out in two areas, optimizing delay size and jitter at low frequencies and reducing packet loss at high frequencies. Also, the frequency of message sending and the size of each sent message can be considered variables to test the performance of different communication systems.

References

1. Cho, J., Sung, J., Yoon, J., Lee, H.: Towards persistent surveillance and reconnaissance using a connected swarm of multiple UAVs. IEEE Access **8**, 157906–157917 (2020). https://doi.org/10.1109/ACCESS.2020.3019963
2. Rosalie, M., Danoy, G., Chaumette, S., Bouvry, P.: Chaos-enhanced mobility models for multilevel swarms of UAVs. Swarm Evol. Comput. **41**, 36–48 (2018)
3. Zhang, Y., Huang, M., Wang, H., Feng, W., Cheng, J., Zhou, H.: A co-verification interface design for high-assurance CPS. Comput. Mater. Continua **1**(58), 287–306 (2019)
4. El-Ferik, S., Almadani, B., Elkhider, S.M.: Formation control of multi unmanned aerial vehicle systems based on DDS middleware. IEEE Access **8**, 44211–44218 (2020). https://doi.org/10.1109/ACCESS.2020.2978008
5. Al-Madani, B., Elkhider, S.M., El-Ferik, S.: DDS-based containment control of multiple UAV systems. Appl. Sci. **10**(13), 4572 (2020)
6. Cho, D.-S., Yun, S., Kim, H., Kwon, J., Kim, W.-T.: Autonomous driving system verification framework with FMI co-simulation based on OMG DDS. In: 2020 IEEE International Conference on Consumer Electronics (ICCE), pp. 1–6 (2020). https://doi.org/10.1109/ICCE46568.2020.9043010
7. Sun, R., Zhou, Z., Zhu, X.: Stability control of a fixed full-wing layout UAV under manipulation constraints. Aerosp. Sci. Technol. **120**, 107263 (2022)
8. Park, S., La, W.G., Lee, W., Kim, H.: Devising a distributed co-simulator for a multi-UAV network. Sensors **20**, 6196 (2020)
9. Martin, J., Nilsson, A., Rhee, I.: Delay-based congestion avoidance for TCP. IEEE/ACM Trans. Netw. **11**, 356–369 (2003)

An Adaptive Low Illumination Color Image Enhancement Method Using Dragonfly Algorithm

Jiang Liu and Shiwei Ma[✉]

Shanghai University, Shanghai, China
masw@shu.edu.cn

Abstract. In order to improve the visual perception of color images under low illumination conditions, an adaptive enhancement method is proposed to enhance the contrast and brightness, enrich the color and avoid over-enhancement. Firstly, the original low illumination image is converted from RGB color space to L*a*b* color space, and a novel gray level transformation function namely piecewise sine function is proposed to improve the brightness of L* channel image. In addition, dragonfly algorithm is utilized to optimize the parameters in piecewise sine function to achieve the best brightness adjustment effect. Then a novel saturation enhancement method is proposed to enrich color information. Subsequently, a fitness function that takes into account both the degree of overall brightness enhancement and the suppression of information loss is applied as the objective function of dragonfly algorithm. Ultimately, the processed L*a*b* color space is transformed back to RGB color space to get the enhanced image. Experimental results verify the effectiveness of the proposed method.

Keywords: Color image enhancement · Low illumination · Piecewise sine function · Dragonfly algorithm · Color space

1 Introduction

With the rapid development of computer vision, optical imaging devices are widely applied on a variety of occasions, such as visual surveillance, medical diagnosis, etc. While influenced by many factors, the quality of images captured by the devices is often degraded. For example, limited by the pixel depth that the devices can record, the detail of objects can hardly be simultaneously captured in both bright regions and dark regions. Also, under low-light conditions, on account of inappropriate device parameters set by users, the contrast and visibility of the output images are usually reduced. As an enormous amount of detail is hidden in the dark regions, it is difficult for human visual system (HVS) to acquire sufficient information from the low illumination images. Thus, enhancing the low illumination images, including improving the brightness and contrast as well as adjusting the color so as to improve the visual perception quality, is demanded.

W. Fan et al. (Eds.): AsiaSim 2022, CCIS 1712, pp. 197–213, 2022.
https://doi.org/10.1007/978-981-19-9198-1_16

Various image enhancement methods have been proposed to improve image quality. In general, the existing methods can be classified into three categories, i.e., the methods based on retinex theory, gray scale transformation, and histogram equalization.

The retinex theory [1–3] proposed by Land and McCann is based on experimental results concerning human visual perception. However, the initial image enhancement methods based on retinex theory aimed at defining a practical implementation of retinex model in image processing, rather than specifically focusing on its validity as a model for human brightness and color perception [4]. The retinex theory-based methods assume that the visible image is produced by illumination and reflection components, thus the enhanced image can be obtained by estimating and eliminating the reflection component. The classical methods include Single-Scale Retinex (SSR) [4], Multiscale Retinex (MSR) [5] and Multi-Scale Retinex with Color Restoration (MSRCR) [5]. They are widely used to enhance low illumination images. Whereas when the original image does not satisfy the grayscale world hypothesis, these methods may cause color distortion. Wang et al. proposed a naturalness-preserved enhancement (NPE) algorithm [6] to enhance image contrast and prevent excessive enhancement of local regions. Fu et al. proposed a weighted variational model for simultaneous reflection and illumination estimation (SRIE) [7] that can preserve the estimated reflectance with more details and suppress noise to some extent.

The methods based on gray scale transformation such as power-law transformation (also known as gamma correction) are practical approaches to enhance low illumination images. In order to more effectively enhance image contrast, Huang et al. proposed an adaptive gamma correction algorithm with weighting distribution (AGCWD) [8]. By calculating the cumulative distribution probability histogram, this method can obtain gamma correction parameters adaptively. However, when the gray distribution of the low illumination image is too concentrated in the low-value range, it is difficult for these methods to effectively improve the image brightness. Al-Ameen proposed an illumination boost algorithm (IBA) [9] to enhance the brightness and contrast of the image. By using several adjustment functions with low computational complexity, this algorithm has the advantages of fast operation speed and easy implementation, but it may over-enhance the low illumination images due to the lack of adaptability.

As a classical image enhancement method, histogram equalization (HE) [10,11] remaps the gray level based on the probability distribution of the gray level of the input image, which can expand the narrow gray level range to a wider range. It has the effect of stretching dynamic range so as to enhance the overall contrast of the image. To avoid excessive enhancement of average brightness, Kim proposed the brightness preserving bi-histogram equalization (BBHE) method [12]. This method firstly separates the histogram of the input image into two parts according to the average brightness. Then histogram equalization is applied for these two parts separately. Inspired by Kim's research, many improved methods were proposed to better overcome the brightness problem, such as dualistic sub-image histogram equalization (DSIHE) [13], recursive sub-image histogram equalization (RSIHE) [14], brightness preserving dynamic histogram equalization (BPDHE) [15], etc. To focus on the enhancement of foreground objects, Arici et al. proposed a method (WAHE) [16] that can reduce large histogram values for smooth areas corresponding to background regions. Contrast limited adaptive histogram equalization (CLAHE) proposed by Pizer et al. [17] and developed by Reza [18], Yadav et al. [19] divides the image into several sub-blocks and enhance each sub-block separately to obtain the enhanced image. Celik and Tjahjadi proposed the contextual and variational contrast (CVC) enhancement algorithm [20], which employs contextual data modeling using the two-dimensional histogram of the input image to perform nonlinear data mapping. Lee et al. proposed a contrast enhancement method based on layered difference representation (LDR) [21,22]. The main purpose of this method is to enhance contrast by amplifying local gray differences. However, it is difficult for the above two methods to effectively improve the brightness of the images.

In recent years, image enhancement is formulated as an optimization problem and solved using nature-inspired optimization algorithms (NIOAs) [23]. Shanmugavadivu et al. proposed a method [24] using particle swarm optimization (PSO) [25] algorithm to optimize histogram in order to enhance the contrast of the image. To enhance the low-illuminance image, Liu et al. proposed an enhancement method [26] based on an optimal hyperbolic tangent curve. The golden section search algorithm was adopted to optimize the mapping parameters of RGB components to maximize the information entropy of the image. In order to avoid color artifacts, the method proposed by Kanmani et al. [27] transforms the low contrast image into the L*a*b* color space and uses PSO algorithm to optimize the gamma correction factor only on the luminance channel. But this method can hardly produce vivid colors.

Dragonfly algorithm (DA) [28] proposed by Mirjalili in 2016 is an NIOA that can be performed in complex search domain to find an optimal solution. It has been proved that DA has great exploration ability and its convergence rate is better than many other NIOAs such as PSO [29]. In order to effectively enhance the brightness and contrast, enrich the color information and reduce the loss of details of the low illumination color images, a novel adaptive enhancement approach using DA is proposed. The main contribution of the proposed method is summarized as follows.

- A new luminance adjustment function namely piecewise sine function is proposed to enhance the brightness of the images.
- DA is used to adaptively find the optimal parameters of piecewise sine function for images under low illumination conditions.
- A function that considers both the degree of overall brightness enhancement and the suppression of information loss in the highlight regions is used as the fitness function of dragonfly algorithm.
- A novel saturation enhancement method is proposed to enrich color information in order to generate vivid images.

The rest of this paper is organized as follows. Principles of DA and color space are described in Sect. 2. The detailed description of the proposed method is presented in Sect. 3. Experimental results and performance analysis are discussed in Sect. 4. Conclusions are drawn in Sect. 5.

2 Principles of DA and Color Space

2.1 Dragonfly Algorithm

DA is designed by modeling the social interaction of dragonflies in navigating, searching for food, and avoiding enemies. To be specific, five main factors are considered to update the position of dragonflies, including separation, alignment, cohesion, attraction towards food sources, and distraction outwards enemies.

The separation is defined as

$$S_i = -\sum_{j=1}^{N} X - X_j \tag{1}$$

where S_i is the separation of the i-th individual, X is the position of the current individual, X_j is the position j-th neighboring individual, and N is the number of neighboring individuals.

The alignment is defined as

$$A_i = \frac{\sum_{j=1}^{N} V_j}{N} \tag{2}$$

where A_i is the alignment of the i-th individual, V_j is the velocity of j-th neighboring individual.

The cohesion is defined as

$$C_i = \frac{\sum_{j=1}^{N} X_j}{N} - X \tag{3}$$

where C_i is the cohesion of the i-th individual.

Attraction towards a food source is defined as

$$F_i = X^+ - X \tag{4}$$

where F_i is the position of food source of the i-th individual, X^+ is the position of the food source.

Distraction outwards an enemy is defined as

$$E_i = X^- + X \tag{5}$$

where E_i is the position of enemy of the i-th individual, X^- is the position of the enemy.

If a dragonfly has at least one neighboring dragonfly, two vectors, i.e., step vector ΔX and position vector X, are considered to update its position. The step vector towards a new position is defined as

$$\Delta X_{t+1} = (sS_i + aA_i + cC_i + fF_i + eE_i) + w\Delta X_t \tag{6}$$

where s is the separation weight, a is the alignment weight, c indicates the cohesion weight, f is the food factor, e is the enemy factor, w is the inertia weight, and t is the current iteration.

Then the updated position vector is calculated as

$$X_{t+1} = X_t + \Delta X_{t+1} \tag{7}$$

If a dragonfly has no neighboring dragonflies, Levy flight will be used to update its position in order to improve the randomness, stochastic behavior, and exploration. Under this condition, the updated position vector is calculated as

$$X_{t+1} = X_t + Levy(d) \times \Delta X_t \tag{8}$$

where d is the dimension of the position vectors.

The Levy flight is defined as

$$Levy(d) = 0.01 \times \frac{r_1 \times \sigma}{|r_2|^{\frac{1}{\beta}}} \tag{9}$$

where r_1, r_2 are two random numbers in $[0, 1]$, β is a constant which equals 1.5, and σ is calculated as

$$\sigma = \left(\frac{\Gamma(1+\beta) \times \sin\left(\frac{\pi\beta}{2}\right)}{\Gamma\left(\frac{1+\beta}{2}\right) \times \beta \times 2^{\left(\frac{\beta-1}{2}\right)}} \right)^{\frac{1}{\beta}} \tag{10}$$

where $\Gamma(x)$ is defined as

$$\Gamma(x) = (x-1)! \tag{11}$$

From the above, it can be seen that DA has the advantages of fewer parameters and simple operations. Also, using of Levy flight makes it possible for DA to search for as many feasible solutions as possible.

2.2 Color Space

Mostly, camera sensors record images with red (R), green (G), and blue (B) components, and monitors display images based on these components accordingly. However, brightness and chroma information is mixed in RGB color space, which is not conducive to the perception and understanding of HVS. L*a*b* and HSV are two color spaces in which luminance component and chrominance components are separated. In L*a*b* color space, the red-green channel (a*) and the yellow-blue channel (b*) are the chrominance components, and the achromatic channel (L*) is the brightness component. The characteristic of this color space is that its L* channel closely matches the human perception of brightness. Therefore, L* channel is selected to enhance the brightness and contrast of the low illumination images. Besides, the evaluation of image brightness in DA is also carried out on this channel. In HSV color space, the value (V) component indicates the luminance, and the chrominance components include a hue (H) component and a saturation (S) component. The chrominance components in HSV color space are more consistent with people's intuitive understanding of color compared with a* and b* in L*a*b* color space. Thus, the chrominance adjustment method in this paper is applied in HSV color space.

3 The Proposed Method

The proposed method for low illumination color image enhancement is as follows. Firstly, the input image in RGB color space is converted to L*a*b* color space and separated into three channels. WAHE is used to enhance the contrast of the L* channel image. Because simply use of WAHE cannot effectively improve the brightness of the image, a gray scale transformation function namely piecewise sine function is proposed. The combination of piecewise sine function and DA makes it possible to adaptively adjust the L* channel image so as to enhance the brightness. Then the three-channel images in L*a*b* color space are merged and converted to HSV color space. In order to obtain a vivid image, a new saturation enhancement approach is used in the S channel image to enrich the color information. Subsequently, the image in HSV color space is converted to L*a*b* color space, and a new evaluation function is proposed to assess the quality of the L* channel image. When the termination condition of DA is satisfied, the image in L*a*b* color space is converted back to RGB color space to acquire the enhanced color image. The overall framework of the proposed method is shown in Fig. 1.

3.1 Piecewise Sine Function

The traditional gamma correction function tends to over-enhance the highlight regions of low illumination images while improving the overall brightness of the images. To overcome this, a gray scale transformation function namely piecewise sine function is proposed to enhance the images. The function is given by

$$I_{en}(x) = \begin{cases} b\sin\left(\frac{\pi}{2a}x\right), & 0 \leq x \leq a \\ (1-b)\sin\left[\left(\frac{\pi}{2(1-a)}\right)(x-1)\right]+1, & a < x \leq 1 \end{cases} \tag{12}$$

where x is the intensity value of each pixel in the input image, $I_{en}(x)$ is the corresponding intensity value in the output image, a, b are two parameters in the range of $[0, 1]$ that control the enhancing degree of the images.

As shown in Fig. 2, various transformation curves can be obtained by adjusting the parameters a and b. It can be seen that the combination of the two parameters represents the position of the piecewise point. Thus, it is easy to figure out to what extent the image can be enhanced by certain parameters. Generally, $a < b$ indicates that the output image will be brighter than the original image, and $a > b$ means that the output image will be darker than the original one.

When it comes to enhancing the brightness of low illumination images, the more b is larger than a, the more the brightness of the image is increased. Figure 3 shows the enhancement results of the piecewise sine function in normalized L* channel according to different parameters a and b.

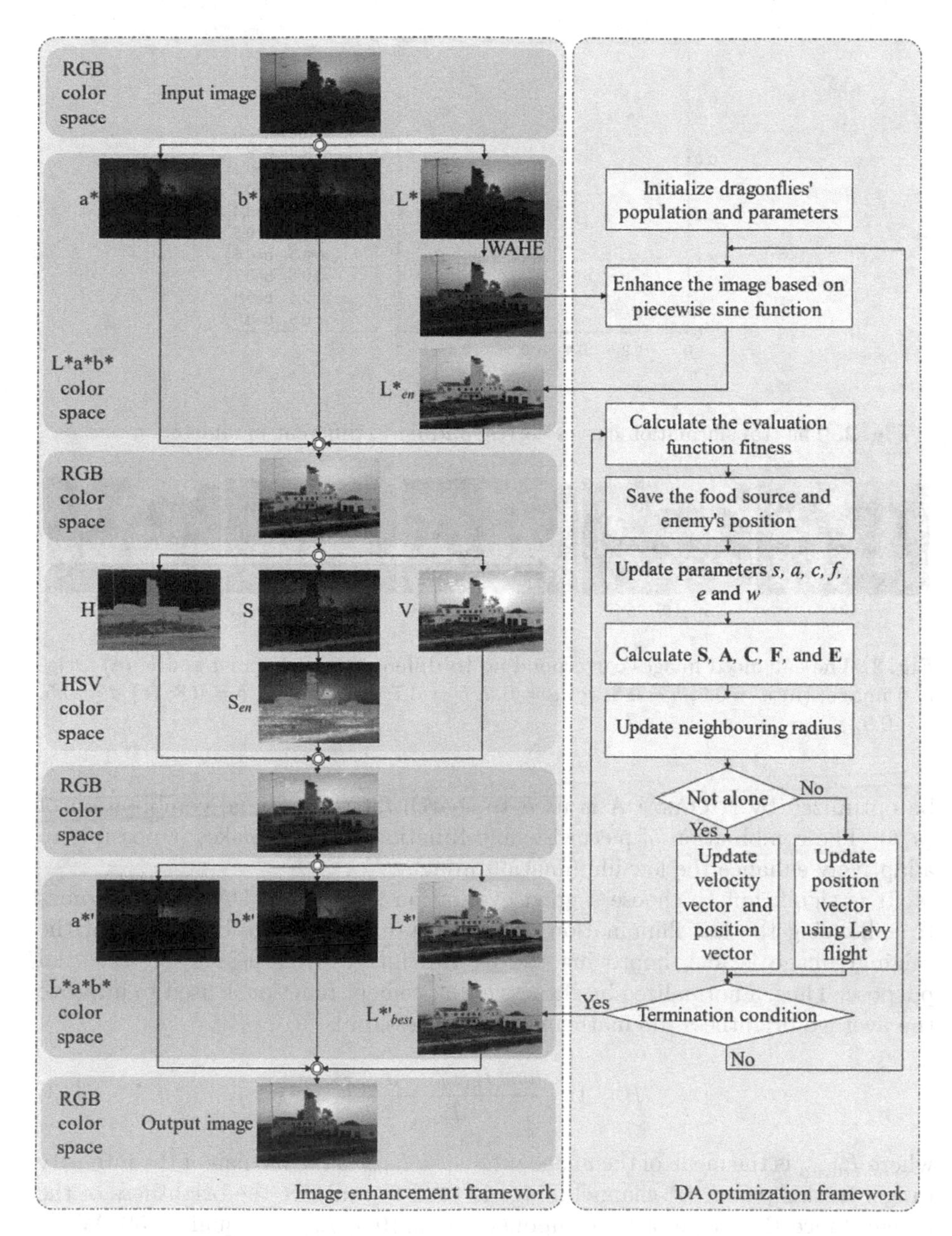

Fig. 1. Overall framework of the proposed method

3.2 Brightness Enhancement Based on DA

In this paper, brightness enhancement is considered as an optimization problem. The proposed piecewise sine function has two parameters a and b which can

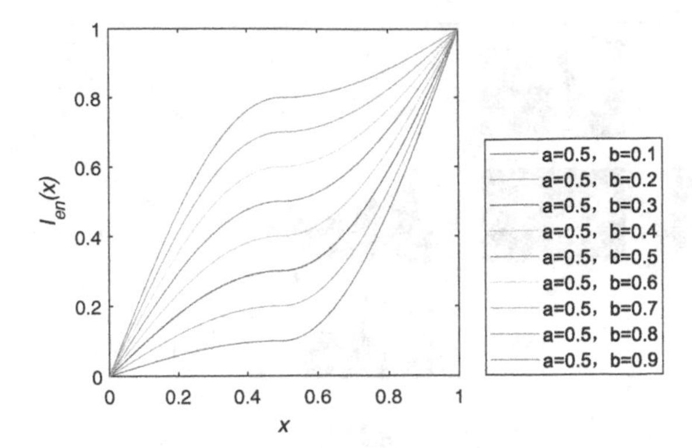

Fig. 2. The transformation curves corresponding to different parameters a and b

Fig. 3. The enhanced images corresponding to different parameters a and b: (a) Original image; (b) $a = 0.5$, $b = 0.6$; (c) $a = 0.5$, $b = 0.7$; (d) $a = 0.5$, $b = 0.8$; (e) $a = 0.5$, $b = 0.9$

be optimized by NIOAs. DA is used to search for the optimal combination of (a, b). The combination of piecewise sine function and DA makes it possible to adaptively enhance the low illumination images.

It is significant to choose a good evaluation function of DA when it comes to enhancing the low illumination images. Since the average brightness of the original image is low, improving the overall image brightness is an essential purpose. Thus, a normalized brightness enhancement function is used to improve the average brightness. Its mathematical expression is

$$B = 1 - \frac{abs(I_{mean} - I_{med})}{I_{med}} \tag{13}$$

where I_{mean} is the mean of the image intensity, I_{med} is the median of the intensity range. In this paper, L* channel image is used to evaluate the brightness of the image. Since the range of L* component is $[0, 100]$, I_{med} is equal to 50. When the value of I_{mean} approaches I_{med}, the value of B will approach 1.

Since highlight regions may exist in some low illumination images, simply improving the overall brightness may lead to the loss of details in these regions. A novel function is utilized to suppress the loss of details while enhancing the brightness of the images. The function can be expressed as

$$P = e^{k-(p-l)} - 1 \tag{14}$$

where k is the penalty parameter, p is the proportion of the maximum intensity value, l is a parameter that limits the acceptable proportion of maximum intensity value. Based on the experience of lots of experiments, k is set as 5 and l is set as 0.07.

The evaluation function of DA is the combination of B and P, which is

$$fitness = B - P \tag{15}$$

By using the fitness function to evaluate the adjusted L* channel image, the brightness can be effectively enhanced while reducing the loss of information in highlight regions.

3.3 Saturation Enhancement

Apart from brightness information, color information also plays an essential role in human visual perception. Even though the chrominance components remain unadjusted, the color of the image that only the brightness of which is enhanced may appear dim to HVS, which can be figured out in Fig. 3. In order to enrich the color information in the images after brightness enhancement, a novel saturation enhancement method is proposed. The method is utilized in HSV color space, and the degree of saturation enhancement depends on the degree of brightness enhancement. Firstly, the expected saturation is calculated by

$$S_{\exp} = \frac{1 - sign(V_{org} - 0.108)}{2} 0.151 + \frac{1 + sign(V_{org} - 0.108)}{2} ({V_{org}}^{0.076} - 0.693) \tag{16}$$

where V_{org} is the pixel intensity in V channel of the original image.

Then the difference of saturation is defined by

$$S_{diff} = S_{org} - S_{\exp} \tag{17}$$

where S_{org} is the saturation of the original image.

Finally, the enhanced saturation is calculated by

$$S_{en} = S_{diff} + \frac{1 - sign(V_{en} - 0.108)}{2} 0.151 + \frac{1 + sign(V_{en} - 0.108)}{2} ({V_{en}}^{0.076} - 0.693) \tag{18}$$

where V_{en} is the pixel intensity in V channel of the image after brightness enhancement.

Figure 4 shows the effect of the proposed saturation enhancement method. The brightness of the original image which is shown in Fig. 3(a) is firstly improved in the normalized L* channel using the proposed piecewise sine function. The parameter a is set as 0.5 and b as 0.9. Then the saturation of the image is enhanced by utilizing the proposed saturation enhancement method. As shown in Fig. 4, it can be illustrated that the saturation of the image is significantly improved and the image becomes more vivid.

Fig. 4. The effect of the proposed saturation enhancement method: (a) Brightness enhanced image; (b) Brightness and saturation enhanced image; (c) Saturation channel of brightness enhanced image; (d) Saturation channel of brightness and saturation enhanced image

4 Experimental Results and Analysis

The experiments in this paper are completed on MATLAB R2020b under Windows 10 operating system. The random access memory is 16 GB. In order to analyze the performance of the proposed method in this paper, the method is compared with nine existing algorithms, including some classical, well-known and state-of-art methods, such as HE [11], CLAHE [19], MSR [5], MSRCR [5], BPDHE [15], CVC [20], LDR [22] and IBA [9].

4.1 Qualitative Evaluation

The images are named 'Road', 'Traffic', 'River', 'Birds', 'Building', 'Indoor', 'Park', 'House', 'Hall' and 'Girl' separately. The experimental results are shown in Figs. 5, 6, 7, 8 and 9.

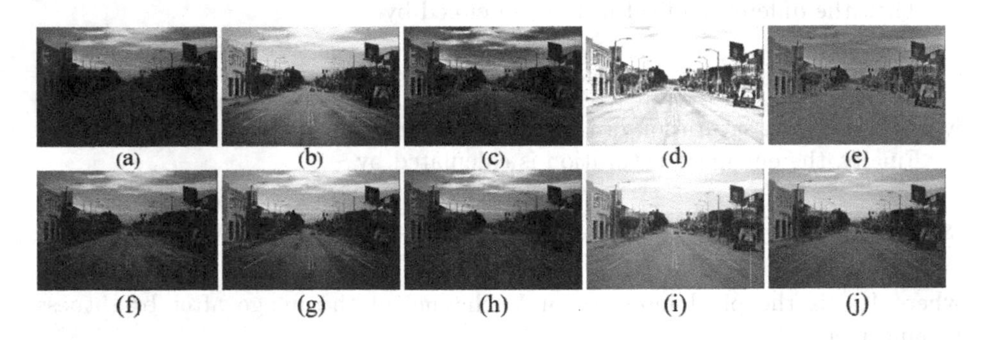

Fig. 5. The enhancement results of image 'Road': (a) Original image; (b) HE; (c) CLAHE; (d) MSR; (e) MSRCR; (f) BPDHE; (g) CVC; (h) LDR; (i) IBA; (j) Proposed method

The enhancement results of image 'Road' are revealed in Fig. 5. Figure 5(a) is the original image in the low illuminance scene. Figure 5(b) is the enhanced image obtained by HE. It can be seen from Fig. 5(b) that the overall brightness is enhanced to some extent, but the color is unnatural. Figure 5(c) is the enhanced image obtained by CLAHE. It can be seen from the image that the brightness is not effectively enhanced, also unnatural shadows appear in the cloud. The enhanced effect of MSR is shown in Fig. 5(d). MSR significantly increases brightness, while the output image appears white on the whole and many details are lost. Figure 5(e) shows the result obtained by MSRCR. This method can effectively improve the brightness of the low illuminance image, but the image is gray and the local contrast is low. Figure 5(f) is the enhanced image obtained by BPDHE. The contrast of the image is enhanced, while the brightness is not improved effectively. Figure 5(g) shows the image enhanced by CVC. The image is like the one enhanced by HE, and also has color problems. In Fig. 5(h), the image enhanced by LDR is like the one enhanced by CLAHE. The brightness of the output image is still low and the details in dark regions are not obvious. As shown in Fig. 5(i), IBA significantly increases the brightness of the low illuminance image. Nevertheless, the sky turns white and many details in highlight regions are lost. Figure 5(j) shows the result obtained by the proposed method in this paper. It can be seen that the brightness and contrast are effectively enhanced. The details of the sky are well preserved, and the color of the enhanced image is natural.

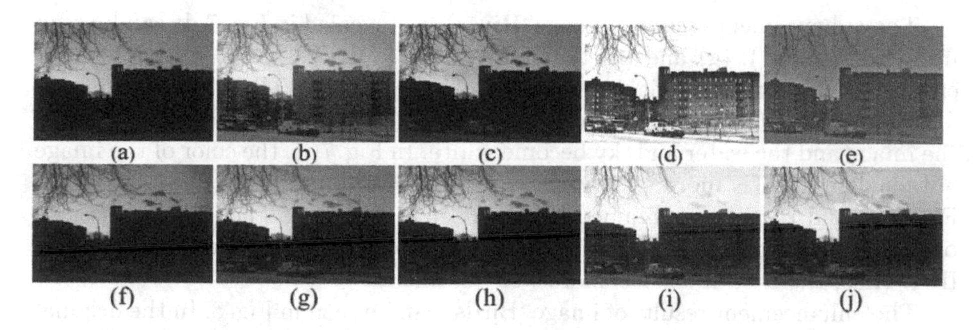

Fig. 6. The enhancement results of image 'Traffic': (a) Original image; (b) HE; (c) CLAHE; (d) MSR; (e) MSRCR; (f) BPDHE; (g) CVC; (h) LDR; (i) IBA; (j) Proposed method

The enhancement results of image 'Traffic' are shown in Fig. 6. In Fig. 6(b), HE effectively enhances the brightness of the original image shown in Fig. 6(a). However, the color of the exterior wall of the building becomes gray. Figure 6(c) shows that CLAHE fails to boost the brightness significantly. Similar problems are shown in Fig. 6(f) enhanced by BPDHE and Fig. 6(g) enhanced by CVC, and the cars in the images are still hard to figure out. Figure 6(d) is the enhanced image obtained by MSR. It is obvious that the upper edge of the exterior wall of the building is black, while the sky is almost pure white. The color distortion is not consistent with human visual perception. The result obtained by MSRCR is shown in Fig. 6(e). The scene seems to be hazy in the image. Figure 6(g) is the image enhanced by CVC, and it lacks color information. The quality of Fig. 6(i) acquired by IBA is significantly improved. However, the enhanced image loses many details of branches. Figure 6(j) shows that the image enhanced by the proposed method has good color and detail retention performance.

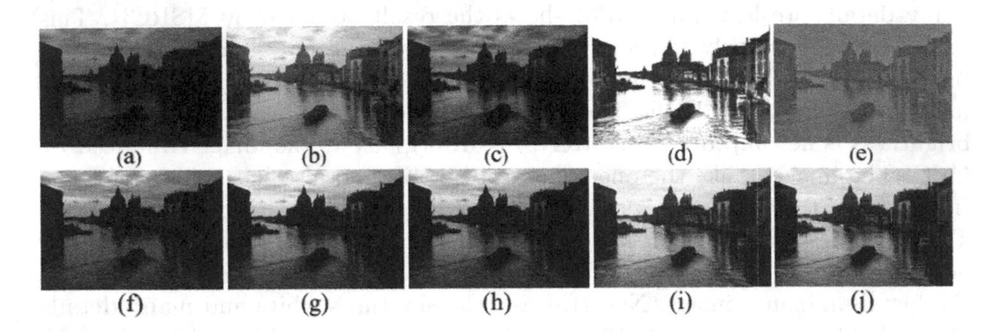

Fig. 7. The enhancement results of image 'River': (a) Original image; (b) HE; (c) CLAHE; (d) MSR; (e) MSRCR; (f) BPDHE; (g) CVC; (h) LDR; (i) IBA; (j) Proposed method

The enhancement results of image 'River' are revealed in Fig. 7. It can be seen from Fig. 7(c), (f), (g) and (h) that CLAHE, BPDHE, CVC and LDR cannot effectively improve the brightness of the original image. The buildings remain dark after being enhanced by these methods. In Fig. 7(d), MSR over-enhances the image and the water and sky become white. In Fig. 7(i), the color of the image obtained by IBA is more saturated than that of the image obtained by HE in Fig. 7(b) and the image obtained by MSRCR in Fig. 7(e). The image enhanced by the proposed method shown in Fig. 7(j) is more vivid than that enhanced by IBA. Also, the overall color is more well retained.

The enhancement results of image 'Birds' can be seen in Fig. 8. In the original image, the birds are in the bright regions, while the background is dark. It is clearly that HE, MSR and IBA over-enhance the image in that the details on the wings are lost, which can be seen in Fig. 8(b), (d) and (i). The color of the image obtained by MSRCR in Fig. 8(e) is gray. In Fig. 8(c), (g) and (h), the branches of the tree remain dark, which means CLAHE, CVC and LDR cannot effectively enhance the image. The image enhanced by BPDHE in Fig. 8(f) has good contrast and is full of details. Compared with that, the image enhanced by the proposed method shown in Fig. 8(j) is more vivid.

The enhancement results of image 'Building', 'Indoor', 'Park', 'House', 'Hall' and 'Girl' are shown in Fig. 9. It can be seen that the enhancement of CLAHE, BPDHE, CVC and LDR is not obvious. HE and MSR tend to over-enhance the original image, so the information in highlight regions may be lost. The images obtained by MSRCR are gray, which is not conducive to human visual perception. Compared with the IBA and the other algorithms, the proposed method can effectively enhance the brightness and contrast, preserve the details in highlight regions and the output images are more vivid.

Fig. 8. The enhancement results of image 'Birds': (a) Original image; (b) HE; (c) CLAHE; (d) MSR; (e) MSRCR; (f) BPDHE; (g) CVC; (h) LDR; (i) IBA; (j) Proposed method

Fig. 9. The enhancement results of image 'Building', 'Indoor', 'Park', 'House', 'Hall' and 'Girl': (a) Original image; (b) HE; (c) CLAHE; (d) MSR; (e) MSRCR; (f) BPDHE; (g) CVC; (h) LDR; (i) IBA; (j) Proposed method

4.2 Quantitative Evaluation

In addition to qualitative evaluation, entropy and average gradient (AG) are used to qualitatively evaluate the effect of the proposed method. Entropy reflects the richness of image information. The greater the entropy of the image, the more information the image contains. The entropy is defined as

$$H = -\sum_{i=0}^{255} p(i) \times \log_2(p(i)) \tag{19}$$

where $p(i)$ is the probability that a certain gray value i appears in the enhanced image.

The quantitative analysis of entropy of processed results by different methods is shown in Table 1. It can be seen that the entropy of the images processed by the proposed method ranks first for six times ('Road', 'Traffic', 'Indoor', 'Park', 'House' and 'Girl') and second for four times ('River', 'Birds', 'Building, 'Hall'). Moreover, the average entropy of all enhanced images of the proposed method is larger than that of any other method, which indicates that the proposed method can enrich more information of low illumination color images.

Table 1. Quantitative analysis of entropy of processed results by different methods

	HE	CLAHE	MSR	MSRCR	BPDHE	CVC	LDR	IBA	Proposed method
Road	7.3199	7.3357	5.5159	6.5979	7.5146	7.2973	7.2689	7.2766	**7.5717**
Traffic	6.8094	6.9056	5.4190	6.5845	6.9897	6.9159	6.8560	5.7230	**7.1680**
River	6.9371	6.9742	5.4826	5.8769	7.0726	7.1059	7.1900	**7.4631**	7.2151
Birds	6.3432	6.5031	6.6105	6.2777	6.9184	6.5795	6.6155	**7.1475**	6.9956
Building	**7.4667**	7.2696	5.2758	6.3525	7.1794	7.3445	7.3883	7.0429	7.4240
Indoor	7.3797	7.4516	5.3100	6.5268	7.1516	7.2725	7.3006	7.6604	**7.6824**
Park	7.4298	7.2411	5.1460	6.1798	7.1060	7.4578	7.4374	5.6196	**7.5025**
House	7.1490	7.1228	5.1176	6.8943	6.9551	7.1442	6.8559	7.1718	**7.4846**
Hall	7.1511	7.2323	6.4042	6.7580	7.0719	7.2200	7.0883	**7.8087**	7.7538
Girl	7.3716	7.4064	5.6274	6.7518	6.9819	7.3333	7.1441	7.5193	**7.6388**
Average	7.1358	7.1442	5.5909	6.4800	7.0941	7.1671	7.1145	7.0433	**7.4437**

AG reflects the sharpness of the outlines and textures in the image. Generally, the larger value of the mean gradient indicates that the outlines and textures are clearer. The AG is defined as

$$AG = \frac{1}{M \times N} \sum_{\alpha=1}^{M} \sum_{\beta=1}^{N} \sqrt{\frac{\left(\frac{\partial f}{\partial x}\right)^2 + \left(\frac{\partial f}{\partial y}\right)^2}{2}} \tag{20}$$

where M is the height of the image, N is the width of the image, α is the row number of the pixel, β is the column number of the pixel, $\frac{\partial f}{\partial x}$ is the horizontal gradient, and $\frac{\partial f}{\partial y}$ is the vertical gradient.

The quantitative analysis of AG of processed results by different methods is shown in Table 2. It is obvious that the AG of the results processed by MSR ranks first for all ten images. Also, the average AG of all enhanced images of MSR is the largest, followed by HE, and the method proposed in this paper is in the third place. However, as is discussed in the qualitative evaluation experiments, although MSR is good at increasing the value of AG, the images tend to be over-enhanced and is not consistent with human visual perception. Besides, color distortion may appear in images processed by HE.

Table 2. Quantitative analysis of AG of processed results by different methods

	HE	CLAHE	MSR	MSRCR	BPDHE	CVC	LDR	IBA	Proposed method
Road	5.9099	4.3712	**7.5011**	4.3593	3.8591	4.6461	3.2254	4.6517	4.4180
Traffic	8.3510	6.2025	**11.1024**	4.1553	5.8955	5.9166	5.5215	6.5600	6.6336
River	6.8057	5.0628	**11.4803**	2.6775	4.3293	5.2591	4.7146	6.9423	7.0238
Birds	12.4130	5.3101	**17.8201**	4.5628	6.2746	7.4456	4.9388	9.0420	8.5675
Building	4.2434	3.0525	**8.9247**	3.7983	2.4862	3.8249	3.8249	4.4838	4.2909
Indoor	16.2583	13.5691	**18.1236**	6.7410	9.8648	11.7839	11.1838	14.0608	16.1052
Park	5.6853	4.7032	**7.6506**	3.4255	3.0283	4.4504	4.1937	4.8084	3.4334
House	4.1164	3.0003	**4.8414**	4.4362	2.1568	3.2389	2.2168	3.0696	3.1827
Hall	8.8156	6.3087	**10.7108**	4.6300	4.8872	5.9437	5.2576	7.1022	7.3232
Girl	6.7401	5.3015	**7.8243**	3.7743	3.2917	5.1186	5.1186	5.3426	6.0031
Average	7.9339	5.6882	**10.5979**	4.2560	4.6074	5.7629	5.0196	6.6063	6.6981

The iterative process of the proposed method for searching fitness values of experimental images is shown in Fig. 10. All ten images are converged within 50 iterations, and good fitness values can be obtained within 20 iterations for some images (see Fig. 10(e), (h), (i) and (j)). The proposed method in this paper can reduce the loss of details in highlight regions. Thus, for the original images that have many details in highlight areas, the converged fitness values are not close to 1 (see Fig. 10(d), (g) and (i)). The iterative curves verify the stability of the proposed method.

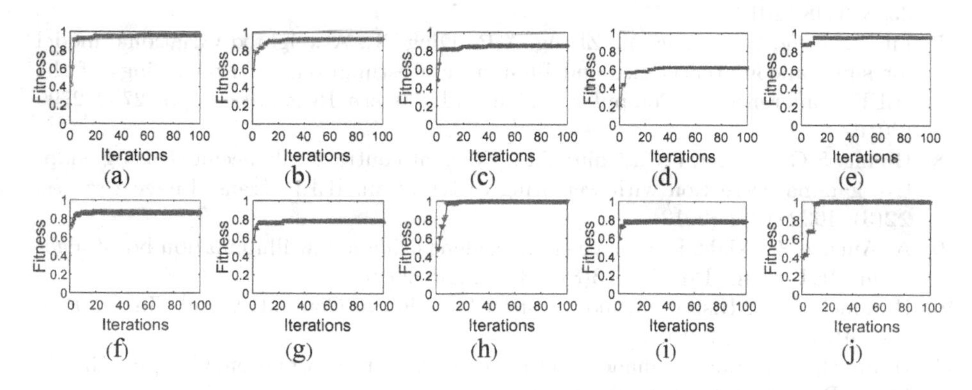

Fig. 10. The iterative process of the proposed method for searching fitness values of experimental images: (a) Road; (b) Traffic; (c) River; (d) Birds; (e) Building; (f) Indoor; (g) Park; (h) House; (i) Hall; (j) Girl

5 Conclusion

In this paper, an adaptive enhancement method is proposed to enhance the low illumination color images. First of all, the original low illumination image is transformed from RGB color space to L*a*b* color space, and the proposed piecewise sine function combined with DA is used to adaptively improve the brightness of the L* channel image. Secondly, a new saturation enhancement method is utilized in the saturation channel of HSV color space to generate a vivid image. Finally, a fitness function that considers both the degree of overall brightness enhancement and the suppression of information loss is proposed as the objective function of DA. Finally, the optimized image is converted to RGB color space. According to the analysis of the experimental results, the proposed method can effectively enhance the contrast and brightness, enrich the color and avoid over-enhancement. The images enhanced by the proposed method are more consistent with human visual perception compared to several classical, well-known and state-of-art algorithms.

References

1. Land, E.H.: The retinex. Am. Sci. **52**(2), 247–264 (1964)
2. Land, E.H.: The retinex theory of color vision. Sci. Am. **237**(6), 108–129 (1977)
3. Land, E.H., McCann, J.J.: Lightness and retinex theory. Josa **61**(1), 1–11 (1971)
4. Jobson, D.J., Rahman, Z.U., Woodell, G.A.: Properties and performance of a center/surround retinex. IEEE Trans. Image Process. **6**(3), 451–462 (1997)
5. Jobson, D.J., Rahman, Z.U., Woodell, G.A.: A multiscale retinex for bridging the gap between color images and the human observation of scenes. IEEE Trans. Image Process. **6**(7), 965–976 (1997)
6. Wang, S., Zheng, J., Hu, H.M., Li, B.: Naturalness preserved enhancement algorithm for non-uniform illumination images. IEEE Trans. Image Process. **22**(9), 3538–3548 (2013)
7. Fu, X., Zeng, D., Huang, Y., Zhang, X.P., Ding, X.: A weighted variational model for simultaneous reflectance and illumination estimation. In: Proceedings of the IEEE Conference on Computer Vision and Pattern Recognition, pp. 2782–2790 (2016)
8. Huang, S.C., Cheng, F.C., Chiu, Y.S.: Efficient contrast enhancement using adaptive gamma correction with weighting distribution. IEEE Trans. Image Process. **22**(3), 1032–1041 (2012)
9. Al-Ameen, Z.: Nighttime image enhancement using a new illumination boost algorithm. IET Image Process. **13**(8), 1314–1320 (2019)
10. Hummel, R.A.: Histogram modification techniques. Comput. Graph. Image Process. **4**(3), 209–224 (1975)
11. Hummel, R.A.: Image enhancement by histogram transformation. Comput. Graph. Image Process. **6**(2), 184–195 (1977)
12. Kim, Y.T.: Contrast enhancement using brightness preserving bi-histogram equalization. IEEE Trans. Consum. Electron. **43**(1), 1–8 (1997)
13. Wang, Y., Chen, Q., Zhang, B.: Image enhancement based on equal area dualistic sub-image histogram equalization method. IEEE Trans. Consum. Electron. **45**(1), 68–75 (1999)
14. Sim, K.S., Tso, C.P., Tan, Y.Y.: Recursive sub-image histogram equalization applied to gray scale images. Pattern Recognit. Lett. **28**(10), 1209–1221 (2007)
15. Ibrahim, H., Kong, N.S.P.: Brightness preserving dynamic histogram equalization for image contrast enhancement. IEEE Trans. Consum. Electron. **53**(4), 1752–1758 (2007)

16. Arici, T., Dikbas, S., Altunbasak, Y.: A histogram modification framework and its application for image contrast enhancement. IEEE Trans. Image Process. **18**(9), 1921–1935 (2009)
17. Pizer, S.M., Johnston, R.E., Ericksen, J.P., Yankaskas, B.C., Muller, K.E.: Contrast-limited adaptive histogram equalization: speed and effectiveness. In: [1990] Proceedings of the First Conference on Visualization in Biomedical Computing, pp. 337–338. IEEE (1990)
18. Reza, A.M.: Realization of the contrast limited adaptive histogram equalization (CLAHE) for real-time image enhancement. J. VLSI Signal Process. Syst. Signal Image Video Technol. **38**(1), 35–44 (2004). https://doi.org/10.1023/B:VLSI.0000028532.53893.82
19. Yadav, G., Maheshwari, S., Agarwal, A.: Contrast limited adaptive histogram equalization based enhancement for real time video system. In: 2014 International Conference on Advances in Computing, Communications and Informatics (ICACCI), pp. 2392–2397. IEEE (2014)
20. Celik, T., Tjahjadi, T.: Contextual and variational contrast enhancement. IEEE Trans. Image Process. **20**(12), 3431–3441 (2011)
21. Lee, C., Lee, C., Kim, C.S.: Contrast enhancement based on layered difference representation. In: 2012 19th IEEE International Conference on Image Processing, pp. 965–968. IEEE (2012)
22. Lee, C., Lee, C., Kim, C.S.: Contrast enhancement based on layered difference representation of 2D histograms. IEEE Trans. Image Process. **22**(12), 5372–5384 (2013)
23. Dhal, K.G., Ray, S., Das, A., Das, S.: A survey on nature-inspired optimization algorithms and their application in image enhancement domain. Arch. Comput. Meth. Eng. **26**(5), 1607–1638 (2019)
24. Shanmugavadivu, P., Balasubramanian, K.: Particle swarm optimized multi-objective histogram equalization for image enhancement. Optics Laser Technol. **57**, 243–251 (2014)
25. Eberhart, R., Kennedy, J.: A new optimizer using particle swarm theory. In: MHS'95. Proceedings of the Sixth International Symposium on Micro Machine and Human Science, pp. 39–43. IEEE (1995)
26. Liu, S., et al.: Enhancement of low illumination images based on an optimal hyperbolic tangent profile. Comput. Electr. Eng. **70**, 538–550 (2018)
27. Kanmani, M., Narasimhan, V.: Swarm intelligent based contrast enhancement algorithm with improved visual perception for color images. Multimed. Tools Appl. **77**(10), 12701–12724 (2018)
28. Mirjalili, S.: Dragonfly algorithm: a new meta-heuristic optimization technique for solving single-objective, discrete, and multi-objective problems. Neural Comput. Appl. **27**(4), 1053–1073 (2016)
29. Rahman, C.M., Rashid, T.A.: Dragonfly algorithm and its applications in applied science survey. Comput. Intell. Neurosci. **2019**, 1–21 (2019)

Target Recognition Method Based on Ship Wake Extraction from Remote Sensing Image

Jun Hong[1], Xingxuan Liu[2], and Hang Dong[1](✉)

[1] Dalian Naval Academy, Dalian, Liaoning, China
504980609@qq.com
[2] Navy Submarine Academy, Qingdao, Shangdong, China

Abstract. Aiming at the difficulty of target recognition in low and medium resolution remote sensing images of maritime mobile ships, in order to accurately match and identify ship types, this paper proposes a ship Kelvin wake extraction method based on Hough transform for matching and identifying ship models. The average distance of Kelvin wake spike wave of different types of ships is obtained through simulation, which can effectively identify the ship attributes in real remote sensing images, and the comparative analysis with the actual image data proves that the algorithm has certain feasibility. Therefore, the Kelvin wake extraction method based on Hough transform can basically achieve the recognition of target ship type under certain conditions and improve the recognition accuracy of maneuvering ship targets at sea by low and medium resolution imaging reconnaissance satellites.

Keywords: Ship wake · Remote sensing image · Kelvin wake · Hough transform

1 Introduction

At present, for the research of imaging reconnaissance satellite target recognition, there are more researches on high-resolution remote sensing images and SAR imaging target recognition, and there are relatively few researches on target recognition of low and medium resolution optical remote sensing images. Literature shows that Academician He You's team conducted in-depth research on signal-level data fusion of multi-source information from satellite images and ship target detection [1, 2]; Yang Guang and others from Aviation University of Air Force composed and analyzed the target detection and recognition in SAR images [3]; Chen Mingrong from Dalian Naval Academy studied the characteristics and extraction method of optical turbulent wake features using a shipboard experimental setup [4], which is more suitable for target detection and not applicable to target identification; Hou Haiping, Gregory Zilman, Du Chong and Chen Jun studied and simulated the Kelvin wake of SAR images based on the electromagnetic scattering characteristics for the SAR imaging mechanism [5–8], but no complete wake feature extraction method has been proposed. In her master's thesis, Bai Haijuan divided the recognition features of ships into ship size features and motion features, etc. Among them, the motion features use the Kelvin wake transverse wave wavelength to project its

W. Fan et al. (Eds.): AsiaSim 2022, CCIS 1712, pp. 214–226, 2022.
https://doi.org/10.1007/978-981-19-9198-1_17

speed, and match the size features and motion features to recognize the maritime ships in different resolution remote sensing images [9], but the object of this article is mainly high-resolution remote sensing images, and the recognition of motion ships in low and medium resolution remote sensing images has not been studied deeply enough; Other units conducted research on machine interpretation of target recognition technology based on big data and neural networks, which has a large theoretical difficulty in terms of data volume requirements for databases containing labels.

Space reconnaissance has become an important source of intelligence for modern warfare because of its advantages such as long reconnaissance distance and not limited by national borders. Especially in the surface ships far sea combat, long-rage intelligence assurance on the increasing reliance on aerospace reconnaissance, wide, low and medium resolution imaging reconnaissance satellite reconnaissance of the enemy's mobile ships at sea has become an important means of maritime combat intelligence assurance. However, low and medium resolution imaging reconnaissance has the problem of difficult target detection and identification, which affects the effectiveness of intelligence use. Research on the identification method of ship targets under low and medium resolution imaging condition is of great value to improve its intelligence support effectiveness.

2 Mathematical Description of the Kelvin Wake of Ships

In 1887, Lord Kelvin proposed the Kelvin wake theory: When a ship is viewed as an ideal point disturbance case, its wake in water contains two distinct waveforms, namely, divergent and transverse, as shown in Fig. 1 [10].

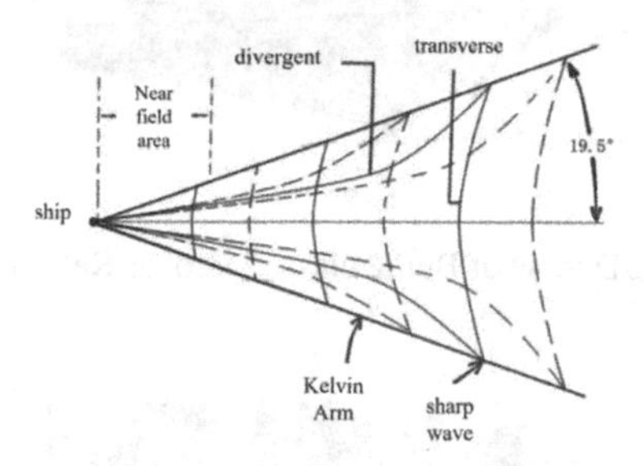

Fig. 1. Schematic diagram of a typical Kelvin wake

Assuming that the ship is sailing along the negative *x-axis* with speed μ, the resulting free surface waves propagate along different angles with the *x-axis* at an angle θ. The wake wave height can be described as [8]:

$$\zeta(x, y) = Re\int_{-\pi/2}^{\pi/2} F(\theta)e^{-iksec^2(xcos\theta+ysin\theta)}d\theta \tag{1}$$

where, $k = g/\mu^2$, $g = 9.8\,\text{m/s}^2$ is the acceleration of gravity and *Re* denotes taking the real part. $F(\theta)$ denotes the characteristic parameters of the ship [5], which leads to:

$$F(\theta) = \frac{2k}{\pi}sec^2\theta\iint\frac{\partial Y(x, z)}{\partial x}e^{kzsec^2\theta+ikzxsec\theta}dxdz \tag{2}$$

where, the integral domain is the surface of the ship and $Y = Y(x, z)$ is the width of the ship at coordinate x at water depth z. With the shape of the ship's side surface approximated as a parabolic shape, when the ship's draught depth is D during navigation, the ship's width calculation formula can be found as:

$$Y(x,z) = \begin{cases} b\left(1-\frac{x^2}{l^2}\right), & -D \le z \le 0, -l \le x \le l \\ 0, & z \le -D \end{cases} \tag{3}$$

where, b is the half-boat width, l is the half-blat length, z is the water depth, and D is the sidewall draft. Substituting Eqs. (2) and (3) into Eq. (1), respectively, the expression for the Kelvin wake can be obtained as:

$$\zeta(x,y) = \frac{4b}{\pi kl}\int_{-\frac{\pi}{2}}^{\frac{\pi}{2}}\left(1-e^{-kDsec\theta}\right)sin\left[ksec^2\theta(xcos\theta+ysin\theta)\right]d\theta \tag{4}$$

Taking the parameters of the Burke class destroyer as an example, the size of the simulation area is 500 × 500 m of infinite depth sea, and the three-dimensional diagram and two-dimensional top view of the Kelvin wake can be derived for the ideal situation at a speed of 10 m/s, as shown in Fig. 2 and Fig. 3.

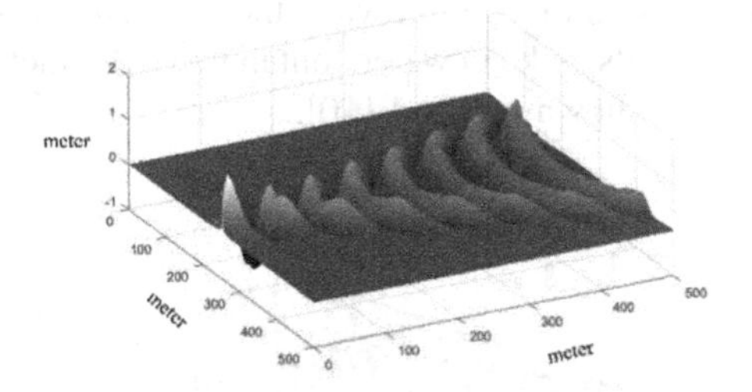

Fig. 2. 3D view of Burke class destroyer Kelvin wake

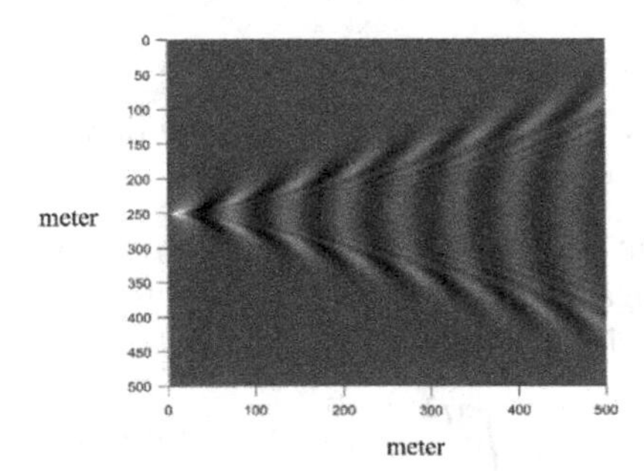

Fig. 3. 2D top view of Burke class destroyer Kelvin wake

The simulation of the ship's Kelvin wake at different speeds without considering the seawater viscosity and the ship's shape is shown in Fig. 4.

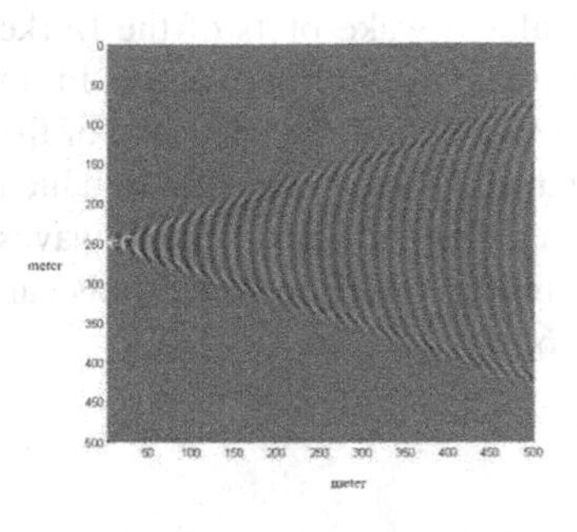

(a) Full draught and Speed 5m/s

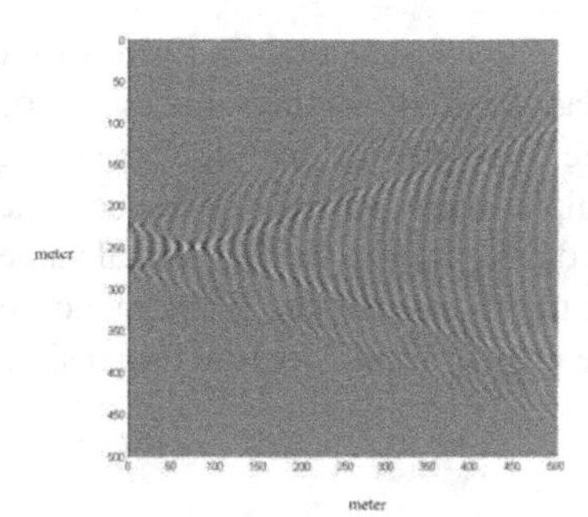

(b) Draught depth 3m and Speed 5m/s

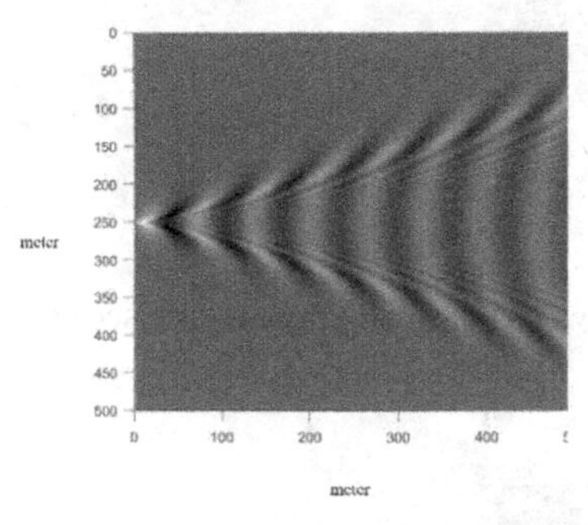

(c) Full draught and Speed 10m/s

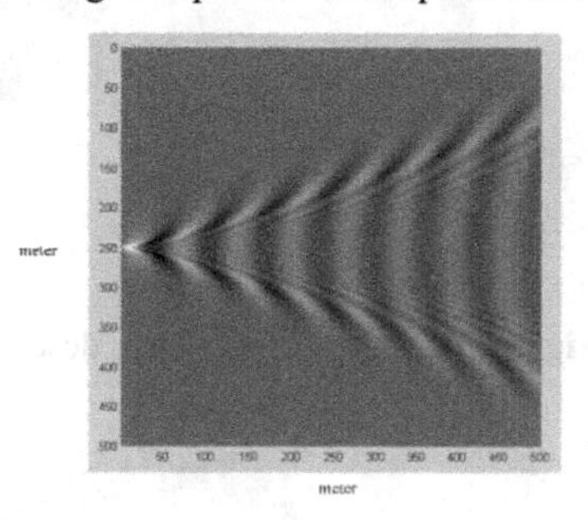

(d) Draught depth 3m and Speed 5m/s

Fig. 4. Kelvin wake variation at different airspeeds

As seen in Fig. 4, the grayscale characteristics of the Kelvin wake enhance with the enhancement of the draft depth for the same ship at the same speed, and the wavelength characteristics do not change significantly. The wavelength of the Kelvin wake sharp wave increases after the speed increases.

3 Kelvin Wake Correction Based on Ship Characteristics

By observing the actual ship images taken by the low and medium resolution imaging satellite, it can be found that the Kelvin wake of a ship during its voyage is different from the theoretical value due to multiple factors such as bow, stern, multi-layer ship wake, waves and seawater viscosity, etc. Therefore, this paper introduces the Kelvin wake correction coefficient based on the ship characteristics, i.e., the viscosity coefficient C.

During the navigation of the ship in the actual ocean background, the ship's shape is integrated by considering two parts, the bow and the stern, and also by considering the influence of seawater viscosity, and the $\zeta(x, y)$ is corrected by the viscosity coefficient C [11]. The expression is:

$$\zeta(x, y) = \zeta(x - l, y) + C\zeta(x + l, y) \tag{5}$$

The coefficient of viscosity C is usually taken as *0.6*.

After introducing the viscosity coefficient, the simulated wake plots of the Burke class destroyer at the same conditions of speed 10 m/s are shown in Fig. 5 and 6. From the three-dimensional diagram of Fig. 5, it can be seen that due to the influence of the ship's shape and the emergence of new wave peaks and troughs, the peak point of which is different from the theoretical value, and the appearance of jagged and covariant waves at the transverse, the phenomenon of superposition of multiple Kelvin wake trails can be clearly seen in its tow-dimensional top view of Fig. 6.

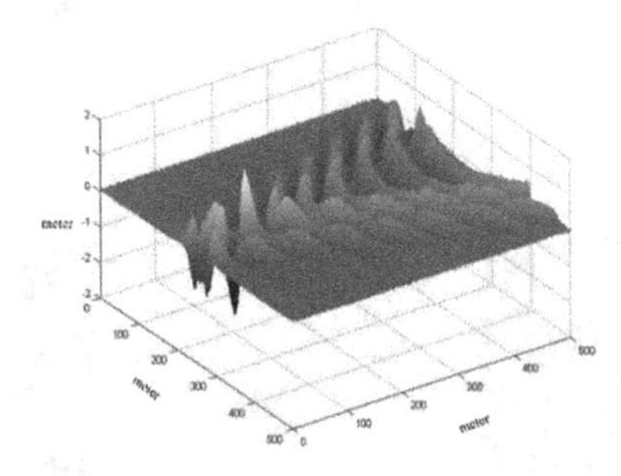

Fig. 5. 3D view of Burke class destroyer Kelvin wake when considering seawater viscosity

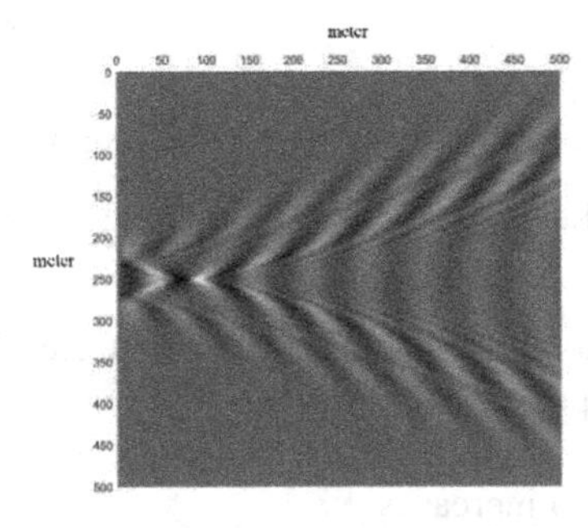

Fig. 6. 2D top view of Burke class destroyer Kelvin wake when considering seawater viscosity

With the above improved method, considering the influence of ship type and seawater viscosity, the Kelvin wake of a certain type of destroyer, a certain type of frigate, a certain type of aircraft carrier and a certain type of dock landing ship in the same speed and full load draft are simulated in this paper, as shown in Figs. 7, 8, 9 and 10.

From Figs. 7, 8, 9 and 10, according to the vessel parameters, it can be seen that the theoretical value of its Kelvin wake is related to the vessel length and width, and the superimposed Kelvin wake formed under the condition of seawater viscosity is different from the theoretical value.

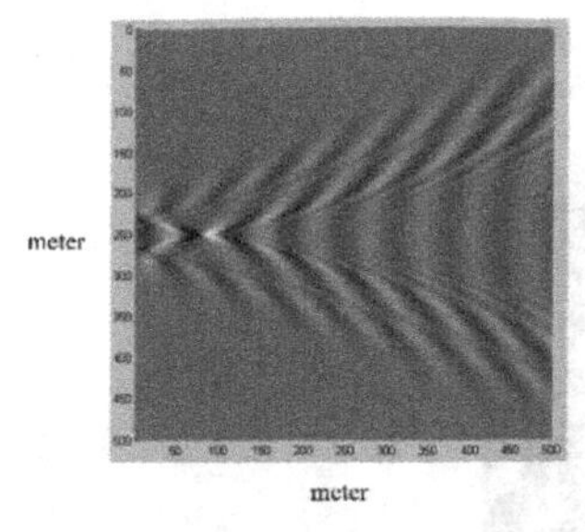

Fig. 7. A destroyer Kelvin wake

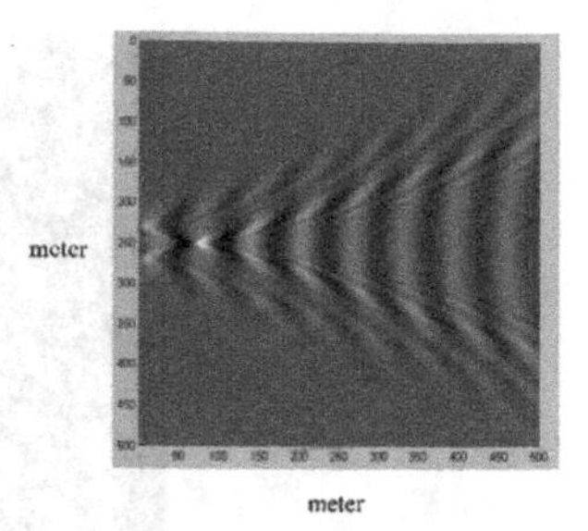

Fig. 8. A frigate Kelvin wake

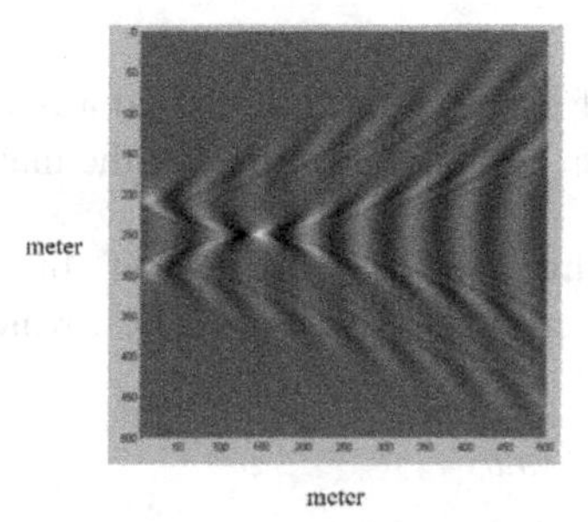

Fig. 9. An aircraft carrier Kelvin wake

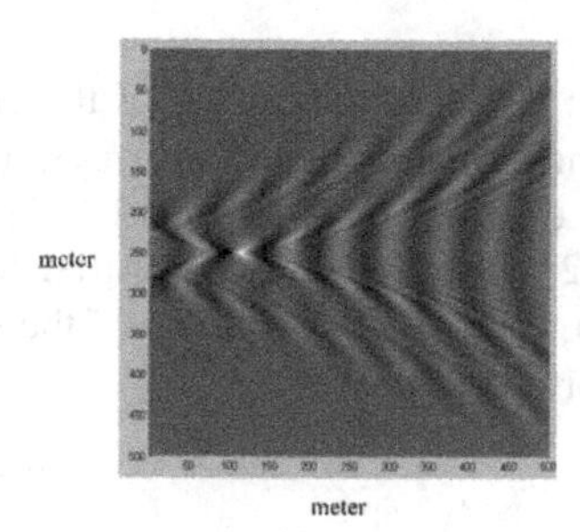

Fig. 10. A dock landing ship Kelvin wake

4 Kelvin Wake Extraction Based on the Hough Transform Algorithm

After the simulation to realize the Kelvin wake of different types of ships, it is also necessary to extract the Kelvin wake of the ships in the actual remote sensing images. The actual ship wake is affected by complex sea conditions, and the wake extraction is difficult, and its sharp wave is easily confused with the marine environment background, so it needs to be carried out by extracting the wave peak point of Kelvin arm after excluding the wave noise, and thus deriving the sharp wave wavelength and matching it with the actual satellite image, so as to realize the target identification through the wake. Therefore, the Kelvin wake algorithm needs to be improved to exclude the noise and get the real Kelvin wake as much as possible. In literature 12, the Hough algorithm is used to identify linear ship formation with V-shaped features, and the analysis of each cusp wave on the V-edge of the wake Kelvin arm is focused, so as to achieve the extraction of ship wake under actual sea conditions.

Taking the detection of the Kelvin wake of a Burke class destroyer as an example, the coordinates, of the center position of the target area $Target(i)$ are obtained by binarizing the wake image and setting the gray value threshold and the area threshold to exclude the image noise points. The coordinates are input into the target matrix $Center(x_i, y_i)$ to obtain the coordinates of the center points of each wave crest region on the Kelvin wake as shown in Fig. 11. The specific algorithm steps are as follows:

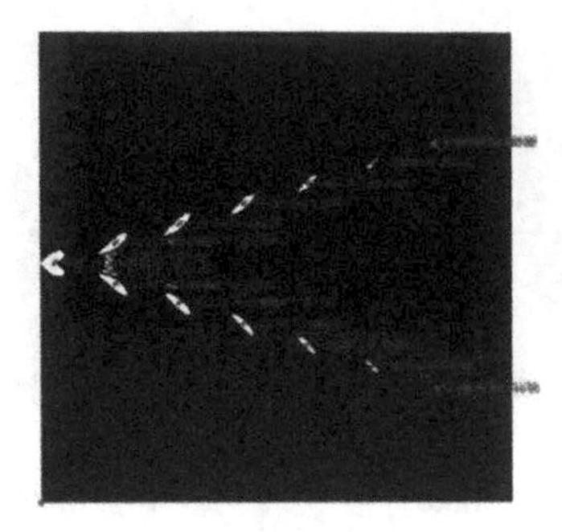

Fig. 11. Coordinates of the center point of the wave crest area

Step 1: Extract the position of the crest point of each region, i.e., the coordinates of the center position, noted as (x_i, y_i), $(i = 1, 2, 3 \ldots\ldots, k)$, and construct the matrix C, noting $C = (x_i, y_i)$, $(i = 1, 2, 3 \ldots\ldots, k)$.
Step 2: Using the Hough transform algorithm, the coordinates of the point (x_i, y_i), $(i = 1, 2, 3 \ldots\ldots, k)$ of the line are converted to polar coordinates, which can be expressed as:

$$\rho = x_i cos\theta + y_i sin\theta \tag{6}$$

ρ is the vertical length from the origin to the line, θ is the angle between the vertical line and the horizontal axis.

Each peak point is transformed into a sine curve in polar coordinates, and the curve is compared between points (θ_i, ρ_i). Establish the matrix P, noting $P = (\theta_i, \rho_i)$, $(i = 1, 2, 3 \ldots\ldots n)$.
Step 3: Create a parameter space accumulator, denoted as $F = (\theta_i, \rho_i)$, $(i = 1, 2, 3 \ldots\ldots m)$. Then divide m intervals of size $\Delta\theta$ and $\Delta\rho$ according to the image pixel size, i.e., the matrix density is $\Delta\theta$ and $\Delta\rho$. The accumulator F is assigned an initial value of 0.
Step 4: If the intersection points in Step 3 are in the same interval (θ_i, ρ_i) of $\Delta\theta$and$\Delta\rho$, then make the accumulator $F\left(\theta_j, \rho_j\right) = F\left(\theta_j, \rho_j\right) + 1$, the accumulation value gradually increases with the number of intersection points, and the local peak in the accumulator F is $F(\theta_1, \rho_1), F(\theta_2, \rho_2), F(\theta_3, \rho_3), \ldots\ldots F(\theta_k, \rho_k)$.
Step 5: Construct the matrix A_{peak}. Store the k local peaks of the matrix F in Step 4 into A_{peak}. Then $A_{peak} = \left(\theta_{peaki}, \rho_{peaki}\right)$, $(i = 1, 2, 3 \ldots\ldots k)$.
Step 6: K-mean clustering of the peak matrix A_{peak} and merging the approximate parameters lead to the parameters of the two Kelvin arms, and two local peaks are obtained from the accumulator local peak matrix plot, the points are the calculated parameters (θ_1, ρ_1) and (θ_2, ρ_2) of the two straight lines, respectively.
Step 7: Substituting the two peak point parameters (θ_1, ρ_1) and (θ_2, ρ_2) into Eq. (6), respectively, two Kelvin arm expressions can be obtained to obtain the sharp wave position point after excluding the wave noise point, as shown in Fig. 12.
Step 8: Substitute the linear equations of the two Kelvin arms into the original image to obtain the wake tip wave spacing parameters, as shown in Fig. 13, to match the ship parameters and navigation characteristics in the image by the tip wave spacing size of the Kelvin arms.

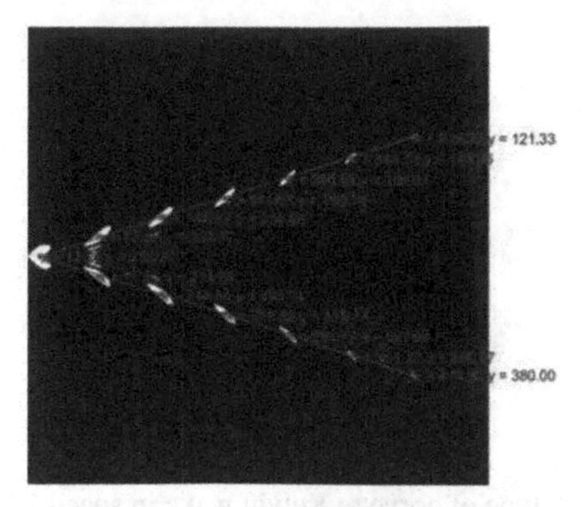

Fig. 12. Effect of Kelvin wake extraction based on the Hough transform algorithm

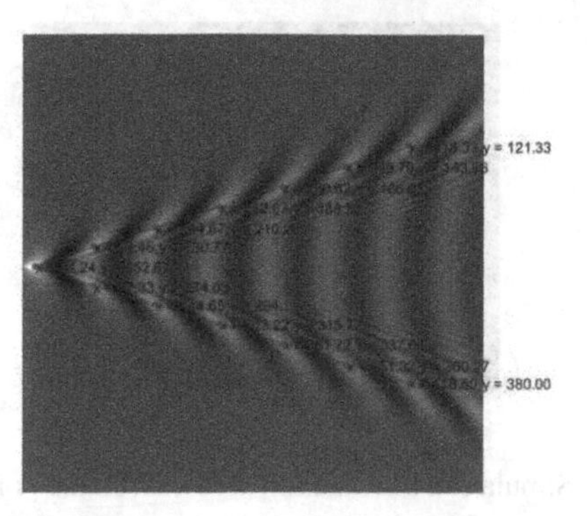

Fig. 13. Kelvin wake extraction based on the Hough transform algorithm

5 Target Identification Example

5.1 Target Wake Simulation

Taking a type of corvette as an example: length 103.2 m, width 10.8 m, draught 3.19 m, speed 7.892 m/s, the Kelvin wake simulation image of the vessel can be obtained according to Eq. (5) as shown in Fig. 14.

The image is binarized to obtain the coordinates of the center position of the target region $Target(i)$. This coordinate is input into the target matrix $Center(x_i, y_i)$, as shown in Fig. 15.

The Kelvin wake is extracted using the Hough transform algorithm, and the parameter space coordinate system is obtained by substituting the coordinate matrix into the polar equation, as shown in Fig. 16, and the parameter space results are saved into the matrix$D = (\theta_i, \rho_i), i = 1, 2, 3, \ldots\ldots, n$.

According to the parameter space characteristics in Fig. 16, the accumulation matrix spacing $\Delta\theta$ and $\Delta\rho$ are set and the peak matrix A is accumulated as shown in Fig. 17. From Fig. 17, it can be seen that the wake image has two local peak points $A\left(\theta_{peak1}, \rho_{peak1}\right)$ and $A\left(\theta_{peak2}, \rho_{peak2}\right)$. These two values are the linear parameters of the Kelvin double-arm identified by the machine, respectively.

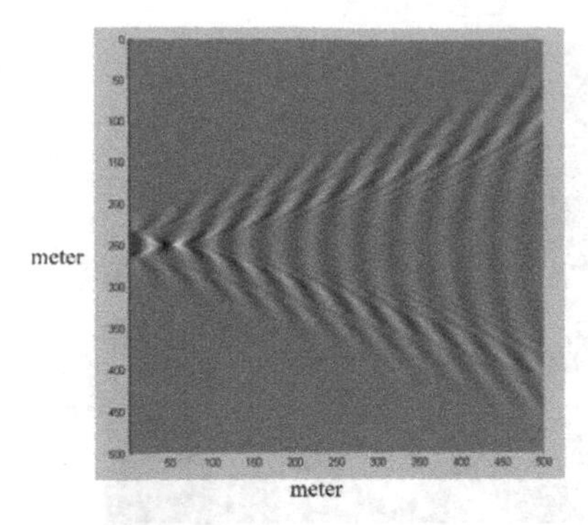

Fig. 14. A type of corvette Kelvin wake in speed 7.892 m/s

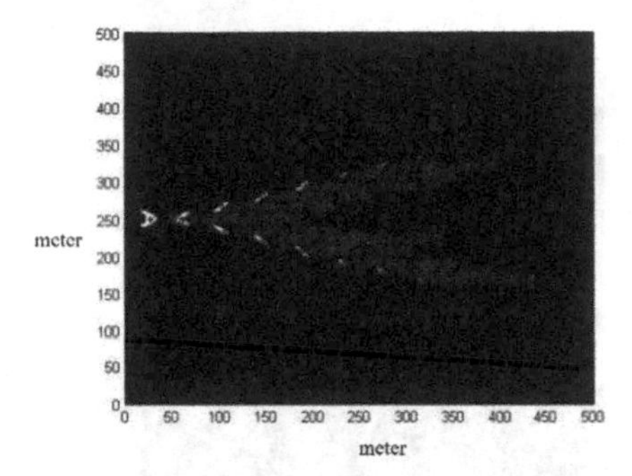

Fig. 15. Simulation binarization area coordinate point map

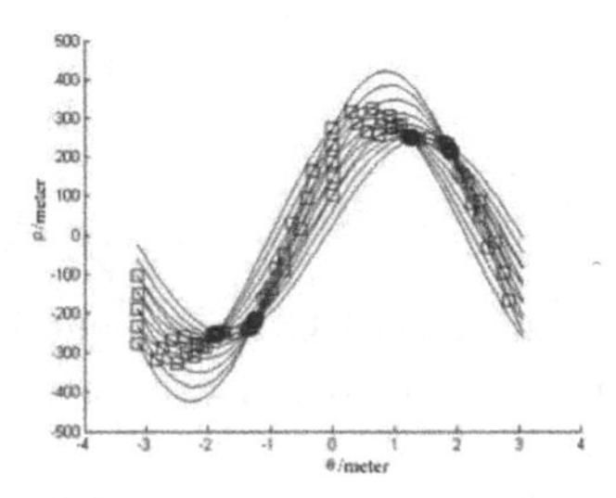

Fig. 16. Wake in parametric spatial coordinate system

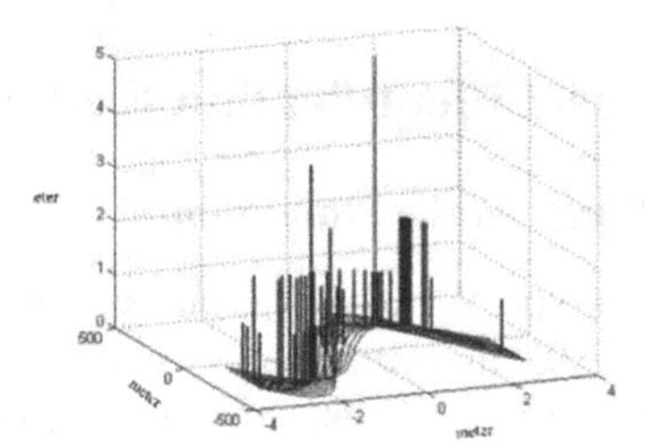

Fig. 17. Parameter space accumulator

The obtained Kelvin double-arm straight line parameters are displayed on the original Fig. as shown in Fig. 18, from which it can be seen that the line can basically satisfy the double-arm wake line, but there are still some deviations. The error can be reduced by extracting a large number of wake sharp wave position points. In the accumulation matrix, the original coordinate point positions of two maximum peak points $A\left(\theta_{peak1}, \rho_{peak1}\right)$ and $A\left(\theta_{peak2}, \rho_{peak2}\right)$ extracted are recorded into the wake target matrices $Target_1(x_i, y_i)$ and $Target_2(x_i, y_i)$, and the average value L_{cusp} of the two matrices is obtained by simulation.

$L_{cusp} \approx 37.245$ (m) is the average spacing value of the Kelvin wake sharp wave at the ship's speed of 7.892 m/s.

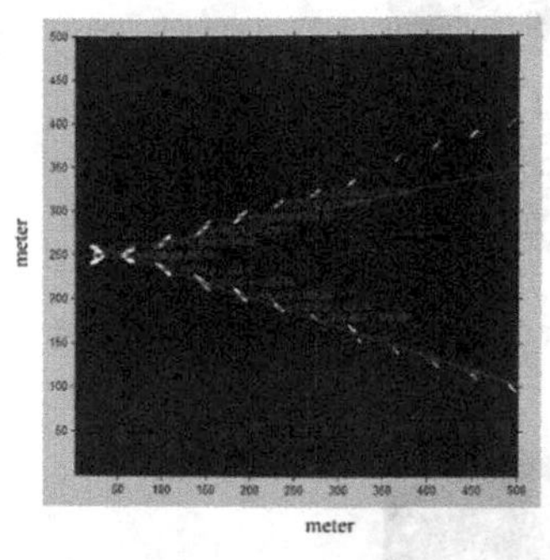

Fig. 18. Simulated Kelvin wake recognition

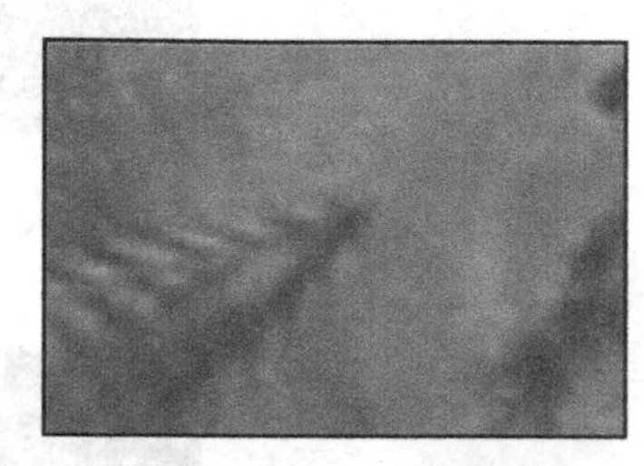

Fig. 19. Actual remote sensing image of the vessel

5.2 Actual Target Kelvin Wake Extraction

Figure 19 shows the actual remote sensing image of a type of ship target and its wake, and the target Kelvin wake is extracted according to the Hough transform algorithm to obtain a gray value three-dimensional stereo image, as shown in Fig. 20. The Kelvin arm in Fig. 20 is obvious on one side, but the middle of the wake is covered by the turbulent wake, and the irregular disturbance noise of the sea surface has more influence. The image binarization process is performed to extract the wave crest position in the image, as shown in Fig. 21.

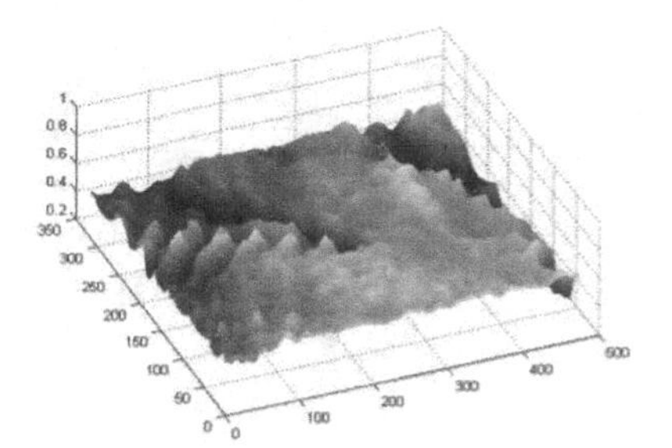

Fig. 20. A gray value 3D stereo image

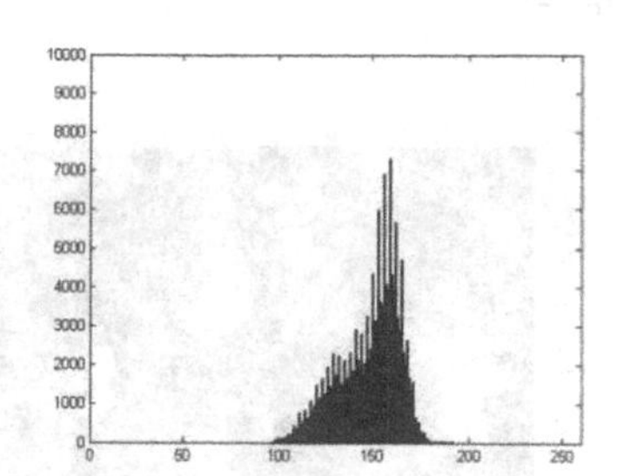

Fig. 21. Grayscale histogram

The grayscale threshold is relatively high in the simulated image because there is no wave background noise, but in the actual image, it is more difficult to distinguish the wake from the wave noise with this threshold. The binarized image under the original threshold condition is shown in Fig. 22. From the Fig., it can be seen that the wave crest at the wake position can form a straight line, and irregular wave points exist in other wave backgrounds, so the threshold range needs to adjusted. Figure 23 shows the binarized image after adjusting the threshold, from which the sharp wave crest of the Kelvin wake can be clearly distinguished as a straight line, and the sea noise is small.

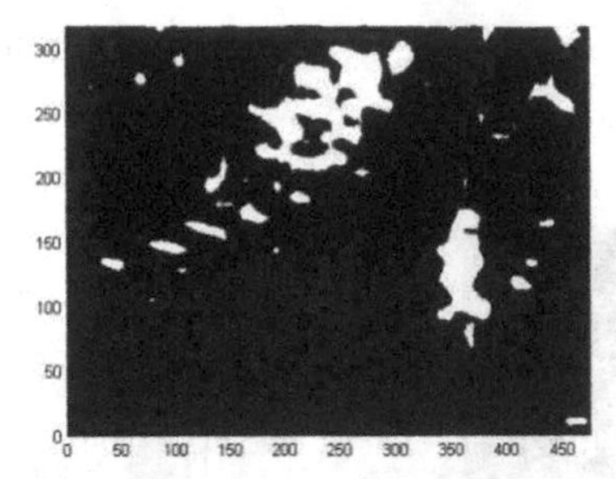

Fig. 22. The location of the wave crest extracted from the simulation diagram

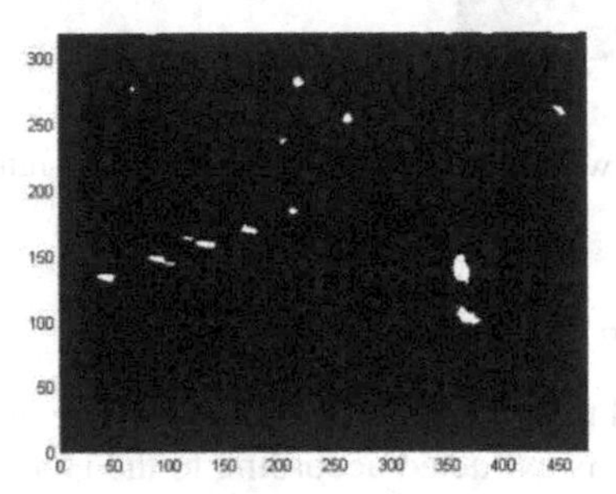

Fig. 23. The location of the wave crest extracted from the actual image

Set the threshold value of the target area to be extracted and the area size of the area, exclude the large cloud noise and other noise in the map, extract the coordinates of the center of the target area, and input them into the target matrix $Center(x_i, y_i)$, as shown in Fig. 24.

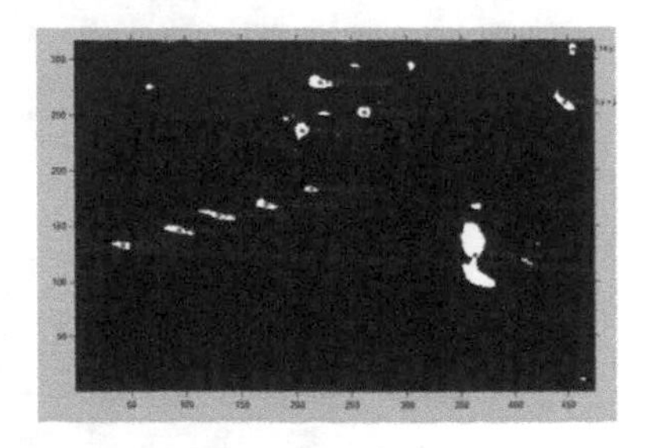

Fig. 24. Target area coordinate location

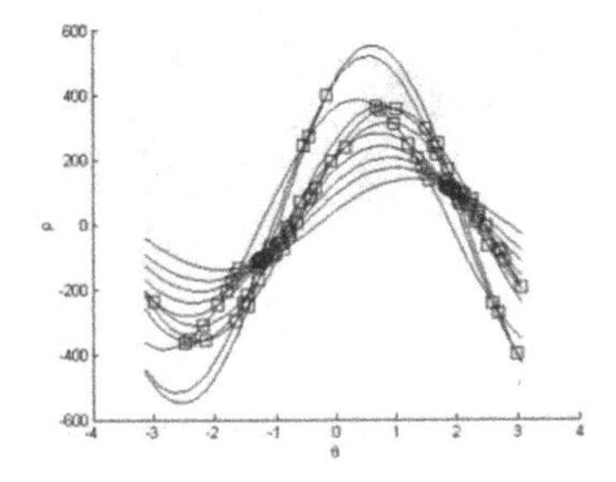

Fig. 25. Parametric spatial coordinate system

The Kelvin wake are extracted using the Hough transform algorithm and substituted into the polar equation to obtain the parameter space values. Figure 25 shows the parameter space coordinate system. The parameter space results are saved into the matrix$D = (\theta_i, \rho_i)$. Set the parameter space accumulation matrix spacing $\Delta\theta$ and $\Delta\rho$, and calculate the accumulation of the peak matrix A to get the local peak point

$A(\theta_{peaki}, \rho_{peaki})$, as shown in Fig. 26. The identified Kelvin wake linear parameters are substituted into the original Fig. to obtain Fig. 27. The ship wake can be clearly identified in the Fig. The coordinate points of the wake sharp wave are substituted into the matrix $Target(x_i, y_i)$ and the average value of the matrix $Target(x_i, y_i)$ is obtained, which is the average spacing value $L_{cusp} \approx 35.569$ of the Kelvin wake sharp wave of this type of ship in the actual remote sensing image.

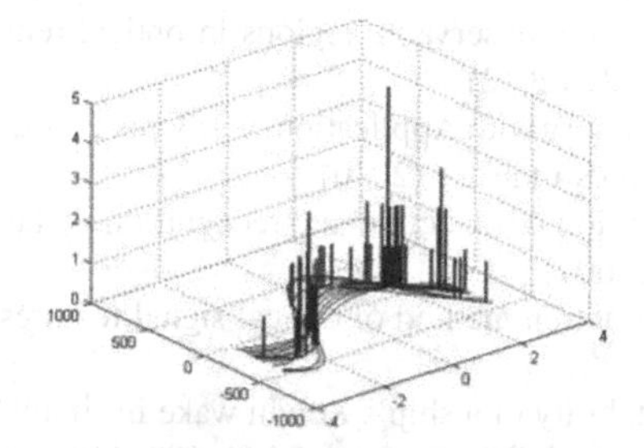

Fig. 26. Spatial accumulation matrix

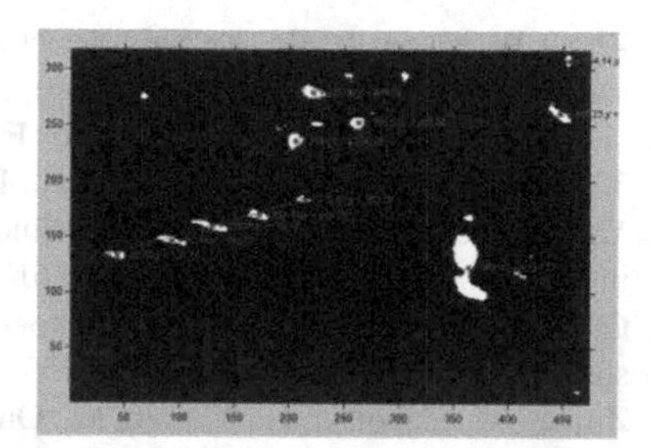

Fig. 27. Algorithm obtained Kelvin wake

6 Conclusion

The average spacing values of Kelvin wake spikes for different types of ships at a speed of 15 knots can be obtained from the simulation, as listed in Table 1.

Table 1. Average distance of Kelvin wake sharp wave for several typical ships at 15 knots

Type	Length (m)	Width (m)	Full draught (m)	Average distance of sharp wave (m)
Corvette 1	103.2	10.8	3.19	37.245
Corvette 2	89	11.14	3.28	20.467
Destroyer 1	157	19	6	29.311
Frigate 1	135	16	4.5	28.180
Nimitz class CV	237.7	40.8	11.3	19.022
Ticonderoga class CG	172.8	16.8	9.5	20.652
Burke class DD	153.8	20.4	6.3	18.925

In the target identification example, the average spacing of Kelvin wake sharp wave of the simulated target is 37.245, while the average spacing of Kelvin wake sharp wave in the real remote sensing image is 35.569, and the error between the simulated data and the real data is 4.712%. From Table 1, it can be seen that the simulation can identify the ship attributes in the real remote sensing image more effectively, which proves that the algorithm has certain feasibility. At the same time, the conditions of using the method in this paper basically do not consider the cloud, wind speed and high speed (over

20 knots) of the target. Therefore, the linear and nonlinear superposition relationship between waves and wake under the influence of high winds and waves, etc. will have a large impact on the results, and the method still needs to be tested and improved by a large amount of actual data.

References

1. Zhou, W., Guan, J., He, Y.: Ship detection from low observable regions in optical remote sensing imagery. J. Image Graph. **17**(9), 1181–1187 (2012)
2. He, Y., Wang, G., Guan, X.: Information Fusion Theory with Application, vol. 3, pp. 239–265, 309–329, 336–370. Publishing House of Electronics Industry (2010)
3. Yang, G., Chen, K., Zhou, M., et al.: Study evolution of detection and recognition on target in SAR image. Prog. Geophys. **22**(2), 617–620 (2007)
4. Chen, M., Chi, W., Xie, T., et al.: Histogram extraction method of optical signal features of ship wake. J. Appl. Opt. **34**(4), 249–253 (2013)
5. Zilman, G., Zapolski, A., Marom, M.: On detectability of a ship's Kelvin wake in simulated SAR images of rough sea surface. IEEE Trans. Geosci. Remote Sens. **53**(2), 609–619 (2015)
6. Hou, H., Chen, B., Liu, C.: Simulation and analysis of Kelvin wake in ocean background. Comput. Integr. Manuf. Syst. **24**(8), 12–14 (2007)
7. Chen J., Zhou Q., Qu C., et al.: SAR imaging simulation of the ship wave. Ship Sci. Technol. **37**(5), 60–62 (2015)
8. Du, C.: Ship wakes modeling and researches on its related electromagnetic methods. Univ. Electron. Sci. Technol. China **12**, 6–10 (2013)
9. Bai, H.: Analysis methodology research on identification characteristics of ship target's remote sensing images. Univ. Electron. Sci. Technol. China **13** (2013)
10. Sun, R.: Research on electromagnetic scattering from ship wakes. Xidian Univ. **5**, 21–22 (2013)
11. Oumansour, K., Wang, Y., Saillard, J.: Multifrequency SAR observation of a ship wake. IEEE Proc. Radar Sonar Navig. **143**(4), 275–280 (1996)
12. Zhang, Y., Dong, S., Bi, K.: Warship formation recognition algorithm based on hough transform and clustering. Acta Armamentarii. **37**(4), 648–655 (2016)

Opacity-Gradation-Based Visualization of Vortices for Large-Scale Ocean Simulation

Soya Kamisaka[1(✉)], Satoshi Nakada[2], Shintaro Kawahara[3], Hideo Miyachi[4], Kyoko Hasegawa[5], Liang Li[5], and Satoshi Tanaka[5]

[1] Graduate School of Information Science and Engineering, Ritsumeikan University, 1-1-1 Noji-higashi, Kusatsu, Shiga, Japan
is0404xs@ed.ritsumei.ac.jp

[2] National Institute for Environmental Studies, 16-2 Onogawa, Tsukuba, Ibaraki, Japan

[3] Japan Agency for Marine-Earth Science and Technology, 3173-25 Showa-machi, Kanazawa-ku, Yokohama, Kanagawa, Japan

[4] Department of Information Systems, Tokyo City University, 3-3-1 Ushikubonishi, Tuzuki-ku, Yokohama, Kanagawa, Japan

[5] College of Information Science and Engineering, Ritsumeikan University, 1-1-1 Noji-higashi, Kusatsu, Shiga, Japan

Abstract. Analysis of vortices is very important from the aspects of marine environment and disaster prevention. With the development of supercomputers, three-dimensional simulations of large-scale objects such as the ocean have been conducted. In ocean current simulations, multi-scale unstructured grids have been employed to precisely reproduce the topography of the seafloor and coastal areas. However, calculating vortices from unstructured grid data using the derivative of the flow velocity requires time-consuming data interpolation. Therefore, in this study, we calculate the covariance matrix using the velocity of each triangular column in the local region. The eigenvalues are used to define vorticity. In order to visualize the obtained vortices, the unstructured grid data are converted into point cloud data. In addition, the point density is adjusted according to the obtained vortices, and the opacity is changed. With the proposed method, we could visualize vortex regions with ambiguous boundaries correctly. During the visualization, we analyzed the causal relationship between the vortices and physical quantities such as flow velocity and salinity obtained from simulation by merging them into the visualization.

Keywords: Visualization · Vortex · Opacity gradation · Ocean simulation

1 Introduction

The analysis of vortices is very important for the marine environment and disaster prevention. For example, garbage such as PET bottles, which are called marine debris, collects at the locations where vortices are generated. By visualizing vortices, it is possible to predict the location and period of vortex generation, which will lead to support for garbage collection. In the Great East Japan Earthquake of 2011, tsunami vortexes

W. Fan et al. (Eds.): AsiaSim 2022, CCIS 1712, pp. 227–238, 2022.
https://doi.org/10.1007/978-981-19-9198-1_18

were generated by the tsunami. The tsunami vortex damaged many vessels. It is considered that the Nankai Trough earthquake will occur with a high probability in the near future. In this situation, we believe that visualization of tsunami vortexes can contribute to disaster prevention by predicting in advance the scale and extent of damage that will occur in the future.

This study aims to develop a visualization method for ocean current data obtained from ocean current simulations. In ocean current simulations for multiscale applications, complex topographic features such as coasts and seafloors must be reproduced precisely. For this reason, unstructured grid type data, which used a mesh other than a cubic grid, is employed. However, derivation of the vortex from the unstructured grid type data by differentiating the flow velocity vector field requires data interpolation, which is computationally time-consuming. Therefore, we use the variance-covariance matrix of the velocity vectors for the calculation of vortex. In the following, we define the covariance matrix of the flow velocity as "velocity variance" and the sum of eigenvalues obtained from the covariance matrix as "vorticity." It means that the local variation of the velocity vector is defined as a vortex. This definition allows us to represent the vortex in the same way as the rotation obtained by differentiating the velocity vector, even if the data are not of cubic grid type. In order to visualize the obtained vortices, we convert the unstructured grid data into point cloud data and visualize them by using our visualization method called stochastic point-based rendering (SPBR) [1, 2]. This method enables fast and highly detailed visualization without data interpolation.

In conventional methods, when visualizing the obtained vortex, a threshold value is set, and only the area where the vorticity exceeds a certain level is visualized with a high opacity, which is called a binary rendering [3]. However, in a natural phenomenon such as a vortex, it is impossible to determine whether or not the area above a certain threshold value is a vortex and to define the boundary clearly. In addition, since the points that are vortices were drawn with equal opacity, the information on the degree of vortex could not be seen. In this study, we propose to control an opacity adaptively according to the value of vorticity when visualizing vortices by adjusting the point density. In other words, a high opacity is given at the points where the vorticity is large, and the image is visualized with a dark emphasis. On the other hand, a low opacity is given at the points where the vorticity is small, and the vortex is visualized thinly. The proposed method generates a gradient-like rendering of the vortex that improves the visibility of vortices and is effective for natural phenomena with ambiguous boundaries such as vortices.

2 Experimental Data

In this study, we treat as the object of visualization the data obtained from a tidal simulation of Osaka Bay and the surrounding sea area [4, 5]. This data was compiled using the prognostic unstructured grid Finite Volume Community Ocean Model (FVCOM) and the ocean prediction system DREAMS (Data assimilation Research of the East Asian Marine System) in a single ocean simulation system using the nesting method. Figure 1 shows the simulation's area of interest. Each data set consists of 298,944 triangle prisms and precisely reproduces the topography of coasts and seafloors. Each triangular prism stores the coordinates of each vertex (x, y, z), the water temperature [°C], the salinity

[‰], the east-west velocity [m/s], the north-south velocity [m/s], the vertical velocity [m/s], the absolute velocity [m/s], and the velocity angle [°] in a mesh. The simulation period is eight months, from March 1, 2015, to October 1, 2015. Numerical calculations are performed every hour, so there are 5,136 steps.

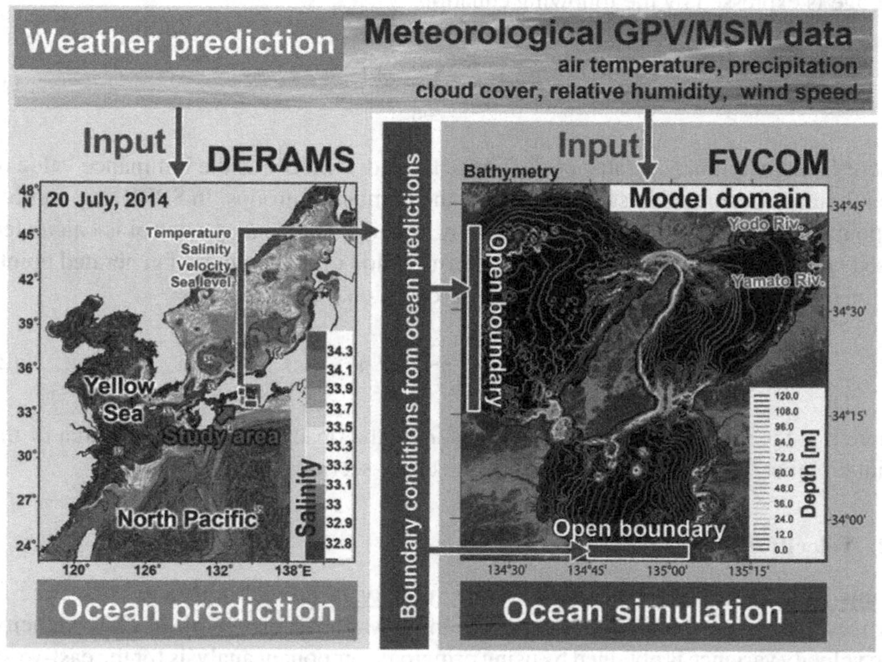

Fig. 1. Integrated system for ocean simulation using domains (left) in the ocean prediction system with salinity and velocity fields (color and arrows) and (right) in the ocean simulation with bathymetry. Domain was resolved with an unstructured grid in the FVCOM. The red square in the left panel indicates our study area. The blue lines and arrows show the data stream from the weather prediction through the ocean prediction to the ocean simulation for the open boundary conditions [3]. (Color figure online)

3 Methods

3.1 Visualization Method

In this study, we use stochastic point-based rendering (SPBR) as a method suitable for visualizing unstructured grid data and fusion visualization of multiple data. SPBR does not require sorting even for large-scale data and thus enables fast and high-definition visualization. SPBR performs rendering in three steps: (1) point generation, (2) point projection, and (3) pixel luminance value determination. The first step of this task is to generate points in the volume space in proportion to the transfer function based on the data to be visualized. In this study, we generate a point cloud with the same physical quantity in a triangular prism with uniform point density. The second step is to project and store the nearest point from the line of sight at each pixel in the image plane. At this time, the generated points are projected in an arbitrary number of groups. In the third step, luminance values are determined by applying an ensemble average to the image group created in step 2. The determination of the luminance value by the ensemble average is expressed by the following equation:

$$B = \frac{1}{L_R} \sum_{i=0}^{L_R - 1} B^{[i]} \tag{1}$$

where B is the luminance value of the image to be drawn, $B^{[i]}$ is the luminance value of the i th image generated in step 2, and L_R is the number of groups. In SPBR, the number of points generated in step 1 is controlled by opacity. The opacity control is calculated based on the binomial distribution. The determination of the number of generated points by the opacity is expressed by the following equation:

$$n = \frac{\ln(1 - \alpha)}{\ln\left(1 - \frac{s}{S}\right)} L_R \tag{2}$$

where n is the number of generation points, α is the opacity, S is the local area of the point cloud on the planar image, and s is the cross-sectional area of the point.

3.2 Velocity Variance

In this study, we analyze the variation of velocity in a local region at each time as vortices. We define "velocity variance" as the covariance matrix of the flow velocity. The velocity variance is obtained by using principal component analysis for the east-west, north-south, and vertical velocities. Let n be the number of points in the neighborhood, v_i, v_j $(i, j = x, y, z)$ be the velocity, and $\overline{v_i}, \overline{v_j}$ $(i, j = x, y, z)$ be the mean of the velocities. Using these values, the velocity variance V_{ij} can be written as:

$$V_{ij} = \frac{1}{n} \sum_{a=0}^{n} \left(v_i^{(a)} - \overline{v_i}\right)\left(v_j^{(a)} - \overline{v_j}\right) \tag{3}$$

Let $\lambda_1, \lambda_2, \lambda_3$ be the eigenvalues obtained by the calculation of the velocity variance V, and define the "vorticity ξ" as follows:

$$\xi \equiv \lambda_1 + \lambda_2 + \lambda_3 \tag{4}$$

Since the vorticity ξ is the sum of the eigenvalues of the covariance matrix, the value of ξ will be larger if the velocity changes rapidly in at least one of the east-west, north-south, and vertical vortex directions.

3.3 Opacity Gradation

Conventional feature area visualization methods use a binary method that gives high opacity to points with feature values above a threshold value. Since the conventional method gives equal opacity to all the features, most of the regions were highlighted depending on the data and threshold, which sometimes reduced the visibility. In addition, in natural phenomena such as vortex, it is difficult to divide the vortex region and the non-vortex region by a simple threshold value. The vorticity is gradually increasing in the vicinity of the vortex. Therefore, it is desirable to adopt some gradation in the expression of vorticity. Therefore, in this study, opacity control is applied to the obtained vorticity for visualization. Opacity control is a method for feature area enhancement visualization of measured points clouds proposed in our previous study [6]. This method allows us to assign opacity according to the vorticity and to represent gradations. The normalized vortex at the point of interest is ξ, the opacity is $\alpha(\xi)$, the maximum and minimum opacity values are $\alpha_{\max}$ and $\alpha_{\min}$ respectively, the two threshold values are $\xi_1, \xi_2(\xi_1 < \xi_2)$, and the dimension of the function is d. The opacity according to the value of the vortex is expressed as follows:

$$\alpha(\xi) = \begin{cases} \frac{\alpha_{\max}-\alpha_{\min}}{(\xi_2-\xi_1)^d}(\xi - \xi_1)^d + \alpha_{\min} & (\xi_1 \leq \xi < \xi_2) \\ \alpha_{\max} & (\xi_2 \leq \xi \leq 1) \end{cases} \quad (5)$$

From Eq. (5), we can see that the value of $\alpha(\xi)$ increases with d when the vortex is between ξ_1 and ξ_2. Furthermore, if the vortex is greater than ξ_2, the opacity is $\alpha_{\max}$. A graph representing Eq. (5) is shown in Fig. 2.

In this study, points with vorticity larger than 0.01 out of the vorticity normalized from 0 to 1 were visualized as vortices. Roughly 20–30% of the visualized area had vorticity larger than 0.01. On the other hand, the points with vorticity larger than 0.15 were near the boundary conditions. These points were considered to be outliers due to simulation. Therefore, the target vortices in this study were distributed in the rage of 0.01 to 0.15. Therefore, $\xi_1 = 0.01$ and $\xi_2 = 0.15$ were used in this study. The maximum and minimum opacity values were set to $\alpha_{\max} = 0.1$ and $\alpha_{\min} = 0.9$, respectively. When $d = 1$, Eq. (5) is linear in the interval ξ_1 to ξ_2. These parameters were fixed in this study.

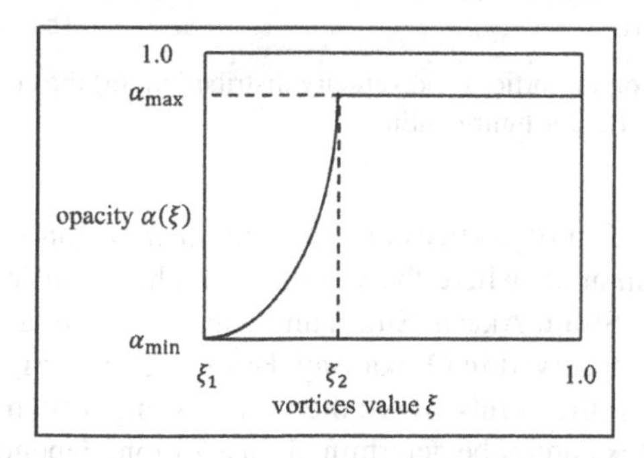

Fig. 2. Relation between vortices value and opacity

4 Visualization Experiments and Considerations

In this study, we perform "fused visualization", in which vortices and other physical quantities are visualized simultaneously. This section describes the results of various experiments and their discussions.

4.1 Vortices and Velocity Distribution

In this section, we show a fused visualization of vortices and velocity distribution. In this study, since the vortices are drawn in red, the visibility will be reduced if a color map including red color is used for the velocity distribution. Therefore, we use the color map without the red color. The visualization results at 10:00 on August 1, 2015, when large vortices were observed, are shown in Fig. 3. Figure 3(a) shows the vortices drawn by the conventional method with a constant opacity of 0.5 for the points where the vorticity is above 0.01. Figure 3(b) shows the vortices of Fig. 3(a) with opacity control applied.

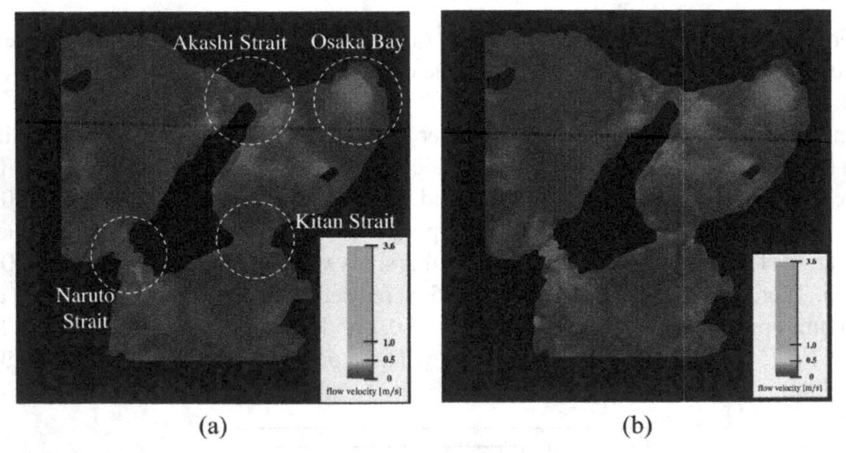

Fig. 3. Fused visualization of vortices and velocity distribution: (a) the conventional method and (b) the proposed method. (Color figure online)

The fused visualization of vortices and velocity distribution indicates that vortices are mainly generated in areas where the velocity is high. In particular, vortices can be observed in the Naruto Strait, Akashi Strait, and Kitan Strait. In addition, vortices with front structures can be observed in Osaka Bay. However, since Fig. 3(a) shows that the same opacity is given to the points where the vortex is larger than the threshold value, the strength of the vortex cannot be determined. In addition, depending on the threshold value, vortices are detected widely. As shown in Fig. 3(b), the vortex is rendered as a gradation by applying the opacity control to the vorticity. The points in high vortices are rendered darker and more pronounced due to the high opacity. In particular, the Naruto Strait and the eastern side of the Akashi Strait are drawn in dark red, indicating that they are strong vortices. On the other hand, the Kitan Strait and Osaka Bay are drawn in light red, indicating weak vortices. Thus, when opacity control is applied, strong vortices are emphasized, and weak vortices are made less noticeable by the gradation expression.

The simulation data used in this study were input from a dataset of precipitation and wind speed created by the Japan Meteorological Agency. Therefore, a large-scale typhoon that hit Japan in August 2015 was also reproduced. The visualization results for 16:00 and 19:00 on August 23, and 4:00 on August 24, when the typhoon hit Japan, are shown in Fig. 4.

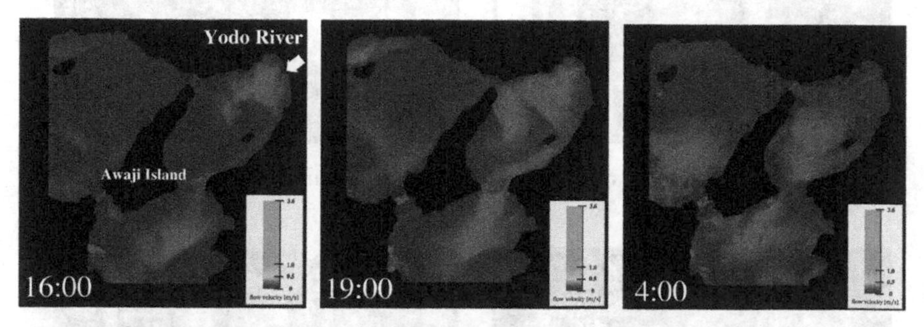

Fig. 4. Fused visualization of vortices and velocity distribution based on typhoon data

The velocity distribution indicates that the flow of the Yodo River was increasing and flowing into Osaka Bay. In conjunction with this, vortices, which do not usually occur in the ocean, can be observed. In particular, strong vortices can be observed in a wide area along the east coast of Awaji Island. These vortices were caused by the direct impact of river water from the Yodo River on Awaji Island. We believe these results will lead to the mitigation of typhoon damage to ship and aquaculture industries.

4.2 Vortices and Salinity Distribution

In this section, we show a fused visualization of vortices and salinity distribution. The visualization result at 7:00 on August 2, 2015, when the relationship between vortices and salinity was evident, is shown in Fig. 5.

The vortex in the Akashi Strait moves eastward and westward with the time. The visualization result shows that the vortex moving from the west to the east of the Akashi Strait extended into Osaka Bay. In conjunction with this vortex, we can confirm that a salt front flows into Osaka Bay. This result allows us to understand the fluctuations of nutrients and plankton, contributing to industrial aspects such as the prediction of fishing grounds.

4.3 Vortices and Water Temperature

In this section, we show a fused visualization of vortices and water temperature distribution. The visualization result at 21:00 on August 6, 2015, when the relationship between vortices and water temperature was evident, is shown in Fig. 6.

From the water temperature distribution, water temperature fronts can be identified in the northwest and southeast of Awaji Island. Vortices were observed along the water

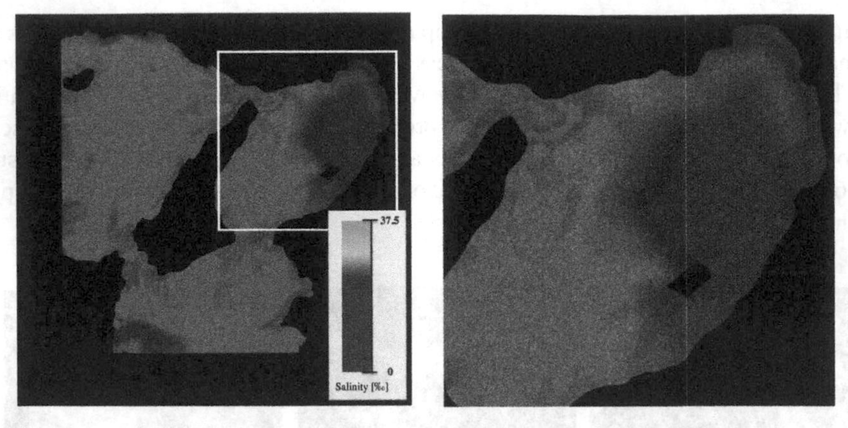

Fig. 5. Fused visualization of vortices and salinity distribution

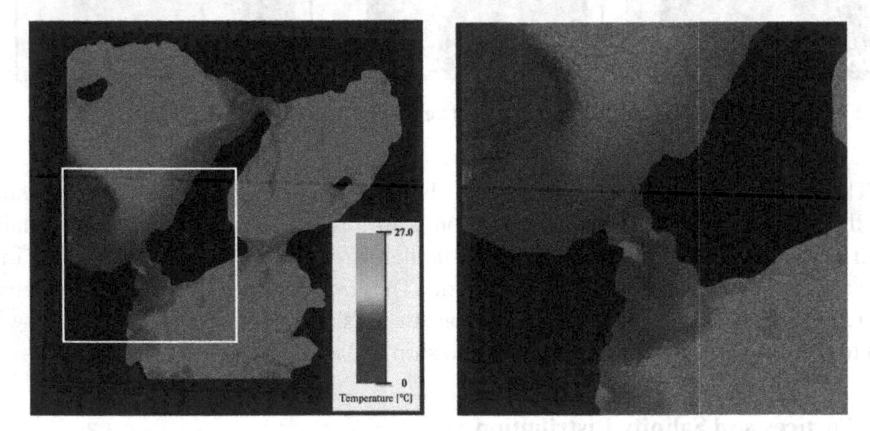

Fig. 6. Fused visualization of vortices and water temperature distribution

temperature fronts. This were caused by the significant changes in the velocity vectors due to the difference in the water temperature.

4.4 Vortices and Tidal Fluctuations

Tidal fluctuation is one of the causes that determine the period of vortex generation. For example, in the Naruto Strait, vortices were generated due to the difference in the height of the sea surface in the strait area caused by the fluctuation of the tide level. In this section, fused visualization was conducted to analyze the relationship between vortices and tide level. In the data used in this study, the tide level was obtained by calculating the width of the z-coordinates of the sea surface and seafloor triangles. The mean of the tide level at each location during a certain period was set to 0, and the difference between the mean and the tide level at each time is expressed as a color map. The visualization results at 4:00, 7:00, and 10:00 on August 2, 2015, are shown in Fig. 7.

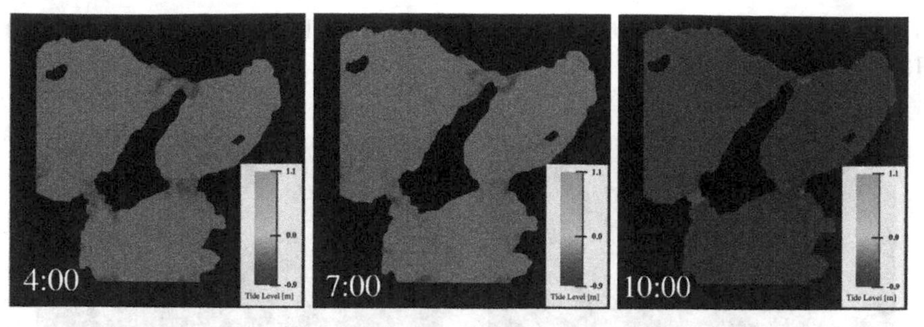

Fig. 7. Fused visualization of vortices and tidal fluctuations

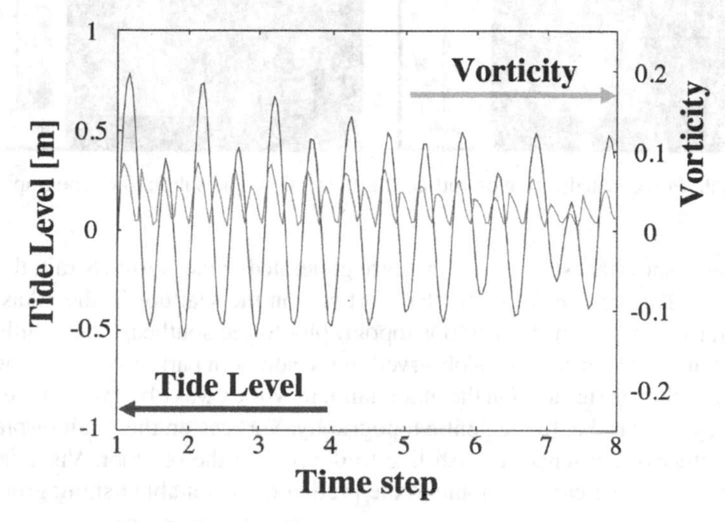

Fig. 8. Relationship between vorticity and tide level

The results show that the vortex in the Naruto Strait was stronger at high and low tides. On the other hand, the vortex was weaker at other times, indicating that the periodicity of the vortex is linked to the tide level. Figure 6(a) also shows that the vortex had a period twice as long as the tide level. The vortex in the Akashi Strait, which was observed to be on the west side at high tide, is found to move to the east side at low tide. This east-west movement is also found to be linked to the tidal cycle. Predicting the vortex generation period will contribute to collect marine debris that were carried by the vortices.

4.5 Vortices on the Seafloor and Submarine Topography

In the previous sections, we visualized vortices at the sea surface. However, vortices also exist on the seafloor and in the sea. In this section, we visualize vortices on the seafloor. The data used in this study were simulated at multiple scale. We calculate the vorticity using the velocity of the triangular columns just above the seafloor. Since vortices on the seafloor are greatly affected by the topography, fused visualization of the vortices and depth distribution was performed. The result of visualization at 13:00 on August 2, 2015, is shown in Fig. 9.

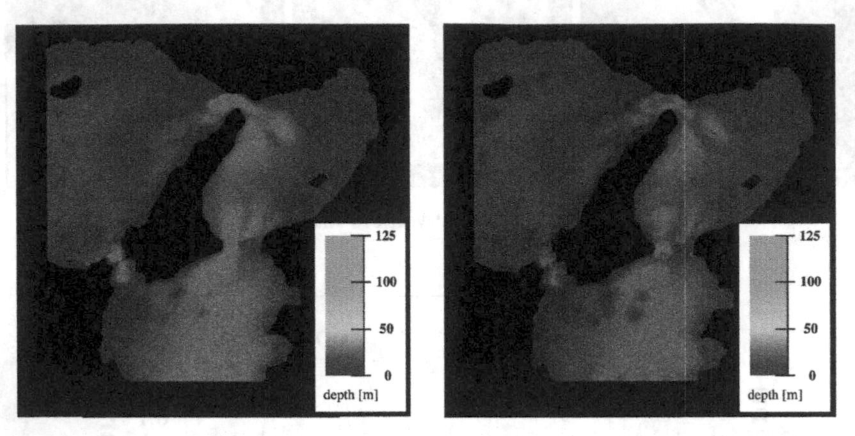

Fig. 9. Fused visualization of vortices of the seafloor and submarine topography

The results show that strong vortices were generated in the Naruto Strait, the Akashi Strait, and the Kitan Strait. In particular, vortices on the seafloor in the Akashi Strait were generated along with the seafloor topography to the southeast and southwest. In addition, small-scale vortices were observed in the southern part of Awaji Island, which did not exist at the sea surface. On the other hand, no vortex was observed on the seafloor of Osaka Bay. This is due to the gentle topography. Vortices on the seafloor play a role in the upwelling of nutrients that fish like to deposit on the seafloor. Visualization of vortices on the seafloor can contribute to the prediction of suitable fishing grounds.

4.6 Vortices Calculated by Normalized Flow Velocity

In this section, we visualized the vortices calculated by using normalized flow velocity. By normalizing the flow velocity, vortices can be detected with emphasis on the direction variation of the flow velocity. The visualization results at 12:00, 13:00, and 14:00 on August 2, 2015, are shown in Fig. 10.

Liner and patchy vortices, which are completely different from the previous visualization results, were confirmed. In particular, the propagation of northwestward vortices was confirmed in Osaka Bay. It is considered that vortices created by the collision of ocean currents can be detected by the proposed method. In the seas around Japan on the Pacific Ocean side, there exist areas where the Kuroshio and the Oyashio collide and plankton gather, which provide good fishing grounds. This method is suitable for predicting such locations.

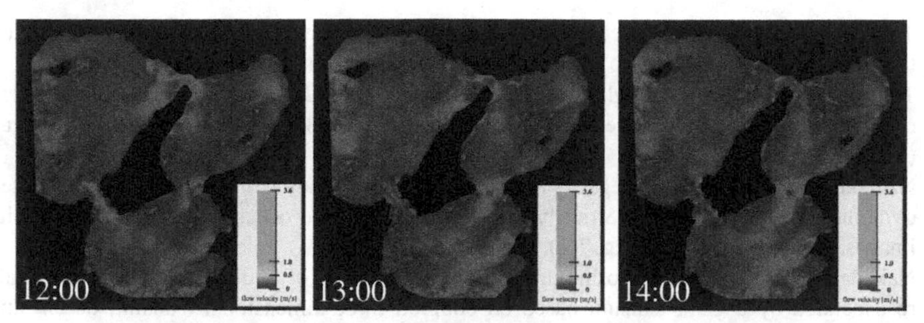

Fig. 10. Fused visualization of vortices calculated by the normalized flow velocity and velocity distribution

5 Conclusion

In this study, we visualized various vortices using simulated tidal current data. In ocean current simulations for multi-scale, unstructured grid type data were employed instead of cubic grid type data to precisely reproduce complex topography such as costs and seafloors. In order to calculate vortices from unstructured grid-type data, we defined vorticity that does not require differentiation. We defined "velocity variance" as the dispersion-covariance matrix constructed using the velocity vectors of triangular columns in the local domain and "vorticity" as the sum of eigenvalues. Opacity control was applied to the visualization of the obtained vortex. This enabled us to represent vortex gradations and faithfully reproduce vortices with ambiguous boundaries. In the actual visualization experiment, fused visualization was performed to analyze the causal relationship between vortex and other physical quantities. The analysis by fused visualization is one of the advantages of the ability to represent vortices in terms of color intensity. Vortex and its relationships to salinity front, temperature front, tidal fluctuation, and submarine topography were elucidated.

There are three prospects. The first is the analysis of longitudinal vortices. In the ocean, depth-dependent currents exist and play a role in transporting nutrients and plankton. However, the proposed method only detects horizontal vortices. This is due to the difference in velocity scales between horizontal and vertical directions. Therefore, it is necessary to define new vorticity that detects longitudinal vortices. The second is to obtain the vortex by differentiating the flow velocity vector field. By using the commonly used "rotation" and visualizing the vortex, the "vorticity" defined in this study is evaluated. The third is to visualize the results of this study in a Virtual Reality (VR) environment. Studies have been conducted to visualize simulation data with VR [7]. By visualizing in VR, it is possible to easily grasp the three-dimensional structure of the ocean and analyze it by interactively manipulating parameters such as threshold value and color maps.

References

1. Tanaka, S., et al.: Particle-based transparent rendering of implicit surfaces and its application to fused visualization. In: EuroVis 2012 (Short Paper), Vienna, pp. 25–29 (2012)
2. Tanaka, S., et al.: See-thorough imaging of laser-scanned 3D cultural heritage objects based on stochastic rendering of large-scale point clouds. In: ISPRS 2016 (Full Paper, Oral), Prague (2016)
3. Nakada, S., Morimoto, I., Hasegawa, K., Tanaka, S.: Visualization of riverine water and vortex dynamics around the Naruto Strait based on high-resolution ocean simulation and satellite images. J. Adv. Simul. Sci. Eng. **7**(1), 214–225 (2020)
4. Nakada, S., Hayashi, M., Koshimura, S.: Transportation of sediment and heavy metals resuspended by a giant tsunami based on coupled three-dimensional tsunami, ocean, and particle-tracking simulations. J. Water Environ. Technol. **16**(4), 161–174 (2018)
5. Nkada, S., Hayashi, M., Koshimura, S.: Salinization by a tsunami in a semi-enclosed bay: tsunami-ocean three-dimensional simulation based on a great earthquake scenario along the Nankai Trough. JSST **3**(2), 206–214 (2016)
6. Kawakami, K., et al.: opacity-based edge highlighting for transparent visualization of 3D scanned point clouds. ISPRS Ann. Photogramm. Remote Sens. Spat. Inf. Sci. **5**, 373–380 (2020)
7. Miyachi, H., Kawahara, S.: Development of VR visualization framework with game engine. JSST **12**(2), 59–67 (2020)

Industrial Metaverse: Connotation, Features, Technologies, Applications and Challenges

Zhiming Zheng[1], Tan Li[1(✉)], Bohu Li[2,3], Xudong Chai[2], Weining Song[4], Nanjiang Chen[1], Yuqi Zhou[1], Yanwen Lin[5], and Runqiang Li[6]

[1] Nanchang University, Nanchang, Jiangxi, China
someone8584@sina.com
[2] China Aerospace Science and Industry Co., Ltd., Haidian, Beijing, China
[3] Beihang University, Haidian, Beijing, China
[4] East China University of Technology, Nanchang, Jiangxi, China
[5] Nanchang Research Institute of Sun Yat Sen University, Nanchang, Jiangxi, China
[6] QuickTech Co., Ltd., Yizhuang, Beijing, China

Abstract. Metaverse expands the cyberspace with more emphasis on human-in-loop interaction, value definition of digital assets and real-virtual reflection, which facilitates the organic fusion of man, machine and material in both physical industry and digital factory. The concept of Industrial Metaverse is proposed as a new man-in-loop digital twin system of the real industrial economy which is capable of man-machine natural interaction, industrial process simulation and industrial value transaction. With the comparison with Metaverse and Digital Twin, the key features of Industrial Metaverse are summarized, which are man-in-loop, real-virtual interaction, process asserts and social network. Key technologies of Industrial Metaverse are surveyed including natural interaction, industrial process simulation, industrial value transaction and large-scale information processing and transmission technologies, etc. Potential application modes of Industrial Metaverse are given at the end as well as the challenges from technology, industry and application.

Keywords: Industrial Metaverse · Metaverse · Man-in-loop simulation · Digital twin · XR · Natural interaction · NFT

1 Introduction

Ever since the first-time literally mentioned in 1992 science fiction "Snow Crash" [1] by Neil Stephenson, Metaverse refers to a fantastic virtual world where people control their avatar to work and compete for higher social position. 2021 as the first year of Metaverse era, numerous world-leading organizations swarm into Metaverse industry in the past year, including Omniverse of Nvidia, Meta of Facebook, Roblox, as well as Enterprise Metaverse of Microsoft and Pico of ByteDance.

Various definition of Metaverse emerges, like in Wikipedia "the metaverse is a 3D virtual space with the characteristics of convergence and physical persistence through

W. Fan et al. (Eds.): AsiaSim 2022, CCIS 1712, pp. 239–263, 2022.
https://doi.org/10.1007/978-981-19-9198-1_19

virtual enhanced physical reality, which is based on the future Internet and has the characteristics of connection perception and sharing" [2] as well as Chen Gang [3] states that "Metaverse is a virtual world linked and created utilizing scientific and technical means to map and interact with the real world, as a digital living space with a new social system". Those concepts from different aspects generally refer to 3 kernel characteristics of Metaverse, which are Human-in-Loop Interaction, Value Definition of Digital Assets and Real-Virtual reflection. Metaverse utilizes Extended Reality (XR) and Natural Language Process (NLP) for immersive interaction with human users, defines and trades the digital valuable assets with Non-Fungible Token (NFT) and Blockchain, while reflecting the real world in virtual space using Digital Twin (DT).

Those features of Metaverse will greatly facilitate the organic fusion of man, machine and material in both physical industry and digital factory, especially in modern intelligent integrated manufacturing systems like Cloud Manufacturing System [4], which are more user-centered, flexible, agile and driven by both physical industrial data and digital product model. From this aspect, this paper proposes Industrial Metaverse as a new man-in-loop digital twin system of the real industrial economy, which simulates all industrial factors including man (workers and engineers), machine (products and equipment), material (raw resources and components) as well as the industrial process (production and maintenance) and the economic activity (pricing and trade).

Compared with Digital Twins and CPS, Industrial Metaverse puts more focus on natural man-machine interaction, industrial process simulation and industrial knowledge valuing, which may hopefully motivate series of exciting innovative technologies and applications in each phase of manufacturing. In this paper, the connotation and features of Industrial Metaverse are proposed comparing with normal Metaverse and Digital Twins, then propose a primary technical architecture and survey key technologies including natural interaction, industrial process simulation, industrial value transaction and large-scale information processing and transmission. Potential application modes of Industrial Metaverse are given at the end as well as the challenges from technology, industry and application.

2 The Connotation and Architecture of Industrial Metaverse

2.1 Connotation

By analyzing the research on related concepts and technical achievements such as digital twin and extended reality, this paper has the following definition of Industrial Metaverse.

Industrial Metaverse is a new man-in-loop digital twin system of the real industrial economy, through the interconnection of industrial digital resources, supported by DT, XR, NFT, NPL, as well as AI technologies, and driven by real-time information in industry. Industrial Metaverse simulates all industrial factors including man (workers and engineers), machine (products and equipment), material (raw resources and components) as well as the industrial process (production and maintenance) and the economic activity (pricing and trade).To generate a virtual-real interactive and full-chain collaborative space, Industrial Metaverse reintegrates industrial data with a variety of new technologies, and it realizes the decentralized sharing of industrial digital resources (including industrial product models, cloud platforms, product process information, etc.). Industrial

Metaverse provides a new solution for intelligent tracking and management of industrial products throughout their life cycle.

2.2 Comparison

According to the connotation of Industrial Metaverse proposed above, it is obvious that Industrial Metaverse has strong relationship with Digital Twin, Simulation, as well as the normal Metaverse. The similarities and differences among those four concepts are summarized as shown in Fig. 1.

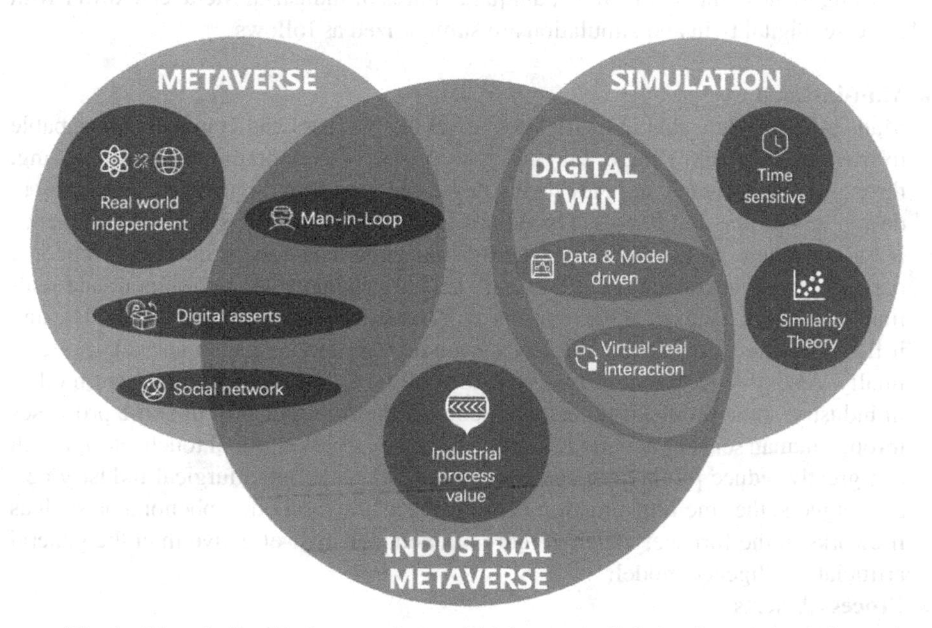

Fig. 1. The relationship between Industrial Metaverse, digital twin and simulation.

Simulation applications first appeared in the early 20th century, which refers to the combination of complete physical mechanisms and deterministic laws to calculate the next state according to the current boundary and state. The concept of digital twin emerged in 2019, which represents a digital replica of a physical entity, and it receive updated data from the physical entity via real-time connection [5]. Using digital twin technology, we can simulate, diagnose and predict on the digital replica. Then, the physical entity can do full analyze and make suitable decisions based on the prediction and simulation results from replica. Therefore, simulation is the core part of digital twin, it's also a feasible support for the creation and operation of twins.

The human-centered Industrial Metaverse is the reproduction of real industrial space scenarios, and digital twins are an important means of restoring the physical world. So, the construction of Industrial Metaverses requires the support of digital twin technology. As the application of the metaverse in the industrial field, it not only focuses on industrial scenarios, but also emphasizes that advantages (like immersive experience, decentralized virtual world, etc.) are reflected in some aspects (such as business management, personnel communication, product services, etc.) other than production. Therefore, the normal operation of Industrial Metaverse needs to be based on digital twin technology.

To sum up, Industrial Metaverse, digital twin and simulation are in a sequential evolution relationship. And it is inseparable from the support of digital twin and simulation technology to build it.

2.3 Features

According to the comparison above, unique features of Industrial Metaverse differ with Metaverse, digital twin and simulation are summarized as follows.

- **Man-in-Loop**
  Man-in-Loop means that industrial personnel like workers and engineers are capable to participate in entire stage, ranging from product development to usage tracking, through digital avatars, and play their roles in the corresponding stages, especially in design and production phases. For AI models used in industry, since its intelligence is based on tens of thousands of trainings on large data sets, it can easily become dependent on specific data sets. Not only because changeable environment and real-time updates of dataset, other reasons will cause deviations in the AI model, and it'll greatly affect production efficiency, even provoke large-scale industrial accidents finally. Also, due to the conservative nature of the application of intelligent models in industry, some production personnel control the start and stop of some processes through human sensory experiences (such as vision, smell, sense of touch, etc.), which can greatly reduce production costs. For example, some metallurgical industry personnel judge the smelting situation in the furnace based on other phenomena such as fireworks in the furnace, which is more efficient and cost-effective than the general artificial intelligence model.
- **Process Asserts**
  In practical applications, process is a very important part of industry, and many companies transform their process into patent. Although the same industrial product may be consistent in appearance, different processes have a great impact on the cost, quality and lifetime of the product. Therefore, in order to create a unique digital stand-in for industrial products in Industrial Metaverse and to software the corresponding industrial knowledge, it is necessary to make the manufacturing process of industrial products as part of the unique identification of the product.
- **Social Network**
  There is a very large industrial system in the real world, and many industrial links often require team participation. Similarly, in Industrial Metaverse, various types of workers communicate and collaborate with each other in the same virtual space through digital avatars to complete their own tasks. A similar connection is also reflected between the client side and the service side. Customers can connect into Industrial Metaverse for an immersive tour of the industrial production line, and then find a suitable party B according to their own needs to achieve precise docking.
- **Virtual-real interaction**

One of the main points in Industrial Metaverse is immersive interact between industrial entities and people in virtual space. Then, the interaction results can be used in the virtual space to guide the process of industrial product design and manufacturing. During the R&D stages of industrial products, workers in different places can enter Industrial Metaverse through their own digital avatars to conduct real-time discussion and modeling with other avatars which solves the real problem of low efficiency of online communication and long product trial period; In the manufacture of industrial products, the use of technologies enables employees to perform tasks(like quality management, manual scheduling, etc.) in Industrial Metaverse, improving worker's operation and management efficiency obviously; At the stages of sales and after-sales of industrial product, employee can reduce the maintenance cost by providing high-quality use guidance and services in cyber space remotely; When managing internal affairs of the company, enterprise personnel can have immersive conversations in internal virtual conference rooms to solve internal contradiction and improve the efficiency of enterprise management.

- **Decentralization**
  Unlike the real industrial world, Industrial Metaverse emphasizes the removal of "central privileges". From a financial perspective, technologies (such as blockchain, NFT, etc.) can improve financial security and transparency, unlock liquidity and growth opportunities, and support a standardized economic system. From an industrial point of view, decentralization can prevent monopoly and enhance the technological innovation capability of enterprises.

2.4 Primary Architecture

In order to build Industrial Metaverse, a primary technical architecture is proposed as shown in Fig. 2 based on the interaction between the human, the industrial physical world and the industrial virtual world, which includes the following main parts.

- **Basic layer.** The five essential elements (which are personnel, equipment, raw material, environment and Principle) are the basis for the construction of Industrial Metaverse. Some data resources (such as motor model, machine tool model, knowledge base, basic user information, etc.), network resources (like routing, optical fiber, etc.), storage resources (For example: hard disk, memory, etc.) and so on, are regarded as part of the basic layer to realize the efficient operation of digital space. So, the basic layer provides all the underlying support for the generation and operation of industrial digital space.
- **Perception layer.** The perception layer is based on factors like requirements and environment in industry. By carrying the IoT technology on equipment, it realizes the perception of all industrial physical elements, ensures that multi-source industrial information can be obtained stably in real time, and provides a basis for the reliable operation of twins. This layer also aims at the features of industrial digital space to realizes the immersive interaction between people and digital space through technologies such as brain-computer interface (BCI). In order to perceive other resources and facilitate the use of the upper layer, this layer also uses technologies such as cloud computing to virtually abstract various resources.

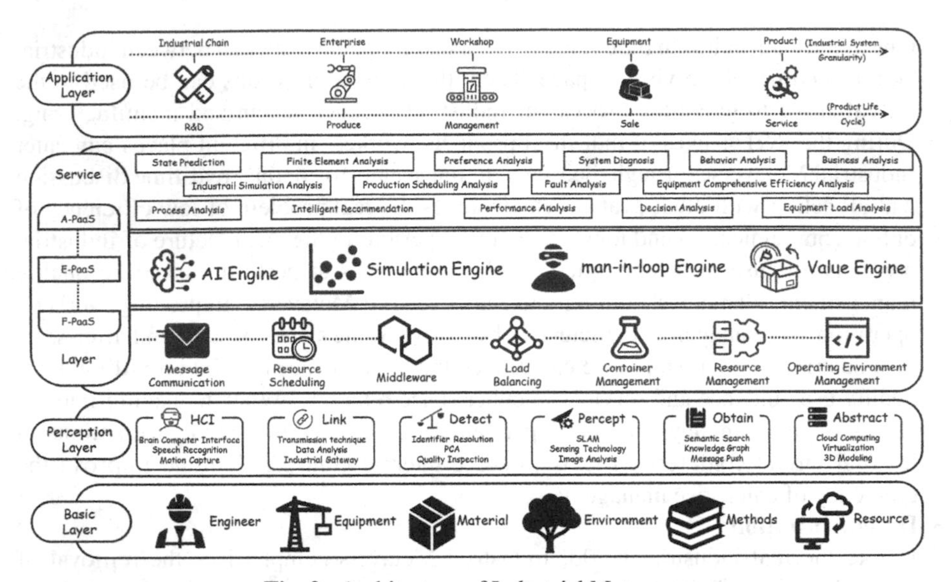

Fig. 2. Architecture of Industrial Metaverse.

- **Service layer.** As the core of the entire architecture, the service layer provides various services for Industrial Metaverse. It makes full use of several advanced technologies, and then provides all packaged basic back-end technologies that encapsulates all basic back-end technologies around the requirements of immersive user experience, high-fidelity industrial environment, and a decentralized social system to users in the form of services. This layer is divided into three sub-layers according to the type of service, which are: F-PaaS (Foundation-PaaS), E-PaaS (Engine-PaaS) and A-PaaS (Analyze-PaaS). F-PaaS provides various basic platform services such as message communication and resource services for the platform; E-PaaS, based on the features and functions of Industrial Metaverse, encapsulates various basic services into an engine for A-PaaS layer is used; A-PaaS, as the direct value embodiment of Industrial Metaverse, will provide users with many analysis functions. Based on real-time perception data, it builds models to realize auxiliary decision-making, state prediction and other functions, so that the twin has the ability to make decisions independently, and practitioners can also provide reference for various decisions during the operation of the corresponding industrial physical entity according to the calculation results of the twin. Not only that, to provide some help for the sales of industrial products, this sub-layer also provides commercial Analyze related functions.
- **Application layer.** The application layer takes the whole life cycle of the product and the granularity of the industrial system as the basic logic, providing industrial personnel with various functions such as immersive digital industrial space, visual management of products, and intelligent scheduling of production. Realizing the requirements of digitization, real-time, and decentralization of Industrial Metaverse.

3 Key Technologies of Industrial Metaverse

The realization of Industrial Metaverse requires various technologies including XR, NFT, AI, etc. Four types of key technologies of Industrial Metaverse are surveyed in this paper, as shown in Fig. 3.

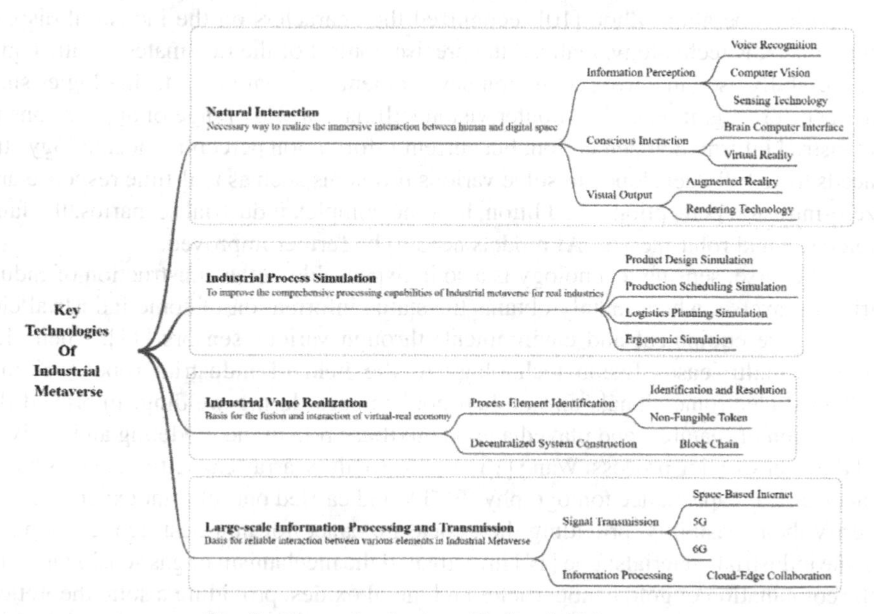

Fig. 3. Key technologies of Industrial Metaverse

3.1 Natural Interaction Technology

Natural Interaction Technology is a necessary way to realize the immersive interaction between human and digital space. The perception of human-factors industrial elements can be realized through some low-intrusive technologies which has the ability to obtain human-machine physical information, and it include Information Perception, Conscious Interaction and Visual Output.

- **Information Perception**
  Intelligent Perception Technology with AI as the main body will be widely used in Industrial Metaverse. At present, technical applications such as speech recognition and computer vision have achieved obvious results in the construction of industrial digital spaces. For example, Olanrewaju [6] used Visual Basic to design a speech recognition interface for a construction quantity assessment system, avoiding the potential safety hazards caused by workers keying in information on the device; Sheldon [7] used speech recognition technology in AR, making the voice user interface (VUI) a viable alternative to device communication modes on construction sites; Pagan [8] installed

a vision positioning system on the gluing robot to enable the robot to locate and reconstruct the shape of the object to be glued, and then applied it to the manufacturing process of footwear, effectively improving the speed and quality of shoe manufacturing; in order to perform low-invasive long-term fatigue monitoring of workers, Yu [9] proposed a 3D motion capture approach and built a fatigue assessment model to alleviate occupational health problems arising from excessive fatigue of workers during operation; Zhou [10] recognized the characters on the industrial display through OCR technology, realized the precise control of the raw material ratios, and successfully used in a live production environment. Not only that, technologies such as speech recognition and computer vision still have a wide range of applications in industrial information perception, but current information perception technology still needs to be fully developed to solve various problems such as real-time response and zero-intrusion perception. In addition, in some complex industrial scenarios, the fault tolerance and robustness of AI models need to be further improved.

Of course, sensing technology is also indispensable in the construction of Industrial Metaverse, which mainly obtains the status information of some industrial elements (like equipment and environment) through various sensors [11]. Chen [12] applied multi-sensor fusion technology to the field of industrial robot welding, obtained electronic signals and molten pool images during welding, improved the dimension of features, and played a good auxiliary role in the modeling and analysis of the laser welding process; Wang [13] developed three array capacitive sensors based on electrical capacitance tomography (ECT), and carried out relevant experiments to verify their feasibility, providing theoretical guidance for non-destructive testing of some industrial materials; Lee [14] investigated the mechanism of gas sensitization by the combination of gold nanoparticles and metal oxides, providing a solid theoretical basis for the fabrication of selective gas sensor arrays commonly used in industrial environmental monitoring; Lin [15] For multi-monitoring tasks, a group-based industrial wireless sensor network (GIWSN) is proposed, which solves the problems of sensor deployment and sleep scheduling, and uses simulation technology to verify the feasibility of the method, successfully overcoming large-scale industrial sensor deployment in scalable. Although the sensing technology has been rapidly developed and widely used in recent years, it still faces various challenges such as energy consumption and reliability, and has been adopted in some domestic precision machining fields. The technological gaps related to sensor perception of ultra-high precision data remain to be filled.

- **Conscious Interaction**

  In real industrial environment, the perception between people and various industrial elements is mainly accomplished through human senses, and corresponding to Industrial Metaverse, people need to be able to experience close to reality infinitely, including hearing, vision, touch, smell, taste and idea. In order to meet the requirements of low-invasive immersive experience, Brain Computer Interface (BCI) technology is a good entry point. Through BIC technology, people can be connected with their own digital avatars, so as to realize the Consciousness Interaction between human and Industrial Metaverse. This technology was first proposed by Vidal et al in 1976 [16], and its main purpose is to convert the activity of the central nervous system into artificial output, thereby optimizing the normal output of the central nervous system. BCI technology can collect and feedback signals through electricity, magnetism, sound, etc., and based on electroencephalography (EEG) is the current mainstream exploration direction. Kwak [17] proposed an error correction regression (ECR) framework and used it in the decoding stage of EEG, which improved the decoding accuracy and the reliability of BCI technology based on ear EEG; Wei [18] acquired EEG on the non-hair scalp region (NHB) and combined it with a classification model to diagnose driver fatigue. This achievement also has good application prospects in the industry; Wang [19] proposed a training method based on both motor imagery and speech imagery, which greatly improves human adaptation to BCI and enables better manipulation of BC. As an emerging field in recent years, Game-BCI has also attracted a lot of attention [20]. It completes the direct control of the game by people through the brain-computer interface and provides a new way of game interaction. For example, Andrew and others [21] developed a game called "Brain Invaders" based on the OpenViBE platform where players can control the elements in the game through BCI, providing theoretical reference for the subsequent application of BCI in the game; Wang [22] combined Motor-Imagery signals with Steady-State Visual Evoked Potentials (SSVEP) to control Tetris games and obtained good feedback from players on the game experience, demonstrating the feasibility of this hybrid approach; Cruz [23], developed a multiplayer cooperative game "KesselRun" based on BCI in 2017. In that game, two players cooperate to control the spacecraft to move left and right to avoid flying objects and persist for enough time to win, and the game has been well-received by players for its realistic and immersive experience. Controlling the interaction of digital humans in Industrial Metaverse through BCI technology can be a good reference for various BCI technologies, but the current development of BCI still has many bottlenecks like low transmission rate, high material cost, and low ease of use [24].

- **Visual Output**
  In Industrial Metaverse, Visual Output refers to the feedback of information in digital space to the human body through vision, which is an important step in the natural interaction process. Commonly used immersive visual output technologies mainly include VR, AR and so on. Burghardt [25] applied VR to the construction process of digital twins for industrial robots, recording human motion in the virtual environment through the VR system, and then reproduced by the real robot; Wolfartsberger [26] addressed the problem arise in engineering design reviews, such as CAD on the PC side not always meeting the functional and ergonomic verification requirements of complex 3D models. To meet the complex 3D model functions and ergonomic verification requirements, a VR-based CAD data visualization software – "VRSmart"- was developed and tested in the real industrial design review process, which improved the accuracy of the review. Yashin [27] proposed a VR-based remote operating system for drones, where operators can manipulate real robots offsite through an immersive virtual environment, making it easy to control the drone. Compared with VR, AR emphasizes the interaction with the physical world, which is essentially an extension of the user's real field of vision. Marino [28] used AR to detect errors in product production and assembly, and combined vision and sensor-based methods to improve the reliability of AR visualization and the quality of product appearance; For a faster speed to deal with unexpected failures, Ariansyah [29], an integrated application based on AR and computerized maintenance management system (CMMS) on the

Hololens platform is appeared, which can automatically highlight some other helpful information (such as abnormal machine parts) in a virtual-real-combined environment. And if the faulty system cannot be handled, the operator can contact the remote technician via Skype for help; To find a good solution to some unpredictable problems in plant layout and improves the efficiency of plant layout, Kokkas [30] used AR in the layout planning of the manufacturing system, allowing users to evaluate layouts in similar applications and skillfully deploy virtual equipment in the field without using markers. Although VR/AR has developed rapidly, there are still many problems [31, 32] like the high price of some corresponding hardware devices, deviation of user experience, poor compatibility, and so on.

It is necessary to build a sufficiently realistic visual-output environment, which cannot be achieved without the underlying support of real-time rendering technology. Many scholars have made great efforts for this. Petsiuk [33] proposed a method for 3D printing anomaly detection based on a physical rendering engine and oriented gradient histogram, which can detect printing errors at an early stage and greatly reduce the raw material cost of 3D printing, and has promising applications in industry; To reduce rendering overhead, Mueller [34] created Temporally Adaptive Shading (TAS) technology, compared with the current mainstream Reverse Reprojection Caching (RRC) technique, it significantly reduces the cost of VR rendering without significant visual quality loss; Accurate estimation of lighting is one of the reasons why AR is realistic enough. Liu [35] proposed a method based on differentiable screen space rendering, providing certain technical support for realizing a sufficiently realistic interactive experience in Industrial Metaverse; Marrinan [36] produced a real time geometry-based omnidirectional stereo rendering method suitable for standard rendering pipelines, which overcomes the problem that in traditional omnidirectional 3D images. The current rendering technology has become more and more mature due to the development in recent years, but various underlying foundations such as computing power, energy consumption, and graphics processing units have a large demand for the use of rendering technology, and are still facing very difficult challenges.

3.2 Industrial Process Simulation Technology

To improve the comprehensive processing capabilities of Industrial metaverse for real industries, it's inseparable from the support of Industrial Process Simulation Technology. As one of the components of current industrial manufacturing, Industrial Process Simulation Technology has been widely used in the industrial field, and it plays an important role in all aspects of the entire life cycle of industrial products, including Product Design Simulation, Production Scheduling Simulation, Logistics Planning Simulation and Ergonomics Simulation, etc.

- **Product Design Simulation**
  As a method of industrial product performance analysis, product design simulation refers to the simulation of various internal and external factors (like structural performance and working principle) in the process of industrial product design and use, so as to provide a basis for the improvement of industrial products. Nowadays, there are also many related technical researches on product design. For example, Chu [37]

studied the virtual maintenance interaction technology based on the general virtual maintenance training platform for engineering equipment based on the systematic analysis, and then developed a semi-physical simulation maintenance training for engineering equipment, which significantly shorten the operation frequency of actual equipment and the training cycle; Shen [38] found a method which uses CFD software to simulate and analyze the pressure distribution of the flow field in the pump to gear life prediction based on virtual simulation; the bending of automobile wheels Fatigue experiment is one of the key experiments to ensure wheel safety. In order to minimize the cost, Chai [39] proposed a bending fatigue experiment simulation method, which used the anisotropic material property data of LGFT wheel to simulate the strength of LGFT wheels, and finally, through the analysis of the experimental results, it is found that the results of the anisotropic material model meet the requirements well; Because the simulation models of complex electromechanical products are often multi-disciplinary and strongly nonlinear, Lin [40] proposed a simulation optimization method for complex electromechanical products based on a deep agent model, which was later applied to the construction of digital prototype of telescopic boom forklift, and the fluctuation and amplitude of the boom motion were strongly reduced. Although the current product design simulation has a deeper application in the industry, there are still various problems such as insufficient underlying computing power, low simulation efficiency, and inability to achieve extremely high fidelity in industrial product simulation.

- **Production Scheduling Simulation**
  As a black-box method, traditional industrial design mode is difficult to obtain the quantitative relationship between design parameters and production equipment. Digital industrial simulation uses the three-dimensional digital prototype of the product to model the product assembly process, and the whole process simulation of the product from parts to finished products can be implemented on the computer, so that the design model which meets the requirements can be transformed more quickly in a low-cost way. Zhang [41] integrated the simulation module into the existing basic process design system and proposed an enhanced ant colony optimization (E-ACO) algorithm for integrated process planning and scheduling (IPPS) in a job shop environment. After that, the original processing technology plan designed was simulated, and the plan was further modified and optimized according to the results; To optimize the water cycle management system in the flotation plant, Michaux [42] not only developed a corresponding process simulation platform that correlates the quality of process water with the performance of the processing plant, but investigates the water recycling scheduling strategy through simulation technology to strengthen the control of process water quality and save the operating costs of flotation equipment; Futáš [43] considered industrial simulation software as a tool to visualize the casting production process, and evaluated simulation results on the improvement of casting quality and reducing shrinkage formation, solving the formation and occurrence of casting defects such as shrinkage cavities in ductile iron castings, Positive impact on improving quality and reducing the cost of product manufacturing; while existing dynamic thermal-hydraulic simulation models have applications in industry, but time-consuming simulation model development increases model generation costs, which limits their use in wider range. Therefore, Martínez [44] proposed a method to

automatically generate thermal-hydraulic process simulation models from 3D plant models and showed by results that the proposed method can generate accurate high-fidelity models that can be accurately predicted. How to strengthen the coordination of process simulation and design and process planning, and how to strengthen the integrity of process simulation in the industrial processing process are the challenges for the construction of simulation systems in Industrial Metaverse.

- **Logistics Planning Simulation**
  In the manufacturing process of real industrial products, engineers develop scheduling plans to coordinate the rapid operation of logistics, but the planning and decision-making upon the past experience can no longer meet the requirements of digital transformation of industrial production processes, and Productive logistics Planning Simulation is a low-cost, high-efficiency solution to the problems we faced. Sun [45] proposed a fusion simulation model based on heuristic algorithm and conflict elimination algorithm in order to shorten the transportation time involved in ladle distribution during steelmaking continuous casting production, which optimized the equipment configuration and determined the optimal production sequence, resulting in minimizing the energy consumption cost; To improve the scheduling efficiency of the job shop, Liao [46] developed a supply chain scheduling optimization simulation system based on particle swarm optimization (PSO), which can overcome the non-convergence in production scheduling; Viharos [47] developed a general discrete event simulation model based on the Siemens Plant Simulation platform, and realized the AGV's robot assembly system simulation and scheduling method, which inputs all product assembly operations as a graph (or tree), and arranges assembly operations on each workstation, greatly reducing the overall manufacturing time and improving manufacturing efficiency; López [48] proposed a simulation framework based on AGV transportation systems, and extended the global framework by replacing on-board control modules and device with event simulators that statistically model AGV behavior, replacing the need for specific simulation tools, replacing the need to use specific simulation tools and improving the generality of the framework. Although the existing Productive Logistics Planning Simulation has many application cases in industry, there is an urgent need to solve various problems such as the variety of current planning simulation platforms, high performance requirements, and operational complexity.
- **Ergonomic Simulation**
  How to deal with the relationship between human, machine and environment is the key to the popularity of industrial products. In Industrial Metaverse, Ergonomic Simulation can be used to simulate the environment of real industrial products, and the simulation results turn back to optimize the design to achieve a positive interaction between the three elements. To make the experience of using exercise bikes better, Zhao [49] took a brand of exercise bike as the research object, used the ergonomics module in CATIA software to conduct virtual simulation analysis on the visibility, accessibility and posture comfort of people in the process of riding, and a corresponding fuzzy comprehensive evaluation system was established. Finally, the simulation results were verified with real experiments, which provided data support for the design of exercise bikes and shortened the product design cycle; Zhou [50] used JACK as the

main simulation tool, combined with The principle of ergonomics analyzes the human-machine-environment system where the driver is located, simulates and analyzes the virtual model of the industrial truck, and gives design optimization suggestions according to the simulation results, which significantly improves the production efficiency and ensures the safety and comfort of the operators. In order to improve the comfort of the health detection integrated machine, by analyzing the physiological and psychological changes of the elderly, to determine the function and module classification of the intelligent health detection integrated machine. Lu [51] used virtual simulation software to simulate the ergonomics of the testing machine, and improved the corresponding equipment through the simulation results, which effectively improved the comfort of the operation of the machine; Wang [52] introduced ergonomic simulation to solve the unreasonable problems in the flexible assembly process of aircraft parts (As show in Fig. 4). Taking the need for rigid assembly of the right wall panel of the pitot tail of a certain aircraft type as an example, the DELMIA human-machine simulation module is used to establish a human model and a simulation evaluation process develop a reasonable assembly process which improves the efficiency of aircraft parts assembly design. In addition, with the complication and strictness of human, machine and environment, there will be higher fidelity requirements for ergonomic simulation in Industrial Metaverse. Building a general comprehensive evaluation system and improving rendering intensity are the key to realizing human-machine engineering simulation. A necessary prerequisite for the mature application of mechanical engineering simulation in Industrial Metaverse.

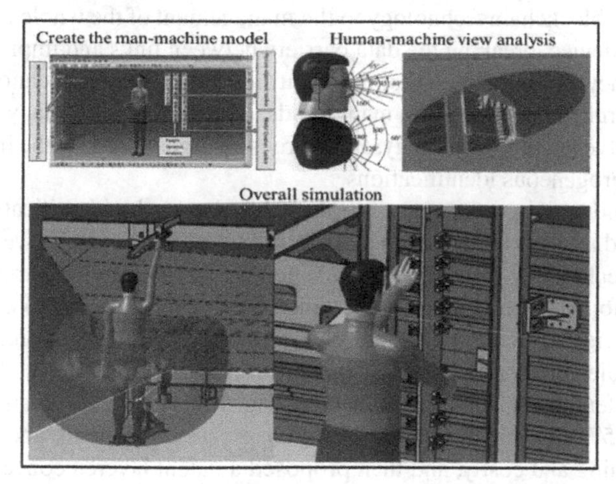

Fig. 4. Application of man-machine engineering simulation technology in aircraft component assembly

3.3 Industrial Value Realization Technology

The value of industrial products is mainly determined by quality, cost and other factors. But in Industrial Metaverse, the value of corresponding digital industrial products is embodied in software-based industrial knowledge and various physical parameters. Therefore, Industrial Value Realization Technology is the basis for the fusion and interaction of virtual-real economy, which mainly includes Process Element Identification and Decentralized System Construction.

- **Process Element Identification**
  Process Element Identification is an important means to measure the value of products in Industrial Metaverse. Through identity resolution technology, all process elements such as industrial equipment and industrial raw materials are connected to build a unique digital stand-in corresponding to industrial products. Xie [53] proposed an industrial IoT identification resolution system based on the Dual-Chord algorithm, which solved the mismatch between logical addresses and physical addresses, improved the resolution rate of the system, and provided some theoretical support to the establishment of Industrial Metaverse identity resolution system; Liu [54] designed a general identity parsing architecture to guide how to develop technical solutions from the joint requirements of services, roles, functions, implementation and security, and security, and verified the feasibility of the architecture through extensive practical applications, which is also applicable in Industrial Metaverse; Wang [55] found a food information traceability method using the relevant technology of identification analysis to clarify the identification relationship between food circulation links and solve the problem, solved the logo nesting problem in traceability, realized the product life cycle traceability information, and established a more reliable quality and safety traceability system in the food industry; Wang [56] applied identification resolution together with blockchain technology to the management of the whole industrial chain of electrical device, breaking the data barriers between links and improving management efficiency. It can be seen that the identification resolution technology has been widely researched and applied in many fields, but there are still many security problems to be solved in this technology, and there is an urgent need for the interoperability of many heterogeneous identifications.

  Digital industrial products are irreplaceable due to the identification of process elements, and in order to this uniqueness in Industrial Metaverse as well, blockchain-based NFT technology plays a key role. When NFT is used in Industrial Metaverse, a product cannot be copied and replaced, which plays a good role in protecting the copyright of the original creator, completely changing the status quo of digital ownership, and it embodies the core features of Industrial Metaverse. Current NFT technology has great potential for application in Industrial Metaverse. For example, Bamakan and others [57] noticed that the application process of patents and trademarks is time-consuming and costly, and then proposed a patent layered conceptual architecture based on NFT. This architecture provides basic elements and guidance for the use of NFT technology in Industrial Metaverse to solve practical problems such as granting patents, funding, and technology; Truby [58] investigate the impact of policy interventions regarding digital currencies on the relationship between NFT and carbon emissions, driving a shift towards digital currency consensus protocols that reduce energy dependence and provide policy guidance for reducing the pressure on carbon emissions in Industrial Metaverse; Karandikar [59] proposed a blockchain-based energy asset trading system, which encapsulates identifiers in the system or unique information and assets of value are modeled as NFTs, which facilitates the construction of peer-to-peer energy trading communities and provides technical references for the trading of virtual industrial products in Industrial Metaverse; Arcenegui [60] blockchain technology is used to manage industrial equipment, and through the physical unclonable function (PUF), smart NFTs are bound to IoT devices, thus establishing a secure communication channel with owners and users, enabling the decentralized management of device, and having good potential for application in the virtual industrial device management module of Industrial Metaverse. However, as NFT technology is more widely used, it has received more and more challenges in terms of reliability and environmental cost [61, 62].

- **Decentralized System Construction**
  Decentralized System Construction refers to construct a decentralized governance system in Industrial Metaverse. And blockchain technology as a distributed ledger, fully guarantees the decentralized operation of Industrial Metaverse. Abidi [63] proposed a privacy protection model for supply chain networks based on blockchain technology and improved the whale optimization algorithm in this model to effectively enhance the security information sharing capability of supply chain networks; Shahbazi [64] used blockchain technology in the creation of distributed smart manufacturing systems (As show in Fig. 5) and improved the transaction based on an integrated edge computing-blockchain-machine learning framework; A layered microgrid architecture is proposed by Hariharasudan [65], which realizes reactive power price management in the electricity market via blockchain technology, and provides certain transparency and security guarantees for both parties; In order to solve the problem of rising prices of industrial products, Hasan [66] a blockchain-based Internet of Things model is proposed to monitor the prices of industrial products, so that the buying and selling between buyers and industrial companies can be monitored, helping the government to control the price market. Familiarity and management. Although the blockchain has many applications in the industrial field, the current blockchain needs to be improved in many aspects such as operation, management, energy consumption and delay [67]. These problems also hinder the rapid expansion of blockchain technology in Industrial Metaverse.

3.4 Large-Scale Information Processing and Transmission Technology

This technology is the basis for reliable interaction between various elements in Industrial Metaverse. To realize rapid calculation and reception of industrial information, some fast and stable network technologies (mainly including Signal Transmission, Information Processing, etc.) can be used.

- **Signal Transmission**
  Some industrial fields with special nature like nuclear industry, mining industry and wind power industry are generally located in remote and harsh environments in real industrial scenarios. If land-based equipment is used to connect the factory to Industrial Metaverse, it will be greatly affected by the environment, which not only makes it difficult to set up land-based communication equipment, but also increases the maintenance difficulty and even leads to disconnection. Therefore, using space-based

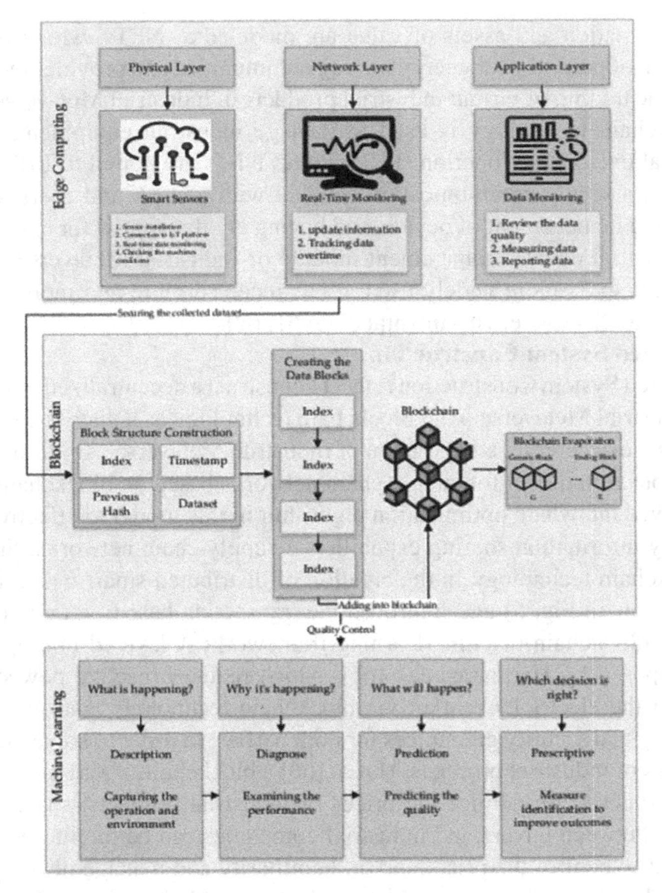

Fig. 5. Smart manufacturing overall architecture based on an integrated system

Internet to ensure access to people and things anytime and anywhere is a reliable way to achieve the barrier-free connection of Industrial Metaverse. Aiming at the problem that the existing random access of Space-based IoT is highly sensitive to load, Fei [68] proposed an intelligent load control-based random access scheme for the S-IoT to maximize network throughput by accurately monitoring and controlling the load. The feasibility of the proposed scheme was verified through simulation experiments, demonstrating the application prospect of the random access scheme in the S-IoT. Although the S-IoT has strong anti-destructive ability and can be accessed anytime and anywhere, the satellite's low bandwidth makes it challenging to provide data transmission services to a massive number of industrial metaverse nodes, so it cannot withstand precise data collection tasks. Fei [69] proposed a sampling-reconstruction (SR) algorithm based on spatiotemporal compressive sensing (ST-CS) technology, which uses the spatiotemporal correlation between the sensory data to accurately reconstruct unacquired sensory data, effectively improving the data collection capability of the S-IoT. Considering the contradiction between periodic activated sporadic transmission of massive user equipment and the short packet communication and limited pilot in uplink S-IoT, Liu [70] proposed a non-orthogonal super imposed pilot grant-free random access (NSP-GFRA) scheme to alleviate pilot collision, and verified the feasibility of the scheme through simulation experiments. Jiao [71], proposed an optimization scheme that combines joint network stability and resource allocation using the KKT(Karush–Kuhn–Tucker) conditions and the particle swarm optimization (PSO) algorithm to maximize the long-term network utility of S-IoT downlink system, which effectively alleviates the problem of communication resource constraints. Currently, the Space-based IoT technology is still in the preliminary application stage. Many of the domestic and foreign technology giants have laid out their plans. Such as the "Xingyun Projet" and "Hongyun Project" of CASIC, the StarLink of SpaceX, the satellite constellation of OneWeb and the Project Kuiper of Amazon, etc. However, there are numerous challenges in the intelligent hybrid NOMA transmission, grant-free SP-NOMA and MSPA random access protocols [72].

When interacting with the industrial digital space in an immersive manner, ultra-reliable and low-latency communication is the most basic requirement, otherwise it will be difficult for people to have an immersive effect, and it will also make people feel uncomfortable. As a new generation of mobile communication technology with high speed, low latency and large connection, 5G plays a very important role in Industrial Metaverse. Rodriguez [73] propose an experimental framework which integrates 5G wireless systems into Industry 4.0 applications and this technology can be used for reliable control of autonomous mobile robots in industry to achieve intelligent path planning and reliable control of the robots. Sasiain [74] combined two critical supportive technologies for 5G (Network function virtualization (NFV) and Software-define networking(SDN)) and proposed a 5G architecture with wireless sensing networks for Industry 4.0, addressing the shortcoming of traditional networks in terms of flexibility, accessibility, and dynamics in real industrial scenarios. Khatib [75] applied 5G technology to better cater for the needs of smart logistics and created a 5G-based optimization system by using an application traffic modeling process, along with a proactive approach to network optimization to effectively aid the logistics industry in reducing management, energy and storage costs. The vertical handover and the seamless connectivity play an important role during the process of industrial smartness. Priya [76] presented a 5G autonomous network selection model which is suitable for Industry 4.0 that effectively improves the unnecessary handover and the ranking abnormality issues. At present, various problems need to be solved such as privacy security protection, standard unification, etc. in the process of pushing 5G to industrial application and it is also an important guarantee to achieve normal and efficient operation of Industrial Metaverse. At present, various problems need to be solved [77–80] such as privacy security protection, standard unification, etc. in the process of pushing 5G to industrial application and it is also an important guarantee to achieve normal and efficient operation of Industrial Metaverse.

With the gradual promotion of 5G technology, a variety of extremely rich multimedia applications such as high-fidelity holograms, immersive virtual-real interaction, and sensory communication are also highly sought after. However, in order to support the smooth operation of such applications, larger system bandwidth and new physical layer technology are required, so 6G will also become the main force supporting the transmission of Industrial Metaverse information, and many applications related to 6G-based Industrial Metaverse will improve the innovation capability of intelligent services through the integration of "communication, sensing and computing" (communication, sensing and computing). At present, many scholars have conducted in-depth research on 6G technology. For example, Mahmood and others [81] are motivated by Machine Type Communication (MTC) and Vertical-specific Wireless Network solution, expressing the prospect that 6G will achieve instant wireless connectivity, and that it will drive the first battier-breaking standard, providing a unified solution that will make all the needs of vertical industries seamlessly connected. Rappaport [82] conducted a more comprehensive study of 6G THz wireless communications and described numerous technical challenges and opportunities for wireless communication and sensing applications above 100 GHz. They present methods that reduce the computational complexity and simplify the adaptive signal processing through analysis of existing results, and discuss the implementation of spatial consistency at THz frequencies, which is expected to play an important role in the construction of 6G. The 6G technology is necessary to support the ubiquitous AI services from the core to the end devices of the network, and Letaief [83] argue that AI powered 6G technology can transform Cognitive-radio (CR) to Intelligent-radio (IR). It will enable AI to play a significant role in designing and optimizing 6G architectures and operations. Basar [84] proposed RIS-space shift keying (RIS-SSK) and RIS-spatial modulation (RIS-SM) schemes to bring the concept of RIS-assisted communication to the realm of index modulation (IM). It will make that RIS-based IM capable of achieving high data rates with remarkably low error rates and can be a potential candidate for future 6G communication systems on the background of beyond multiple-input multiple-output (MIMO) solutions. At present, the research on 6G is still in the theoretical stage. Also, high security, confidentiality and privacy as key features of 6G, should receive special attention from relevant researchers [85, 86].

- **Information Processing**

  The diversity and real-time nature of industrial data put forward higher requirements for information processing in Industrial Metaverse, while cloud-edge collaboration technology provides a good solution to the challenges faced by cloud computing such as load and latency, and the limitations brought by local terminals in edge computing. This technology can tightly combine cloud computing and edge computing to achieve a balance between cloud and edge, and improve the capability of information processing in Industrial Metaverse through reasonable allocation and scheduling of resources in the cloud and edge. There are many cases of applying cloud-edge collaboration technology in industry. For example, Jing [87] proposed a Cloud-Edge collaboration framework with deep learning-based which can achieve highly accurate remaining useful life (RUL) of machinery prediction. It can satisfy the requirements of intelligent manufacturing services for rapid response, and is widely used in industry with potential. Yang [88] proposed a cloud-edge-device collaboration framework, which provides excellent suggestions for improving the collaboration mechanism in Industrial Metaverse. Jian [89] proposed a Cloud-Edge based two-level hybrid scheduling learning model and improved bat scheduling algorithm with interference factors and variable step size (VSSBA), and finally predict the scheduling results by Neural Network model. As the experimental results show, the model can improve the performance of the cloud manufacturing platform in real-life applications efficiently. In order to improve the performance of cloud-side collaboration, Laili [90] developed a practical model for task scheduling considering two cloud-side collaboration models, which reduces the overall task computation time and saves energy to a certain extent. Because of the ultra-large scale of data generated by industry, it is challenging for Industrial Metaverse to achieve real-time response. Therefore, it is crucial to break through the difficult design of the underlying computing architecture at both ends of the cloud edge and the security issues brought about by the cloud edge collaboration, etc., to drive the data-driven Industrial Metaverse.

4 Applications and Challenges of Industrial Metaverse in the Future

4.1 Applications

Industrial Metaverse complements the industrial ecology of the future intelligent integrated manufacturing system in three aspects of man-machine interaction, industrial process simulation and industrial economic system, giving birth to numbers of Industrial Metaverse primordial enterprises with the essence of digital industrial social networks, which will fully support all kinds of digital industrial applications and virtual economic systems under the mode of natural interaction. Industrial Metaverse will promote series of innovative applications at different domain in the vertical and horizontal levels of digital industry. Part of the predictable future applications of Industrial Metaverse are shown in Fig. 6.

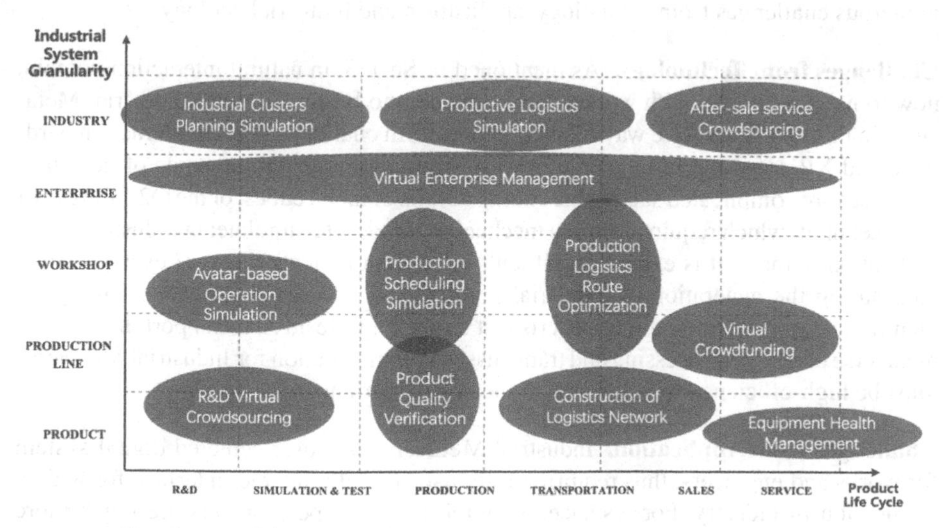

Fig. 6. Part of the predictable future applications of Industrial Metaverse.

Future applications of Industrial Metaverse can be sorted into two categories as follow,

- Vertical level (coarse and fine granularity of manufacturing system): industrial chain level (digital industrial cluster/Park simulation and planning based on industrial brain), factory/manufacturing enterprise level (virtual organization based on meta universe enterprise/virtual enterprise management and collaboration), workshop/production line level (production line simulation and process scheduling based on digital twins), equipment/product level (digital delivery and virtual training of industrial products), etc.;
- Horizontal level (the whole life cycle of Intelligent Manufacturing): R&D links (R & D crowdsourcing based on industrial meta universe and user participation in design), testing (simulation of product operation friendliness and operation safety based on virtual human), production and manufacturing (handicraft process optimization and capacity evaluation based on standard man hour data) Logistics transportation (productive logistics collaboration based on working hours, logistics network simulation planning based on Swarm Intelligence), marketing (virtual trial and crowdfunding marketing based on digital products), and after-sales service (equipment health management and fault diagnosis based on digital twins, product application and skill training based on skill meta universe), etc.

4.2 Challenges

Lots of "so-called" Industrial Metaverse applications are reviewed by the authors' team, which lead to a conclusion that most of the applications are actually digital twins, CPS or merely XR system in essence, lack of supports to man-in-loop interaction and industrial value transaction. As a new digital industry approach, Industrial Metaverse still faces numerous challenges from technology, application and industrial ecology.

Challenges from Technology. As mentioned in Sect. 3, in natural interaction domain, how to natural interact with workers and engineers to bring them into Industrial Metaverse in minimally invasive way remains a long-term challenge to both interaction hardware and XR software; in industrial process simulation domain, amounts of industrial processes are complicated such as the cutting details of metal cubes, or the O2-blowing of hot steel melt, which require accurate mechanism models to simulate; in industrial value realization domain, it is extremely difficult to "holographically" record every detailed data during the generation of industrial product and knowledge, as well as uniquely identify and pricing those digital asserts just and sound. Besides, to support the massive metaverse users, the processing and transmission of information for Industrial Metaverse must be high-efficiency to ensure the immersive experience.

Challenges from Application. Industrial Metaverse is a user-centered digital system for works and engineers, thus requiring fully user-friendly interact interface for widely application in industry. For instance, automobile prototype designers are much more familiar with burin and chisel rather than the PICO game handles to build a sludge model for their latest vehicles. How to design the easiest but most accurate operating procedure to make the manufacturers willing to enter Industrial Metaverse to implement virtual industry.

Challenges from Ecology. Industrial Metaverse is a complicated meta-synthesis information system integrated with high-performance ICT resources, immersive man-machine interaction devices, as well as the industrial models and data. The Industrial Metaverse ecology concludes suppliers from industrial sensors, wearable devices, CAD/CAE software, digital content producers, digital asserts exchange, etc.., which are far from mature world widely.

5 Conclusion

Human-in-Loop Interaction, Value Definition of Digital Assets and Real-Virtual reflection are important features that enable Metaverse possibly to support the organic fusion of man, machine and material in both physical industry and digital factory.

Industrial Metaverse fuses technologies from Metaverse, Digital Twins and Simulation to build a new man-in-loop virtual intelligent manufacturing system which is capable of man-machine natural interaction, industrial process simulation and industrial value transaction, which is featured with man-in-loop, real-virtual interaction, process collection and social network.

Besides traditional DT and Metaverse technologies, natural interaction, industrial process simulation, industrial value transaction and large-scale information processing and transmission technologies are considered key technologies of Industrial Metaverse.

With the development of technology and ecology, Industrial Metaverse will promote series of innovative applications at different domain in the vertical and horizontal levels of digital industry. However, challenges from technology, application and industrial ecology remain to be solved for Industrial Metaverse.

References

1. Stephenson, N.: Snow Crash. Bantam (1992)
2. Wikipedia - Metaverse. https://en.wikipedia.org/wiki/Metaverse
3. Peking University scholars released the START map of metaverse characteristics and attributes [EB/.OL], 19 November 2021. https://www.sohu.com/a/502061675_162758. Accessed 25 Apr 2022
4. Li, B.H., Chai, X.D., Zhang, L.: Cloud Manufacturing: a new service-oriented networked manufacturing model. CIMS **6**(01), 7+16 (2010). https://doi.org/10.13196/j.cims.2010.01.3.libh.004
5. Zhang, L.: Cold thinking about digital twin and the modeling and simulation technology behind it. J. Syst. Simul. **32**(4), 744 (2020)
6. Olanrewaju, O.I., Sandanayake, M., Babarinde, S.A.: Voice assisted key-in building quantities estimation system. J. Eng. Proj. Prod. Manag. **10**(2), 114–122 (2020)
7. Sheldon, A., Dobbs, T., Fabbri, A., et al.: Putting the AR in (AR)chitecture-Integrating voice recognition and gesture control for Augmented Reality interaction to enhance design practice (2019)

8. Pagano, S., Russo, R., Savino, S.: A vision guided robotic system for flexible gluing process in the footwear industry. Robot. Comput. Integr. Manuf. **65**, 101965 (2020)
9. Yu, Y., Li, H., Yang, X., et al.: An automatic and non-invasive physical fatigue assessment method for construction workers. Autom. Constr. **103**, 1–12 (2019)
10. Zhou, M., Liu, Z.Y., Lu, G.P., et al.: Application of OCR-based automatic identification of digital instruments in industrial field. Instrument User (2021)
11. Javaid, M., Haleem, A., Singh, R.P., et al.: Significance of sensors for Industry 4.0: roles, capabilities, and applications. Sens. Int. **2**, 100110 (2021)
12. Chen, B., Wang, J., Chen, S.: Modeling of pulsed GTAW based on multi-sensor fusion. Sens. Rev. **29**, 223–232 (2009)
13. Wang, W., Zhao, K., Zhang, P., et al.: Application of three self-developed ECT sensors for monitoring the moisture content in sand and mortar. Constr. Build. Mater. **267**, 121008 (2021)
14. Lee, J., Jung, Y., Sung, S.H., et al.: High-performance gas sensor array for indoor air quality monitoring: the role of Au nanoparticles on WO_3, SnO_2, and NiO-based gas sensors. J. Mater. Chem. A **9**(2), 1159–1167 (2021)
15. Lin, C.C., Deng, D.J., Chen, Z.Y., et al.: Key design of driving Industry 4.0: joint energy-efficient deployment and scheduling in group-based industrial wireless sensor networks. IEEE Commun. Mag. **54**(10), 46–52 (2016)
16. Vidal, J.J.: Toward direct brain-computer communication (1973)
17. Kwak, N.S., Lee, S.W.: Error correction regression framework for enhancing the decoding accuracies of ear-EEG brain–computer interfaces. IEEE Trans. Cybern. **50**(8), 3654–3667 (2019)
18. Wei, C.S., Wang, Y.T., Lin, C.T., et al.: Toward drowsiness detection using non-hair-bearing EEG-based brain-computer interfaces. IEEE Trans. Neural Syst. Rehabil. Eng. **26**(2), 400–406 (2018)
19. Wang, L., Huang, W., Yang, Z., et al.: A method from offline analysis to online training for the brain-computer interface based on motor imagery and speech imagery. Biomed. Signal Process. Control **62**, 102100 (2020)
20. Cattan, G.: The use of brain–computer interfaces in games is not ready for the general public. Front. Comput. Sci. **3**, 628773 (2021)
21. Andreev, A., Barachant, A., Lotte, F., et al.: Recreational applications of OpenViBE: brain invaders and use-the-force (2016)
22. Wang, Z., Yu, Y., Xu, M., et al.: Towards a hybrid BCI gaming paradigm based on motor imagery and SSVEP. Int. J. Hum. Comput. Interact. **35**(3), 197–205 (2019)
23. Cruz, I., Moreira, C., Poel, M., Ferreira, H., Nijholt, A.: Kessel Run - a cooperative multiplayer SSVEP BCI game. In: Chisik, Y., Holopainen, J., Khaled, R., Luis Silva, J., Alexandra Silva, P. (eds.) INTETAIN 2017. LNICSSITE, vol. 215, pp. 77–95. Springer, Cham (2018). https://doi.org/10.1007/978-3-319-73062-2_6
24. Xu, M.P., He, F., Jung, T.P., Gu, X.S., Ming, D.: Current challenges for the practical application of electroencephalography-based brain-computer interfaces. Engineering **7**(12), 1710–1712 (2021)
25. Burghardt, A., Szybicki, D., Gierlak, P., et al.: Programming of industrial robots using virtual reality and digital twins. Appl. Sci. **10**(2), 486 (2020)
26. Wolfartsberger, J.: Analyzing the potential of Virtual Reality for engineering design review. Autom. Constr. **104**, 27–37 (2019)
27. Yashin, G.A., Trinitatova, D., Agishev, R.T., et al.: AeroVr: virtual reality-based teleoperation with tactile feedback for aerial manipulation. In: 2019 19th International Conference on Advanced Robotics (ICAR), pp. 767–772. IEEE (2019)
28. Marino, E., Barbieri, L., Colacino, B., et al.: An augmented reality inspection tool to support workers in Industry 4.0 environments. Comput. Ind. **127**, 103412 (2021)

29. Ariansyah, D., Rosa, F., Colombo, G.: Smart maintenance: a wearable augmented reality application integrated with CMMS to minimize unscheduled downtime. Comput. Aid. Des. Appl. **17**(4), 740–751 (2020)
30. Kokkas, A., Vosniakos, G.-C.: An Augmented Reality approach to factory layout design embedding operation simulation. Int. J. Interact. Des. Manuf. (IJIDeM) **13**(3), 1061–1071 (2019). https://doi.org/10.1007/s12008-019-00567-6
31. Damiani, L., Demartini, M., Guizzi, G., et al.: Augmented and virtual reality applications in industrial systems: a qualitative review towards the Industry 4.0 era. IFAC-PapersOnLine **51**(11), 624–630 (2018)
32. Baroroh, D.K., Chu, C.H., Wang, L.: Systematic literature review on augmented reality in smart manufacturing: collaboration between human and computational intelligence. J. Manuf. Syst. **61**, 696–711 (2021)
33. Petsiuk, A., Pearce, J.M.: Towards smart monitored AM: open source in-situ layer-wise 3D printing image anomaly detection using histograms of oriented gradients and a physics-based rendering engine. Addit. Manuf. **52**, 102690 (2022)
34. Mueller, J.H., Neff, T., Voglreiter, P., et al.: Temporally adaptive shading reuse for real-time rendering and virtual reality. ACM Trans. Graph. (TOG) **40**(2), 1–14 (2021)
35. Liu, C., Wang, L., Li, Z., et al.: Real-time lighting estimation for augmented reality via differentiable screen-space rendering. IEEE Trans. Vis. Comput. Graph. **01**, 1 (2022)
36. Marrinan, T., Papka, M.E.: Real-time omnidirectional stereo rendering: generating 360 surround-view panoramic images for comfortable immersive viewing. IEEE Trans. Vis. Comput. Graph. **27**(5), 2587–2596 (2021)
37. Chu, W., He, X., Zhu, Z.: Research on the semi-physical simulation maintenance training system of one engineering equipment. In: 2019 International Conference on Modeling, Analysis, Simulation Technologies and Applications (MASTA 2019), pp. 34–39. Atlantis Press (2019)
38. Shen, H., Li, Z., Qi, L., et al.: A method for gear fatigue life prediction considering the internal flow field of the gear pump. Mech. Syst. Signal Process. **99**, 921–929 (2018)
39. Chai, W., Liu, X., Shan, Y., et al.: Research on simulation of the bending fatigue test of automotive wheel made of long glass fiber reinforced thermoplastic considering anisotropic property. Adv. Eng. Softw. **116**, 1–8 (2018)
40. Lin, J.L.: Simulation and optimization method and application of complex electromechanical products based on deep surrogate model. Guangdong University of Technology (2021)
41. Zhang, S., Wong, T.N.: Integrated process planning and scheduling: an enhanced ant colony optimization heuristic with parameter tuning. J. Intell. Manuf. **29**(3), 585–601 (2014). https://doi.org/10.1007/s10845-014-1023-3
42. Michaux, B., Hannula, J., Rudolph, M., et al.: Study of process water recirculation in a flotation plant by means of process simulation. Miner. Eng. **148**, 106181 (2020)
43. Futáš, P., Pribulová, A., Fedorko, G., et al.: Failure analysis of a railway brake disc with the use of casting process simulation. Eng. Fail. Anal. **95**, 226–238 (2019)
44. Martínez, G.S., Sierla, S.A., Karhela, T.A., et al.: Automatic generation of a high-fidelity dynamic thermal-hydraulic process simulation model from a 3D plant model. IEEE Access **6**, 45217–45232 (2018)
45. Sun, L., Yu, Y., Jin, H., et al.: An optimised steelmaking-continuous casting scheduling simulation system with unity 3D. Int. J. Simul. Process Model. **15**(3), 213–224 (2020)
46. Liao, J., Lin, C.: Optimization and simulation of job-shop supply chain scheduling in manufacturing enterprises based on particle swarm optimization. Int. J. Simul. Model. **18**(1), 187–196 (2019)
47. Viharos, A.B., Németh, I.: Simulation and scheduling of AGV based robotic assembly systems. IFAC-PapersOnLine **51**(11), 1415–1420 (2018)

48. López, J., Zalama, E., Gómez-García-Bermejo, J.: A simulation and control framework for AGV based transport systems. Simul. Model. Pract. Theory **116**, 102430 (2022)
49. Zhao, C.Y., Li, J.L., Ren, J.J., et al.: Ergonomic simulation and evaluation of upright program-controlled exercise bike based on CATIA. Mech. Des. **4**, 140–144 (2019)
50. Zhou, A., Zhang, J.M., Yang, Q., et al.: Simulation analysis of ergonomics of industrial truck cab based on JACK. Mech. Des. **37**(1), 26–34 (2020)
51. Lu, N., Zhu, D.X., Li, F.Y.: Ergonomic design and simulation analysis of intelligent health monitoring integrated machine for the elderly. Mech. Des. **37**(10), 128–133 (2020)
52. Wang, W., Gao, X.S., Mu, Z.G., et al.: Application of ergonomic simulation technology in aircraft component assembly. Aviat. Manuf. Technol. **60**(9), 92–96 (2017)
53. Xie, R., Wang, Z., Yu, F.R., Huang, T., Liu, Y.: A novel identity resolution system design based on Dual-Chord algorithm for industrial Internet of Things. Sci. China Inf. Sci. **64**(8), 1–14 (2021). https://doi.org/10.1007/s11432-020-3016-x
54. Liu, Y., Chi, C., Zhang, Y., et al.: Identification and resolution for industrial internet: architecture and key technology. IEEE Internet Things J. **9**, 16780–16794 (2022)
55. Wang, Z., Ye, T., Xiong, A.: Research of food traceability technology based on the Internet of Things name service. In: 2016 IEEE International Conference on Internet of Things (iThings) and IEEE Green Computing and Communications (GreenCom) and IEEE Cyber, Physical and Social Computing (CPSCom) and IEEE Smart Data (SmartData), pp. 100–106. IEEE (2016)
56. Wang, H., Sun, Z.: Research on multi decision making security performance of IoT identity resolution server based on AHP. Math. Biosci. Eng. **18**(4), 3977–3992 (2021)
57. Bamakan, S.M.H., Nezhadsistani, N., Bodaghi, O., et al.: Patents and intellectual property assets as non-fungible tokens; key technologies and challenges. Sci. Rep. **12**(1), 1–13 (2022)
58. Truby, J., Brown, R.D., Dahdal, A., et al.: Blockchain, climate damage, and death: policy interventions to reduce the carbon emissions, mortality, and net-zero implications of non-fungible tokens and Bitcoin. Energy Res. Soc. Sci. **88**, 102499 (2022)
59. Karandikar, N., Chakravorty, A., Rong, C.: Blockchain based transaction system with fungible and non-fungible tokens for a community-based energy infrastructure. Sensors **21**(11), 3822 (2021)
60. Arcenegui, J., Arjona, R., Román, R., et al.: Secure combination of IoT and blockchain by physically binding IoT devices to smart non-fungible tokens using PUFs. Sensors **21**(9), 3119 (2021)
61. Chohan, U.W.: Non-fungible tokens: blockchains, scarcity, and value. Critical Blockchain Research Initiative (CBRI) Working Papers (2021)
62. Pinto-Gutiérrez, C., Gaitán, S., Jaramillo, D., et al.: The NFT hype: what draws attention to non-fungible tokens? Mathematics **10**(3), 335 (2022)
63. Abidi, M.H., Alkhalefah, H., Umer, U., et al.: Blockchain-based secure information sharing for supply chain management: optimization assisted data sanitization process. Int. J. Intell. Syst. **36**(1), 260–290 (2021)
64. Shahbazi, Z., Byun, Y.C.: Improving transactional data system based on an edge computing–blockchain–machine learning integrated framework. Processes **9**(1), 92 (2021)
65. Hariharasudan, A., Otola, I., Bilan, Y.: Reactive power optimization and price management in microgrid enabled with blockchain. Energies **13**(23), 6179 (2020)
66. Hasan, M.K., Akhtaruzzaman, M., Kabir, S.R., et al.: Evolution of industry and blockchain era: monitoring price hike and corruption using BIoT for smart government and Industry 4.0. IEEE Trans. Ind. Inform. **18**, 9153–9161 (2022)
67. Chen, Y., Lu, Y., Bulysheva, L., et al.: Applications of blockchain in Industry 4.0: a review. Inf. Syst. Front., 1–15 (2022). https://doi.org/10.1007/s10796-022-10248-7
68. Fei, C., Jiang, B., Xu, K., et al.: An intelligent load control-based random access scheme for space-based Internet of Things. Sensors **21**(4), 1040 (2021)

69. Fei, C., Zhao, B., Yu, W., et al.: Towards efficient data collection in space-based Internet of Things. Sensors **19**(24), 5523 (2019)
70. Liu, Z., Jiao, J., Wu, S., et al.: Non-orthogonal superimposed pilot grant-free random access scheme in satellite-based IoT. In: 2022 IEEE Wireless Communications and Networking Conference (WCNC), pp. 1407–1412. IEEE (2022)
71. Jiao, J., Sun, Y., Wu, S., et al.: Network utility maximization resource allocation for NOMA in satellite-based Internet of Things. IEEE Internet Things J. **7**(4), 3230–3242 (2020)
72. Jiao, J., Wu, S., Lu, R., et al.: Massive access in space-based Internet of Things: challenges, opportunities, and future directions. IEEE Wirel. Commun. **28**(5), 118–125 (2021)
73. Rodriguez, I., Mogensen, R.S., Fink, A., et al.: An experimental framework for 5G wireless system integration into Industry 4.0 applications. Energies **14**(15), 4444 (2021)
74. Sasiain, J., Sanz, A., Astorga, J., et al.: Towards flexible integration of 5G and IIoT technologies in Industry 4.0: a practical use case. Appl. Sci. **10**(21), 7670 (2020)
75. Khatib, E.J., Barco, R.: Optimization of 5G networks for smart logistics. Energies **14**(6), 1758 (2021)
76. Priya, B., Malhotra, J.: 5GAuNetS: an autonomous 5G network selection framework for Industry 4.0. Soft. Comput. **24**(13), 9507–9523 (2019). https://doi.org/10.1007/s00500-019-04460-y
77. Li, S., Da Xu, L., Zhao, S.: 5G Internet of Things: a survey. J. Ind. Inf. Integr. **10**, 1–9 (2018)
78. Wang, N., Wang, P., Alipour-Fanid, A., et al.: Physical-layer security of 5G wireless networks for IoT: challenges and opportunities. IEEE Internet Things J. **6**(5), 8169–8181 (2019)
79. Chettri, L., Bera, R.: A comprehensive survey on Internet of Things (IoT) toward 5G wireless systems. IEEE Internet Things J. **7**(1), 16–32 (2019)
80. Xu, L., Collier, R., O'Hare, G.M.P.: A survey of clustering techniques in WSNs and consideration of the challenges of applying such to 5G IoT scenarios. IEEE Internet Things J. **4**(5), 1229–1249 (2017)
81. Mahmood, N.H., Alves, H., López, O.A., et al.: Six key features of machine type communication in 6G. In: 2020 2nd 6G Wireless Summit (6G SUMMIT), pp. 1–5. IEEE (2020)
82. Rappaport, T.S., Xing, Y., Kanhere, O., et al.: Wireless communications and applications above 100 GHz: opportunities and challenges for 6G and beyond. IEEE Access **7**, 78729–78757 (2019)
83. Letaief, K.B., Chen, W., Shi, Y., et al.: The roadmap to 6G: AI empowered wireless networks. IEEE Commun. Mag. **57**(8), 84–90 (2019)
84. Basar, E.: Reconfigurable intelligent surface-based index modulation: a new beyond MIMO paradigm for 6G. IEEE Trans. Commun. **68**(5), 3187–3196 (2020)
85. Dang, S., Amin, O., Shihada, B., et al.: What should 6G be? Nat. Electron. **3**(1), 20–29 (2020)
86. Tataria, H., Shafi, M., Molisch, A.F., et al.: 6G wireless systems: vision, requirements, challenges, insights, and opportunities. Proc. IEEE **109**(7), 1166–1199 (2021)
87. Jing, T., Tian, X., Hu, H., et al.: Cloud-Edge collaboration framework with deep learning-based for remaining useful life prediction of machinery. IEEE Trans. Ind. Inform. (2021)
88. Yang, C., Wang, Y., Lan, S., et al.: Cloud-edge-device collaboration mechanisms of deep learning models for smart robots in mass personalization. Robot. Comput. Integr. Manuf. **77**, 102351 (2022)
89. Jian, C., Ping, J., Zhang, M.: A cloud edge-based two-level hybrid scheduling learning model in cloud manufacturing. Int. J. Prod. Res. **59**(16), 4836–4850 (2021)
90. Laili, Y., Guo, F., Ren, L., et al.: Parallel scheduling of large-scale tasks for industrial cloud-edge collaboration. IEEE Internet Things J. (2021)

Analysis and Suppression for Shaft Torsional Vibrations in Wind Energy Conversion System with MPPT Control

Hongfei Zhang[1](✉), Yue Xia[1], Xu Liu[2], Songhuai Du[1], Juan Su[1], and Huapeng Sun[1]

[1] China Agricultural University, Haidian District, Beijing 100083, China
xiayuexiayue@163.com

[2] Technische Universität Berlin, Straße Des 17, 10623 Berlin, Germany

Abstract. An improved maximum power point tracking (MPPT) control algorithm is proposed based on a doubly-fed induction generator wind energy conversion system. The control objective below rated wind speed is to maximize the extracted energy from the wind while reducing mechanical loads. The existing MPPT control method do rarely not take into account the influence of control method and control parameters on the mechanical parts of wind turbines such as transmission shafts. In order to bring some improvement the applicability of the control method, an improved method of MPPT control is proposed, which superimposes an additional torsional vibration suppression command on the basis of the original method, and suppresses the torsional vibration of the drive shaft of the wind power generation system while ensuring the maximum power tracking. Test studies substantiate the claims made and demonstrate the application of the methodology.

Keywords: Maximum power point tracking · Drive shaft · Torsion vibration

1 Introduction

Wind power systems are widely used in power systems as a typical representative of renewable energy generation technologies. According to the annual report data of Global Wind Energy Council (GWEC), the total installed capacity of wind power worldwide is up to 837GW by 2021 [1]. With the increasing size of wind turbines, the flexibility of the drive shaft keeps increasing [2]. In the actual operation process, the drive shaft vibrates for a long time, which not only increases the fatigue load of the wind turbine [3], but also endangers the stability of the power grid in serious cases [4].

The wind power generation system involves various physical systems such as aerodynamic subsystems, mechanical transmission subsystems, and electrical subsystems. Different subsystems are coupled with each other. At the same time, the system stability and performance are affected by the controller and its parameters, and the dynamic characteristics of the system are very complex [4, 5]. In the control system of wind power generation system, maximum power point tracking (MPPT) control plays an important role. Its use can effectively improve the conversion and utilization efficiency of wind

W. Fan et al. (Eds.): AsiaSim 2022, CCIS 1712, pp. 264–272, 2022.
https://doi.org/10.1007/978-981-19-9198-1_20

energy in wind power systems, increase the annual power generation capacity of the units, and reduce the cost of wind power generation [5]. Maximum power point tracking (MPPT) algorithms have speed control, power control [6]. Due to wind turbine characteristics, small changes in generator speed can lead to large changes in the torque applied to the mechanical drive between the wind turbine and the generator [7]. However, most wind turbine manufacturers usually give little consideration to the impact of the control strategy and control parameters on the mechanical parts of the wind turbine (e.g., the drive shaft) when selecting the control strategy for the wind turbine, which leads to an increase in the fatigue load of the turbine, greatly reducing the service life of the turbine and even making the control less effective. Therefore, under the premise of maximum power tracking, it is important to analyze the mechanism of control algorithm and control parameters on the mechanical parts, and study a set of wind turbine control methods that take into account the mechanical losses of the turbine.

The rest of this paper is structured as follows. In the second part, the mathematical model of the doubly-fed wind power system under consideration of drive shaft flexibility is described. In the third part, the conventional speed control method and the improved speed control method. In the fourth part, the correctness of the control method is verified in Matlab/Simulink. In the fifth part, the work done in this paper is summarized.

2 Wind Turbine Modelling

2.1 Aerodynamic Modelling

The capture of wind energy by the wind turbine is reflected as the process in which the wind acts on the blades to form mechanical torque. According to Betz's proof, the power absorbed by a wind turbine from wind energy can be expressed as [8]:

$$P_a = \frac{1}{2}\rho\pi R^2 C_p(\lambda, \beta)v^3 \tag{1}$$

With the tip speed ratio

$$\lambda = \frac{\omega_r R}{v} \tag{2}$$

where ρ is the air density, β is the blades pitch angle, R is the blade length, V is the wind velocity, ω_t is the rotating speed of the wind turbine, and C_p is the power coefficient. The $C_p(\lambda, \beta)$ curve is displayed in Fig. 1.

When the system works below the rated wind speed, the main control objective is to make the system capture as much wind energy as possible under the premise of stable operation.

In order to achieve this objective, the system generally keeps β unchanged. When the wind speed changes, the electromagnetic torque is adjusted to adjust the rotor speed, so that the rotor speed is to track the optimal tip speed ratio that ensure the best power efficiency.

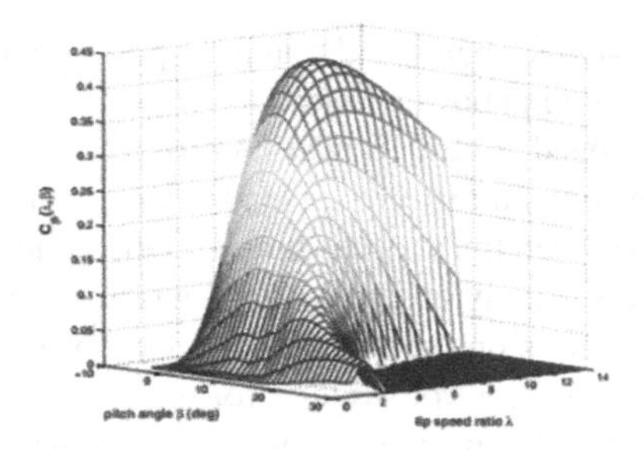

Fig. 1. Power coefficient as a function of blade tip speed ratio and propeller pitch angle

2.2 Drive Shaft Mechanical Modelling

In order to accurately reflect the influence of the flexibility of the transmission system on the dynamic characteristics of the wind power generation system, in the transient analysis of the power system, the two-mass model is generally used to describe the transmission system.

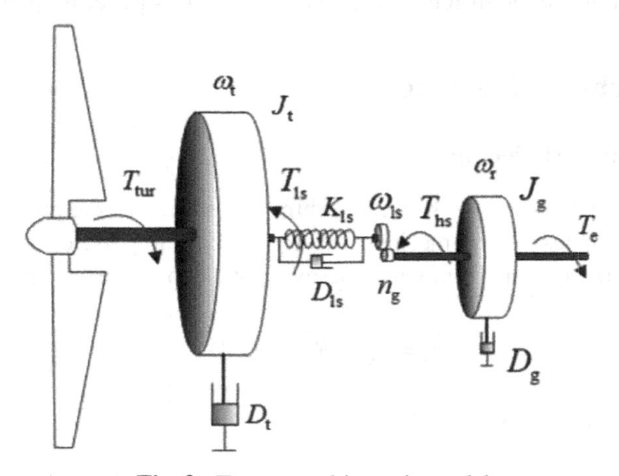

Fig. 2. Two-mass drive train model

Assuming a two-mass drive train model with a flexible low-speed shaft shown in Fig. 2, the turbine inertia J_t is driven at a speed ω_t by the aerodynamic torque T_{tur} and the generator inertia J_g is driven by the high-speed shaft torque T_{hs} and decelerated by the electromagnetic torque T_e.

Dynamics of the two-mass drive train model are described by [6]:

$$J_t \frac{d\omega_t}{dt} = T_{tur} - T_{ls} - D_t \omega_t \tag{3}$$

$$J_g \frac{d\omega_r}{dt} = T_{hs} - T_e - D_g \omega_r \tag{4}$$

$$T_{ls} = K_{ls}(\theta_t - \theta_{ls}) + D_{ls}(\omega_t - \omega_{ls}) \tag{5}$$

With

$$\omega_t = \frac{d\theta_t}{dt}, \omega_{ls} = \frac{d\theta_{ls}}{dt} \tag{6}$$

where ω_t is the rotating speed of the wind turbine, ω_r is the mechanical angular velocity of the generator, ω_{ls} is the low-speed shaft speed, D_t is the turbine damping coefficient, D_{ls} is the shaft damping coefficient, D_g is the generator damping coefficient, θ_t is the turbine-side rotor position, θ_{ls} is the gearbox-side low-speed shaft position, T_{ls} is the low-speed shaft torque. The aerodynamic torque T_{tur} of the wind turbine is given by [8, 9]:

$$T_{tur} = \frac{1}{2}\pi \rho r^3 V_w^2 C_T(\lambda) \tag{7}$$

where $C_T(\lambda)$ is the torque coefficient. Both coefficients are related by [10]:

$$C_T(\lambda) = \frac{C_P(\lambda)}{\lambda} \tag{8}$$

Due to the gearbox, the generator experiences a torque that is reduced ng times and a speed that is increased ng times. Then, the following relation holds [10]:

$$n_g = \frac{T_{ls}}{T_{hs}} = \frac{\omega_r}{\omega_{ls}} \tag{9}$$

3 Influence of Traditional Control Method on Torsional Vibration of Drive Shaft

When the system works below the rated wind speed, the traditional control method only considers the tracking of the maximum output power, but does not consider the torsional vibration suppression of the drive shaft.

This section will introduce two commonly used MPPT control methods and analyze their influence on the torsional vibration of the drive shaft.

3.1 Speed Control

The speed of the generator is controlled according to the wind speed, so that the two are in line with the relationship between the optimal tip speed ratio.The system takes the error between the measured ω_e and the reference command ω_{e_ref} as the input of the speed controller, and the output of the speed controller is the electromagnetic torque command $T_{e\_ref.}$ The reference ω_{t_ref} is calculated by the measured wind speed for the given optimal tip speed λ_{opt} ratiousing (11) [8]. A block diagram interpretation of the speed control is depicted in Fig. 3.

This control method can be applied if the optimal value of the tip speed ratio, λ_{opt}, is known.

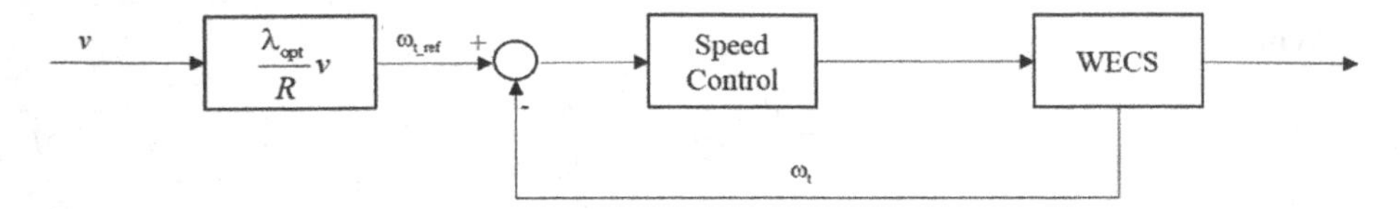

Fig. 3. A block diagram interpretation of the speed control

The turbine operates on the ORC if

$$\lambda(t) = \lambda_{opt} \tag{10}$$

Which supposes that the shaft rotational speed is closed-loop controlled such that to reach its optimal value:

$$\omega^* = \omega_{t\_ref} = \frac{\lambda_{opt}}{R} v \tag{11}$$

3.2 Improved Speed Control

The design idea of this method is as follows: by detecting the electromagnetic torque T_e of the wind turbine, if T_e changes slowly, it means that the mechanical stress of the system is within a reasonable range at this time, and no changes are needed. If the T_e change distance is sharp, it means that the mechanical stress of the system is large at this time, and the mechanical loss is large, so the change of T_e needs to be gentle. The smooth change of T_e can be achieved by making the change of w_m smooth.

The specific implementation process is as follows: Detect the electromagnetic torque T_e of the wind turbine and the real-time wind speed v, and calculate the generator speed reference value ω^* through the formula (11); judge the change speed of the electromagnetic torque T_e, if the change is slow, then directly output the speed Reference value ω^*; otherwise, a lag compensation device is introduced to output the lag-compensated speed reference value ω_{lc}^*, but if the signal of ω_{lc}^* is directly output, chopping will occur, that is, the final feedback to the system will occur. The speed reference signal will jump, causing the system to become unstable. However, through the use of formula (12), a smooth transition of switching between the two modes can be effectively achieved, thereby avoiding system oscillation.

$$\omega^*(n+1) = \alpha\omega_{lc}^* + (1-\alpha)\omega^*(n) \tag{12}$$

When the electromagnetic torque changes gently, the system adopts the traditional speed control for maximum power control. If the electromagnetic torque changes speed or its amplitude is too large, the system switches to the lag compensation mode, and the parameter α in formula (12) is set in 1s Gradient from 0 to 1, that is, to achieve a smooth transition of the system within 1 s. Then the system continues to run in the lag compensation mode for 3 s, and detects the change speed of the electromagnetic torque of the system again. If the change speed of the electromagnetic torque of the system is still very large, the output speed reference value ω_{lc}^{*} is maintained in the lag compensation mode, otherwise, it switches back to the traditional speed The control mode outputs the speed reference ω^{*}.

4 Case Studies

In order to verify the influence of the proposed improved MPPT control method on the shafting torsional vibration suppression, this section will compare and verify two examples. In Sect. 4.1, comparative tests are carried out under step wind speed conditions. In Sect. 4.2, the effectiveness of the method is verified under random wind speed conditions. The parameters of the wind power generation system are shown in Table 1.

Table 1. System parameters for the simulation.

Parameters	Symbol	Value
Rated power	P_{tur}	1.5 MW
Rated wind speed	V_{w}	12 m/s
Optimal tip speed ratio	λ_{opt}	7
Maximum power coefficient	C_{P_opt}	0.374
Gear box ratio	n	72
Inertia costant of turbine	H_{t}	4.32 s
Inertia costant of generator	H_{g}	0.685 s

4.1 Step Test

In the initial stage, the wind power generation system is in a stable working state, and the wind speed is 9 m/s. At t = 5 s, the wind speed will be 8 m/s. It can be seen from Fig. 5 that, compared with the traditional speed control method, the improved algorithm proposed in this paper produces a smaller change in the electromagnetic torque generated by the wind power generation system under the step wind speed. The effectiveness of the improved algorithm constructed in this paper is further verified (Fig. 4).

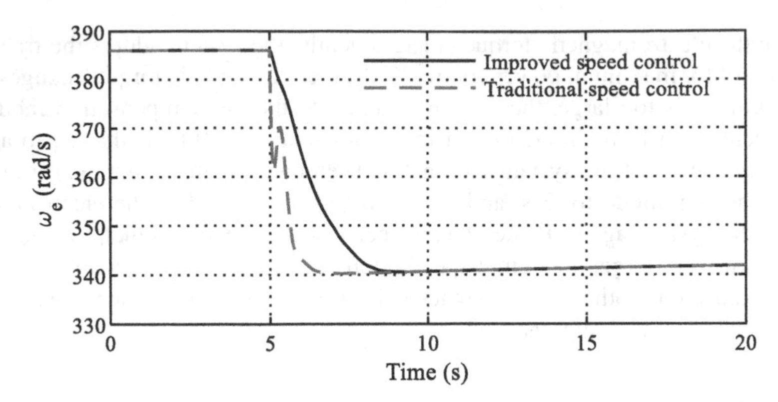

Fig. 4. Rotational speed step response

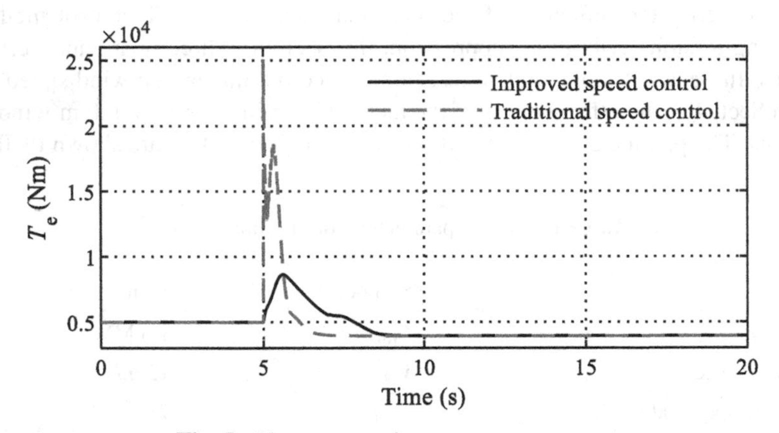

Fig. 5. Electromagnetic torque step response

4.2 Random Wind Speed Test

This section further analyzes the simulation results of the model under variable wind speed conditions. The wind speed curve is shown in Fig. 6. The average wind speed is 9 m/s. The wind speed curve is obtained from the IEC standard von Karman spectrum with a disturbance strength of 0.12. The specific modeling method of the wind speed model can be found in the literature [7].

It can be seen from Fig. 7 and Fig. 8 that under random wind conditions, applying the improved algorithm proposed in this paper, the output speed and electromagnetic torque of the wind power generation system change more smoothly. That is, under this algorithm, the driveshaft torque variation of the system is lower.

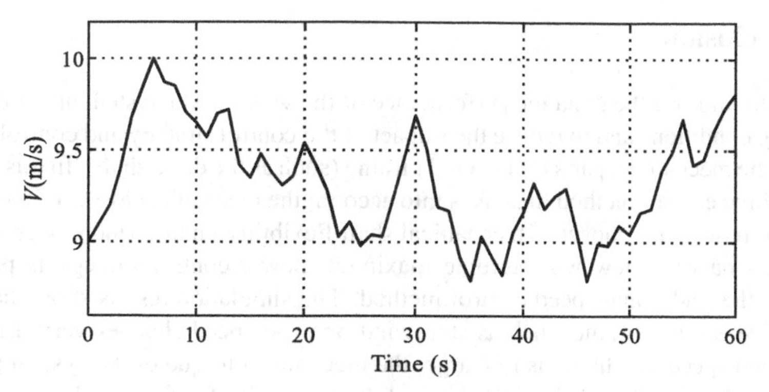

Fig. 6. Wind speed curve

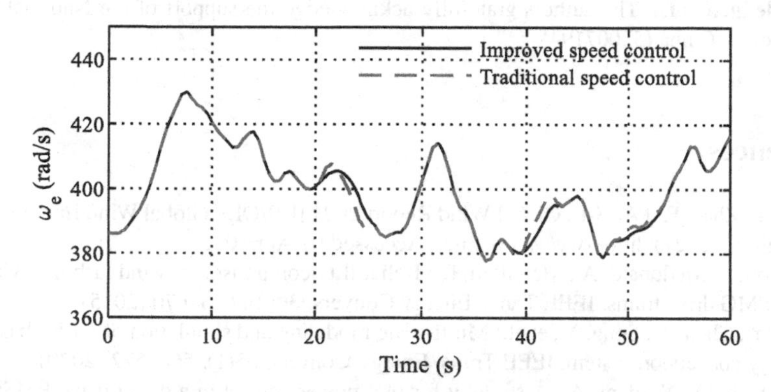

Fig. 7. Electrical angular velocity comparison

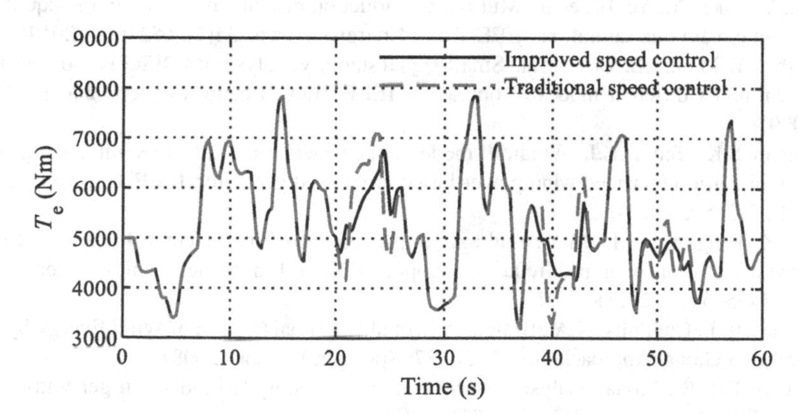

Fig. 8. Electromagnetic torque comparison

5 Conclusion

In order to improve the dynamic performance of the wind power system under different operating conditions and to reduce the impact of the control strategy and control parameters on the mechanical parts of the wind turbine (such as the drive shaft). In this paper, a wind turbine control method that takes into account the mechanical losses of the turbine is used to reduce the impact of mechanical shaft flexibility on the wind power system.

In this paper, a new wind turbine maximum power control strategy is proposed based on the traditional speed control method. The simulation results show that when the wind speed is extreme, such as step wind or wind speed changes very drastically or the wind speed amplitude is too large, the mechanical torque of the system changes more smoothly, i.e. the mechanical stress of the system is effectively reduced.

Acknowledgements. The authors gratefully acknowledge the support of the National Science Foundation of China (52007194).

References

1. Lee, J., Zhao, F.: PACS-L: Global Wind Report 2021 [EB/OL]. Global Wind Energy Council, Belgium: (2021). http://www.Gwec.net. Accessed 01 Apr 2021
2. Carroll, J., Mcdonald, A., Mcmillan, D.: Reliability comparison of wind turbines with DFIG and PMG drive trains. IEEE Trans. Energy Convers. **30**(2), 663–670 (2015)
3. Xia, Y., Chen, Y., Song, Y., et al.: Multi-scale modeling and simulation of DFIG-based wind energy conversion system. IEEE Trans. Energy Convers. **35**(1), 560–572 (2020)
4. Alizadeh, O., Yazdani, A.: A strategy for real power control in a direct-drive PMSG-based wind energy conversion system. IEEE Trans. Power Deliv. **28**(3), 1297–1305 (2013)
5. Xia, Y., Chen, Y., Ye, H., et al.: Multi-scale induction machine modeling in the dq0 domain including main flux saturation. IEEE Trans. Energy Convers. **34**(2), 652–664 (2019)
6. Mishra, Y., Mishra, S., Li, F., et al.: Small-signal stability analysis of a DFIG-based wind power system under different modes of operation. IEEE Trans. Energy Convers. **24**(4), 972–982 (2009)
7. Salman, S.K., Teo, A.L.J.: Windmill modeling consideration and factors influencing the stability of a grid-connected wind power-based embedded generator. IEEE Trans. Power Syst. **18**(2), 793–802 (2003)
8. Xie, Z., Feng, Y.T., Ma, M.Y., et al.: An improved virtual inertia control strategy of DFIG-based wind turbines for grid frequency support. IEEE J. Emerg. Sel. Topics Power Electron. **9**(5), 5465–5477 (2021)
9. Munteanu, I., Cutululis, N.A., Bratcu, A.I., et al.: Optimal Control of Wind Energy Systems: Towards a Global Approach, pp. 30–36, 77. Springer, London (2008)
10. Mei, F., Pal, B.: Modal analysis of grid-connected doubly fed induction generators. IEEE Trans. Energy Convers. **22**(3), 728–736 (2007)

Networked Modeling and Simulation

Design and Implementation of Gigabit Ethernet Traffic Integer Module Based on ZYNQ

Kang Lei[1], Mei Haihong[1(✉)], Li Weihao[2], and Ren Xuchao[1]

[1] School of Computer Science, Xi'an Shiyou University, Xi'an 710065, China
1770695421@qq.com

[2] School of Communication and Information Engineering, Xi'an University of Posts and Telecommunications, Xi'an 710100, China

Abstract. To solve the problem that a large number of services such as video, voice and multimedia need to be transmitted on the network, and these services are extremely sensitive to bandwidth, jitter and delay, the traffic integer is proposed. The traffic integer is used to classify different types of data, cache the data, and periodically send the data in the cache. This ensures that the high-speed data can be stably output on the link to avoid data loss, and ensures data continuity without data mutation. It adopts an open-loop control mode, according to different types of data system will be given the corresponding input, there is no interaction between input and output, can improve the utilization rate of bandwidth, reduce transmission delay, improve processing speed. Zynq7000 is used as the main control chip to realize data receiving and data processing on a single chip, which can improve the stability of work. Compared with the implementation of traffic integer in software, implementation in hardware has a faster processing speed.

Keywords: Gigabit ethernet · ZYNQ · Traffic integer

1 Introduction

With the continuous development of the Internet, the development of data storage and transmission technology is driven, and at the same time, there are higher requirements for real-time, reliability, speed and other aspects of data transmission [1]. With the continuous progress of aerospace scientific research, aircraft structures and functions have become increasingly complex, and their reliability and safety need to be further improved [2]. With the development of semiconductor and integrated circuit technology, data acquisition and storage technology has made great breakthroughs in acquisition and storage speed, data storage capacity, reliability and integration, improving the reliability and safety of aircraft [3].

In the aerospace environment, a large amount of data needs high-speed network transmission. Gigabit Ethernet technology, as a new generation of high-speed Ethernet technology, can provide 1 Gbps communication bandwidth, bringing users an effective solution to improve the core network [4]. In this paper, the MAC controller is used as an independent IP core to connect to the PHY chip, which has better portability and

W. Fan et al. (Eds.): AsiaSim 2022, CCIS 1712, pp. 275–289, 2022.
https://doi.org/10.1007/978-981-19-9198-1_21

controllability compared with the PHY layer and MAC layer integrated into one chip [5]. Zynq-7000 chip from Xilinx Company is used as the core device. The chip integrates ARM processing system (PS) and FPGA programmable logic (PL) effectively. The two are connected through AXI bus to realize the high bandwidth communication between PL and PS. PHY chip combined with Tri_Mode_Ethernet MAC soft core can achieve gigabit transmission rate [6].

The algorithms for traffic integer mainly include leaky bucket algorithm and token bucket algorithm, TCR algorithm proposed by Shrikrishna et al. and PostACK proposed by Huan Yun Wei, Ying Dar Lin et al. [7]. Around the token bucket integer, many integer methods have been proposed to control the burst volume and bandwidth of the output traffic [8]. Greenberg and Roxford combined token bucket integer and scheduling mechanism to realize fair token bucket integer and so on. The token bucket algorithm is widely used, and the kernel integer module adopts the token bucket algorithm [9]. Now, many foreign institutions and organizations have done a lot of research on packet scheduling and traffic integer. Some manufacturers also integrate traffic integer, monitoring and packet scheduling modules to form their own products, such as Packeteer. However, there are few such products in China, and the research in this area needs to be strengthened, so the research in this area is of great significance.

The design idea of this paper is that the traffic integer module can receive different types of data, classify and process each type of data, and also set up a certain buffer area, plus the token algorithm to complete the integer function to reduce data loss. Possibility. Sending the received data periodically and periodically is to ensure the continuity and security of the data and avoid data loss caused by data congestion. The main contribution is to use the set buffer, token bucket algorithm and round-robin scheduling to complete the function of traffic integer. This function is completed on the ZYNQ development board, which realizes hardware acceleration so that data can be transmitted at high speed.

2 Traffic Integer and Related Technologies

2.1 Overview of Traffic Integer

The traffic integer is a measure to proactively adjust the output rate of the traffic and control the rate of the output data so that the data can be sent out at a uniform rate [10]. The traffic integer is used to match the transmission rate of downstream devices. When data is transferred from a high-speed link to a low-speed link, if burst traffic occurs, the bandwidth bottleneck at the egress of the low-speed link will cause serious data loss. In this case, the traffic integer is required. The traffic integer is usually implemented using the buffer and token bucket algorithm. If the data is sent too fast, the data is cached in the buffer first. Under the control of the token bucket, the data of the buffer can be evenly sent out, so that the speed limit of different types of data can be realized separately. Buffering improves bandwidth utilization and reduces data retransmission.

2.2 Related Algorithms

2.2.1 Leaky Bucket Algorithm

The principle of the leaky bucket algorithm is that the request enters the leaky bucket first, which is irregular data flow with unrestricted flow, and then the leaky bucket processes the request at a constant rate. Out of the bucket is the regular data flow after flow limiting. When the water above is added too quickly, it can cause overflow, which means denial of the request. The specific principle is shown in Fig. 1. A leaky bucket algorithm can change irregular data flow into regular data flow [11], process request smoothing, and restrict data transmission rate forcibly.

However, the leaky bucket algorithm has obvious shortcomings, which can not accurately predict the network bandwidth. Suppose the transmission rate is less than the real network bandwidth. In this case, if some burst traffic is generated on the link, the real network bandwidth allows the transmission to be completed more quickly. Since the leaky bucket algorithm is still transmitted at a constant transmission rate, it cannot make full use of network resources when burst traffic occurs, resulting in transmission delay [12]. To solve this problem, token bucket algorithm is proposed.

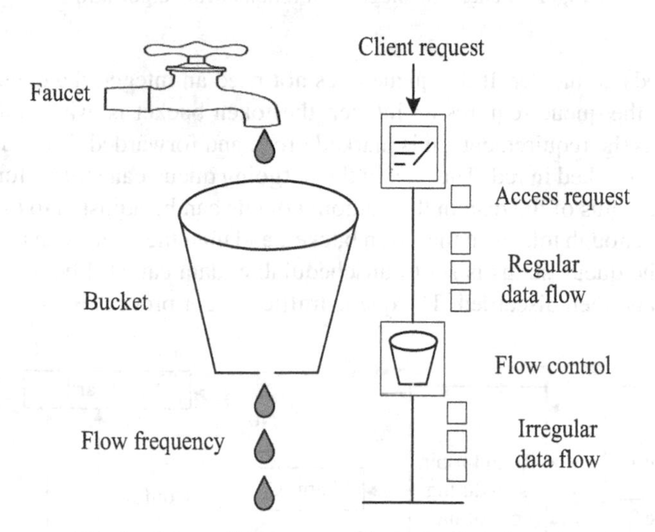

Fig. 1. Schematic diagram of leaky bucket algorithm

2.2.2 Token Bucket Algorithm

The principle of token bucket algorithm is that the system adds tokens to the bucket in a unit time. If there is a request to be processed, the request will be processed only after the token is obtained from the token bucket [13]. Reject the request if there is no redeeming token in the bucket. The specific principle is shown in Fig. 2.

The token bucket algorithm can solve the problem of burst data flow by allowing burst data flow as long as there are tokens in the token bucket until the maximum limit is reached [14].

2.2.3 Queue Traffic Integer

The queue traffic integer refers to the traffic integer for each queue of the outgoing interface [15]. When a queue is dispatched after polling, you need to check whether

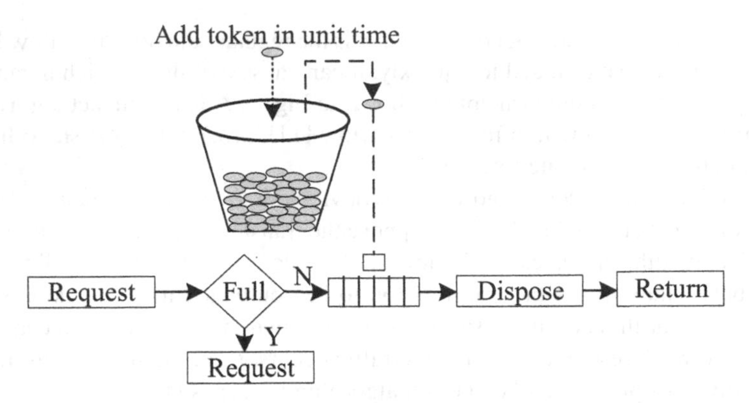

Fig. 2. Schematic diagram of token bucket algorithm

the queue needs an integer. If the queue does not need an integer, the queue is directly forwarded. If the queue requires an integer, the token bucket is evaluated. If the data flow rate meets the requirements, it is marked green and forwarded. If the data flow rate is too fast, it is marked in red. The data in the outgoing queue can still be forwarded out, and the queue status of the data in the outgoing queue can be adjusted to unschedulable until there are enough tokens in the token bucket, and then the queue can be rescheduled [16]. When the queue status is set to unschedulable, data can still be queued until the queue is full and then discarded. The queue traffic integer process is shown in Fig. 3.

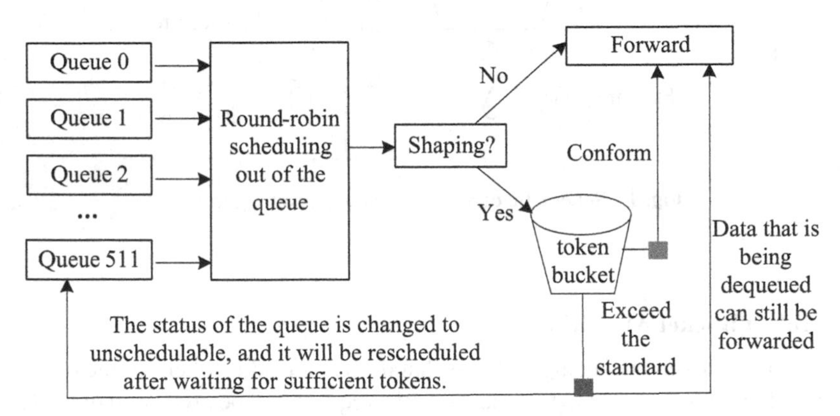

Fig. 3. Queue traffic integer process diagram

3 Control System

The control system can be classified into the closed-loop control system and the open-loop control system according to the structural characteristics of the system [17].

3.1 Closed-Loop Control System

The closed-loop control system is equivalent to the feedback control system, and its principle is shown in Fig. 4. The controller and the control object can be connected in both positive and reverse directions. The output of the control object is fed back to the input and compared with the input, and the comparison result is the deviation signal [18]. The deviation signal applied to the controller makes the input more approximate to the expected value.

The closed-loop control system has negative feedback regulation, which can adjust the output automatically. It has a certain anti-interference ability to the disturbance outside the system. However, the closed-loop operation of the system may produce some instability, so the problem of poor stability exists [19]. In addition, the interaction between the input and output ends of the closed-loop control system will bring certain transmission delays, occupy bandwidth resources and reduce bandwidth utilization.

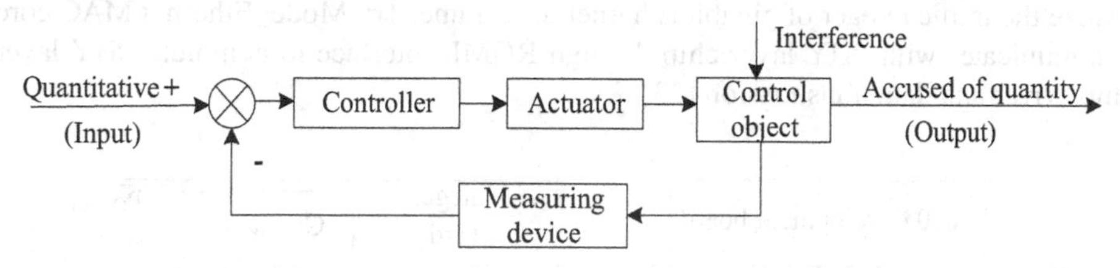

Fig. 4. Schematic diagram of closed-loop control system

3.2 Open-Loop Control System

The principle of the open-loop control system is shown in Fig. 5. There is only a positive connection between the controller and the control object, and the system does not need to measure the output or compare the output with the input [20]. The input of the system is calculated based on the estimated output.

There is no feedback link in the open-loop control system and its structure is simple. Open-loop operation has high stability [21]. The most important is that there is no need for input and output to interact, which reduces transmission delay and improves bandwidth utilization.

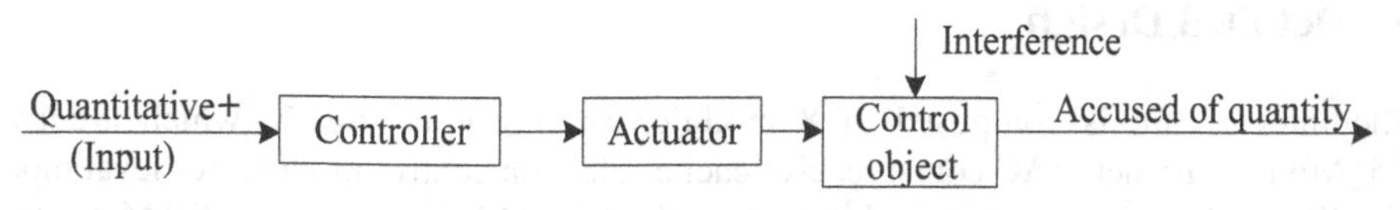

Fig. 5. Schematic diagram of open-loop control system

The open-loop control system is used in this paper. The token bucket algorithm is a typical open-loop control, which only needs quantitative input according to the system and does not need feedback from the output end. Of course, the input needs to transmit different types of data. For different types of data, the system can adjust the amount of data to make the output closer to the expected value.

4 Overall Design

The overall architecture of the traffic integer is shown in Fig. 6. The P2020 development board on the left is the dual-core communication processor of QorIQ P2 platform under PowerPC architecture [22]. The ZYNQ development board on the right is collectively referred to as the integer card. The optical fiber data is packaged and converted into Ethernet frames through the P2020 development board, and sent to the integer card through the Ethernet interface. The P2020 development board communicates with the PS side processor of ZYNQ device through SPI interface, and then it can configure the relevant registers in the logic of the GIGABit traffic integer card of PL side, and realize the traffic integer of gigabit Ethernet data frame. Tri_Mode_Ethernet MAC core communicates with PHY layer chip through RGMII interface to complete PHY layer chip driver and data transmission [23].

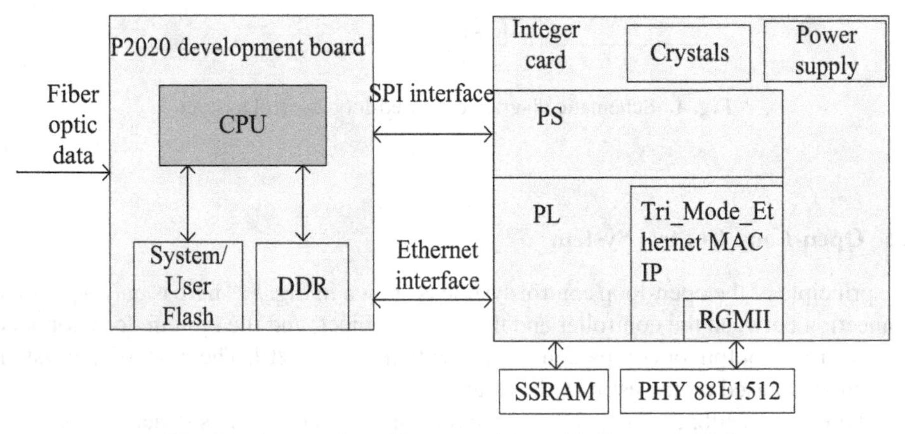

Fig. 6. The overall architecture of the traffic integer module

5 Detailed Design

The integer card is composed of 8 modules as shown in Fig. 7, which are two Tri_Mode_Ethernet MAC cores, receive cache selection control module, retrieval module, SSRAM_Ctrl module, scheduling module, SSRAM_if module (SSRAM control interface module), and send cache module respectively. The integer card is externally connected to the SSRAM chip. The integer card retrieves the incoming Ethernet frames and stores them in SSRAM, and then the scheduling module periodically sends out Ethernet frames.

The Tri_Mode_Ethernet MAC core enables transmission of Ethernet frames at gigabit rates. The receiving cache selection control module mainly stores Ethernet frames temporarily to determine whether the frames need to be retrieved. If the retrieved frames need to be retrieved, the module extracts the keyword IENA_ID and sends it to the retrieval module. The retrieval module retrieves the configuration unit based on IENA_ID and returns the result. If the retrieval is successful, the corresponding queue number is returned. If the retrieval fails, the frame is discarded. The SSRAM_Ctrl module is mainly responsible for sending and receiving scheduling requests, and stores data frames to the corresponding address of SSRAM according to the address allocated by the scheduling module. SSRAM_if module is responsible for data interaction between SSRAM memory chip and SSRAM_Ctrl as well as polling and arbitration access to SSRAM. The sending cache module is responsible for reading Ethernet frames and sending them to RGMII controller according to the interface timing of RGMII controller to complete data transmission.

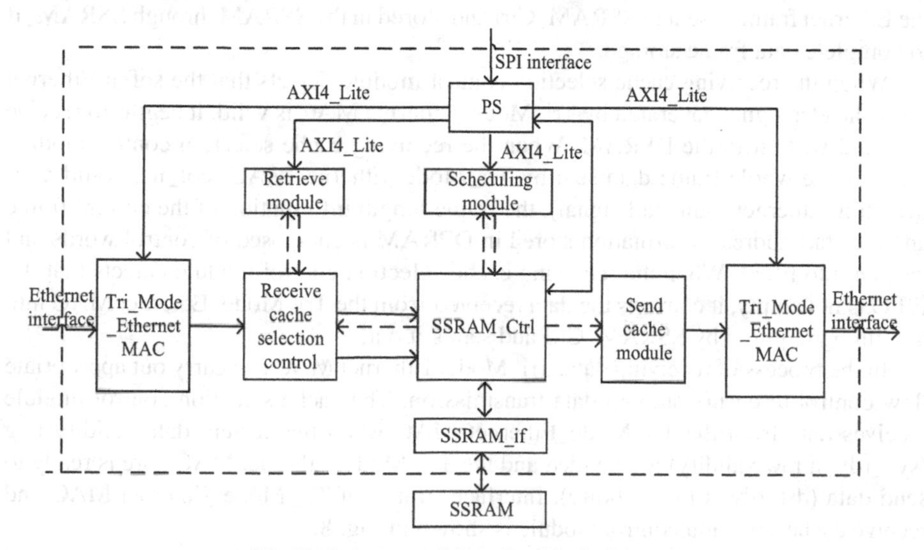

Fig. 7. Integer card function block diagram

5.1 Tri_Mode_Ethernet MAC

Tri_Mode_Ethernet MAC is the IP core provided by Vivado, which needs to be configured. PS uses the AXI4_Lite interface to initialize the MAC core. The configuration includes receiving configuration, sending configuration, flow control configuration, speed configuration, MDIO configuration, and the rate configuration is thousands of mbit/s. Ethernet frames are transmitted through Ethernet interfaces, stored temporarily in FIFO, converted into AXI_Stream and sent out.

5.2 Receive Cache Selection Control Module

The receiving cache selection control module mainly stores Ethernet frames temporarily to determine whether the frames need to be retrieved. If the retrieved frames need to be retrieved, the module extracts the keyword IENA_ID (different IENA_ID is configured according to different types of data) and sends it to the retrieval module. This includes receiving DPRAM and receiving FIFO. The DPRAM space is set as 4K space, divided into two caches, storing data frames as Ethernet frames of regrouping packets, and adding custom information traffic integer queue (IENA_ID) for identification of retrieval. Receive the control word information of each frame stored in the FIFO. When the cache data is valid, the module sends a retrieval request to the retrieval module. When the retrieval is successful, rxctrl signal will be sent to SSRAM_Ctrl module, requesting to receive the data; otherwise, the current frame will be discarded. If the protocol type of the current frame is 8'h01 or 16'h0806, the frame is directly sent to the queue with the queue number 8'hFE without retrieval. After receiving data from the RGMII controller, the Ethernet frame is sent to SSRAM_Ctrl and stored in the SSRAM through SSRAM_if to complete data frame storage.

When the receiving cache selection control module detects that the sof_n Ethernet frame header signal generated by Tri_Mode_Ethernet MAC is valid, it begins to receive data and write it to the DPRAM. When the receiving cache selection control module receives the whole frame data sent by Tri_Mode_Ethernet MAC (eof_n is valid after detecting Ethernet frame tail signal), the frame length information of the current frame and the start address information stored in DPRAM is composed of control words and written into FIFO. When the receiving cache selection control module detects that the FIFO is not empty, it converts the data received from the Tri_Mode_Ethernet MAC into the timing required by SSRAM_Ctrl and sends it out.

In the process of receiving data, Tri_Mode_Ethernet MAC can carry out appropriate flow control to ensure smooth data transmission. The cache selection control module receives data from the Tri_Mode_Ethernet MAC when the current data validity flag (syc_rdy_n low validity) is provided and the Tri_Mode_Ethernet MAC core is ready to send data (dst_rdy_n low validity). Interface timing of Tri_Mode_Ethernet MAC and receive cache selection control module is shown in Fig. 8.

Syc_rdy_n low valid indicates that the MAC of Tri_Mode_Ethernet is ready to send the data flag. Syc_rdy_n and dst_rdy_n are both valid. Data is valid and is responsible for the insert waiting period.

5.3 Retrieval Module

The retrieval module retrieves the configuration unit based on IENA_ID and returns the result. If the retrieval is successful, the corresponding queue number is returned. If the retrieval fails, the frame is discarded. The retrieval module consists of 512 32-bit configuration units, which are made up of 4-block 64x64 DPRAM. Operations include initialization, configuration, and retrieval.

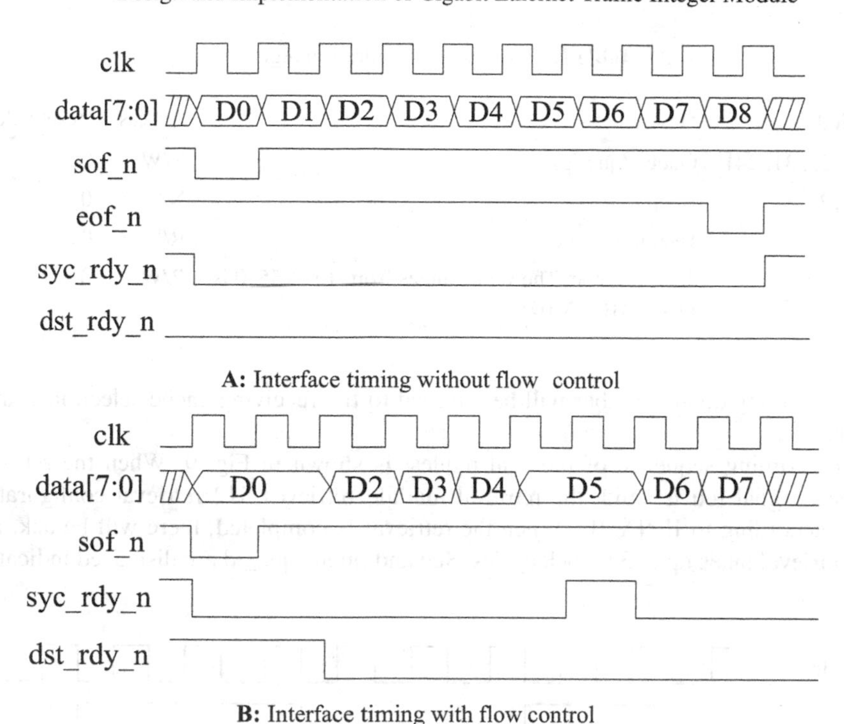

Fig. 8. Tri_Mode_Ethernet MAC core and receive cache selection control module interface timing

5.3.1 Initialization

Initialization is the initialization of the retrieval configuration space (all zeros written) within 32 cycles after the system is reset. To speed up the initialization of the 4-block 64 × 64 DPRAM, dual-port simultaneous initialization is designed, with port A and port B starting retrieval at the same time, so that 512 32-bit configuration units can be initialized in 32 clock cycles.

5.3.2 Configuration

Configuration is PS through the host interface to complete the configuration of the retrieval unit. The common configuration operation is completed before the retrieval, and the PS CPU completes the configuration of 512 configuration units. Each configuration unit is 32 bits, where bit15-bit0 indicates 16-bit IENA_ID, bit16 indicates valid flag, bit23-bit17 indicates reserved, and bit31-bit23 indicates the queue number. The retrieval configuration register definitions are shown in Table 1.

5.3.3 Retrieve

The receiving cache selection control module sends the IENA_ID of the Ethernet frame to be retrieved to the retrieval module. The retrieval module matches the IENA_ID with the retrieval table when the retrieval is enabled. If the match is successful, the

Table 1. Retrieving configuration registers

Period	Describe	Access	Reset value
{23, 22, 31: 24}	Queue (que-id)	R/W	0
21: 17	Keep	NA	0
16	Useful indicator	R/W	0
15: 0	Traffic queue. The value ranges from 1 to 255. 0 is invalid.(IENA-ID)	R/W	0

corresponding queue number will be returned to the receiving cache selection control module.

The timing sequence of retrieval request is shown in Fig. 9. When the retrieval request signal req is valid, the retrieval module retrieves 512 retrieval configuration units according to IENA_ID. When the retrieval is completed, there will be ack, and the retrieval takes up to 32 clock cycles. Suc and queue que_id are displayed indicating success.

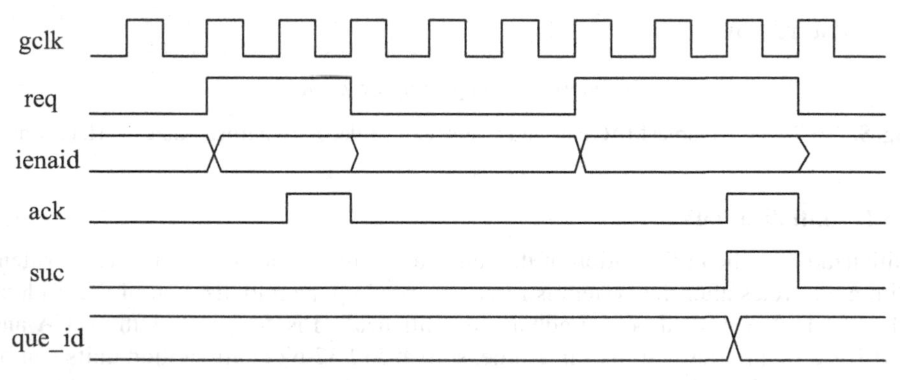

Fig. 9. Retrieval request timing

5.4 SSRAM_Ctrl Module

The SSRAM_Ctrl module is mainly responsible for sending and receiving scheduling requests, and stores data frames to the corresponding address of SSRAM according to the address allocated by the scheduling module. The module communicates with the receiving cache selection control module, the sending cache module, the SSRAM_if module and the scheduling module to complete the storage and forwarding control of receiving frames.

5.5 Scheduling Module

The scheduling module is responsible for the time integer and Ethernet frame scheduling of 512 queues (each queue stores the same type of data). PS configures the sending interval of each queue before Ethernet frame transmission. The format of the queue configuration register is shown in Table 2. Allocate a corresponding storage address for the received data frame and start queuing time, the polling scheduling arbiter decides that the frame in this queue needs to be sent, and sends a dispatching request to SSRAM_Ctrl module.

Table 2. Queue configuration registers

Period	Describe	Access	Reset value
31: 16	Keep	NA	0
15	Sampling mode is disabled. When the value is 1, the BAG queue is empty and will not request to send	R/W	1
14	Free to transmit, out of BAG's control	R/W	1
13: 0	BAG interval, unit: 1ms, initial value: 10 ms	R/W	A

5.5.1 BAG Timing Period Control

The host accesses the configuration register through the register interface to configure BAG, sampling enable, and free_to_trans functions for one of the 512 queues. The BAG timing clock uses the unit of 1ms. The default BAG power-on reset of each queue is 10 ms. The host can be reconfigured.

For each queue, when a frame is queued, the VL_cnt count value increases by 1, and when dispatching is completed, the count value decreases by 1, and the VL_cnt reset value is 0. When there is data in the queue, that is, VL_cnt is not 0, BAG time is up, or free_to_trans is 1, the queue sends a scheduling request to the scheduling module. However, if a frame in the queue is not sent after it is scheduled, the next frame in the queue will not send a dispatch request until the frame is sent by the port.

5.5.2 Queue Scheduling and Arbitration

Queue scheduling and arbitration use the polling algorithm for scheduling requests of 512 queues. When a queue is scheduled, the system outputs scheduling signals and obtains the queue number.

Firstly, 512 sending scheduling requests are divided into 16 groups on average, and 16 groups are detected synchronously. According to the design requirements, 512 requests may reach multiple channels at the same time. When a valid request is detected in one channel, the request signal corresponds to position 1. At the same time, the scheduler starts polling schedule. The scheduler is divided into two levels. The first level searches from 16 groups; if the first level polls to a valid queue, the second level continue to search within the group. According to the value of the two-level scheduling counter, the decoding gets the queue number of the scheduling. After the scheduling, the message word of the queue will be taken out of the memory according to the number of the scheduled queue, and dispatching request will be sent to the scheduling module of the output queue. After receiving the response signal, the state machine will return to an idle state.

5.6 SSRAM_if Module

SSRAM_if is responsible for data interaction between SSRAM memory chip and SSRAM_Ctrl and polling and arbitrating SSRAM access. Because the SSRAM is a single-port chip, each module needs to use the SSRAM_if module to perform polling and arbitration before accessing the SSRAM [24].

5.7 Send Cache Module

When the sending cache module detects Ethernet frames that need to be sent, it is responsible for reading Ethernet frames and sending them to the RGMII controller according to the interface timing of the RGMII controller to complete data transmission. Send cache module includes send DPRAM and send FIFO. The space of DPRAM is 4 KB, which stores the data information of each frame, and the FIFO stores the control word of each frame.

6 The Test Results

The XC7Z100FFG900-1 development board of XILINX ZynQ-7000 series is selected in this paper. The circuit design uses Verilog hardware description language, and the function simulation is completed in Questasim environment, and the simulation waveform is observed. The functional simulation of Tri_Mode_Ethernet MAC core is shown in Fig. 10. Tri_Mode_Ethernet MAC core can realize data receiving and sending. The functional simulation results of the retrieval module are shown in Fig. 11. PS can configure the retrieval configuration unit of the retrieval module through AXI bus. The functional simulation results of the scheduling module are shown in Fig. 12. Similarly, PS can configure the configuration register of the scheduling module through AXI bus. The functional simulation of SSRAM memory is shown in Fig. 13. SSRAM memory can normally complete the read and write function. Thus, this circuit can realize the function of data transmission.

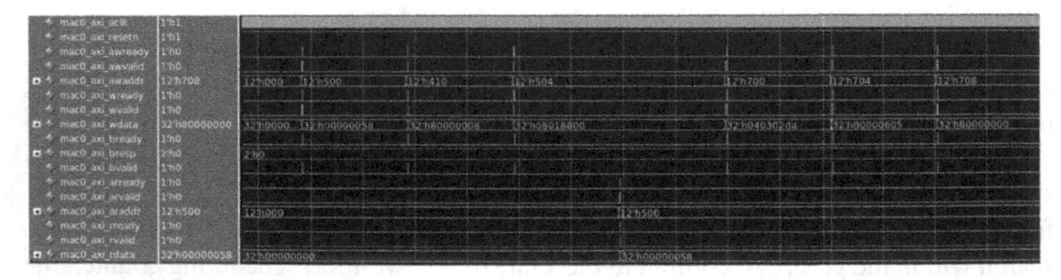

Fig. 10. Tri Mode Ethernet MAC core function simulation results

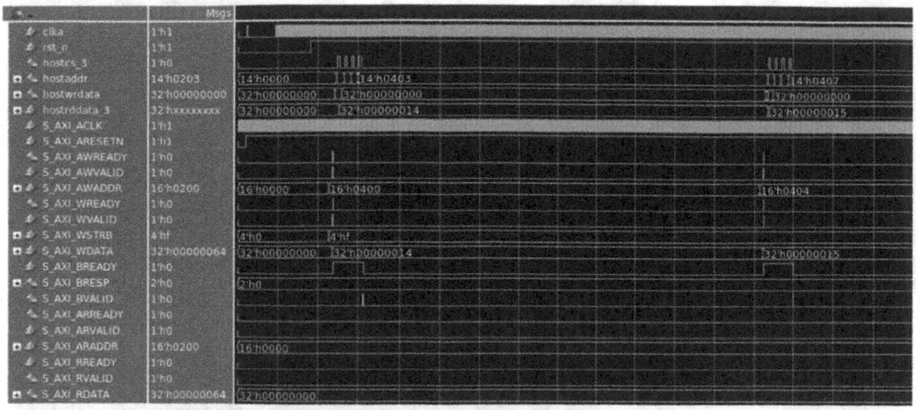

Fig. 11. Retrieve module function simulation result

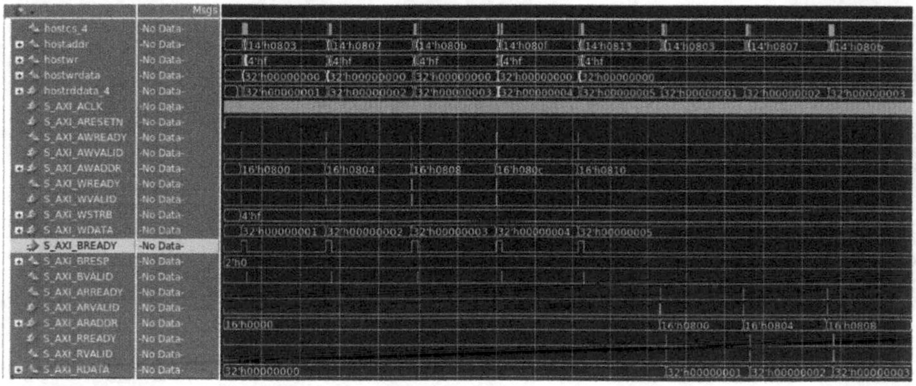

Fig. 12. Scheduling module function simulation result

The FPGA testing is completed by connecting PC, P2020 development board and the integer card. The host sends optical fiber data to the P2020 development board, which converts the received optical fiber data into Ethernet frames and then sends them to the integer card for traffic integer, and then the integer card sends Ethernet frames to the host. The host uses Wireshark software to capture Ethernet frames in real time. The host can also capture signal verification by vivado software. The result is shown in Fig. 14. A total of 411 frames are received, including 411 correct frames (corresponding to signal 3 in the figure), 0 error frames (corresponding to signal 4 in the figure), and 0 overflow frames (corresponding to signal 5 in the figure), Ethernet frames are stored in SSRAM (corresponding to signal 1 in the figure) and sent out (corresponding to signal 2 in the figure), and finally 411 frames are sent out (corresponding to signal 6 in the figure). The test results show that Ethernet frames can be transmitted smoothly over the link, avoiding frame loss.

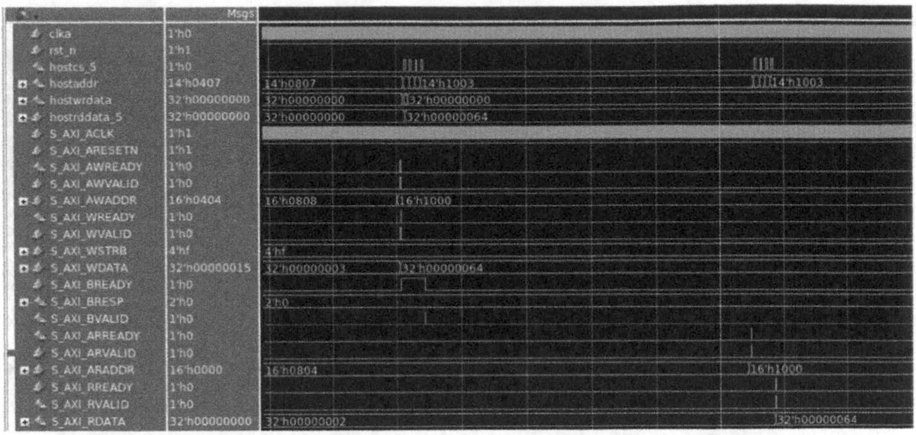

Fig. 13. SSRAM memory function simulation result

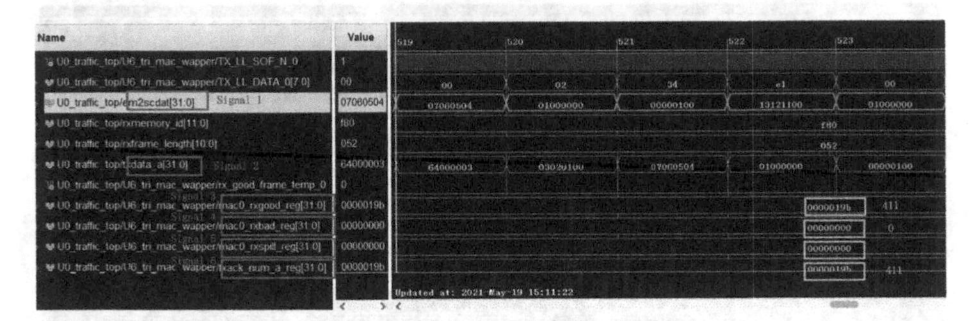

Fig. 14. Integer card test results

7 Conclusion

Aiming at the large number of diversified services appearing on the current network, it is necessary to ensure high-speed transmission, and to make the data not lost as much as possible, so this paper proposes a traffic integer module. First, the Tri_Mode_Ethernet MAC soft core is used in combination with the PHY chip to achieve a gigabit transmission rate, and the traffic integer function is completed by setting the buffer and token bucket algorithm, reducing the transmission delay and improving the bandwidth utilization. Periodically send data out to ensure data persistence. Experiments have shown that data can be stably transmitted on the link, and data packet loss rarely occurs. Traffic integer modules will be better developed in the future in applications such as network monitoring, data communications, and data recorders for aircraft.

References

1. Gao, Y., Huang, J.: J. Beijing Inst. Electron. Sci. Technol. **16**(2), 68–69 (2008)
2. Zhen, G., Wang, Q., Jiao, X., Chu, Q.: Instrum. Tech. Sens. (10), 40–44 (2019). (in Chinese)

3. Zhang, C., Luo, F.: Design of high-speed data transmission system based on gigabit Ethernet. Electron. Sci. Technol. **24**(01), 44–46 (2011)
4. Wang, Q.: Analysis of Custom Protocol and Design of Interface IP Core Based on Gigabit Ethernet. North University of China (2019)
5. Chen, H.: Design and Implementation of High-Speed Data Recorder Based on Gigabit Ethernet. Sun Yat-sen University (2021)
6. He, B.: Xilinx All-programmable ZynQ-7000 SoC Design Guide. Tsinghua University Press, Beijing (2013)
7. Wei, H.-Y., Tsao, S.-C., Lin, Y.-D.: Assessing and improving TCP rate shaping over edge gateways. IEEE Trans. Comput. **53**(3), 259–275 (2004)
8. Boyer, P.E., Guillenmin, F.M., et al.: Spacing cells protects and enhances utilization of ATM network links. IEEE Netw. **6**(5), 38–49 (1992)
9. Sidi, M., Liu, W.Z., Cidon, I: Congestion control through input-rate regulation. IEEE Trans. Communi. **41**, 471–477 (1993)
10. Liu, Y.: Research and Implementation of Network Traffic Integer Strategy Based on Leak-bucket Theory and Token Bucket Algorithm. Northeast Normal University (2008)
11. Gao, L.: Research and Implementation of IP Network Traffic Integer System. University of Electronic Science and Technology of China, Chengdu (2016)
12. Chen, B.: Research on QoS Oriented Packet Scheduling and Traffic Integer Types in IP Access Networks. Zhejiang University (2003)
13. Miao, N., Lin, J.: Research and design of multi-token bucket flow integer algorithm. Microelectron. Comput. **28**(11) (2011)
14. Lin, N.: Suo Girls Middle School: Hierarchical Dynamic Flow Integer Algorithm based on Token Bucket. Mod. Comput. (Prof. Ed.) (12) (2011)
15. Liu, Y.: Design of Flow Integral Circuit Module in Network Processor. Southeast University, Jiangsu (2017)
16. Jiang, W.: Research on Multi-priority Queue Packet Scheduling. Hunan Normal University (2009)
17. Chen, K.: Sci. Consult. (Sci. Technol. Manag.) (06), 73–74 (2010)
18. Dong, S.: China Mod. Educ. Equip. (20), 38–40 (2020). (in Chinese)
19. Xu, J., Li, J., Jin, J., Huang, Z.: Speed closed-loop control system based on microcontroller. Autom. Instrum. (04), 21–21 (2005)
20. Liang, Y., Hu, Y., Zhu, L.: Research on reactive power compensation control strategy of power supply system based on open-loop control. Metall. Power (08), 1–4 (2014)
21. Fang, Y., Hang, B.: Open-loop control system of stepper motor based on microcontroller. Electr. Mach. Control Appl. (04), 61–64 (2006)
22. Chen, W., Xia, K., Yuan, J.: Design of P2020 processor display interface based on FPGA. Microprocessors **35**(02), 8–10 (2014)
23. Wen, F., Han, Y.: Electron. Meas. Technol. **201**, 44(1), 150–154. (in Chinese)
24. Ji, A., Wang, S., Wang, H.: Design of external memory interface for high-end SOC chips. Microelectron. Comput. (01), 71–73 (2006)

[illegible]

Flight Simulation, Simulator, Simulation Support Environment, Simulation Standard and Simulation System Construction

Flight Control of Underwater UAV Based on Extended State Observer and Sliding Mode Method

Canhui Tao(✉), Zhiping Song, and Baoshou Wang

China Ship Scientific Research Center, Wuxi 214082, China
taocanhui@cssrc.com.cn

Abstract. Disturbance rejection control of underwater UAV is a key technology. First of all, In this paper, the model of underwater UAV will be established, and the closed-loop control system will be constructed according to the virtual control quantity. Then, sliding mode controllers and extended state observer are designed to realize the attitude and position control. Finally, the simulation is performed to show the robustness of UAV. It can be proved that under the influence of inhomogeneous media, UAV can still have good robustness and compensate for interference.

Keywords: Underwater UAV · Sliding mode control · Extended state observer

1 Introduction

The underwater UAV has become a research hotspot in recent years due to its ability to operate both underwater and above water. The UAV is a typical nonlinear, strongly coupled and multi-variable underactuated system [1]. In order to reduce the impact of disturbance, the actuator of underwater UAV will be more than four rotors. This paper studies an underwater UAV with six rotor which is called hex-rotor UAV. so the design of its controller has become a key point in the application research of aircraft [2].

There are two main categories of flight controller design methods [3, 4], one is linear control method, the other is nonlinear control method. In addition, with the increasing research interest of intelligent algorithms, some self-learning control methods, such as the use of artificial neural networks and fuzzy control methods, have also been applied in this field [5].

Among the linear control methods, PID controller [3] is widely used in engineering practice due to its advantages of low cost and easy implementation. However, due to the tedious control parameter tuning of PID controller, it does not have the ability of batch transplantation, resulting in its poor robustness to different hex-rotor models. LQR controller is easy to adjust the flight system to achieve the optimal state by considering the flight performance indexes. Therefore, it is widely used in the flight control of the hex-rotor.

The use of nonlinear control method for aircraft controller design has been the focus of research in recent years [6]. For underactuated system, Lyapunov function can

W. Fan et al. (Eds.): AsiaSim 2022, CCIS 1712, pp. 293–304, 2022.
https://doi.org/10.1007/978-981-19-9198-1_22

be deduced by backward construction through backstepping method to ensure system stability, which has been widely concerned in trajectory control of hex-rotor aircraft. Sliding mode control is also a control method with strong robustness. However, for the actual hex-rotor flight control system, the chattering of sliding mode controller is difficult to achieve, so the performance of the system will be affected.

Observers-based and disturbance estimation methods are one of the most common methods for random fault diagnosis and characterization. In reference [5], the Extended State Observer (ESO) was used to estimate the uncertainty of the system online, and good control effect was achieved, but the amplitude of attitude Angle changed greatly. Literature [8, 9] respectively used sliding mode Thau observer and radial basis function observer to observe blade damage faults, and verified the fault-tolerant control effect through numerical simulation.

In this paper, the sliding mode method combined with the extended state observer is proposed to design the flight control system of the hex-rotor. This method can overcome the disturbance and parameter perturbation problems well, and the output of the controller is limited to a reasonable range, which can be used as a reference for practical systems.

2 System Model

The structural feature of hex-rotor UAV is shown in Fig. 1, which is a kind of aircraft with six degrees of freedom. The distance to the center of gravity is l.

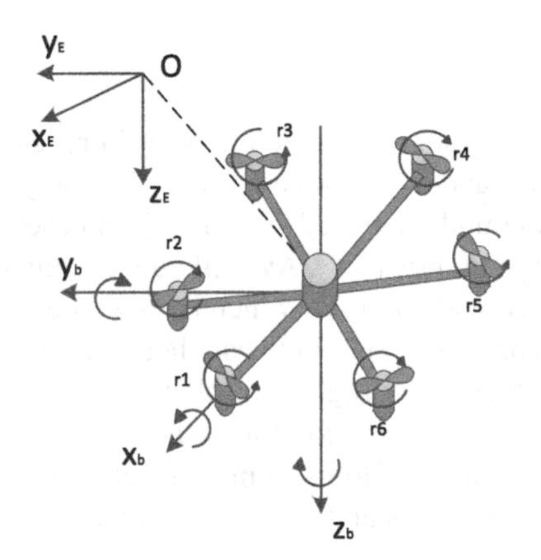

Fig. 1. The hex-rotor UAV structure diagram

Let the body coordinate system be and the ground coordinate system be, then the relationship between the two coordinate systems is as follows:

$$R = \begin{bmatrix} c_\theta c_\psi & s_\phi s_\theta c_\psi - c_\phi s_\psi & c_\phi s_\theta c_\psi + s_\phi s_\psi \\ c_\theta s_\psi & s_\phi s_\theta s_\psi + c_\phi c_\psi & c_\phi s_\theta s_\psi - s_\phi c_\psi \\ -s_\theta & s_\phi c_\theta & c_\phi c_\theta \end{bmatrix} \tag{1}$$

where $c_{\cdot} = \cos(\bullet), s_{\cdot} = \sin(\bullet)$.

According to the American standard, set up the body fixed frame $Ox_by_bz_b$ and inertial frame $Ox_Ey_Ez_E$. $\mathbf{Ox_b}$ is set in the line of $r1$ and $r4$, and points to $r1$, which is set as the right ahead of the hex-rotor UAV. The attitude angles that refer to Euler angles ($\phi\ \theta\ \psi$), are determined by the relationship between $Ox_Ey_Ez_E$ and $Ox_by_bz_b$.

R is the transformation matrix from $Ox_by_bz_b$ to $Ox_Ey_Ez_E$,

The speeds of $r1$ to $r6$ are $(\omega_1\ \omega_2\ \omega_3\ \omega_4\ \omega_5\ \omega_6)$, and $(\omega_1\ \omega_2\ \omega_3\ \omega_4\ \omega_5\ \omega_6) = (-\Omega_1\ \Omega_2\ -\Omega_3\ \Omega_4\ -\Omega_5\ \Omega_6)$, where the minus in front of Ω_i shows that the rotation of the rotor acts in the opposite direction to $\mathbf{Oz_b}$. The attitude and the position of hex-rotor UAV can be changed by changing the value of Ω_i. Increase (or decrease) the value of Ω_1, and decrease (or increase) the value of Ω_4 at the same time to achieve the pitch movement and the translation along the diction of $\mathbf{Ox_E}$. Increase (or decrease) the value of Ω_5 and Ω_6, and decrease (or increase) the value of Ω_2 and Ω_3 at the same time to achieve the roll movement and the translation along the diction of $\mathbf{Oy_E}$. Increase (or decrease) the value of Ω_1, Ω_3 and Ω_5, and decrease (or increase) the value of Ω_2, Ω_4 and Ω_6 at the same time to achieve the yaw movement. Increase or decrease the six rotations at the same time to achieve the translation along the diction of $\mathbf{Oz_E}$.

Set up the line movement and angular motion equations in the inertial frame as:

$$\sum \vec{F} = m\frac{d\vec{V}}{dt}$$
$$\sum \vec{M} = \frac{d\vec{H}}{dt} \tag{2}$$

where $\vec{F}$ represents force and $\vec{V}$ is the absolute speed of the mass relative to the ground coordinate system. m is the mass of the hex-rotor UAV. $\vec{H}$ is angular momentum. $\vec{M}$ is resultant moment.

Next, the motion of the hex-rotor UAV is discussed respectively from line movement and angular motion.

a) Line Movement

In the process of actual flight, the hex-rotor UAV is affected by the lift force F_T that produced by the rotors, gravity G and the air resistance F_D. Ignore the air resistance effects when the hex-rotor UAV in hovers and flies slowly. Show F_T and G in the inertial frame as follows:

$$F_T^E = RF_T^b = \begin{pmatrix} c_\phi s_\theta c_\psi + s_\phi s_\psi \\ c_\phi s_\theta s_\psi - s_\phi c_\psi \\ c_\phi c_\theta \end{pmatrix} \left(-\sum_{i=1}^{6} F_i\right) \tag{3}$$

$$F_D^E = \begin{bmatrix} 0 \\ 0 \\ mg \end{bmatrix} \tag{4}$$

where F_i is the lift produced by ri, $i = \{1, 2, 3, 4, 5, 6\}$.

According to Newton-Euler's first law, the equation of line movement is:

$$m\begin{bmatrix}\ddot{x}\\ \ddot{y}\\ \ddot{z}\end{bmatrix} = F_T^E + F_D^E \tag{5}$$

Take (3) and (4) into (5), and the equations of line movement of the hex-rotor UAV in the inertial frame are as follows:

$$\begin{cases} m\ddot{x} = -\left(c_\phi s_\theta c_\psi + s_\phi s_\psi\right)\sum_{i=1}^{6} F_i \\ m\ddot{y} = -\left(c_\phi s_\theta s_\psi - s_\phi c_\psi\right)\sum_{i=1}^{6} F_i \\ m\ddot{z} = -c_\phi c_\theta \sum_{i=1}^{6} F_i + mg \end{cases} \tag{6}$$

b) Angular Motion

During flight, the hex-rotor rotates around the center of mass under the effect of torque. Ignore the gyroscopic effect, because the rotor mass and the size of the hex-rotor UAV are small. The inertia matrix is as follows:

$$I = \begin{bmatrix} I_{xx} & I_{xy} & I_{xz} \\ I_{yx} & I_{yy} & I_{yz} \\ I_{zx} & I_{zy} & I_{zz} \end{bmatrix} = \begin{bmatrix} I_x & 0 & 0 \\ 0 & I_y & 0 \\ 0 & 0 & I_z \end{bmatrix} \tag{7}$$

where, I_x, I_y and I_z are the rotational inertias correspond to each axis respectively.

Establish angular motion equation in body fixed frame as:

$$\sum \vec{M}^b = \mathrm{I}\dot{\mathrm{W}}^b + \mathrm{W}^b \times \left(I W^b\right) \tag{8}$$

where

$$\mathrm{W}^b = \left(p\ q\ r\right)^T \tag{9}$$

and p, q, r are angular velocities in body fixed frame separately.

$$\sum \vec{M}^b = \left(L\ M\ N\right)^T \tag{10}$$

Here, L, M, N are components of $\sum \vec{M}^b$ in body fixed frame.

Substitute (9) and (10) into (8), and $\sum \vec{M}^b$ can be expressed as follows:

$$\sum \vec{M}^b = \begin{bmatrix} L \\ M \\ N \end{bmatrix} = \begin{bmatrix} \dot{p} I_x + qr\left(I_z - I_y\right) \\ \dot{q} I_y + pr\left(I_x - I_z\right) \\ \dot{r} I_z + pq\left(I_y - I_x\right) \end{bmatrix} \tag{11}$$

Angular motion equation of six rotor unmanned aerial vehicle in body fixed frame is obtained:

$$\begin{cases} \dot{p} = \left(L + qr\left(I_y - I_z\right)\right)/I_x \\ \dot{q} = \left(M + pr(I_z - I_x)\right)/I_y \\ \dot{r} = \left(N + pq\left(I_x - I_y\right)\right)/I_z \end{cases} \tag{12}$$

The relationship between $\left(p\, q\, r\right)$ and $\left(\dot{\theta}\, \dot{\phi}\, \dot{\psi}\right)$ is established according to the relation between body fixed frame and inertial frame:

$$\begin{cases} \dot{\phi} = p + \left(rc_{\phi} + qs_{\phi}\right)\tan\theta \\ \dot{\theta} = qc_{\phi} - rs_{\phi} \\ \dot{\psi} = \left(rc_{\phi} + qs_{\phi}\right)/c_{\theta} \end{cases} \tag{13}$$

Here, θ and ϕ are assumed to be very small when six rotor unmanned aerial vehicle is flying, so the relationship between $\left(\mathrm{p\, q\, r}\right)$ and $\left(\dot{\theta}\, \dot{\phi}\, \dot{\psi}\right)$ is come down to:

$$\begin{cases} \dot{\phi} = p \\ \dot{\theta} = q \\ \dot{\psi} = r \end{cases} \tag{14}$$

So (12) can be expressed in the inertial frame as:

$$\begin{cases} \ddot{\phi} = \left(L + \dot{\theta}\dot{\psi}\left(I_y - I_z\right)\right)/I_x \\ \ddot{\theta} = \left(M + \dot{\phi}\dot{\psi}(I_z - I_x)\right)/I_y \\ \ddot{\psi} = \left(N + \dot{\phi}\dot{\theta}\left(I_x - I_y\right)\right)/I_z \end{cases} \tag{15}$$

According to [8], the lift force F_i and moment M_i is proportional to Ω_i^2, and the scaling factors are k_L and k_M. So

$$\begin{aligned} &\sum_{i=1}^{6} F_i = k_L \sum_{i=1}^{6} \Omega_i^2 \\ &L = \left(\sqrt{3}/2\right)Lk_L\left(\Omega_5^2 + \Omega_6^2 - \Omega_2^2 - \Omega_3^2\right) \\ &M = Lk_L\left(\Omega_1^2 - \Omega_4^2\right) \\ &N = k_M\left(\Omega_1^2 - \Omega_2^2 + \Omega_3^2 - \Omega_4^2 + \Omega_5^2 - \Omega_6^2\right) \end{aligned} \tag{16}$$

For simple research, take the rotations as the control input, and divide the whole system into four control system: pitch, roll, yaw and up-and-down movement. U_1, U_2, U_3 and U_4 are defined as the control input of each movement:

$$\begin{aligned} &U_1 = \left(\sqrt{3}/2\right)Lk_L\left(\Omega_5^2 + \Omega_6^2 - \Omega_2^2 - \Omega_3^2\right) \\ &U_2 = Lk_L\left(\Omega_1^2 - \Omega_4^2\right) \\ &U_3 = k_M \sum_{i=1}^{6} (-1)^{i+1}\Omega_i^2 \\ &U_4 = \sum_{i=1}^{6} F_i = k_L \sum_{i=1}^{6} \Omega_i^2 \end{aligned} \tag{17}$$

Above all, in the inertial frame, the simplified model of the hex-rotor UAV is as follows:

$$\begin{cases} \ddot{x} = -(c_\phi s_\theta c_\psi + s_\phi s_\psi)U_4/m \\ \ddot{y} = -(c_\phi s_\theta s_\psi - s_\phi c_\psi)U_4/m \\ \ddot{z} = (-c_\phi c_\theta U_4 + mg)/m \\ \ddot{\phi} = U_1/I_x + \dot{\theta}\dot{\psi}(I_y - I_z)/I_x \\ \ddot{\theta} = U_2/I_y + \dot{\phi}\dot{\psi}(I_z - I_x)/I_y \\ \ddot{\psi} = U_3/I_z + \dot{\phi}\dot{\theta}(I_x - I_y)/I_z \end{cases} \quad (18)$$

3 Control System Design

Based on the hex-rotor aircraft model described above, the following block diagram of the flight control loop system can be obtained. Since the attitude of the hex-rotor can form an independent sub-model, it can be set as the inner loop of the control system. After the attitude controller with good performance is designed, the trajectory tracking controller will be designed for the system.

Definition of state variables is:

$$\begin{aligned} &[x_1\ x_2\ x_3\ x_4\ x_5\ x_6\ x_7\ x_8\ x_9\ x_{10}\ x_{11}\ x_{12}] \\ &= [\phi\ \dot{\phi}\ \theta\ \dot{\theta}\ \psi\ \dot{\psi}\ z\ \dot{z}\ x\ \dot{x}\ y\ \dot{y}] \end{aligned} \quad (19)$$

System models in the form of state equation is as follows:

$$\begin{bmatrix} \dot{x}_1 \\ \dot{x}_2 \\ \dot{x}_3 \\ \dot{x}_4 \\ \dot{x}_5 \\ \dot{x}_6 \\ \dot{x}_7 \\ \dot{x}_8 \\ \dot{x}_9 \\ \dot{x}_{10} \\ \dot{x}_{11} \\ \dot{x}_{12} \end{bmatrix} = \begin{bmatrix} x_2 \\ b_1U_1 + a_1x_4x_6 \\ x_4 \\ b_2U_2 + a_2x_2x_6 \\ x_6 \\ b_3U_3 + a_3x_2x_4 \\ x_8 \\ g - \cos x_1 \cos x_3 U_4/m \\ x_{10} \\ -u_xU_4/m \\ x_{12} \\ -u_yU_4/m \end{bmatrix} \quad (20)$$

with,

$$b_1 = 1/I_x, b_2 = 1/I_y, b_3 = 1/I_z, a_1 = (I_y - I_z)/I_x, a_2 = (I_z - I_x)/I_y;$$
$$a_3 = (I_x - I_y)/I_z, u_x = c_\phi s_\theta c_\psi + s_\phi s_\psi, u_y = c_\phi s_\theta s_\psi - s_\phi c_\psi.$$

In the previous, the model of the three channels of attitude was established, which are very close in form and relatively independent. Therefore, we will choose any channel for controller design. Taking the ϕ as an example, the control law design and stability proof are given in the following.

The model of ϕ channel is:

$$\begin{cases} \dot{x}_1 = x_2 \\ \dot{x}_2 = b_1 U_1 + a_1 x_4 x_6 + d_\phi \end{cases} \tag{21}$$

The extended state observer is designed according to the system model

$$\begin{cases} \dot{\xi}_1 = \xi_2 - \frac{\alpha_1}{\tau}(\xi_1 - x_1) \\ \dot{\xi}_2 = b_1 U_1 + a_1 \sigma_1 + \hat{d}_\phi - \frac{\alpha_2}{\tau^2}(\xi_1 - x_1) \\ \dot{\hat{d}}_\phi = -\frac{\alpha_3}{\tau^3}(\xi_1 - x_1) \end{cases} \tag{22}$$

where ξ_1 is the estimated value of x_1, ξ_2 is the estimated value of x_2, $\hat{d}_\phi$ is the estimated disturbance. α_1, α_2, and α_3 need to satisfy Hurwitz's condition. τ is a time constant that can be set to adjust the bandwidth. σ_1 is the value of $x_4 x_6$, which may be obtained by state observer or sensors.

$$e_1 = x_1 - x_{1c} \tag{23}$$

A linear sliding surface is introduced

$$s_1 = e_1 + c_1 \dot{e}_1 \tag{24}$$

The sliding mode variable s can be estimated by the extended state observer

$$\hat{s}_1 = \xi_1 - x_{1c} + c_1(\xi_2 - \dot{x}_{1c}) \tag{25}$$

Set the following control laws

$$U_1 = \frac{1}{b_1}(-a_1 \sigma_1 + \ddot{x}_{1c} - \hat{d}_\phi) + \frac{1}{b_1 c_1}(\dot{x}_{1c} - \xi_2 - k\hat{s}_1) \tag{26}$$

Let the Lyapunov function $V_1 = \frac{1}{2}s_1^2$.

Then,

$$\begin{aligned} \dot{V}_1 &= s_1 \dot{s}_1 = s_1(\dot{e}_1 + c_1 \ddot{e}_1) \\ &= s_1(\dot{x}_1 - \dot{x}_{1c} + c_1(\ddot{x}_1 - \ddot{x}_{1c})) \\ &= s_1(x_2 - \dot{x}_{1c} + c_1(b_1 U_1 + a_1 \sigma_1 + d_\phi - \ddot{x}_{1c})) \\ &= s_1(x_2 + c(-\hat{d}_\phi + \frac{1}{c_1}(-\xi_2 - k\hat{s}_1) + d_\phi)) \\ &= s_1(-k(s_1 - \tilde{s}_1) + \tilde{x}_2 + c_1 \tilde{d}_\phi) \\ &= -ks_1^2 + s_1(k\tilde{s}_1 + \tilde{x}_2 + c_1 \tilde{d}_\phi) \end{aligned}$$

where $\tilde{s}_1 = s_1 - \hat{s}_1$, $\tilde{x}_2 = x_2 - \xi_2$, $\tilde{d}_\phi = d_\phi - \hat{d}_\phi$

If the estimated error value of the extended state observer is a sufficiently small value, then there exists a constant Δ_ϕ satisfying

$$\left|k\tilde{s}_1 + \tilde{x}_2 + c_1\tilde{d}_\phi\right| < \Delta_\phi \tag{27}$$

So $\dot{V}_1$ satisfies that

$$\begin{aligned}\dot{V}_1 &\le -ks_1^2 + s_1\Delta_\phi \\ &\le -ks_1^2 + \frac{1}{2}s_1^2 + \frac{1}{2}\Delta_\phi^2 \\ &= -(2k-1)V_1 + \frac{1}{2}\Delta_\phi^2\end{aligned}$$

Then V_1 satisfies the following inequality

$$\begin{aligned}V_1 &\le e^{-(2k-1)(t-t_0)}V_1 + \frac{1}{2}\Delta_\phi^2\int_{t_0}^{t} e^{-(2k-1)(t-\tau)}d\tau \\ &= e^{-(2k-1)(t-t_0)}V_1 - \frac{1}{2}\Delta_\phi^2\int_{t_0}^{t} e^{-(2k-1)(t-\tau)}d(t-\tau) \\ &= e^{-(2k-1)(t-t_0)}V_1 - \frac{1}{2}\Delta_\phi^2\int_{t_0}^{t} e^{-(2k-1)n}dn \\ &= e^{-(2k-1)(t-t_0)}V_1 + \frac{1}{2(2k-1)}\Delta_\phi^2(1 - e^{-(2k-1)(t-t_0)})\end{aligned}$$

Hence, when $k > \frac{1}{2}$,

$$\lim_{t\to\infty} V \le \frac{1}{2(2k-1)}\Delta_\phi^2 \tag{28}$$

By adjusting $\alpha_1, \alpha_2, \alpha_3$ of the extended state observer to meet Hurwitz's condition and adjusting ε to a small value, the observer can quickly track the state, that is, the observation error is small, and the state of the system can reach an equilibrium point, and the motion can be kept approximately on the sliding mode surface. So the error will converge to zero according to the law of exponential decay. The larger the value of k is, the faster the system state converges to the sliding mode surface.

Similarly, the other two control input can be expressed as:

$$U_2 = \frac{1}{b_2}(-a_2\sigma_2 + \ddot{x}_{3c} - \hat{d}_\theta) + \frac{1}{b_2c_2}(\dot{x}_{3c} - \xi_4 - k\hat{s}_2) \tag{29}$$

$$U_3 = \frac{1}{b_3}(-a_3\sigma_3 + \ddot{x}_{5c} - \hat{d}_\psi) + \frac{1}{b_3c_3}(\dot{x}_{5c} - \xi_6 - k\hat{s}_3) \tag{30}$$

The model of z channel is:

$$\begin{cases}\dot{x}_7 = x_8 \\ \dot{x}_8 = g - \frac{\cos x_1 \cos x_3}{m}U_4 + d_z\end{cases} \tag{31}$$

The extended state observer is designed according to the system model

$$\begin{cases} \dot{\xi}_7 = \xi_8 - \frac{\beta_1}{\tau}(\xi_7 - x_7) \\ \dot{\xi}_8 = b_4 U_4 + g + \hat{d}_z - \frac{\beta_2}{\tau^2}(\xi_7 - x_7) \\ \dot{\hat{d}}_z = -\frac{\beta_3}{\tau^3}(\xi_7 - x_7) \end{cases} \tag{32}$$

where $b_4 = -\frac{\cos x_1 \cos x_3}{m}$.

The linear sliding surface of z channel is introduced

$$s_4 = e_4 + c_4 \dot{e}_4 \tag{33}$$

The sliding mode variable s can be estimated by the extended state observer

$$\hat{s}_4 = \xi_7 - x_{7c} + c_4(\xi_8 - \dot{x}_{7c}) \tag{34}$$

Set the following control laws

$$U_4 = \frac{1}{b_4}(-g + \ddot{x}_{7c} - \hat{d}_z) + \frac{1}{b_4 c_4}(\dot{x}_{7c} - \xi_8 - k\hat{s}_4) \tag{35}$$

Similarly, The Lyapunov stability of the system can be proved.

4 Simulation

In this section, the math model of the UAV, the sliding mode controller and extended state observer are validated. The parameters of UAV are shown in the Table 1 [9].

Table 1. Parameters of the simulation model

m (kg)	3.96	I_x (N·m^2)	0.363
g (N/kg)	9.8	I_y (N·m^2)	0.363
k_L (kg·m^2)	1.91×10^{-3}	I_z (N·m^2)	0.651
k_M (kg·m^2)	4.21×10^{-5}	L (m)	0.45

The initial state of the UAV is

$$\begin{pmatrix} x\ y\ z\ \phi\ \theta\ \psi \end{pmatrix} = \begin{pmatrix} 0\ 0\ 0\ 0\ 0\ \frac{\pi}{6} \end{pmatrix}$$

The simulation time is 20 s and the parameters of the controllers are as follows:

$$\begin{aligned} &c_1 = 5, k_1 = 1, c_2 = 5, k_2 = 1, \\ &c_3 = 5, k_3 = 1, c_4 = 8, k_4 = 2, \\ &\beta_1 = 3, \beta_2 = 3, \beta_3 = 1, \\ &\alpha_1 = 3, \alpha_2 = 3, \alpha_3 = 1, \end{aligned}$$

To verify this control system from two cases: spot hover and robustness. Besides that, compare the effect with PID controller and extended state observer.

4.1 Spot Hover

Hovering is the key mode of underwater UAV operation. It is necessary to change the position and attitude frequently when working on the surface and underwater. This group of experiments will simulate the effect of fixed-point hovering.

If setting the command

$$\left(x_c\ y_c\ z_c\ \phi_c\ \theta_c\ \psi_c\right) = \left(5\ 5\ -5\ 0\ 0\ \tfrac{\pi}{6}\right)$$

The simulation results of PID controller are shown in the Fig. 2. The simulation results of the proposed method (sliding mode) controller are shown in the Fig. 3.

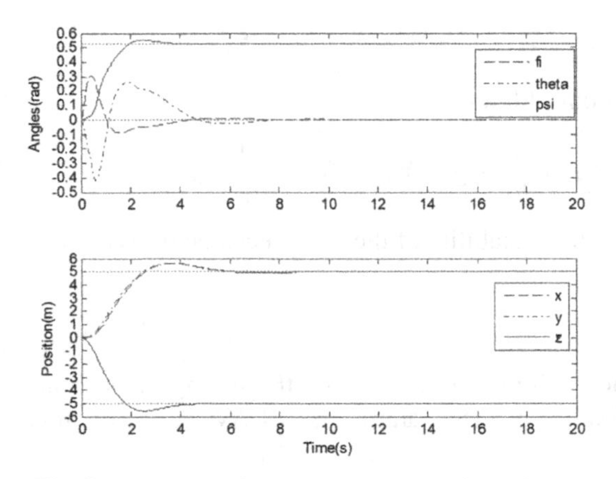

Fig. 2. PID controller response curves of spot hover

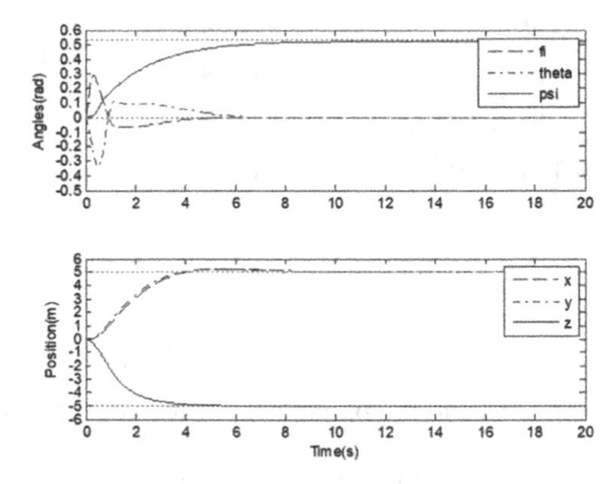

Fig. 3. Sliding mode controller response curves of spot hover

It can be seen from the simulation that the two kinds of control strategies can ensure the hex-rotor UAV arrive at the target location in about 8 s, and steadily hovering. But the sliding mode controller ensures the UAV response progress more smoothly than the PID controller.

4.2 Robustness

Whether operating on the surface or underwater, the fluctuation of the current will be particularly severe, so the robustness of the UAV system needs to be verified.

Considering the uncertainty of the model, we set the parameters of nominal model shown in the Table 2.

Table 2. Parameters of the nominal model

m (kg)	3.96	I_x (N·m^2)	0.5
g (N/kg)	9.8	I_y (N·m^2)	0.5
k_L (kg·m^2)	1.91×10^{-3}	I_z (N·m^2)	0.5
k_M (kg·m^2)	4.21×10^{-5}	L (m)	0.4 s 5

At 10 s, add a distraction (with amplitude value of 0.3 rad, lasting 0.1 s) to the feedback of θ, and add a distraction (with amplitude value of 2 m, lasting for 0.2 s) to the feedback of *y*. The effects of the two controllers are shown in the Fig. 4 and Fig. 5.

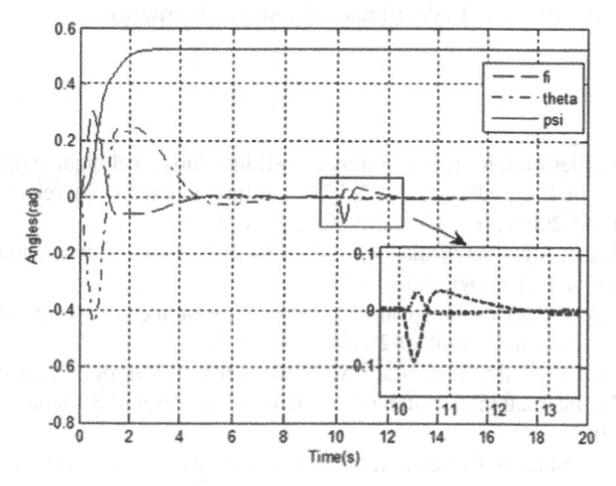

Fig. 4. The control effect of PID controller

From Fig. 4, using PID controller, the amplitude fluctuate of pitch and roll is about 0.02 rad and lasts 2 s after adding uncertainty and disturbance. From Fig. 5, the amplitude fluctuate is about 0.09 rad and lasts 0.5 s. This can prove that sliding mode controller with extended state observer shows better robustness.

5 Conclusion

The kinematic equations of the hex-rotor UAV are set up in this paper, and then considering the uncertainty of the math model and the external disturbance, the sliding mode

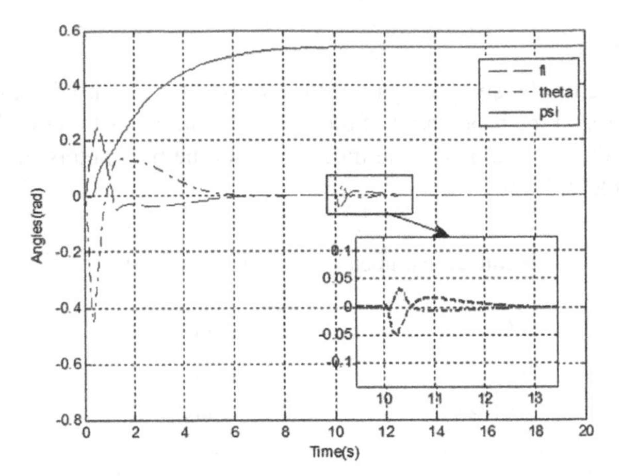

Fig. 5. The control effect of sliding mode controller

controller with extended state observer is designed to control the attitude and the position of the hex-rotor UAV. The simulation experiments show that this kind of controller not only can ensure the hex-rotor UAV spot hover stably, but also show better robustness than PID controller when the UAV affected by some distraction.

References

1. Bouabdallah, S., Siegwart, R.: Backstepping and sliding-mode techniques applied to an indoor micro quadrotor. In: Proceedings of the 2005 IEEE International Conference on Robotics and Automation, ICRA 2005, pp. 2247–2252. IEEE (2005)
2. Suresh, K.K., Kahn, A.D., Yavrucuk, I.: GTMARS - flight controls and computer architecture. Georgia Institute of Technology, Atlanta (2000)
3. Amin, R., Li, A.: A review of quadrotor UAV: control methodologies and performance evaluation. Int. J. Autom. Control **10**(2), 87–103 (2016)
4. Rosales, C., Soria, C., Carelli, R., et al.: Adaptive dynamic control of a quadrotor for trajectory tracking. In: International Conference on Unmanned Aircraft Systems (ICUAS), Miami, pp. 143–149 (2017)
5. De Crousaz, C., Farshidian, F., Neunert, M., et al.: Unified motion control for dynamic quadrotor maneuvers demonstrated on slung load and rotor failure tasks. In: IEEE International Conference on Robotics and Automation (ICRA), Seattle, pp. 2223–2229 (2015)
6. Yang, H., Lee, Y., Jeon, S., et al.: Multi-rotor drone tutorial: systems, mechanics, control and state estimation. Intell. Serv. Robot. **10**(2), 79–93 (2017)
7. Ali, N., Tawiah, I., Zhang, W.: Finite-time extended state observer based nonsingular fast terminal sliding mode control of autonomous underwater vehicles. Ocean Eng. **218**, 108179 (2020)
8. Nie, B.: Study on the modeling and control of a micro quadrotor. National University of Defense Technology (2006)
9. Ling, J.: Study of flight control algorithm for quad-rotor aircraft. Nanchang University (2013)

Analysis of Autonomous Take-Off and Landing Technology of Shipborne Unmanned Helicopter

Qibing Zhao[1](✉), Zhongyuan Yang[1], Chenfan Zhu[1], Jia Zhu[2], and Wan Sun[3]

[1] China Shipbuilding and Marine Engineering Design and Research Institute, Shanghai, China
zhaoqb@nuaa.edu.cn
[2] Wuhu Shipyard Co., Ltd., Wuhu, China
[3] Beijing Zhonghangzhi Technology Co., Ltd., Beijing, China

Abstract. The take-off and landing of shipborne unmanned helicopters are the most dangerous parts in the process of its dispatch and recovery. By comparing and analyzing the advantages and disadvantages of surface aviation support system for helicopter among the navies of different countries, the key technologies of the take-off and landing process of shipborne unmanned helicopter are summarized. In order to solve problems such as the difficulties of autonomous take-off and landing of shipborne UAVs, a three-stage strategy in landing process is proposed for a 300-tons ship and a 350 kg-class unmanned helicopter, and a real-time forecasting algorithm is used to predict the motion of the deck. Simulation results verify the feasibility of the proposed strategy and the effectiveness of the forecasting algorithm based on FlightGear software, which lays a theoretical foundation to carry out the ship-aircraft adaptability test of shipborne unmanned helicopter and dramatically reducing the risk of the test.

Keywords: Autonomous take-off and landing technology · Shipborne unmanned helicopter · Simulation

1 Introduction

The shipborne helicopter, as a kind of important equipment carried by aircraft carriers [1], amphibious ships [2], destroyers [3] and other ships, can accompany the mother carrier into every corner of the ocean to carry out various combat missions, provide real-time, flexible, and relatively accurate battlefield perception information for maritime operations, enhance the combat capability of single ship/ship formation, and further enrich and optimize the ship combat mode. Considering that modern local maritime conflicts are characterized with unclear targets, numerous uncertain factors, and a wide range of activities, shipborne UAV [4] has the outstanding advantages of low risk, easy deployment, easy accompanying, high maneuverability, and high cost-effectiveness ratio in comparison with the traditional manned aircraft. It could perform important tasks such as battlefield awareness, target locking, relay communication, damage assessment, with increasing importance in the process of strengthening air power at sea. Therefore, it is more and more necessary for shipboard unmanned aerial vehicles to be equipped on ships, and it has become a hot area for development for navies of various countries.

W. Fan et al. (Eds.): AsiaSim 2022, CCIS 1712, pp. 305–320, 2022.
https://doi.org/10.1007/978-981-19-9198-1_23

In the training and actual combat, the armed forces of various countries have found that the probability of accidents in the take-off and landing stages is the highest in the whole process carrying out the mission of carrier-borne unmanned helicopter. In the process of dispatch and recovery, unmanned helicopters need to overcome many interference factors, such as sea wind and waves, errors of its flight control system, and errors of ship guidance system and so on. Thus, it is necessary to break through key technologies, and only in this way can realize the fully automatic take-off/landing on the ship.

This paper mainly summarizes the recent research progress in regard to the key technologies of carrier-based aircraft take-off and landing. A scheme for fully automatic take-off and landing of the unmanned helicopter using the LSM-AIC-AR(n) [5] real-time prediction method with limited memory is implemented to predict the ship deck motion. After the construction of an unmanned helicopter take-off and landing scenario based on Flight Gear, the simulation is carried out to verify the feasibility of autonomous takeoff and landing at sea.

2 Shipborne UAV Take-Off and Landing Technology

According to the current situation of engineering, many experts and scholars are devoted to the research of autonomous take-off and landing technology of shipborne UAVs. Without additional sensor equipment during the autonomous take-off, the unmanned aerial vehicle uses its own navigation sensor to assist, while the autonomous landing needs additional guidance, landing assistance equipment and others. Therefore, the related research is focused on how to realize the safe and stable landing direction of UAV, among which the key technologies include ship guidance, auxiliary landing, and ship deck motion prediction.

2.1 Landing Guidance Technology

It is abroad for UAV landing guidance [5] and positioning technology of unmanned aerial vehicles, mainly including secondary radar guidance, ship-to-machine optical tracking guidance, differential satellite positioning guidance, analog satellite positioning guidance, etc.

(1) Secondary radar guidance

U.S. UCARS-V1 landing guidance system uses shipborne millimeter-wave secondary radar to locate the unmanned aerial vehicle, which has been applied to guide MQ-8B "Fire Scout" unmanned helicopter landing (Fig. 1).

As shown in Fig. 2, the automatic landing system of the French Navy D2AD carrier-based UAV has realized the relative positioning of the ship to the carrier-based UAV by arranging and installing radar arrays on the flight deck and radar beacons on the carrier-based UAV.

Fig. 1. UCARS-V1 landing guidance system

Fig. 2. UAV landing guidance system shipboard radar

(2) Ship-to-machine optical tracking and guidance

SADA system, another landing guidance system developed in France, has utilized shipboard infrared tracking equipment to accurately track the UAV with infrared sensing technology and send out flight instructions to adjust the flight route until the harpoon of the UAV is aligned with the center of the landing aid grille. As early as in 2008, the system was installed on the destroyer Montcalm and the completed the joint debugging test with a certain type of UAV.

(3) Differential satellite positioning guidance

The United States Precision Approach and Landing System (JPALS) [6] adopts high-precision differential satellite positioning technology and has been verified by landing X-47B UAVs and F/A-18C UAVs. The technology is mature and suitable for aircraft carriers and land airstrips.

(4) Simulated satellite positioning and guidance

The pseudo-satellite positioning system (Deck Finder) developed by AIRBUS in Europe provides 3D images of the relative position of RPAS based on six simulated satellite transmitters on the take-off and landing operation deck to assist the aircraft land safely (Fig. 3).

Fig. 3. European deck finder pseudo-satellite positioning system

2.2 Auxiliary Landing Technology

(1) Landing Assistance, Fixation and Traction System (RAST)

As shown in Fig. 4, Canada has developed the landing assistance, fixation, and traction system (RAST), which is the world's mainstream ship-borne helicopter landing assistance system. It has been equipped with the navies of the United States, Canada, Japan, India, Argentina, Australia, Spain, and many other countries. The system is mainly composed of pull-down cable, winches, clamping trailer, console, track, and other equipment. In the process of a helicopter landing, the ship pulls down the rope and clamps the trailer to realize safe landing and ship surface transfer.

(2) Helicopter anti-skid net

As shown in Fig. 5, in order to prevent Ka-27 series of ship-borne helicopters from rolling over during landing, the special anti-skid nets are equipped in Russia and China. The structure and principle of the anti-skid net are simple, and the four wheels of the landing gear of the clamping machine will be caught by the nylon net once they touch the ship, reducing the possibility of the helicopter sliding on the flight deck.

Fig. 4. Helicopter landing aid system RAST

Fig. 5. Ka-27 helicopter surface traction

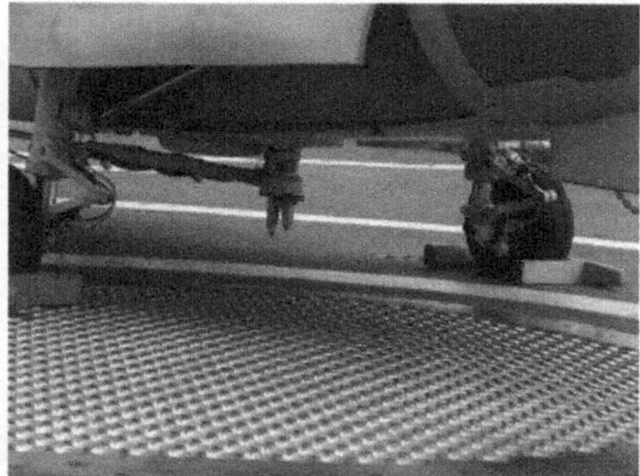

Fig. 6. Z-9 helicopter harpoon

(3) Helicopter Landing grille

As shown in Fig. 6, Chinese Navy has developed a special landing grille on the deck to improve the sea state adaptability of Z-9 series helicopters. Once the helicopter touches the ship, it is inserted into the circular grille and locked by its fork-shaped locking structure, which could provide a rigid connection between the helicopter and the flight deck.

2.3 Ship Deck Motion Prediction

Deck motion prediction mainly comes from the landing demand of aircraft carriers. In the 1960s, some countries began to study the extremely short-term forecast method of aircraft carrier movement [7] After the 1970s, modern control theory was applied to the online real-time prediction of ship motion. The U.S. Navy Hawaii Laboratory and Ames Research Center of NASA have conducted in-depth research on the feasibility of ship attitude prediction and accurately predicted ship movement information in the next five to six seconds or even longer in real-time. In the 1980s, the United States began to install equipment and procedures for real-time and extremely short-term forecasts of aircraft carrier movement on aircraft carriers. Ship motion prediction is included in the flight management of carrier-based aircraft.

For non-aircraft carrier ship platforms, British Airways and General Instruments (AGI) developed the Landing Period Designator, LPD) [8], which can measure the deck motion information in real-time, predict the deck rest period and aircraft landing energy index, and provide the best time window indication for manned aircraft and unmanned aircraft on board. This product is mainly equipped with destroyer-type ships in the United States, Britain, Canada, Israel, and other countries, used for take-off/launch and recovery of manned helicopters and unmanned aerial vehicles, and is also a part of the verification plan of American "Fire Scout" unmanned helicopter [9]. The LPD consists of the deck motion sensor installed below the deck, the landing aid indicator above the deck, and the display and control unit in the bridge, as shown in the Fig. 7.

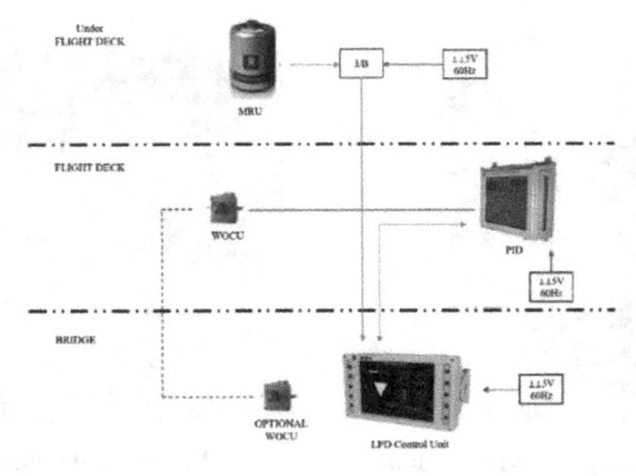

Fig. 7. Composition and cross-linking relationship of LPD system

3 Autonomous Take-Off and Landing Program on the Flight Deck

The target ship has been in a six-degree-of-freedom motion state [10] at sea. In this paper, the landing guidance scheme is selected, and the take-off and landing process of the UAV are planned to provide input for numerical simulation.

3.1 Landing Guidance Scheme

Compared with secondary radar guidance, ship-to-machine optical tracking guidance, differential satellite positioning guidance, simulated satellite positioning guidance, and other guidance methods, differential satellite positioning technology has outstanding advantages such as high guidance accuracy, long working distance, all-weather and all-time guidance, etc. Differential satellite positioning guidance [11] is used as a landing guidance method in simulation.

3.2 Take-Off Process

The take-off process of unmanned helicopter carrier surface is similar to the preparation process of land take-off. Jet fuel is injected after checking the aircraft's state. To reduce the interference of hull movement before the vertical climb, the unmanned helicopter takes off in attitude holding mode. When the absolute speed of the forwarding flight disappears to 0 after leaving the ship, it means that the take-off mission is completed. The take-off process is shown in the Fig. 8.

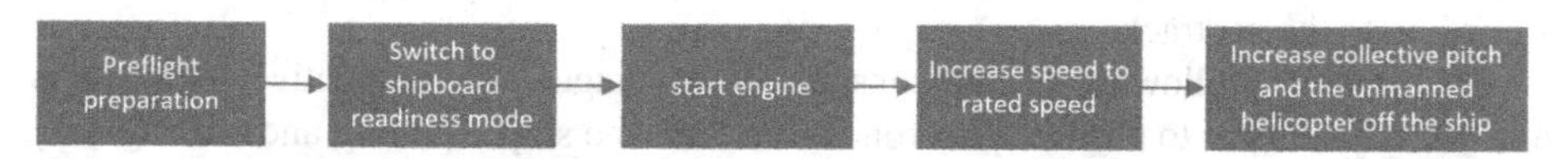

Fig. 8. Take-off process of an unmanned helicopter on board

3.3 Landing Process

The landing process is mainly divided into four stages. As shown in the Fig. 9, 0–1 indicates the return trajectory, 1–2 indicates the descent approach trajectory, 2–3 indicates the hovering follow-up trajectory and 3–4 indicates the rapid landing.

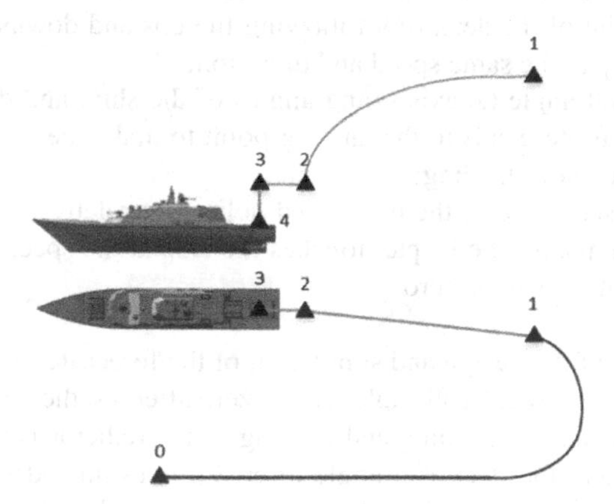

Fig. 9. Flow chart of landing segment trajectory

(1) Design of return approach scheme

The schematic diagram of the return approach is as follows: (1) Collecting current information on ship heading, speed, and relative position (2) Planning an ideal descent trajectory (3) Guiding the unmanned helicopter to the rear of the ship (4) Keeping a certain relative distance and speed with the ship.

The return approach design consists of three stages. The respective steps are as follows: constant height and deceleration; constant speed and falling height; constant height and deceleration.

(2) Hover follow-up scheme design

The schematic diagram of the hovering follow-up is s as follows: adjusting the unmanned helicopter's position, speed, and heading to keep pace with the ship in the transitional stage; guiding the unmanned helicopter to move above the landing point of the ship according to the set track.

The hovering follow-up design consists of two stages. The respective steps are as follows: adjust itself to ensure synchronization with the ship; opening and closing.

(3) Design of rapid landing scheme

The schematic diagram of rapid landing is as follows: (1) Guiding the unmanned helicopter to descend steadily (2) Catching the ideal landing opportunity when it reaches a safe altitude (3) Controlling the unmanned helicopter to descend rapidly to complete the landing.

The rapid landing phase control process includes the following three parts:

(1) Control the unmanned helicopter to descend and hover at a fixed safe height above the landing point of the deck (not following the ups and downs of the deck), and fly with the ship at the same speed and direction;
(2) Estimate the roll angle (Z-axis swing angle) of the ship, and determine the time when the aircraft descends to the landing point to make the roll angle of the ship zero at the moment of landing;
(3) Control the descent rate of the unmanned helicopter relative to the landing point, at which the unmanned helicopter touches the ship at the specified time point and the roll angle of the ship is zero.

The importance of the design and simulation of the forecaster is highlighted. If the predictor predicts that the deck roll angle will be zero after 8 s, the unmanned helicopter will be triggered to start descending and landing. The predictor is updated 2 s before touching the ship, and the deck roll angle after 2 s is estimated so that the descent suspension control system can decide whether to give up landing or not. The predictor is only used when the wave motion is obvious. If the sea is calm and the ship's roll motion is less than 1° to 2°, then the start time for the descent does not depend on the estimated value of the ship.

4 Real Time Prediction of Ship Deck Motion

4.1 Sea State Simulation Model

The flight deck motion is assumed to be a stationary random process with a narrow frequency band, and the sinusoidal wave combination is used to describe the motion.

The rolling, pitch, and heave models are defined in Eqs. 1, 2 and 3, and their motion response curves are shown in Fig. 10(a), 10(b) and 10(c).

Rolling simulation model:

$$theta(t) = 0.005sin(0.46t) + 0.00946sin(0.58t) + 0.00725sin(0.7t) + 0.00845sin(0.82t) \quad (1)$$

Pitch simulation model:

$$phi(t) = 0.021sin(0.46t) + 0.0431sin(0.54t) + 0.029sin(0.62t) + 0.022sin(0.67t) \quad (2)$$

Heave simulation model:

$$h(t) = 0.2172sin(0.4t) + 0.417sin(0.5t) + 0.3592sin(0.6t) + 0.2227sin(0.7t) \quad (3)$$

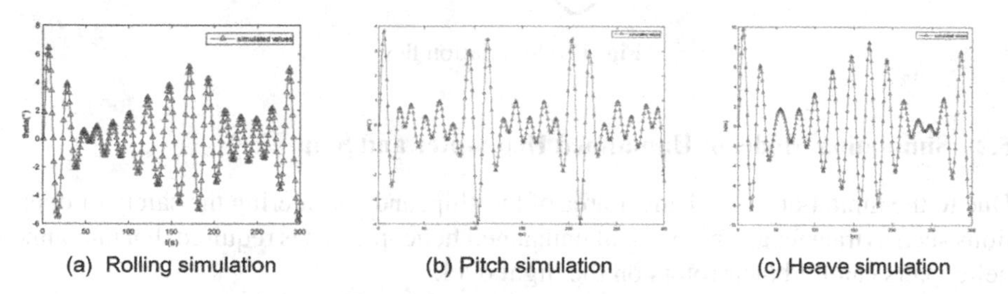
(a) Rolling simulation (b) Pitch simulation (c) Heave simulation

Fig. 10. Motion response curve diagram

4.2 LSM-AIC-AR(n) Real-Time Prediction Algorithm

LSM-AIC-AR(n) real-time prediction algorithm contains three main parts. The estimation of the model coefficients was performed by the Least Squares Method and the Akaike Information Criterion is used for estimating the model order. Real-time estimation method based on limited memory method is used for estimation in the next steps. Detailed simulation flow is shown in detail in Fig. 11.

5 Simulation and Analysis

5.1 Simulation Software and Scenarios

Building the simulation environment at the basis of Flight Gear software [12] and the control law of the external unmanned helicopter can realize the real feedback of the unmanned helicopter and visually observe the process of the unmanned helicopter taking off and landing on the ship from a three-dimensional visual perspective.

The geographical background of the sea area near Bohai Bay is set as a simulation scene. The ship is about 10 nautical miles away from the coast, with a speed of 25 knots and a heading of 45° northeast. At the simulated ship roll angle of $-7.5 \leq \varphi \leq +7.5$, pitch angle of $-2 \leq \theta \leq +2$, the corresponding period is 8 s, the heaving amplitude is ± 2.2 m, the corresponding period is 8 s, and the motion curve contains a rest period with small motion amplitude.

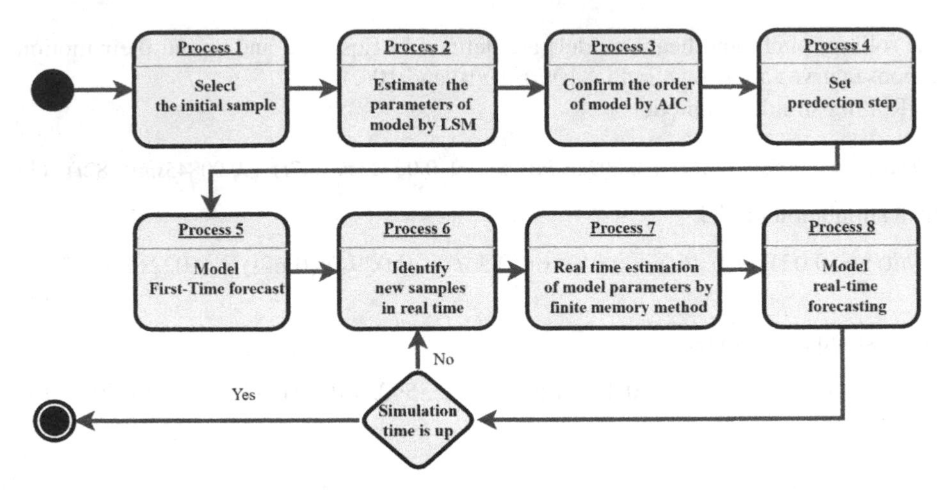

Fig. 11. Simulation flow

5.2 Simulation Model of Unmanned Helicopter and Ship

Due to the limitation of the hangar area of the ship, and considering the safety of operations such as transfer and mooring of unmanned helicopters, it is required that unmanned helicopters can fold their rotors on the flight deck.

A certain type of unmanned helicopter is a highly reliable universal unmanned helicopter developed by Beijing Zhonghangzhi Technology Co., Ltd., which adopts a coaxial double-rotor configuration. It is characterized with small fuselage size, convenient transportation and transshipment, flexible use, vertical take-off, landing and hovering in the air, low requirements for take-off and landing sites, strong wind resistance, tailless propeller failure, and other safety risks. It has great advantages in automation degree, adaptability to a complex environment, multi-tasking ability, safety, reliability, operability, simple maintenance, economical use, etc., and can meet various needs and product demands of different types of users.

A certain type of ship has a full displacement of 300 tons, a total length of about 80 m, and a width of about 23 m, as shown in Fig. 11. It has a wide flight deck and multi-type helicopters can take off or land on the deck. In this paper, assuming the ship as a typical ship-borne platform, a three-dimensional model is established and imported into simulation platform to carry out the relevant analysis of UAV take-off and landing. To focus on the whole process of UAV take-off or landing, and weaken the detailed operation of landing, the lighting and facilities for landing assistance are not considered in the simulation process.

5.3 Real-Time Rest Period Forecast of Deck

LSM-AIC-AR(n) real-time prediction model based on finite memory method is used to predict the flight deck motion of carrier-borne aircraft. The flight deck motion model established above was adopted, and the period T = 0.5 s was set, and the simulation model time was set to 150 s. The flight deck motion predictor starts to calculate at 50 s and predicts deck motion after 8 s. Blue * is the sampling point of the simulated wave model, and green one is the estimated data point.

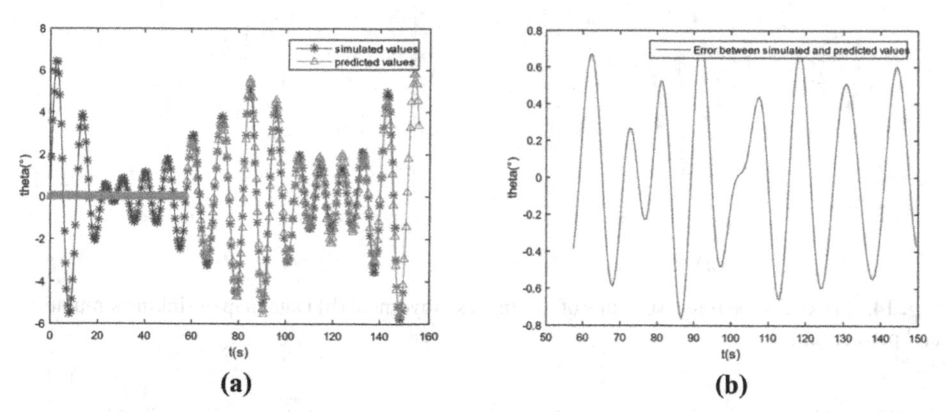

Fig. 12. (a) Real-time prediction results of rolling 8 s motion (b) Rolling simulation and prediction error data (Color figure online)

The real-time tracking curve and tracking error curve of the estimator for rolling motion at the ideal landing point are shown in Fig. 12(a) and (b), respectively.

It can be seen from Fig. 12(a) and (b) that the estimator can accurately track the roll motion of the ideal landing point after 8 s, and the estimation error is less than $\pm 0.8°$, indicating a relatively accurate estimation.

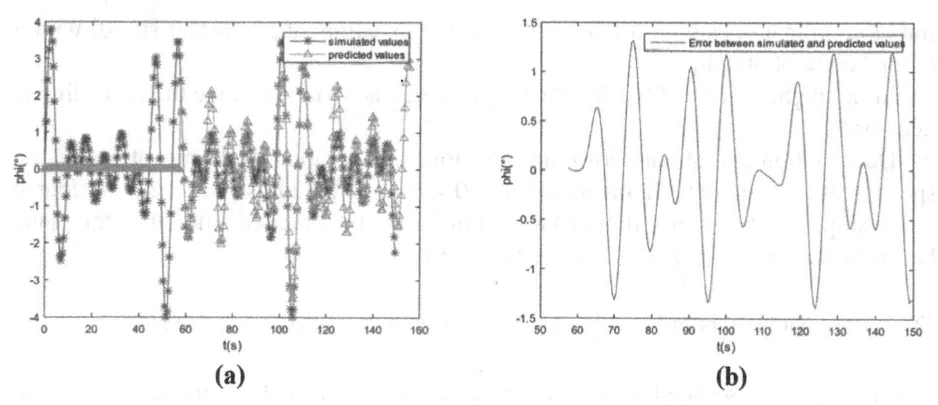

Fig. 13. (a) Real-time prediction results of pitch 8 s motion (b) Pitch simulation and prediction error data

The real-time tracking curve and tracking error curve of the estimator for pitch motion at the ideal landing point are shown in Fig. 13(a) and (b), respectively.

It can be seen from Fig. 13(a) and (b) that the estimator can accurately track the pitch motion of the ideal landing point after 8 s, and the estimation error is less than $\pm 1.5°$, indicating a relatively accurate estimation.

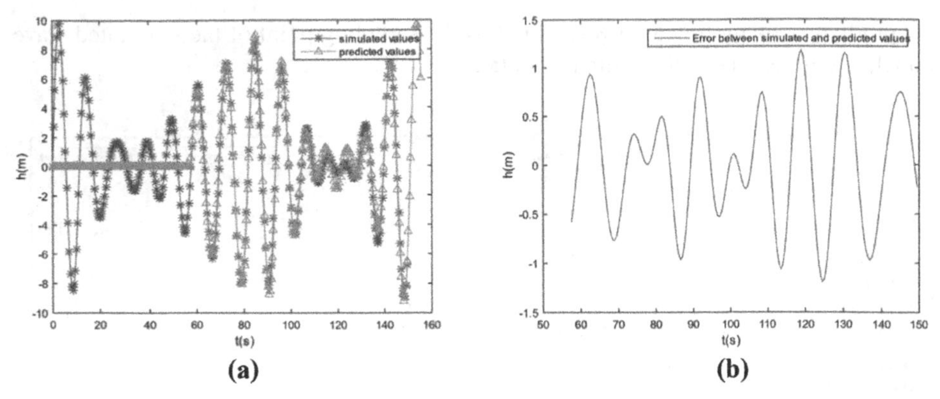

Fig. 14. (a) Real-time forecast results of sinking 8 s movement (b) Data map of sinking simulation and prediction error

The real-time tracking curve and tracking error curve of the predictor for the heave movement at the ideal landing point are shown in Fig. 14(a) and (b), respectively.

It can be seen from Fig. 14(a) and (b) that the estimator can accurately track the heave movement of the ideal landing point after 8 s, and the estimation error is less than ± 1.5 m, indicating a relatively accurate estimation.

5.4 Description of Landing and Recovery Process

(1) Flight parameters of an unmanned helicopter in initial appearance state

Initial coordinate position: 8 km south of the ship sailing at a constant speed, with a flying height of 500 m;

Initial flight speed and heading: the flight speed is 30 m/s, and the flight heading is due north;

Deceleration and altitude lowering position: Deceleration to 25 m/s at a cruising speed of 30 m/s, and after flying north for 250 s, the aircraft adjusted to the same course as the ship, and the altitude dropped to 100 m during the process. After 20 s, the plane began to descend to 20 m, as shown in Fig. 15(a).

(2) Flight parameters of unmanned helicopter entering guidance phase state

When the plane descended to 20 m, it formally entered the landing guidance stage, and the relative positioning equipment began to work formally.

The mother ship starts to send the movement parameters of the ship to the unmanned helicopter, including heading, speed, real-time position information, deck roll Angle, deck pitch Angle, deck heave amplitude and wind direction and speed on deck, to guide the aircraft to chase the ship, as shown in Fig. 15(b).

After receiving the relevant data sent by the carrier, the unmanned helicopter adjusted its flight speed and altitude and lowered its landing gear to guide the aircraft to chase the ship. At 60 s, the aircraft reached directly above the ship deck and kept the same

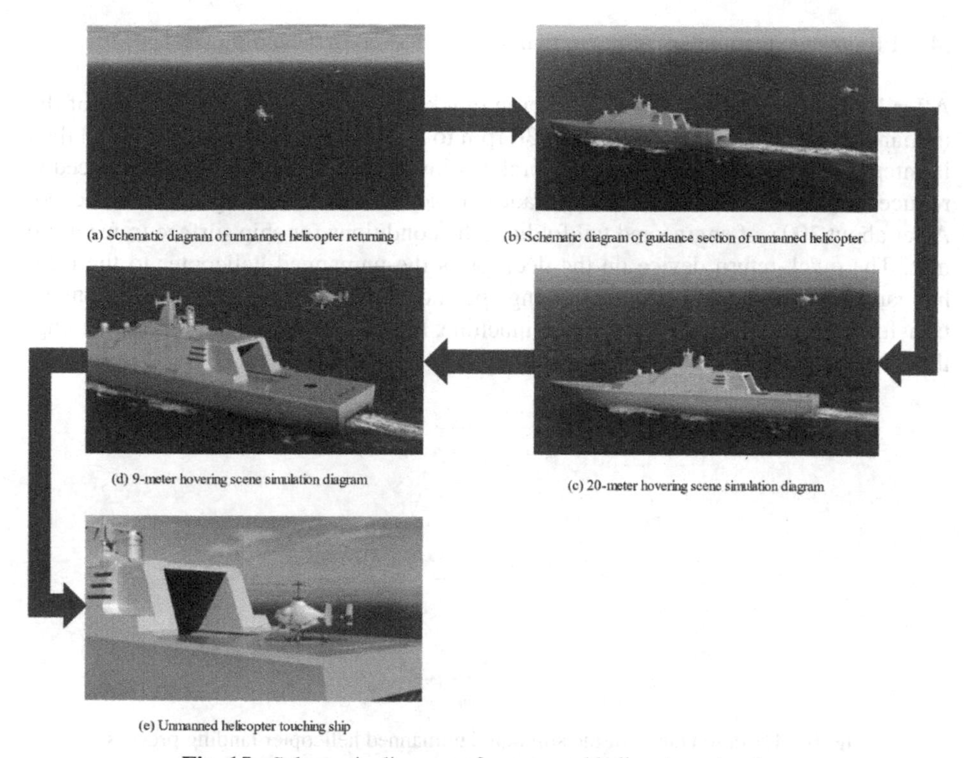

(a) Schematic diagram of unmanned helicopter returning
(b) Schematic diagram of guidance section of unmanned helicopter
(c) 20-meter hovering scene simulation diagram
(d) 9-meter hovering scene simulation diagram
(e) Unmanned helicopter touching ship

Fig. 15. Schematic diagram of unmanned helicopter returning

speed as the ship (25 knots 12.86 m/s) to achieve a relatively static flight with the mother ship. At this moment, the static gap period is predicted for the movement data of the carrier-carrying deck and the landing is carried out waiting for the opportunity, as shown in Fig. 15(c).

(3) Flight parameters of unmanned helicopter entering landing phase

Adjusting the approach course of the landing ship according to the wind direction of the deck, the plane continues to land to 9 m. According to the definition of the 6-m-high deck model, the plane is 3 m away from the deck at the time. Allowing for the heave and pitch height of the ship, the helicopter needs to keep a safe height and avoid collision with the deck, as shown in Fig. 15(d).

After the clearance period is determined, the landing gear descends at a speed of 0.6 m/s. Finally, the landing gear is achieved and the airborne rapid recovery device is combined with the surface rapid recovery device to complete the landing process, as shown in Fig. 15(e).

(4) Flight state parameters of the unmanned helicopter after landing

After landing and fixing with a ship-borne quick return device, the rotor speed of the unmanned helicopter is reduced from 730 rpm to 300 rpm lasting for the 60 s, and then it enters the idle state. Then, the rotor brake is involved in the work, the rotor speed is reduced to 0 rpm within 10 s, and the blades are adjusted to feathering state and locked. After about 300 s of engine and tail folding, the conditions for ship surface transfer are met. The quick-return device on the deck pulls the unmanned helicopter to the right hangar for parking at a horizontal moving speed of 1 m/s, to complete the autonomous transfer task of the autonomous landing machine. In the process of autonomous landing, the elevation curve of the unmanned helicopter is shown in Fig. 16.

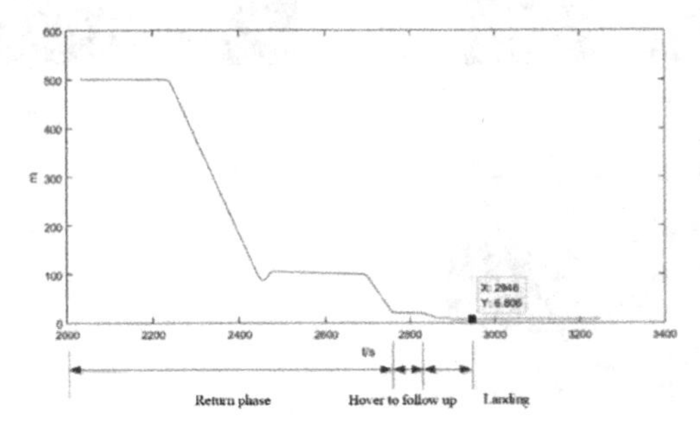

Fig. 16. Elevation curve of the simulated unmanned helicopter landing process

The motion tracking of the unmanned helicopter and warship is shown in Fig. 17. The blue Heli curve is the flight track of an unmanned helicopter, and the red ship is the motion track of a warship. The horizontal and vertical coordinates respectively represent the longitude and latitude of the simulated sea environment. The speed of the ship is 12.8 m/s, the altitude of the simulated deck is 6.0 m, and the center of the aircraft is 1.2 m above the landing gear.

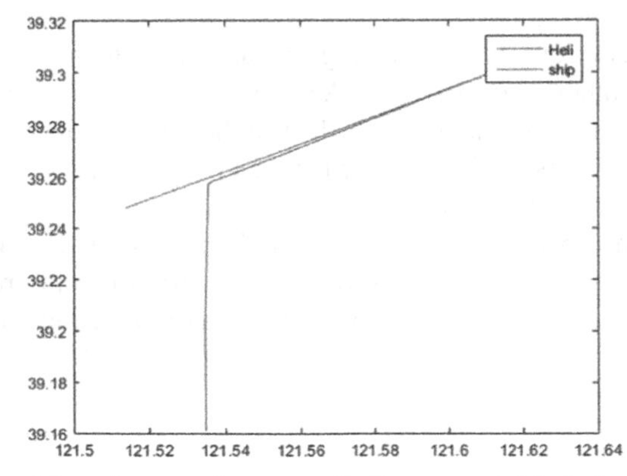

Fig. 17. Simulation of the trajectory of unmanned helicopter and ship during landing (Color figure online)

6 Conclusion

In this paper, the key technologies and scheme design of autonomous take-off and landing of the unmanned helicopter are analyzed, and the simulation of flight deck motion prediction based on the LSM-AIC AR(N) model is carried out by using Flight Gear software, and its feasibility is verified. The simulation analysis based on computer software can economically, efficiently, and intuitively verify the feasibility of unmanned helicopters taking off and landing at sea, but it is difficult to simulate the complex environmental conditions such as crustal effect in the process of taking off and landing. Subsequently, the take-off and landing test of an unmanned helicopter on the platform will be carried out based on the land-based swing platform (as shown in Fig. 18). The swing platform will be used to simulate the rolling, pitch, and depth of the flight deck, and the risk of equipment on the ship will be further reduced through the test measurement data and optimization of the flight control system.

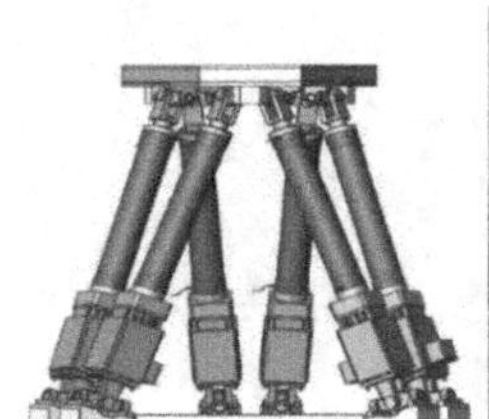

Fig. 18. Diagram of swing table

References

1. Zhu, Y., Xiong, Z., Hu, Y.: On the development trends of aircraft carriers. Chin. J. Ship Res. **11**(1), 1 (2016). https://doi.org/10.3969/j.issn.1673-3185.2016.01001
2. Ferreiro, L., Autret, G.: A comparison of French and U.S. amphibious ships. Naval Eng. J. **107**(3), 167–184 (1995)
3. O'Reilly, P.: Aircraft/deck interface dynamics for destroyers. Mar. Technol. **24**(1), 15–25 (1987)
4. Webb, K., Rogers, J.: Adaptive control design for multi-UAV cooperative lift systems. J. Aircr. **4**, 1–21 (2021)
5. Huang, Y.: Research on key technology of automatic carrier landing for unmanned helicopter. Northwestern Polytechnical University (2015)
6. Zhen, Z.: Research development in autonomous carrier-landing/ship-recovery guidance and control of Unmanned Aerial Vehicles. Acta Automatica Sinica **45**(4), 669 (2019). https://doi.org/10.16383/j.aas.2018.c170261
7. Pervan, B., Chan, F.C., Gebre-Egziabher, D., et al.: Performance analysis of carrier-phase DGPS navigation for shipboard landing of aircraft. NAVIG. J. Inst. Navig. **50**(3), 181–192 (2003)
8. Yakimenko, O.A., Kaminer, I.I., Lentz, W.J., et al.: Unmanned aircraft navigation for shipboard landing using infrared vision (S0018-9251). IEEE Trans. Aerosp. Electron. Syst. **38**(4), 1181–1200 (2002)

9. Ferrier, B.D.: Experiences with the landing period designator (LPD) helicopter recovery aid in high sea states (1998)
10. Theodore, C.R., Colbourne, J.D., Tischler, M.B.: Rapid frequency-domain modeling methods for unmanned aerial vehicle flight control applications. J. Aircr. **41**(4), 735–743 (2004)
11. Xu, D., Liu, X., Wang, L.: Influence of carrier motion on landing safety for carrier-based airplanes. J. Beijing Univ. Aeronaut. Astronaut. **37**(3), 289–294 (2011)
12. Shi, L.: Influence of disturbances on ship-based aircraft's landing process. Comput. Simul. **26**(12), 46 (2009). https://doi.org/10.3969/j.issn.1006-9348.2009.12.014
13. Zhang, J., Xu, H., Zhang, D., et al.: Safety modeling and simulation of multi-factor coupling heavy-equipment airdrop. Chin. J. Aeronaut. **27**(5), 1062–1069 (2014)

Research on UAV State Estimation Method Based on Variable Structure Multiple Model

JianWei Chen[1,2], Yu Wang[3](✉), Siyang Chen[3], Wei Lu[3], and Cheng Ma[3]

[1] Northwestern Polytechnical University, Xi'an, China
[2] China Academy of Launch Vehicle Technology, Beijing, China
[3] Beijing Institute of Astronautical Systems Engineering, Beijing, China
wangyu_hitsa@hit.edu.cn

Abstract. In order to provide accurate UAV state estimation information for UAV monitoring and control, the UAV state estimation method based on variable structure multi model is studied. Firstly, the state model set is established for the motion forms of UAV, such as smooth flight, lateral turning maneuver and longitudinal jumping maneuver, and the measurement model is established based on the radar measurement principle; Then, based on the variable structure multi model framework, a model set adaptive strategy is proposed, which can solve the problems of too many threshold parameters and complex adaptive strategy; Finally, the simulation scenarios of two radar tracking UAVs are built, and the superiority of the proposed method in the accuracy of state estimation is verified by mathematical simulation.

Keywords: Variable structure multi model · UAV · State estimation

1 Introduction

With the rapid development of UAV technology, more and more UAVs are emerging in the sky of cities for transportation, reconnaissance and other tasks. UAV not only brings us convenience, but also increases the threat to our life security [1]. In order to realize the monitoring and control of UAV, how to carry out high-precision state estimation for UAV has attracted more and more attention and research. Compared with the ground target with limited movement form, UAV can carry out horizontal maneuver, vertical jump maneuver and other maneuvers, with strong maneuverability and uncertain maneuvering timing, which makes it very difficult to estimate its flight state with high precision. In order to improve the accuracy of UAV state estimation, the key is to establish the target motion model and propose a high-precision state estimation method.

At present, there are two main modeling ideas in the establishment of target motion model: 1) Establish a parametric dynamic model [2–6]. Based on the force analysis, the expressions of engine thrust, aerodynamic force and earth gravity are derived respectively, which are converted to the same coordinate system and the dynamic model is obtained simultaneously. The model contains unknown parameters such as overall parameters and aerodynamic parameters. The overall parameters are assumed to be

W. Fan et al. (Eds.): AsiaSim 2022, CCIS 1712, pp. 321–333, 2022.
https://doi.org/10.1007/978-981-19-9198-1_24

known as prior information in most literatures. If the prior information is sufficient and the unknown parameters are estimated accurately, it has the characteristics of high accuracy. However, according to the results of the current literature, the accuracy of aerodynamic parameter identification is not high. In addition, the parameterized dynamic model is complex and has the disadvantage of large amount of calculation. 2) Establish kinematic model [7–10]. Common kinematic models include CV model, CA model, CT model, singer model, current statistical model, Sine wave model, etc. This kind of model does not involve coordinate transformation and force analysis, so the model form is relatively simple and easy to apply, but there is a problem of limited accuracy.

Scholars at home and abroad have proposed many models based on the above two ideas. Literature [11] proposed a nonlinear Markov acceleration model from the perspective of statistics. In reference [12], the aerodynamic acceleration is modeled as a second-order time-dependent model, which can describe the periodic and attenuation changes of acceleration. In order to cover a variety of maneuvers, reference [13] designed a singer model with adaptive frequency parameters. With the rapid development of multi model methods and the enhancement of computer capabilities, scholars at home and abroad have turned their attention to multi model methods [14, 15]. From the above analysis, it can be seen that the current mainstream solution is to use the multi model method to describe the target motion for UAV, which has a variety of maneuvering forms.

The early multi model method is static multi model method. Each filter operates independently and is not connected with each other. Until the 1980s, Markov chain process was used to represent the dynamic process of the system and the transformation process between models [16]. According to different model inputs, dynamic multi model methods are divided into generalized pseudo Bayesian (GPB) [17] and interacting multiple model (IMM) [18]. IMM method has higher trajectory tracking accuracy because it considers the matching degree between the model and the actual motion. At present, it has been widely used in the field of target trajectory tracking [19–21].

In static multi model method, GPB method and IMM method, the model is always the same, which is also collectively referred to as fixed structure multi model method. The more models used in this kind of method, the more comprehensive the target maneuver coverage will be. However, the amount of calculation will inevitably increase. At the same time, there is a problem of poor accuracy caused by model competition. In order to solve this problem, Professor Li XR proposed variable structure multiple model (VSMM) [22, 23] in his series of articles. The tracking of different maneuver forms is realized through two key steps: model set adaptation and model set based adaptive estimation. Model set adaptation is the core of VSMM algorithm, and it is also the difficulty of the algorithm.

It can be seen from the above analysis that the main problem of variable structure multi model method is how to realize model set adaptation. The existing methods have the problems of too many threshold parameters and complex adaptive strategies, so it is necessary to further study the model set adaptive strategies.

In this paper, the different motion forms of UAV are modeled separately to form the state model set, and then the state estimation method based on variable structure multi model is studied, and the model set adaptive switching strategy is proposed.

2 Construction of UAV Tracking Model

2.1 State Model Set

In this section, a model set covering different maneuver forms will be established for the UAV's smooth flight (uniform speed, uniform acceleration, uniform deceleration), lateral maneuver, longitudinal jump maneuver and other motion forms.

When the UAV flies steadily or turns laterally, the trajectory is relatively flat. It can be assumed that the acceleration at the next moment is within the neighborhood of the acceleration at the current moment. Assume that the "current" statistical probability of maneuvering acceleration obeys the modified Rayleigh distribution, and the mathematical form is

$$f(a) = \frac{a_{\max} - a}{\sigma^2} \exp\left[-\frac{(a_{\max} - a)^2}{2\sigma^2}\right], \ a_{\max} - a \geq 0 \tag{1}$$

where, a denotes acceleration; $a_{\max}$ denotes the maximum of acceleration; σ^2 denotes the variance.

According to the property of Rayleigh distribution, the expected acceleration of UAV is

$$E[a] = a_{\max} - E[a_{\max} - a] = a_{\max} - \sigma\sqrt{\frac{\pi}{2}} \tag{2}$$

The variance of UAV acceleration is

$$\sigma_a^2 = \frac{4 - \pi}{2}\sigma^2 \tag{3}$$

Therefore, when the acceleration expectation $E[a]$ is known, we can obtain

$$\sigma = \sqrt{\frac{2}{\pi}}(a_{\max} - E[a]) \tag{4}$$

Substitute into Eq. (3)

$$\sigma_a^2 = \frac{4 - \pi}{\pi}(a_{\max} - E[a])^2 \tag{5}$$

Establish the non-zero mean time correlation model of acceleration, we have

$$\ddot{X}(t) = \overline{a}(t) + a(t) \tag{6}$$

where, $\overline{a}(t)$ denotes the "current" mean of acceleration, which is assumed to be constant in the sampling period. $a(t)$ is the noise of acceleration, which is modeled as a stochastic processes with zero mean exponential decay autocorrelation

$$R_a(\tau) = E[a(t)a(t + \tau)] = \sigma_a^2 \cdot e^{-\alpha|\tau|} \tag{7}$$

where, σ_a^2 denotes the variance of acceleration noise; α denotes the maneuver frequency. The stronger the maneuver, the smaller the value α.

Take Fourier transform to obtain power spectral density, and denotes the Laplacian operator as $s = wj$, we have

$$S_a(s) = \frac{-2\alpha\sigma_a^2}{(s+\alpha)(s-\alpha)} \tag{8}$$

The transfer function

$$H(s) = \frac{1}{s+\alpha} \tag{9}$$

The acceleration noise $a(t)$ satisfies

$$\dot{a}(t) = -\alpha \cdot a(t) + w(t) \tag{10}$$

where, $w(t)$ is Gaussian noise with zero-mean and variance $2\alpha\sigma_a^2$.

Derive formula (10), we can obtain

$$\dddot{X}(t) = -\alpha \cdot \ddot{X}(t) + \alpha \cdot \overline{a}(t) + w(t) \tag{11}$$

And the discrete form model

$$\boldsymbol{X}_{k+1} = \boldsymbol{\Phi}_k \boldsymbol{X}_k + \boldsymbol{U}_k \overline{a}_k + \boldsymbol{w}_k \tag{12}$$

where,

$$\boldsymbol{\Phi}_k = e^{AT} = \begin{bmatrix} 1 & T & \frac{\alpha T - 1 + e^{-\alpha T}}{\alpha^2} \\ 0 & 1 & \frac{1-e^{-\alpha T}}{\alpha} \\ 0 & 0 & e^{-\alpha T} \end{bmatrix} \tag{13}$$

$$\boldsymbol{U}_k = \int_t^{t+T} e^{\boldsymbol{A}(t+T-\tau)} \boldsymbol{U}_1 d\tau = \begin{bmatrix} \frac{-T+\alpha T^2/2+\left(1-e^{-\alpha T}\right)/\alpha}{\alpha} \\ T - \frac{1-e^{-\alpha T}}{\alpha} \\ 1-e^{-\alpha T} \end{bmatrix} \tag{14}$$

where, $\overline{a}_k$ is the mean of "Current" acceleration, T is the sampling period, α is the maneuver frequency, the process noise w_k is zero-mean Gaussian noise, the variance $\boldsymbol{Q}_k$ is

$$\boldsymbol{Q}_k = E\left[\boldsymbol{w}_k \boldsymbol{w}_k^T\right] = 2\alpha\sigma_a^2 \boldsymbol{\Gamma}_k \boldsymbol{\Gamma}_k^T \tag{15}$$

where,

$$\boldsymbol{\Gamma}_k = \int_t^{t+T} e^{\boldsymbol{A}(t+T-\tau)} \boldsymbol{B} = \begin{bmatrix} \frac{\left(1-e^{-\alpha T}\right)/\alpha + 0.5\alpha T^2 - T}{\alpha^2} \\ \frac{T-\left(1-e^{-\alpha T}\right)/\alpha}{\alpha} \\ \frac{1-e^{-\alpha T}}{\alpha} \end{bmatrix} \tag{16}$$

The variance of "Current" acceleration σ_a^2 is

$$\sigma_a^2 = \begin{cases} \frac{4-\pi}{\pi}[a_{\max} - \overline{a}]^2, & \overline{a} \geq 0 \\ \frac{4-\pi}{\pi}[a_{-\max} + \overline{a}]^2, & \overline{a} < 0 \end{cases} \tag{17}$$

where, $a_{\max}$ and $a_{-\max}$ are the maximum positive and negative acceleration, respectively.

During the longitudinal jump maneuver of UAV, the acceleration has the characteristics of periodicity, attenuation and time-varying mean value of acceleration. In this section, the non-zero mean attenuation oscillation sine wave model (NZSW) is derived to represent the acceleration of UAV and meet the characteristics of periodicity, attenuation and time-varying mean value of acceleration. It is assumed that the acceleration $a(t)$ is composed of acceleration mean $\overline{a}(t)$ and acceleration disturbance $a_0(t)$, i.e.

$$a(t) = \overline{a}(t) + a_0(t) \tag{18}$$

The autocorrelation function of acceleration $a(t)$ is expressed by sine wave autocorrelation time function model. In addition, the exponential attenuation term is introduced to reflect the attenuation characteristics of acceleration, i.e.

$$R_{a_0}(\tau) = E(a_0(t)a_0(t+\tau)) = \sigma_a^2 \cos(\omega_0 \tau) \cdot e^{-\mu|\tau|} \tag{19}$$

where, τ denotes time interval; σ_a^2 denotes the variance of acceleration; ω_0 denotes the oscillation frequency; μ denotes attenuation coefficient.

The power spectral density of $R_a(\tau)$ is obtained by Fourier transform

$$S_{a_0}(w) = F\big(R_{a_0}(\tau)\big) = \sigma_a^2 \left[\frac{\mu}{\mu^2 + (w+\omega_0)^2} + \frac{\mu}{\mu^2 + (w-\omega_0)^2}\right] \tag{20}$$

where, w denotes angular rate.

Therefore

$$\begin{aligned} S_{a_0}(w) = {} & \frac{\sigma_a^2}{2}\left[\frac{1}{\mu - (\omega_0 - w)j} + \frac{1}{\mu + (\omega_0 + w)j}\right] \\ & + \frac{\sigma_a^2}{2}\left[\frac{1}{\mu + (\omega_0 - w)j} + \frac{1}{\mu - (\omega_0 + w)j}\right] \end{aligned} \tag{21}$$

and

$$\begin{aligned} S_{a_0}(w) = {} & 2\mu\sigma_a^2 \frac{\sqrt{\mu^2 + \omega_0^2} + wj}{[\mu - (\omega_0 - w)j][\mu + (\omega_0 + w)j]} \\ & \cdot \frac{\sqrt{\mu^2 + \omega_0^2} - wj}{[\mu - (\omega_0 + w)j][\mu + (\omega_0 - w)j]} \end{aligned} \tag{22}$$

Define Laplacian operator $s = wj$, we can obtain

$$H(s) = \frac{s + \sqrt{\mu^2 + \omega_0^2}}{s^2 + 2\mu s + \mu^2 + \omega_0^2} \tag{23}$$

The second-order Markov model of acceleration is

$$\ddot{a}_0(t) + 2\mu\dot{a}_0(t) + \left(\mu^2 + \omega_0^2\right)a_0(t) = \dot{w}(t) + \sqrt{\mu^2 + \omega_0^2}\,w(t) \tag{24}$$

The Eq. (24) can be further simplified as:

$$\ddot{a}_0(t) = -\left(\mu^2 + \omega_0^2\right)a_0(t) - 2\mu\dot{a}_0(t) + w_c(t) \tag{25}$$

where, $w_c(t)$ is zero-mean Gaussian noise with variance $2\mu\sigma_a^2$.

Therefore

$$\ddot{a}(t) = \ddot{\overline{a}}(t) - \left(\mu^2 + \omega_0^2\right)[a(t) - \overline{a}(t)] - 2\mu\left[\dot{a}(t) - \dot{\overline{a}}(t)\right] + w_c(t) \tag{26}$$

Define the state vector as follows

$$\boldsymbol{X}(t) = \left[x(t)\ v(t)\ a(t)\ \dot{a}(t)\right]^{\mathrm{T}} \tag{27}$$

Then the non-zero mean attenuation oscillation sine wave model can be written as

$$\dot{\boldsymbol{X}}(t) = \boldsymbol{A}\cdot\boldsymbol{X}(t) + \boldsymbol{U}_1\overline{a}(t) + \boldsymbol{U}_2\dot{\overline{a}}(t) + \boldsymbol{U}_3\ddot{\overline{a}}(t) + \boldsymbol{B}w_c(t) \tag{28}$$

where, $\boldsymbol{A}$ is the transition matrix; $\boldsymbol{U}_1$, $\boldsymbol{U}_2$, $\boldsymbol{U}_3$ are non-zero mean compensation term, which is used to compensate the influence of the mean value of acceleration, acceleration and acceleration derivative on the state quantity, respectively; $\boldsymbol{B}$ is noise vector.

Neglecting the influence of the mean value of the acceleration derivative on the state quantity, and assume that the sampling time is T, the discrete form of the model is

$$\boldsymbol{X}_{k+1} = \boldsymbol{F}_k\boldsymbol{X}_k + \boldsymbol{U}_{1k}\overline{a}_k + \boldsymbol{U}_{2k}\dot{\overline{a}}_k + \boldsymbol{W}_k \tag{29}$$

where,

$$\boldsymbol{F}_k = \begin{bmatrix} 1\ T\ f_1(T)\ f_5(T) \\ 0\ 1\ f_2(T)\ f_6(T) \\ 0\ 0\ f_3(T)\ f_7(T) \\ 0\ 0\ f_4(T)\ f_8(T) \end{bmatrix} \tag{30}$$

Acceleration compensation term is

$$\boldsymbol{U}_{1k}\overline{a}_k = \int_t^{t+T} e^{\boldsymbol{A}(t+T-\tau)}\boldsymbol{U}_1\overline{a}(t)d\tau = \left(\mu^2 + \omega_0^2\right)\overline{a}(t)\int_t^{t+T}\begin{bmatrix} f_5(t+T-\tau) \\ f_6(t+T-\tau) \\ f_7(t+T-\tau) \\ f_8(t+T-\tau) \end{bmatrix}d\tau \tag{31}$$

Acceleration derivative on the state quantity item

$$\boldsymbol{U}_{2k}\dot{\overline{a}}_k = \int_t^{t+T} e^{\boldsymbol{A}(t+T-\tau)}\boldsymbol{U}_2\dot{\overline{a}}(t)d\tau = 2\mu\dot{\overline{a}}(t)\int_t^{t+T}\begin{bmatrix} f_5(t+T-\tau) \\ f_6(t+T-\tau) \\ f_7(t+T-\tau) \\ f_8(t+T-\tau) \end{bmatrix}d\tau \tag{32}$$

The process noise matrix $\boldsymbol{Q}_k$ is

$$\boldsymbol{Q}_k = E\left(\boldsymbol{W}_k\boldsymbol{W}_k^T\right) = q\begin{bmatrix} q_{11}\ q_{12}\ q_{13}\ q_{14} \\ q_{21}\ q_{22}\ q_{23}\ q_{24} \\ q_{31}\ q_{32}\ q_{33}\ q_{34} \\ q_{41}\ q_{33}\ q_{43}\ q_{44} \end{bmatrix} \tag{33}$$

2.2 Radar Measurement Function

The measurements of the radar is usually expressed in the spherical coordinate system, which is composed of the elevation angle γ, azimuth angle η and the distance ρ between the radar and the target.

The measurenment function is

$$\begin{bmatrix} \rho \\ \gamma \\ \eta \end{bmatrix} = \boldsymbol{h}(\boldsymbol{x}) + \boldsymbol{v} = \begin{bmatrix} \sqrt{X_L^2 + Y_L^2 + Z_L^2} \\ \arcsin \frac{Y_L}{\sqrt{X_L^2+Y_L^2+Z_L^2}} \\ \arctan 2(-Z_L,\ X_L) \end{bmatrix} + \boldsymbol{v} \tag{34}$$

where,$\boldsymbol{v}$ is zero-mean Gaussian measurement noise

$$\begin{bmatrix} X_L \\ Y_L \\ Z_L \end{bmatrix} = C_I^o(\begin{bmatrix} x \\ y \\ z \end{bmatrix} - \begin{bmatrix} x_r \\ y_r \\ z_r \end{bmatrix}) \tag{35}$$

where, $\boldsymbol{C}_I^o$ is transformation matrix from geocentric inertial system to radar system, $[x_r, y_r, z_r]^T$ is the coordinates of the radar in the geocentric inertial system.

3 Variable Structure Multi Model Filtering Method

3.1 Variable Structure Multi Model Filtering Framework

The main idea of interactive multi model filtering (IMM) is to sum the estimation results of multiple filters matching different motion models to obtain the state estimation. The weight of each filter estimation result depends on the model matching probability and is calculated from the filter residual and prior information.

The precise Bayesian inference framework for interactive multi model filtering can be divided into five steps:

(1) Mode Mixing

Evolution of model probability from time $k-1$ to k

$$P\left(m_{k-1} \mid y^{k-1}\right) \overset{\text{mixing}}{\rightarrow} P\left(m_k \mid y^{k-1}\right) \tag{36}$$

where, m_{k-1} and m_k are motion mode of time $k-1$ and k, respectively; $\boldsymbol{y}^{k-1} = \{y_i\}_{i=1}^{k-1}$ is measurement sequence.

Define model set Θ, $\forall i, j \in \Theta$, we have

$$P\left(m_k = j|\boldsymbol{y}^{k-1}\right) = \sum_{i\in\Theta} P\left(m_k = j|m_{k-1} = i, \boldsymbol{y}^{k-1}\right) P\left(m_{k-1} = i|\boldsymbol{y}^{k-1}\right) \tag{37}$$

where, $P\left(m_k = j|m_{k-1} = i, \boldsymbol{y}^{k-1}\right)$ is model transfer probability from model i to model j.

(2) State interaction

State interaction step generates initial state probability density

$$p\left(x_{k-1} \mid m_{k-1}, y^{k-1}\right) \xrightarrow{\text{interaction}} p\left(x_{k-1} \mid m_k, y^{k-1}\right) \tag{38}$$

According to the theory of conditional probability and full probability, $\forall i, j \in \Theta$, we have

$$p\left(x_{k-1}|m_k = j, y^{k-1}\right) = \frac{\sum_{i \in \Theta} P\left(m_k = j|m_{k-1} = i, y^{k-1}\right)p\left(x_{k-1}|m_{k-1} = i, y^{k-1}\right)}{P\left(m_k = j|, y^{k-1}\right)} \tag{39}$$

(3) State propagation

The state propagation step propagates the state probability density from time $k-1$ to time k

$$p\left(x_{k-1} \mid m_k, y^{k-1}\right) \xrightarrow{\text{evolution}} p\left(x_k \mid m_k, y^{k-1}\right) \tag{40}$$

According to state transfer function, $\forall i, j \in \Theta$, we have

$$p\left(x_k|m_k = j, y^{k-1}\right) = \int p\left(x_k|x_{k-1}, m_k = j, y^{k-1}\right)p\left(x_{k-1}|m_k = j, y^{k-1}\right)dx_{k-1} \tag{41}$$

(4) State update

The state update step obtains the likelihood equation, corrects the prior information

$$p\left(x_k \mid m_k, y^{k-1}\right) \xrightarrow{\text{correction}} p\left(x_k, m_k \mid y^k\right) \tag{42}$$

Define likelihood $p\left(y^k|x_k\right)$, the posterior distribution form is

$$p\left(x_k, m_k = j|y^k\right) \propto P\left(m_k = j|y^{k-1}\right)p\left(x_k|m_k = j, y^{k-1}\right)p\left(y^k|x_k\right) \tag{43}$$

(5) Fusion

Calculate of model probability at time k

$$P\left(m_k = j \mid y^{k-1}\right) \xrightarrow{\text{fusion}} P\left(m_k = j \mid y^k\right) \tag{44}$$

According to Bayesian rule, the state posterior probability density function can be obtained

$$p\left(x_k|y^k\right) = \sum_{j \in \Theta} p\left(x_k|m_k = j, y^k\right)P\left\{m_k = j|y^k\right\} \tag{45}$$

The framework of variable structure multi model filtering adaptive the model set based on IMM algorithm, in which the real-time variable model set is used. The model set is the combination of different models. At each filtering time, the model set is optimized according to the specific switching criteria, and then the state estimation is obtained according to the optimized model set. The main steps are as follows:

(1) Run IMM

Define current model set Θ_k, run IMM;

(2) Model set activation

Set the activation criteria for the model set. If the activation conditions are met, activate the new model subset that matches the target motion pattern better, and form a new model set with the original model set, where; If not, the original model set will be maintained.

Set the activation criteria for the model set. If the activation conditions are met, activate the new model subset Θ_n that matches the target motion pattern better, and form a new model set Θ_a with the original model set, where $\Theta_a = \Theta_n \cup \Theta_k$; If not, the original model set Θ_k will be maintained.

(3) Model set termination

Set the model set termination criteria. If the termination conditions are met, the model subset Θ_n or Θ_k that does not match the target motion mode will be terminated, the reserved model subset will be used as a new model set; If the termination conditions are not met, the original model set is maintained.

(4) Recursive computation

Define $k = k + 1$ and return to step (1).

3.2 Model Set Adaptive Strategy Design

Model set adaptive strategy is the core of variable structure multi model algorithm. A good model set adaptive strategy can select the smallest optimal model set from the candidate model set to achieve efficient and high-precision state estimation. The current mainstream adaptive strategies based on model set probability and model set likelihood ratio mainly have the following problems:

Problem 1: There are too many threshold parameters to be set. Model set adaptation requires preset 2 model set activation threshold parameters and 2 model set termination threshold parameters.

Problem 2: The adaptive strategy is complex. Model set activation and model set termination are required. The activation and termination of model sets cause changes in the number of models in the model set, and the algorithm time also changes, which is not conducive to practical engineering applications.

In order to avoid the problems of too many threshold parameters to be set and complex adaptive strategies, this paper adopts the idea of combining model set activation and model set termination operations, that is, to build a model set with a clear transformation relationship, and directly complete the switching of model sets according to the probability of edge models, without activating and terminating model sets respectively. In addition, the model set with a fixed number of models is adopted to transform the process of model set adaptation into the process of continuously selecting fixed size model sets from the complete model set, so as to increase the number of models without increasing the amount of computation. In order to avoid the problem of model switching by mistake, a model switching criterion with memory depth is designed. Specific adaptive strategies are as follows:

Use M to represent the model; Θ to represent a model set; $\boldsymbol{S}$ to represent all models, which is called a complete model set. For an example, define $\boldsymbol{S}$ as a complete model composed of 13 models, i.e., $M_1, M_2, \ldots, M_{13}$. Assume that the current model set is $\Theta[M_7] = \{M_3, M_6, M_7, M_8, M_{11}\}$, where, M_7 is the central model, M_3, M_6, M_8, M_{11} are marginal models. Define the model probability of M_j at time k as μ_k^j, which is

$$\mu_k^j = P\left\{m_k = M_j | \boldsymbol{y}^k\right\} \tag{46}$$

Calculate the probability of each model in the model set $\Theta[M_7]$. When the probability of the edge model within the memory range satisfies

$$\mu_T^j = \max_{M_l \in \Theta[M_7]} \mu_T^l \tag{47}$$

where, $T = \{k - d, k - d + 1, \ldots, k\}$ is the memory interval, d is the memory depth. The selection of d is related to the target maneuvering frequency.

In addition, the model probability also needs to meet that when

$$\mu_T^j > \mu_{th} \tag{48}$$

the model set switches to $\Theta\left[M_j\right]$. Where, μ_{th} is the preset probability threshold, and $\mu_{th} > 1/N$. N is the model number in $\Theta[M_7]$. The larger the μ_{th}, the lower the switching frequency.

4 Simulation

4.1 Simulation Scene

Assume that the initial longitude and latitude coordinates of the UAV are (135E, 35N), the initial flight altitude is 10 km, the initial flight speed is 80 m/s, and the initial heading angle is 60°. For the first 300 s, the UAV flies steadily. For 300–600 s, the UAV perform lateral maneuver. For 600–900 s, the UAV fly in jump maneuver mode.

Two active radars are used to track the trajectory of UAV, and the deployment positions are (165E, 40N) and (170E, 45N) respectively. Assuming that the UAV is visible to both radars, the radar ranging accuracy is 10 m (1σ), the angle measurement accuracy is 1 mrad (1σ), and the measurement frequency is 5 Hz.

4.2 Simulation Parameter

The proposed method is used to track the UAV. The "current" statistical model (CSM) is used to describe the steady flight and lateral turning maneuver of the UAV, and the NZSW model is used to describe the jumping maneuver of the UAV.

Set the initial model set as $\Theta[M_4]$, the initial model probabilities are $\left[\ 1/3\ 1/3\ 1/3\ \right]$. The transfer probabilities are $\pi_{ij} = l$. If $i = j, l = 0.95$; If $i \neq j, l = 0.025$.

4.3 Simulation Result

In order to verify the superiority of the proposed algorithm in model switching, the new method is compared with the single model method (CSM) and the structured multi model algorithm (IMM). The CSM adopts the "current" statistical model with a maneuver frequency of 1/60. The three models in the IMM are the "current" statistical model with a maneuver frequency of 1/60, the "current" statistical model with a maneuver frequency of 1/40, and the NZSW model with a oscillation frequency of 0.5. The full range position and velocity estimation results obtained from 500 Monte Carlo simulations are shown in Fig. 1 and Fig. 2. See Figs. 3, 4, 5, 6, 7 and 8 for the enlarged figures.

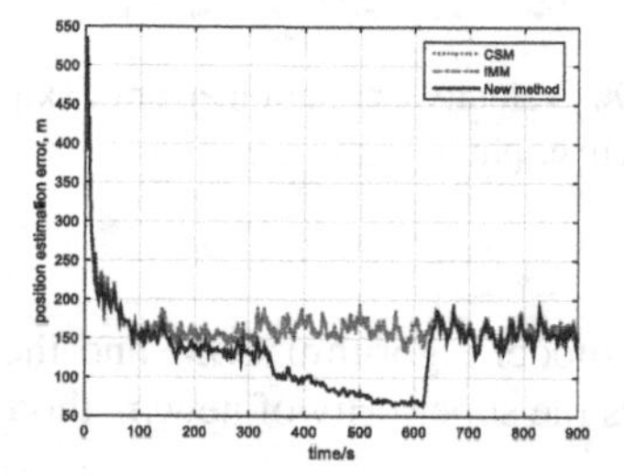

Fig. 1. Position estimation error of the whole phase

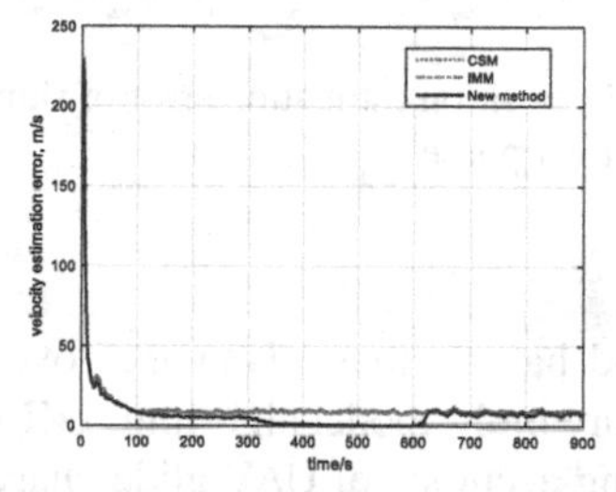

Fig. 2. Velocity estimation error of the whole phase

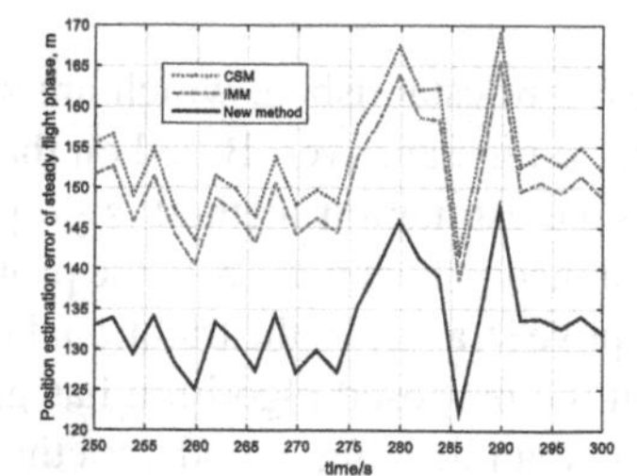

Fig. 3. Position estimation error of steady flight phase

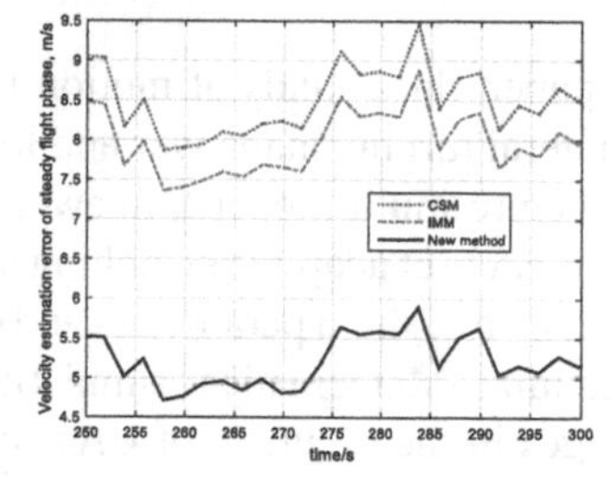

Fig. 4. Velocity estimation error of steady flight phase

From the above results, it can be seen that the new method proposed in this paper has the highest state estimation accuracy in the steady flight phase, lateral maneuver phase and jump maneuver phase compared with the single model algorithm CSM and the fixed structure multi model algorithm IMM. Especially in the lateral maneuver phase, the new

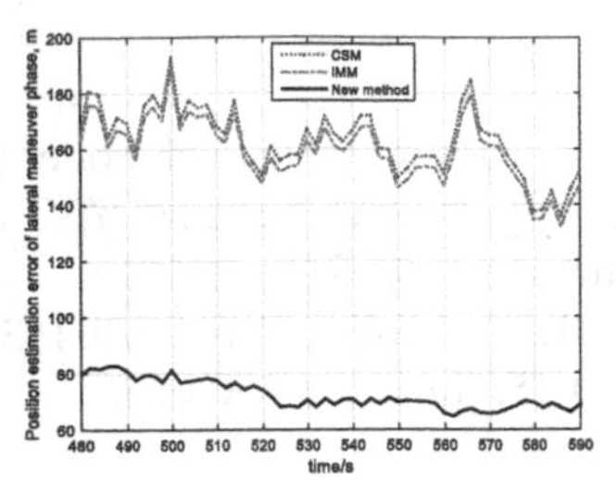

Fig. 5. Position estimation error of lateral maneuver phase

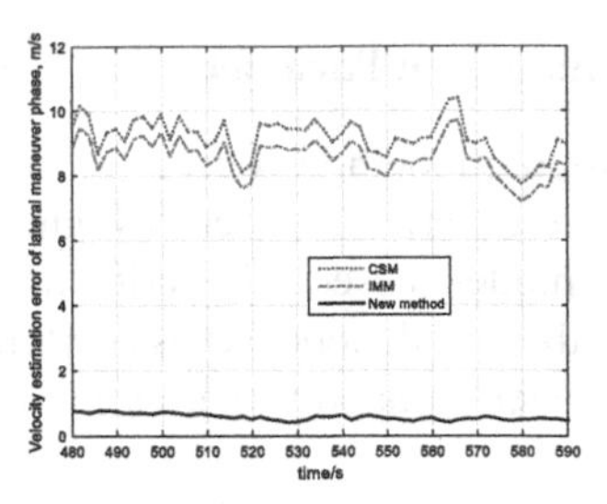

Fig. 6. Velocity estimation error of lateral maneuver phase

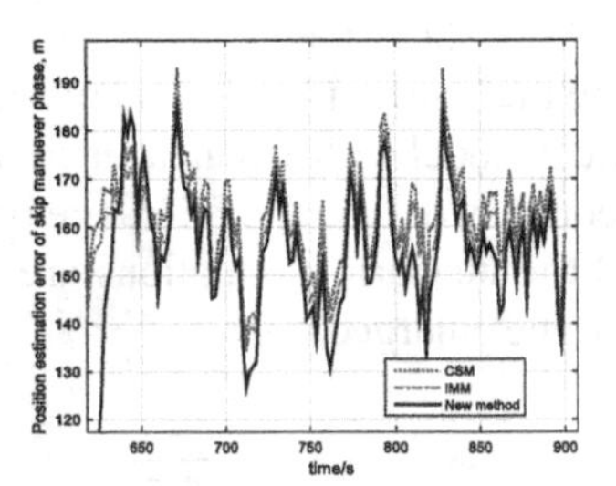

Fig. 7. Position estimation error of skip maneuver phase

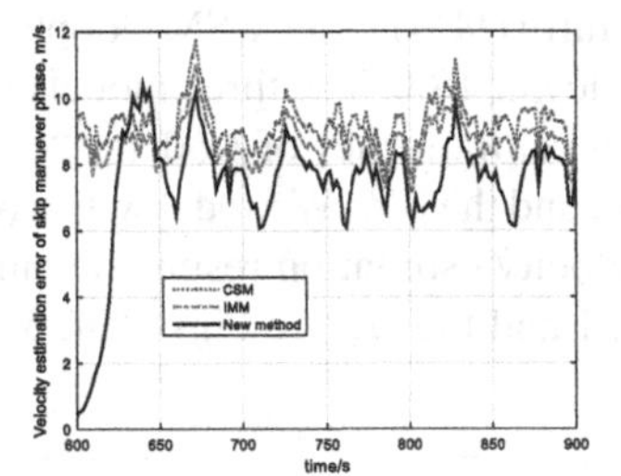

Fig. 8. Velocity estimation error of skip maneuver phase

method has obvious advantages over the single model algorithm CSM and the fixed structure multi model algorithm IMM. This shows the superiority of new method in the tracking accuracy of UAV glide phase.

5 Conclusion

In this paper, three kinds of motion models of UAV are established, which are steady flight, lateral turning maneuver and longitudinal jumping maneuver; Based on the variable structure multi model framework, the UAV state estimation method is proposed, and the model set adaptive switching strategy is designed, which can avoid the problems of too many threshold parameters and complex adaptive strategy in the traditional model set switching. Mathematical simulation shows that the proposed algorithm has greater advantages in the accuracy of UAV state estimation than the single model method and the fixed structure multi model method.

References

1. Wang, Y., Wang, X., Cui, N.: Hybrid consensus-based distributed pseudomeasurement information filter for small UAVs tracking in wireless sensor network. IET Radar Sonar Navig. **14**(4), 556–563 (2020)

2. Huang, J., Zhang, H., Tang, G., et al.: Radar tracking for hypersonic glide vehicle based on aerodynamic model. In: 2017 29th Chinese Control and Decision Conference (CCDC), Chongqing, China, pp. 1080–1084 (2017)
3. Kim, J., Vaddi, S.S., Menon, P.K., et al.: Comparison between nonlinear filtering techniques for spiraling ballistic missile state estimation. IEEE Trans. Aerosp. Electron. Syst. **48**(1), 313–328 (2012)
4. Zhang, K., Xiong, J., Han, C., Lan, X.: A tracking algorithm of hypersonic glide reentry vehicle via aerodynamic model. J. Astronaut. **38**(02), 123–130 (2017)
5. Zhang, K., Xiong, J., Fu, T., et al.: Aerodynamic parameter modeling of hypersonic gliding missile for adaptive tracking. J. Natl. Univ. Def. Technol. **041**(001), 101–107 (2019)
6. Lu, Z., Cai, Y., He, J., et al.: A tracking algorithm for near-space hypersonic vehicles based on aerodynamic acceleration model. In: Chinese Association of Automation, Wuhan, China, pp. 381–388 (2017)
7. He, S., Wu, P., et al.: A tracking algorithm for near space hypersonic reentry glide target. J. Astronaut. **41**(05), 553–559 (2020)
8. Zhou, H.: A "Current" statistical model and adaptive tracking algorithm for maneuvering targets. Acta Aeronautica et Astronautica Sinica **4**(1), 73–86 (1983)
9. Wang, G., Li, J., Zhang, X., et al.: A tracking model for near space hypersonic slippage leap maneuvering target. Acta Aeronautica et Astronautica Sinica **36**(7), 2400–2410 (2015)
10. Singer, R.A.: Estimating optimal tracking filter performance for manned maneuvering targets. IEEE Trans. Aerosp. Electron. Syst. **6**(4), 473–483 (1970)
11. Cheng, Y., Tang, S., Lyu, S., et al.: Aerodynamic statistics-based trajectory estimation of hypersonic maneuvering target. IEEE Access **8**, 227642–227656 (2020)
12. Li, F., Xiong, J., Qu, Z., et al.: A damped oscillation model for tracking near space hypersonic gliding targets. IEEE Trans. Aerosp. Electron. Syst. **55**(6), 2871–2890 (2019)
13. Huang, J., Li, Y., Tang, G., et al.: Adaptive tracking method for hypersonic glide target. Acta Aeronautica et Astronautica Sinica **41**(9), 323786 (2020)
14. Shan, F., Tan, Q., Li, Q., et al.: Trajectory tracking for hypersonic glide vehicles based on improved sine-AIMM. In: Chinese Control Conference, Dalian, China, pp. 5475–5480 (2017)
15. Zhang, X., Wang, G., et al.: Tracking of hypersonic boost-to-glide trajectory target in near-space. J. Astronaut. **36**(10), 1125–1132 (2015)
16. Tugnait, J.K.: Adaptive estimation and identification for discrete systems with markov jump parameters. IEEE Trans. Autom. Control **27**(5), 1054–1065 (1981)
17. Ackerson, G., Fu, K.: On state estimation in switching environments. IEEE Trans. Autom. Control **15**(1), 10–17 (1970)
18. Blom, H.A.P., Bar-Shalom, Y.: The interacting multiple model algorithm for systems with Markovian switching coefficients. IEEE Trans. Autom. Control **33**(8), 780–783 (1988)
19. Hernandez, M., Farina, A.: PCRB and IMM for target tracking in the presence of specular multipath. IEEE Trans. Aerosp. Electron. Syst. **56**(3), 2437–2449 (2020)
20. Nadarajah, N., Tharmarasa, R., Mcdonald, M., et al.: IMM forward filtering and backward smoothing for maneuvering target tracking. IEEE Trans. Aerosp. Electron. Syst. **48**(3), 2673–2678 (2012)
21. Ho, T.J.: A switched IMM-Extended Viterbi estimator-based algorithm for maneuvering target tracking. Automatica **47**(1), 92–98 (2011)
22. Li, X.R., Zhi, X., Zhang, Y.: Design and evaluation of a model-group switching algorithm for multiple-model estimation with variable structure. IEEE Trans. Autom. Control **45**(11), 2047–2060 (2002)
23. Li, X.R.: Multiple-model estimation with variable structure-Part II: model-set adaptation. IEEE Trans. Autom. Control **45**(11), 2047–2060 (2000)

A Study of Self-position Estimation Method by Lunar Explorer by Selecting Corresponding Points Utilizing Gauss-Newton Method

Mitsuki Itoh and Hiroyuki Kamata(✉)

School of Science and Technology, Meiji University, Tokyo, Japan
kamata@meiji.ac.jp

Abstract. The JAXA/ISAS SLIM project aims to land a small unmanned spacecraft on the Moon with pinpoint accuracy at its destination. This research aims to realize a method for estimating the flight position of a lunar explorer using image matching technology. The flight position is estimated by high-precision image matching between the lunar surface image taken by the probe and the lunar surface map image. However, the lunar surface image taken by the spacecraft is assumed to be a degraded low-resolution image due to various disturbances. This causes positional errors in the corresponding points between the captured image and the map image, which hinders the improvement of the accuracy of the transformation matrix. Therefore, an optimization method of transformation matrices based on the Gauss-Newton method is used to improve the accuracy of spacecraft flight position estimation. Besides, this method cannot eliminate the position error of the corresponding points themselves, thus limiting the improvement of position estimation accuracy. Then this study proposes a method for selecting corresponding points with small position error by utilizing the transformation matrix optimized by the Gauss-Newton method. Compared to the conventional method, the proposed method improved the number of successful estimates and the estimated average error from the true value. In particular, the proposed method was found to be effective for brightness, contrast, and noise disturbances, for which the conventional method had low position estimation accuracy.

Keywords: Self-location estimation · Gauss-Newton method · Image matching

1 Introduction

We are participating in the SLIM (Smart Lander for Investigating Moon) project of JAXA/ISAS as co-investigators. We are aiming to demonstrate high-precision landing technology on the Moon by a small unmanned lunar explorer [1]. The project aims to achieve a pinpoint landing of a small lunar explorer within 100 m of the target site. Currently, the lunar surface points to be explored have been narrowed down from years of space research. From this, a technology for pinpoint landing at a target point on the lunar surface is required [2].

W. Fan et al. (Eds.): AsiaSim 2022, CCIS 1712, pp. 334–346, 2022.
https://doi.org/10.1007/978-981-19-9198-1_25

The goal of this research is to realize a highly accurate flight position estimation technology for lunar explorers by image matching. This technology matches lunar surface images taken by a spacecraft while cruising over the Moon with lunar surface map images built into the spacecraft as a database. High-precision image matching enables accurate estimation of the spacecraft's flight position. In this research, image matching technology based on the feature point detection method, which assumes backup processing on the earth, is also being considered [3].

However, the position estimation from captured images accompanied by disturbance is a major challenge. Lunar images taken by spacecraft are considered to be low-resolution images due to various disturbances and degradation caused by Jpeg compression and restoration. In a low-resolution image, the feature values obtained from the luminance gradient will change, resulting in misalignment and incorrect matching of feature point detection positions between the captured image and the map image. Therefore, we believe that the accuracy of image matching will be reduced and errors will be created in the estimated flight position. For this reason, we are aiming for an image matching method with high accuracy even for captured images accompanied by disturbance. In this project, we are studying accurate correspondence point search, outlier exclusion by robust estimation method, and highly accurate correspondence point selection method [4, 5].

In this study, we propose a method for optimizing the transformation matrix utilizing the Gauss-Newton method [6]. Here, the feature points that correspond between the captured image and the map image are called correspondence points. The deviation of the detected positions of the feature points between the correspondence points is called the position error of the correspondence points. In the proposed method, the transformation matrix is adjusted so that the position error of the correspondence points is small. This enables the estimation of a transformation matrix with less discrepancy between images and improves the position estimation accuracy. However, since this is a method to adjust the transformation matrix so as to reduce the position error of the corresponding points, the corresponding point error cannot be eliminated, and a certain amount of deviation remains in the flight position estimation. For this reason, there is a limitation in optimizing the transformation matrix using the Gauss-Newton method [6].

This study proposes a correspondence point selection method using a transformation matrix based on the Gauss-Newton method. The objective is to achieve more accurate flight position estimation than before by increasing the accuracy of the correspondence points used in the transformation matrix estimation. For this purpose, we evaluate the correspondence points using the transformation matrix obtained from the Gauss-Newton method, and aim to increase the number of correspondence points effective for improving the accuracy of position estimation, thereby achieving more accurate transformation matrix estimation.

2 Principles

2.1 Flow of Flight Position Estimation

In this study, the flight position of a spacecraft is estimated by image matching based on the feature point detection method. This is an estimation method that assumes backup processing on the earth. In backup processing, the images taken by the lunar explorer are likely to be degraded low-resolution images. There are two main reasons for this. The first is degradation due to disturbances associated with the captured images. It is assumed that disturbances such as brightness change, contrast change, blurriness, fluctuation, noise, etc. are associated with the captured image. The second is image degradation caused by compression and restoration to Jpeg images. Considering communication costs, this study assumes that the captured images will be compressed into a Jpeg image of approximately 8 kB and sent to the ground. In this case, it is assumed that mosaic noise caused by image compression and restoration will be added to the image when various disturbances are mixed in [3, 4].

Considering these factors, this project is studying a flight position estimation method based on the feature point detection method [3, 4]. The flow of this method is shown in Fig. 1.

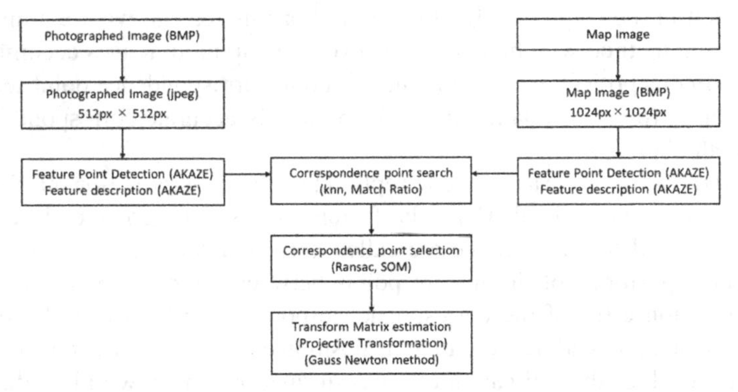

Fig. 1. Flight position estimation method based on feature point detection method

The location estimation method shown in Fig. 1 can be divided into four steps.

- **Step.1: Feature point detection and feature description**

AKAZE features are used for feature point detection and feature description in captured and map images. It is tolerant of image scale and rotation [7]. Based on previous research, we believe that this is an effective feature point detection and feature description method for lunar images with disturbance [4].

- **Step.2: Correspondence point search**

  k-NN (k-Nearest Neighbor) method is used to search for candidate correspondence points from feature points in the captured image and the map image [8]. The distance between feature vectors of feature points in both images is calculated by brute force; in Match Ratio, a combination of correspondence points with high confidence is selected as correspondence points using the Ratio Test [9]. In this case, let α be $0 < \alpha < 1$.

$$\text{(Closest distance feature)} < \text{(Second nearest distance feature)} * \alpha \quad (1)$$

- **Step.3: Correspondence point selection and outlier exclusion**

Outlier removal exclusion of corresponding points using RANSAC [10]. A model of the transformation matrix is created, and the corresponding points are evaluated and excluded for removal. Next, using SOM (Self Organizing Map), correspondence points that are effective for improving accuracy are selected [5]. From the correspondence points, candidate estimated coordinates are created. Since the flight position estimation seeks a certain location on the lunar surface, the candidate estimated coordinates are considered to be densely distributed in some parts. Therefore, we considered that the correspondence points used for the candidate estimated coordinates for densely populated locations would be highly accurate. In addition, using SOM, correspondence points are selected from the dense locations by weighting based on the positional relationship of the candidate estimated coordinates.

- **Step.4: Transformation matrix estimation**

Using singular value decomposition, a transformation matrix is created from the selected corresponding points. This allows the coordinates of the captured image to be transformed onto the map image, and the flight position can be estimated. In this case, the transformation matrix can be optimized by using the Gauss-Newton method [6].

This study further examines a method of repeating Step 3. And Step 4. Using the Gauss-Newton method. By doing so, we aim to improve the accuracy of the transformation matrix by selecting correspondence points with fewer positional errors.

2.2 Transform Matrix Estimation by Projection Transform

A transformation matrix is estimated from the selected correspondence points. This transformation matrix transforms the center coordinates of the captured image into the coordinates on the map image. In this study, the projective transformation is employed as the transformation matrix estimation method [11]. This is to accommodate the scale, rotation, and tilt of the image due to the acquisition attitude of the spacecraft. Let the feature point coordinates of the captured image be (x_i, y_i), the feature point image of the map image be (X_i, Y_i), and the transformation matrix be $\boldsymbol{H}$ with elements h_{00}~h_{22}. These can be expressed by the following relation.

$$\begin{bmatrix} \overline{X}_i \\ \overline{Y}_i \\ M_i \end{bmatrix} = \begin{bmatrix} h_{00} & h_{01} & h_{02} \\ h_{10} & h_{11} & h_{12} \\ h_{20} & h_{21} & h_{22} \end{bmatrix} \begin{bmatrix} x_i \\ y_i \\ 1 \end{bmatrix} \quad X_i = \frac{\overline{X}_i}{M_i},\ Y_i = \frac{\overline{Y}_i}{M_i} \tag{2}$$

Next, find the transformation matrix H. From Eq. (2), assemble the simultaneous equations.

$$\begin{bmatrix} -x_i & -y_i & -1 & 0 & 0 & 0 & X_i \\ 0 & 0 & 0 & -x_i & -y_i & -1 & Y_i \end{bmatrix} \begin{bmatrix} h_{00} \\ h_{01} \\ h_{02} \\ h_{10} \\ h_{11} \\ h_{12} \\ h_{20} \\ h_{21} \\ h_{22} \end{bmatrix} = \begin{bmatrix} 0 \\ 0 \end{bmatrix} \tag{3}$$

Equation (3) is then expressed as $\boldsymbol{A} \cdot \boldsymbol{H} = 0$. Since the matrix $\boldsymbol{A}$ is not regular, the solution of $\boldsymbol{H}$ can be found by the least-squares method or other methods.

2.3 SVD (Singular Value Decomposition)

SVD is utilized as a method to find the optimal solution $\boldsymbol{H}$ from the homogeneous simultaneous equations $\boldsymbol{A} \cdot \boldsymbol{H} = 0$ [12]. For a matrix $\boldsymbol{A}$ whose rank is R and M × N, SVD can be decomposed into three matrices.

$$\boldsymbol{A} = \boldsymbol{VSU}^{\boldsymbol{T}} \tag{4}$$

The optimal solution $\boldsymbol{H}$ of $\boldsymbol{A} \cdot \boldsymbol{H} = 0$ is then found. In the method using SVD, the optimal solution $\boldsymbol{H}$ is found when $\boldsymbol{VSU}^{\boldsymbol{T}}$ is minimized; the optimal solution $\boldsymbol{H}$ can be calculated by finding the minimum singular value λ_{min} that minimizes $\boldsymbol{VSU}^{\boldsymbol{T}}$ and the corresponding component of $\boldsymbol{V}$, the right singular vector V_{min}. In this case, $\boldsymbol{H}$ is obtained as a unit-length solution with the condition that $\boldsymbol{H} \neq 0$, $|\boldsymbol{H}| = 1$. From the matrix $\boldsymbol{A}$ multiplied by its transpose matrix, we obtain the following equation.

$$\boldsymbol{AA}^{\boldsymbol{T}} V_i = \lambda_i^2 V_i \quad (\mathrm{i} = 0, 1, \ldots \mathrm{R} - 1) \tag{5}$$

For Eq. (5), the inverse iteration method can be used to find the minimum singular value λ_{min} and the corresponding right singular vector V_{min}. The optimal solution $\boldsymbol{H}$ can be obtained and the transformation matrix can be estimated. The program uses Oopencv's solvez function, an image processing library [4].

2.4 Optimization of Transformation Matrices by Gauss-Newton Method

To optimize the transformation matrix estimated by SVD, we use the Gauss-Newton method, a nonlinear optimization technique [13]. This aims to improve the estimation accuracy of the estimated transformation matrix. In transformation matrix estimation by SVD, the transformation matrix is estimated from multiple corresponding point coordinates. However, when the estimated transformation matrix H and the corresponding point coordinates are applied to the left side of Eq. (3), it does not equal 0. Because of this, an optimization calculation of the transformation matrix is performed using the Gauss-Newton method so that the errors between all correspondence points used in the transformation matrix estimation are reduced. This allows us to obtain a transformation matrix with high estimation accuracy with reduced influence of the position of correspondence point errors and calculation accuracy. The formulas for the error function e and Jacobian J are shown in Eqs. (6) and (7).

1. Compute JJ^T and $-J^T\boldsymbol{e}$ from the corresponding points and the transformation matrix **H** used in 1.
2. Solve $JJ^T\delta H = -J^T$ and calculate δ **H**.
3. Update the transformation matrix **H** as **H** = **H** + δ **H**
4. Convergence is assumed when the change in the transformation matrix, δ**H**, turns from a decreasing trend to an increasing trend.
5. repeat 1 ~ 5 to optimize the transformation matrix **H**.

$$e = \frac{1}{2}\sum_{i=0}^{N}(X_i - x_i)^2 \tag{6}$$

$$J = \begin{bmatrix} \frac{\partial X_1}{\partial h_{00}} & \frac{\partial X_1}{\partial h_{01}} & \frac{\partial X_1}{\partial h_{02}} & \frac{\partial X_1}{\partial h_{10}} & \frac{\partial X_1}{\partial h_{11}} & \frac{\partial X_1}{\partial h_{12}} & \frac{\partial X_1}{\partial h_{20}} & \frac{\partial X_1}{\partial h_{21}} & \frac{\partial X_1}{\partial h_{22}} \\ \frac{\partial Y_1}{\partial h_{00}} & \frac{\partial Y_1}{\partial h_{01}} & \frac{\partial Y_1}{\partial h_{02}} & \frac{\partial Y_1}{\partial h_{10}} & \frac{\partial Y_1}{\partial h_{11}} & \frac{\partial Y_1}{\partial h_{12}} & \frac{\partial Y_1}{\partial h_{20}} & \frac{\partial Y_1}{\partial h_{21}} & \frac{\partial Y_1}{\partial h_{22}} \\ \vdots & \vdots & \vdots & \vdots & \vdots & \vdots & \vdots & \vdots & \vdots \\ \frac{\partial X_N}{\partial h_{00}} & \frac{\partial X_N}{\partial h_{01}} & \frac{\partial X_N}{\partial h_{02}} & \frac{\partial X_N}{\partial h_{10}} & \frac{\partial X_N}{\partial h_{11}} & \frac{\partial X_N}{\partial h_{12}} & \frac{\partial X_N}{\partial h_{20}} & \frac{\partial X_N}{\partial h_{21}} & \frac{\partial X_N}{\partial h_{22}} \\ \frac{\partial Y_N}{\partial h_{00}} & \frac{\partial Y_N}{\partial h_{01}} & \frac{\partial Y_N}{\partial h_{02}} & \frac{\partial Y_N}{\partial h_{10}} & \frac{\partial Y_N}{\partial h_{11}} & \frac{\partial Y_N}{\partial h_{12}} & \frac{\partial Y_N}{\partial h_{20}} & \frac{\partial Y_N}{\partial h_{21}} & \frac{\partial Y_N}{\partial h_{22}} \end{bmatrix} \tag{7}$$

One drawback of the Gauss-Newton method is that it is prone to local solutions. However, the transformation matrix estimated by SVD is considered to be less prone to local solutions because it is highly accurate to a certain extent, but still yields high solution candidates. For this reason, we believe that the Gauss-Newton method is a good match for high-precision location estimation algorithms based on feature point detection methods [6].

2.5 Correspondence Point Selection by Gauss-Newton Method

The transformation matrix estimated from the Gauss-Newton method is used for correspondence point selection. Although the Gauss-Newton method can be expected to improve the accuracy of the transformation matrix obtained from SVD, it has limitations in improving accuracy. Essentially, these methods optimize the transformation matrix to reduce the position error of the correspondence points based on all the correspondence points to be used. At this time, the originally occurring position error of the corresponding points cannot be eliminated, and a certain amount of position estimation error remains. If combinations with large position errors of corresponding points are included or if the number of corresponding points is small, the accuracy of the transformation matrix cannot be improved. For this reason, to achieve highly accurate transformation matrix estimation, it is necessary to increase the number of corresponding points with small position errors.

The accuracy of the transformation matrix itself estimated from the Gauss-Newton method is expected to be high. We expected to improve the accuracy of position estimation by using this transformation matrix to evaluate and select corresponding points. The procedure of this method is shown below.

1. Select correspondence points where the correspondence point error is less than a set threshold.
2. determine the position error of the corresponding point, which is the distance between the feature point coordinates of the captured image converted into a map image and the feature point coordinates of the map image of the corresponding point.
3. If the position error of the corresponding point is less than or equal to the set threshold, the point is selected as the corresponding point.
4. Find the transformation matrix using SVD for the selected correspondence points
5. Repeat 1.~4. to improve the accuracy of the transformation matrix.

In this case, the number of repetitions and threshold parameters are set in advance.

3 Verification Method

This section examines the method of estimating row positions by selecting corresponding points using the Gauss-Newton method. First, the parameters necessary for selecting corresponding points using the Gauss-Newton method are set.

The lunar surface images used in the verification will be described. A lunar surface image taken by the lunar orbiter KAGUYA is used for the map image. A 512 px × 512 px lunar surface image is randomly cropped from this map image. Scale change, rotation, tilt, and disturbance are added to this lunar surface image, and the compressed jpeg image is used as the captured image. The disturbances at this time are shown below. For each disturbance, 1000 captured images are prepared.

Table 1. Type of disturbance

Disturbance	Image Overview	Disturbance	Image Overview
Nominal	No disturbance	**Noise**	Add sesame salt noise to assume radiated noise
Lightness	Add variation in brightness	**Aberration**	Add distortion to assume lens distortion
Contrast	Add variation to contrast	**Camera shake**	Add blur associated with high-speed movement
Blur	Add blur with a Gaussian filter	**Vignetting**	Add vignetting due to caliper eclipse
Tremor	Add brightness fluctuations		

In the verification, the estimation is considered successful when the estimated mean error is within 3 px ($\approx$100 m) from the true value.

3.1 Threshold Setting for Correspondence Point Selection

In order to select the corresponding points from the transformation matrix estimated by the Gauss-Newton method, a threshold value is set based on the position error of the corresponding points. 0.5 px~.0 px was used, changing the threshold value in 0.5 times increments. The position error of the correspondence point is evaluated by the mean and standard deviation of the position error of the correspondence point for each disturbance, the estimated number of successes, and the estimated mean error from the true value. The corresponding points to be evaluated were those after the Match Ratio and after the k-NN method. The disturbances were brightness, contrast, and noise, which had low accuracy in the conventional method.

3.2 Setting the Number of Iterations in Selecting Correspondence Points

Set the number of iterations for selecting corresponding points using the Gauss-Newton method. The number of iterations was verified in increments of 2 times from 1 ~15. The estimated number of successes and the estimated average error are used to evaluate the results. The disturbances were brightness, contrast, and noise, which had low accuracy in the conventional method.

3.3 Comparison of Conventional and Proposed Methods

The estimation results from the conventional method and the proposed method were compared and evaluated. The conventional method is a position estimation method using the Gauss-Newton method. The number of successful estimations and the average estimation error are used to evaluate the results. The parameters obtained in the validation were set for the proposed method. In addition, the method was evaluated on all disturbance images.

4 Results

4.1 Threshold Setting for Correspondence Points

Table 1 Shows the mean and standard deviation of the corresponding point error for each disturbance. Figures 2 and 3 show the estimated number of successes and the estimated average error from the true value by selecting the correspondence point after Match Ratio and after k-NN method when the threshold is varied.

As Figs. 2 and 3 show, the estimation accuracy was higher when the correspondence point selection by threshold was done after k-NN method. For brightness, the estimated number of successful images and the estimated mean error became smaller when the position error of the correspondence point was small. For contrast, the estimated number of successful images is more stable than for other disturbances, regardless of the error of the correspondence point. In addition, the estimated mean error became smaller when the correspondence point error was small. For noise, the estimated number of successful images and the estimated mean error became smaller when the error of the correspondence point was large. Since the appropriate correspondence point threshold

Table 2. Position error of the corresponding point for each disturbance

Disturbance Pattern	Average Correspondence point error	Standard deviation Correspondence point error	Average + Standard deviation
Lightness	**1.3165 [px]**	**0.8381 [px]**	**2.1546 [px]**
Contrast	**1.3606 [px]**	**0.8688 [px]**	**2.2294 [px]**
Noise	**2.1408 [px]**	**1.1398 [px]**	**3.2806 [px]**

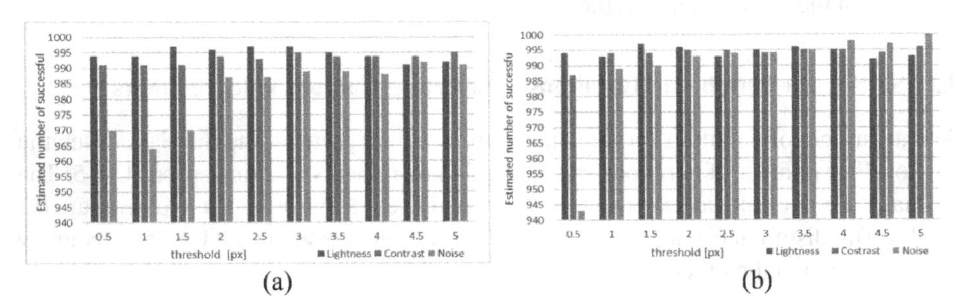

(a) (b)

Fig. 2. Estimated number of successes per threshold (a): Correspondence point selection after Match Ratio (b): Correspondence point selection after k-NN method

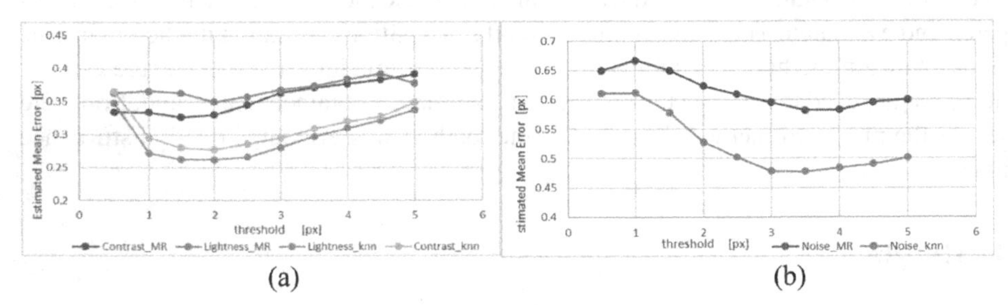

(a) (b)

Fig. 3. Estimated average error per threshold (a): Correspondence point selection after Match Ratio (b): Correspondence point selection after k-NN method

differs depending on the disturbance, we considered it necessary to divide the cases according to the error of the correspondence point, and from Table 2 and Fig. 3 we can see that the threshold of the error of the correspondence point at which the sum of the mean and standard deviation of the error of the correspondence point for brightness and contrast and the estimated mean error are minimized takes a value around 2.0 px. In addition, the noise takes a large value compared to the two disturbances. Therefore, the case is divided by 2.2 px, which is the sum of the mean and standard deviation of the error of the correspondence point that the contrast has. We thought that this would allow us to separate brightness, contrast, and noise. When the error of the correspondence point is smaller than 2.2 px, the threshold is set at 2.0 px, where the total number of successful estimates of brightness and contrast is the largest. When the error of the correspondence point was larger than 2.2 px, the threshold was set at 5.0 px, which was the maximum number of successful noise images.

4.2 Setting the Number of Iterations in Selecting Correspondence Points

Figures 4 and 5 shows the estimated number of successes and estimated average error by selecting correspondence points after Match Ratio and after k-NN method when the number of iterations of correspondence point selection is varied.

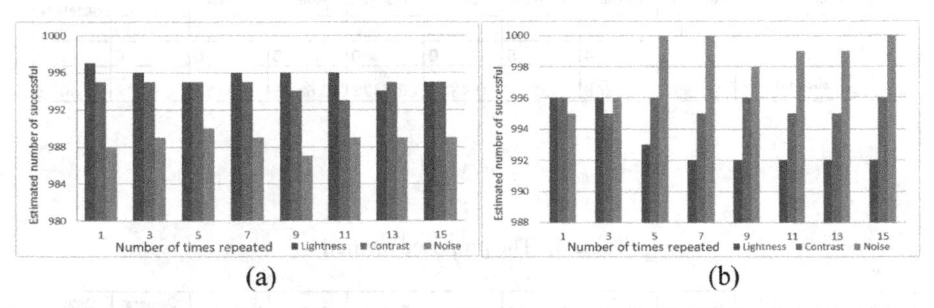

Fig. 4. Estimated number of successes per Number of repeated (a): Correspondence point selection after Match Ratio. (b): Correspondence point selection after k-NN method.

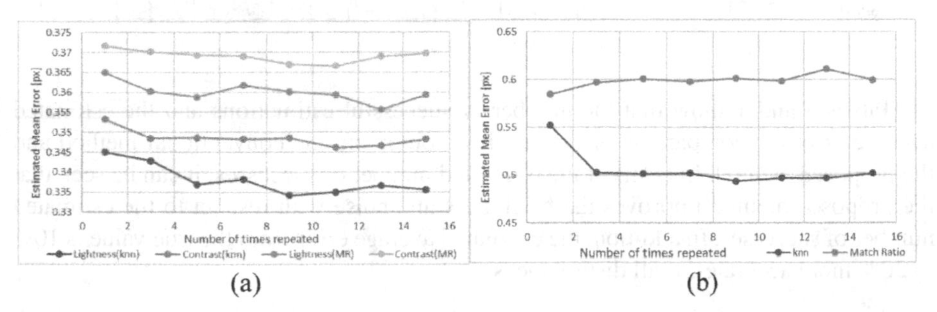

Fig. 5. Estimated average error per Number of repeated (a): Correspondence point selection after Match Ratio (b): Correspondence point selection after k-NN method

Figure 4 (a) shows that the number of successful copies did not change significantly for the three disturbances. In Fig. 4 (b), the estimated number of successful images for brightness decreases with increasing number of iterations. For contrast, no trend can be seen in the number of successful images as the number of iterations increases. In Fig. 5 (a), the range of change is about 0.01 px for all patterns, even when the number of iterations is increased. In Fig. 5 (b), the values were almost constant. The number of iterations was determined by the sum of the mean and standard deviation of the correspondence point errors in the proposed method (2.2 px), and when it was smaller than 2.2 px, the number of iterations was set to the one time when the number of successfully estimated images became large in terms of brightness. When the number of iterations is larger than 2.2 px, the number of iterations is set to 7, which is the maximum number of successes estimated for the noise.

4.3 Comparison of Conventional and Proposed Methods

The parameters determined from validation 4.1, 4.2 and 4.3 were set Table 3 shows the estimation results of the conventional method Table 4 shows the estimation results of the proposed method.

Table 3. The conventional method

Disturbance pattern		Nominal	Lightness	Contrast	Blur	Tremor	Noise	Aberration	Camera shake	Optical Vignetting
Estimation result	success	1000	996	995	1000	1000	995	1000	1000	1000
	failure	0	4	5	0	0	5	0	0	0
Estimation error[px]	success ave	0.3593	0.3886	0.3982	0.4339	0.4423	0.6435	0.3682	0.3629	0.3462

Table 4. The proposed method

Disturbance pattern		Nominal	Lightness	Contrast	Blur	Tremor	Noise	Aberration	Camera shake	Optical Vignetting
Estimation result	success	1000	998	995	1000	1000	998	1000	1000	1000
	failure	0	2	5	0	0	2	0	0	0
Estimation error[px]	success ave	0.2914	0.3102	0.3291	0.3824	0.3647	0.4989	0.3017	0.2952	0.2835

Tables 3 and 4 show that the number of successful estimations and the estimated average error are comparable or better when comparing the conventional method and the proposed method. In terms of the estimated number of successes, it can be seen that the proposed method improves the brightness and noise with respect to the estimated number of successes. In addition, the estimated average error from the true value is 10% to 20% more accurate for all disturbances.

5 Conclusions

The proposed method of flight position estimation by selecting corresponding points using the Gauss-Newton method is considered to be more effective than the conventional method in improving the accuracy of position estimation. For brightness, contrast, and noise, where the conventional method had failed to estimate images, the number of successfully estimated images and the estimated average error from the true value were equal to or greater than the values obtained by the conventional method. In particular, the position estimation accuracy was improved by using the correspondence points after the k-NN method, because the number of correspondence points is the largest in the selection of correspondence points after the k-NN method, and it was possible to leave a sufficient number of correspondence points after the selection of correspondence points. Therefore, it is thought that the threshold value and the number of iterations enabled the use of more accurate correspondence points in the transformation matrix estimation than before. In

Figs. 3 and 4, the threshold values of the position error of the corresponding points that gave the best results in position estimation accuracy were different for brightness, contrast and noise. The reason for this is that images with noise are consider to have a large shift in the position of feature point detection. For this reason, it was necessary to increase the number of corresponding points by increasing the threshold value. This improved the accuracy of flight position estimation in images taken with disturbance, which had been a problem with the conventional method.

On the other hand, for brightness, contrast, and noise, the results of parameter estimation were sometimes more accurate than the results of the proposed method. One possible reason for this is the effect of case separation due to the positional error of the corresponding points. While some images were more accurate due to case separation, others could not be estimated with the appropriate parameter settings. To further improve the accuracy of location estimation, it is necessary to improve the method so that the parameters can be set based on the information possessed by each image.

Furthermore, the number of successful contrast estimation images was not different in the proposed method from that of the conventional method. This is thought to be due to the fact that many incorrect matches remain at the point of correspondence point search. On, it is necessary to consider an accurate correspondence point search method even for images taken with disturbance.

References

1. Yoshikawa, S., et al.: Conceptual study on the guidance, navigation and control system of the Smart Landing for Investigating Moon(SLIM). In: Proceedings of Global Lunar Conference 2010, pp. 59–62 (2010)
2. Ishida, T., et al.: crater-based optical navigation technologies for lunar precision landing in SLIM project. In: Proc. 16th IPPW-2019 (2019)
3. Komatsubara, Y., et al.: Crater detection and backup processing based on the feature extraction method. In: Proceedings of 62th Space Sciences and Technology Conference, 1D10 (2018)
4. Ohara, S., et al.: Study on high precision matching between captured image with disturbance and map image. In: Proceedings of 63th Space Sciences and Technology Conference, 1B08 (2019)
5. Itoh, M., Kamata, H.: Increasing the accuracy of self-positioning by selection corresponding points using self-organizing maps. In: The 34th Workshop on Circuits and Systems Speech, pp. 143–148 (August 28, 2021)
6. Itoh, M., Kamata, H.: A study on self-positioning by optimizing the transformation matrix using Gauss-Newton method. In: RISP International Workshop on Nonlinear Circuits, Communications and Signal Processing 2022, pp. 109–112 (February 28, 2022)
7. Alcantarilla, P.F., Nuevo, J., Bartoli, A.: Fast explicit diffusion for accelerated features in nonlinear scale spaces. In: British Machine Vision Conf. (BMVC) (2013)
8. Dasarathy, B.V.: Nearest neighbor (NN) norms: NN pattern classification techniques. IEEE Computer Society Press; IEEE Computer Society Press Tutorial (1991)
9. Lowe, D.: Distinctive image features from scale-invariant keypoints, cascade filteringapproach. IJCV **60** pp. 91–110 (2004)
10. Raguram, R., Chum, O., Pollefeys, M., Matas, J., Frahm, J.: Usac: a universal framework for random sample consensus. IEEE Trans. Pattern Anal. Mach. Intell. (2013)
11. Hirotaka, N, Rangarajan, P., Kenichi, K.: High accuracy homography computation without iterations. IPSJ SIG Technical Report, Vol. 2010-CVIM-170, No. 55 (2010)

12. Masashi, I., Shinya, S., Yoshisama, N.: Accurate twisted factorization of real symmetric tridiagonal matrices and its application to singular value decomposition. Trans. Japan Soc. Ind. Applied Math. **15**(3), 461–481 (2005)
13. Hartley, H.O.: The modified Gauss-Newton method for the fitting of non-linear regression functions by least squares. Am. Stat. Assoc. Am. Soc. Qual. **3**(2), 269–280 (1961)

Influence of Wave Parameters on Taxiing Characteristics of Seaplane

Qing Wen[1(✉)], Zhihang Cheng[1], Rui Deng[2], and Kangzhi Yang[1]

[1] Research and Development Centre, AVIC General HUANAN Aircraft Industry Co., Ltd., Zhuhai 519040, Guang Dong, China
wenqing0601@163.com

[2] School of Marine Engineering and Technology, Sun Yat-Sen University, Zhuhai 519040, Guang Dong, China

Abstract. Sailing speed and ocean wave environment have obvious effects on the motion characteristics of seaplane on waves. Using the numerical simulation method, based on the VOF method and the dynamic overlapping grid technology, the dynamic characteristics of the seaplane sailing on the wave surface are simulated, and the effects of the sailing speed, wavelength and wave height on the resistance characteristics, heave and pitch motion are investigated respectively. The numerical simulation results show that with the increase of navigation speed, the interaction between the body and the wave is stronger, and the peak resistance, the double amplitude values of heave and pitch are increasing; With the increase of wavelength, the wave steepness decreases, the peak value and average value of resistance curve decrease gradually, and the motion characteristics of heave and pitch and acceleration also have the same trend; With the increase of wave height, the wave steepness of the wave increases, the wave energy encountered by amphibious aircraft increases, the peak value and average value of the resistance curve of the aircraft in the wave gradually increase, and the heave, pitch motion characteristics and acceleration characteristics also have the same trend.

Keywords: Sailing speed · Wave surface · Motion response · Water plane

1 Introduction

The biggest difference between the take-off and landing of a seaplane on the water and the take-off and landing on the land is that it has to experience the high-speed taxiing state on the water surface during the take-off and landing process [1]. In this state, the speed of the aircraft is very high, and the maximum speed is close to the speed of air flight, but it also experiences the load of water and air two-phase flow. At this time, the free surface is impacted and torn due to the high Froude motion of the aircraft, resulting in strong splash. At the same time, the wave parameters of the water surface also has a great impact on the pitch and heave motion response of the aircraft [2–5], and the density of the water is about 800 times that of the air, which leads to the characteristics of large amplitude and rapid change of the load borne by the seaplane in this process, It

W. Fan et al. (Eds.): AsiaSim 2022, CCIS 1712, pp. 347–358, 2022.
https://doi.org/10.1007/978-981-19-9198-1_26

brings great challenges to the prediction of seaplane in the state of high-speed taxiing on the wave surface [6, 7]. Since the 1960s, many pool experiments have been carried out at home and abroad on seaplane taxiing on the wavy water surface [2, 6–9]. With the development of computer technology, the numerical simulation technology has also made great progress in this area, including simulation accuracy research [10], wave water surface landing load [11].

The research on seakeeping performance of seaplane is mainly to study the motion law of seaplane on waves. Seaplane moving on waves will have significant degree of freedom motion on waves due to the disturbance of waves. The influence of different speed and wave parameters on the taxiing stability of seaplane is studied by studying the variation law of aircraft angle of attack amplitude and heave amplitude in waves.

2 Numerical Simulation Method

2.1 Simulation Method

In seakeeping of seaplane, the water velocity and air velocity are both less than Mach 0.3. Therefore, the compressibility of the medium is not considered in the water phase and air phase, and the related flow is incompressible three-dimensional unsteady turbulent flow, which is described by Reynolds averaged NS equation [12]. The realizable k-epsilon two layer model is selected as the turbulence model [13]. VOF method is used to capture the water vapor mixed free surface.

In the numerical simulation of the motion of seaplane in waves, the generation of wave environment and the construction of boundary wave reflection elimination method are very important. The wave making method of the numerical wave pool adopts the velocity boundary method, that is, the wave is directly simulated at the velocity inlet boundary, and the wave making type is the first-order linear wave of VOF wave; When the numerical wave propagates to the physical boundary, the boundary wave reflection usually occurs. The boundary reflection of the wave will interfere with the calculation and the wave quality. The construction of the elimination method of the wave boundary reflection in this study usually has the wave resistance wave elimination method. In the numerical study, the wave damping is generated by applying resistance to the vertical motion at the outlet boundary of the calculation domain.

The heave and pitch motions of a seaplane in waves are solved by a six degree of freedom motion model. In the calculation of seakeeping performance of seaplane, two coordinate systems are used to solve the six degree of freedom equation, one is called the initial coordinate system (earth coordinate system), and the other is called the non initial coordinate system (body coordinate system).

2.2 Simulation Model

The simulation of the wave motion of the seaplane is carried out at the model scale. The model scale is consistent with the physical waveform in the test. According to the wave test model of the whole seaplane model, the three-dimensional digital model of the aircraft is treated as the calculation model for numerical simulation. The calculation model components for numerical simulation mainly include the fuselage, wing, flap, wing fairing, main lift fairing, nacelle and propeller cover, pontoons and some pontoon struts, tail wings Elevator, wave suppression plate and water rudder. Flap deflection 20 degrees, elevator -12 degrees.

The seakeeping calculation model under the model scale is scaled in equal proportion, with the scale of 1:10. Taking the similarity of Froude number (FR) as the criterion, the size, mass, speed, moment of inertia, model test speed and other parameters of the model meet the scale relationship corresponding to Table 1 (representing the scale of the model, = 1:10).

Table 1. Scale relation of model test

Name	Full scale value	Scale	Model value
Length	L	λ	λ L
Force	F	λ^3	λ^3 F
Mass	m	λ^3	λ^3 m
Velocity	V	$\sqrt{\lambda}$	$\sqrt{\lambda}$ V
Moment of inertia	I	λ^5	λ^5 I
Acceleration	a	1	a
Angular acceleration	α	λ^{-1}	λ^{-1} α

2.3 Computational Domain and Boundary Conditions of Wave Motion Simulation

In the numerical study of seakeeping of seaplane in waves, the seaplane is axisymmetric geometry, and the flow characteristics have symmetry, so the calculation domain of seakeeping problem is a half model, and the symmetry plane is processed at the mid longitudinal section.

In order to ensure the full propagation of numerical waves and the full development of free surface waveform wake in seakeeping calculation of seaplane, sufficient seaplane free surface interaction and fluid development areas are reserved during the establishment of calculation area. The axial range of the overall calculation area is -2.15loa $\leq$ x $\leq$ 7.3loa, the spanwise range is 0.0lpp $\leq$ y $\leq$ 2.7loa, and the vertical range is -2.7loa $\leq$ Z $\leq$ 2.7loa, Where loa is the total length of the seaplane. The numerical simulation domain of seakeeping performance of seaplane is shown in Fig. 1.

The inlet boundary is the velocity inlet. The inlet condition of the first-order waveform free surface is established based on the water vapor composition of VOF. The speed of the seaplane encounter is set by specifying the incoming velocity of the inlet free surface. The outlet boundary is the pressure outlet. The outlet condition of the first-order waveform free surface is established based on the water vapor component of VOF, and the height of the free surface is maintained by specifying the hydrodynamic pressure of the free surface at the outlet. The upper and bottom boundaries of the computational domain are consistent with the velocity inlet boundary conditions. The lateral and symmetric surfaces of the computational domain are symmetric boundary conditions, and the aircraft surface is a non slip wall surface.

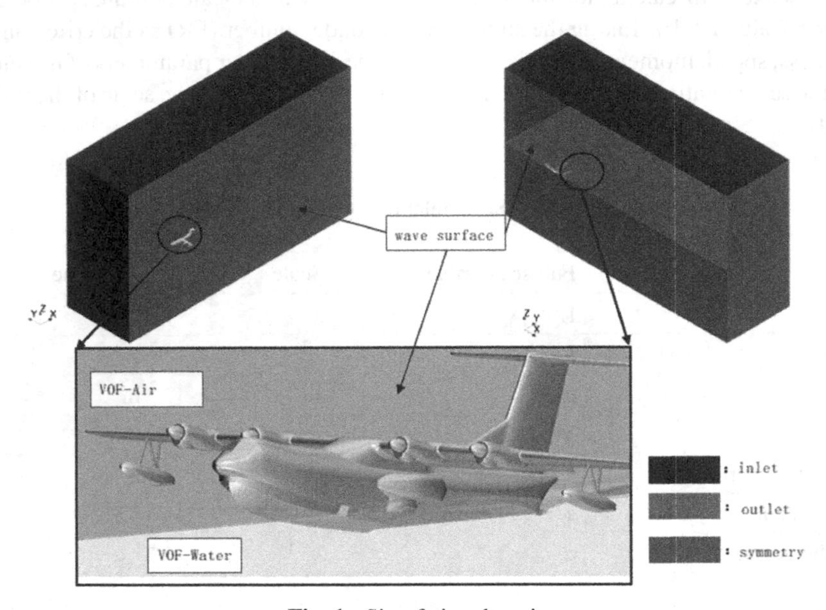

Fig. 1. Simufation domain

2.4 Mesh Generation

The cutting volume rectangular grid is used to divide the calculation area, and the prism boundary layer grid is used to divide the boundary layer at the fuselage. The corresponding encryption for physical phenomena is carried out for the free surface of the seaplane, the area around the aircraft fuselage and the tail area [14]. In the process of boundary layer mesh generation, different regions are divided according to the different media of the underwater part and the above water part of the seaplane. In the model scale hydrodynamic calculation, the y+ of the bottom part is about 200 and the y+ of the air part is about 30. The high-precision simulation of numerical waves generated based on boundary conditions is the most important part. The capture of wave shape has strict grid requirements. Generally speaking, at least 18–20 grids are required to capture the wave height shape within the wave height range, and at least 100–120 grids are required to capture the wavelength shape within the wavelength range. During the whole calculation process, it is also necessary to ensure that the Coloane number of the free surface is below 0.5, The calculation time format is second-order. The calculation grid is shown in Fig. 2.

The large heave and pitch motions of seaplane in waves need to be captured by overlapping mesh model and DFBI degree of freedom model. The overlapping grid can well realize the simulation of six degrees of freedom motion of heave and pitch of seaplane in waves in the seakeeping calculation of seaplane [15]. The VOF method is used to track the free surface, and the motion capture of the degree of freedom model can well realize the study of the impact of waves on the taxiing stability of seaplane.

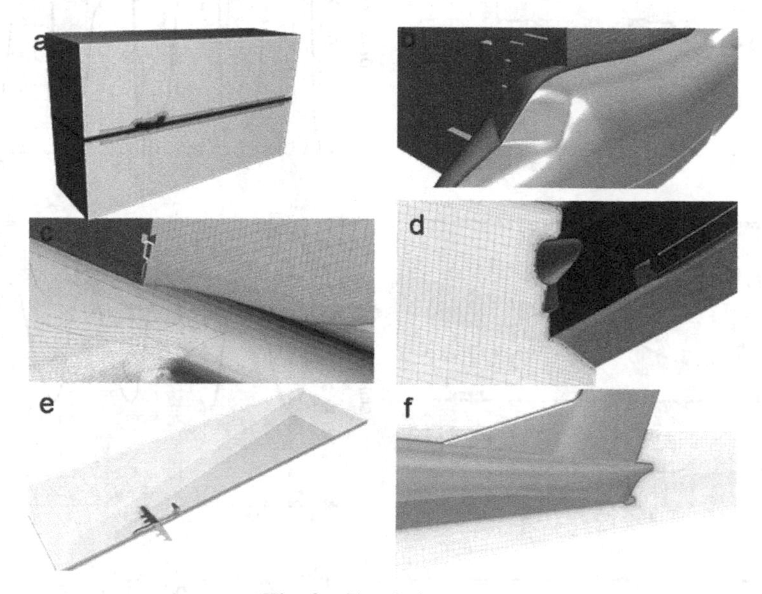

Fig. 2. Simulation mesh

3 Seakeeping Simulation of Seaplane in Waves

3.1 Influence of Speed Factor on Taxiing Stability

In the study of the influence of speed factors on the sliding stability of seaplane in waves, the wave height of 0.05 M, the wavelength of 8 m, and the speed of 4 m/s and 9 m/s were selected as the study conditions. See Fig. 3 for resistance curve, heave and pitch curve of seaplane at different speeds.

With the increase of speed, the curves of different speeds have obvious periodicity. At the same time, the number of waves encountered by seaplane increases gradually, and the interaction with waves is stronger, which shows that the peak value of resistance curve increases with the increase of speed. In addition, the average resistance of seaplane in waves also increases with the increase of speed.

Like the resistance curve of seaplane in waves, the motion characteristics of seaplane in waves also have significant periodicity. Due to the different speed, the encounter frequency between the body and the wave is also different, and the periodicity of the curve under different working conditions is also significantly different. With the increase of speed, the double amplitude and average values of heave and pitch of seaplane in waves increase with the increase of speed.

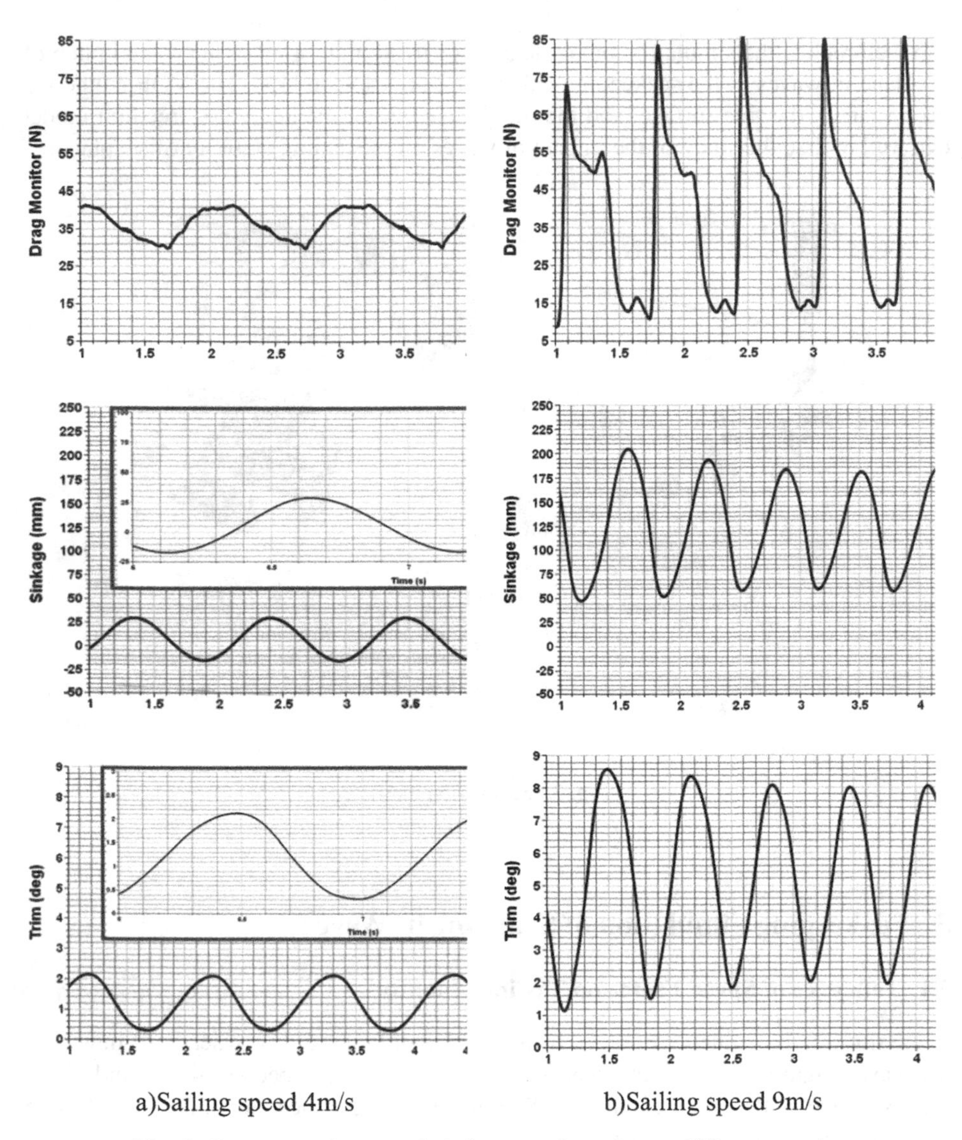

a)Sailing speed 4m/s b)Sailing speed 9m/s

Fig. 3. Drag curve, heave and pitch curve of seaplane at different speeds

See Fig. 4 for contour distribution diagram of seaplane waveform under different speed conditions. The free wave surface distribution of the interaction between the seaplane and the waves can be clearly observed from the waveform contour distribution diagram. The numerical calculation strategy in this study has better captured and reproduced the numerical waves. During the wave transfer from the entrance to the far rear of the seaplane, the parameters such as the shape and amplitude of the waves are well maintained. With the increase of speed, the encounter period of seaplane and wave is different. This description is further confirmed by the phase of wave and the relative position of aircraft in the waveform contour map. In addition, with the increase of the speed, the Kelvin angle of the free surface waveform decreases gradually, and the interaction of the waves presents a discontinuous free surface wake with periodic peaks and troughs.

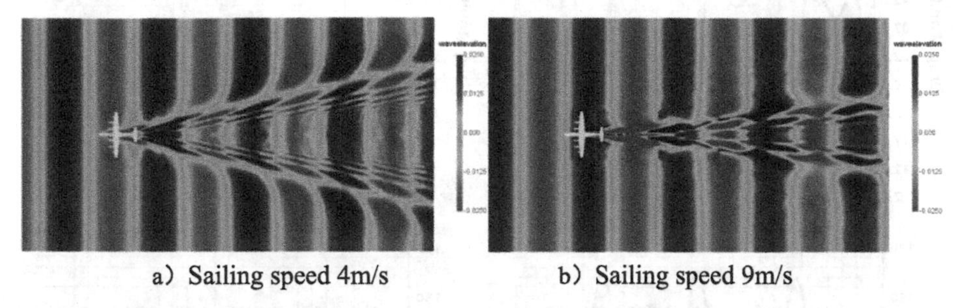

a) Sailing speed 4m/s b) Sailing speed 9m/s

Fig. 4. Contour distribution of seaplane waveform at different speeds

3.2 Influence of Wavelength Factor on Sliding Stability

In the research on the influence of wavelength factor on the taxiing stability of seaplane in waves, the wave height of 0.05 M, the wavelength of 8 m and 16 m, and the speed of 8 m/s are selected as the research conditions. The resistance curve, heave curve and pitch curve of seaplane at different wavelengths are shown in Fig. 5.

The curves at different wavelengths have obvious periodicity. At the same wave height, with the increase of wavelength, the wave steepness becomes smaller, and the wave intensity encountered by the seaplane decreases, which shows that the peak value of the resistance curve gradually decreases with the increase of wavelength. In addition, the average resistance of seaplane in waves also decreases with the increase of wavelength. For the heave and pitch curves of seaplane in waves, the periodicity of the curves is more significant, and the amplitude and mean value also decrease with the increase of wavelength.

According to the waveform contour distribution diagram of seaplane under different wavelength conditions in Fig. 6, the free wave surface distribution of seaplane and wave can be clearly observed. With the increase of wavelength, the encounter period of seaplane and wave is different. The phase of wave in the waveform contour diagram and the relative position of aircraft further confirm this description. Under the three wavelength States, the speed of the seaplane is the same, and the Kelvin angle of the free surface waveform is basically the same. However, due to the different wavelengths, in the small wavelength state. Due to the mutual interference of waves, the discontinuous effect of periodic wave peaks and troughs on the free surface wake is more significant. In addition, under the condition of larger wavelength, the tearing phenomenon of aircraft to waves is more significant.

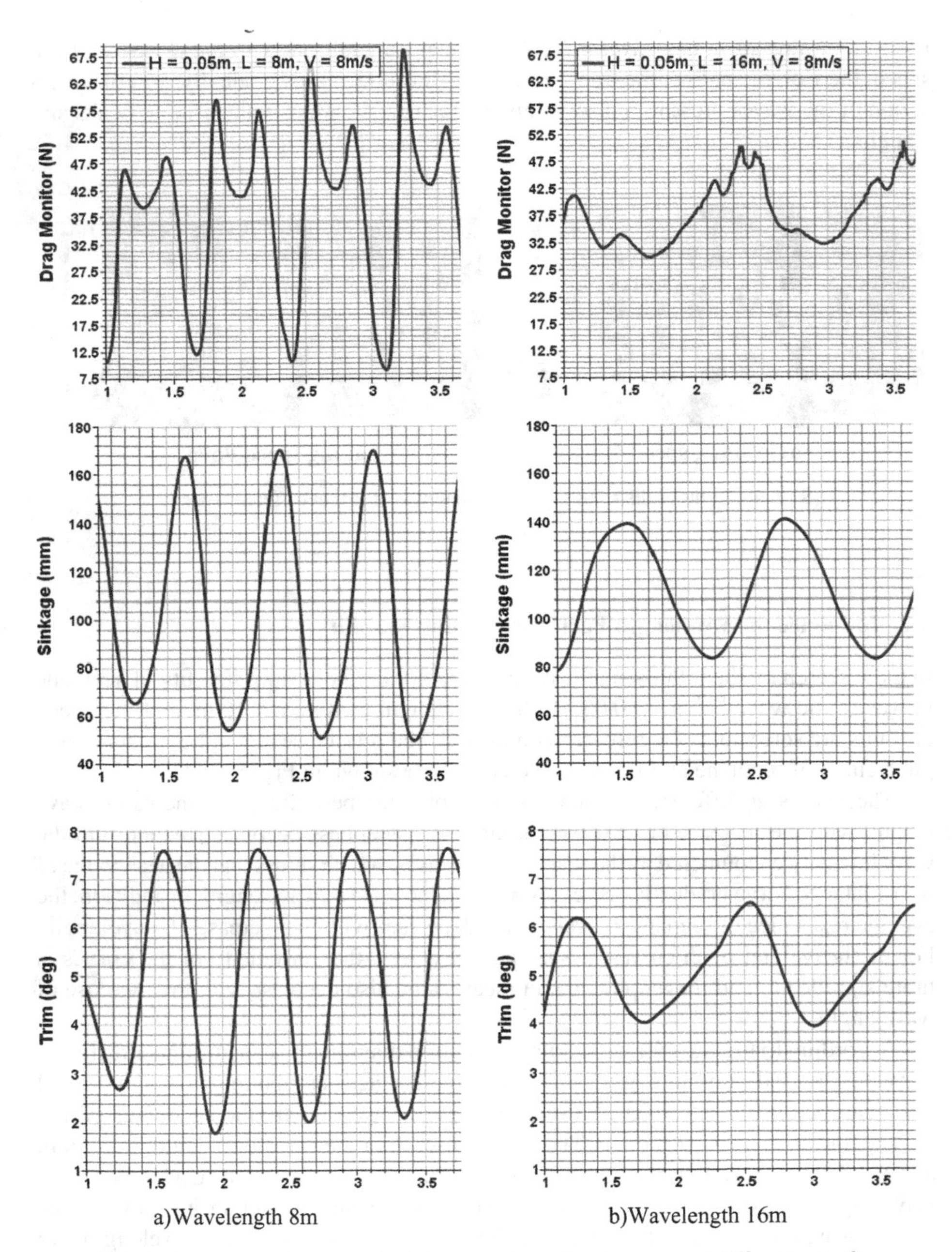

a)Wavelength 8m b)Wavelength 16m

Fig. 5. Drag curve, heave and pitch curve of seaplane in waves at different speeds

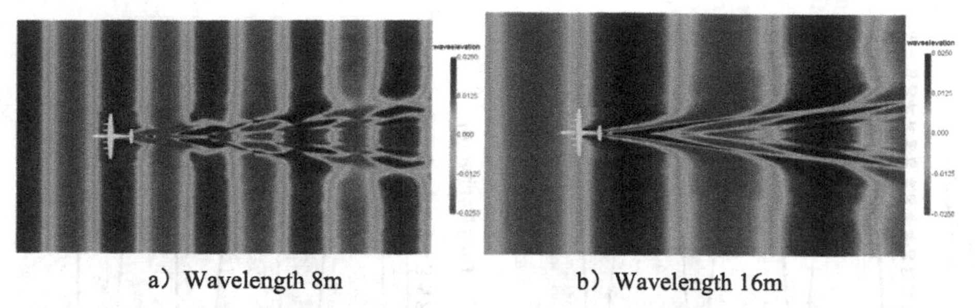

a) Wavelength 8m b) Wavelength 16m

Fig. 6. Contour distribution diagram of seaplane waveform under different wavelength conditions

3.3 Influence of Wave Height on Sliding Stability

In the study of the influence of wave height on the sliding stability of seaplane in waves, the wave height of 0.05 and 0.1 M, the wavelength of 8 m and the speed of 8 m / s are selected as the study conditions. See Fig. 7 for resistance curve, heave curve and pitch curve of seaplane at different wavelengths. The curves under different wave heights have obvious periodicity. At the same wavelength, with the increase of wave height, the wave steepness becomes larger, and the wave strength encountered by the seaplane increases, which shows that the peak value of the resistance curve increases with the increase of wave height. In addition, the average resistance of seaplane in waves also increases with the increase of wave height. For the heave and pitch curves of seaplane in waves, the periodicity of the curves is more significant, and the amplitude and mean value also increase with the increase of wave height.

From the contour distribution diagram of seaplane waveforms at different wave heights in Fig. 8, it is obvious to observe the free wave surface distribution of seaplane and waves under different wave height conditions. Since the wavelength remains the same, the encounter period of seaplane and waves is the same. The phase of waves in the contour diagram of waveforms and the relative position of aircraft further confirm this description. Because the speed of the seaplane is the same under the two wave heights, the Kelvin angle of the free surface waveform is basically the same. When the wave length is the same, the wave steepness and the wave energy increase with the increase of wave height. In wavelet high state. The interaction between the body and the waves is weaker than that under the condition of large wave height. Due to the mutual interference of the waves, the discontinuous effect of the periodic wave peaks and troughs, the free surface wake is more significant under the condition of large wave height. Under the condition of larger wave height, the pitch angle of the aircraft is larger, and the peak value of heave is larger.

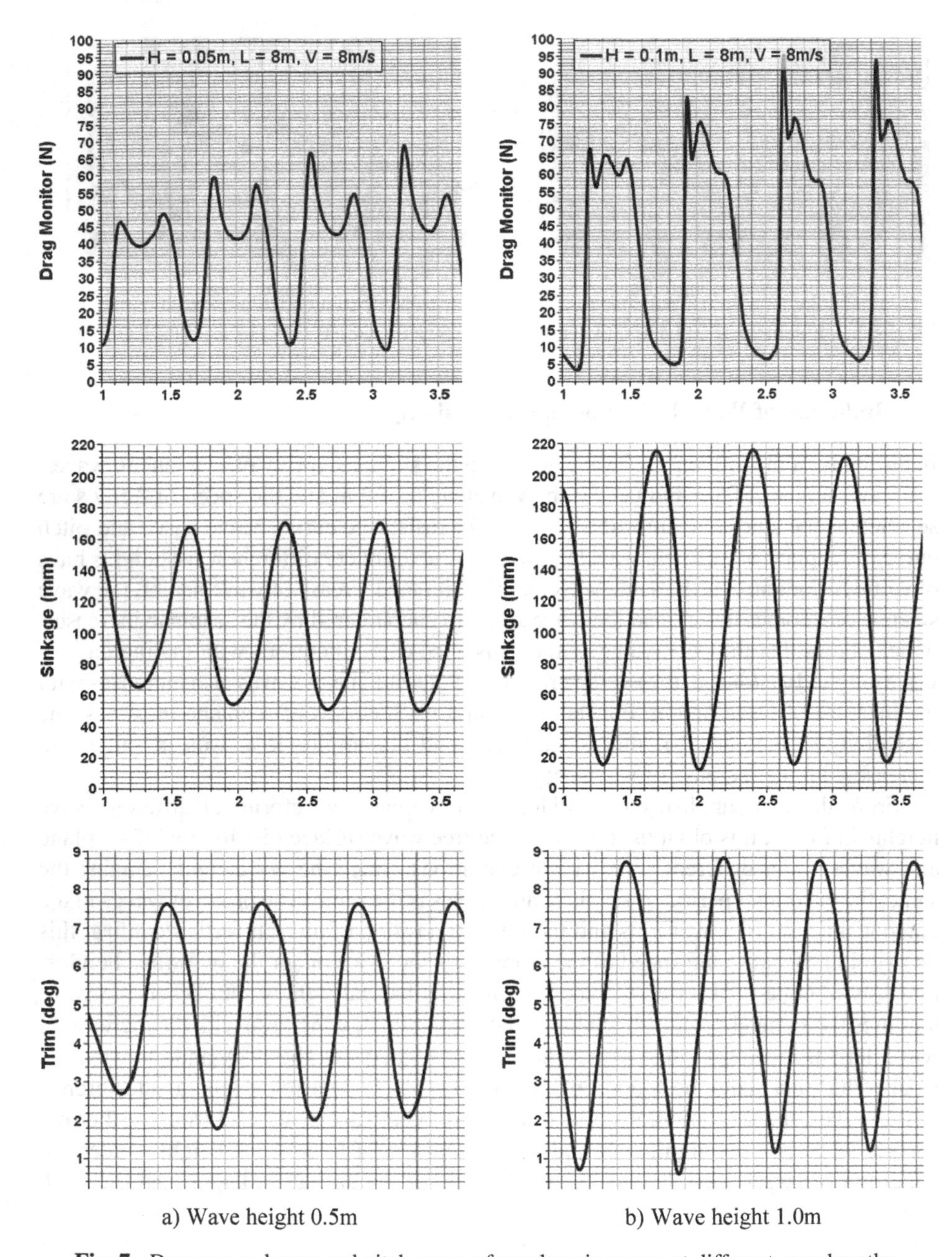

a) Wave height 0.5m

b) Wave height 1.0m

Fig. 7. Drag curve, heave and pitch curve of seaplane in waves at different wavelengths

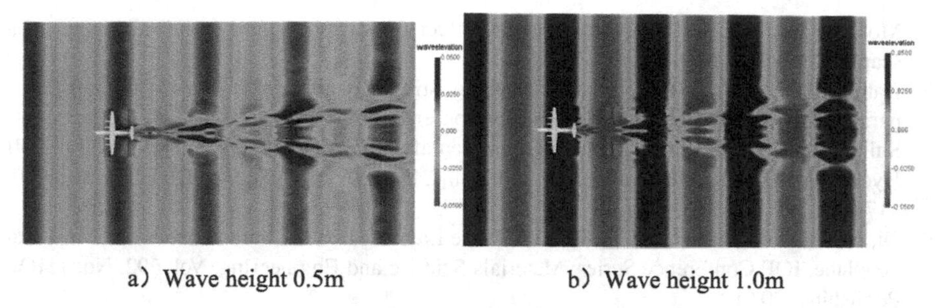

a) Wave height 0.5m b) Wave height 1.0m

Fig. 8. Waveform contour distribution under different wave heights

4 Conclusion

The effects of different speed factors, different wavelength factors and different wave height factors on the taxiing stability of seaplane in waves are simulated and analyzed. The numerical results show that:

Under the wave condition with wave height of 0.05 m and wave length of 8 m, the interaction between the body and the wave becomes stronger with the increase of the speed of the seaplane, which shows that the peak value and average value of the resistance curve increase with the increase of the speed, and the double amplitude and average values of the heave and pitch in the waves of the seaplane and the peak value of the acceleration curve also increase with the increase of the speed.

When the wave height is 0.05 M, the wave length is 8, 16 m, and the speed is 8 m/s, the wave steepness decreases with the increase of the wave length, and the peak value and average value of the resistance curve of the seaplane in the wave gradually decrease. The heave and pitch motion characteristics and acceleration characteristics also have the same trend.

When the wave height is 0.05, 0.1 m, the wavelength is 8 m, and the speed is 8m/s, with the increase of wave height, the wave steepness of the wave increases, the wave energy encountered by the seaplane increases, the peak value and average value of the resistance curve of the aircraft in the wave gradually increase, and the heave, pitch and acceleration characteristics also have the same trend.

References

1. Leyva, C., Leonel, A.: Seaplane conceptual design and sizing. Diss. University of Glasgow (2012)
2. Gao, X., et al.: Research on wave motion response characteristics of a seaplane. J. Phys.: Conf. Ser. **1985**(1). IOP Publishing (2021)
3. Huang, M., Wu, B., Li, C., et al.: The influence of aerodynamic on stability performance of amphibian on rough water. J. Phys.: Conf. Ser. IOP Publishing, **2012**(1), 012007 (2021)
4. Nebylov, V., Nebylov, A.: Seaplane landing smart control at wave disturbances. IFAC Proc. Vol. **44**(1), 3021–3026 (2011)
5. Du, H., Fan, G.-L., Yi, J.-Q.: Nonlinear longitudinal attitude control of an unmanned seaplane with wave filtering. Int. J. Autom. Comput. **13**(6), 634–642 (2016). https://doi.org/10.1007/s11633-016-0962-x

6. Mottard, E.J.: A brief investigation of the effect of waves on the take-off resistance of a seaplane. NASA TN D-165 (1959)
7. Dathe, I., Deleo, M.: Hydrodynamic characteristics of seaplanes as affected by hull shape parameters. In: Advanced Marine Vehicles Conference (1989)
8. Shihua, J.M., Seth, A., Li, Y., et al.: Experimental and Computational Analyses of Take-off Hydrodynamics of an Amphibian Aircraft Hull[C]//AIAA AVIATION 2020 FORUM. 2020: 3174
9. Qi, H. U., et al. Numerical Simulation of Wave Landing Loads Characteristics of Twin-Float Seaplane. IOP Conference Series: Materials Science and Engineering. Vol. 692. No. 1. IOP Publishing, 2019
10. Feng, Sun, et al. Numerical simulation method for wave surface landing of seaplane. IOP Conference Series: Materials Science and Engineering. Vol. 751. No. 1. IOP Publishing, 2020
11. Qu, Q., Liu, C., Liu, P., et al.: Numerical simulation of water-landing performance of a regional aircraft. J. Aircr. **53**(6), 1680–1689 (2016)
12. Lind, S.J., et al.: Numerical predictions of water–air wave slam using incompressible–compressible smoothed particle hydrodynamics. Appl. Ocean Res. **49**, 57–71 (2015)
13. Chen, J., et al.: Numerical study of wave effect on aircraft water-landing performance. Appl. Sci. **12**(5), 2561 (2022)
14. Wang, S., Guedes Soares, C.: Experimental and numerical study of the slamming load on the bow of a chemical tanker in irregular waves. Ocean. Eng. **111**, 369–383 (2016)
15. Mousaviraad, S.M., Wang, Z., Stern, F.: URANS studies of hydrodynamic performance and slamming loads on high-speed planing hulls in calm water and waves for deep and shallow conditions. Appl. Ocean. Res. 51, 222–240 (2015)

Target Tracking and Motion Estimation of Fixed Wing UAVs Based on Vision Measurement

Nian Danni(✉), Zhang Sibo, and Zhu Ma

Beijing Institute of Spacecraft System Engineering, Beijing 100094, China
13811990096@163.com

Abstract. In view of the target tracking control and state estimation of unmanned aerial vehicle (UAV), a fixed wing unmanned aerial vehicle with a front fixed camera is used as the research object. The target is estimated by visual measurement and extended Kalman filter, and the 3D local coordinate system research guidance law. The dynamic equation of UAV and the state equation and observation equation of the target are set up, and the Kalman filter is designed to estimation the state of the ground moving target. The simulation results show that motion estimation based on EKF can achieve effective tracking of targets.

Keywords: Fixed wing UAV · Target tracking · EKF · Motion estimation

1 Introduction

In recent years, the application of UAV has been in a growing trend. They can be applied in many aspects, such as intelligence, reconnaissance and surveillance systems, search and rescue, border control, natural disasters or battlefield damage assessment [1]. Generally speaking, these applications have very strict requirements for the ground environment. Many large UAV systems can meet the application requirements, but the cost is also huge. For such missions, compared with the rotor UAV, the small fixed wing UAV can quickly achieve high-speed flight under the condition of low energy consumption, and its dynamics increases the visibility of ground targets. Usually, we install the PTZ and its control system on the UAV, use the visual sensor (camera) fixed with the PTZ to obtain the motion information of the target, automatically adjust the speed, heading and rotation angle of the UAV, so that the tracked target is always in the center of the camera's field of vision, and maintain a certain relative position relationship between the UAV platform and the target, So as to realize the tracking of the target. The coupling between UAV and camera requires complex navigation, guidance and control algorithms and robust motion estimation to compensate the motion of unknown targets in time. However, additional load and control strategy will increase the energy consumption of UAV and limit the flight time of UAV, especially small and micro UAV [2]. How to effectively track the ground moving target on the premise of reducing energy consumption has become a necessary ability for the development of UAV [3]. In this paper, we will use a small fixed wing UAV to track moving targets through a fixed camera at the front (Fig. 1).

W. Fan et al. (Eds.): AsiaSim 2022, CCIS 1712, pp. 359–369, 2022.
https://doi.org/10.1007/978-981-19-9198-1_27

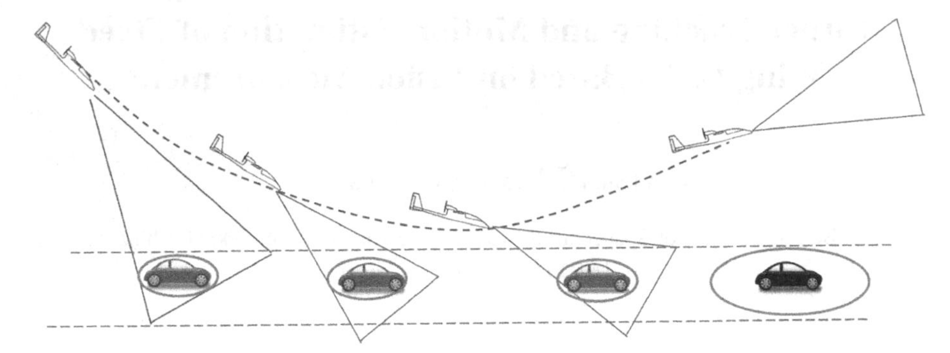

Fig. 1. Schematic diagram of front camera fixed wing UAV tracking ground target

2 Coordinate System

When studying the motion of small unmanned helicopter, in order to accurately describe its motion state information, a reference coordinate system is needed. When the small unmanned helicopter is tracking the moving target on the ground, the relative motion between the UAV, the target and the PTZ camera is more complex. In order to facilitate the analysis and model establishment, it is necessary to analyze the motion relationship between each part in different coordinates.

The guidance law of UAV tracking moving target can be divided into two categories: global guidance and local guidance. Global guidance law generally requires complete environmental knowledge, and the target coordinates are usually represented by the global inertial coordinate system. The path planning algorithm is generally developed by aligning the camera line of sight with the target vector, as shown in the Fig. 2 below [4]. In this paper, we use a local inertial system, such as camera coordinate system, to estimate the relative motion of the target. The advantage of using local guidance law is that it can greatly improve the computational efficiency, because it involves less coordinate system and inherent attributes that can be operated without global environment knowledge. The coordinate systems used in this paper mainly include inertial coordinate system, unmanned body coordinate system, camera coordinate system, image coordinate system and pixel coordinate system. The specific definitions of each coordinate system are as follows [5] (Fig. 2):

(1) Inertial coordinate system F^i

In the process of motion, we need a reference coordinate system to describe the position of UAV and moving target in the actual environment. This reference coordinate system is the inertial coordinate system. The origin of the inertial coordinate system can be selected at any point O^s on the ground, and the axes of the inertial coordinate system can be selected according to the geographical direction of North East ground.

(2) Body coordinate system 0 F^v

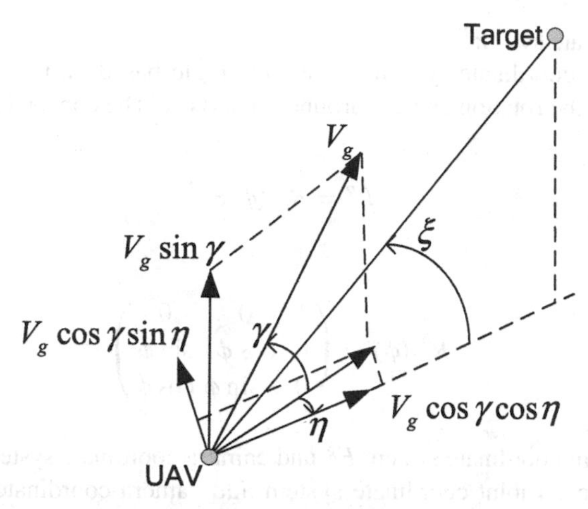

Fig. 2. Spherical coordinate system

The origin of the body coordinate system 0 is selected as the centroid of the UAV, and the unit vectors, i^v, j^v, k^v, points to the north, East and geocenter respectively.

(3) Body coordinate system 1 F^{v1}

The body coordinate system 1 is the yaw angle based on the body coordinate system 0 and the rotation angle ψ around the axis Z. The conversion formula from F^v to F^{v1} is:

$$P^{v1} = R_v^{v1}(\psi)P^v$$

where,

$$R_v^{v1}(\psi) = \begin{pmatrix} \cos\psi & \sin\psi & 0 \\ -\sin\psi & \cos\psi & 0 \\ 0 & 0 & 1 \end{pmatrix}$$

(4) Body coordinate system 2 F^{v2}

The body coordinate system 2 is the yaw angle based on the body coordinate system 1 and the rotation angle θ around the axis Y. The conversion formula from F^{v1} to F^{v2} is:

$$P^{v2} = R_{v1}^{v2}(\theta)P^{v1}$$

where,

$$R_{v1}^{v2}(\theta) = \begin{pmatrix} \cos\theta & 0 & -\sin\theta \\ 0 & 1 & 0 \\ \sin\theta & 0 & \cos\theta \end{pmatrix}$$

(5) Body coordinate system F^b

The body coordinate system is the yaw angle based on the body coordinate system 2 and the rotation angle ϕ around the axis X. The conversion formula from F^{v2} to F^b is:

$$P^b = R_{v2}^b(\phi)P^{v2}$$

where,

$$R_{v2}^b(\phi) = \begin{pmatrix} 1 & 0 & 0 \\ 0 & \cos\phi & \sin\phi \\ 0 & -\sin\phi & \cos\phi \end{pmatrix}$$

(6) Universal joint coordinate system F^g and camera coordinate system F^c

The universal joint coordinate system and camera coordinate system can be converted from the body coordinate system, and the conversion matrix is:

$$R_b^g = \begin{pmatrix} \cos\alpha_{el}\cos\alpha_{az} & \cos\alpha_{el}\sin\alpha_{az} & -\sin\alpha_{el} \\ -\sin\alpha_{az} & \cos\alpha_{az} & 0 \\ \sin\alpha_{el}\cos\alpha_{az} & \sin\alpha_{el}\sin\alpha_{az} & \cos\alpha_{el} \end{pmatrix}$$

$$R_g^c = \begin{pmatrix} 0 & 1 & 0 \\ 0 & 0 & 1 \\ 1 & 0 & 0 \end{pmatrix}$$

where, α_{el} and α_{az} are the elevation and azimuth of the universal joint respectively. For cameras without universal joints, the conversion matrix R_b^g and R_g^c are fixed matrices [6].

3 Control System

Let $\left(P_n\ P_e\ P_d\right)^T$ be the coordinates of the UAV in the north-east-down coordinate system, V_g is the ground speed, χ is the heading angle, γ is the track angle, ϕ is the rolling angle and g is the sea level gravity acceleration, then the UAV dynamic equation can be written as:

$$\dot{p}_n = V_g \cos\chi \cos\gamma$$

$$\dot{p}_e = V_g \sin\chi \cos\gamma$$

$$\dot{p}_d = -V_g \sin\gamma$$

$$\dot{\chi} = \frac{g}{V_g}\tan\phi$$

$$\dot{V}_g = \alpha_{Va}(V_g^c - V_g)$$

$$\dot{\gamma} = \alpha_{\gamma a}(\gamma^c - \gamma)$$

$$\dot{\phi} = \alpha_\phi(\phi^c - \phi)$$

where, $\alpha_{Va}, \alpha_{\gamma a}, \alpha_\phi$ is the constant positive control gain. It is assumed that the wind speed is zero and the ground speed when the UAV approaches the target V_g is a fixed value, i.e. $\dot{V}_g = 0$. Let r is the distance from the UAV to the target point, η, ξ is the azimuth and elevation of the relative target in the body 1 coordinate system. The kinematic equation of the target relative to the UAV can be expressed as:

$$\dot{r} = -V_g(\cos\gamma\cos\eta\cos\xi + \sin\gamma\sin\xi)$$

$$\dot{\eta} = \frac{V_g}{r}\frac{\cos\gamma\sin\eta}{\cos\xi} - \dot{\chi}$$

Let $\rho = \frac{V_g}{r}$, Then the above formula can be written as:

$$\dot{\rho} = \rho^2(\cos\gamma\cos\eta\cos\xi + \sin\gamma\sin\xi)$$

$$\dot{\eta} = \rho\frac{\cos\gamma\sin\eta}{\cos\xi} - \dot{\chi}$$

$$\dot{\xi} = \rho(\cos\gamma\cos\eta\sin\xi - \sin\gamma\cos\xi)$$

The control goal is to seek ϕ^c and γ^c to make η and $\gamma - \xi$ close to zero. Assuming that the speed direction of the UAV is on the same line as the x-axis of the body coordinate system, the pitch angle of the UAV is equal to the inclination angle of the flight channel, i.e. $\theta = \gamma$.

Suppose f is the focal length, M is the detector pixel size (square), and υ is the camera field angle. The relationship among the three is as follows:

$$f = \frac{M}{2\tan\left(\frac{\upsilon}{2}\right)}$$

Given the position coordinate $\left(\varepsilon_x, \varepsilon_y\right)$ of the target in the image, the target position in the camera coordinate system P^c is $\left(\varepsilon_x, \varepsilon_y, f\right)$, the target position in the body 1 coordinate system is $P^{v1} = R_{v2}^{v1}R_b^{v2}R_g^bR_c^gP^c$, and η, ξ is

$$\eta = \mathrm{a}\tan 2\left(P^{v1}(2), P^{v1}(1)\right)$$

$$\xi = -\tan^{-1}\left(\frac{P^{v1}(3)}{\sqrt{P^{v1}(1)^2 + P^{v1}(2)^2}}\right)$$

4 State Estimation

Extended Kalman filter EKF [7]: select system state vector $x = (\rho, \eta, \xi)^T$, system output vector $y = (\eta, \xi)^T$ and system input vector $u = (\dot{\chi}, \xi)^T$, then the system state equation can be expressed as:

$$\begin{aligned} \dot{x} &= f(x,u) + q \\ y &= h(x) + w \end{aligned}$$

where, $f(x,u) = \begin{pmatrix} \rho^2(\cos\gamma\cos\eta\cos\xi + \sin\gamma\sin\xi) \\ \rho\frac{\cos\gamma\sin\eta}{\cos\xi} - \dot{\chi} \\ \rho(\cos\gamma\cos\eta\sin\xi - \sin\gamma\cos\xi) \end{pmatrix}$, $h(x) = \begin{pmatrix} \eta \\ \xi \end{pmatrix}$,

q is process noise, representing model error and system disturbance, W is measurement noise, and represents detector noise. q and W are zero mean Gaussian random noise, and the covariance matrices are Q and R respectively.

Let the state estimation of state variable x be $\hat{x}$ and the covariance estimation be P, then the Kalman filter estimation is

$$\begin{aligned} \dot{\hat{x}} &= f(\hat{x}, u) \\ \dot{P} &= AP + PA^T + Q \end{aligned}$$

where

A is defined as $\frac{\partial f}{\partial x} = \begin{pmatrix} A_{11} & A_{12} & A_{13} \\ A_{21} & A_{22} & A_{23} \\ A_{31} & A_{32} & A_{33} \end{pmatrix}$,

$$\begin{aligned} A_{11} &= 2\rho(\cos\gamma\cos\eta\cos\xi + \sin\gamma\sin\xi) \\ A_{12} &= -\rho^2\cos\gamma\sin\eta\cos\xi \\ A_{13} &= \rho^2(-\cos\gamma\cos\eta\cos\xi + \sin\gamma\sin\xi) \\ A_{21} &= \frac{\cos\gamma\sin\eta}{\cos\xi} \\ A_{22} &= \frac{\rho\cos\gamma\cos\eta}{\cos\xi} \\ A_{23} &= \frac{\rho\cos\gamma\sin\eta\sin\xi}{\cos^2\xi} \\ A_{31} &= \cos\gamma\cos\eta\sin\xi - \sin\gamma\cos\xi \\ A_{32} &= -\rho\cos\gamma\sin\eta\sin\xi \\ A_{33} &= \rho(\cos\gamma\cos\eta\cos\xi + \sin\gamma\sin\xi) \end{aligned}$$

Assuming that the Kalman gain at time k is L_k, when the measurement starts, the Kalman filter is updated to [8]

$$L_k = P_k^- C^T \left(C P_k^- C^T + R \right)^{-1}$$
$$P_k^+ = (I - L_k C) P_k^-$$
$$\hat{x}^+ = \hat{x}^- + L_k \left(y_k - h\left(\hat{x}^-\right) \right)$$

where, superscript + and − represent the variable values respectively before and after updating, and I is the identity matrix, C is defined as $\frac{\partial h}{\partial x} = \begin{pmatrix} 0 & 1 & 0 \\ 0 & 0 & 1 \end{pmatrix}$.

5 Simulation Result

In this section, we assume a scene tracking the target through the above algorithms. Assume that the UAV is 100 m above the ground, whose speed is 13m/s, and the target starts from location (100, 0) m in the North-East coordinate frame, whose speed is 10m/s. Figure 3 shows the top-down view of the trajectories of the UAV and the target. The blue solid line represents the trajectory of UAV and the red dotted line represents the trajectory of the target.

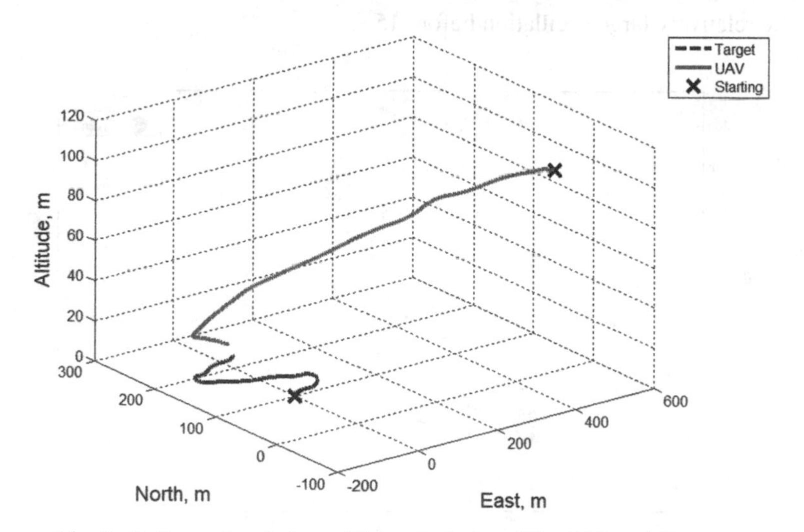

Fig. 3. 3-dimensional view of the trajectories of the UAV and the target

We can see that the target is moving in a curve. Figure 4 is the trajectory tracking diagram of UAV Based on EKF, in which the red dotted line is the theoretical trajectory of the target, the star point is the measured value, and the blue solid line is the estimated trajectory value. It can be seen that although the target trajectory is not linear, the UAV can track the trajectory well.

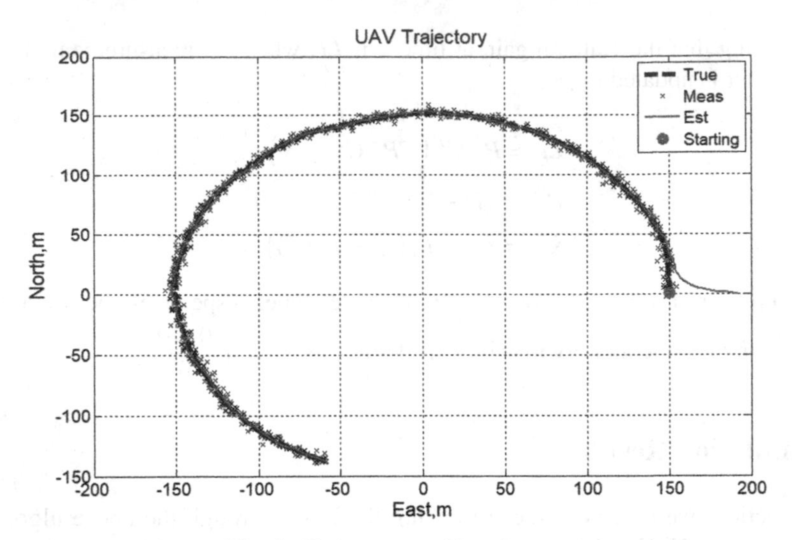

Fig. 4. Trajectory tracking diagram

Figure 5 shows the target trajectory in the camera view and Fig. 6 shows the corresponding evolution of the target motion and its distance to the center of the image. It can be seen that the target trajectory in the camera view is close to the center of the image except the relatively large oscillation before 15 s.

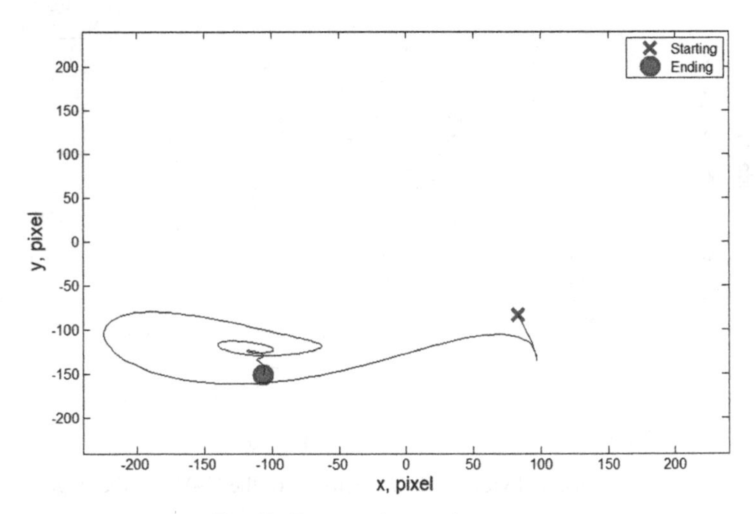

Fig. 5. Target trajectory in camera

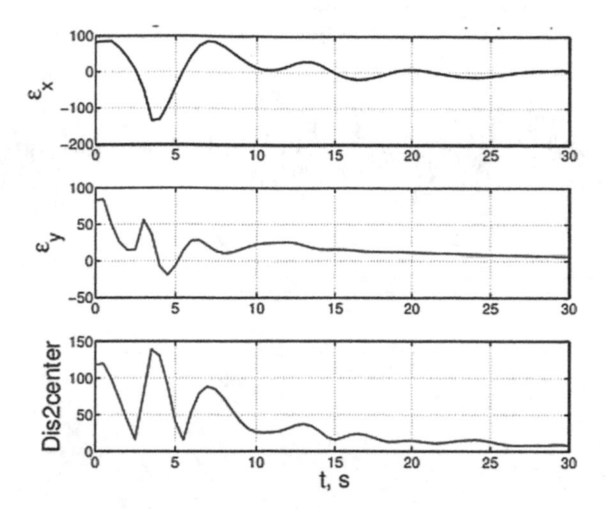

Fig. 6. Target coordinate evolution

The following Fig. 7 shows the specific deviation values of position coordinates (North and East) and heading angle and rolling angle.

We can see from the dynamic deviation diagram, the heading angle and rolling angle and position coordinates EAST and NORTH, each of them tends to zero when compared with the real value. Which indirectly indicates that the tracking effect is good.

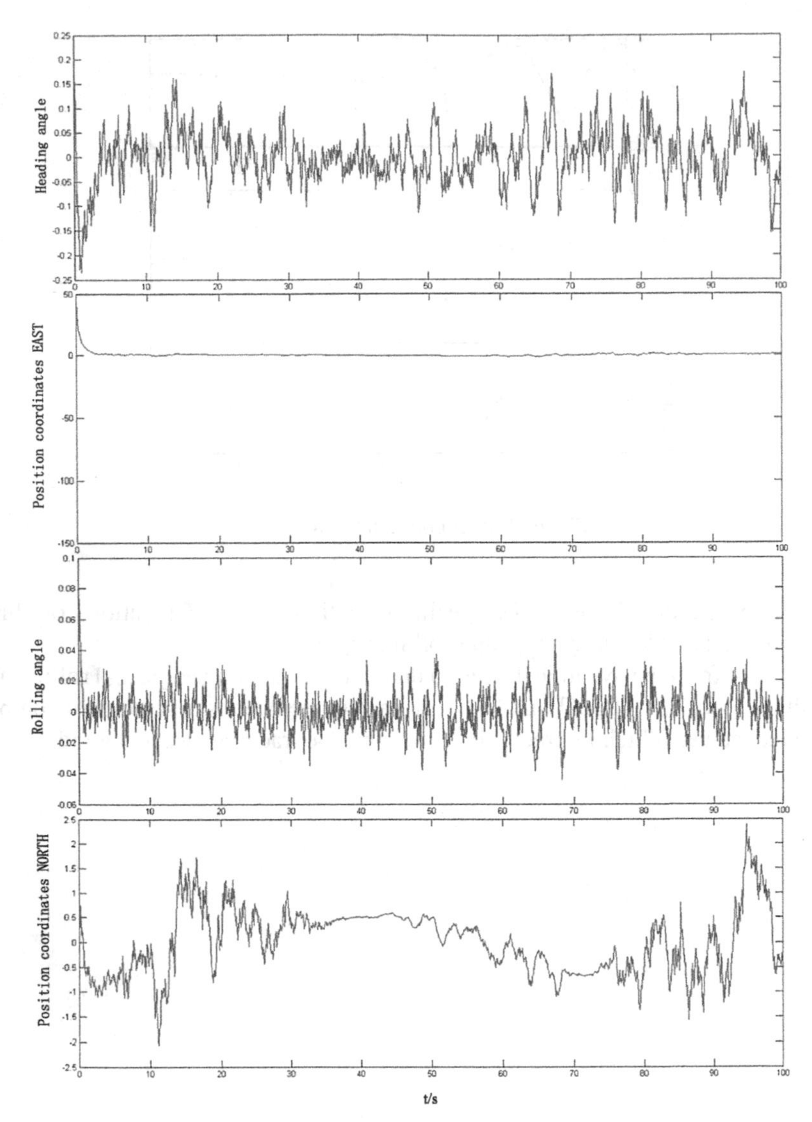

Fig. 7. Dynamic deviation diagram

6 Conclusions

The research of UAV target tracking technology based on vision has received extensive attention from researchers all over the world and has become one of the research hotspots in the field of UAV [9]. In this paper, the problem of target tracking and motion estimation of fixed wing UAV with front camera is analyzed and simulated. The three-dimensional local coordinate system is used to study the guidance law, the UAV dynamic equation and the state equation and observation equation of the target are established, and the extended Kalman filter is used to estimate the state of the ground moving target. The simulation results show that the extended Kalman filter method can effectively realize the motion state estimation of the UAV.

References

1. Farmani, N., Sun, L., Pack, D.: Tracking multiple mobile targets using cooperative Unmanned Aerial Vehicles, pp. 395–400 (2015), Unmanned Aircraft Systems (ICUAS), 2015 International Conference on Unmanned Aircraft Systems
2. He, R., Bachrach, A., Achtelik, M., et al.: On the design and use of a micro air vehicle to track and avoid adversaries. Int. J. Robot. Res. **29**(5), 529–546 (2010)
3. Hou, Z., Han, C.: A survey of visual tracking. Acta Automatica Sinica **32**(4), 603–617 (2006)
4. Liu, S., Hu, C., Zhu, J.: Study on computer vision-based approaches to estimate position and orientation of unmanned helicopter. Comput. Eng. Des. **25**(4): 564–568 (2004)
5. Li, Z., Xu, Y., Yang, G.: Simulation of UAV flight control for tracking ground targets. Sci. Technol. Eng. **8**(5), 1374–1378 (2008)
6. Zhang, G.: Machine Vision, pp. 24–27. Science Press, Beijing (2005)
7. Liu, J., Jiang, H., Shi, X.: Study and application of Kalman filtering for target tracking. Inf. Technol. (10): 174–177 (2011)
8. Wang, X., Zeng, Q., Xiong, Z., et al.: Development and research analysis on vision-based navigation technologies. Inf. Control. **39**(5), 607–613 (2010)
9. Zhu, S., Wang, D., Low, C.B.: Ground target tracking using UAV with input constraints. J. Intell. Robot. Syst. **69**(1–4), 417–429 (2013)

Wind Aided Aerodynamic Characteristics in the Quadcopter UAV Control Modeling

Wei Cao[1(✉)], Xiaoyi Liu[2], Pinggang Yu[1], Xiao Xu[1], and Jie Zhang[3]

[1] College of Joint Operation, National Defense University, Beijing 100091, China
cassanocao@hotmail.com
[2] No. 96942 Unit of PLA, Beijing 102208, China
[3] Graduate School, National Defense University, Beijing 100091, China

Abstract. Quadcopter UAVs are always exposed to time-varying winds in the flight process, which makes it much more challenging to be detected and recognized by radar systems. Conventional solutions are to estimate and reject the wind aided adverse impacts, while in this paper we regard the wind aided aerodynamic characteristics as prior environmental information rather than disturbance. In particular, we establish a Matlab Simulink based quadcopter UAV control model in wind environment which is composed of wind shear, turbulence and gusts, and investigate the wind aided velocity and Doppler frequency fluctuations in a modeling and simulation way since the wind aided aerodynamic characteristics of the quadcopter UAV are difficult to be described in mathematical expressions. The simulation results reveal that: (1) the wind gusts are the primary components for the velocity and Doppler frequency fluctuations; (2) the impacts of wind turbulence are negligible due to the quad-copter UAVs' wind resistance capability; (3) the impacts of the wind shear can be transformed between horizontal and vertical dimensions as a consequence of the coupling effect.

Keywords: Quad-copter UAV · Aerodynamic characteristics · Wind velocity

1 Introduction

Quadcopter UAVs, i.e. mini drones, are widely used in a plethora of industria and commercial areas [1], such as medical rescue, weather forecast, aerial remote sensing, movie shooting, live events broadcasting, goods delivery, etc. While in the meantime, the increasing popularity brings about serious concerns of the potential and probable threats upon public safety [2]. Therefore it is very important to develop countermeasures to detect, localize, track, recognize, control and kill the quad-copter UAVs.

A lot of efforts have been made to detect the quad-copter UAVs as the first step to deal with their low, slow, small target characteristics [3–8]. Radar systems, in mono-static or multi-static configuration, are among the best candidates for anti-quad-copter UAVs, and are assumed to outperform all other candidates

W. Fan et al. (Eds.): AsiaSim 2022, CCIS 1712, pp. 370–381, 2022.
https://doi.org/10.1007/978-981-19-9198-1_28

such as infrared and images under bad illumination conditions [4]. In particular, both active and passive radar systems can be placed near the place to be protected and used to detect the drones. Once the drone is detected, its neutralization can be carried out by launching a series of countermeasures.

However, the detection performance is still likely to be improved theoretically since not all the prior information has been utilized. During the actual flights, quad-copter UAVs are always exposed to various disturbances and uncertainties, such as wind gusts, which results in difficulties in achieving robust performance and accurate flight paths because of their small structure and light weight. In spite of the various linear and nonlinear control techniques to maintain stability against the varying wind, quad-copter UAVs present sudden velocity and Doppler frequency fluctuations as aerodynamic characteristics [9–12]. In [9], a disturbance observer based backstepping control algorithm was designed to realize the trajectory following control for UAVs with the wind disturbance. While in [10], the authors analyzed the influence of horizontal wind on the drone and built a model on a multi-rotor UAV in the wind field. The paper [11] calculated the maximum wind speed that the UAV can keep hovering state under the compound wind field by analyzing the variation of the unstable wind and the movement state of the quadrotor UAV. Finally in [12], the authors extensively studied the effect of wind on the connectivity and safety of a large scale swarm with one leader and multiple follower UAVs. In particular, they examined the relationship between different parameters including UAV speed, UAV mass, wind speed, drag force, and number of UAVs maintaining the desired safety and connectivity requirement in a swarm.

Generally, wind is regarded as the disturbance that may degrade the detection performance, so researchers seek to develop several methods to estimate and then mitigate the adverse effects of winds on quad-copter UAVs [13]. However, the wind aided aerodynamic characteristics may be used to improve the detection performance. To the best of our knowledge, not much attention has been addressed to the use of the wind aided aerodynamic characteristics as prior information, which is probably due to the lack of detailed process description between the wind field and flight control model of quad-copter UAV.

The main contributions of this paper are summarized as follows:

(1) We are the first to investigate the wind aided aerodynamic characteristics of the quad-copter UAVs in a modeling and simulation way, so that the velocity and Doppler frequency fluctuations are likely to be employed to improve the detection performance rather than being treated as disturbance in radar signal processing.
(2) We figure out that the wind gusts are the primary components for the Doppler frequency fluctuations, while the wind turbulence is negligible due to the quad-copter UAVs' wind resistance capability. Besides, the impacts of the wind shear can be transformed between horizontal and vertical dimensions as a consequence of the coupling effect.

The remainder of this paper is organized as follows. The wind velocity model is introduced in Sect. 2. A quad-copter UAV control modeling platform in wind environment is set up in Sect. 3 and the corresponding simulation and analytic results are presented in Sect. 4. The concluding remarks are provided in Sect. 5.

2 Wind Velocity Model

The flight altitude of quad-copter UAV is generally below 600 m, while in some limited areas it is required to be below 120 m. And in the low altitude zones, wind is assumed as a more common weather condition in the flight phase of quad-copter UAV compared with rain, fog, sleet. etc. Noted that the wind field is comprised of wind direction and wind velocity. In this paper, we only investigate the target characteristics considering the wind velocity while ignoring the wind direction. Therefore, the wind velocity at low altitude is modeled as a linear summation of wind shear, turbulence and gusts

$$v_{wind} = v_{shear} + v_{turbulence} + v_{gust} \tag{1}$$

In particular, wind shear and gusts represent the expectation of the wind velocity, while turbulence represents a random perturbation change.

Wind shear is a change in wind velocity over a short distance, which is most often associated with strong temperature inversions or density gradients. It can occur either horizontally or vertically, while vertical wind shear is the most commonly described shear. Empirically at the low altitude, it can be formulated as

$$v_{shear} = v_{20} \frac{\ln\left(\frac{h}{z_0}\right)}{\ln\left(\frac{20}{z_0}\right)} \tag{2}$$

where v_{shear} is the mean wind shear velocity, v_{20} is the measured wind shear velocity at an altitude of 6 m/20 ft, h is the altitude, and z_0 is a constant equal to 0.15 ft for Category C flight phases and 2.0 ft for all other flight phases according to the Military Specification MIL-F-8785C [14]. We can see from formula (2) that vertical wind shear is expected to be constant at the fixed altitude.

Wind turbulence generally refers to rapid and irregular fluctuations in wind velocity, which is often described by Dryden model. In particular, the wind turbulence velocity is calculated by using the parameter v_{20}, i.e. the wind speed at 20 ft altitude. The larger v_{20} is, the more significant the wind turbulence is. Typically for light turbulence, the wind speed at 20 ft is 15 knots; for moderate turbulence, the wind speed is 30 knots, and for severe turbulence, the wind speed is 45 knots.

Wind gust is a sudden, seconds-long burst of high-speed wind which is followed by a lull. It represents a variation over time or space relative to a baseline constant wind. Generally wind gust can be modeled as the standard '1-cosine' shape, and the mathematical representation of the discrete wind gust is

$$v_{gust} = \begin{cases} 0 & x < 0 \\ \frac{v_m}{2}\left(1 - \cos\left(\frac{\pi x}{d_m}\right)\right) & 0 \leqslant x \leqslant d_m \\ v_m & x > d_m \end{cases} \quad (3)$$

where v_{gust} is the gust velocity, v_m is the gust velocity amplitude, d_m is the gust length, and x is the distance traveled.

3 Quadcopter UAV Control Modeling in Wind Environment

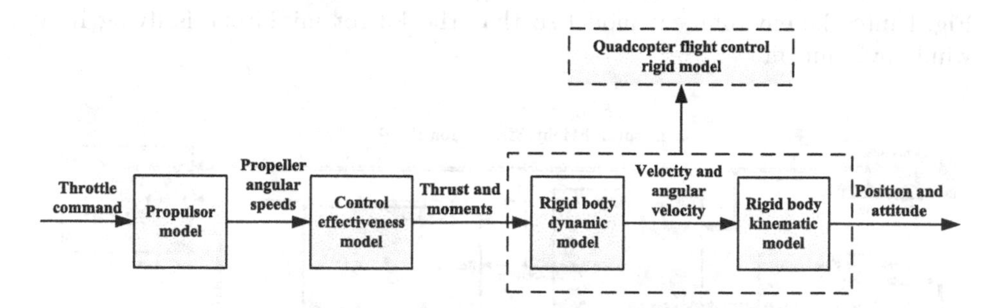

Fig. 1. Quadcopter UAV control model structure

Quadcopter UAV is a nonlinear, under-actuated, and highly-coupled complex system. According to the book [15], the Quadcopter UAV control model is a cascade of four models as shown in Fig. 1, i.e. propulsor model, control effectiveness model, rigid body dynamic model and rigid body kinematic model. And the Quadcopter flight control rigid model, which consists of the rigid body dynamic model and rigid body kinematic model, can be described mathematically as

$$\begin{cases} {}^{\mathrm{e}}\dot{\mathbf{p}} = {}^{\mathrm{e}}\mathbf{v} \\ {}^{\mathrm{e}}\dot{\mathbf{v}} = \mathrm{g}\mathbf{e}_3 - \frac{f}{m}\mathbf{R}\mathbf{e}_3 \\ \dot{\theta} = \mathbf{W} \cdot {}^{\mathrm{b}}\omega \\ \mathbf{J} \cdot {}^{\mathrm{b}}\omega = -{}^{\mathrm{b}}\omega \times \left(\mathbf{J} \cdot {}^{\mathrm{b}}\omega\right) + \mathbf{G_a} + \tau \end{cases} \quad (4)$$

where the left superscript ^{e} and ^{b} represent the geodetic and body coordinates, respectively, ${}^{\mathrm{e}}\mathbf{p}$ is the gravity vector, ${}^{\mathrm{e}}\mathbf{v}$ is the flight speed, g is the acceleration of gravity, m is the mass, f is the total pulling force owing to the four rotors, $\mathbf{R}$ is the rotation matrix from the body coordinates to the geodetic coordinates, $\theta = [\phi, \theta, \psi]^{\mathrm{T}}$ is the Euler angle that stands for the Roll, Pitch, and Yaw angles, $\mathbf{W}$ is the transform matrix which is illustrated in details in formula (5), ${}^{\mathrm{b}}\omega$ is the body angular velocity, $\mathbf{J}$ is the moment of inertia, $\mathbf{G_a}$ is the gyroscopic torque, τ is the torque upon the body axis caused by four rotors.

$$\mathbf{W} = \begin{bmatrix} 1 & \tan\theta\sin\phi & \tan\theta\cos\phi \\ 0 & \cos\phi & -\sin\phi \\ 0 & \sin\phi/\cos\theta & \cos\phi/\cos\theta \end{bmatrix} \tag{5}$$

Since the wind aided dynamic characteristics of the quadcopter UAV are very hard to be described in mathematical expressions, we establish a quadcopter simulation model platform to investigate the aforementioned issue in a black-box way by modifying the Matlab Quadcopter Flight Simulation Model as demonstrated in Fig. 2 which is from Matlab Quadcopter Project. In particular, the quadcopter UAV is a Parrot minidrone as illustrated in Fig. 3. And the environment model is extended by adding the wind module which is shown in Fig. 4 into the environment model so that the Parrot minidrone is flying in a wind environment.

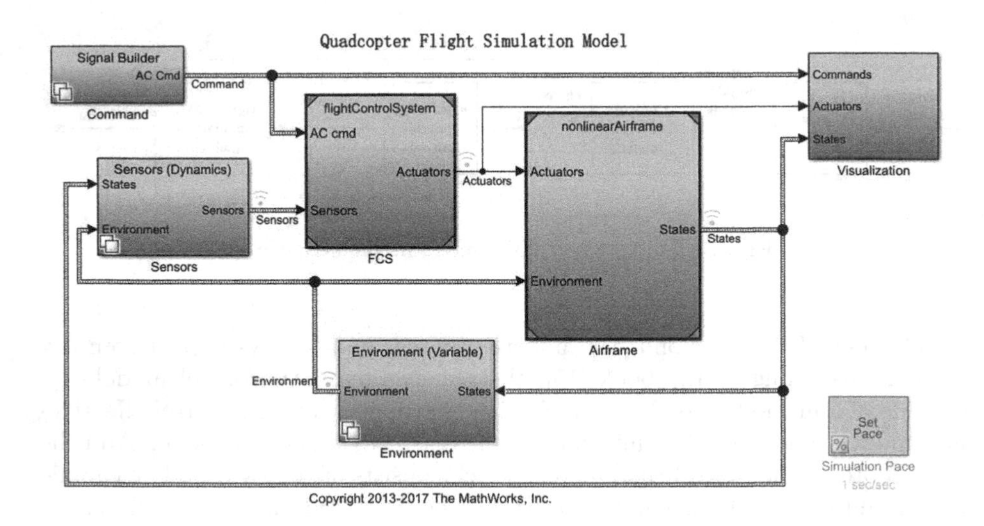

Fig. 2. Quadcopter UAV flight simulation model

4 Simulation Results

The simulation scenario is designed as a flight process that the Parrot minidrone lifts off from ground to the expected altitude and then hovers in a wind environment. The wind is set as the combination of wind shear, turbulence and three gusts where the three wind gusts start at 25, 45, 65 s, respectively. The total simulation time length is 100 s.

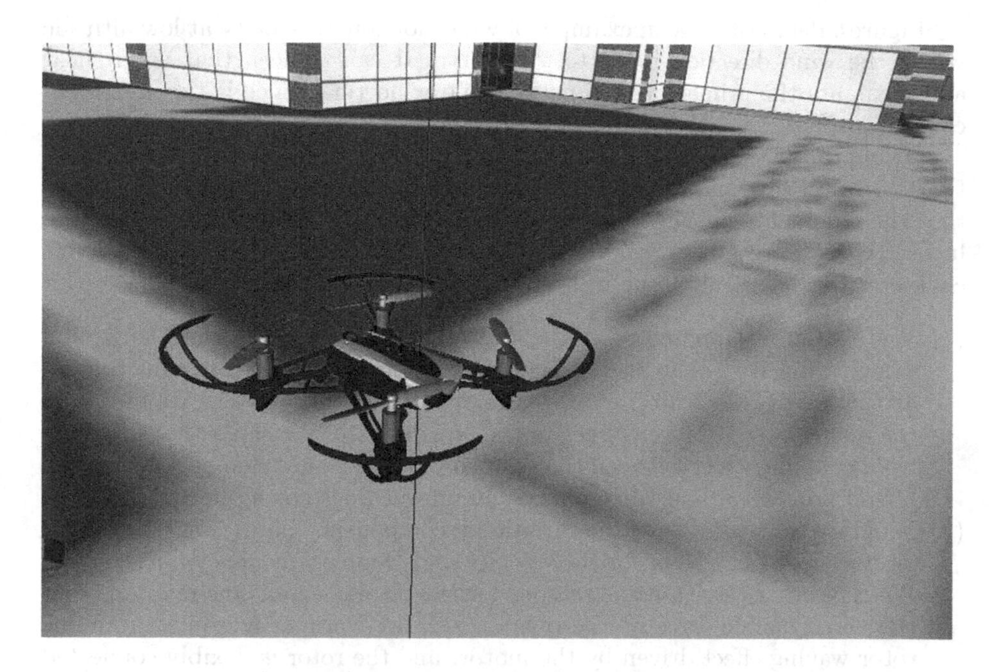

Fig. 3. Parrot minidrone model

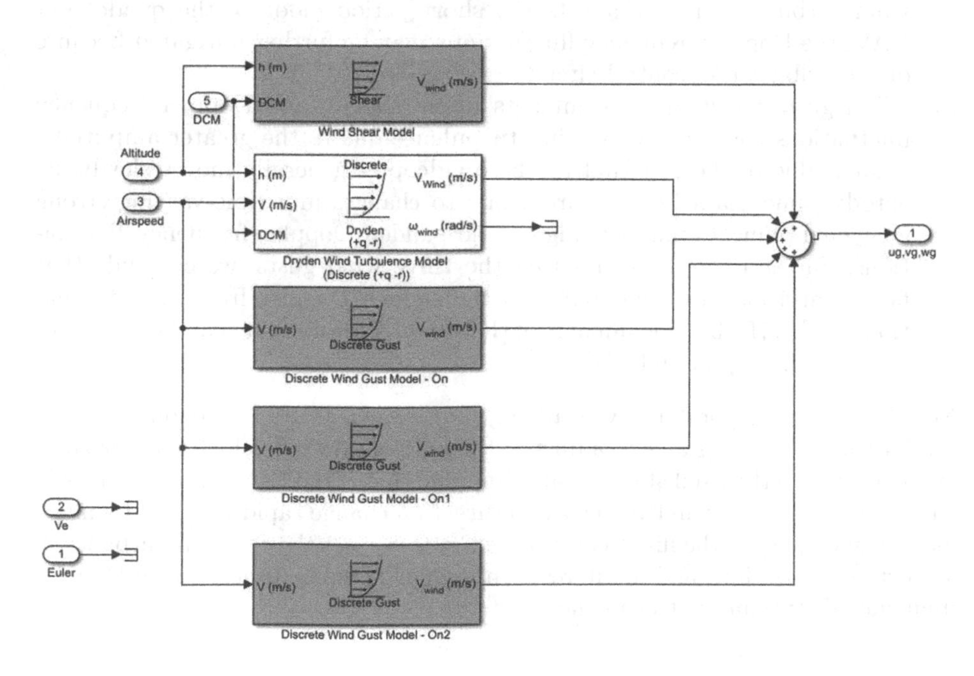

Fig. 4. Wind simulation module

Figure 5 demonstrates an example of wind horizontal velocity at low altitude where the wind direction is set to the North. It can be seen that wind shear and gusts are the primary components while wind turbulence is the secondary component.

Owing to the time-varying horizontal wind velocity, we can see that the fluctuations of velocity and Doppler frequency are shown in Fig. 6 and Fig. 7. Noted that the nominal frequency is set to be the carrier frequency of Long Term Evolution (LTE) telecommunication signals, i.e. 2.6GHz, so that the aerodynamic characteristics are likely to be utilized by LTE passive radars.

(1) Wind shear is supposed to be the steady component in the wind horizontal velocity, but it leads to a sudden velocity and Doppler frequency fluctuation at the start of the lift-off process as shown in Fig. 6b and Fig. 7b, and then the quadcopter UAV converges to the steady state. We claim that it results from coupling effect between the vertical and horizontal channels during the lift-off process, although the Parrot minidrone does not fly horizontally.
(2) Wind turbulence is one of the variational components, but it is the secondary or even negligible factor for the velocity and Doppler frequency fluctuations. We claim that there are two reasons for the phenomenon. One reason is that the aerodynamic force of the quadcopter UAV primarily comes from the rotor waving effect driven by the motor, and the rotor is flexibly connected to the fuselage, so the rotational movement of the rotor mitigates the impact of turbulence on the quadcopter UAV; while the other one reason is that as wind turbulence mainly affects the short period mode of the quadcopter UAV, the Doppler frequency fluctuations may be further mitigated because of the stabilization control circuits.
(3) Wind gusts result in more impacts upon velocity and Doppler frequency fluctuations compared with wind turbulence due to the greater amplitude. That is due to the fact that as the quadcopter generally moves slowly, its aerodynamic characteristics are prone to changes in a time-varying strong gust environment, thus leading to the sudden Doppler frequency fluctuations. Furthermore, by comparing the three wind gusts, we conclude that faster variation rate contributes to the sudden Doppler frequency fluctuations in Fig. 7f while the impacts of the slower variation rate can be negligible as shown in Fig. 7d and Fig. 7e.

Noted that the temporal intervals among the three gusts are long enough so that the Parrot minidrone is able to adjust to the horizontal wind velocity disturbance and converge to the fixed steady state. Imagine that if the Parrot minidrone is in a environment where wind velocity and direction change rapidly, it will be much more challenging for the flight control system to correct the speed disturbance in a short period of time, and it will result in a more complex time-varying Doppler frequency disturbance effect in the spectrum.

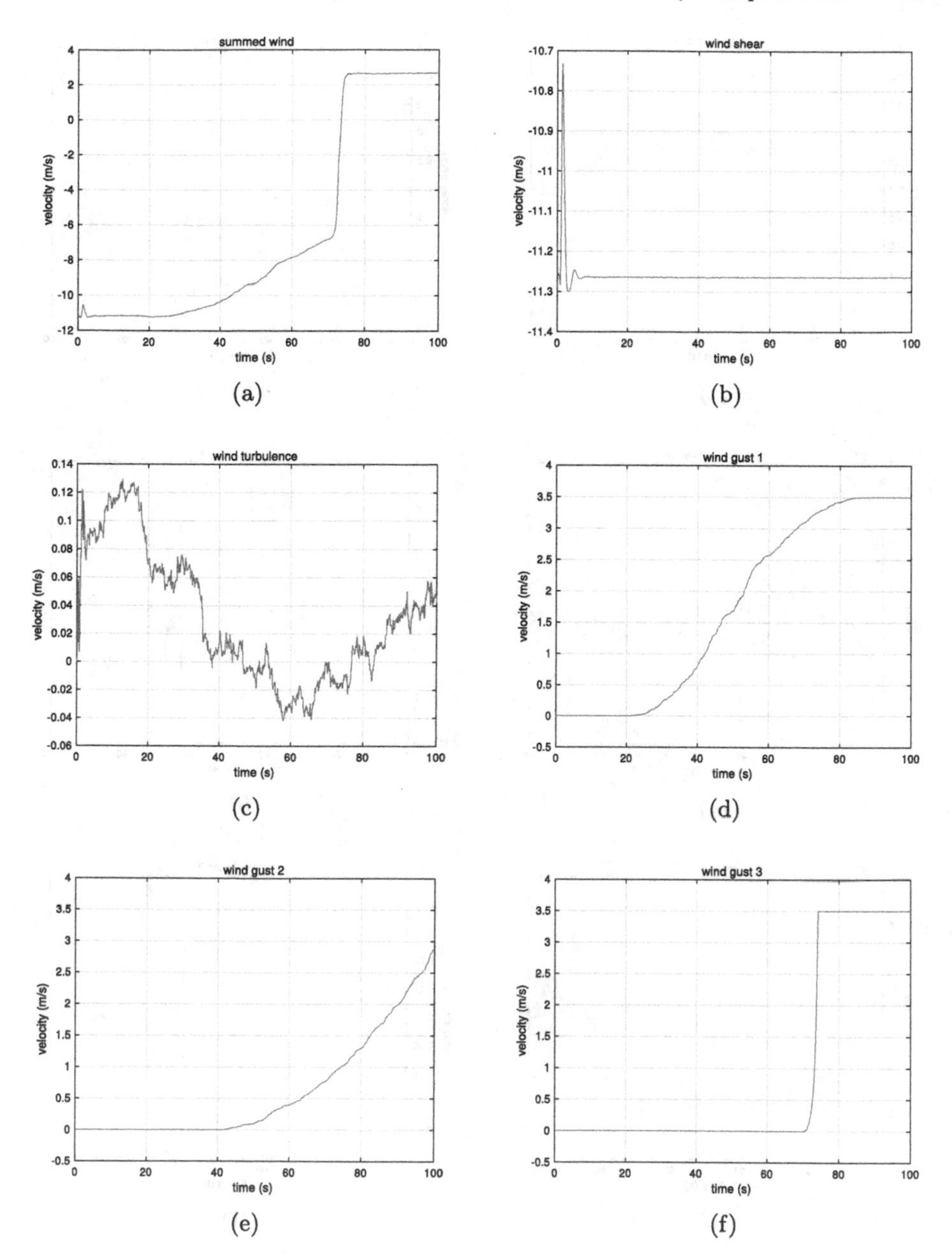

Fig. 5. An example of wind horizontal velocity at low altitude: (a) summed wind; (b) wind shear; (c) wind turbulence; (d) wind gust 1; (e) wind gust 2; (f) wind gust 3

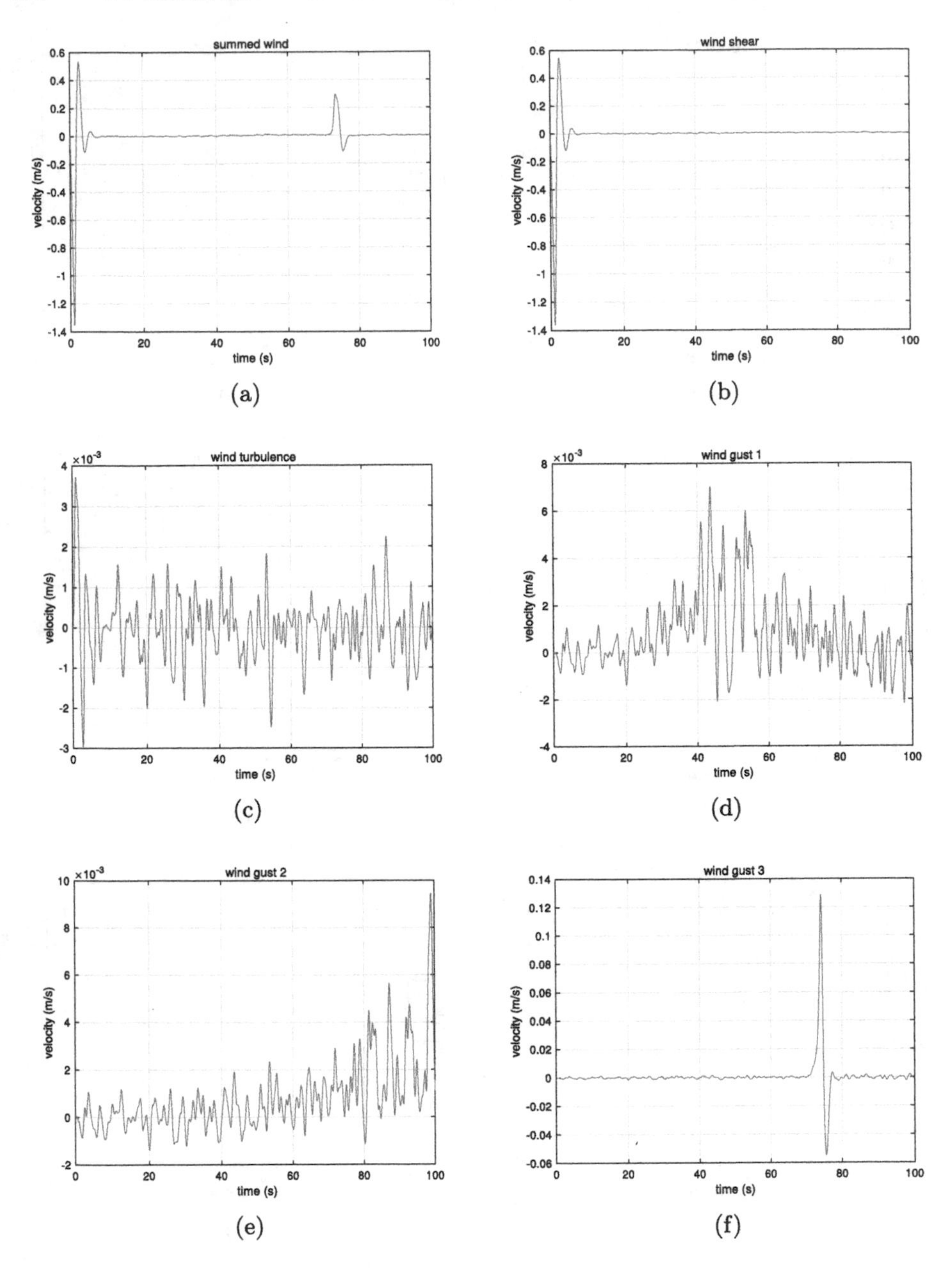

Fig. 6. Parrot minidrone velocity fluctuations in wind environment: (a) summed wind; (b) wind shear; (c) wind turbulence; (d) wind gust 1; (e) wind gust 2; (f) wind gust 3

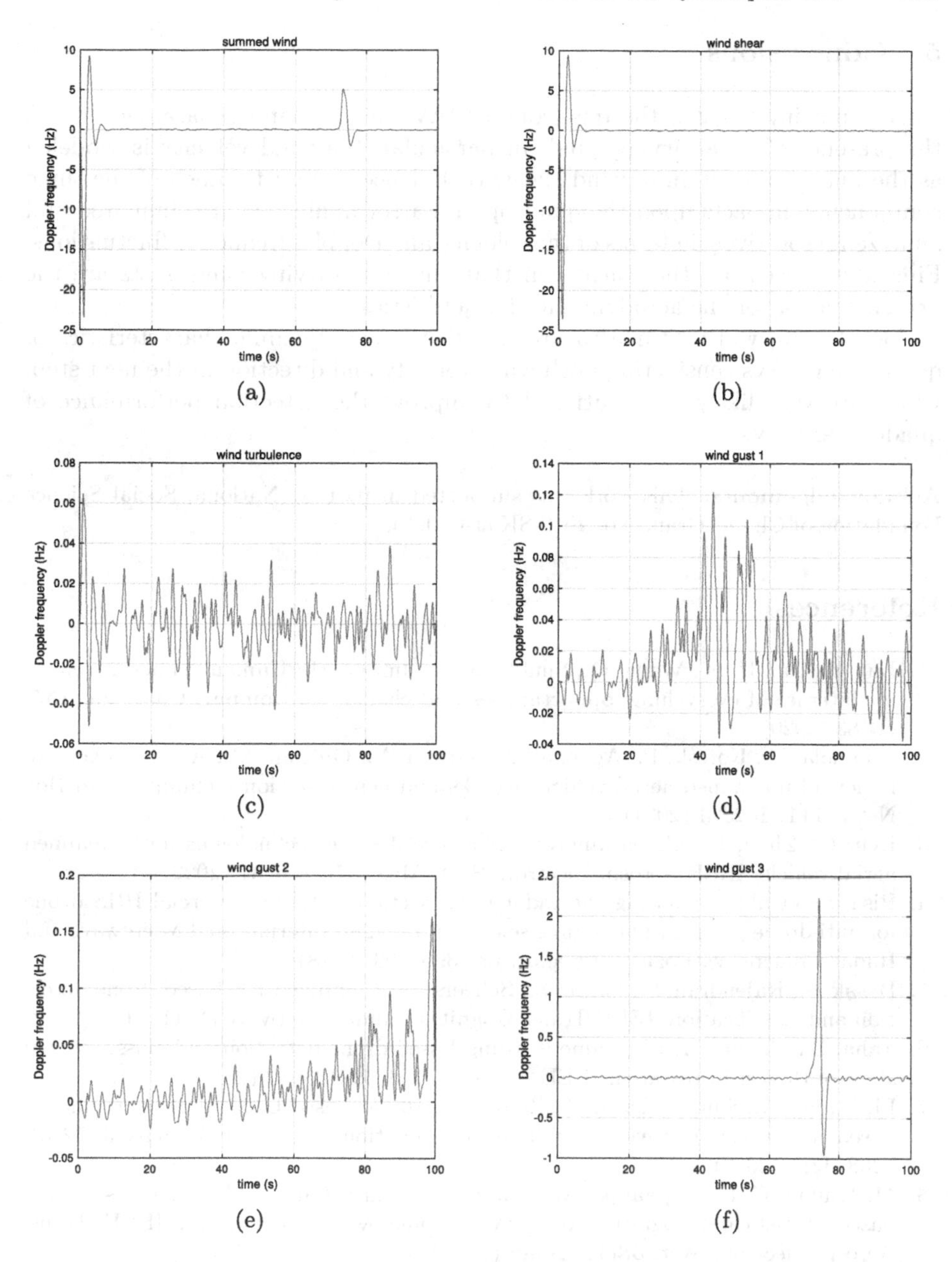

Fig. 7. Parrot minidrone Doppler frequency fluctuations in wind environment: (a) summed wind; (b) wind shear; (c) wind turbulence; (d) wind gust 1; (e) wind gust 2; (f) wind gust 3

5 Conclusions

This paper investigates the quadcopter UAVs' aerodynamic characteristics in the presence of time-varying wind. In particular, the wind velocity is modeled as the linear summation of wind shear, turbulence and gusts. Each of the three components' impacts upon the quadcopter Parrot minidrone are simulated and analyzed respectively in terms of the velocity and Doppler frequency fluctuations. Finally we reach out the conclusion that the time-varying wind gusts are the primary factor for the aerodynamic characteristics.

Besides, we will continue to investigate the aerodynamic characteristics of quadcopter UAVs considering both wind velocity and direction in the next step, which are very likely to be utilized to improve the detection performance of quadcopter UAVs.

Acknowledgements. This work was supported in part by National Social Science Foundation of China (Grant No. 2019-SKJJ-C-063).

References

1. Liu, Y., Dai, H.-N., Wang, Q., Shukla, M.K., Imran, M.: Unmanned aerial vehicle for internet of everything: opportunities and challenges. Comput. Commun. **155**, 66–83 (2020)
2. Chamola, V., Kotesh, P., Agarwal, A., Gupta, N., Guizani, M.: A comprehensive review of unmanned aerial vehicle attacks and neutralization techniques. Ad Hoc Netw. **111**, 102324 (2021)
3. Lyu, C., Zhan, R.: Global analysis of active defense technologies for unmanned aerial vehicle. IEEE Aerosp. Electron. Syst. Mag. **37**(1), 6–31 (2022)
4. Pisa, S., et al.: Evaluating the radar cross section of the commercial IRIS drone for anti-drone passive radar source selection. In: 22nd International Microwave and Radar Conference, Poznan, Poland , pp. 699–703 (2018)
5. Basak, S., Rajendran, S., Pollin, S., Scheers, B.: Combined RF-based drone detection and classification. IEEE Trans. Cognit. Commun. Netw. **8**(1), 111–120 (2022)
6. Taha, B., Shoufan, A.: Machine learning-based drone detection and classification: state-of-the-art in research. IEEE Access **7**, 138669–138682 (2019)
7. Li, Y., Fu, M., Sun, H., Deng, Z., Zhang, Y.: Radar-based UAV swarm surveillance based on a two-stage wave path difference estimation method. IEEE Sens. J. **22**(5), 4268–4280 (2022)
8. M. Ezuma, C. K. Anjinappa, M. Funderburk and I. Guvenc: Radar cross section based statistical recognition of UAVs at microwave frequencies. IEEE Trans. Aerosp. Electron. Syst. **58**(1), 27–46 (2022)
9. Wu, K., Fan B., Zhang, X.: Trajectory following control of UAVs with wind disturbance. In: 36th Chinese Control Conference, Dalian, China , pp. 4993–4997 (2017)
10. Hou, Y., Huang, W., Zhou, H., Gu, F., Chang, Y., He, Y.: Analysis on wind resistance index of multi-rotor UAV. In: Chinese Control And Decision Conference, Nanchang, China, pp. 3693–3696 (2019)

11. Haidong, Z., Qiuyu, C., Chongfa, Z., Yajie, D., Yufeng, M., Jun, Y.: Stability research of quadcopter UAV under unstable wind. In: IEEE 7th International Conference on Control Science and Systems Engineering, Qingdao, China, pp. 114–118 (2021)
12. Tegicho, B., Geleta, T., Bogale, T., Eroglu, A., Edmonson, W., Bitsuamlak, G.: Effect of wind on the connectivity and safety of large scale UAV swarms. In: IEEE International Black Sea Conference on Communications and Networking, Bucharest, Romania, pp. 1–6 (2021)
13. Azid, S.I., Kumar, K., Cirrincione, M., Fagiolini, A.: Robust motion control of nonlinear quadrotor model with wind disturbance observer. IEEE Access **9**(1), 149164–149175 (2021)
14. U.S. Military Specification MIL-F-8785C, 5 November (1980)
15. Quan, Q.: Introduction to Multicopter Design and Control. Springer, Singapore (2017)

A Trajectory Re-planning Simulation Design of Multistage Launch Vehicle

Haolei Ma(✉), Xuefeng Li, and Tianliang Zhang

Beijing Aerospace Automatic Control Institute, Beijing 100854, China
mahaolei@163.com

Abstract. When a thrust anomaly occurs in the flight, launch vehicles should have the ability to deal with faults and deviations autonomously, predict trajectory reachable range and re-plan a new flight trajectory rapidly. The entry orbit elements are determined by flight status, and reachable range of aircraft is predicted. According to state equations and performance indexes, Hamilton function of the problem is constructed. Adaptive dynamic adjustment and control command re-planning can be adopted to ensure orbit entry. On the other hand, original problem can be re-described, and degradation scheme can be planned. According to mission partition, the optimal control problem is constructed and solved via sequence convex programming. The simulation results show that this method can re-plan trajectories rapidly in different situations and remedy abnormal states to a certain extent.

Keywords: Launch Vehicle · Orbital Mechanics · Flight Deviation · Trajectory Planning · Flight Simulation

1 Introduction

The diversification of space missions and complexity of structure are launch vehicles' future development trends. Launch vehicles must have the ability to handle faults independently [1]. It is an urgent problem for the design of launch vehicle control system to ensure flight safety and to maximize reachable mission profile. The thrust system has become a sensitive system because of its complex working environment and harsh working conditions. The thrust system fault will lose control ability of launch vehicles, and eventually lead catastrophic accidents. The flight failure caused by thrust system has shown an increasing trend in recent years [2]. In 2016, the second stage engine of a Falcon-9 launch vehicle of SpaceX exploded during its flight. In 2019, the third stage engine of a CZ launch vehicle decreased abnormally, resulting in mission failure, etc.

In a flight mission, the thrust system of launch vehicle breaks down and deviates from original designed nominal trajectory. It will be difficult to complete the mission if the guidance and control scheme under nominal trajectory continues to be used [3]. It is necessary to evaluate flight capability, make independent decisions and plan to

W. Fan et al. (Eds.): AsiaSim 2022, CCIS 1712, pp. 382–394, 2022.
https://doi.org/10.1007/978-981-19-9198-1_29

complete capability construction of different mission modes. Countermeasures of trajectory re-planning and control reconfiguration are taken to maximize flight reliability and ensure completion of the flight mission. In recent years, more complex three-dimensional motion models are used in trajectory re-planning [4]. The optimization algorithm needs to consider multi-path constraints [5] and multi-terminal constraints [6], and try to solve the real-time problem [7]. With the development of modern computer technology, it is possible to solve the trajectory planning problem rapidly based on numerical method [8]. The convex programming method [9] can effectively reduce complexity of solving the problem, and can achieve planning on the premise of ensuring accuracy rapidly. This paper designs a mission partition scheme, reconstructs constraints and performance indicators of the optimization problem, solves optimization problem under different conditions by using sequential convex programming method [10]. The flight trajectory is reprogrammed that meets conditions. If flight failure or large deviation makes it impossible to enter scheduled orbit, it will enter the suboptimal orbit through mission degradation.

2 Model and Orbit Parameters

In ECI (Earth Centered Inertial) system $O_E\text{-}X_iY_iZ_i$, state equations of launch vehicle is described. The O_E is located at the Earth centroid. The O_EX_i points to the longitude line of the launch point in equatorial plane of the Earth. The plane of $O_EX_iY_i$ is equatorial plane. The O_EZ_i is the Earth's axis of rotation and points to North Pole.

The continuous optimal control problem is transformed into a standard nonlinear programming problem with unknown coefficients of state and control variables. A general optimal control problem of nonlinear system includes three parts: state equations, constraint conditions and performance index. The whole optimal control problem requires that performance index reach the minimum value under the condition of satisfying state equations and constraints. The standard format can be written as

$$\begin{aligned} &\min \ J \\ &\text{s.t.} \quad \frac{\mathrm{d}\boldsymbol{x}(\mathrm{t})}{\mathrm{dt}} = f[\boldsymbol{x}(t), \boldsymbol{u}(t), t], t \in [t_0, t_\mathrm{f}] \\ &\qquad E[\boldsymbol{x}(t), \boldsymbol{u}(t), t] = 0 \\ &\qquad I[\boldsymbol{x}(t), \boldsymbol{u}(t), t] \le 0 \end{aligned} \tag{1}$$

where E represents equality constraints and I represents inequality constraints respectively. State equations can be determined according to kinematic and dynamic equations of launch vehicle.

It is assumed that the thrust direction of launch vehicle is always along the body axis. According to Newton's Second law, motion equations of launch vehicle in ECI system are as follows

$$\begin{cases} \frac{\mathrm{d}\boldsymbol{r}(t)}{\mathrm{d}t} = \boldsymbol{v}(t) \\ \frac{\mathrm{d}\boldsymbol{v}(t)}{\mathrm{d}t} = -\frac{\mu}{[r(t)]^3}\boldsymbol{r}(t) + \frac{\mathbf{T}(t)}{m(t)} + \frac{\boldsymbol{F}_{\text{Aero}}}{m(t)} \\ \frac{\mathrm{d}m(t)}{\mathrm{d}t} = -\frac{T(t)}{g_0 \cdot I_{\text{sp}}} \end{cases} \tag{2}$$

where $\boldsymbol{r} = [x, y, z]^{\mathrm{T}}$ is the projection of displacement vector in ECI system. $\boldsymbol{v} = [v_x, v_y, v_z]^{\mathrm{T}}$ is the projection of velocity vector. $\boldsymbol{T}$ is engine thrust. $\boldsymbol{F}_{\boldsymbol{Aero}}$ is aerodynamics. I_{SP} is engine specific impulse.

When launch vehicle runs on orbit of central gravitational field without power, there are some invariant characteristic variables, namely momentum constant, energy constant and integral constant vector.

The momentum moment of launch vehicle in orbit is

$$\boldsymbol{H} = \boldsymbol{r} \times \boldsymbol{v} = r \cdot v \cdot \begin{bmatrix} \sin\Omega \sin i \\ -\cos\Omega \sin i \\ \cos i \end{bmatrix} = H \cdot \begin{bmatrix} \sin\Omega \sin i \\ -\cos\Omega \sin i \\ \cos i \end{bmatrix} \tag{3}$$

where i represents the orbit inclination of the target orbit, and Ω represents the right ascension point of the target orbit.

The energy constant of launch vehicle in orbit can be expressed as

$$E = \frac{1}{2}v^2 - \frac{\mu}{r} = -\frac{\mu}{2p}(1 - e^2) = -\frac{\mu}{2a} \tag{4}$$

where E represents the energy of launch vehicle when it is running on the orbit, and a represents the semi major axis of the orbit.

The cross multiplication of motion equation and momentum moment is

$$\left(\frac{\mathrm{d}^2\boldsymbol{r}}{\mathrm{d}t^2} + \frac{\mu}{r^3}\boldsymbol{r}\right) \times \boldsymbol{H} = \frac{\mathrm{d}^2\boldsymbol{r}}{\mathrm{d}t^2} \times \boldsymbol{H} + \frac{\mu}{r^3}\boldsymbol{r} \times \boldsymbol{H} = 0 \tag{5}$$

According to the triple cross multiplication of vectors, it can be substituted into Eq. (5), then integral

$$\boldsymbol{v} \times \boldsymbol{H} = \frac{\mu}{r}(\boldsymbol{r} + r\boldsymbol{c}) \tag{6}$$

where $\boldsymbol{c}$ is the integral constant vector, from which we can get the orbit constant vector when the vehicle is running on the target orbit

$$\boldsymbol{C} = \boldsymbol{v} \times \boldsymbol{H} - \frac{\mu}{r}\boldsymbol{r} \tag{7}$$

3 Mission Partition

3.1 Estimation of True Proximal Angle

For a certain time, the acceleration of the launch vehicle's coordinate system relative to the inertial coordinate system is determined according to measurement results of accelerometers. The axial apparent acceleration of launch vehicle is approximately equal to the ratio of engine thrust to rocket mass, i.e.

$$W_{bx}(t) = \frac{T(t)}{m(t)} = \frac{T(t)}{m_0 - \dot{m} \cdot t} \tag{8}$$

Equation (8) variable conversion

$$W_{bx}(t) = \frac{g_0 I_{SP} \cdot \dot{m}}{m_0 - \dot{m} \cdot t} = \frac{g_0 I_{SP}}{\frac{m_0}{\dot{m}} - t} = \frac{U}{M - t} \tag{9}$$

where $U = g_0 I_{SP}, M = m_0/\dot{m}$.

Integrate Eq. (9) to estimate the apparent velocity increment under the action of thrust in a subsequent certain period of time

$$V_T = \int_0^{t_k} W_{bx}(t)\mathrm{d}t = \int_0^{t_k} \frac{U}{M - t}\mathrm{d}t = U \ln \frac{M}{M - t_k} \tag{10}$$

Conduct integral calculation to estimate the range increment under the action of thrust in a subsequent certain period of time

$$S_T = \int_0^{t_k} \int_0^{t} W_{bx}(t)\mathrm{d}t\mathrm{d}t = \int_0^{t_k} \int_0^{t} \frac{U}{M - t}\mathrm{d}t\mathrm{d}t = U \ln \frac{M}{M - t_k}(t_k - M) + Ut_k \tag{11}$$

If the rotational angular velocity of the optimal orbital maneuver thrust vector is known, the gravitational loss can be estimated

$$V_G = \frac{1}{24} \cdot t_k^2 \cdot V_T \cdot \left[\omega_s^2(1 - 3\sin^2 \overline{\gamma}) - \omega_{opt}^2 + (\omega - \omega_{opt})^2\right] \tag{12}$$

where ω is the actual rotation angular velocity of the thrust, $\omega_s = \sqrt{\mu/r^3}$ is the angular velocity of the circular orbit at the corresponding altitude, and $\overline{\gamma}$ is the average included angle between the thrust vector direction on the target orbit plane and the orbit tangent. If the attitude of the aircraft can ideally track the optimal rotational angular velocity, the range loss caused by gravity can be estimated as

$$S_G = \int_0^{t_k} V_G \mathrm{d}t = \frac{1}{432}\left[S_{GA} \cdot \ln \frac{M}{M - t_k} + 6M^3 \cdot \ln(M - t_k) + S_{GB}\right] \tag{13}$$

The specific expressions of S_{GA} and S_{GB} are

$$S_{GA} = 6U \cdot \left[\omega_s^2(1 - 3\sin^2 \overline{\gamma}) - \omega_{opt}^2\right] \cdot t_k^3 \tag{14}$$

$$S_{\mathrm{GB}} = 3Mt_k^2 + 6M^2 t_k + 2t_k^3 \tag{15}$$

The increment of geocentric angle from the current position to the orbit entry point can be expressed as

$$\Delta\Phi = \frac{\cos\overline{\gamma}}{r_{yk}}(v_0 t_k + S_{\mathrm{T}} - S_{\mathrm{G}}) \tag{16}$$

The proximal point angle of the mission target orbit is ω_f, The true proximal angle of orbit entry point can be estimated by using integral operation which calculates speed increment and range increment under the action of thrust and gravity in a subsequent certain period of time

$$f = \Phi_0 + \Delta\Phi - \omega_f \tag{17}$$

3.2 Estimation of Apogee Height and Orbit Inclination

Under the action of thrust, the differential form of apogee height can be expressed by orbital elements

$$\frac{\mathrm{dH_a}}{\mathrm{dt}} = A(1+e)\sin f \cdot u_r + A(2 + 2e\cos f + B - eB) \cdot u_\theta \tag{18}$$

where

$$A = \frac{a^{3/2}(1+e)^{1/2}}{\mu^{1/2}(1-e)^{1/2}}, B = \frac{e\cos^2 f + 2\cos f + e}{1 + e\cos f}$$

In the subsequent flight, the apogee altitude increment required for the orbit entry point can be expressed as

$$\Delta H_{aK} = \int_0^{t_k} [A(1+e)\sin f \cdot u_r + A(2 + 2e\cos f + B - eB) \cdot u_\theta] \cdot \mathrm{dt} \tag{19}$$

On the premise of meeting other constraints, the maximum apogee height of the rocket shall be ensured as far as possible. The performance index can be described as

$$J = -\Delta H_{aM} = -\int_0^{t_f} F(x,u)\,\mathrm{dt} \tag{20}$$

where

$$F(x,u) = [A(1+e)\sin f \cdot u_r + A(2 + 2e\cos f + B - eB) \cdot u_\theta] \tag{21}$$

According to the state equation and performance index, the Hamilton function of the problem is constructed. When the flight time of subsequent missions reaches the maximum value t_M, the apogee altitude increment that the launch vehicle can obtain is

$$\Delta H_{a\mathrm{M}} = \int_0^{t_{\mathrm{M}}} [A(1+e)\sin f \cdot u_{r\mathrm{M}} + A(2 + 2e\cos f + B - eB) \cdot u_{\theta\mathrm{M}}] \cdot \mathrm{dt} \tag{22}$$

For the out of plane orbit transfer mission, in addition to determining the farthest reachable range according to the apogee altitude, it is also necessary to analyze the deviation of orbit inclination. Similarly, the increment of orbit inclination that the rocket can obtain is

$$\Delta i_{\mathrm{M}} = \int_{0}^{t_{\mathrm{M}}} \left[\frac{a^{1/2}}{\mu^{1/2}(1-e^2)^{1/2}} \left[\frac{(1-e^2)\cos(\omega+f)}{1+e\cos f} \right] \cdot u_{h\mathrm{M}} \right] \cdot \mathrm{dt} \tag{23}$$

3.3 Orbit Entry Conditions

When flight deviation of launch vehicle is within a small range, the original control command can be adjusted adaptively, and shutdown point can be reasonably selected to enter the scheduled mission orbit.

In the process of adaptively adjusting and completing orbit entry, it is advisable to set the correction amount of shutdown point to meet

$$\Delta t_{\mathrm{N1}}/\Delta t = \Delta i/n_a,\ \Delta t_{\mathrm{N2}}/\Delta t = \Delta H_a/n_c$$

where, Δt_{N1} is correction amount of shutdown time required to reach predetermined orbit inclination and Δt_{N2} is correction amount of shutdown time required to reach predetermined apogee height. The time constraint is constructed

$$\sqrt{\Delta t_{\mathrm{N1}}^2 + \Delta t_{\mathrm{N2}}^2} \le \Delta t \tag{24}$$

It can be determined that the apogee height and orbit inclination boundary conditions for the selected orbit mission are

$$\sqrt{(\Delta t_{\mathrm{N1}}/\Delta t)^2 + (\Delta t_{\mathrm{N2}}/\Delta t)^2} \le 1 \tag{25}$$

According to the definition of shutdown point correction, Eq. (25) is equivalent to

$$\sqrt{(\Delta i/n_a)^2 + (\Delta H_a/n_c)^2} \le 1 \tag{26}$$

Similarly, it can be determined that the apogee height and orbit inclination boundary conditions for the scheduled re-planning orbit mission are

$$\begin{cases} \sqrt{(\Delta i/n_a)^2 + (\Delta H_a/n_c)^2} > 1 \\ \sqrt{(\Delta i/n_b)^2 + (\Delta H_a/n_d)^2} \le 1 \end{cases} \tag{27}$$

It can be determined that the apogee height and orbit inclination boundary conditions for the sub-optimal re-planning orbit mission are

$$\begin{cases} \sqrt{(\Delta i/n_b)^2 + (\Delta H_a/n_d)^2} > 1 \\ F_{H_a - i}(\Delta H_{a\mathrm{N}}, \Delta i_{\mathrm{N}}) \ge 0 \end{cases} \tag{28}$$

The derivation process can refer to the author's relevant works [11].

When flight deviation occurs, the residual flight capability of launch vehicle needs to be analyzed. According to current flight status, it can be judged whether to re-plan the trajectory or not. As an example, the orbital inclination and apogee height deviation can be judged, as shown in Fig. 1.

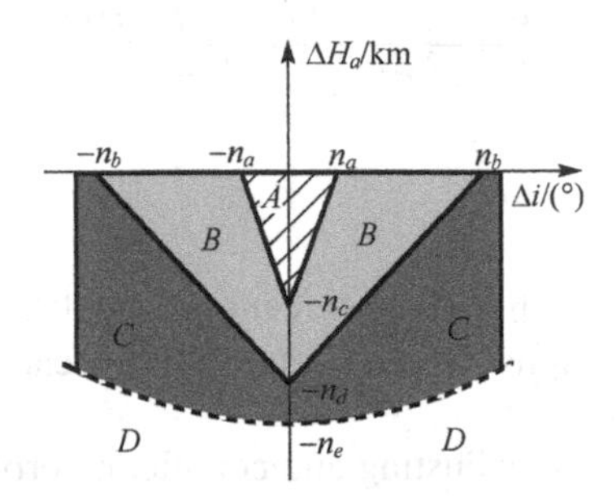

Fig. 1. Mission partition under abnormal flight conditions

Table 1. Mission partition range and strategy

Fault	Deviation range	Mission strategy
A	Within small deviation	Adjust to orbit adaptively (Not Re-planning)
B	Beyond deviation range can be recovered	Trajectory re-planning and enter scheduled orbit
C	Beyond deviation range cannot get into scheduled orbit	Mission downgrade and enter sub-optimal orbit
D	Irreparable catastrophic failure	—

Table 1 shows Mission Partition under flight abnormalities. Different mission strategies can be selected according to the flight deviation range of launch vehicle. When the flight deviation of launch vehicle is within allowable range (Fig. 1, Area A), payloads can be put into the target orbit by adjusting adaptability. When the deviation exceeds Area A, but the deviation range is not large (Fig. 1, Area B), the trajectory planning design can be used. Considering position, velocity, target and constraints of launch vehicle, the trajectory satisfying mission requirements of the flight terminal can be obtained by fast planning algorithm, and payloads can be put into the target orbit. When the deviation range of launch vehicle is too large, but it has not caused fatal failure, payloads cannot enter the target orbit (Fig. 1, Area C). Launch vehicle needs to carry out trajectory re-planning, and payloads are put into sub-optimal orbit through mission decision and trajectory planning to remedy abnormal flight state.

4 Trajectory Planning

Launch vehicle trajectory re-planning problem can be described as an optimal control problem. The optimal control model takes motion model under the action of limited thrust as the state equation.

In this paper, the circular orbit is taken as a transfer orbit, and an online trajectory re-planning problem is described. In the autonomous planning scheme design of downgraded orbit with circular orbit as the target orbit, it is necessary to simplify the control problem model and complete the design of optimization problem.

When ignoring the influence of aerodynamics, this problem is modeled as an optimal control problem when it is solved according to control theory

$$\begin{aligned}
\min\ & J = -m(t_f) \\
\text{s. t.}\ & \dot{\boldsymbol{r}} = \boldsymbol{v} \\
& \dot{\boldsymbol{v}} = \boldsymbol{g} + \frac{\mathbf{T}}{m} \\
& \dot{m} = -\frac{\|\boldsymbol{T}(t)\|}{I_{sp} g_0} \\
& \boldsymbol{s}(t_0) = \boldsymbol{s}_0 \in \mathbf{R}^7,\quad \boldsymbol{s}_0 = [\boldsymbol{r}_0^{\mathrm{T}}, \boldsymbol{v}_0^{\mathrm{T}}, \mathrm{m}]^{\mathrm{T}} \\
& \|\boldsymbol{r}(t_f)\| = r_{set} \in \mathbf{R} \\
& \boldsymbol{r}^{\mathrm{T}}(t_f)\, \boldsymbol{h} = 0,\ \boldsymbol{v}^{\mathrm{T}}(t_f)\, \boldsymbol{h} = 0 \\
& \boldsymbol{r}^{\mathrm{T}}(t_f) \boldsymbol{v}(t_f) = 0 \\
& \|\boldsymbol{r}(t_f)\| \|\boldsymbol{v}(t_f)\|^2 = \mu \\
& 0 \le \|\boldsymbol{T}(\mathrm{t})\| \le \mathrm{T}_{\max},\quad \boldsymbol{m}(\mathrm{t_f}) \ge \boldsymbol{m}_{low}
\end{aligned} \tag{29}$$

The idea of degraded trajectory design is to ensure the suboptimal orbit close to the schedule trajectory as much as possible. In order to implement this strategy in trajectory planning and design, the performance index function can be selected as

$$\min J = \Gamma_r \cdot \left\| \frac{\boldsymbol{r}(t_f)}{r_f} - \mathbf{1}_{3\times 1} \right\| + \Gamma_v \cdot \left\| \frac{\boldsymbol{v}(t_f)}{v_f} - \mathbf{1}_{3\times 1} \right\| \tag{30}$$

where Γ_r,Γ_v respectively represent the coefficients of terminal displacement deviation and velocity deviation. The physical meaning of this performance index is to make the terminal state close to the target point state in an equal proportion, so as to ensure that terminal displacement and speed of the degraded planning are close to terminal flight state of the original target orbit. Using this performance index can ensure the transfer range as small as possible with the original target orbit to a certain extent, so as to facilitate the subsequent orbit maneuver adjustment of spacecraft.

The control variables of the original problem are reconstructed and transformed into a form that is easy to be solved by convex optimization method.

In convex optimization, the expression of integral is not generally used, and discretization is transformed into matrix multiplication. Through transformation, the optimization problem can be transformed into a standard second-order cone optimization problem:

$$\begin{aligned} &\min \ \hat{\boldsymbol{c}}^{\mathrm{T}} \hat{\boldsymbol{x}} \\ &\text{s. t.} \ \hat{\boldsymbol{A}} \hat{\boldsymbol{x}} = \hat{\boldsymbol{b}} \\ &\hat{\boldsymbol{x}} \in \boldsymbol{K} \end{aligned} \tag{31}$$

A primal-dual interior point method [12] is used to solve trajectory re-planning problems.

5 Simulation and Analysis

Taking launch vehicle as an example, orbit constraints of the transfer orbit are convex analyzed. There is deviation in the flight process, the real-time trajectory re-planning scheme is used to determine constraints of mission scheduled orbit, and the corresponding implementation strategy is adopted.

Consider the failure of launch vehicle's secondary engine. Under different flight deviations, different transfer orbits and final orbits are planned respectively, as shown in Fig. 2.

Parameters of launch vehicle: total mass is 187200 kg, load mass is 7000 kg, engine specific impulse is 318 s.

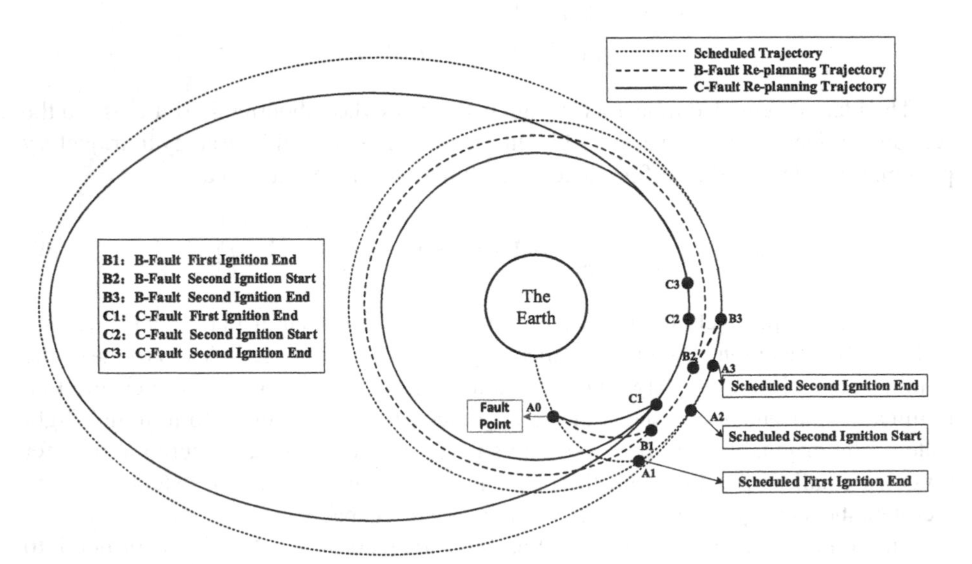

Fig. 2. Trajectory Re-planning Scheme under Propulsion Failure

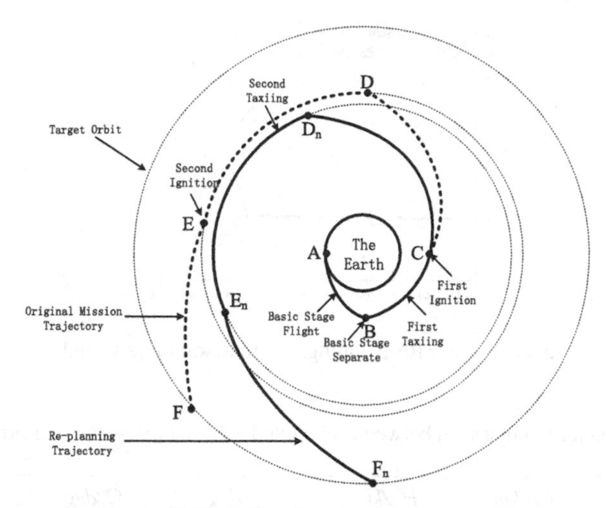

Fig. 3. B-Fault Re-planning into target orbit

5.1 Re-planning into B-fault Orbit

When thrust of the launch vehicle is deviated due to failure, the mission cannot be completed according to original plan. Under the condition of thrust deviation, engines can be shut down and re-ignited normally, and the specific impulse of engine remains unchanged. The remaining fuel can be fully burned. According to the orbit entry capability evaluation of the launch vehicle (in Area B), combined with the current requirements of the remaining fuel and parameters of target orbit, considering flight constraints, the mission is realized through trajectory re-planning, as shown in Fig. 3.

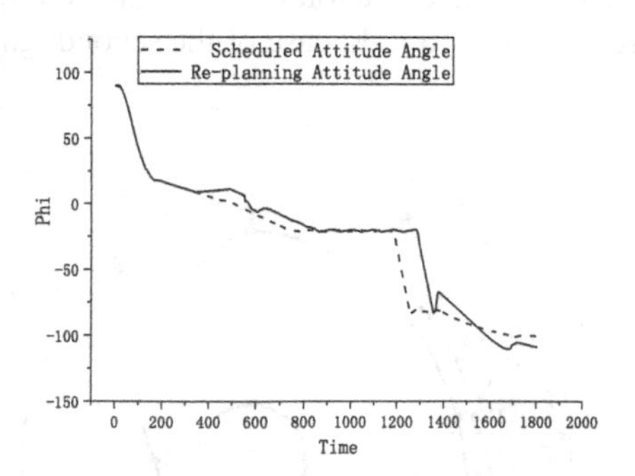

Fig. 4. B-Fault Re-planning: comparison of pitch angle

When thrust decreases by 20% (in 350 s). The comparison of attitude control angles between re-planning trajectory and original target trajectory is shown in Figs. 4 and 5, where the dotted line is the change of the original target trajectory and the solid line is the change of the re-planning trajectory.

The flight control parameters after trajectory re-planning are shown in Table 2. The schedule orbit can be reached by re-planning trajectory.

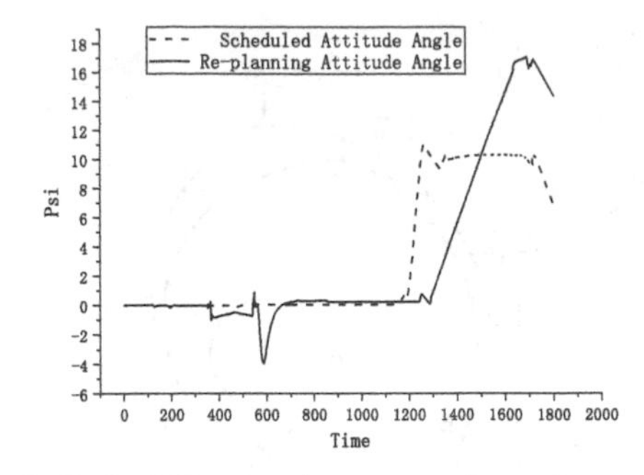

Fig. 5. B-Fault Re-planning: comparison of yaw angle

Table 2. Parameter Comparison between scheduled orbit and Re-planning orbit in B-Fault

Transfer Orbit	*Hp/km*	*Ha/km*	*i/deg*	*Ω/deg*	*ω/deg*
Scheduled	160.14	267.53	19.51	17.80	130.17
Re-planning	139.59	168.98	19.50	17.65	−45.17
Target Orbit	*Hp/km*	*Ha/km*	*i/deg*	*Ω/deg*	*ω/deg*
Scheduled	200.04	46025.92	17.00	13.89	179.01
Re-planning	200.05	46026.10	17.00	13.90	179.01

5.2 Re-planning into C-Fault Orbit

Select the same launch vehicle parameters as simulation case in 5.1, and engines meet parameter requirements. When the deviation of secondary ignition flight is large, the launch vehicle cannot reach target orbit. In case of the second ignition, thrust decreases

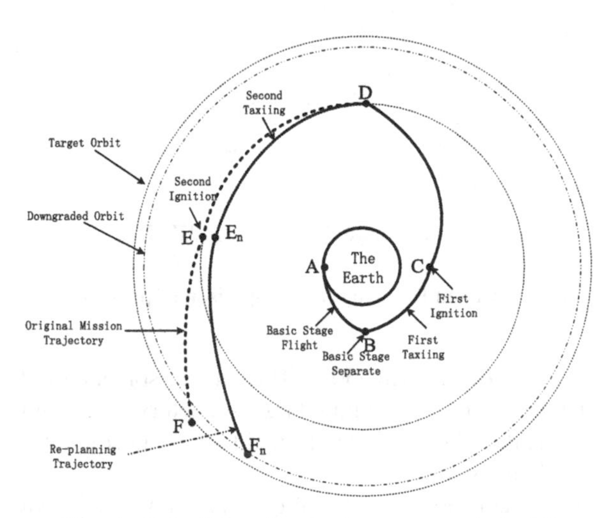

Fig. 6. C-Fault Re-planning into downgraded orbit

by 40%. Considering flight constraints, switch to the downgraded mission mode (in Area C). Downgraded orbit entry is achieved by re-planning trajectory, as shown in Fig. 6.

The comparison of attitude control angles between re-planning trajectory and original target trajectory is shown in Figs. 7 and 8, where the dotted line is the change of the original target trajectory and the solid line is the change of the re-planning trajectory.

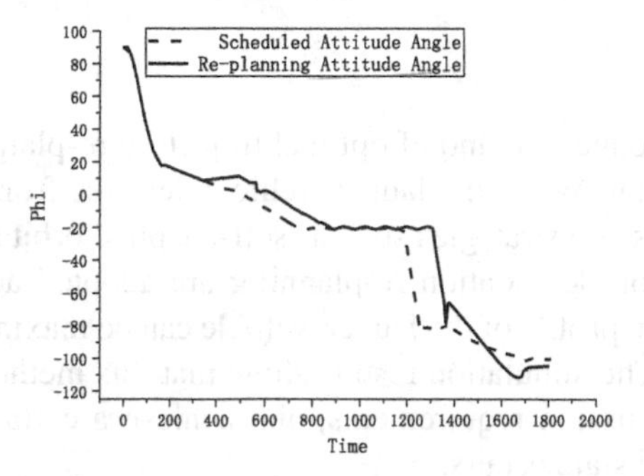

Fig. 7. C-Fault Re-planning: comparison of pitch angle

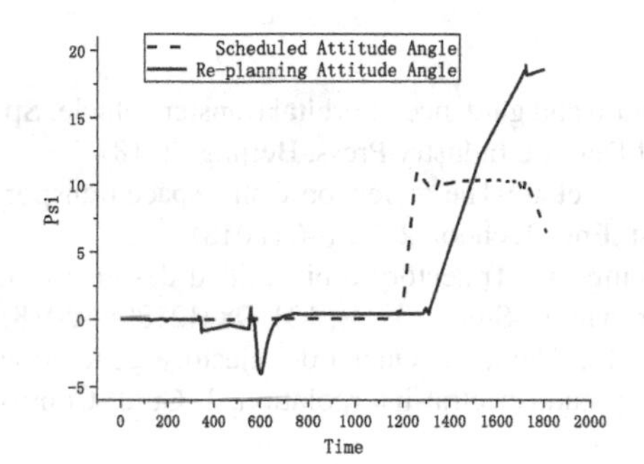

Fig. 8. C-Fault Re-planning: comparison of yaw angle

The flight control parameters after trajectory re-planning are shown in Table 3.

Table 3. Parameter comparison between scheduled orbit and Re-planning orbit in C-Fault

Transfer Orbit	*Hp/km*	*Ha/km*	*i/deg*	*Ω/deg*	*ω/deg*
Scheduled	160.14	267.53	19.51	17.80	130.17
Re-planning	137.24	151.55	19.50	17.64	-26.67
Target Orbit	*Hp/km*	*Ha/km*	*i/deg*	*Ω/deg*	*ω/deg*
Scheduled	200.04	46025.92	17.00	13.89	179.01
Re-planning	199.88	45428.26	17.01	13.92	178.98

The downgrade orbit can be reached by re-planning trajectory.

It should be noted that results of real-time planning are not unique, so it is necessary to deal with relaxation to ensure the solution and convergence as much as possible. Real-time planning can adapt to different fault conditions. The planning scheme can re-plan trajectory of the mission independently, and has strong adaptability for the emergent mission.

6 Conclusion

This paper mainly introduces a kind of optimal trajectory re-planning design of launch vehicle mission partition. When the launch vehicle deviates from the scheduled flight trajectory, different mission strategies such as self-adaptive orbit adjustment, trajectory re-planning and mission degradation re-planning are adopted according to abnormal conditions. The mission profile of the launch vehicle can be maximized by trajectory re-planning technology. The simulation results show that this method can re-plan a flight trajectory that meets mission requirements, and achieve a certain degree of recovery when a flight abnormal state occurs.

References

1. Li, X., Li, C.: Navigation and guidance of orbital transfer vehicle. Springer Nature Singapore Pte Ltd. and National Defense Industry Press, Beijing (2018)
2. Long, L., Li, P., Qin, X., et al.: The review on China space transportation system of past 60 years. Astronaut. Syst. Eng. Technol. **2**(2), 1–6 (2018)
3. Fang, Q., Liu, Y., Wang, X.: Trajectory-orbit unified design for aerospace vehicles. Acta Aeronauticaet et Astronautica Sinica. **39**(4), 121398–121398 (2018)
4. Sagliano, M., Mooij, E., Theil, S.: Onboard trajectory generation for entry vehicles via adaptive multivariate pseudospectral interpolation. J. Guid. Control. Dyn. **40**(2), 466–476 (2017)
5. Harris, M.W., Akmee, B.: Maximum divert for planetary landing using convex optimization. J. Optim. Theory Appl. **162**(3), 975–995 (2013)
6. Liu, X., Shen, Z., Lu, P.: Entry trajectory optimization by second-order cone programming. J. Guid. Control. Dyn. **39**(2), 227–241 (2016)
7. Scharf, D.P., Acikmese, B., Dueri, D., et al.: Implementation and experimental demonstration of onboard powered-descent guidance. J. Guid. Control. Dyn. **40**(2), 213–229 (2017)
8. Xie, Y., Zhou, W., Du, D., et al.: Rapid computation and analysis of finite thrust maneuver capability. Missiles Space Veh. **3**(1), 14–17 (2017)
9. Liu, X.: Autonomous Trajectory Planning by Convex Optimization. Iowa State University (2013)
10. Cheng, X., Shang, T., Xu, F., et al.: Online trajectory planning for launch vehicles with successive convex programming. J. Astronaut. **42**(2), 202–210 (2021)
11. Li, X., Xu, F., Zhou, W., Ma, H.: Redundant Inertial Measurement Unit Reconfiguration and Trajectory Replanning Technology of Launch Vehicle. Science Press, Beijing (2021)
12. Mattingley, J., Boyd, S.: CVXGEN: A Code Generator for Embedded Convex Optimization. Optim. Eng. **13**(1), 1–27 (2012)

Research on Cooperative Multi-constraint Guidance Law for Leader-Follower Multi-aircraft

Du Xin[1](✉), Diao Guijie[1,2], Liu Zhe[1], and Gong Ningbo[1]

[1] Beijing Electro-Mechanical Engineering Institute, Beijing 100074, China
260126131@qq.com
[2] Science and Technology On Complex System Control and Intelligent Agent Cooperation Laboratory, Beijing 100074, China

Abstract. In this paper, the design and optimization of multi-aircraft cooperative guidance law with multi-constraint is studied for high-speed and large maneuvering targets. Firstly, the motion of the aircraft and the target are analyzed in the inertial coordinate system. The relative motion model and the corresponding state equation are established in a plane. Then, based on the sliding mode control method, the multi-constrained ballistic forming guidance law of the lead aircraft is designed to make it hit the target in multiple dimensions. Finally, the aircraft-to-target range and line-of-sight angle of the aircraft are selected as the state variables to be controlled, and the multi-constrained cooperative guidance law is designed for the follow aircraft. So that the motion of follow aircraft can track the lead aircraft accurately, the controlled state variables converges quickly, and hit the target at the same time until the end. The simulation results show that the leader-follower cooperative guidance law designed in this paper is effective for intercepting high-speed and large maneuvering targets.

Keywords: Cooperative guidance · Multi-constrained · Leader-follower · Large maneuvering · Sliding mode control

1 Introduction

Multi-aircraft cooperation can fuse the information detected by radar, infrared and other sensors, realize information sharing through highly reliable inter-aircraft communication, and complete mutual tactical and technical cooperation. Multi-aircraft cooperative attack has the advantages of high efficiency, high fault tolerance and intrinsic parallelism. It can not only greatly improve penetration capability, but also accomplish tasks impossible for a single aircraft, such as achieving tactical stealth, enhancing the capability of electronic countermeasures and identifying and searching moving targets [1]. At this time, on the one hand, with the development of data link technology, the time delay of inter-aircraft communication is greatly reduced; on the other hand, with the development of computer technology on the aircraft, the computing speed and storage capacity on the aircraft are growing rapidly, and the online real-time computing ability

W. Fan et al. (Eds.): AsiaSim 2022, CCIS 1712, pp. 395–404, 2022.
https://doi.org/10.1007/978-981-19-9198-1_30

is greatly increased. These technological breakthroughs have promoted the emergence and development of multi-aircraft cooperative attack technology [1]·

At present, there are still some important and typical problems in the research process of multi-aircraft cooperative attack, such as: (1) the accuracy of aircraft against large maneuvering targets is very poor, far less than that of stationary targets; (2) the traditional aircraft guidance law, such as proportional guidance law, generally only takes miss distance as a constraint, which will reduce the interception efficiency of aircraft. The multi-constraint cooperative guidance law can solve these problems successfully.

At present, some progress has been made in the research of multi-constraint guidance law for multi-aircraft cooperation. Reference [3] takes the coordination variable as the center. The lower layer of each aircraft adopts impact-time control guidance (ITCG), and the upper layer adopts centralized or distributed coordination strategy, which can accurately hit the target. This method can dynamically adjust the weight coefficients of each aircraft, and the topological network is scalable and has certain generality. In reference [4], on the basis of considering modeling errors and disturbances, a cooperative guidance law based on dynamic surface is designed, which realizes the double constraints of attack time and miss distance. This method designs an integrated guidance and control system with strong robustness. Based on the dynamic inverse control theory, a guidance strategy of leader-follower is designed in Reference [5] to make the follow aircraft converge to the lead aircraft quickly. References [4, 5] do not need to estimate the remaining flight time, and the aircraft speed is variable during flight. However, the above methods are all aimed at stationary or low-speed targets, and the targets have no maneuverability; Only time constraint is considered in the terminal guidance process, which is relatively single.

To solve the above problems, an improved distributed leader-follower cooperative guidance law is proposed in this paper. By designing the corresponding guidance law and efficient information exchange, the motion state of follow aircraft can track the lead aircraft accurately, and the control variables converge quickly. This method considers a variety of constraints, can intercept large maneuvering targets, and the amount of information interaction between aircraft is little.

2 Problem Description

The case of multiple aircraft attacking a single target in a two-dimensional plane is considered. Suppose a total of n aircraft intercept one target. The aircraft with the largest initial distance is chosen as the lead aircraft, while the other $(n - 1)$ aircraft are the follow aircraft. The target makes high-speed and large maneuvering motion in the whole process. As shown in Fig. 1, M1 stands for the lead aircraft and M_i stands for the follow aircraft, where $i = 2 \ldots N$. T stands for the target. r_1 and r_i stand for the relative distances between the lead aircraft and the i-th follow aircraft, σ_1 and σ_i stand for their respective line of sight angles, γ_1 and γ_i stand for their respective flight path angles, θ_1 and θ_i stand for their respective aircraft leading angles, a_{m1} and a_{mi} stand for their respective normal accelerations, (x_1, y_1), (x_i, y_i) stand for their coordinates in the coordinate system, and V_{m1} and V_{mi} stand for their respective velocities. V_t stands for the target speed, a_t stands for the target normal acceleration, and γ_t stands for the target flight path angle. For all angles in the figure, the counterclockwise direction is the positive direction.

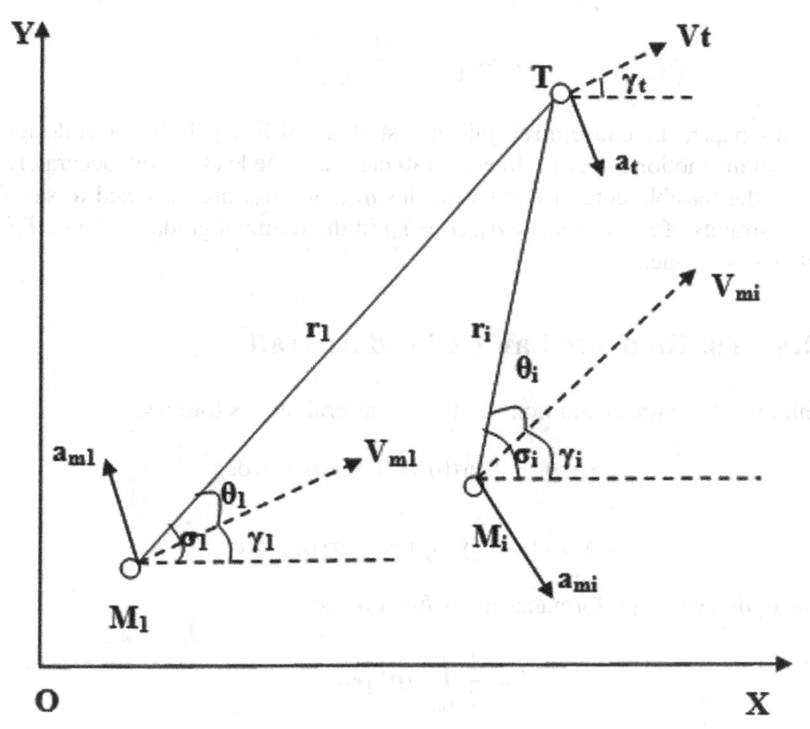

Fig. 1. Geometry of aircraft and target

In order to simplify the calculation, the following assumptions are made in this paper: (1) In this paper, the velocity of the aircraft used to intercept the target is constant; (2) When the target moves in high-speed and large maneuver, the velocity is also constant; (3) The delay effect of aircraft guidance & control system and inter-aircraft communication is ignored.

As can be seen from Fig. 1, the relative motion equations of the lead aircraft, the follow aircraft and the target are as follows:

$$\dot{r}_1 = V_t\cos(g_t - s_1) - V_{m1}\cos(g_1 - s_1) \tag{1}$$

$$\dot{\sigma}_1 = \frac{V_t\sin(\gamma_t - \sigma_1)}{r_1} - \frac{V_{m1}\sin(\gamma_1 - \sigma_1)}{r_1} \tag{2}$$

$$\dot{\gamma}_1 = \frac{a_m 1}{V_m 1} \tag{3}$$

$$\dot{\gamma}_t = \frac{a_t}{V_t} \tag{4}$$

$$\dot{r}_i = V_t\cos(\gamma_t - \sigma_i) - V_{mi}\cos(\gamma_1 - \sigma_i) \tag{5}$$

$$\dot{\sigma}_i = \frac{V_t\sin(\gamma_t - \sigma_i)}{r_i} - \frac{V_{m_i}\sin(\gamma_l - \sigma_i)}{r_i} \tag{6}$$

$$\dot{\gamma}_i = \frac{a_{m_i}}{V_{m_i}} = \frac{a_{i_1} + a_{i_2}}{V_{mi}} \tag{7}$$

In this paper, the cooperative guidance strategy of leader-follower is designed to ensure that the motion state of follow aircraft can track the lead aircraft accurately. That is to say, the feasible control input variables a_{m1} and a_{mi} are designed to satisfy the three constraints of $r_1 = r_i, \sigma_1 = \sigma_i, t_f = T_d$ in the terminal guidance stage. T_d is the expected attack time.

3 Design of Guidance Law for Lead Aircraft

The initial and terminal conditions of the lead aircraft are as follows:

$$x_1(0) = 0, y_1(0) = 0, \sigma_1(0) = 0\,\text{deg} \tag{8}$$

$$x_1(t_f) = 0, y_1(t_f) = 0, \sigma_1(t_f) = \sigma_f \tag{9}$$

The optimization performance index function is:

$$J = \frac{1}{2}\int_{t_0}^{t_f} (a_{m1}^{\mathrm{T}} \bullet a_{m1})dt \tag{10}$$

In reference [8], a new attack time control guidance law ITCG is proposed, which divides the aircraft normal acceleration into two terms. The first item is the control variable input obtained from the pure proportional guidance law, which is used to control the aircraft to hit the target accurately; The second item is the offset item input, which is used to control the aircraft to hit the target at the expected time. As shown in formula (11)–(15).

$$\dot{\gamma}_1 = \frac{a_m 1}{V_m 1} = \frac{a_{B1} + a_{F1}}{V_{m1}} \tag{11}$$

$$a_{B1} = NV_m 1\dot{\sigma}_1 \tag{12}$$

$$a_{F1} = \frac{3}{2}V_{m1}\dot{\sigma}_1(1 - \sqrt{1 + \frac{4(N+1)(2N-1)V_{m1}^3 e_{T_1}}{3r_1^3\dot{\sigma}_1^2}}) \tag{13}$$

$$e_{T1} = T_{god} - T_{goPNG1} = T_d - t - T_{goPNG1} \tag{14}$$

$$T_{goPNG1} = \frac{r_1}{V_{m1}}(1 + \frac{\sin^2(\gamma_1 - \sigma_1)}{2(2N-1)}) \tag{15}$$

Among them: N is proportional navigation coefficient, $N = 3 + m$ and $m > -1$. T_{gopng1} is the remaining flight time of the lead aircraft. This method adds time control term on the basis of proportional guidance law, so as to achieve the purpose of attack time constraint. The object of interception in reference [8] is stationary target, which is extended to high speed and large maneuvering target in the following.

It is assumed that the target continues to make large maneuvers. The velocity and the normal acceleration of target are constant. The simulation result can be obtained according to the normal acceleration control input derived from Eqs. (11)–(13). It shows that although it can hit the target in a certain time, it is quite different from the expected attack time T_d due to the approximate estimation and small angle linearization in the above calculation. So the sliding mode control is used to compensate for it in the following.

In this section, a new sliding mode guidance law is designed considering the constraints of attack time and attack angle. R stands for ideal relative distance. The value of R is $V_{m1}(T_d - t)$. And the terminal attack angle is σ_f. As t approaches T_d, $r1$ converges to R *and* zero, σ_1 converges to σ_f. As long as a feasible controller can be designed to make r_1&σ_1 track R&σ_f in the expected time, T_d、σ_f can be set reasonably.

Select the following state variables:

$$\begin{cases} x_1 = R - r_1 + \sigma_1 - \sigma_f \\ x_2 = -V_{m1} - \dot{r_1} + \dot{\sigma_1} \end{cases} \tag{16}$$

Combination formula (1)–(3) and derivation of formula (16). In order to meet the requirement of convergence speed of state variable r_1, the following sliding surface is selected:

$$s = x_1 + x_2 \tag{17}$$

In order to suppress the jitter of the control quantity near the sliding surface, the law of power approach is selected:

$$\dot{s} = -k|s|^a \text{sign}(s) \tag{18}$$

where k and a are normalization parameters, which require k > 0 and 1 > a > 0. The Eq. (19) can be obtained by (18) and derivative of Eq. (17).

$$\hat{a}_{m1} = \frac{k|s|^a sign(s) - V_{m1} - \dot{r}_1 - r\dot{q}^2 + a_t(\sin(g_t - s_1) + \cos(g_t - s_1)/r_1) + (1 - 2\dot{r}_1/r_1)\dot{s}_1}{\sin(g_1 - s_1) + \cos(g_1 - s_1)/r_1} \tag{19}$$

In Eq. (19), a_{s1} represents the sliding mode guidance law bias term. And it is a component of the input of the control variable. So the above formula is the new guidance law based on sliding mode control.

4 Design of Cooperative Guidance Law

Based on the multi-constraint guidance law of lead aircraft, a new cooperative guidance law for follow aircraft is designed in this section. According to reference [9], the second-order nonlinear system can keep stable under the action of specific control input. So this paper designs the following state variables.

$$\begin{cases} x_1 = r_1 - r_i + \sigma_1 - \sigma_i \\ x_2 = \dot{x_1} \end{cases} \tag{20}$$

The system equation is obtained by deriving the upper formula.

$$\begin{cases} \dot{x_1} = x_2 \\ \dot{x_2} = D_1 + D_2 + D_3 + D_4 + D_5 + D_6 + D_7 + D_8 \end{cases} \tag{21}$$

Among them, $D_1 \sim D_8$ is as follows:

$$D_1 = -V_t \sin(\gamma_t - \sigma_1)(\dot{\gamma}_t - \dot{\sigma_1}) \tag{22}$$

$$D_2 = V_{m1} \sin(\gamma_1 - \sigma_1)(\dot{\gamma}_1 - \dot{\sigma}_1) \tag{23}$$

$$D_3 = V_t \sin(\gamma_t - \sigma_i)(\dot{\gamma}_t - \dot{\sigma_i}) \tag{24}$$

$$D_4 = V_{mi} \sin(\gamma_i - \sigma_i)(\dot{\gamma}_i - \dot{\sigma}_i) \tag{25}$$

$$D_5 = Vt\frac{\cos(\gamma_t - \sigma_1)(\dot{\gamma}_t - \dot{\sigma_1})r_1 - \sin(\gamma_t - \sigma_1)\dot{r_1}}{r_1^2} \tag{26}$$

$$D_6 = -V_{m_1}\frac{\cos(\gamma_1 - \sigma_1)(\gamma_1 - \sigma_1)r_1 - \sin(\gamma_1 - \sigma_1)\dot{r}_1}{r_1^2} \tag{27}$$

$$D_7 = -Vt\frac{\cos(\gamma_t - \sigma_i)(\dot{\gamma}_t - \dot{\sigma_i})r_i - \sin(\gamma_t - \sigma_i)\dot{r_i}}{r_i^2} \tag{28}$$

$$D_8 = V_{m_i}\frac{\cos(\gamma_i - \sigma_i)(\dot{\gamma}_i - \dot{\sigma}_i)r_i - \sin(\gamma_i - \sigma_i)\dot{r}_i}{r_i^2} \tag{29}$$

According to reference [9], the input form of specific control variable is as follows:

$$\dot{x_2} = u = k_1|x_1|^{b_1}\text{sign}(x_1) + k_2|x_2|^{b_2}\text{sign}(x_2) \tag{30}$$

In this equation, k_1 and k_2 are both greater than 0. The value range of b_1 is 0–1 and $b_2 = 2b_1/(1 + b_1)$. a_{i2} represents the additional control input variable. It can be obtained by solving the system of equations. The system consists of Eqs. (7), (21), and (30). The expression for the variable a_{i2} is as follows:

$$a_{i2} = \frac{u - D_1 - D_2 - D_3 - V_{mi}\sin(\gamma_i - \sigma_i)(\frac{a_{i1}}{V_{m1}} - \dot{\sigma}_i) - D_5 - D_6 - D_7 - D_9}{\sin(\gamma_i - \sigma_i) + \cos(\gamma_i - \sigma_i)/r_i} \tag{31}$$

$$D_9 = V_{mi}\frac{\cos(\gamma_i - \sigma_i)(\frac{a_{i1}}{V_{mi}} - \dot{\sigma}_i)r_i - \sin(\gamma_i - \sigma_i)\dot{r}_i}{r_i^2} \tag{32}$$

In this formula, the pure proportional guidance law a_{i1} is used to guide the aircraft to hit the target. Because of $\cos(\gamma_i - \sigma_i) < < r_i$, this term can be regarded as zero. The simplified formula (31) can make the dynamic system variable r_i converge r_1 more uniformly and rapidly. This can achieve the effect of coordinated strike. The final cooperative guidance law a_{mi} can be obtained.

5 Numerical Simulation

In this chapter, two guidance laws proposed in this paper are simulated and analyzed. Suppose there are two aircraft attacking the target. One of them is the lead aircraft, which is equipped with a high-precision seeker; The other one is a follow aircraft. Both of them have the same velocity. The initial position of the target is (31000,0)m. The initial trajectory deflection angle is 0. Velocity is 1600 m/s. The normal overload of the target is 9 g.

The data listed in Table 1 are the initial parameters of two aircraft.

Table 1. Initial parameters of aircraft

Aircraft	Speed/(m/s)	Line of sight angle/(°)	Flight path angle/(°)	Aircraft- target distance/(km)
M1	1800	0	30	31
M2	1800	0	30	30

In order to make the follow aircraft traceable to the lead aircraft, it is usually that the aircraft with the largest initial value of T_{goPNG} is selected as the lead aircraft. That is M_1. And M_2 is the follow aircraft. The lead aircraft adopts the multi-constraint guidance law designed in Sect. 3, and takes k equal to 0.1 and a equal to 0.7. The cooperative

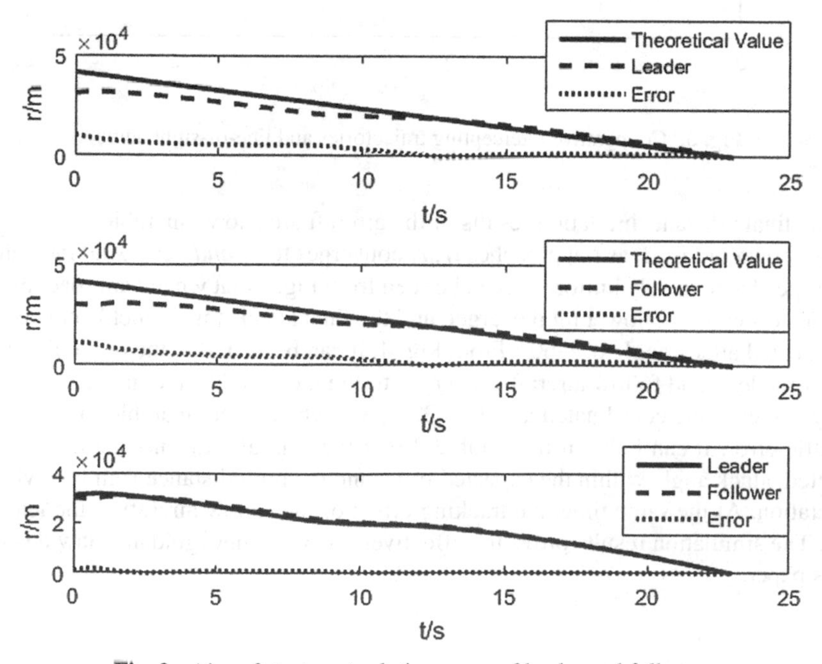

Fig. 2. Aircraft-to-target relative range of leader and follower

guidance law designed in Sect. 4 is adopted for the follow aircraft. Take k_1 equal to 20, $k_2 = 20$, b_1 equal to 0.5 and b_2 equal to 2/3. Set T_d equal to 23 s, σ_f equal to $-6\ 0°$, and N equal to 3. The fourth-order Longo-Kutta method is adopted. The simulation step dt is 0.001 s. The simulation stopping condition is that the relative distance between the aircraft and the target is less than 2 m. The simulation results are as follows (Fig. 2).

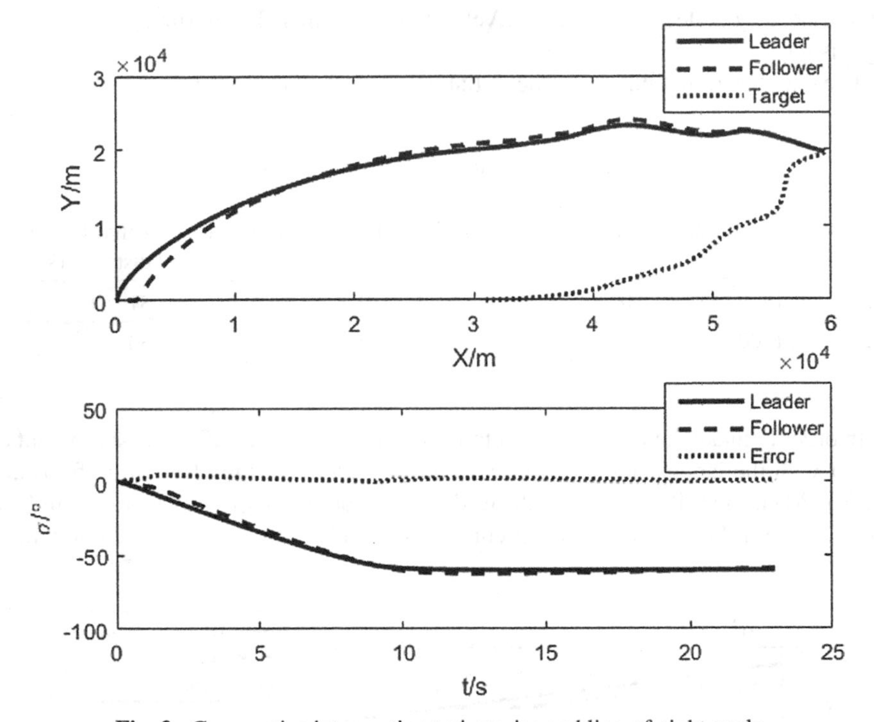

Fig. 3. Cooperative intercepting trajectories and line-of-sight angle

The final relevant simulation results of the aircraft are shown in Table 2:

As shown in Fig. 2, as t approaches t_f, r_i converges to r_1 *and* zero. And the convergence speed is relatively uniform. It can be seen from Fig. 3 that when t approaches 23 s, both the leader and follower hit the target, and the final attack angle quickly converge to the expected attack angle of $-60°$. From Fig. 4, it can be seen that the remaining flight time of the lead and follow aircraft converges to the ideal flight time in about 10 s, and finally achieves the coordinated attack in 22.993 s, which is comparable to the T_d and has little error. It can be seen from Table 2 that the lead aircraft hits the target at the expected attack angle within the expected time, and the miss distance is in line with the expectation. At the same time, the tracking effect of the follow aircraft to the leader is great. The simulation results prove the effectiveness of the new guidance law designed in this paper.

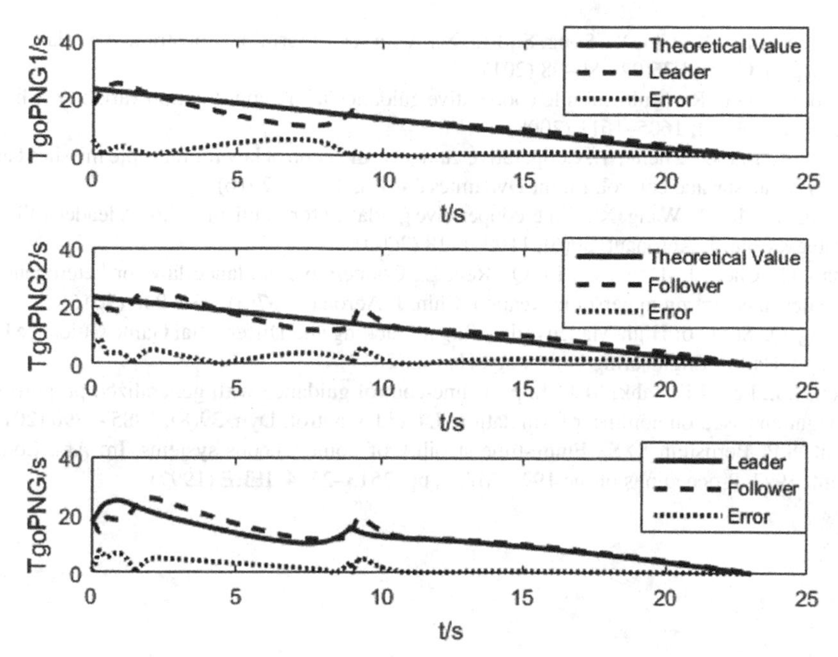

Fig. 4. Time-to-go and error of cooperative guidance

Table 2. Simulation results of aircraft

Aircraft	Remaining Flight Time/(s)	Terminal line-of-sight angle/(°)	Aircraft- target distance/(m)
M1	22.993	−60	9.428
M2	22.993	−59.13	53.06

6 Conclusion

Aiming at high-speed and large maneuvering targets, this paper presents a new guidance model of leader-follower. A multi-constraint guidance law based on sliding mode control is designed for the lead aircraft, and a tracking cooperative guidance law is designed for the follow aircraft. This method has certain generality to solve the problem of intercepting large maneuvering targets. How to popularize this method considering the time-varying velocity of aircraft and target, the delay of inter-aircraft communication and the limitation of aircraft overload is the work to be done in the next step.

References

1. Zhang, K., Liu, Y., Guan, S.: An investigation into the issues of penetration and cooperative engagement for aerodynamic missile under the condition of systems countermeasures. Tact. Missile Technol. **2**, 1–7 (2005) (in Chinese)

2. Zhao, Y., Liao, X., Chi, X., Song, S., Liu, X.: A survey of terminal constrained guidance law. Aerospace Control **35**(02), 89–98 (2017)
3. Zhao, S., Zhou, R.: Multi-missile cooperative guidance using coordination variables. Chin. J. Aeronaut. **29**(06), 1605–1611 (2008)
4. Wang, X., Liu, D., Zheng, Y.: Cooperative guidance and control law for multiple missiles based on dynamic surface control. Flight Dynamics **34**(03), 48–52 (2016)
5. Zhang, Y., Ma, G., Wang, X.: Time-cooperative guidance for multi-missiles: A leader-follower strategy. Chin. J. Aeronaut. **30**(6), 1109–1118 (2009)
6. Zhao, Q., Chen, J., Dong, X., Li, Q., Ren, Z.: Cooperative guidance law for heterogeneous missiles intercepting hypersonic weapon. Chin. J. Aeronaut. **37**(3), 936–948 (2016)
7. Zhang, Q.: Study of High-Maneuvering Target Tracking and Differential Game Guidance Law Design. Harbin Engineering University, Harbin (2017)
8. Jeon, I.S., Lee, J.I., Tahk, M.J.: Impact-time-control guidance with generalized proportional navigation based on nonlinear formulation. J. Guid. Control. Dyn. **39**(8), 1885–1890 (2016)
9. Bhat, S.P., Bernstein, D.S.: Finite-time stability of homogeneous systems. In: Am. Control Conf. 1997. Proceedings of the 1997, Vol. 4, pp. 2513–2514. IEEE (1997)

High Performance Computing, Parallel Computing, Pervasive Computing, Embedded Computing and Simulation

FPGA-Based Hardware Modeling on Pigeon-Inspired Optimization Algorithm

Yu Zhao, Chun Zhao(✉), and Yue Liu

Beijing Information Science and Technology University,
School of Computer Science, Institution of Smart Manufacturing & Complex System,
BISTU, Beijing 100101, China
zhaochun@bistu.edu.cn

Abstract. By learning behavioral characteristics and biological phenomena in nature, such as birds, ants, and fireflies, intelligent optimization algorithms (IOA) is proposed. IOA shows feasibility in solving complex optimization problems in reality. Pigeon-inspired optimization (PIO) algorithm, which belongs to intelligent optimization algorithms, is proposed by the pigeons homing navigation behavior inspired. PIO is superior to other algorithms in dealing with many optimization problems. However, the performance of PIO processing large-scale complex optimization problems is poor and the execution time is long. Population-based optimization algorithms (such as PIO) can be optimized by parallel processing, which enables PIO to be implemented in hardware for improving execution times. This paper proposes a hardware modeling method of PIO based on FPGA. The method focuses on the parallelism of multi-individuals and multi-dimensions in pigeon population. For further acceleration, this work uses parallel bubble sort algorithm and multiply-and-accumulator (MAC) pipeline design. The simulation result shows that the implementation of PIO based on FPGA can effectively improve the computing capability of PIO and deal with complex practical problems.

Keywords: Intelligent optimization algorithm · Pigeon-inspired optimization · FPGA

1 Introduction

Intelligent optimization algorithm (IOA), also known as meta-heuristic algorithm, is suitable for parallel processing, with strong versatility, good robustness. IOA gets good results in a wide range of scientific and practical problems. In recent years, IOA attracts a lot of research interest worldwide. Li et al. propose a simulation design optimization method of IOA based on MATLAB, which

Supported by the National Key R&D Program of China (No. 2018YFB1701600).

W. Fan et al. (Eds.): AsiaSim 2022, CCIS 1712, pp. 407–423, 2022.
https://doi.org/10.1007/978-981-19-9198-1_31

effectively shorten the simulation design cycle and improve the accuracy of the design to a certain extent [1]. Li et al. propose an extended Kalman filter (EKF) algorithm based on particle swarm optimization (PSO), which can get smaller filtering deviation for aircraft trajectory tracking [2].

Inspired by the homing behavior of pigeons, Duan and Qiao propose pigeon-inspired optimization (PIO) in 2014 [3]. PIO shows good performance and receives extensive research. Typically, Duan and Qiu propose a multi-objective PIO method, which is suitable for solving multi-objective optimization problems and successfully applies to parameter design of brushless direct current motor [4]. Alazzam et al. propose a discrete pigeon-inspired optimizer to solve the multiple traveling salesmen Problem [5]. Zhang and Duan propose a modified PIO model adopting Gaussian strategy and verify the feasibility and effectiveness in solving orbital spacecraft formation reconfiguration problems [6]. Moreover, PIO is also used to solve other problems, such as parameter design of the controller for small unmanned helicopters [7], fuzzy energy management strategy for parallel hybrid electric vehicle [8], Underwater Wireless Sensor Networks [9], the target detection task for Unmanned Aerial Vehicles (UAVs) at low altitude [10], feature selection by improving binary pigeon-inspired optimization [11], active disturbance rejection attitude controller for quadrotors [12] etc.

However, there are complex scenarios, large scale and complex computing problems in the real world, such as multi-task assignment, mixed resource scheduling, complex control parameters. In application scenarios dealing with large-scale and complex problems, the IOA, including PIO, execution efficiency decreases. As the calculation complexity increases, the execution time of the PIO increases rapidly, which degrades the performance of PIO. Due to the inherent parallelism of the PIO, PIO can be designed on parallel acceleration platforms to solve the above problems. Currently, studies on parallel design of IOA are mainly divided into the following three patterns: multi-core (open multi-processing), distributed (MapReduce), and heterogeneous computing-based parallel platforms (graphics processing unit-GPU, FPGA, Application Specific Integrated Circuit-ASIC). The first two methods take a general-purpose computer as the hardware carrier and divide tasks into several parts for parallel execution, but still serial in nature. Heterogeneous computing refers to the use of hardware with different architectures to speed up the task of computing. GPU as a parallel acceleration platform is widely used, but GPU has the disadvantages of high power consumption and insufficient flexibility. Using AISC to accelerates IOA algorithm is also feasible, but ASIC is expensive and Not reconfigurable. FPGA is a reconfigurable chip with flexibility and low energy consumption. Parallel design technology can effectively accelerate IOA on FPGA platform. therefore, this paper presents a PIO hardware modeling method based on FPGA, including parallelization analysis, state machine design, and IP core scheduling.

The rest of the paper is structured as follows. Section 2 reviews the typical hardware modeling and implementation of IOA. Section 3 introduces the proposed method. Result and analysis are carried out in Sect. 4. Section 5 concludes this paper.

2 Related Work

Intelligent optimization algorithms encounter challenges in the face of complex calculations. In order to solve the above problems, researchers design various acceleration methods on various platforms. There are some studies on parallel design based on classic algorithms such as genetic algorithm (GA), particle swarm optimization (PSO) and ant colony algorithm (ACO), while few research on PIO. Zou et al. accelerate GA and PSO algorithm by using FPGA, open multi-processing and Compute Unified Device Architecture (CUDA) [13]. Zhou et al. execute PSO in parallel on GPU by using the general-purpose computing ability of GPU and based on the software platform of CUDA from NVIDIA [14]. Menezes et al. propose a parallel design for ACO GPU-based. in which distinct parallelization strategies are compared and analyzed [15]. Juang et al. propose a parallel design of ACO and apply to a fuzzy controller [16]. Moreover, Djenouri et al. improve bees swarm optimization algorithm by using GPU parallelism [17]. Jiang et al. accelerate Whale Optimization Algorithm (WOA) by using OpenCL-based FPGA parallel design [18]. Sadeeq et al. implement the firefly optimization algorithm based on FPGA [19].

3 Hardware Modeling of PIO Based on FPGA

This section introduces the hardware modeling of PIO, first introduces briefly PIO, then conducts modeling analysis from the perspective of acceleration, and finally the detailed design method is given.

3.1 Pigeon-Inspired Optimization Algorithm

The PIO is proposed by mimicking the special navigation behavior of pigeons in the homing process. In PIO, two operators are designed to mimic the mechanism which pigeons use different navigation tools at different stages of finding a target.

Map and Compass Operator
Pigeons can sense the earth's magnetic field using magnetic objects and form a mental map. Pigeons use the sun's altitude as a compass to steer flight, and as pigeons approach target, pigeons rely less on the sun and magnetic field.

Landmark Operator
The landmark operator is used to mimics the influence of landmarks on pigeons in navigation tools. The closer the pigeons get to destination, the more the pigeons rely on landmarks. Pigeons familiar with landmarks fly straight to destination. And pigeons unfamiliar with the landmark follow pigeons familiar with the landmark.

Suppose the search space is n-dimensional, the i-th pigeon can be represented by an n-dimensional vector, $X_i = (x_{i,1}, x_{i,2}, ..., x_{i,n})$. The velocity of the pigeon, which represents the change in position of the pigeons, can be represented by another n-dimensional vector, $V_i = (v_{i,1}, v_{i,2}, ..., v_{i,n})$. The global best position obtained by comparing the positions of all pigeons after each iteration is $X_g = (x_{g,1}, x_{g,2}, ..., x_{g,n})$. Then, each pigeon updates velocity and position according to the following two equations:

$$V_i(t) = V_i(t-1)e^{-Rt} + rand(X_g - X_i(t-1)) \tag{1}$$

$$X_i(t) = X_i(t-1) + V_i(t) \tag{2}$$

where t is the current number of iterations; R is the map and compass factor, with a range of $[0, 1)$, which is used to control the impact of the latest velocity on the current velocity; $rand$ is a random number, uniformly distributed in $[0, 1)$; Eq. (1) is used to update velocity of the pigeon according to latest velocity of the pigeon and the distance of current position of the pigeon from the global best position. Then the pigeon updates position with new velocity of the pigeon according to Eq. (2). As the number of iterations reaches the requirement, stop the work of map and compass operator and continue to work in landmark operator.

In the landmark operator, pigeons depend on landmarks for flight. After each iteration, the number of pigeons decrease by half by Eq. 3. Pigeons far from destination are unfamiliar with the landmarks and can not discern the path, such pigeons are discarded. X_c is the center position of the remaining pigeons, which be used as a landmark and reference for flying. The equation in the landmark operator is given as follow:

$$N_p(t) = \frac{N_p(t-1)}{2} \tag{3}$$

$$X_c(t) = \frac{\sum_{n=1}^{N_p(t)} X_i(t) fitness(X_i(t))}{N_p \sum_{n=1}^{N_p(t)} fitness(X_i(t))} \tag{4}$$

$$X_i(t) = X_i(t-1) + rand(X_c(t) - X_i(t-1)) \tag{5}$$

where N_p is the size of the population; $fitness$ is an evaluation function calculating the fitness of each pigeon. Equation (4) is used to calculate the center value of pigeons. Then the pigeon flies toward a new position according to Eq. (5). As the number of iterations of the landmark operator reaches the requirement, the landmark operator stops working and the algorithm is finished.

3.2 Hardware Modeling Analysis of PIO

In order to improve the execution efficiency of the algorithm, finding the acceleration in hardware modeling is necessary. Four kinds of acceleration are proposed as follows.

Multi-individual Parallelism
The intelligent optimization algorithm based on population have good parallelism because the individual in each iteration is independent from each other. Specifically, velocity updates, position updates, and individual evaluations of PIO can all be processed in parallel.

Multi-dimensional Parallelism
Inside each operation of PIO, the dimensions are independent of each other. Thus, multi-dimensional parallel operations can be implemented. Assuming a solution space of 10 dimensions, the speed of multidimensional operations can theoretically be increased by nearly 10 times.

Pipeline Design of Multiply-and-Accumulator (MAC) Circuit
To calculate the center value of the pigeons by Eq. (4), MAC operation is required. This work can carry out pipeline design of MAC (see Fig. 1) to accelerate PIO. Pipeline design enables adder and multiplier to be fully utilized in limited time.

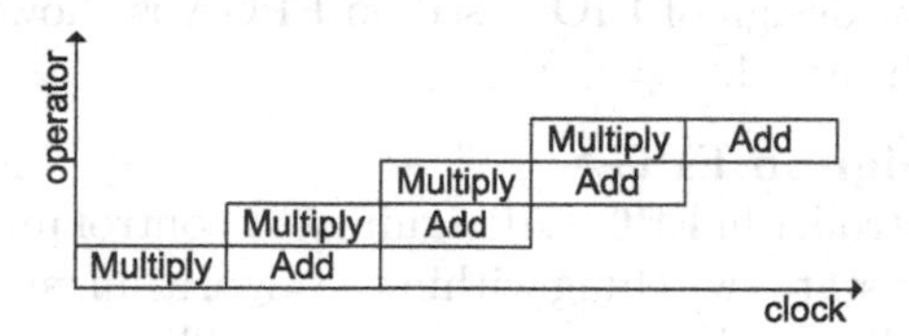

Fig. 1. Pipeline design of MAC circuit.

Sort Algorithm Parallelism
In the landmark operator, sorting is required before calculate the center value of better pigeons. This work exploits parallelism to implement bubble sort algorithm. Parallel bubble sort allows parallel comparison and swapping of independent data [20]. The Odd-even transposition technique is well suited for this case. Let

$$A =< a_1, a_2, a_3, ..., a_i, a_j, ..., a_n > \tag{6}$$

is a list of n elements. Any pair of adjacent elements is called an element pair, such as a_i, a_j. If the a_i is odd, the element pair is an odd pair, and if the a_i is even, the element pair is an even pair.

The descending order of 10 numbers is taken as an example to illustrate parallel bubble sort. A basic operator (see Fig. 2) including a phase of odd pair and a phase of even pair which work sequentially is defined. But each phase is compared and swapped in parallel. The parallel bubble sort algorithm is given below (see Algorithm. 1).

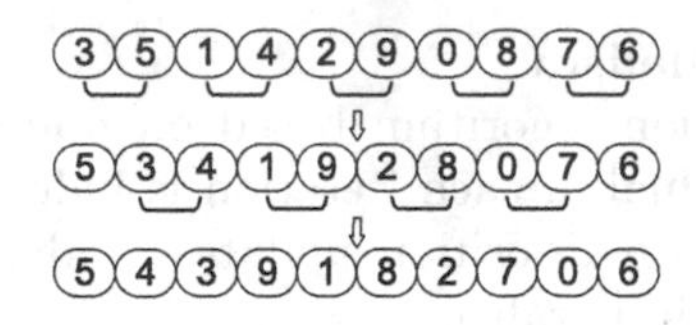

Fig. 2. Base operation of parallel bubble sort.

3.3 Parallel Design of PIO Based on FPGA

A new FPGA can be treated as a blank processor. Except specific logical resources and other auxiliary computing resources, the hardware circuit can be automatically generated after the HDL code is downloaded to the FPGA chip.

In this work, the pigeon population is stored in the top-level module's signal. And the PIO parameters are declared as follows: solution space dimension $D=10$, the population size $N_p = 10$, map and compass factor $R = 0.2$, the number of iteration $iter1 = 15$ and $iter2 = 15$ for two operators.

The program flow design of PIO based on FPGA is shown in Fig. 3. And the data flow is shown in Fig. 4.

Control Unit Design in FPGA

Since there is no controller in FPGA, designing the control unit (CU) is necessary. CU is responsible for state switching within the algorithm, such as timing control, communication, reading and writing memory, etc. There is also a CU inside each module to improve the parallelism of the algorithm. In the top-level module design, multiple Arithmetic logical unit (ALU) modules are used to carry out multi-individual parallelism through CU.

Arithmetic Logical Unit (ALU) Design in FPGA

There is no fixed computing architecture and ALU in FPGA. Considering the mathematical formula of PIO, designing ALU is necessary. Since there is no decimal point calculation in FPGA, the design of ALU in this work chooses floating point operation in accordance with IEEE754 international standard. ALU can be used in parallel if there are sufficient logical resources in the FPGA.

Ten ALU modules are designed in this work, which are evaluation module, sorting module, updating best fitness module, exponent module, random module, updating velocity module in compass operator, updating position module in compass operator, MAC module, center value module, and updating position module in landmark operator. To make the description clearer, some control signals, such as reset signal and enable signal, are hidden and each module is described as follows:

Algorithm 1. Parallel bubble sort algorithm (descending sort)

Input: A(an unsorted array of length n)
Output: A(an sorted array of length n)

```
i ← 0
while i < n/2 do
    j ← 0
    while j < n − 1 do                    /* Do in parallel */
        if A[j] < A[j + 1] then
            temp ← A[j]
            A[j] ← A[j + 1]
            A[j + 1] ← temp
        end if
        j ← j + 2
    end while
    j ← 1
    while j < n − 1 do                    /* Do in parallel */
        if A[j] < A[j + 1] then
            temp ← A[j]
            A[j] ← A[j + 1]
            A[j + 1] ← temp
        end if
        j ← j + 2
    end while
    i = i + 1
end while
```

Evaluation Module

In this work, using a general benchmark function [21] as fitness function to carry out hardware modeling of PIO based on FPGA, and the benchmark function is:

$$fitness(X_i) = \sum_{i=1}^{n} [x_i + 0.5]^2 \tag{7}$$

where X_i is the i-th pigeon, x_i is i-th dimension of the pigeon. By comparative experiments, PIO gets good performance while x_i is $[0, 15]$. Hardware modeling of evaluation module is shown in Fig. 5. The module operates according to Eq. (7). The module uses multi-dimensional parallelism and receives a position and then outputs a fitness value.

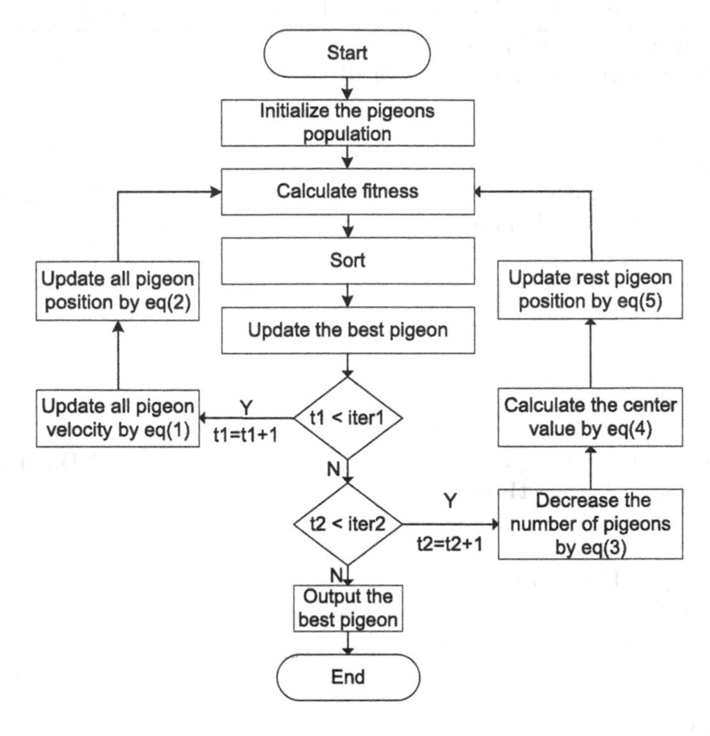

Fig. 3. Program flow design of PIO.

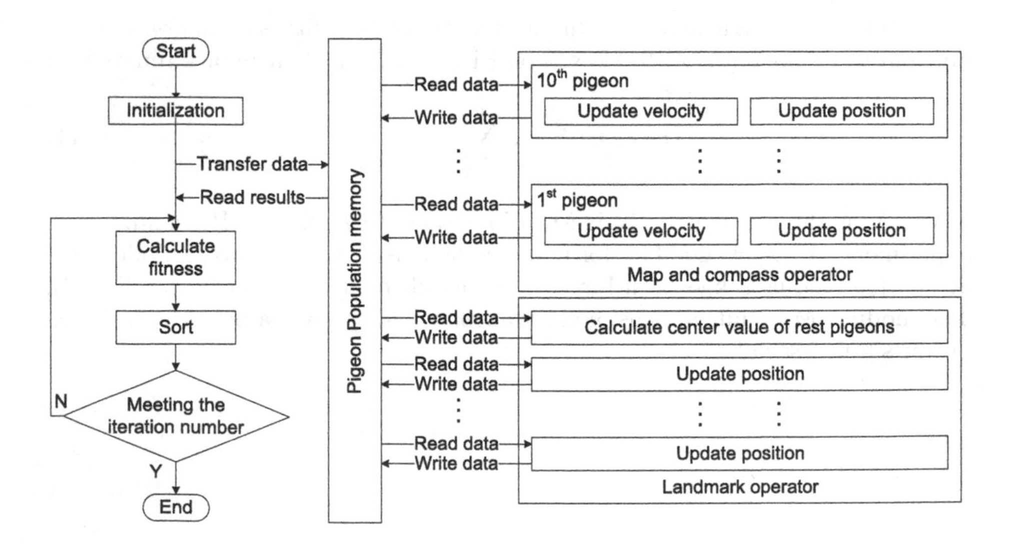

Fig. 4. Data flow design of PIO.

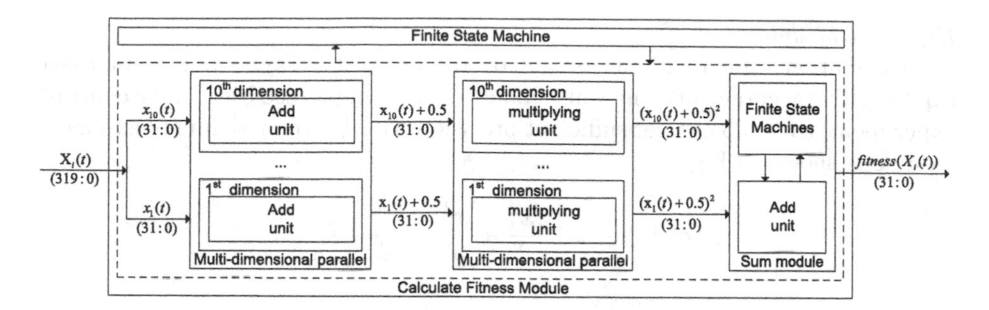

Fig. 5. Hardware modeling of evaluation.

Sorting Module
The sorting module sorts the pigeons according to fitness by using the above parallel bubble sorting algorithm, and the hardware modeling of sorting is shown in Fig. 6. The swapped module in the figure is used to compare and swap two pigeons. Considering module reuse, the sorting module uses only five swapped modules. More swapped modules are shown in the figure for ease of description.

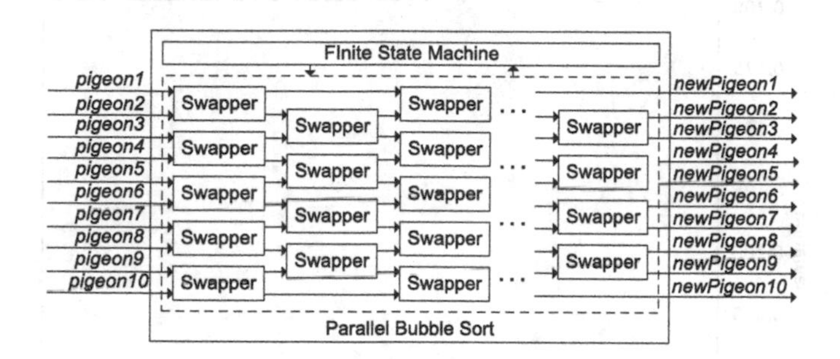

Fig. 6. Hardware modeling of Sorting.

Updating Best Fitness Module
After the sorting is finished, by comparing two fitness values, the X_g may be updated. Hardware modeling of updating best fitness module is shown in Fig. 7.

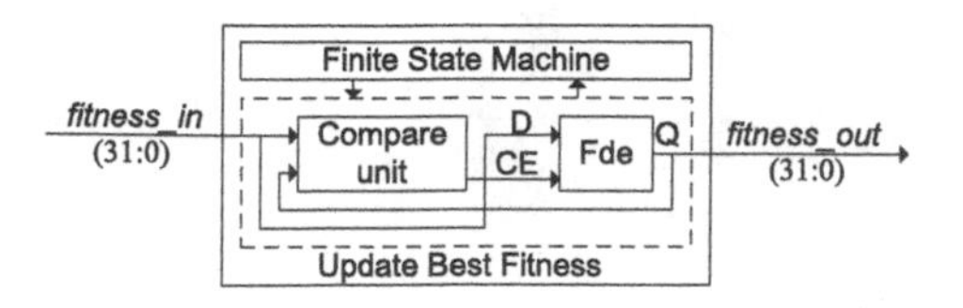

Fig. 7. Hardware modeling of updating best fitness

Exponent Module
Approximation value of e^x can be got by the kth-order Taylor polynomial (see Eq. (8)). The accuracy of e^x module changes as k changes (see Fig. 8). By contrast experiment, let k = 5 can get sufficient precision. Hardware modeling of exponent module is shown in Fig. 9.

$$e^x = 1 + x + \frac{x^2}{2!} + \frac{x^3}{3!} + \frac{x^4}{4!} + \frac{x^5}{5!} \tag{8}$$

Random Module
This work design random number generator by linear Feedback Shift Registers (LFSR) [22]. A 8-bit random binary sequences can be got by 8-bit LFSR (see Fig. 10). To get the random number, uniformly distributed in [0, 1), the "8-bit random binary sequences" is defined as a 9-bit fixed point number. The fixed point number consists of 1-bit sign digits, 3-bit integer digits and 5-bit fractional digits. The value of 1-bit sign digits is 0. Obviously, the ranges of fixed point number is [0, 8). And then the random number uniformly distributed in [0, 1) can be got by divide the fixed point number by eight.

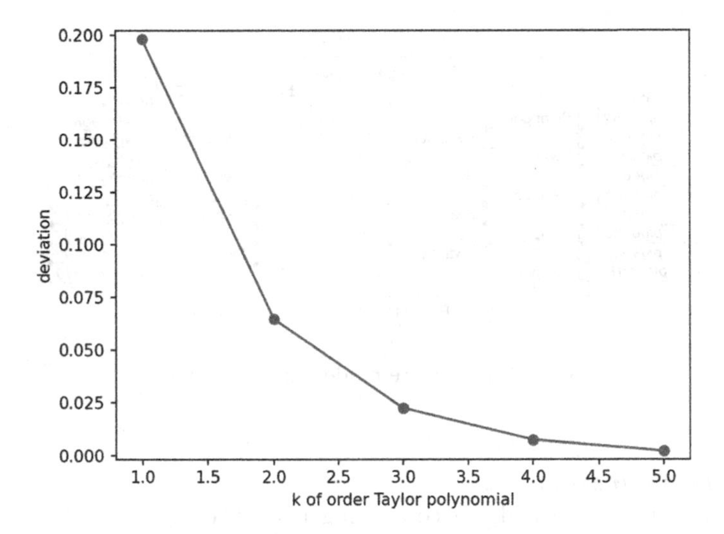

Fig. 8. Precision of exponent module,which x in range [0, 15].

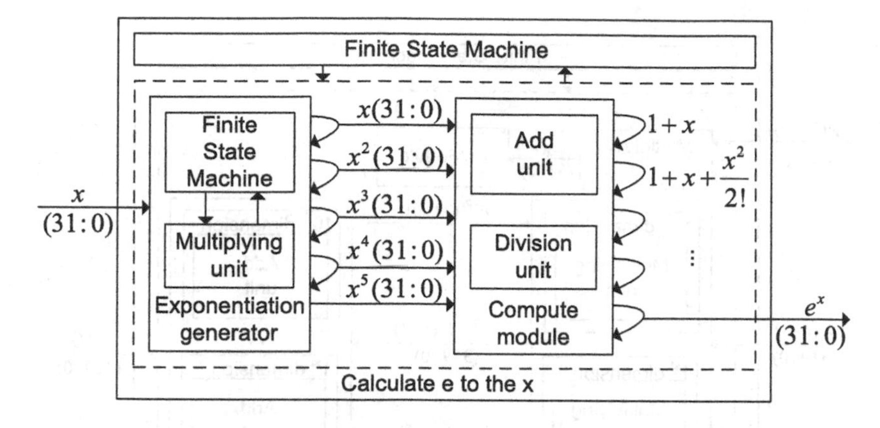

Fig. 9. Hardware modeling of exponent.

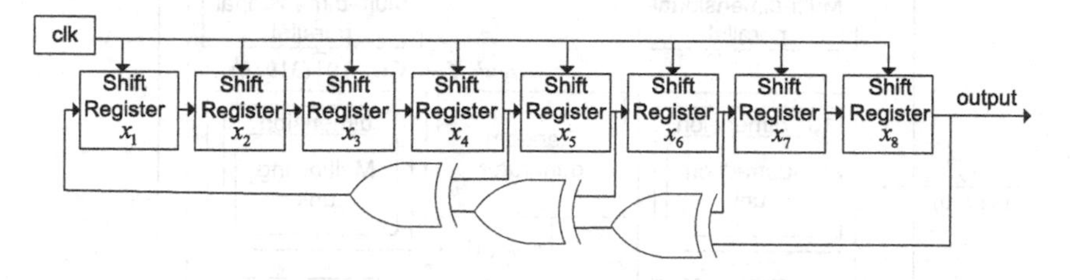

Fig. 10. 8-bit LFSR design.

Updating Velocity Module in Compass Operator
The updating velocity module in compass operator is used to update the velocity of pigeon by Eq. (1). Hardware modeling of updating velocity module in compass operator is shown in Fig. 11, which uses multi-dimensional parallelism.

Updating Position Module in Compass operator
The updating position module in compass operator is used to update the position of pigeon by Eq. (2). Hardware modeling of updating position module in compass operator is shown in Fig. 12, which uses multi-dimensional parallelism.

MAC Module
To calculate X_c by Eq. (4), MAC circuit is required. Hardware modeling of MAC is shown in Fig. 13. Through the control of the finite state machine, the module uses pipeline design.

Center Value Module
In the landmark operator, pigeons of low fitness value are discarded by Eq. (3), and then the center value of remaining pigeons are calculated. To get the center value (X_c), a center value module using multi-dimensional parallelism is designed(see Fig. 14).

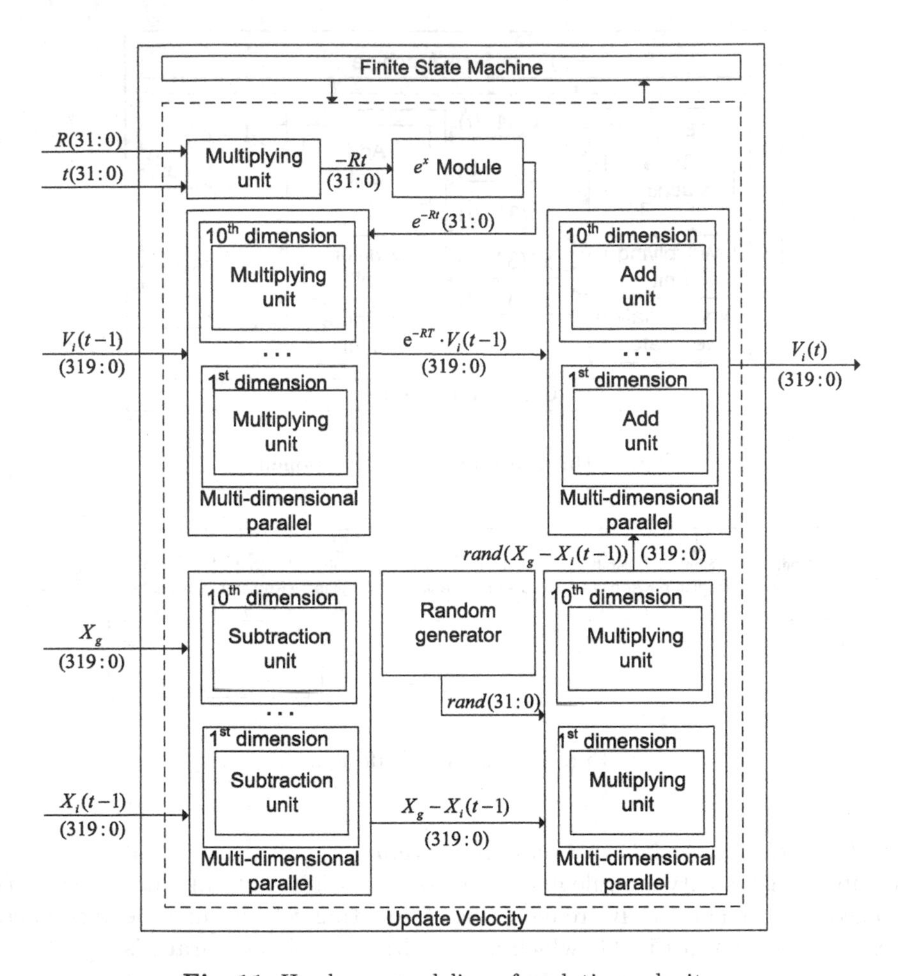

Fig. 11. Hardware modeling of updating velocity.

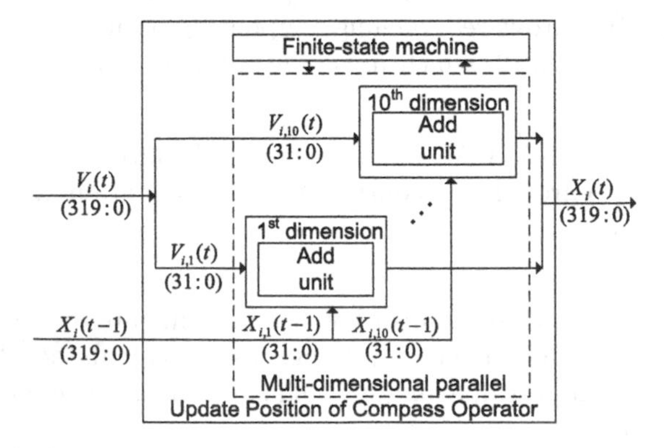

Fig. 12. Hardware modeling of updating position in the compass operator.

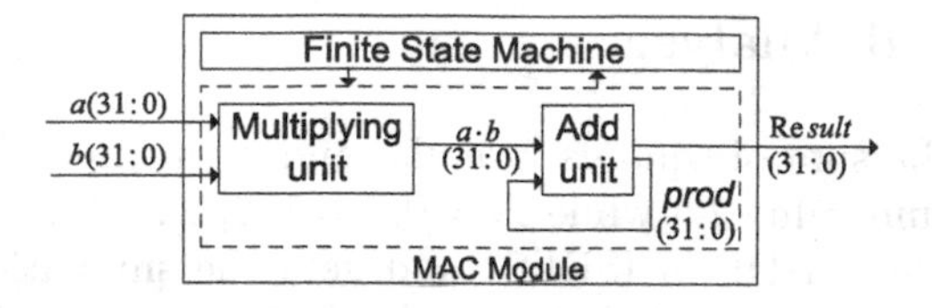

Fig. 13. Hardware modeling of MAC

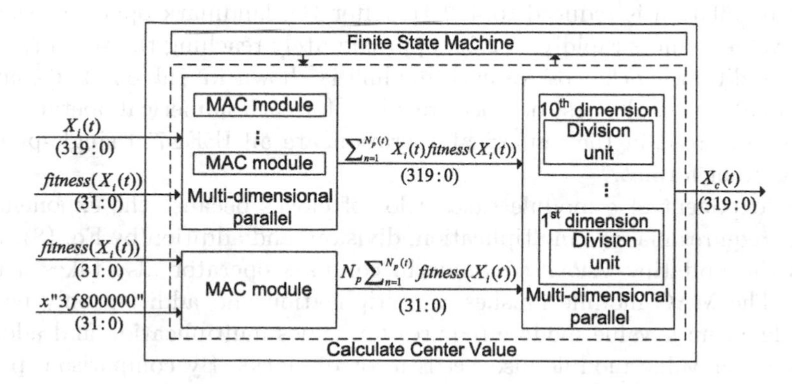

Fig. 14. Hardware modeling of center value

Updating Position module in landmark operator

The updating position module in landmark operator is used to update the position of pigeon by Eq. (5). Hardware modeling of updating position module in landmark operator is shown in Fig. 15, which uses multi-dimensional parallelism.

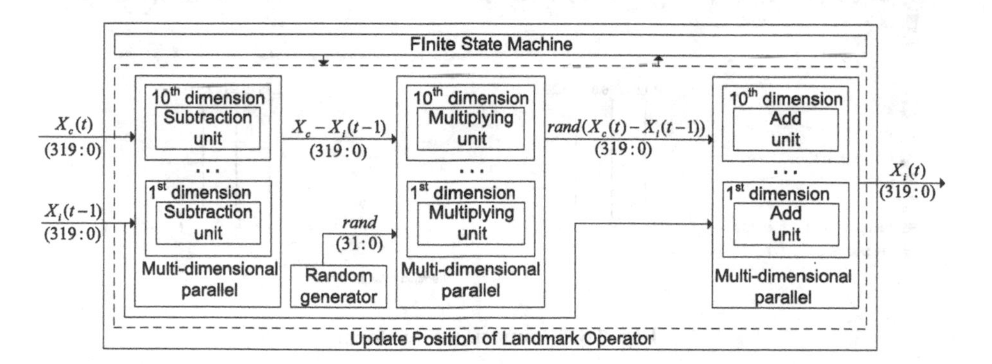

Fig. 15. Hardware modeling of updating position in the landmark operator

4 Results and Analysis

Equation (7) is the sum of squares formula, and x_i in range $[0, 15]$. Obviously, fitness get minimum value 2.5 while $x_i = 0, i = 1, 2, 3, ..., 10$.

This work codes VHDL in ISE14.7 and gets the simulation waveform (see Fig. 16). The best fitness of the pigeon population can be obtained after initialization is 562.14. After the compass and map operator, the best fitness of pigeon population is reduced to 452.41. After the landmark operator, the best fitness value reduces rapidly to 2.54, approximately reaching the true best value 2.5. In addition, clocks cost of each module is shown in Table 1. For comparison, Table 2 shows the cost of clocks required for each numerical operation. The numbers involved in the numerical operation are all IEEE754 single-precision floating-point format.

The exponent of e module takes a lot of clocks because the exponent of e module requires massive multiplication, division, and addition by Eq. (8), which causes the updating velocity module in compass operator also takes a lot of clocks. The MAC module finishes a multiplication and addition only need 16 clock. The center value module requires ten times multiplication and addition, so the center value module also needs a lot of clocks. By comparison, parallel bubble sort is 6.95 times faster than serial bubble sort, and the more the number of sorts, the more obvious the acceleration effect.

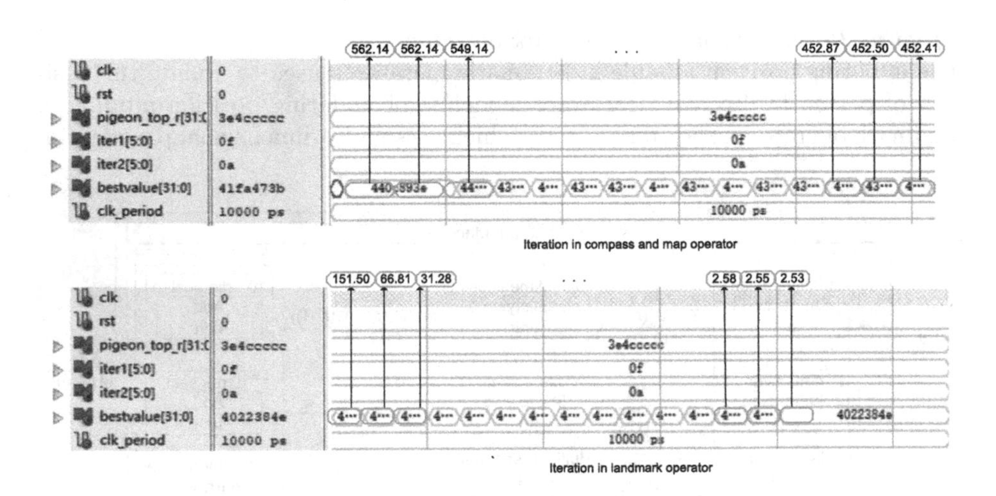

Fig. 16. Simulation waveform

Table 1. Clock cost of each module.

Module name	Clocks
Evaluation module	93
Sorting module	148
Updating best fitness module	4
Exponent module	259
Random module	43
Updating velocity module in compass operator	370
Updating position module in compass operator	34
MAC module	16
Center value module	242
Updating position module in landmark operator	88

Table 2. Clock cost of each numeric operation.

Operation	Clocks
Add	12
Subtraction	12
Multiplication	8
Division	28

5 Conclusion

This paper presents a method of hardware modeling of PIO based on FPGA, and improve the compute performance by parallel design. Firstly, modeling analysis is carried out from the perspective of acceleration, and then the program flow design and data flow design are presented. In addition, the research is a reference for hardware modeling of other intelligent optimization algorithms.

In this work, state machine control is used in the whole process of hardware modeling, including timing control, communication, reading and writing memory, etc. The ALU module uses multi-dimensional parallelism to accelerate PIO. Through CU, multiple ALUs can be used in parallel to accelerate PIO.

In the future, parallel design of PIO can map to a specific FPGA chip to validate PIO performance by comparing with other implementation methods. In addition, PIO based on FPGA can be applied to practical engineering problems.

Acknowledgment. This work is supported by the National Key R&D Program of China (No. 2018YFB1701600).

References

1. Li, W., et al.: A simulation design and optimization method based on MATLAB and intelligent optimization algorithm. In: Proceedings of 2020 China Simulation Conference, pp. 396–402 (2020)
2. Li, L., et al.: Improved EKF aircraft trajectory tracking algorithm based on PSO. In: Proceedings of the 33rd China Simulation Conference, pp. 64–69 (2021)
3. Duan, H., Qiao, P.: Pigeon-inspired optimization: a new swarm intelligence optimizer for air robot path planning. Int. J. Intell. Comput. Cybern. **7**(1), 24–37 (2014)
4. Qiu, H.X., Duan, H.B.: Multi-objective pigeon-inspired optimization for brushless direct current motor parameter design. Sci. China Technol. Sci. **58**(11), 1915–1923 (2015). https://doi.org/10.1007/s11431-015-5860-x
5. Alazzam, H., Alsmady, A., Mardini, W.: Solving multiple traveling salesmen problem using discrete pigeon inspired optimizer. In: 2020 11th International Conference on Information and Communication Systems (ICICS). IEEE (2020)
6. Zhang, S., Duan, H.: Gaussian pigeon-inspired optimization approach to orbital spacecraft formation reconfiguration. Chin. J. Aeronaut. **28**(1), 200–205 (2015)
7. Zhang, D., Duan, H., Yang,Y.: Active disturbance rejection control for small unmanned helicopters via Levy flight-based pigeon-inspired optimization. Aircraft Engineering and Aerospace Technology (2017)
8. Pei, J.Z., YiXin, S., Zhang, D.H.: Fuzzy energy management strategy for parallel HEV based on pigeon-inspired optimization algorithm. Sci. China Technol. Sci. **60**(3), 425–433 (2017)
9. Yu, S., et al.: Node self-deployment algorithm based on pigeon swarm optimization for underwater wireless sensor networks. Sensors **17**(4), 674 (2017)
10. Li, C., Duan, H.: Target detection approach for UAVs via improved pigeon-inspired optimization and edge potential function. Aerosp. Sci. Technol. **39**, 352–360 (2014)
11. Pan, J.-S., et al.: Improved binary pigeon-inspired optimization and its application for feature selection. Appl. Intell. **51**(12), 8661–8679 (2021)
12. Yuan, Y., Duan, H.: Active disturbance rejection attitude control of unmanned quadrotor via paired coevolution pigeon-inspired optimization. Aircraft Engineering and Aerospace Technology (2021)
13. Zou, X., et al.: Parallel design of intelligent optimization algorithm based on FPGA. Int. J. Adv. Manuf. Technol. **94**(9), 3399–3412 (2018)
14. Zhou, Y., Tan, Y.: GPU-based parallel particle swarm optimization. In: 2009 IEEE Congress on Evolutionary Computation. IEEE (2009)
15. Menezes, B.A.M., et al.: Parallelization strategies for GPU-based ant colony optimization solving the traveling salesman problem. In: 2019 IEEE Congress on Evolutionary Computation (CEC). IEEE (2019)
16. Juang, C.-F., et al.: Ant colony optimization algorithm for fuzzy controller design and its FPGA implementation. IEEE Trans. Ind. Electron. **55**(3), 1453–1462 (2008)
17. Djenouri, Y., et al.: Exploiting GPU parallelism in improving bees swarm optimization for mining big transactional databases. Inf. Sci. **496**, 326–342 (2019)
18. Jiang, Q., et al.: Improving the performance of whale optimization algorithm through OpenCL-based FPGA accelerator. In: Complexity 2020 (2020)
19. Sadeeq, H., Abdulazeez, A.M.: Hardware implementation of firefly optimization algorithm using FPGAs. In: 2018 International Conference on Advanced Science and Engineering (ICOASE). IEEE (2018)

20. Lipu, A.R., et al.: Exploiting parallelism for faster implementation of Bubble sort algorithm using FPGA. In: 2016 2nd International Conference on Electrical, Computer & Telecommunication Engineering (ICECTE). IEEE (2016)
21. Yao, X., Liu, Y., Lin, G.: Evolutionary programming made faster. IEEE Trans. Evol. Comput. **3**(2), 82–102 (1999)
22. Babitha, P.K., Thushara, T., Dechakka, M.P.: FPGA based N-bit LFSR to generate random sequence number. Int. J. Eng. Res. General Sci. **3**(3), 6–10 (2015)

CAD/CAE/CAM/CIMS/VP/VM/VR/SBA

An End-to-End Edge Computing System for Real-Time Tiny PCB Defect Detection

Kehao Shi[1], Zhenyi Xu[2(✉)], Yang Cao[2,3], and Yu Kang[2(✉)]

[1] Department of Automation, University of Science and Technology of China, Hefei 230026, China

[2] Institute of Artificial Intelligence, Hefei Comprehensive National Science Center, Hefei 230088, China

xuzhenyi@mail.ustc.edu.cn, {forrest,kangduyu}@ustc.edu.cn

[3] Institute of Advanced Technology, University of Science and Technology of China, Hefei 230088, China

Abstract. In this paper, a low cost real-time monitoring and automatic detection system for PCB defect detection is studied. Two types of algorithms (single-stage algorithm and two-stage algorithm) TDD-NET algorithm and YOLO v5s algorithm in the field of deep learning object detection are selected. First, the research looks for data enhancement methods for problems with small scale of existing public data sets. Then, among at the problem of tiny target detection, feature fusion is used to improve the detection rate of tiny targets. Then, the performance of the dataset on the two algorithms is tested, and the performance indexes of the two algorithms are compared, and the real-time monitoring using YOLO v5s is realized. Finally, YOLO v5s is deployed in NVIDIA embedded hardware Jetson Nano. In addition, TensorRT and DeepStream are used to accelerate the model while invoking the CSI camera for real-time detection. After the hardware deployment, the algorithm can not only obtain the image data from the video and send it to the model for reasoning, but also maintain high accuracy and reasoning speed.

Keywords: PCB defect detection · Feature fusion · Model compression · Edge deployment

1 Introduction

Printed Circuit Board (PCB) has been developed rapidly, but it also faces the problem of quality assurance at the same time. If the defects in PCB cannot be detected in time, it will affect the use of the PCB, which may bring huge economic loss. Therefore, it is necessary to carry out defect detection in PCB. But because of the rapid development of PCB, it makes the current PCB defect

This work was supported in part by the National Natural Science Foundation of China (62103124), Major Special Science and Technology Project of Anhui, China (202003a07020009), China Postdoctoral Science Foundation (2021M703119).

W. Fan et al. (Eds.): AsiaSim 2022, CCIS 1712, pp. 427–440, 2022.
https://doi.org/10.1007/978-981-19-9198-1_32

detection methods difficult to work, so we should develop an automatic detection system for PCB defects.

We apply deep learning algorithm to realize PCB defect detection. The research belongs to the field of object detection, which the common algorithms are divided into twp-stage detection and one-stage detection. We select one of the two types of algorithms: the two-stage detection TDD-Net (Tiny Defect Detection Network) and the single-stage detection YOLO v5s (You Only Look Once v5s), then we will compare performances in terms of complexity and speed and so on. In actual manufacturing, it is always hoped that rapid detection can be performed on the premise of ensuring accuracy. Therefore, we finally deploys YOLO v5s on the NVIDIA embedded hardware Jetson Nano to realize real-time detection with a CSI camera. The research framework of the research is shown in Fig. 1.

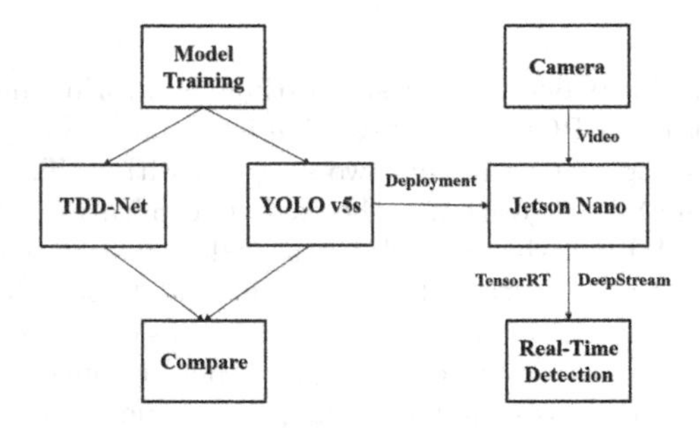

Fig. 1. The research frame. We first train the two algorithms, and then compare their performance. By comparison, YOLO v5s is deployed on the Jetson nano. The Jetson nano calls the camera to obtain real-time pictures, and then combined with TensorRT and Deepstream to realize real-time PCB defect detection.

2 Object Detection Overview

Object detection includes both recognition (e.g. image classification) and localization (e.g. position regression) tasks [10], which require distinguishing the target to be detected from the background in the image and exploiting boundaries boxes for precise localization and correct classification label prediction.

2.1 Common Methods of Object Detection

The use of deep learning technology to complete object detection is gradually becoming the mainstream method. In the field of deep learning, object detection algorithms can be divided into two types: two-stage detection and single-stage detection [7]. The algorithm pipeline is shown in Fig. 2. As can be seen from it, the idea of the two-stage detection is to first search for several candidate regions that may contain objects of interest in the input image, then classify and judge these regions, and finally output the results. The core idea of the single-stage algorithm is to transform localization into regression problem to solve.

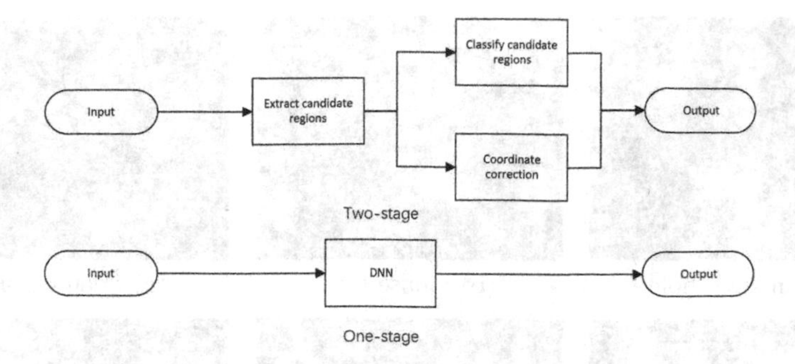

Fig. 2. The pipeline of the one-stage algorithm and the two-stage algorithm

2.2 Evaluation Metrics

For a detector, it is necessary to design corresponding indicators to evaluate its performance so as to select the required detector. The following are commonly used indicators [7]:

1) *IoU* (Intersection Over Union) is often used to measure the degree of overlap between two regions. *IoU* represents the ratio of the overlapping area to the total area of two bounding boxes (prediction box B_p and ground truth box B_{gt}) in the object detection, which is defined as follows:

$$IoU = \frac{area(B_p \cap B_{gt})}{area(B_p \cup B_{gt})}. \tag{1}$$

2) *mAP* (mean Average Precision) is used to measure the accuracy of the predicted box category and location. Its definition is as follows:

$$mAP = \frac{\sum_{q=1}^{Q} AveP(q)}{Q}, \tag{2}$$

where Q represents numbers of categories; $AveP(q)$ represents the average accuracy under a certain category.

3 Methodology

In this section, we first introduce the processing of the dataset and algorithms, then display and compare the results of the algorithms.

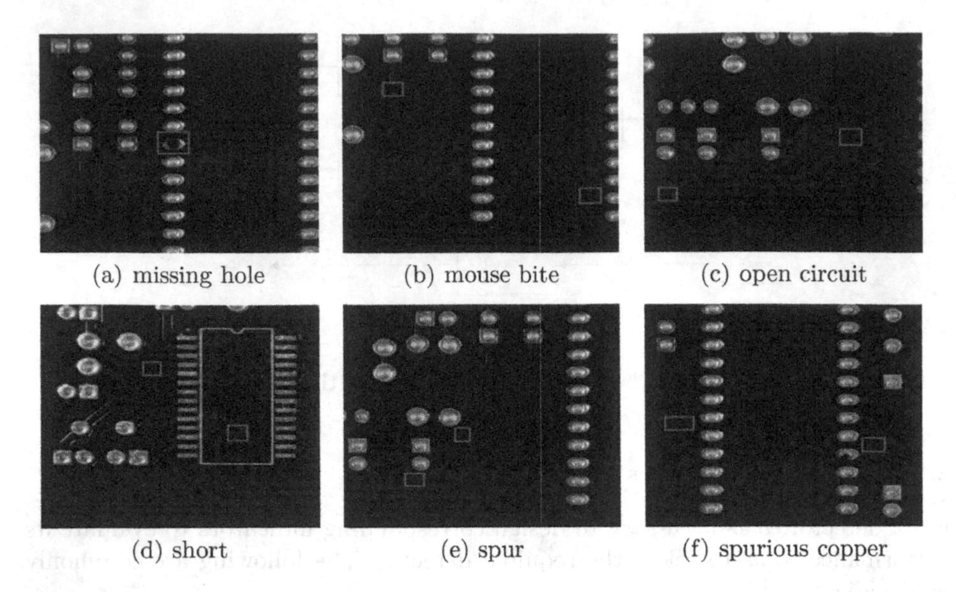
(a) missing hole (b) mouse bite (c) open circuit
(d) short (e) spur (f) spurious copper

Fig. 3. Examples of defects. From (a) to (f), the defects are missing hole, mouse bite, open circuit, short, spur and spurious copper respectively.

3.1 Data Preprocessing

The training dataset comes from the Intelligent Robot Open Laboratory of Peking University, which includes missing hole, mouse bite, open circuit, short, spur and spurious copper. The size of each image is 2777 × 2138, and may contain several defects. Some examples are shown in Fig. 3. The detailed description of the dataset refers to Table 1. There are only 693 images in this dataset. Because using this dataset the for training model directly may cause serious overfitting and cannot accurately predict new data, it needs to be expanded to a certain extent. To train the model, the dataset uses the following techniques for data augmentation: changing image brightness, image cropping, and random rotation, etc. In addition, image processing techniques such as median filtering, salt and pepper noise, and Gaussian filtering are often used for data augmentation [6]. The augmented dataset is described in Table 2.

Table 1. Experiment dataset details

Defect types	Number of images	Number of defects
Missing hole	115	497
Mouse bite	115	492
Open circuit	116	482
Short	116	491
Spur	115	488
Spurious cropper	116	503
Total	693	2953

Table 2. Augmented dataset size

Defect types	Number of images	Number of defects
Missing hole	1832	3612
Mouse bite	1852	3684
Open circuit	1740	3548
Short	1732	3508
Spur	1752	3636
Spurious cropper	1760	3676
Total	10668	21664

3.2 Tiny Object Detection Problem

The targets to be detected are all tiny defects, and the main reason for the difficulty in detecting tiny targets is that the tiny target labeling area accounts for a small proportion. As shown in Fig. 4, in a neural network, the shallow layer feature map has a strong structure and contains more detailed information, while the high layer will retain most of the semantic information [6]. Therefore, it is necessary to fuse low-level semantic information and high-level detailed information, which will be discussed later.

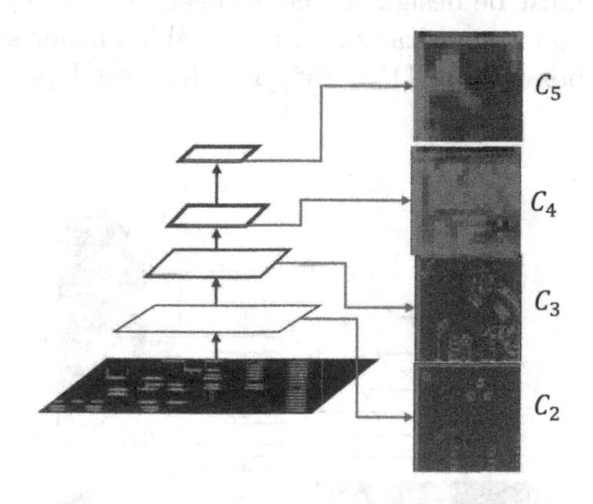

Fig. 4. Schematic diagram of the feature pyramid. The outputs of blocks are denoted as C_2 to C_5 [4].

3.3 TDD-Net

The regional convolutional neural network (Region-CNN, R-CNN) series of algorithms lack certain technical support for the detection of tiny targets. Therefore, it needs to be improved. The two-stage TDD-Net used in this algorithm is based on Faster R-CNN and next the algorithm principle will be introduced.

1) Review of Faster R-CNN Algorithm

The main workflow of Faster R-CNN is as follows: for the input image, the model can search for areas where there may be objects of interest on the image, and then judge in these areas, finally output the selected object and the confidence of the object. Figure 5 shows the whole process, which shows that the Region Proposal Network (RPN) is a multi-task, which needs to predict the object to be detected or the background in the proposed region and regress the predicted box.

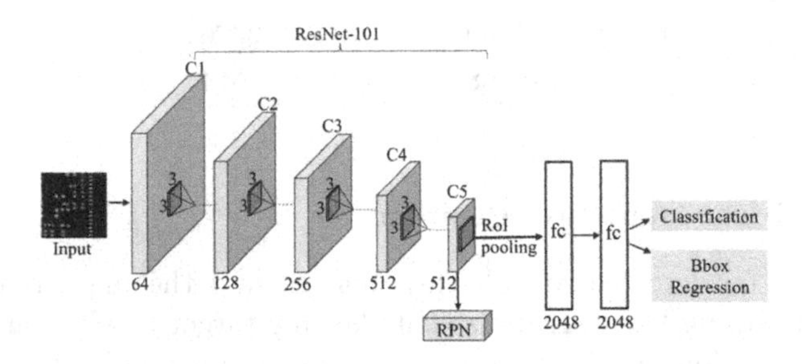

Fig. 5. Faster R-CNN pipeline [4]

2) Anchor design

PCB defects are tiny objects that may only exist in a small area. For tiny defects, reasonable anchors must be designed. Use k-means clustering on PCB training set bounding boxes to automatically find reasonable anchor scales. The scale of the final bounding box area is $\{15^2, 25^2, 40^2, 60^2, 80^2\}$ pixels with an aspect ratio of {2, 3, 4, 5}.

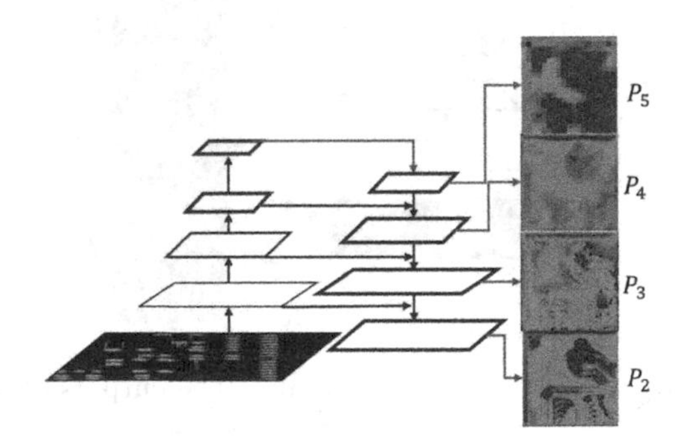

Fig. 6. Schematic diagram of feature fusion. The outputs of merged maps are denoted as P_2 to P_5 [4].

3) Multi-scale feature fusion

TDD-Net combines low-resolution features that are semantically strong with high-resolution features that are semantically weak and structurally strong through a feature pyramid. The process of generating the feature map is continuously up-sampling the upper layer feature map, and then the up-sampled feature map is merged with low-level feature map. Finally, a 3×3 convolutional layer is appended to each merged map to generate the final feature map [2,4]. Denote the feature atlas as $\{P_2, P_3, P_4, P_5\}$. In addition, for feature fusion, TDD-Net adopts the simplest nearest neighbor up-sampling while there are often new technologies such as normalized fusion and attention mechanism and so on [2]. Feature fusion results on different convolutional layers are shown in Fig. 6.

4) Online hard example mining

The online hard example mining strategy is applied to the PCB defect detection, and the Region of Interest (RoI) network runs a forward pass on the feature map and all RoIs. The hard RoI module then uses these RoI losses to automatically select examples.

5) Loss function

The total loss of the algorithm is a weighted sum of 4 different losses, which are rpn_loss_bbox, rpn_loss_cls, fast_rcnn_loss_bbox and fast_rcnn_loss_cls. A Fast R-CNN network has two output layers (classification score and bounding box score), where the first output layer outputs a discrete probability distribution while the second layer outputs bounding box regression offsets [4].

3.4 YOLO v5s

We select YOLO v5s to complete model training and inference. The structure of YOLO v5s consists of the input, the backbone network, the neck network and the output prediction network [9]. The input includes data augmentation and adaptive anchor design for enriching the dataset and targeting the best anchor box values for different targets and different datasets; the backbone network includes the Focus module and the Cross Stage Partial Networks (CSPNet) to reduce the cost of computation; the neck network includes the Feature Pyramid Networks (FPN) and Path Aggregation Networks (PAN) structure for feature fusion to improves tiny object detection rate; the output is used to generate bounding boxes and predict object classes [1].

4 Experiments and Hardware Deployment

This section will show about experimental results and then talk about hardware deployment.

4.1 Experimental Results

1) TDD-Net experimental details and results
The environment is Tensorflow 1.2, and the GPU is GTX 1080 Ti. The implementation of the algorithm is based on Tensorflow. At the same time, in order to prevent overfitting and local minima, the regularization coefficient 0.0001 the learning rate 0.001 and the momentum 0.9 are used. Figure 7 shows examples of the algorithm results. TDD-Net annotates the existing defects with anchor boxes, categories and confidences. Figure 8 shows the average precision (Average Precision, AP) of different categories of TDD-Net on the test set so that the mAP of TDD-Net in the test set is calculated to be 0.989.

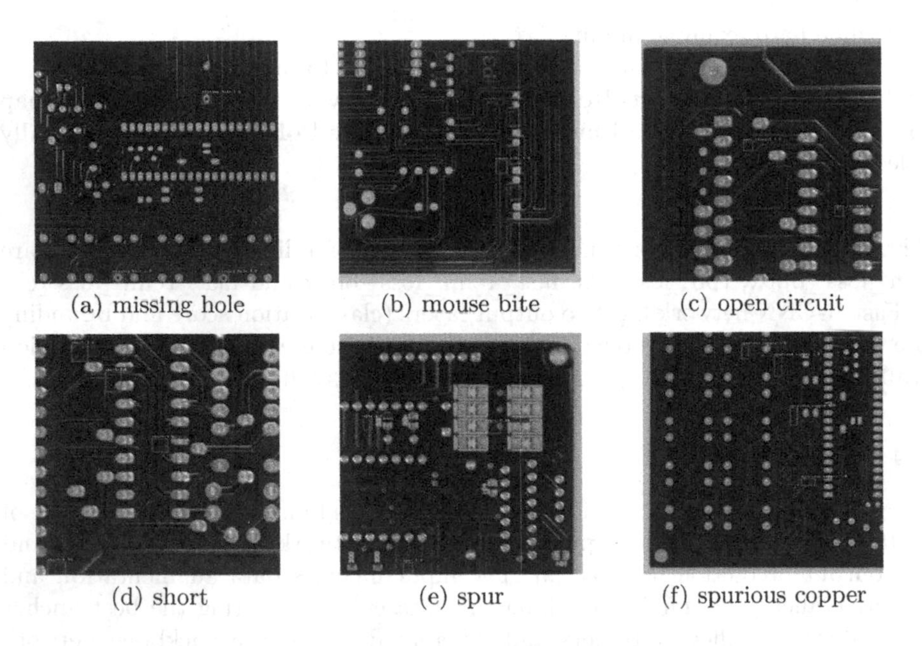

Fig. 7. TDD-Net algorithm detection results.

2) YOLO v5s experimental details and results

Table 3. YOLO v5s evaluation

Type of defects	Number of labels	Precision	Recall	AP
Missing hole	770	0.98	0.986	0.982
Mouse bite	748	0.986	0.988	0.983
Open circuit	642	0.994	0.997	0.993
Short	667	0.981	0.986	0.982
Spur	743	0.98	0.988	0.984
Spurious cropper	749	0.98	0.999	0.983

The environment for YOLO v5s is Pytorch 1.7, the CUDA version is 10.2, and the GPU is GTX 1080 Ti. The learning rate 0.01, the momentum size 0.937 and the weight decay 0.0005 are used. The entire training time is approximately 7 h and some examples of the detection effect of YOLO v5s is shown in Fig. 9. Like the evaluation of the TDD-Net algorithm, the algorithm mAP is 0.986. What's more, the accuracy metrics of the YOLO v5s on the test set are shown in Table 3.

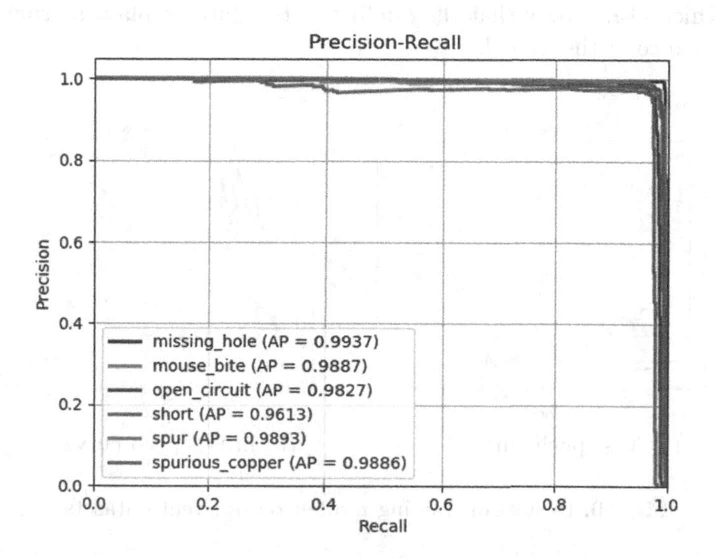

Fig. 8. Precision-recall curve of TDD-Net.

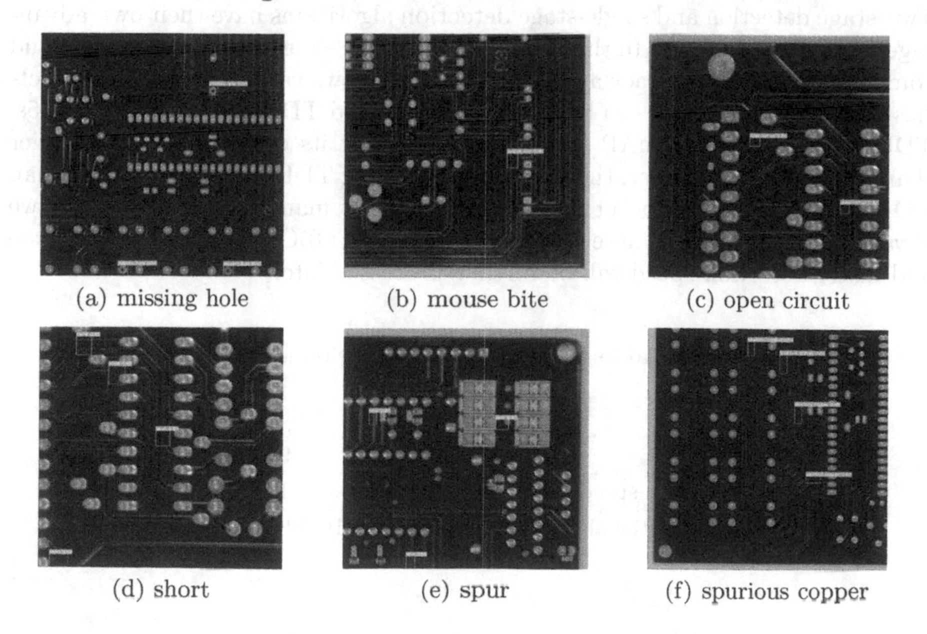

(a) missing hole (b) mouse bite (c) open circuit

(d) short (e) spur (f) spurious copper

Fig. 9. YOLO v5s detection results

Training the algorithm used both the original dataset without data augmentation and the new dataset with data augmentation. The training results are shown in Fig. 10. It can be seen that using the new data-augmented dataset can play an important role in training the model, which the loss function and the mAP can converge better in a small number of training epochs. However, if you use the dataset without data augmentation to train the model, not only a lot of epochs are required, but the final algorithm is not as effective as using the new dataset, which also proves that the quality of the dataset plays a crucial role in the performance of the model.

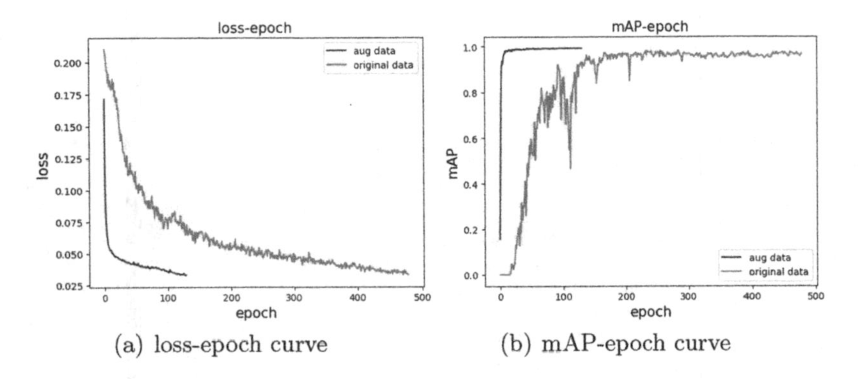

(a) loss-epoch curve (b) mAP-epoch curve

Fig. 10. Results of training models on different datasets

3) Algorithm comparison

Two-stage detection and single-stage detection algorithms have their own advantages and disadvantages. In this study, the selected algorithms are analyzed and compared from the accuracy and speed. Table 4 shows various performance metrics of two algorithms. It can be seen that compared TDD-Net with YOLO v5s, TDD-Net has a better mAP than YOLO v5s, but its detection time is longer than YOLO v5s. Moreover, the space complexity of TDD-Net is also larger than YOLO v5s. In the context of the actual intelligent manufacturing industry, we always hope that can achieve fast inspection, so YOLO v5s with smaller scale and faster detection speed will be further processed later.

Table 4. Comparison of algorithms

	TDD-Net	YOLO v5s
mAP	0.989	0.986
Detection speed	143.5 ms	9.1 ms
Computational complexity	495.26 MB	54.52 MB

4.2 Hardware Deployment

NVIDIA embedded hardware supports deep learning frameworks and model compression techniques, which can greatly optimize learning models. Considering the need to reduce the actual production cost as much as possible, YOLO v5s is deployed into the embedded hardware. We choose Jetson Nano b01 as the embedded deployment hardware.

1) TensorRT model acceleration principle
TensorRT is a model compression technology supported by NVIDIA embedded hardware that allows efficient deployment of deep learning models. There are two main methods for TensorRT to optimize the model: reducing parameter accuracy and inter-layer fusion. Usually an uncompressed deep learning network owns high precision parameters such as 32-bit floating point numbers. TensorRT can reduce the parameters precision to 16-bit floating point numbers or 8-bit integer numbers, which can speed up inference. In addition, TensorRT can decompose and reconstruct neural network networks, merge and optimize compatibility layers and achieve model compression. Figure 11 shows how to optimize: first, it eliminates unused layers and avoids unnecessary calculation. Compatible layers are then merged into one (vertical merge), and similar layers of the same inputs can also be merged (horizontal merge). After that, a highly compressed neural network model is obtained [8]. In addition to the methods mentioned above, current research methods also include the use of the Open Neural Network Exchange (ONNX) to further compress the model, which the space complexity of the model will be smaller [3].

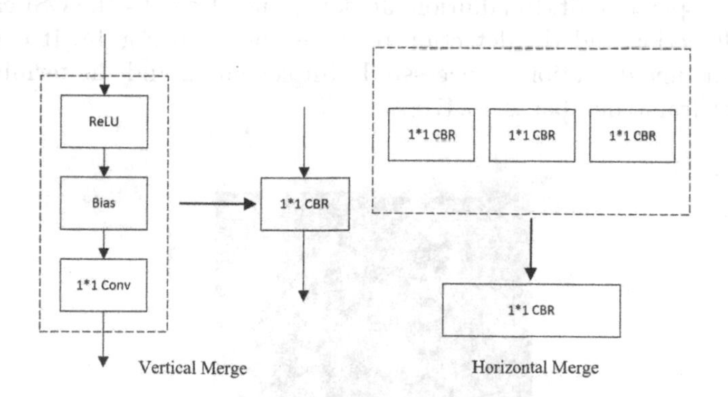

Fig. 11. TensorRT optimization strategy

2) DeepStream video stream processing
If you wait for a camera to take pictures and pass it into the model for detection, it will take much time. Therefore, combined with the actual situation, we use Jetson Nano to call the CSI camera to obtain real-time images and continuously

transmit them into the model for inference and judgment, which can not only speed up the detection speed but also reduce the time cost.

The DeepStream core consists of several hardware accelerator plugins. DeepStream uses the open source GStreamer framework to build an optimized graph architecture. Figure 12 shows a typical application: analysis from input video to output video. Plugins are used for each individual block. The bottom is the different hardware engines used in the application and the independence between plugins ensures the quickness of use [5].

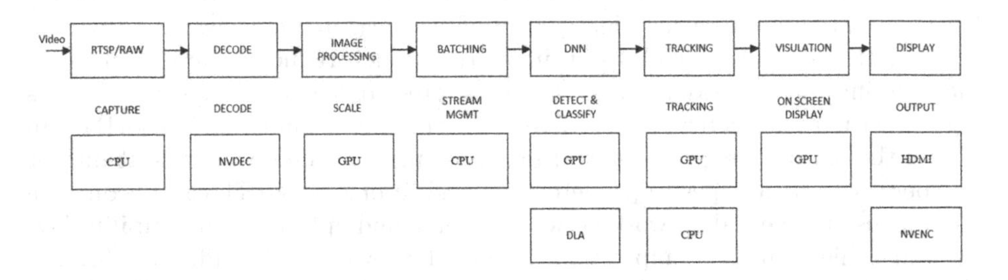

Fig. 12. DeepStream video analysis [5]

3) Details and results of hardware deployment experiments
The hardware environment is the Jetson Nano b01, which the operating system is Linux Ubuntu 18.04, the version of CUDA is 10.2, the version of TensorRT is 8.2, the version of DeepStream is 6.0.

Figure 13 is a example of defect detection by Jetson Nano, where 2 represents the detected open circuit. In addition, Jetson Nano also calls the CSI camera for real-time detection and the detection result is shown in Fig. 14. It can be seen that the real-time detection is successfully implemented and the terminal shows that present frame number is 13.47.

Fig. 13. Jetson Nano deployment detection results

Because TensorRT is used to compress the model, the inference speed of the model will change. Table 5 shows the difference between the algorithm before model compression and after model compression. It can be seen that after the acceleration with TensorRT, the model has a faster inference speed and is closer to the actual production scenario.

Table 5. Model compression effect

	Before compression	After compression
YOLO v5s	247 ms	82.6 ms

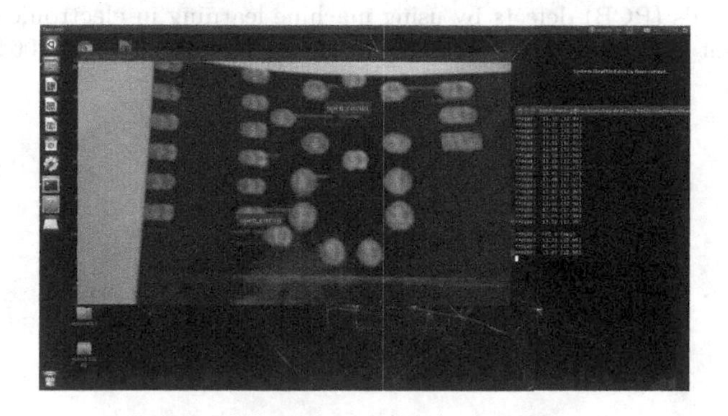

Fig. 14. Real-time detection display

5 Conclusion

According to the requirements of PCB defect detection, we firstly expound the necessity of data augmentation to expand the dataset. Then the related technology to solve the difficulty of tiny target detection is explained, and then two kinds of algorithms are used to realize the defect detection of PCB. In addition, because of the needs of actual production manufacturing, YOLO v5s is deployed into the embedded hardware Jetson Nano, which combines with TensorRT to compress the model and DeepStream to process the video stream. Finally, we realize the real-time detection of PCB defects.

References

1. Chen, B.: Intelligent detection of printed circuit board defects based on YOLOv5. Electronic Test (2022)
2. Chen, H.: Small target detection algorithm based on convolutional neural network (2021)
3. Chen, H.: Design and implementation of video target tracking in mobile terminal based on YOLO. Ph.D. thesis, Heilongjiang University (2021)
4. Ding, R., Dai, L., Li, G., Liu, H.: TDD-Net: a tiny defect detection network for printed circuit boards. CAAI Trans. Intell. Technol. **4**(2), 110–116 (2019)

5. Liu, M.: Domestic traffic sign recognition and embedded implementation based on YOLOv3. Hainan University (2020)
6. Sun, C.: PCB defect detection based on image processing and deep learning (2021)
7. Tan, Z.: Target detection and recognition technology based on deep learning. Chemical lndustry Press (2021)
8. Wang, C.: Research on fast detection algorithm of vehicle exhaust target and embedded device deployment. Ph.D. thesis, University of Science and Technology of China (2021)
9. Wang, X.: Design of intelligent control device of traffic signal light based on YOLO v5 object detection algorithm. Technol. Highw. Transp. **38**(1), 7 (2022)
10. Zakaria, S.S., Amir, A., Yaakob, N., Nazemi, S.: Automated detection of printed circuit boards (PCB) defects by using machine learning in electronic manufacturing: current approaches. IOP Conf. Ser. Mater. Sci. Eng. **767**, 012064 (2020)

Simulation and Analysis of the Scattering Minifying Function of Electromagnetic Wave Expander

Xudong Pang[1](✉), Beibei Cao[1], Liping Wu[1], Xiaolong Zhang[2], and Shouzheng Zhu[3]

[1] Shanghai Publishing and Printing College, Shanghai, China
pxd210@163.com, bbc@sppc.edu.cn
[2] Shanghai Spaceflight Precision Machinery Institute, Shanghai, China
[3] East China Normal University, Shanghai, China

Abstract. Transformation electromagnetics, which provides a new way to manipulate electromagnetic waves with coordinate transformation methodology, has become an important inspiration in wave propagation theories. In this paper, a two-dimensional cylindrical electromagnetic wave expander is discussed with its two basic variables. The variations of each constitutive parameter extremum are analyzed in detail with these two variables respectively. The different curves that vary with the basic variables can provide some guidance for the production of wave expanders. The scattering minifying functions of the wave expander are further studied through numerical simulations of two-dimensional embedded aircraft models and the quantitative calculations of their scattering width values, along with which the scattering width of the expander itself is also studied. A conclusion is drawn that the minification degree of the scattering cross-section of the embedded object inside the wave expander depends on the reduction factor of the wave expander.

Keywords: Transformation electromagnetics · Wave expander · Scattering characteristics

1 Introduction

Transformation optics/electromagnetic can provide a new way to manipulate electromagnetic waves in a designed manner[1], where the coordinate transformation methodology eventually leads to the redesign of the media constitutive parameter tensors, of which is also named as metamaterials[2]. The methodology and the ideal of metamaterials inspired innovations in many other branches of physics researches [3, 4]. Although the electromagnetic devices [5] such as the wave expander have been studied before, there is still a lack of detailed media parameter analysis work and a further study of the scattering characteristics of the wave expander with embedded models.

In this paper, the 2-D cylindrical model of the wave expander is introduced, where two basic variables of this metamaterial shell of the wave expander are defined as the reduction factor and the relative thickness, with which the variations of each constitutive

W. Fan et al. (Eds.): AsiaSim 2022, CCIS 1712, pp. 441–448, 2022.
https://doi.org/10.1007/978-981-19-9198-1_33

parameter extremums are analyzed with detailed illustrations. The parameter analysis work here gives a further study for the constitutive parameters of the wave expander, which can also be used as guidance for the production of the metamaterial-assisted devices of wave expanders. A comparative study for the scattering characteristics of the wave expander with embedded object inside is performed using the Comsol Mutiphysics finite-element-based (FEM) electromagnetics solver, along with which the scattering width of the expander itself is also studied.

2 Basic Model

As can be seen from Fig. 1(a), the electromagnetic wave expander can be classified as one of the polygonal-line coordinate transformation devices [5]. The simulation model of the wave expander in Fig. 1(b) uses 2-D cylindrical shape where the inner radius is *a* and the outer radius is *b*. In Fig. 1(a), *r* represents the cylinder radius in the original system while *r'* represents the cylinder radius in the transformed system. Line 1 which is drawn in solid line, indicates that $0 \leq r \leq s$ is expanded to $0 \leq r' \leq a$ $(0 < s < a)$ and $s \leq r \leq b$ is compressed to $a \leq r' \leq b$ accordingly. Such a coordinate transformation as line 1 leads to wave expander. Line 2 in dashed line, represents the wave concentrator [5]-[6]. Line 1 and 2 are respectively located at the two sides of the $k = 1$ base line. Here a conclusion can be drawn that: $k > 1$ means the region in this space is expanded while $k < 1$ means the region is compressed.

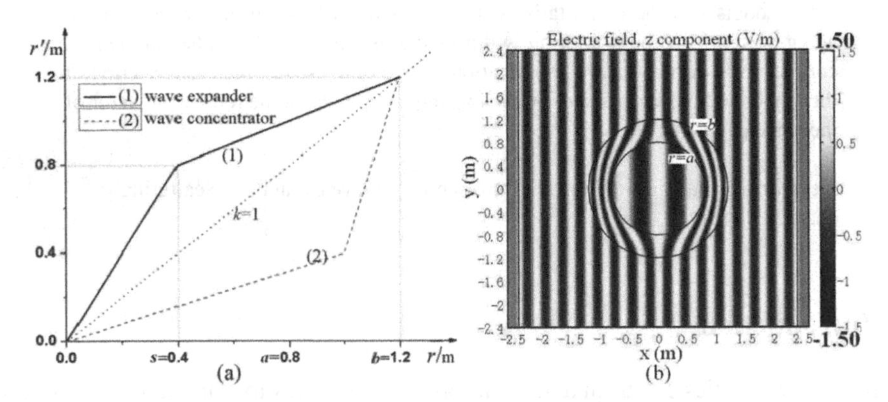

Fig. 1. Comparative schematic of wave concentrator and the wave expander with a brief verification (a) polyline coordinate transformation schematic (b) simulation result of the cylindrical wave expander where $a = 0.8$ m, $b = 1.2$ m, $s = 0.4$ m, $\beta = 1/2, f = 1$ GHz.

The formula of the coordinate transformation used in this paper is given as [6]:

$$r' = r/\beta,\ \theta' = \theta,\ z' = z(0 \leq r \leq ss = \beta a, 0 < s < a) \tag{1}$$

$$r' = a + \frac{b-a}{b-s}(r-s), \theta' = \theta,\ z' = z(s \leq r \leq b) \tag{2}$$

$$r' = r, \theta' = \theta, z' = z(r > b) \tag{3}$$

Here $\beta = s/a$ $(0 < \beta < 1)$. This device has a dispersing effect to the incident wave, thus leads to a scattering minifying function. Here the reduction factor $\tau = 1/\beta$ $(\tau > 1)$ is defined to represent the minifying capability of the wave expander.

The expressions of the permittivity components are shown below as [6]:

$$\varepsilon_r = \frac{(b-s)r - b(a-s)}{(b-s)r} \tag{4}$$

$$\varepsilon_\theta = \frac{(b-s)r}{(b-s)r - b(a-s)} = \frac{1}{\varepsilon_r} \tag{5}$$

$$\varepsilon_z = \varepsilon_r \times \left(\frac{b-s}{b-a}\right)^2 \tag{6}$$

$$\varepsilon_{xx} = \varepsilon_r \times \cos^2\theta + \varepsilon_\theta \times \sin^2\theta \tag{7}$$

$$\varepsilon_{xy} = (\varepsilon_r - \varepsilon_\theta) \times \sin\theta \times \cos\theta \tag{8}$$

$$\varepsilon_{yy} = \varepsilon_r \times \sin^2\theta + \varepsilon_\theta \times \cos^2\theta \tag{9}$$

Here $\sin\theta = y/r$, $\cos\theta = x/r$. (4)-(9) used the ideal parameter sets, so the expander itself does not causes any obvious scatterings. Figure 1(b) shows the simulation result of the electromagnetic wave expander in Fig. 1(a) with a transverse electric polarized (TE) plane-wave illuminated from left to right, where $a = 0.8$ m, $b = 1.2$ m, $s = 0.4$ m, $\beta = s/a = 1/2, f = 1$ GHz.

3 Parameter Analysis

In this part, the curves of the constitutive parameter extremums vary with the basic variables such as the reduction factor τ or the relative thickness D are shown below respectively. The relative thickness of the wave expander in Fig. 1(b) is defined as $D = (b\text{-}a)/a$.

First $\boldsymbol{\varepsilon}$-τ is studied while D remains constant. A series of data are collected and drawn to represent the quantitative relationship between each component of permittivity parameter $\boldsymbol{\varepsilon}$ and the reduction factor τ. The size of the expander model is $a = 0.8$ m, $b = 1.2$ m $(0 < s < a, \tau > 1, D = 0.5)$.

Figure 2, which is drawn with (a) and (b) due to the different ranges of the permittivity parameter $\boldsymbol{\varepsilon}$, shows the curves of the extremums for each component of $\boldsymbol{\varepsilon}$ vary with the reduction factor τ. Here the component directions of $\boldsymbol{\varepsilon}$ include r, θ, z, xx, xy and yy. In Fig. 2(a), curve 1 and 2 represent the variations of $\varepsilon_{\theta\max} = \varepsilon_{xx\max} = \varepsilon_{yy\max}$ and $|\pm\varepsilon_{xy}|_{\max}$ respectively, which all increase linearly with τ. The extremums in xy direction are a pair of plus-minus numbers with equal absolute value, both of which are represented by $|\pm\varepsilon_{xy}|_{\max}$ here. In Fig. 2(b), the curves 1, 2, 3 and 4 with different symbols represent the variations of $\varepsilon_{\theta\min} = \varepsilon_{z\max}$, $\varepsilon_{z\min}$, $\varepsilon_{r\max}$ and $\varepsilon_{r\min} = \varepsilon_{xx\min} = \varepsilon_{yy\min}$ respectively, where $\varepsilon_{\theta\min}$ and $\varepsilon_{z\max}$ increase gradually with τ while $\varepsilon_{r\max}$ $\varepsilon_{r\min}$ $\varepsilon_{xx\min}$ and $\varepsilon_{yy\min}$ decrease gradually with τ, but $\varepsilon_{z\min}$ has a tiny increase in the very beginning and then decrease gradually with τ.

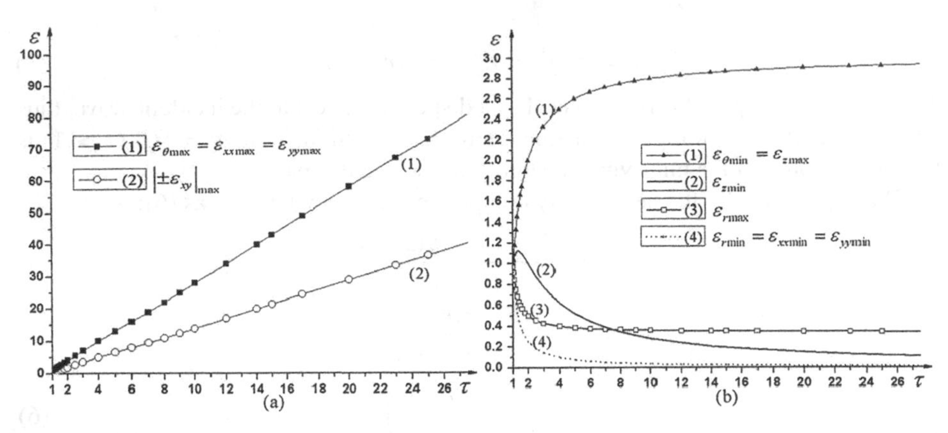

Fig. 2. Curves of the extremums for each component of $\boldsymbol{\varepsilon}$ vary with τ (a) curves of $\boldsymbol{\varepsilon}$-τ with larger range of $\boldsymbol{\varepsilon}$ (b) curves of $\boldsymbol{\varepsilon}$-τ with smaller range of $\boldsymbol{\varepsilon}$

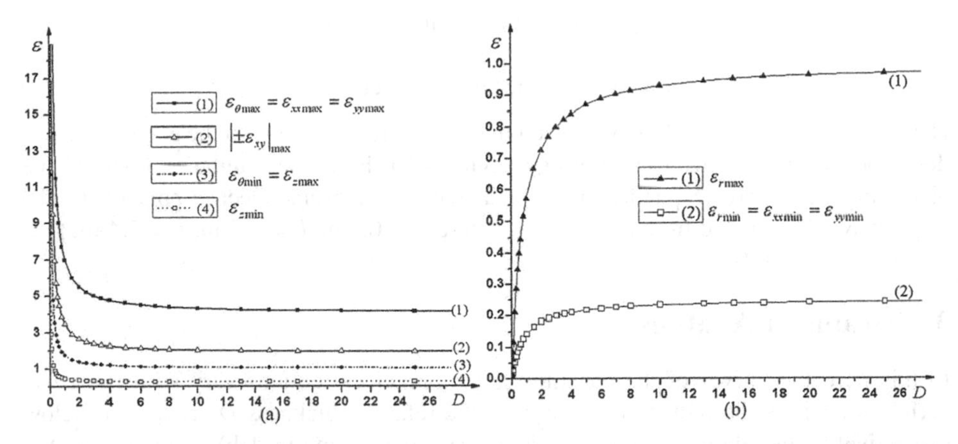

Fig. 3. Curves of the extremums for each component of $\boldsymbol{\varepsilon}$ vary with D (a) curves of $\boldsymbol{\varepsilon}$-D with larger range of $\boldsymbol{\varepsilon}$ (b) curves of $\boldsymbol{\varepsilon}$-D with smaller range of $\boldsymbol{\varepsilon}$

Then $\boldsymbol{\varepsilon}$-D is studied while τ remains constant. A similar method is used here to study the quantitative relations between each component of $\boldsymbol{\varepsilon}$ and the relative thickness D. The size of the model used here is $a = 0.8$ m, $s = 0.2$ m ($\tau = 4$), and the outer radius b starts from 0.8 m. Figure 3, which is also drawn with (a) and (b) due to the different ranges of $\boldsymbol{\varepsilon}$, shows the curves of the extremums for each component of ε vary with the relative thickness D. In Fig. 3(a), the curves 1, 2, 3 and 4 represent the variations of $\varepsilon_{\theta max} = \varepsilon_{xxmax} = \varepsilon_{yymax}$, $|\pm\varepsilon_{xy}|_{max}$, $\varepsilon_{\theta min} = \varepsilon_{zmax}$ and ε_{zmin} respectively, which all decrease gradually with D. In Fig. 3(b), curve 1 and 2 represents the variations of ε_{rmax} and $\varepsilon_{rmin} = \varepsilon_{xxmin} = \varepsilon_{yymin}$, which all increase gradually with D.

The different curves that vary with the basic variables can provide some guidance for the production of wave expanders.

4 Scattering Minifying Function

Figure 1(b) shows that the field distribution in $0 \leq r' \leq a$ appears to be more dispersed after the coordinate transformation. Such function reminds us of the characteristic of concave lens, which also has a dispersing effect to the incident waves. Associated with the minifying imaging function of the concave lens, the following theoretical deductions can be proposed that [7]: (1) electromagnetic wave expander has a dispersing function to the incidence wave, hence a minifying function to the scattering cross-section (SCS) of the embedded object; (2) its reduction degree depends on $\beta = 1/\tau$ ($\tau > 1$).

In the following part, a validation to the scattering minifying function of electromagnetic wave expander is performed with numeric simulations and quantitative calculations. Since there are few papers about the experimental test and the implementability[7] of the electromagnetic wave expanders in this paper, the following numerical simulation method is more valuable for theoretical research. The method used here is to embed a 2-D object model in the core inside the wave expander in Fig. 1(b), and then simulate the scattering pattern using the FEM solver. Here a 2-D simulation model using the geometric cross-section of the Northrop B-2 spirit bomber is chosen as the embedded object, which is a 2-D conformal shrinking model of the practical aircraft. The size of the embedded model is wingspan $w = 1.4$ m, body length $h = 0.56$ m. Here the metal boundary is applied to the embedded B-2 shaped object. The size of the wave expander here is $a = 0.8$ m, $b = 1.2$ m and an $f = 1$ GHz cylinder wave is used as the wave source with a distance $d = 2.53$ m away. For a 2-D model, the SCS of the object is actually the scattering width [6] of its cross-section. So a quantitative calculation for the scattering width value of each 2-D B-2 shaped model is made according to its simulation result. The Huygens principle is required here to convert the near field data from the FEM solver into the far field data which will be used to calculate the scattering width value. The formula used here is: $\sigma = 2\pi r |\boldsymbol{E}^s(\varphi) 2| \boldsymbol{E}^i|^2$.

The validation to the scattering minifying function of the electromagnetic wave expander with $\beta = 1/4$ is carried out through simulations with a cylinder wave illuminated from up to down, as shown in Fig. 4(a)–(c). Figure 4(a) is the simulation result of the original size model ($w = 1.4$ m, $h = 0.56$ m); In Fig. 4(b), the embedded model with original size is placed in the core inside the $\beta = 1/4$ wave expander; Fig. 4(c) is the simulation result of a conformal model with 1/4 original size ($w_2 = 0.35$ m, $h_2 = 0.14$ m). The scattering width values of the B-2 shaped models in Fig. 4(a)–(c) are shown in Fig. 4(d)–(f) accordingly, where Fig. 4(d) is the scattering width result of the model in Fig. 4(a); Fig. 4(e) is the scattering width result of the embedded model in Fig. 4(b); Fig. 4(f) is the scattering width result of the model in Fig. 4(c). In Fig. 4(d)–(f), the y-axis represents the results of scattering width value with a same range of 0 –0 m for further comparison. As can be seen from Fig. 4 that the SCS of the original size embedded model which is wrapped by the $\beta = 1/4$ wave expander is mainly equal to that of the 1/4 size conformal model, both of which are obviously smaller than the SCS of the original size model in the air.

Theoretically, the wave expander itself does not have any scattering from the outside, which makes it an invisible device as shown in Fig. 1(b). But, a more precise calculation of its scattering width shows a different result. Here the scattering width values of the 2-D cylindrical metamaterial shell of the wave expander in Fig. 4(b) is also studied (τ

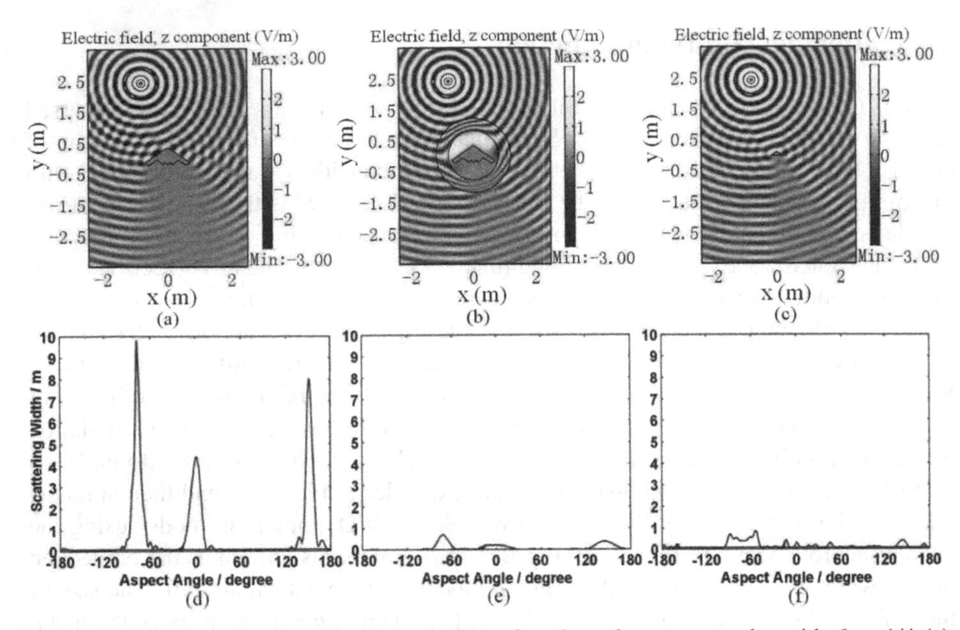

Fig. 4. Verifications for the scattering minifying function of wave expander with $\beta = 1/4$ (a) scattering pattern of the original size model; (b) overall scattering result of the $\beta = 1/4$ wave expander with the original size embedded model inside; (c)scattering pattern of a conformal model with 1/4 original size; (d)–(f) scattering width value of the model in (a)–(c).

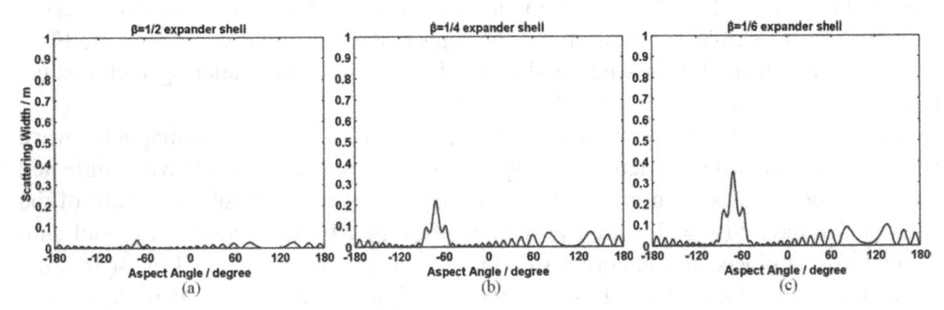

Fig. 5. The scattering width values of the cylindrical metamaterial shell of the 2-D wave expander with (a) $\tau = 2$ (b) $\tau = 4$ (c) $\tau = 6$, the calculation model uses the expander in Fig. 4(b).

= 4), the result of which is shown in Fig. 5(b). In Fig. 5, we also calculate the scattering width values of the expander shell with other reduction factors ($\tau = 1/\beta$), where $\tau = 2$ is shown in Fig. 5(a) and $\tau = 6$ is shown in Fig. 5c). From Fig. 5, we can see that the scattering width values of the expander shell increase with the reduction factor τ yet remains a similar curve shape, and when τ reaches the value of 6 or more, the scattering width value would be comparable with that of the shrinked B-2 shaped model and can't be ignored. From this point of view, we can also suggest that a wave expander with the reduction factor τ less than 6 or 7 is feasible, which is also less difficult to fabricate. Generally speaking, the overall scattering width values of the expander shell is tiny and also somehow depends on the mesh number of the simulation model, so the result in Fig. 5 don't contradict with the basic theory before and can be used to explain the tiny difference between Fig. 4(e) and (f).

The work above can prove that the reduction degree of the scattering cross-section of the 2-D embedded model inside the wave expander depends on $\beta = 1/\tau$. In a real case, a stealth aircraft is often coated with absorbing materials to reduce its Radar Cross Section (RCS), so a simulation for an absorbing boundary condition of the aircraft model is also carried out with the same conclusion drawn. The work in Fig. 4 shows that the SCS of an object placed inside the wave expander is equivalent to that of a conformal object with a smaller geometric cross section in the air.

In military field, this scattering minifying function can be further applied to camouflage real aircrafts to drones and can also avoid radar detections. And the device of electromagnetic wave expander alone is an extension for concave-lens-like devices in the electromagnetic spectrum, which deserves a more profound study.

5 Conclusion

In this paper, A two-dimensional cylindrical electromagnetic wave expander was discussed with varies of reduction factor τ or the relative thickness D, and the scattering minifying functions of the wave expander are studied. The variations of each constitutive parameter extremums are analyzed in detail. A comparative study is made to the scattering minifying function of the wave expander which verifies the aforementioned theoretical deductions that the minification degree of the scattering cross-section of the embedded object inside the wave expander depends on the reduction factor of the wave expander and the scattering width values of the expander shell increase with the reduction factor τ yet remains a similar curve shape, and when τ reaches the value of 6 or more, the scattering width value would not be ignored. From this point of view, we can also suggest that a wave expander with the reduction factor τ less than 6 or 7 is feasible, which is also less difficult to fabricate.

References

1. Zhang, P.-F., Yan, Y.-K., Liu, Y., Mittra, R.: A look at field manipulation and antenna design using 3D transformation electromagnetics and 2D surface electromagnetics. Frontiers of Information Technology & Electronic Engineering **21**(3), 351–365 (2020). https://doi.org/10.1631/FITEE.1900489
2. Cui, T.J., Qi, M.Q., Wan, X., et al.: Coding metamaterials, digital metamaterials and programmable metamaterials. Light Sci. Appl. **3**(10), 218 (2014)
3. Zhu, S., Zhang, X.: Metamaterials: artificial materials beyond nature. Natl. Sci. Rev. **5**(2), 131 (2018)
4. Dai, G.-L.: Designing nonlinear thermal devices and metamaterials under the Fourier law: A route to nonlinear thermotics. Front. Phys. **16**(5), 1–36 (2021). https://doi.org/10.1007/s11467-021-1048-y
5. Cheng, F., Yang, M., Huang, J., Jing, Y., et al.: Arbitrarily shaped homogeneous concentrator and its layered realization, Opt. Commun. **435**, 150–158 (2019)

6. Pang, X., Tian, Y., Wang, L., et al.: Simulation and analysis of a new electromagnetic wave concentrator with reduced parameter Sets. AsiaSim SCS AutumnSim **2016**, 300–307 (2016)
7. Pang, X., Zhu, S.: The implementability of electromagnetic wave expander based on metamaterials, In: 2013 7th International Congress on Advanced Electromagnetic Materials in Microwaves and Optics, pp. 193–195 (2013)

The Study on Flow Characteristics of Inlet Flow Field of Compressor Experiment

Bobo Jia(✉) and Zhibo Zhang

AECC Shenyang Engine Research Institute, Shenyang 100015, China
nianming126@163.com

Abstract. The main function of the intake system of the compressor test rig is to ensure that the air flow into a compressor is uniform and consistent. Firstly, the dimensionless total pressure on the experimental measurement section from the numerical is almost agreed with that from the experimental, except that there is only 5% deviation at 10% channel height. Therefore, the numerical simulation method is correct and the result is reliable. Secondly, the characteristics of pressure field, velocity field and boundary layer thickness before the variable inlet guide vanes (VIGV) and the first stage rotor (R1) are studied. The results show that the closed area formed by the intake stabilizing chamber and the inlet bell mouth of the test article has not affected the flow around the bell-mouth, especially the evolution and development of boundary layer. Two vortices directly behind the support plate wake are observed at the hub, which is caused by the flow channel falls. These two vortices have opposite rotational directions and axial vorticity direction. Throttle does not affect the dimensionless velocity and dimensionless pressure distribution at the inlet of the test article, nor does it affect the thickness of the boundary layer. With the increase of rotational speed and mass flow, the radial distribution becomes increasingly uniform. Meanwhile, the thickness of boundary layer at the hub and shroud decreases slightly, but the range of variation is within 1% of the channel height.

Keywords: Compressor experiment · Intake system · Thickness of boundary layer · Throttle ratio · Wake of support plate

1 Introduction

By adjusting the geometric parameters, the variable cycle engine (VCE) can change its aerodynamic thermodynamic cycle, and meet different requirements of the plane under different flight conditions. Therefore, VCE can provide relatively high thrust and low fuel consumption in overall flight envelope, and possesses the advantageous properties of both the turbofan engine and the turbojet engine. As a consequence, it has gradually become main research interests in the field of aero-engine. Core engine drive fan (CDFS), high-pressure compressor and the forward variable area bypass injector between them are the key geometric adjustable components for the VEC to realize the variable cycle. Mastering the aerodynamic characteristics of the variable cycle compressor components

W. Fan et al. (Eds.): AsiaSim 2022, CCIS 1712, pp. 449–462, 2022.
https://doi.org/10.1007/978-981-19-9198-1_34

can provide data support for the overall characteristic adjustment of VCE. Therefore, it is necessary to carry out the aerodynamic performance test for the variable cycle compressor on the compressor component test rig.

In the aerodynamic design of the compressor, uniform and stable ideal gas is used as the inlet working medium, and there are strict requirements for the inlet gas flow boundary layer thickness, distortion and turbulence. In order to ensure the consistency with the designed air inlet conditions, the compressor test article usually does not directly absorb air from the surrounding atmosphere, but indirectly absorbs air through the air inlet system of the test equipment. After the compressor test article and the compressor test equipment are installed and connected, the test article and the test equipment together constitute a complex dynamic operation system, and there is a strong interaction between them [1]. Due to the three-dimensional unsteady inverse pressure flow characteristics of axial-flow compressor, the compressor is more sensitive to the changes of inlet and outlet boundary conditions, especially the variable cycle compressor with high aerodynamic load and compact structure layout. The valve position opening and flow capacity of the inlet / outlet throttle device of the test equipment will directly affect the flow range of the compressor; The evolution and development process of the boundary layer in the inlet flow field of the compressor will directly affect the inlet flow angle and size relative to the compressor rotor tip, and then affect its work capacity. Therefore, this paper will focus on the research of the flow field quality of the inlet system.

Some studies [2] on the intake system mainly focus on the influence of the rectifying device in the surge tank on the intake air quality. However, these studies focus on the macro impact of the inlet flow field quality on the compressor aerodynamic performance, and less on the flow details of the inlet flow field. This paper studies influence of the relatively closed area near the bell mouth of the test article on the mainstream flow; This paper studies the formation and flow mechanism of two vortices with opposite directions induced by the support plate at the hub; This paper studies the characteristics of pressure field, velocity field and boundary layer thick-ness before VIGV and R1; This paper studies the effects of inlet throttle ratio and rotational speed on the pressure, velocity and boundary layer thickness before R1; These factors directly affect the quality of the flow field at the inlet of the test article; These factors directly determine the measurement layout design at the inlet of the test article. In addition, this paper pays special attention to the flow mechanism revelation of the inlet boundary layer development. Therefore, the research results of this paper will play an important role in improving the design ability of intake system and experimental scheme.

2 Experimental Requirements

The variable cycle compressor test platform is shown in Fig. 1. The test platform is mainly composed of high-precision variable frequency power control system, gear system for multiplying physical rotational speed of the power control system, exhaust volute, exhaust throttle, variable cycle compressor test articles, gas collecting chamber and its connected suction system for front bypass exhaust of compressor, air intake system, etc. The air intake system mainly includes bell mouth, flow pipe, transition section, diffuser section, air inlet throttle device, air inlet pressure intake stabilizing chamber, etc. The

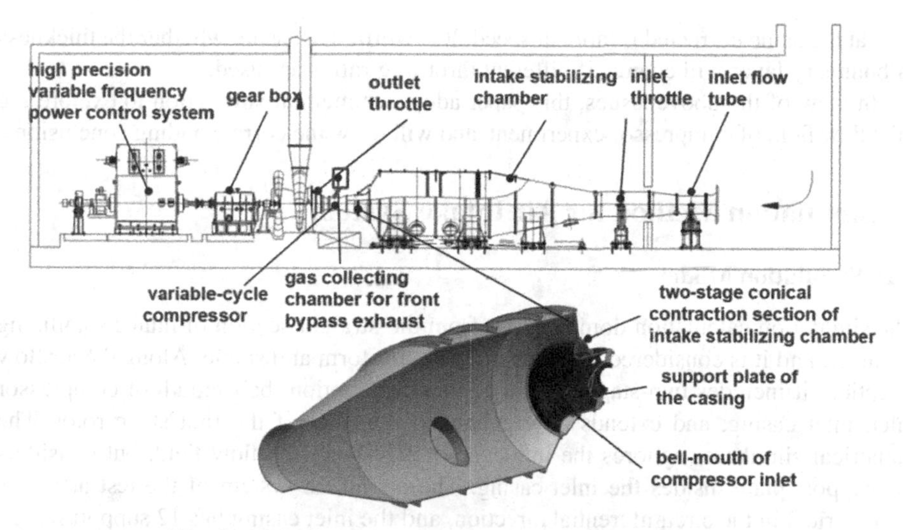

Fig. 1. Variable-cycle compressor experimental platform.

main function of the intake system is to ensure that the inlet flow field of the test article is uniform and the intake throttling during the compressor aerodynamic performance test.

The intake stabilizing chamber is the key component of the air inlet system, and its main function is to ensure that the air flow into the test article is uniform and stable. The intake stabilizing chamber mainly includes diffusion section, straight section and contraction section. There is perforated plate, rectifier net, honeycomb device, etc. inside the box to reduce the turbulence of air flow. The intake stabilizing chamber adopts two-stage conical contraction section to connect with the bell mouth of the test article and air inlet casing through sealing rubber ring and pressure ring, without flow leakage.

The inlet bell mouth of the compressor needs to be inserted into the conical contraction section of the intake stabilizing chamber. Due to the size limitation of the conical contraction, the inlet bell mouth is small relative to the standard. The influence of the relatively closed area between the bell mouth, the contraction section of the intake stabilizing chamber, and the inlet casing of the compressor on the flow field of the inlet bell mouth is worth studying. The influence of the wake after the support plate of the air inlet casing on the flow field at the hub and shroud of the inlet passage is worth studying.

The measuring position of the boundary layer at the inlet of the test article is behind the support plate and in front of VIGV. Due to size limitations, the boundary layer probe cannot be placed in the front section of R1. The boundary layer before R1 plays key role in the work capacity of the rotor. Whether the thickness of the boundary layer in front of R1 is the same as that in front of VIGV is worth discussing.

During the compressor component test, considering the power and torque of the test article at different rotational speeds, as well as the power and torque range of key components of the test rig, such as motor and torque measuring instrument, it is usually necessary to throttle the inlet of the compressor, and reduce the inlet pressure and mass flow at the same corrected rotational speed. It is worth discussing whether the thickness of boundary layer will change if different throttling ratios are used.

In view of the above issues, this paper adopts numerical simulation to explore the inlet flow field of compressor experiment, and will draw the corresponding conclusions.

3 Simulation Method for Air Intake System

3.1 Simulation Model

The simulation calculation domain starts from the straight section of intake stabilizing chamber, and it is considered that the air flow is uniform and stable. Along the air flow direction, it includes two-stage conical contraction section, bell mouth of compressor inlet, inlet casing, and extends downstream to the inlet of the first stage rotor. The numerical simulation ignores the influence of VIGV on the flow field, but considers the support plate insides the inlet casing. The air intake system of the test article is symmetrical in the circumferential direction, and the inlet casing has 12 support plates, so the calculation domain of the numerical simulation is set as a single channel, and the circumferential angle is 30 degrees.

3.2 Mesh Generation

The structured hexahedral mesh of calculation domain is shown in Fig. 2. From the bell mouth of test article to the front of the first stage rotor, the local grid at the wall is refined, and total number of grid nodes is 691080. In order to improve the mesh quality of the bending section of the bell mouth, the O-grid scheme is adopted twice in the domain before bell mouth and once after bell mouth. The Y-grid scheme around axis and O-grid scheme around support plate are adopted. Taking the determinant and angle as two evaluation criteria, the aspect ratio (determinant) of grid is more than 0.59 and the angle is more than 25°. In terms of a curved geometric structure at the inlet bell mouth of the test article, the grid quality is excellent.

In order to obtain the boundary layer thickness in front of the first stage rotor, the dimensionless thickness of the first layer of grid Yplus at the wall is particularly important [3, 4]. Yplus is defined as follows:

$$y^+ = \frac{\sqrt{\tau_\omega / \rho}\,\Delta n}{\nu} \tag{1}$$

where Δn is the distance between the first and second mesh points off the wall, τ_ω is wall-shear-stress, ρ is density, and ν is kinematic viscosity coefficient. Turbulent boundary layer is divided into inner and outer region. Inertial force is dominant in the outer region, and the upper limit depends on Reynolds number. The inner is divided into three layers [5, 6]: laminar sub layer (Yplus < 5), buffer region ($5 <$ Yplus $+ < 30$), and Log-law region ($30 <$ Yplus). In the laminar sub layer, viscosity plays a leading role and the velocity parallel to the wall is linear with the distance from the wall. The maximum Yplus value is 2.44 at the hub and 2.10 at the shroud, both less than 5, as shown in Fig. 3, and therefore the first layer of grid captures the laminar sub layer.

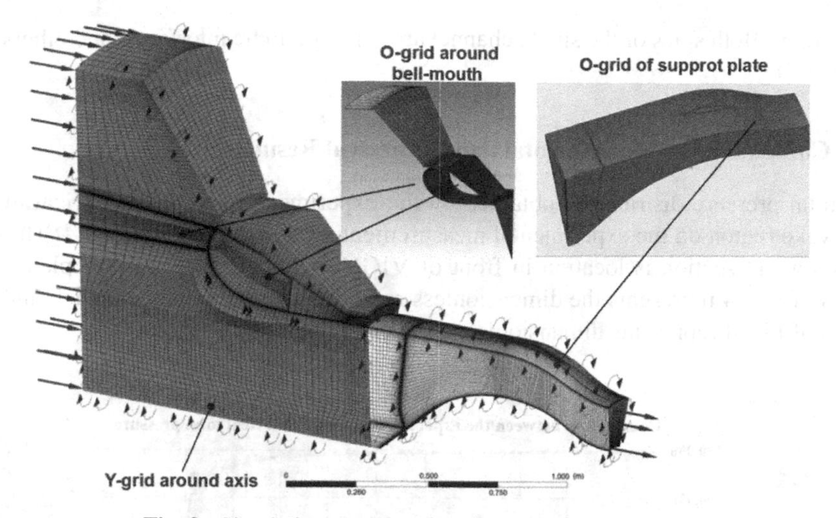

Fig. 2. Simulation model and mesh for air intake system.

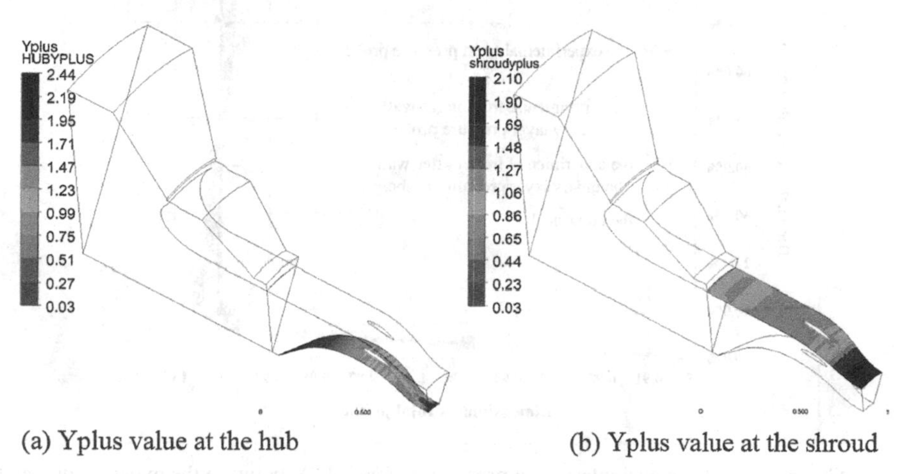

(a) Yplus value at the hub (b) Yplus value at the shroud

Fig. 3. Yplus value at the hub and shroud.

3.3 Turbulence Model and Boundary Condition Setting

ANSYS CFX 19.2 is used for simulation calculation and result processing. In order to better simulate the possible flow separation, the turbulence model of shear stress transport (SST) is adopted. This model uses the robustness of the model near the wall to capture the motion of the viscous bottom layer, and uses the model in the mainstream region to avoid the disadvantage that the model is too sensitive to the inlet turbulence parameters.

The reference pressure in the calculation domain is set to 1 atmosphere, so the pressure of the calculation result represents the gauge pressure. The inlet is set to the total pressure of 0 Pa and the total temperature of 288.15 K; The outlet is set as mass flow outlet. Both sides of the single channel are set as periodic interfaces, and others are set as wall.

3.4 Comparison of Experimental and Numerical Results

The total pressure distribution obtained by the experimental and the numerical in the non-wake region on the experimental measurement section is shown in Fig. 4. The test measurement section is location in front of VIGV and behind the support plate. The y-axis of Fig. 4 represents the dimensionless radial height in the inlet channel, and the x-axis of Fig. 4 represents dimensionless total pressure.

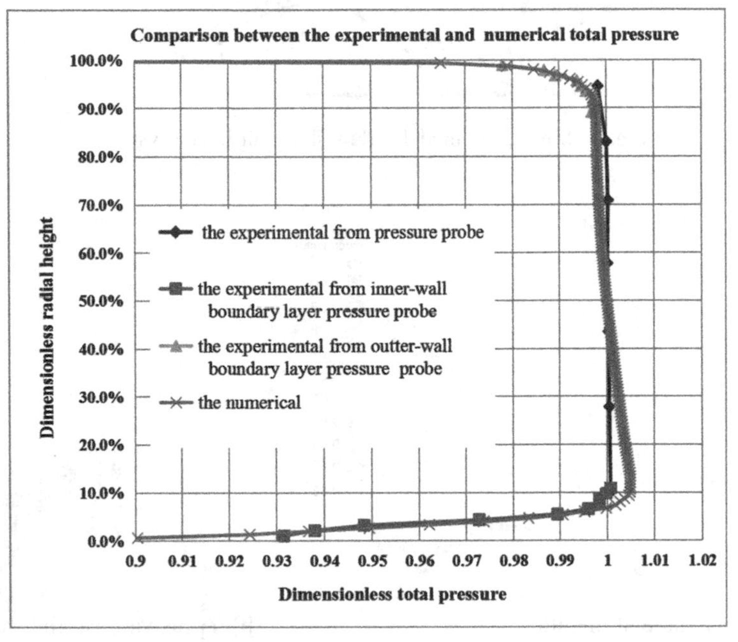

Fig. 4. Comparison of dimensionless total pressure before VIGV between the experimental and the numerical.

In order to quantitatively analyze the influence of the wake behind the support plate on the sections in front of VIGV and R1, Fig. 9 shows the dimensionless pressure in the wake region and main flow region. The y-axis of Fig. 9 represents the dimensionless radial height in the inlet channel, and the x-axis of Fig. 9 represents dimensionless total pressure. The dimensionless total pressure represents the total pressure at a certain position in the radial direction divided by the total pressure in the main flow region at 50% of the channel height. The experimental total pressure at 1%–10% and 89%–99% of channel height is measured by the boundary layer probe, and the value in the main flow region is measured by the total pressure probe. The value calculated by numerical simulation almost coincides with the experimental, and there is only 0.5% discrepancy at 10% channel height, which is completely within the acceptable range. Therefore, it can be concluded that the simulation method for air intake system is correct and the numerical results are reliable.

4 Simulation Results and Analysis

4.1 Flow Field Characteristics Around Bell-Mouth

The flow field characteristics near the bell-mouth are shown in Fig. 5. The velocity increases gradually from the straight section of intake stabilizing chamber to the front of the first stage rotor (R1). In front of R1, the speed at the shroud is greater than that at the hub. The velocity vector at the bell-mouth indicates that the flow completely conforms to the bell-mouth direction, and there is no flow separation.

The boundary layer begins to form near the bell-mouth and is very thin here. It becomes thicker and thicker with the flow into the straight section. After this straight section and when the flow goes out of the chamber, the flow area gradually decreases, the velocity gradually increases and the pressure gradually decreases. Because the pressure distribution in the boundary layer along the flow direction is the same as that in the mainstream, the boundary layer becomes thin again with favorable pressure gradient. On the other hand, the gradual increase of the mainstream velocity here strengthens its energy supply capacity to the fluid in the boundary layer and the velocity gradient in the boundary layer increases and it will gradually become thinner.

In the relatively closed area formed by the conical contraction of the intake stabilizing chamber, the bell-mouth of the test article, and the inlet casing, the pressure is slightly higher compared with that outside this closed area around the bell mouth. Therefore,

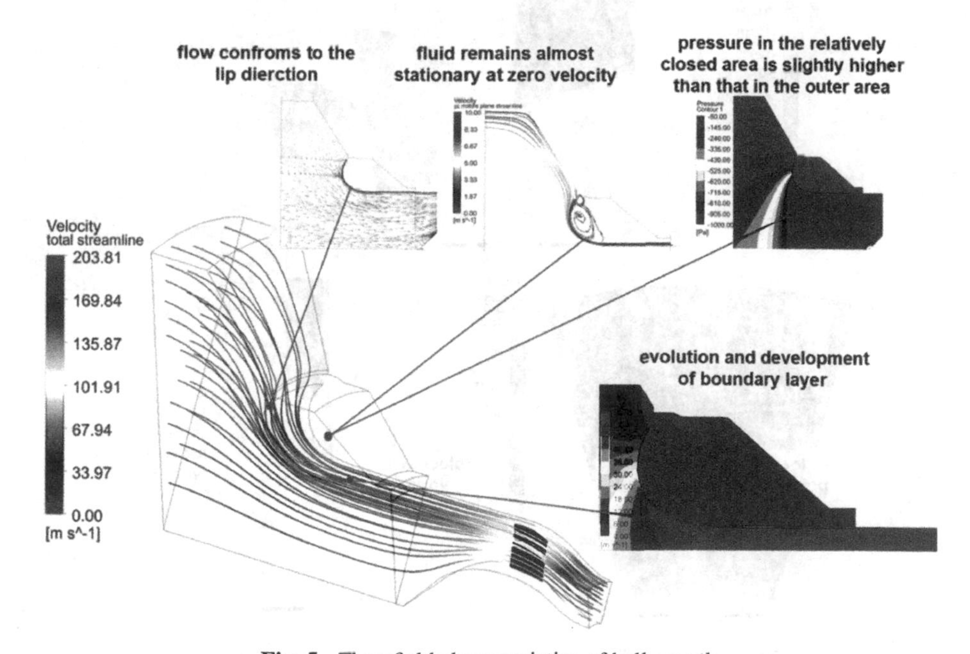

Fig. 5. Flow field characteristics of bell-mouth.

the air flow in this closed area tends to flow outward through the small gap. The flow outside this small gap is accelerating and its velocity is high compared with that inside the gap. Therefore, near the gap inside the closed area, a small counterclockwise vortex occurs, which is observed from the view that the flow moves from left to right. Under this small anticlockwise vortex near the gap, A relatively large anticlockwise vortex is formed. This vortex is also induced by the downward movement of air flow around the gap. However, the flow velocity in this closed area induced by these two vortices is low, no more than 0.4 m/s. Compared with the flow outside the gap, the flow velocity in this closed area is negligible, which has not affected the inlet of the bell-mouth, especially the evolution and development of boundary layer.

4.2 Flow Field Characteristics Around Support Plate

The measurement section before VIGV is the total pressure and boundary layer thickness measurement section at the compressor inlet. The comparison of its flow characteristics with that before R1 is the focus of the following analysis. Figure 6 shows the comparison of total pressure between the measured section before VIGV and the section before R1. The wake directly behind the support plate has dissipated most in front of VIGV and basically completely before R1. However, on these two sections, there are

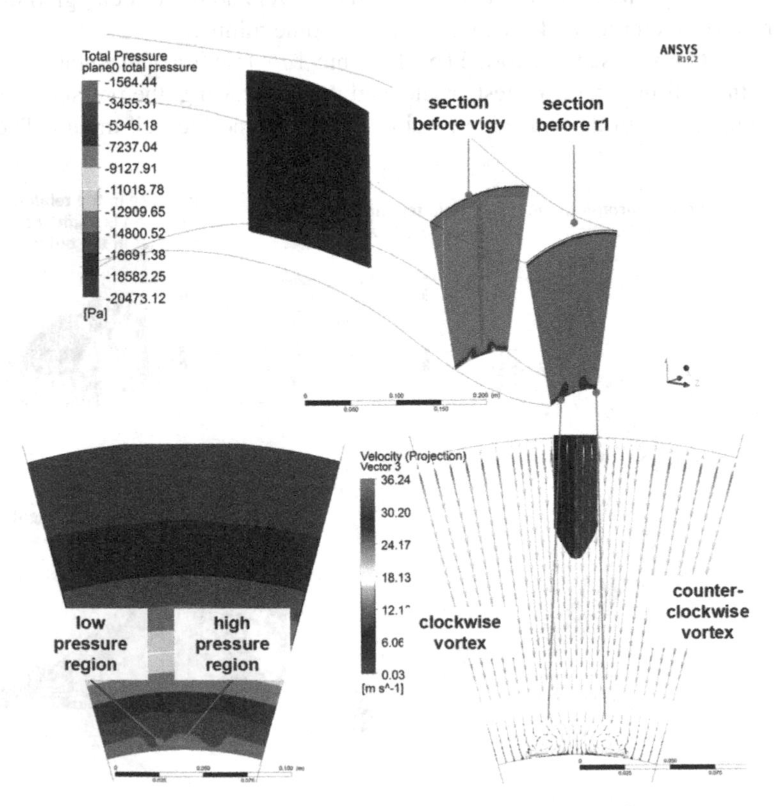

Fig. 6. Comparison of total pressure before VIGV and R1.

two low-pressure vortex regions near the hub directly behind the support plate wake. Looking against the airflow direction, the left vortex is clockwise, and the right vortex is counterclockwise. This section focuses on the formation process and flow mechanism of these two vortices.

Figure 7 shows the vorticity contour in the x-, y- and z-axis direction on different sections perpendicular to the support plate from the front of it to the back of it, which is used to represent the development of the fluid boundary layer near the wall. The z-axis direction is the flow direction. The definition of vorticity (curl of velocity) is shown in formula 2.

$$\begin{aligned} \text{Curl X} &= \frac{\partial u_z}{\partial y} - \frac{\partial u_y}{\partial z} \\ \text{Curl Y} &= \frac{\partial u_x}{\partial z} - \frac{\partial u_z}{\partial x} \\ \text{Curl Z} &= \frac{\partial u_y}{\partial x} - \frac{\partial u_x}{\partial y} \end{aligned} \tag{2}$$

In Fig. 7 (a), the vorticity in the x-axis direction at different sections is positive at the hub and negative at the shroud. Because the velocity in the mainstream area is greater than that in the boundary layer, the velocity gradient $\frac{\partial u_z}{\partial y}$ is just opposite at the hub and shroud, and the velocity gradient $\frac{\partial u_y}{\partial z}$ can be ignored. After passing through the support plate, there is already vorticity in the x-axis direction near both sides of the wake at the hub, which is personally believed to be caused by the three-dimensional effect of the vortex and vorticity in the z-axis direction.

In Fig. 7 (b), the vorticity in the y-axis direction is mainly caused by the wake behind the blade. The velocity is the wake is less than that in the mainstream region. The velocity gradient $\frac{\partial u_z}{\partial x}$ on both sides of the wake along the x-axis direction is opposite, and the velocity gradient $\frac{\partial u_x}{\partial z}$ can be ignored. Therefore, the vorticity direction on the left and right sides of the wake is opposite. The vorticity direction in the y-axis direction before the support plate is almost 0, while after passing through the support plate, the vorticity in the y-axis direction is the strongest. With the gradual dissipation of the wake, the vorticity in the y-axis direction gradually weakens.

In Fig. 7(c), the vorticity in the z-axis direction does not appear in front of the support plate, nor is it obvious at the trailing edge of the support plate, but it is obvious in the measurement section and the front section of the first stage rotor. However, as the flow passage decreases in the radial direction, the fluid will have a velocity component from the shroud to the hub, and it is along the negative direction of the y-axis, which is shown in Fig. 6. This velocity component rolls up two vortices in opposite directions on both sides of the support plate wake at the hub. Observed in the direction of anti-z axis, the vortex on the left is clockwise (vorticity value is negative), and the vortex on the right is counterclockwise (vorticity value is positive). From the pressure contour in Fig. 6, it can be seen that the pressure at the hub directly behind the support plate is higher than that at both sides, so the flow flows from high pressure to low pressure. Because the vortex causes energy dissipation, it is observed in Fig. 6 that these two vortices are low-pressure regions. These two low-pressure regions are observed at the hub, not

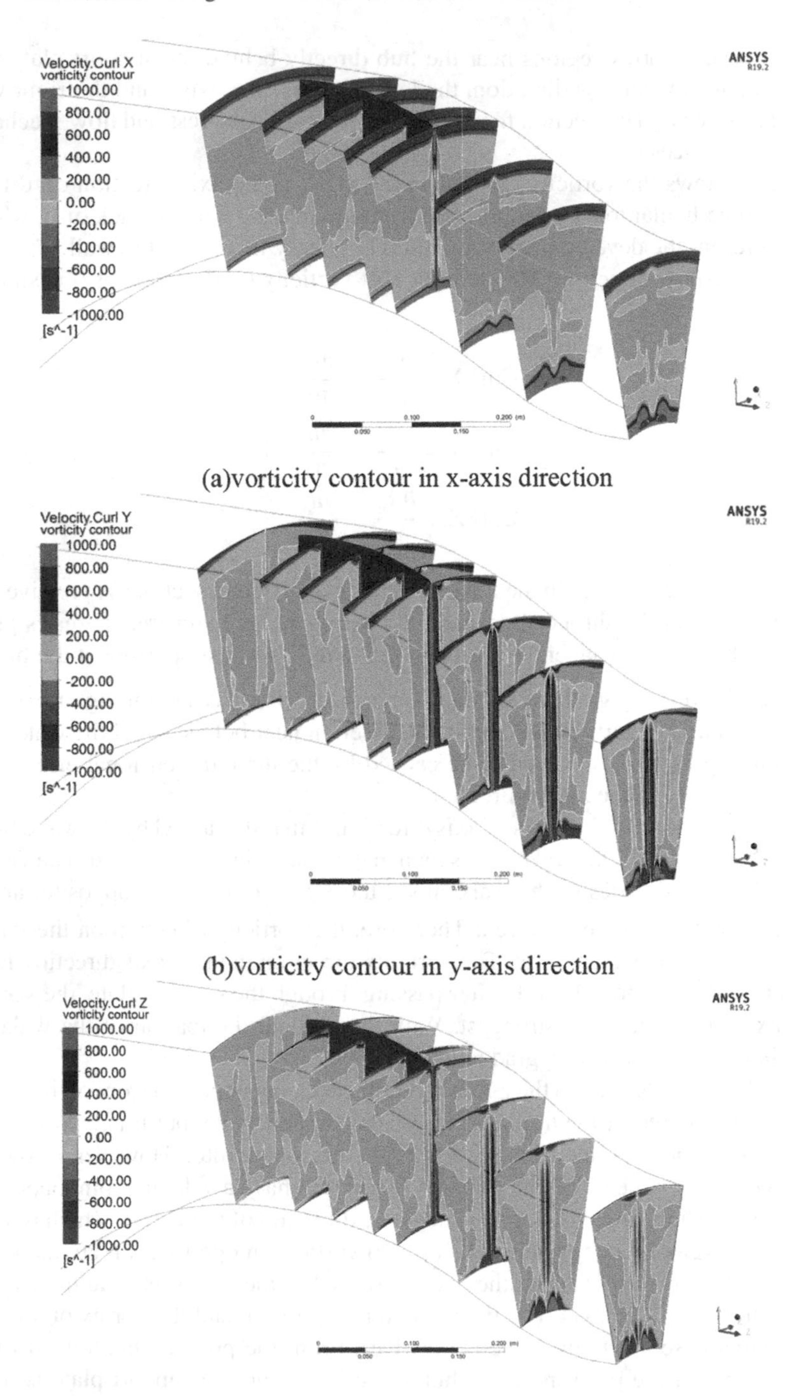

(a)vorticity contour in x-axis direction

(b)vorticity contour in y-axis direction

(c)vorticity contour in z-axis direction

Fig. 7. Vorticity contour around support plate

at the shroud, because the flow channel falls, and the hub is expanding the shroud is contracting.

In order to quantitatively analyze the influence of the wake behind the support plate on the sections in front of VIGV and R1, Fig. 8 shows the dimensionless pressure in the wake region and main flow region. The y-axis of Fig. 8 represents the dimensionless radial height in the inlet channel, and the x-axis of Fig. 8 represents dimensionless total pressure. The dimensionless method is the same as that shown in Fig. 4. The dimensionless total pressure in the wake region is 0.98 in front of VIGV and 0.99 in front of R1. The value of 0.98 and 0.99 indicates that the support plate wake is almost dissipated at the inlet of the compressor and has little effect on the flow field.

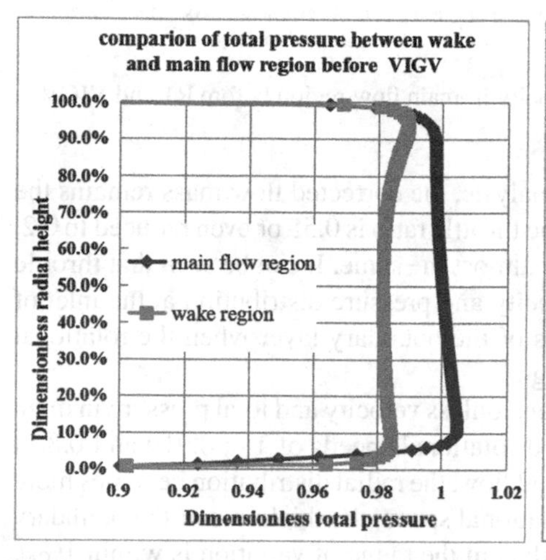

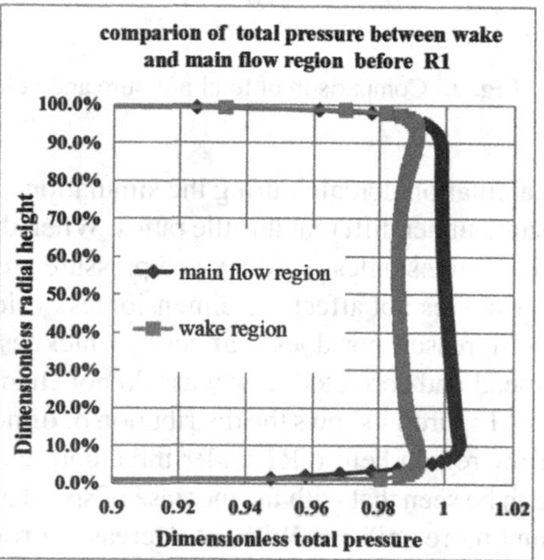

Fig. 8. Comparison of total pressure in wake region and mainstream region before R1 and VIGV

Figure 9 shows the comparison of dimensionless velocity and pressure in the main flow region before VIGV and R1. Dimensionless velocity is the velocity at a certain position in the radial direction divided by the velocity in the main flow region at 50% of the channel height. The dimensionless velocity before R1 is 1.02 at the shroud and 0.94 at the hub, while the dimensionless velocity before VIGV is 1.09 at the shroud and 0.83 at the hub. The velocity distribution before R1 is more uniform than that before VIGV. The dimensionless total pressure distribution before R1 and VIGV is almost identical. The thickness of the boundary layer is about 6% at the shroud and 7% at the hub, one percent higher.

4.3 The Influence of Throttle Ratio and Rotational Speed on the Boundary Layer

Figure 10 shows the distribution of dimensionless velocity and total pressure in main flow region before R1 under the inlet throttle ratio of 1.0 (standard intake condition), 0.55, 0.40, 0.30 and 0.20. The inlet throttle ratio represents the ratio of total pressure before the compressor to the standard atmospheric pressure (101325 Pa). By adjusting the matching relationship between the inlet total pressure and the outlet flow mass in the

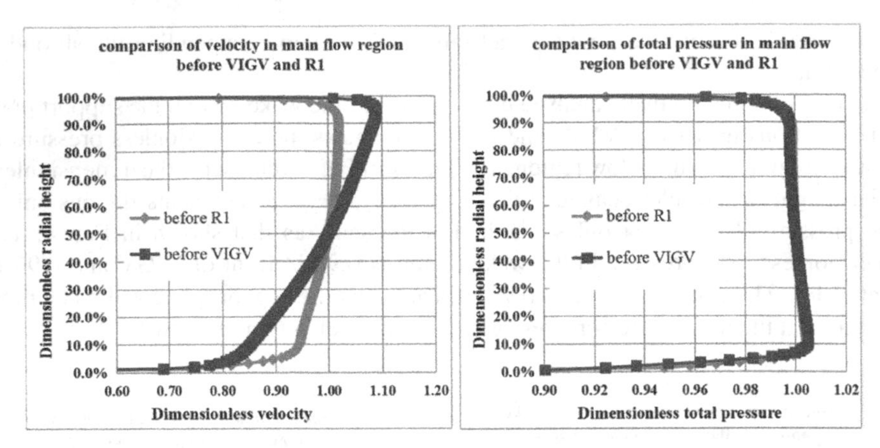

Fig. 9. Comparison of total pressure and velocity in main flow region before R1 and VIGV

calculation domain during the simulation analysis, the corrected flow mass remains the same under different throttle ratios. When the throttle ratio is 0.55 or even reduced to 0.2, the dimensionless velocity and pressure are almost the same. It can be seen that throttle ratio does not affect the dimensionless velocity and pressure distribution at the inlet of compressor, nor does it affect the thickness of the boundary layer when the rotational speed and corrected flow mass do not change.

Figure 11 shows the distribution of dimensionless velocity and total pressure in main flow region before R1 under three corrected rotational speeds of 1.0, 0.919 and 0.8. It can be seen that with the increase of speed and flow, the radial distribution becomes more and more uniform. With the decrease of rotational speed, the thickness of the boundary layer at the hub and shroud increases slightly, but the range of variation is within 1%.

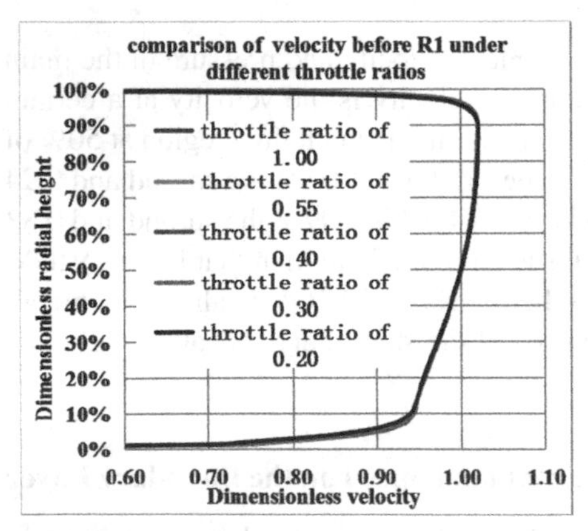

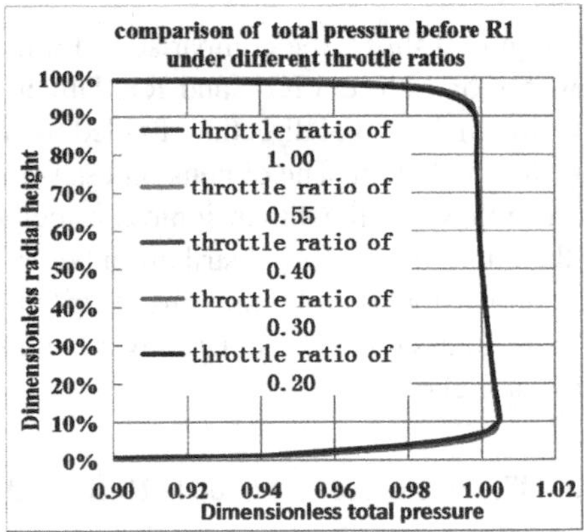

Fig. 10. Comparison of total pressure and velocity in main flow region before R1 under different throttle ratios

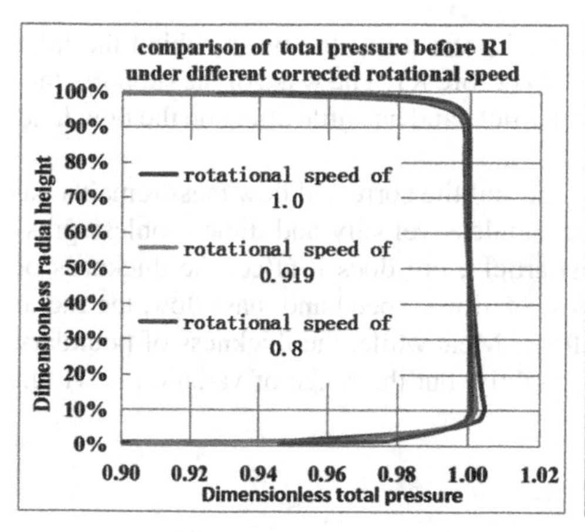

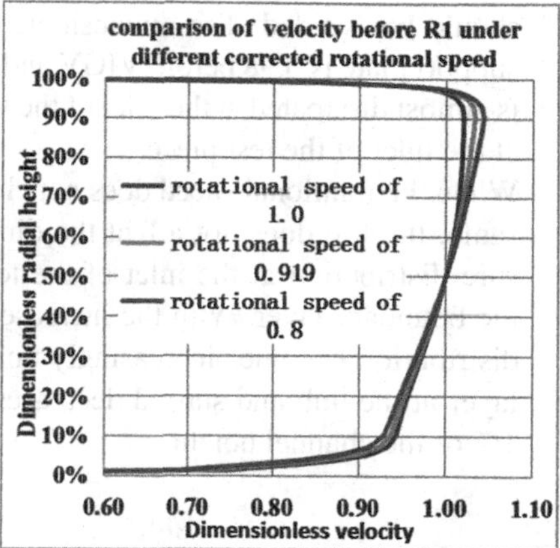

Fig. 11. Comparison of total pressure and velocity in main flow region before R1 under different rotational speeds.

5 Conclusion

In this paper, the flow field characteristics in the intake system of a variable cycle compressor test article are analyzed by numerical simulation. The simulation calculation domain starts from the straight section of the intake stabilizing chamber to the two-stage conical contraction section; From the inlet bell-mouth of the compressor test article, the inlet casing, and finally to the front of the first stage rotor. The influence of the relatively closed area near the bell mouth of the test article on the mainstream flow is studied; The formation and flow mechanism of two vortices with opposite directions induced by the support plate at the hub are studied; The characteristics of pressure field, velocity field and boundary layer thickness before VIGV and R1 are studied; The effects of inlet throttle ratio and rotational speed on the pressure, velocity and boundary layer thickness before R1 are studied. The conclusions are as follows:

1. In the mesh generation scheme, the first layer of mesh at the hub and shroud captures the viscous bottom layer of the boundary layer, and the mesh generation is reasonable. The dimensionless total pressure before VIGV calculated by numerical simulation almost coincides with the experimental, and there is only 0.5% discrepancy at 10% channel height, which is completely within the acceptable range. The numerical simulation method is correct and the result is reliable;
2. The airflow enters the bell mouth from the intake stabilizing chamber, and the flow completely conforms to the bell mouth direction. There is no flow separation. The closed area formed by the intake stabilizing chamber and the inlet bell mouth of the test article has not affected the inlet of the bell-mouth, especially the evolution and development of boundary layer. The air inlet of the test article is in good condition;
3. Two vortices directly behind the support plate wake are observed at the hub, which is caused by the flow channel falls. These two vortices have opposite rotational directions and axial vorticity direction. When measuring the thickness of the boundary layer in the experimental test, the region twice the thickness of the support plate

should be avoided. The dimensionless total pressure of the wake behind the inlet support plate is 0.98 before VIGV and 0.99 before R1. The wake of the support plate is almost dissipated at the inlet of the test article and has little effect on the flow field at the inlet of the test piece.

4. When the rotational speed does not change and the corrected flow mass remains the same, throttle does not affect the dimensionless velocity and dimensionless pressure distribution at the inlet of the test article, nor does it affect the thickness of the boundary layer. With the increase of rotational speed and mass flow, the radial distribution becomes increasingly uniform. Meanwhile, the thickness of boundary layer at the hub and shroud decreases slightly, but the range of variation is within 1% of the channel height.

References

1. Xiang, H., Wu H, Gao J, Liu Z: GE Ning. Preliminary study of two aerodynamic coupling problems in axial compressor test system. Gas Turbine Experiment and Research 30(2), 22–31(2017)
2. Xiang H; Hou M; Ge N; Liuzhi G; Yang R: Characterization and mechanism analysis of several abnormal aerodynamic phenomena of axial flow compressor experiment. Journal of Aerospace Power 31(3), (2016.)
3. Ma, X., Bai, J.: Small unmanned aerial vehicle aerodynamic design of low reynolds number. Aeronautical Computing Technique **42**(6), 17–24 (2012)
4. Wang, W., Wang, Z., Zhang, H., Nie, C.: Numerical simulation and experiment of laminar separation bubble transition and corner separation of compressor stator cascade. Journal of Aerospace Power **32**(9), 2273–2282 (2017)
5. Chen, M.: Fundamentals of viscous fluid dynamics(M). Higher education press, Beijing (2004)
6. Jing S, Zhang M: Fluid Dynamics. Xi'an: Xi'an Jiaotong University Press, 2001

Real-Time Ski Jumping Trajectory Reconstruction and Motion Analysis Using the Integration of UWB and IMU

Xuan Li[1], Yanfei Shen[2], Yi Qu[2], Xie Wu[3], and Yu Liu[3](✉)

[1] Beijing Bingfeng Technology Co., LTD, Beijing 100084, China
[2] Beijing Sport University, Beijing 100084, China
[3] Shanghai University of Sport, Shanghai 200438, China
yuliu@sus.edu.cn

Abstract. To satisfy an increasing demand to reconstruct an athlete's motion for performance analysis, this paper proposes a new method for reconstructing the position, acceleration, velocity and angle of skis in the context of ski jumping trajectories. Therefore, a real-time Measurement System was used. The system consisted of wearable devices attached to the athletes and fixed ultra-wide band (UWB) antennas next to the jumping hill. To determine the accuracy and precision of the method, six athletes of the China A or B National Team performed 25 measured ski jumps. The method was used to measure the trajectory, velocity, acceleration and skis angles during the jump. The measurements are compared with camera measurements of a Markerless Human Movement Automatic Capture System to assess their accuracy. The test results demonstrate that the method has sufficient accuracy and reliability for ski jumping. Thus, the system can be used as a tracking system during training and competitions for coaches and sports scientists.

Keywords: Ski jumping · Tracking · Trajectory · Ultra-wideband (UWB) · Inertial measurement unit (IMU) · Wearable sensors

1 Introduction

With the advancements in technology and miniaturization, systems with wearable devices are being increasingly used in recreational activities, physical rehabilitation, and sports. Synergetic use of sensor, communication, and computer technology, i.e., ubiquitous computing, can now be successfully applied in sports without obstructing or distracting the athletes during the performed actions [1]. Ubiquitous computing produces many new types of systems in sport, which are used by athletes, coaches, and staff.

Accurate sports localization and tracking with the possibility of real-time feedback has gained much interest during the last couple of years [2]. Localization and tracking systems can be applied during training and competition to assist the evaluation of the performance of the athletes. Nowadays, there are many commercially-available radio frequency (RF)-based solutions for sports localization and tracking [3].

W. Fan et al. (Eds.): AsiaSim 2022, CCIS 1712, pp. 463–478, 2022.
https://doi.org/10.1007/978-981-19-9198-1_35

Ski jumping consists of a complex sequence of two dimensional (i.e., in-run and take-off) and three-dimensional (i.e., early-flight, stable-flight, and landing) movements [4], and it is recommended analyzing this sequence in its entirety when studying the kinematics or evaluating performance [5]. Acquiring accurate position, velocity, and acceleration of the athlete during a ski jump has always been essential to athletes, coaches, and sports researchers to analyze the movement for improving the jumping performance. Consequently, there is a need for systems that can measure the entire jump sequence in 3D without time-consuming set-up or post-processing [6].

Traditional solutions to meet this demand are video analysis techniques based on camera recorded videos [7] and wearable sensors based on a global navigation satellite system (GNSS). The former solution has the disadvantage that it generally only covers limited parts of the jumping area [8]. To record the entire jump, a large number of systems merely provide information about the athlete's position, while velocity and acceleration have to be estimated via numeric differentiation, which may be subject to large errors. The second solution, i.e., wearable sensors, has the disadvantage that a global navigation satellite system (GNSS) provides low position accuracy and low sample rate. For instance, Global Positioning System (GPS) can only provide 3 m circular error probable (50%) and up to 10 Hz sample rate [9].

In recent years, the UWB (ultra-wideband) technology has attracted considerable attention in the field of indoor positioning, where GPS signal is not available [10]. In addition to indoor positioning, UWB offers a good alternative for a confined space outdoor positioning, where the frequency and accuracy of satellite systems are insufficient for the needs of application. The UWB technology can be used in many game sports. According to the previous study, some basic general positioning requirements for such sports are that the positioning error should not exceed 15–20 cm and the sampling rate should not be below 10–15 Hz [2].

Trajectory reconstruction techniques are a class of methods to achieve an accurate state estimation of a moving object by properly combining information from the kinematic model with measurement data from sensors, including but not limited to IMUs, magnetometers, and GNSS receivers [9]. Trajectory reconstruction techniques are also termed flight path reconstruction in the aerospace field. A systematic overview of such methods for an aircraft can be found in a previous study [11]. Similar techniques can be also applied to ski jumping applications.

This paper presents a new real-time method to reconstruct trajectory and track motion, including the position, velocity, acceleration and angle of skis, of a ski jumper by the UWB and IMU integration. Additionally, to fully utilize all available information to improve the estimation quality, the geometric shape of the ski jumping hill is modeled as a set of additional soft constraints and included in the algorithm framework. Furthermore, the measurements are compared with camera measurements of a Markerless Human Movement Automatic Capture System to assess the accuracy.

2 Related Work

UWB localization is implementable in many indoor and outdoor scenarios. Our interest lies in its use in sport applications. As already mentioned, some basic general localization requirements for localization in sports are: the positioning error should not exceed 15–20 cm and the sampling rate should not be lower than 10–15 Hz [2]. In the following paragraphs we list some of the related work in localization in sport.

The accuracy of UWB positioning depends on several technological factors as well as on the way the position is acquired and calculated. An RTLS (real-time localization system) algorithm can be based on different operation schemes [12]. Most of the UWB positioning systems implement time measurement in order to get parameters such ToF (Time of Flight) or TDOA (Time Difference of Arrival), which can be used for position calculation.

All ToF (Time of Flight) based systems work on the basis of determining the time it takes for a radio signal to propagate from a transmitter to a receiver. Once this time is known accurately then the distance between the transmitter and the receiver can be determined since the speed of propagation of radio waves in air is known. It is extremely difficult to make sure all elements of the system to be time synchronized. Besides, there are also various physical effects such as clock drift from which this simple scheme suffers and that can be corrected by a more advanced scheme known as Symmetric Double-Sided Two-Way Ranging (SDS-TWR). Three UWB messages were exchanged to calculate a range between the tag and anchor: poll, reply, final [3]. The message exchange is shown in Fig. 1. This way of ranging has the advantage of eliminating clock drift. The range can be calculated using: range = ToF * c, with c equal to the speed of light. The time of flight (ToF) was calculated with Formula (1), where the variables are related to the timestamps used in Fig. 1.

$$ToF = \frac{(t4 - t1)(t6 - t3) - (t3 - t2)(t5 - t4)}{(t4 - t1) + (t6 - t3) + (t3 - t2) + (t5 - t4)} \tag{1}$$

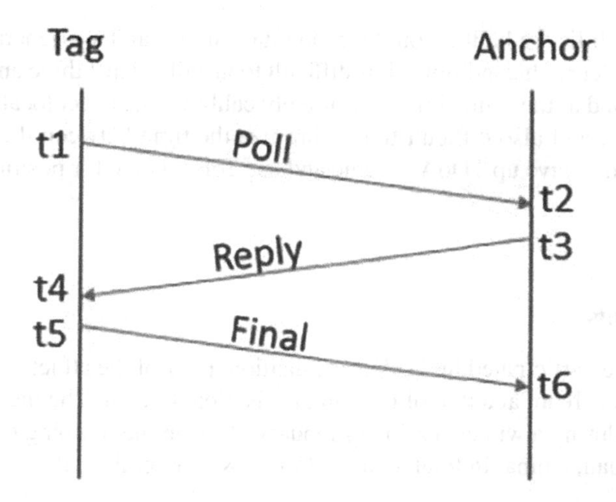

Fig. 1. SDS-TWR scheme.

Multi-lateration is an alternative scheme based on measuring the time difference of arrival (TDOA) of the signal. In this scheme three or more anchors are positioned in known locations around the area in which tagged items are to be located. Each of these anchors is time synchronized to the others. Since all anchors are time synchronized, the tag only transmits and does not receive, this is also known as One Way ranging. But is not possible with only one anchor since it relies on the difference in the arrival times at several anchors to calculate the location (Fig. 2).

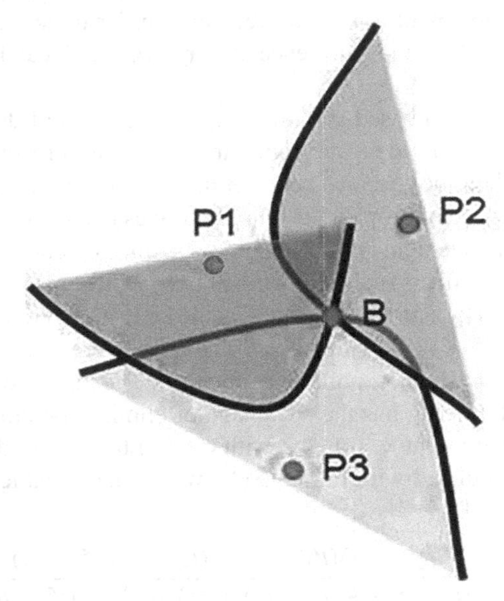

Fig. 2. TDOA scheme.

Ski jumping hills are built on outdoor mountain slopes and are generally more than 100 m wide. Under such conditions, it is difficult to install at least three anchors to cover the whole hill, and at the same time to accurately calibrate the exact location and height of these anchors, and also difficult to synchronize the time between these anchors. In this case, we had to give up TDoA scheme and use ToF scheme for position calculation.

3 Methods

3.1 Participants

Six athletes were participated in the data acquisition. Four of the athletes was part of the China A National Team and two of the China B National Team. The measurement data were collected during a winter session in January 2021 on the jumping hill (Hill Size = 108 m) in Laiyuan, China. In total, data of 25 jumps were collected.

3.2 Measurement Systems

The first main component of the tracking system comprises UWB antennas with fixed positions along the jumping hill. Figure 3 shows the positions of the UWB antennas. The

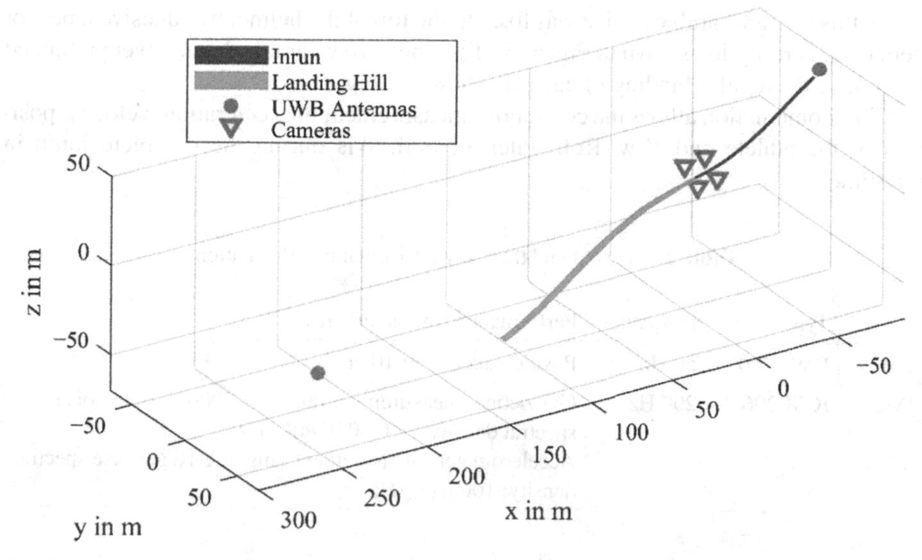

Fig. 3. Definition of the local coordinate system of the ski jumping hill. This includes the positions of the UWB antennas, the cameras, and the 3D models of the ski jumping in-run and landing hill.

chosen position allows minimal packet loss in wireless communication and practically 100% line-of-sight (LOS) conditions between the anchor and the tags.

The used UWB radio settings are shown in Table 1. The low data rate and high preamble length resulted in a long range. The highest possible pulse repetition frequency was chosen to obtain higher accuracy on the first path time stamp. The downside of this choice was that this required greater power consumption.

Table 1. UWB radio settings.

Parameter value	Value
Channel	2
Data rate	850 kbps
Pulse repetition frequency (PRF)	64 MHz
Preamble length	512 symbols

The other main component is the wearable devices that are attached to the athletes. The wearable device consists of the following sensors: a Decawave DW1000 and an InvenSense ICM-20648 IMU. Table 2 summarizes the performance characteristics of the sensors provided by the respective manufacturers. The UWB sensor DecaWave DW1000 chip provides position measurements with a 50 Hz sample rate and 10 cm accuracy [13]. The IMU measures triaxial rotational rates as well as specific forces (i.e., non-gravitational accelerations) at 200 Hz frequency [14].

In this test, a wearable device was fixed to the top of the helmet by adhesive tapes for better signal reception shown in Fig. 4. And another two wearable devices were mounted in front of the regular binding of each ski shown in Fig. 5.

This combination allows the continuous measurement of acceleration, velocity, position of the athlete and Yaw, Roll, Pitch of both skis during the complete jump in real-time.

Table 2. Sensors and their key performance information.

Sensor	Type	Frequency	Performance Characteristics
UWB	DW 1000	50 Hz	Position accuracy: 10 cm
IMU	ICM-20648	200 Hz	Gyroscope: measurement range: $\pm$ 2000°/s, rate noise spectral density: $\pm$ 0.004(°/s)/$\sqrt{}$Hz; Accelerometer: measurement range: $\pm$ 16 g, noise spectral density: 100 μg/$\sqrt{}$Hz

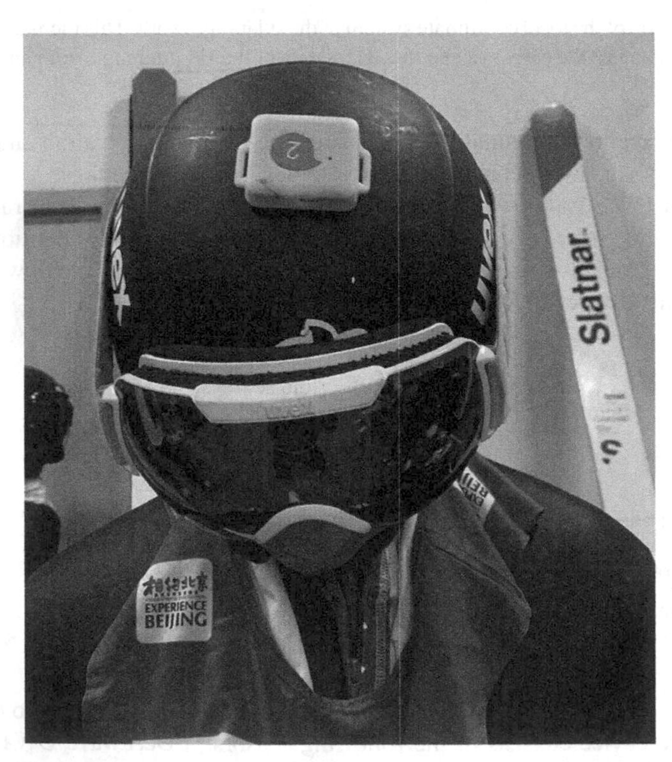

Fig. 4. Attachment of the wearable device on the top of the helmet.

Fig. 5. Attachment of the wearable device in front of the binding of the skis.

3.3 Evaluation System

In addition, in order to verify the accuracy of the system, we used a Markerless Human Movement Automatic Capture System to automatically track human motion and estimate posture [15–18]. The system was calibrated using Direct Linear Transformation(DLT) method to automatically obtain the three-dimensional coordinates of 21 joint points of the human body in motion. Four high-speed cameras (120 Hz, SONY FDR-AX700, SONY, Japan) were installed in the take-off area, conducted synchronous video shooting in the take-off area (Fig. 6), and quickly identified the athlete's posture through AI technology to form the calculation and output of three-dimensional spatial kinematic characteristic indicators. The output indicators integrated the research results of relevant literature, including the athletes takeoff speed, start stretching position, hip angle, knee angle, angle between trunk and horizontal axis, angle between shoulder and knee joint connection and horizontal axis, and hip and knee joint extension angular velocity, angle between trunk and horizontal axis, angle between shoulder and knee joint connection and horizontal axis in take-off stage [19, 20], and display three-dimensional feature map and its key characteristic values. The effectiveness of the system has been validated in a recent study [21].

3.4 Algorithm

The hill profile used in this study was the Laiyuan hill size (HS)-140 m Olympic ski jumping hill representing the modern large hill construction (Fig. 7). The hill consists of an inrun area (A ~ T) and a landing area (T ~ U). The inrun area is composed of a

Fig. 6. The installation position and coverage of the four cameras.

straight section(A ~ E1) with γ gradient, followed by a curve (E1 ~ E2) with the radius r_1 at the beginning of the takeoff and a straight takeoff table (E2 ~ T) with length t and gradient α.

The landing area from P to L is of a circular shape which is determined by the radius r_1. This radius starts at the P point with the tangent angle β_P. At the K point and at L the tangent angles are β and β_L.

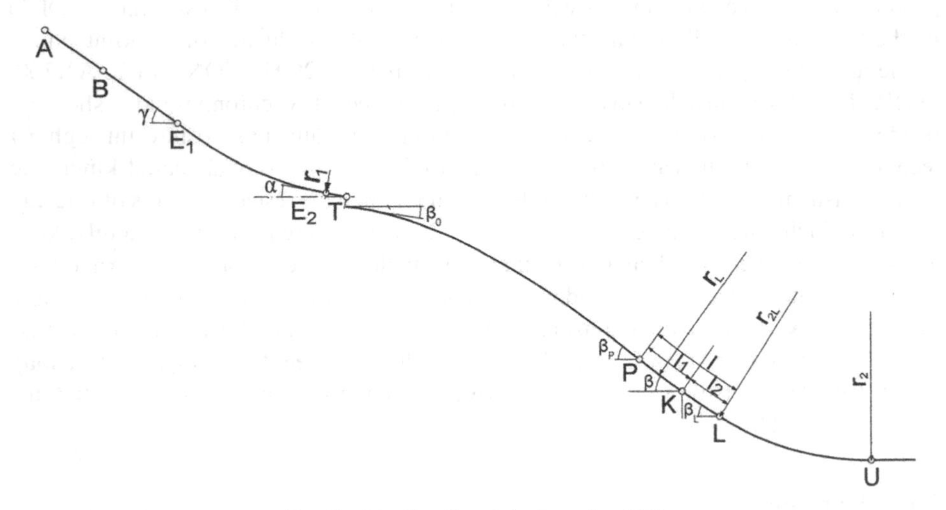

Fig. 7. The Profile of the Jumping Hill.

There are five sections of ski jumping:

1. inrun linear section (A ~ E1);
2. inrun curve section (E1 ~ E2);
3. linear take-off table section at the end of the inrun where the actual take-off takes place (E2 ~ T);
4. flight section;
5. landing section.

In the first section, each time position coordinates were calculated by the following manner:

$$\begin{cases} (x - x_0)^2 + (y - y_0)^2 = d^2 \\ y = \gamma x + b \end{cases} (A_x \le x \le E_{1x}, E_{1y} \le y \le A_y) \tag{2}$$

where (x_0, y_0) is the coordinate of the UWB anchor, d is the distance between the anchor and the tag which is calculated by the UWB SDS-TWR scheme, γ is the gradient of the straight section of inrun, b is the intercept.

In the second section, each time position coordinates were calculated by the following manner:

$$\begin{cases} (x - x_0)^2 + (y - y_0)^2 = d^2 \\ (x - x_1)^2 + (y - y_1)^2 = r_1^2 \end{cases} (E_{1x} \le x \le E_{2x}, E_{2y} \le y \le E_{1y}) \tag{3}$$

where (x_0, y_0) is the coordinate of the UWB anchor, d is the distance between the anchor and the tag which is calculated by the UWB SDS-TWR scheme, $(x_1, y_1)$is the coordinate of the center of the curve section of inrun with the radius r_1.

In the third section, each time position coordinates were calculated by the following manner:

$$\begin{cases} (x - x_0)^2 + (y - y_0)^2 = d^2 \\ y = \alpha x + e \end{cases}, (E_{2x} \le x \le T_x, T_y \le y \le E_{2y}) \tag{4}$$

where (x_0,y_0) is the coordinate of the UWB anchor, d is the distance between the anchor and the tag which is calculated by the UWB SDS-TWR scheme, α is the gradient of the take-off table section of inrun, e is the intercept.

In the fourth section, it's difficult to calculate the flight trajectory. But we can calculate the vacuum flight distance based on the velocity of take-off, and then compare with the jumping distance to analyze the effect of the flying postures.

In the fifth section, each time position coordinates were calculated by the following manner:

$$\begin{cases} (x - x_0)^2 + (y - y_0)^2 = d^2 \\ (x - x_2)^2 + (y - y_2)^2 = r_L^2 \end{cases}, (P_x \le x \le L_x, L_y \le y \le P_y) \tag{5}$$

where (x_0,y_0) is the coordinate of the UWB anchor, d is the distance between the anchor and the tag which is calculated by the UWB SDS-TWR scheme, (x_2,y_2) is the coordinate of the center of the landing curve from P to L with the radius r_L.

We can get the whole trajectory by combining the tracks of the above five sections. Figure 8 shows an example trajectory of one jump.

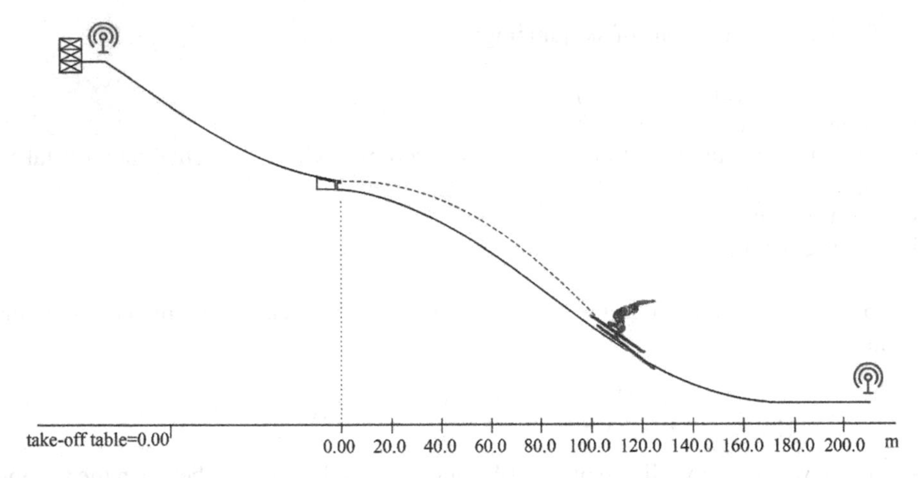

Fig. 8. Illustrative trajectory figure of one jump.

4 Results and Discussion

With the trajectory of the whole jumping course, we can further calculate the velocity curve of the whole course. Then, with the IMU's inertial data, we can get the Velocity & acceleration – distance curve of one jump. The key action points can be matched with the positions. Figure 9 shows an example trajectory of one jump.

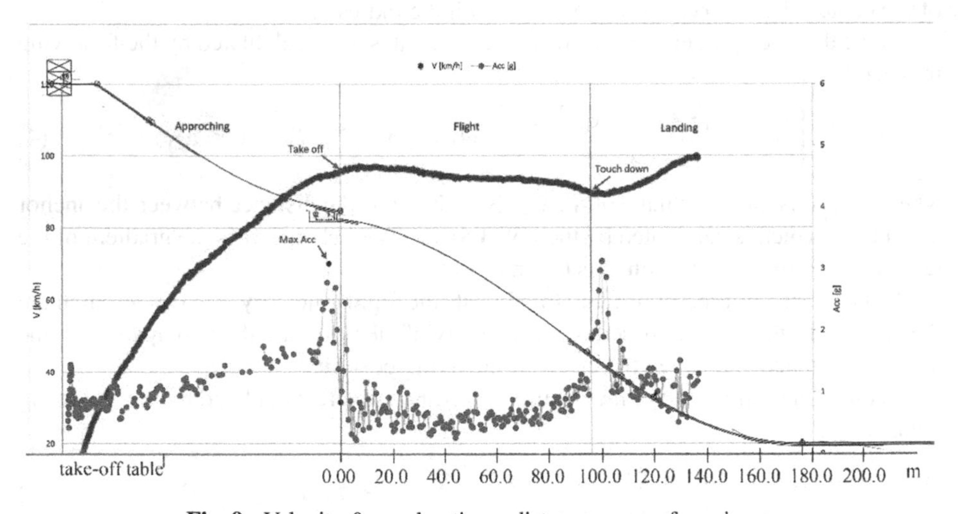

Fig. 9. Velocity & acceleration – distance curve of one jump.

4.1 Jumping Distance

The jumping distance of the recorded jumps was determined with the official video-based measurement system that has been used in Fédération Internationale de Ski (FIS) competitions for over 25 years. The system operator was certified by FIS for video distance measurements. To determine the jumping distance, the system operator determines the first camera frame where both skis are flat on the landing hill [22]. Through the camera calibration, the respective jumping distance is determined and rounded down to a resolution of 0.5 m.

From Fig. 7 we can get the jumping distance is 93.4 m. The official distance is 93.5 m. The bias is 0.1 m. We analyzed all the jumping distance of 35 ski jumps. The jumping distances range from 55 m to 106 m. The average bias of this system is 0.3 m.

4.2 Velocity and Acceleration

In Ski Jumping only the jumping velocity can be measured by the photocell beam located 10 m before the edge of the takeoff and 0.2 m above the snow profile [22]. The system operator was certified by FIS for jumping velocity measurements. But the velocity in other position cannot be measured. Furthermore, the acceleration of inrun, taking-off, flying, landing cannot be measured neither.

From the figure we can know the take-off velocity is 94.7 km/h, the touch-down velocity is 86.6 km/h, the starting position of extension is at 5.1 m before take-off point, the max acceleration in inrun section is 3.08 G at 3.1 m before take-off point, the Elapsed Time of extension is 0.2 s.

Then we compare the results to the camera measurements. We used four high-speed cameras to take synchronous video shots of the athletes in the take-off area, as shown in Fig. 10. At the same time, we use AI technology to quickly identify athlete posture and form three-dimensional space kinematics features, as shown in Fig. 11. From Fig. 11 we can get the angle of lower-limb knee extension. The lowest point means that athlete started the extension movement. And then we can get the starting position of extension is at 5.15 m before take-off point, the Elapsed Time of extension is 0.22 s. This result is consistent with our previous analysis.

The official jumping velocity of this jump is 89 km/h, which is lower than the speed measured by our method. This is because the official speed measurement method is through the photocell beam located 10 m before the edge of the takeoff. This means that the official take-off speed is calculated as the average speed of 10 m before the take-off point. The speed obtained by our method is the instantaneous speed of the athlete in 0.02 s at the take-off point. Therefore, it is reasonable that the take-off speed obtained by our method is slightly higher than the official speed. Moreover, the instantaneous speed obtained by our method can better reflect the subtle speed changes of athletes, which is more helpful to analyze the subtle impact of athletes' different take-off actions on speed.

Furthermore, from Fig. 7, we can also do the following calculations.

$F_r - m*g = m*a$.

From Fig. 7, we know $a_{max} = 3.08$ g; When starting the extension movement, a = 1.8 g.

So $\Delta a = 3.08\text{ g} - 1.8\text{g} = 1.28\text{ g}$;

Fig. 10. Perspective Scheme for Four Cameras to Capture Take-off Movements synchronously.

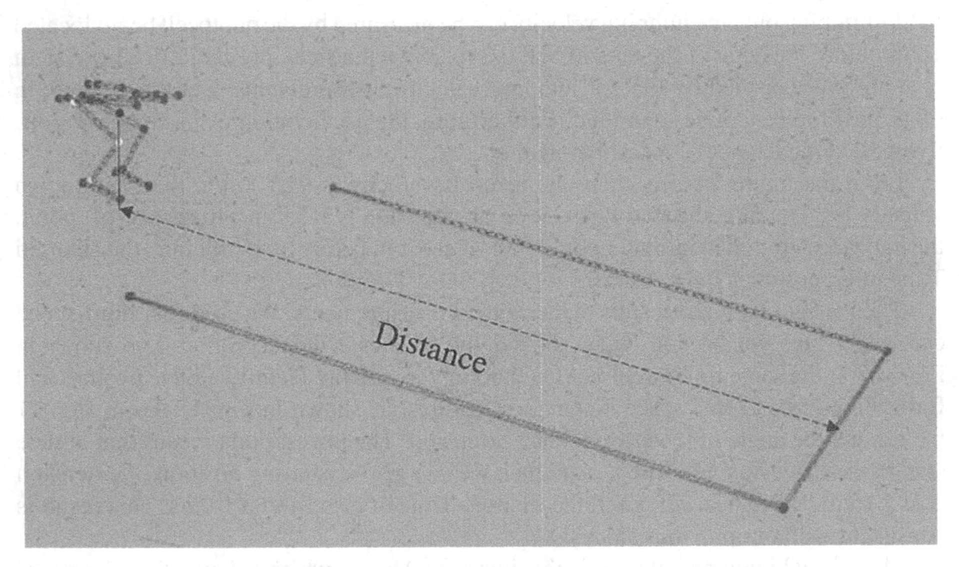

Fig. 11. 21 joint points of the human body in jumping area.

So F_{rmax} = m*g + m*(1.28 g) = m*g*(1 + 1.28) = 2.28*m*g = 2.28 BW;
This means that the maximum normal force of extension is 2.28 times Body Weight.

4.3 Angle Measurements of Skis

How to get the Angle of the two skis in three directions in the air is always a problem. Some schemes use the camera to get the Angle of the skis at certain positions. For example, the pitch Angle of the ski can be obtained by the camera on the landing slope side. Or use the pan–tilt–zoom (PTZ) camera mounted at the upper end of the in-run to get the projection of the V-opening-angle of the two skis which is determined on the PTZ camera image plane [23].

However, either method requires manual marking and has considerable measurement error. Additionally, camera-based tracking methods are used [19], which are unobtrusive but have the disadvantage of only covering a part of the jumping hill, or many cameras are needed and must be combined. Furthermore, the post-processing of video data has high computational costs and does not work during bad weather conditions.

In the cameras at the slope side of the landing hill, when the jumper is near the center of the camera picture, the jumper appears large, and thus the unbent part of the skis is identified more precisely. But when the jumper moved to the edge of the picture, the measurement accuracy decreased significantly. When the jumper moved out of the frame, it was impossible to measure.

The determination of the V-angle has a worse precision compared to the angle determined from the cameras next to the landing hill. One reason for this is that the V-angle is calculated using the orientation of both skis, which both introduce a measurement error. Apart from this, due to the missing calibration of the PTZ camera, no 3D vectors could be used to measure the V-angles. Instead, they were taken directly from the image. The angles measured by the PTZ are projected onto the image plane, but effects such as radial distortion through the camera lens are not corrected. This may introduce a bias in the comparison of both methods. Another reason for the bias and worse precision may be the less accurate labeling of the PTZ camera images. This is introduced through the bending of the skis in conjunction with the jumper appearing relatively small in the image.

Our solution, using IMU combined with UWB, can solve these problems well. The sampling rate of IMU is high enough and the angular precision of $-0.2 \pm 0.8°$[13]. Through the combination with UWB, the angular data and position data can be well combined, and the real-time data transmission ability can be achieved through UWB.

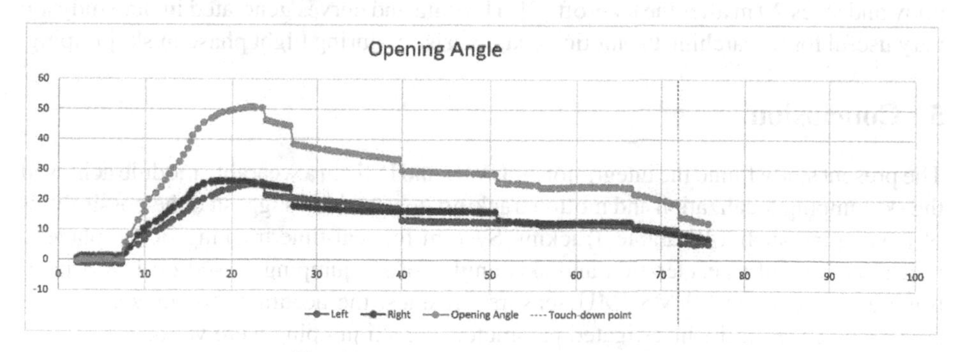

Fig. 12. Opening angle of the two skis.

Figures 12, 13, 14 show opening angle, roll angle, pitch angle of the two skis during the flight phase. The rate of change of these angles is fast in the early flight phase, but after the stable phase of flight, the rate of change of these angles slows, and there is a considerable mutation till landing. This is consistent with several studies about flight phase in ski jumping [20]. Early flight is considered as a crucial phase for length of the jump in ski jumping since it also reflects the take-off action. Mistakes during the take-off cannot be corrected during the flight phase, but the benefits of a successful take-off

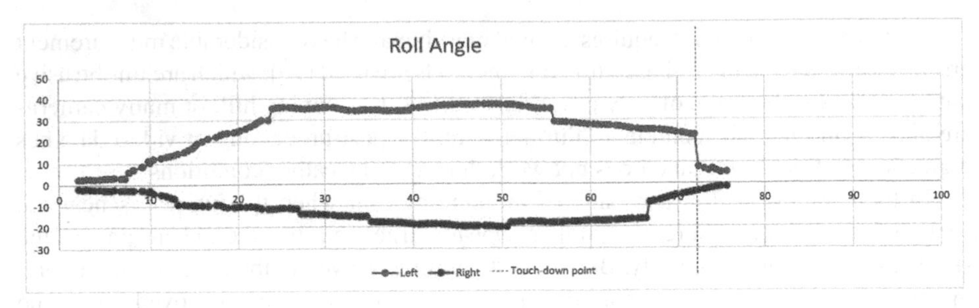

Fig. 13. Roll angle of the two skis.

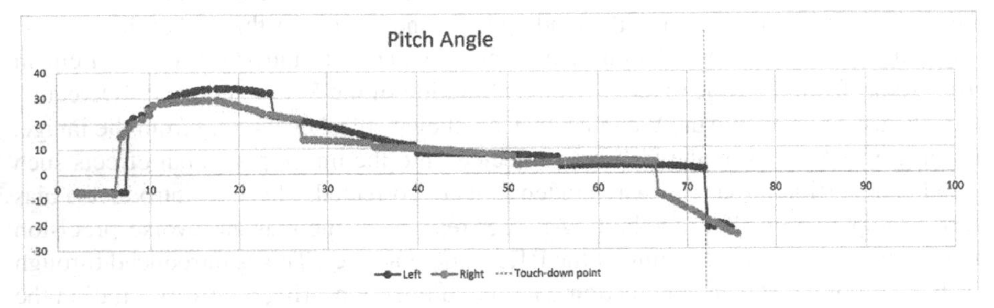

Fig. 14. Pitch angle of the two skis.

action can be destroyed by mistakes during the flight [7]. According to Schwameder and Muller [24] a long jumping distance could be obtained by the optimal combination of high vertical take-off velocity, big rotation impulse at take-off and small angle between body and skies 20 m after the take-off [7]. The data and curves generated in our study are very useful for researching the motion characteristics during flight phase of ski jumping.

5 Conclusion

The present study found the integration of UWB and IMU in a wearable module achieved the ski jumping localization and motion tracking goals, which suggested the possibilities of developing such a Wearable Tracking System for real-time tracking of the player's trajectory, velocity, acceleration and skis angles of ski jumping, based on UWB positioning chipsets and MEMS IMU sensors. Besides, the accuracy of this system and correspondence for the investigated parameters for ski jumping were valid.

6 Application

Looking at the possible applications of such a tracking system, it had great potential to support coaches and sports scientists to improve further the technique of the athletes to jump even further. Furthermore, for the live application of this system during TV broadcasting, it also had great potential for the live application of this system during TV broadcasting, since information such as live-speed or trajectory comparison might be fascinating for the spectators.

Acknowledgments. The work was supported by the National Key R&D Program of China (2019YFF0301803).

Data Availability Statement: The data is available from the authors on reasonable request.

Conflicts of Interest: The authors declare no conflict of interest.

References

1. Baca, A., Dabnichki, P., Heller, M., et al.: Ubiquitous computing in sports: A review and analysis. J. Sports Sci. **27**, 1335–1346 (2009)
2. Leser, R., Baca, A., Ogris, G.: Local Positioning Systems in (Game) Sports. Sensors **11**, 9778–9797 (2011)
3. Minne, K., Macoir, N., Rossey, J., et al.: Experimental evaluation of UWB indoor positioning for indoor track cycling. Sensors **19**, 2041 (2019)
4. Schwameder, H.: Biomechanics research in ski jumping, 1991–2006. Sports Biomechanics **7**, 114–136 (2008)
5. Denoth, J., Luethi, S.M., Gasser, H.H.: Methodological problems in optimization of the flight phase in Ski jumping. Int. J. Sport Biomech. **3**, 404–418 (1987)
6. Chardonnens, J., Favre, J., Cuendet, F., et al.: A system to measure the kinematics during the entire ski jump sequence using inertial sensors. J. Biomech. **46**, 56–62 (2013)
7. Virmavirta, M., Isolehto, J., Komi, P., et al.: Characteristics of the early flight phase in the Olympic ski jumping competition. J. Biomech. **38**, 2157–2163 (2005)
8. Groh, B.H., Warschun, F., Deininger, M., et al.: Automated ski velocity and jump length determination in ski jumping based on unobtrusive and wearable sensors. Proc. ACM Interact. Mob. Wearable Ubiquitous Technol. **1**, 1–17 (2017)
9. Fang, X., Grüter, B., Piprek, P., et al.: Ski Jumping Trajectory Reconstruction Using Wearable Sensors via Extended Rauch-Tung-Striebel Smoother with State Constraints. Sensors **20**, 1995 (2020)
10. Umek, A., Tomažič, S., Kos, A.: Application for impact position evaluation in tennis using UWB localization. Proc. Comput. Sci. **147**, 307–313 (2019)
11. Mulder, J.A., Chu, Q.P., Sridhar, J.K., et al.: Non-linear aircraft flight path reconstruction review and new advances. Prog. Aerosp. Sci. **35**, 673–726 (1999)
12. Umek, A., Kos, A.: Validation of UWB positioning systems for player tracking in tennis. Pers. Ubiquit. Comput. 1–11 (2020). https://doi.org/10.1007/s00779-020-01486-0
13. Decawave DW1000 Chip. https://www.decawave.com/product/dw1000-radio-ic/ (accessed on 20 April 2022)
14. InvenSense ICM-20648 6-Axis IMU Datasheet. https://invensense.tdk.com/download-pdf/icm-20648-datasheet/ (accessed on 30 March 2022)
15. Mathis, A., Mamidanna, P., Cury, K.M., et al.: DeepLabCut: markerless pose estimation of user-defined body parts with deep learning. Nat. Neurosci. **21**, 1281–1289 (2018)
16. Sun K, Xiao B, Liu D, et al. Deep High-Resolution Representation Learning for Human Pose Estimation. In: 2019 IEEE/CVF Conference on Computer Vision and Pattern Recognition (CVPR), pp. 5686–5696. IEEE, Long Beach, CA, USA (2019).
17. Nie X, Feng J, Zhang J, et al. Single-Stage Multi-Person Pose Machines. In: 2019 IEEE/CVF International Conference on Computer Vision (ICCV), pp. 6950–6959. IEEE, Seoul, Korea (South) (2019)
18. Wojke N, Bewley A, Paulus D. Simple online and realtime tracking with a deep association metric. In: 2017 IEEE International Conference on Image Processing (ICIP), pp. 3645–3649. IEEE, Beijing (2017)

19. Virmavirta, M., Isolehto, J., Komi, P., et al.: Take-off analysis of the Olympic ski jumping competition (HS-106m). J. Biomech. **42**, 1095–1101 (2009)
20. Arndt, A., Brüggemann, G.-P., Virmavirta, M., et al.: Techniques Used by Olympic Ski Jumpers in the Transition from Takeoff to Early Flight. J. Appl. Biomech. **11**, 224–237 (1995)
21. Liu, H., Li, H., Qu, Y.: Validity of an artificial intelligence system for markerless human movement automatic capture. J. Beijing Sport Univ. **44**, 125–133 (2021)
22. FIS International Competition Rules (ICR) Ski Jumping. https://www.fis-ski.com/en/inside-fis/document-library/ski-jumping-documents (accessed on 10 April 2022)
23. Link, J., Guillaume, S., Eskofier, B.M.: Experimental Validation of Real-Time Ski Jumping Tracking System Based on Wearable Sensors. Sensors **21**, 7780 (2021)
24. Schwameder H, Müller E. Biomechanical description and analysis of the V-technique in ski-jumping. Spectrum der Sportwissenschaften 71, 5–36

Numerical Simulation Analysis of Flow Field in Intake System of a Core Engine Test

Wang Anni(✉), Zhang Zhibo, and Chen Yehui

AECC Shenyang Engine Research Institute, Shenyang 100015, China
17741346055@163.com

Abstract. Taking the core engine test air intake system as the research object, through the numerical simulation calculation of the test intake device and the compressor inlet of a core engine, the influence of the selection of the test section of the intake device on the accuracy of the air flow field test is explored, and the influence factors of the compressor air inlet struts and probe on the test accuracy are analyzed. The results show that: the selection of different test sections of the air intake device, the uniformity of static pressure distribution, the unevenness of static pressure (no more than 1%) and flow coefficient are different, but the inlet flow pipe, the inlet probe of test section I and the inlet support plate have a certain impact on the inlet flow field of the compressor. The test uniformity should be considered when arranging the circumferential positions of the probes of the main test sections of the compressor, and the influence of the wake area of the probe of section I and the air inlet struts should be considered in space.

Keywords: Core engine · Air intake device · Flow tube · Compressor · Flow field · Numerical simulation

1 Introduction

The core engine is an important part of aero-engine. The mastery of core engine performance is the key to the success of engine modification or new engine development [1]. The research on the core engine design technology is inseparable from the core engine test and verification platform. Through the use of the whole process parameter measurement technology, the aerodynamic performance and thermal characteristics data of the core engine can be obtained, and the matching design technology of each component of the core engine can be further mastered [2].

A core engine consists of a compressor, a main combustion chamber, a high-pressure turbine, a transmission and various systems. During the test, it is equipped with an air intake device, a retractor and diffuser nozzle, etc. [3]. The air intake device and compressor inlet are collectively referred to as the core engine test inlet system. The air inlet flow pipe in the air inlet device is a test equipment used to accurately measure the air flow at the core engine inlet and ensure that the core engine inlet has good flow field quality [4, 5]. Since the air flow is an indirect measurement parameter, it is generally calculated by measuring the total pressure, static pressure and other physical parameters.

W. Fan et al. (Eds.): AsiaSim 2022, CCIS 1712, pp. 479–493, 2022.
https://doi.org/10.1007/978-981-19-9198-1_36

Therefore, it is necessary to find a suitable test section to reduce the test error. The total inlet temperature and total inlet pressure of the core engine participate in the calculation of the core engine pressure ratio, temperature ratio, unit cycle work and other important performance parameters, while the inlet flow pipe, air inlet struts and inlet probe have a certain impact on the measurement of the total inlet pressure and temperature in space.

In this paper, the core engine test air intake system, the intake device and the compressor inlet, are taken as the research objects for numerical simulation calculation. Referring to the test results of a certain core engine, the influence of the test section of the intake device on the air flow is explored, and the influence of the wake loss of the air inlet struts and the angular position of the inlet probe on the test accuracy of the compressor inlet parameters is evaluated to improve the validity of the test data, to achieve accurate measurement and performance evaluation of the core engine intake system. The results of this study provide technical support for the test performance verification and engineering application of the core engine.

2 Research Object

The composition of the test intake system of a certain core engine is shown in Fig. 1. The intake flow pipe and the intake pressure stabilization box form an intake device, and the compressor inlet is connected with the intake flow pipe and the pressure stabilization box, so as to play the role of diversion and meet the flow continuity principle under the condition of strict airtightness.

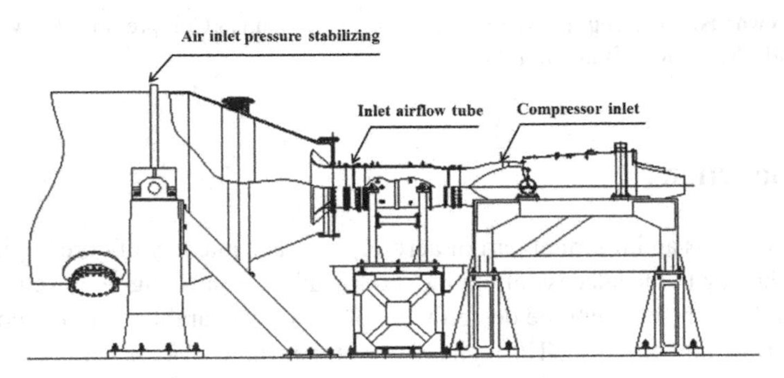

Fig. 1. Structure composition of core engine test air intake system.

This numerical simulation adopts UG modeling. In order to simulate the actual test conditions of the core engine, the simulation calculation model of the intake device is composed of part of the pressure stabilizing box and complete flow pipe, and the simulation calculation model of the compressor inlet is composed of part of the flow pipe, probe and air inlet struts. In the modeling process, the wall thickness of the flow pipe is ignored and replaced by a rotating surface (Figs. 2 and 3).

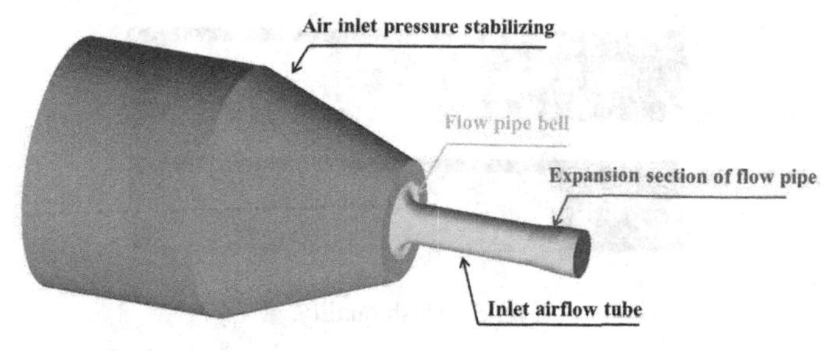

Fig. 2. Simulation model of core engine test air intake device.

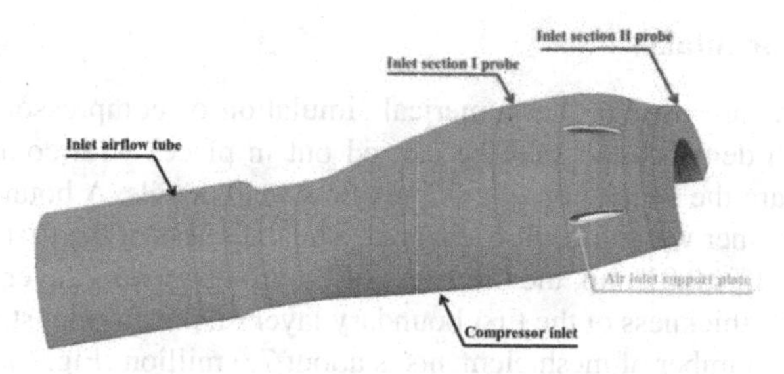

Fig. 3. Simulation model of core engine inlet.

3 Meshing

3.1 The Test Intake Device

The three-dimensional numerical simulation calculation domain grid is topologically divided by ICEM. Using the symmetrical characteristics of the air intake device, the grid is divided by O grid and C grid. The grid quality is good. In order to reduce the calculation error, a boundary layer is added on the wall of the flow pipe. The thickness of the first boundary layer on the wall is 0.1 mm. Finally, the total number of grid elements is 2.88 million. The schematic diagram of the flow pipe network is as follows (Figs. 4 and 5).

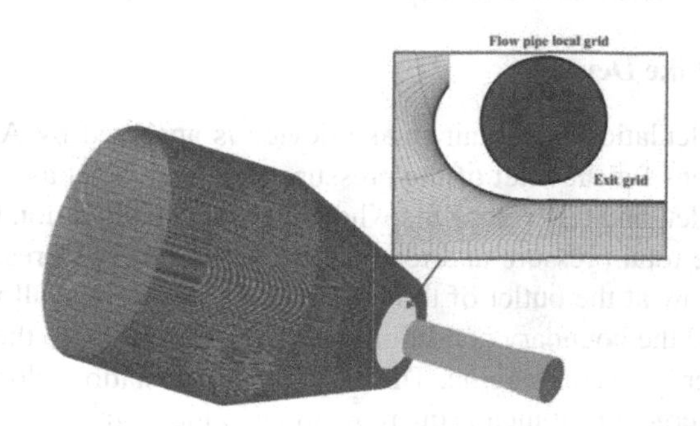

Fig. 4. Grid model of air intake device.

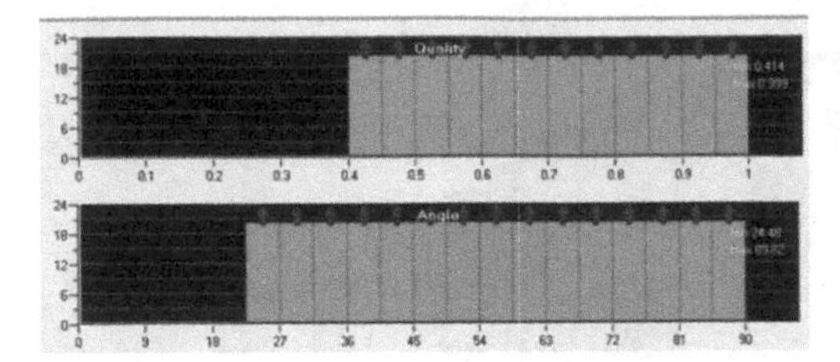

Fig. 5. Mesh quality.

3.2 Compressor Intake

Tetrahedral grids are used in the numerical simulation of compressor intake system. Automatic mesh densification shall be carried out in places with complex geometric structure to ensure the simulation of complex flow field details. A boundary layer shall be added on the inner wall of the flow channel. The thickness of the first boundary layer on the casing wall is 0.01 mm, the thickness of the first boundary layer on the probe is 0.01 mm, and the thickness of the first boundary layer on the air inlet struts is 0.01 mm. Finally, the total number of mesh elements is about 7.6 million (Fig. 6).

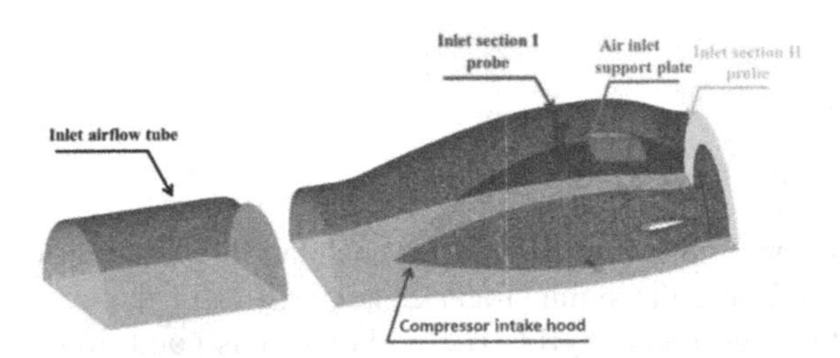

Fig. 6. Grid model of compressor inlet.

4 Boundary Condition Setting

4.1 The Test Intake Device

The numerical calculation of the air intake device is analyzed by ANSYS CFX. The boundary conditions for the inlet of the pressure stabilizer is set as inlet, the flow pipe outlet is set as outlet, and the wall of the whole pressure stabilization box and flow pipe is set as wall. The total pressure and temperature at the inlet of pressure stabilizer are given, the mass flow at the outlet of the flow tube is given, the wall is set as a nonslip adiabatic wall, and the boundary conditions are given according to the 1.0 design speed working parameters of a core engine. This numerical calculation adopts steady calculation, and the convergent root mean square residual value is given as 10–7. Considering the convergence and reliability of the calculation results, RNG k-ε Model is selected as the turbulence model. The RNG model adds a condition to the ε equation. RNG theory provides an analytical formula considering the viscosity of low Reynolds number flow, making the RNG k-ε model has higher reliability and accuracy in a wider range of flows, and is more suitable for simulating turbulent flows [6]. The simulated material library is air ideal gas without considering the heat transfer and the temperature change of working medium (Table 1).

Table 1. Boundary condition setting of each section of air intake device.

Section name	Boundary condition
The inlet of pressure stabilizer	Total temperature and total pressure
Wall surface of the pressure stabilizer	Non slip wall
Wall surface of flow pipe	Non slip wall
Flow pipe outlet	Outlet, given mass flow
Calculation type	Steady state calculation
Turbulence model	Standard k- ε Model
Material library	Air ideal gas

4.2 Compressor Intake

The numerical calculation of compressor intake is analyzed by ANSYS CFX. The flow pipe inlet boundary condition is set as inlet, the imported test section II is set as outlet, the wall of the whole casing wall, probe and air inlet struts are set as wall, and the middle position of the compressor is set as symmetry. The inlet total pressure and temperature are given, the static pressure is given at the outlet, and the boundary conditions are given according to the 1.0 design speed operating parameters of a core engine. This numerical calculation adopts steady state calculation, and the convergent root mean square residual value is given as 10–7. Considering the convergence and reliability of the calculation results, turbulence model chosen for k-ε model, the simulated material library is air ideal gas without considering the heat transfer and the temperature change of working medium (Table 2).

Table 2. Boundary condition setting of compressor inlet sections.

Section name	Boundary condition
Flow pipe inlet	Total temperature and total pressure
Casing wall, probe and air inlet struts	Non slip wall
Model middle	Symmetry
Model exit	Outlet, given mass flow
Calculation type	Steady state calculation
Turbulence model	Standard k- ε Model
Material library	Air ideal gas

5 Results and Analysis

5.1 The Test Intake Device

In this section, the streamline distribution, static pressure nephogram, Mach number nephogram, static pressure nonuniformity and flow coefficient of the inlet test section under the typical state points of the inlet device in the test inlet system of a core engine are analyzed, to select the appropriate inlet test section, so as to calculate the more accurate air flow.

Streamline Distribution. Figure 7 shows the three-dimensional limit streamline distribution of the air inlet device. It can be seen that the streamline distribution in the air inlet pressure stabilizer is relatively uniform. Due to the contraction effect of the air inlet flow pipe flare position on the air flow, the air flow begins to accelerate before entering the flow pipe. The acceleration map is in a fan shape. The flow velocity gradually becomes uniform after entering the flat and straight section, reaches the maximum flow velocity at the handover position of the flow pipe flare, and tends to be stable in the straight section. After the fluid enters the expansion section, the flow velocity decreased significantly. The air velocity in this area changed significantly from the wall to the main flow area. From the limit streamline, it can be seen that the flow in the flow tube is relatively good, there is no obvious backflow, the air flow is mixed evenly at the flow pipe flare, and the overall flow is relatively stable.

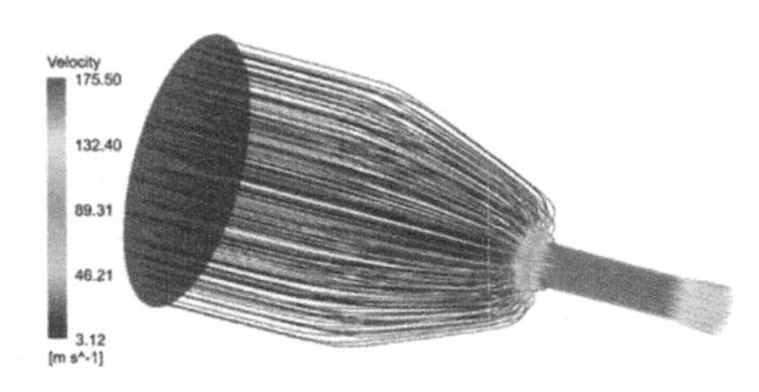

Fig. 7. Three dimensional streamline distribution of flow tube.

Static Pressure Cloud Distribution. Figure 8 shows the static pressure cloud diagram of the meridional section of the air inlet device. It can be seen that the static pressure uniformity of the air inlet pressure stabilizer is good, and the static pressure uniformity at the flow pipe flare decreases. As shown in Fig. 9, there is a very low pressure at the junction of the flow pipe flare and the straight section, resulting in a sharp change in the static pressure of the air flow from the wall to the mainstream area, and the pressure distribution in the straight section of the flow pipe tends to be uniform. After the fluid enters the expansion section, the static pressure gradually increases.

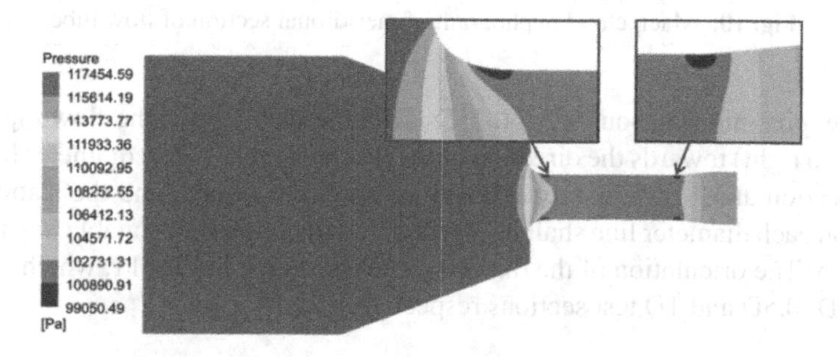

Fig. 8. Static pressure nephogram of meridional section of flow tube.

Fig. 9. Static pressure nephogram of flow tube.

Mach Cloud Distribution. The distribution of Mach number cloud diagram is similar to that of static pressure cloud diagram. As shown in Fig. 10, the flow rate in the air inlet pressure stabilizer is low and uniform. When the fluid enters the flow pipe flare position, the Mach number gradually increases, and the flow rate begins to increase. When the fluid enters the flow pipe flat and straight section, the Mach number reaches the maximum, and after entering the expansion section, the Mach number gradually decreases. Affected by the boundary layer, the Mach number near the wall of the flow tube is low.

Static Pressure Curve Distribution at Different Test Position. As the lowest pressure occurs at the junction of flow pipe flare and the straight section, the static pressure changes sharply along the mainstream direction and the pressure distribution is uneven, so it is not suitable to select the wall static pressure measurement section before this position. The test section shall be selected at the position where the inlet of the straight section

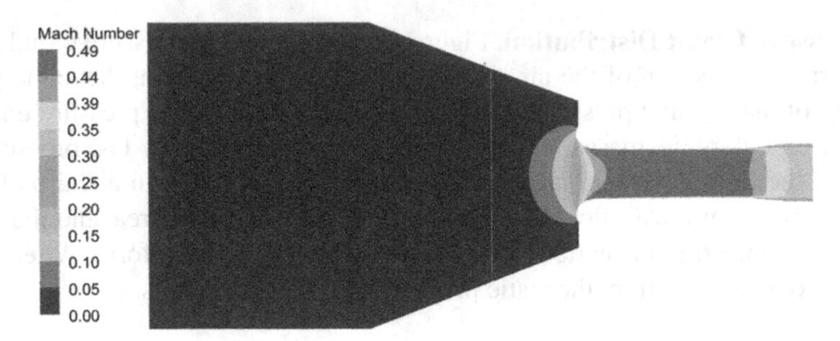

Fig. 10. Mach cloud nephogram of meridional section of flow tube.

of the flow pipe moves about 0.2 ~ 0.25D (D is the diameter of the flow pipe) axially (from left to right) towards the direction of the test piece [6, 7]. The diameter lines of the straight section along different axial positions shall be defined, and the static pressure changes on each diameter line shall be investigated to obtain a reasonable position of the test section. The orientation of the diameter lines is shown in Fig. 11, which are 0.25D, 0.3D, 0.4D, 0.5D and 1D test sections respectively.

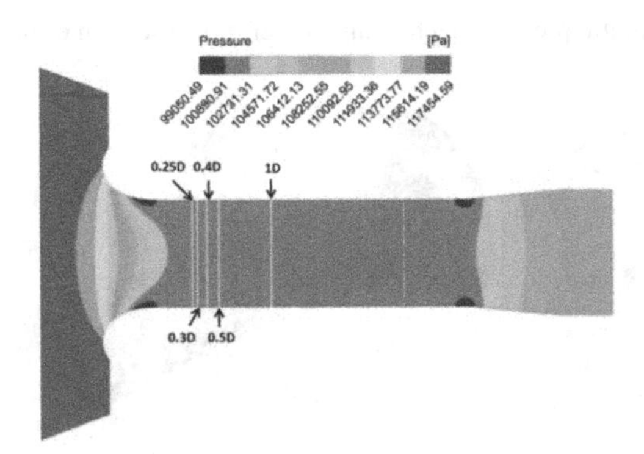

Fig. 11. Schematic diagram of test section position.

Figure 12 shows the distribution of static pressure curves of different test sections. It can be seen that the uniformity of static pressure curve of 0.25D test section is low. As the test section moves backward, the uniformity of static pressure gradually becomes better, and the static pressure distribution of 1D test section is relatively uniform. In practical engineering applications, it is recommended to select the position around 0.2–0.25D [7, 8]. Since the uniformity of 0.25D is low, it is recommended to select the position after 0.25D as the measurement position.

Analysis of Static Pressure Non-uniformity. On the measuring section, more than four wall static pressure measuring points shall be uniformly distributed in the circumferential direction to measure the static pressure of each wall respectively, and the static pressure nonuniformity shall be calculated according to formula (1). It is required that the static

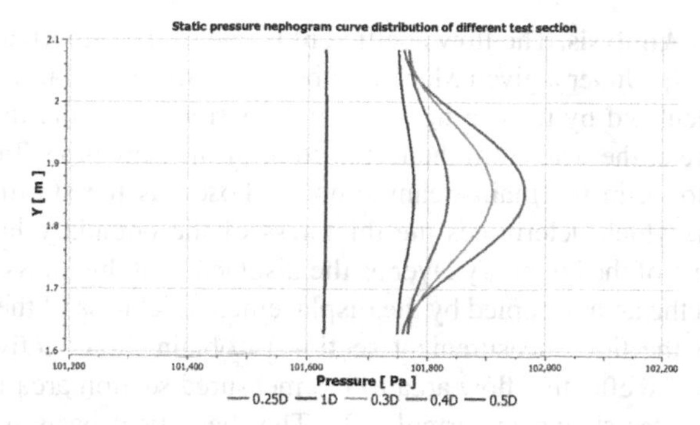

Fig. 12. Static pressure nephogram curve distribution of different test section.

pressure nonuniformity of the inlet flow pipe shall not be greater than 1% [8].

$$D_{ps,av} = \frac{P_{s,max} - P_{s,min}}{P_{t,av} - P_{s,av}} \tag{1}$$

where: $P_{s,max}$ is the maximum static pressure;
$P_{s,min}$ is the minimum static pressure;
$P_{t,av}$ is the average value of total pressure;
$P_{s,av}$ is the average static pressure.

By analyzing the static pressure nonuniformity of the 0.25D ~ 1.0D test section, as shown in Fig. 12, the static pressure nonuniformity of the 0.25D test section is 1.33%, the static pressure nonuniformity of the 0.3D test section is 0.92%, the 1.0D static pressure nonuniformity is 0.02%, and the static pressure nonuniformity is not greater than 1%. Therefore, it is recommended to select the 0.3d ~ 1.0d position for the test section.

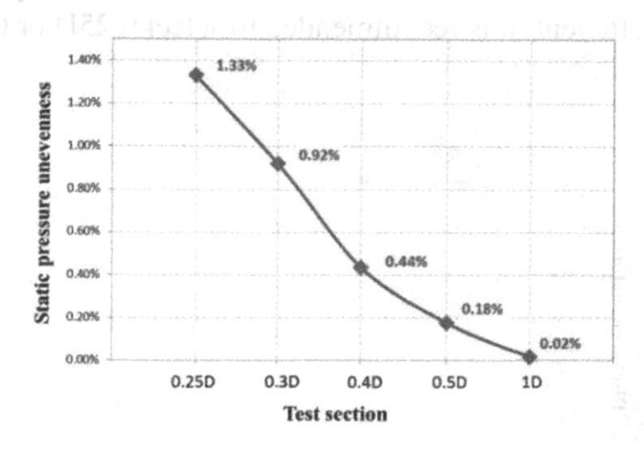

Fig. 13. Distribution of static pressure unevenness in different test sections.

Flow Coefficient Analysis. The flow coefficient is one of the important parameters for flow calculation [8]. Under a given Mach number, the radial distribution of the velocity coefficient is calculated by measuring the radial distribution of the total pressure and the static pressure at the wall from the test section of the flow pipe. Taking 0.99 times the velocity factor λ in the mainstream zone is chosen as the starting point for the attachment layer, which determines the thickness of the boundary layer. Take 0.125 times the thickness of the boundary layer as the displacement thickness of the boundary layer, and deduct the area occupied by the displacement thickness of the boundary layer from the area of the flow measurement section to obtain the effective flow area i.e. A_{eff}. The ratio of the effective flow area to the measured section area i.e. A is the flow coefficient i.e. K_w, as shown in formula (2). The theoretical basis of this method is based on the fact that the static pressure gradient in the mainstream and boundary layer of the flow measurement section is zero, which satisfies Eq. (3). The smaller the flow coefficient, the larger the flow measurement error caused by the boundary layer.

$$K_w = \frac{A_{eff}}{A} \tag{2}$$

$$\frac{\partial p}{\partial r} = 0 \tag{3}$$

In the above formula, the determination of the flow coefficient is based on the assumption that the static pressure is uniformly distributed along the radial direction, that is, the flow loss is only caused by the boundary layer. However, in the actual test process, the static pressure radial distribution of the measured section may be non-uniform, resulting in the non-uniform distribution of velocity, which will lead to the flow calculation error.

Calculate the flow coefficient of the five test sections of 0.25D ~ 1.0D through formula (2), as shown in Fig. 13, the flow coefficient of the test section of 0.25D is 0.9925. As the test section moves backward, the flow coefficient gradually decreases, and the minimum flow coefficient of the test section of 1.0D is 0.9865. The smaller the flow coefficient, the greater the flow error caused by the boundary layer of the intake flow pipe. Therefore, the test section of the intake device should try to select the section with a large flow coefficient, it is recommended to select 0.25D or 0.3D (Fig. 14).

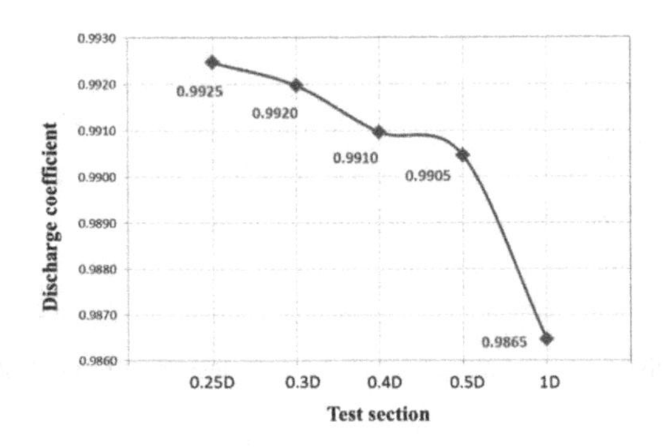

Fig. 14. Distribution of discharge coefficient in different test sections.

To sum up, based on the above analysis of static pressure distribution, static pressure nonuniformity and flow coefficient of different test sections of air intake device from 0.25D to 1.0D, the test section with uniform static pressure distribution, static pressure nonuniformity not greater than 1% and large flow coefficient shall be selected. Therefore, the air intake device selects 0.3D measurement section as the test section of inlet total pressure, inlet total temperature and inlet static pressure, which has high accuracy and small error.

5.2 Compressor Intake

In this section, the streamline distribution and total pressure nephogram at the typical state point of the compressor inlet in the test intake system of a certain core engine are analyzed, and the influence of the inlet flow pipe connecting the compressor transition step, the inlet probe of the test section I, and the air inlet struts on the flow field of the compressor test section II (the main test section of the compressor) is analyzed, so as to find out the influencing factors on the compressor inlet parameter test.

Limit Streamline Cloud Diagram. Through CFX post-processing analysis, the limit streamline is obtained, as shown in Fig. 15. It can be seen that the overall distribution of the limit streamline at the inlet of the core engine is relatively uniform, and the flow condition is good. The flow rate increases from the outlet of the inlet flow pipe to the inlet of the core engine due to the transfer steps. When it flows to the air inlet struts of the compressor, the flow rate decreases, the flow rate increases again after passing through the air inlet struts. The inlet section I probe and the air inlet struts have little effect on the inlet flow line of the core engine, and there is no flow field separation.

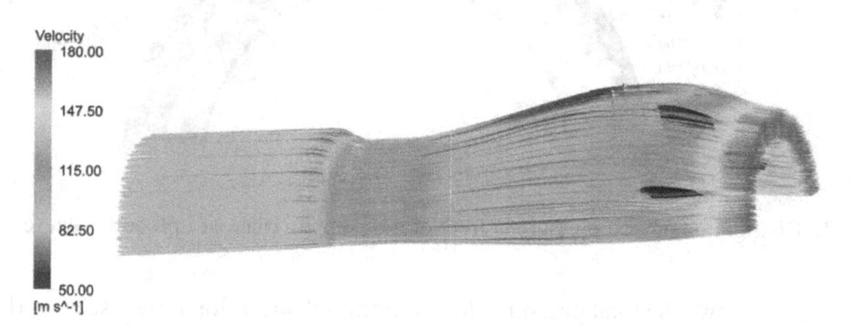

Fig. 15. Cloud diagram of inlet limit streamline.

Total Pressure Nephogram Distribution. Figure 16 shows the distribution of total pressure in the meridian section of the compressor inlet section. It can be seen that the overall distribution of total pressure in the main flow area of the compressor inlet is relatively uniform. Affected by the connecting steps between the inlet flow pipe and the compressor, a low-pressure wake area appears in the flow field at the compressor inlet casing, and the wake area extends to the test section II, indicating that the connecting steps between the inlet flow pipe and the compressor have a great impact on the flow field near the compressor inlet casing. During the design of the inlet device for the core engine test, a smooth transition should be ensured at the connection with the compressor inlet, so as to reduce the influence of the flow field.

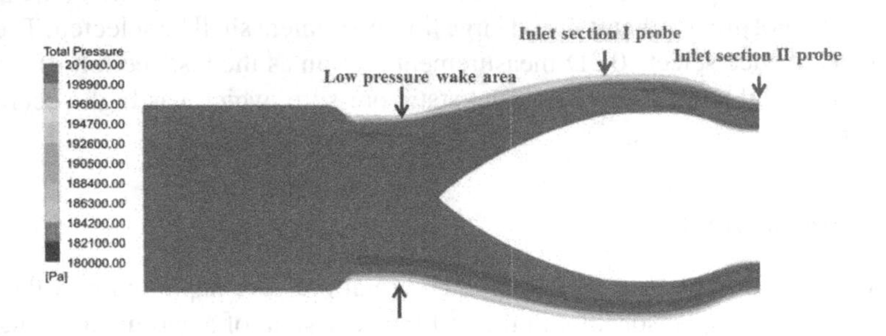

Fig. 16. Distribution of total pressure cloud diagram at inlet meridian section.

Figure 17 shows the distribution of total pressure cloud after the I probe at the inlet section of the compressor and before the air inlet struts, in order to analyze the impact of the I probe at the inlet section on the inlet flow field of the compressor. From the figure, it can be seen that the flow field presents a low-pressure wake area after passing through the probe, and the total pressure is significantly reduced.

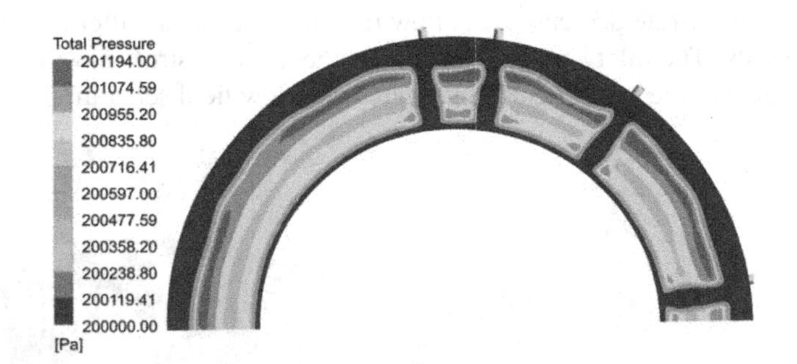

Fig. 17. Distribution of total pressure in front of rear support plate of I probe at inlet section.

Figure 18 shows the total pressure cloud diagram distribution of test section II after the compressor air inlet struts. As shown in the figure, wake areas appear after the air flow passes through the compressor inlet section I probe and the air inlet struts, indicating that the flow field is affected after the air flow passes through the compressor inlet section I probe and the air inlet struts. Therefore, the relative positions of the test section II probe and the compressor inlet section I probe and the air inlet struts in space should be considered in the circumferential layout, avoid the wake area of up-stream probe and air inlet struts.

Mach Cloud Distribution. Figure 19 shows the distribution of Mach number cloud of test section II behind the compressor air inlet struts, similar to the total pressure cloud map. As shown in the figure, the Mach number decreases and the air velocity decreases after the air flows through the compressor inlet section I probe and the air inlet struts.

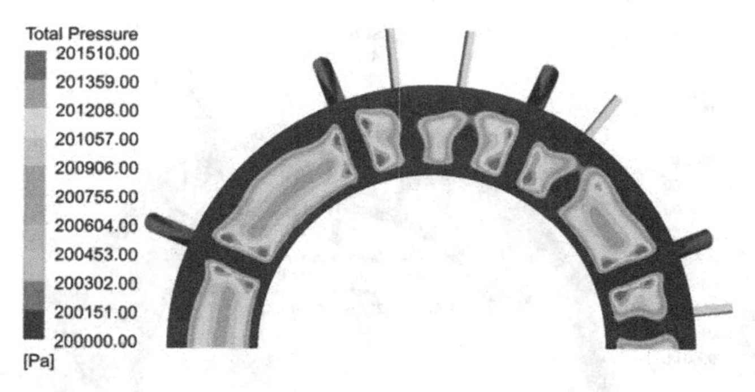

Fig. 18. Distribution of total pressure cloud diagram of test section II behind air inlet support plate.

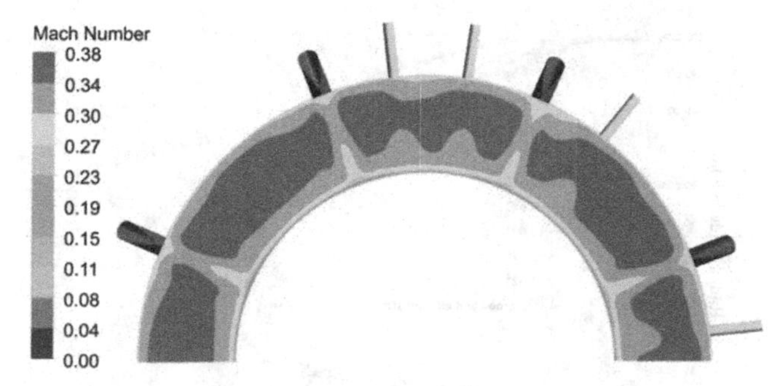

Fig. 19. Mach number cloud diagram distribution of test section II behind air inlet support plate.

Total Pressure Distribution Curve. Through the above analysis, the compressor inlet section I probe and the air inlet struts have a certain impact on the flow field of test section II, so the spatial position should be explored when arranging the probe on test section II. Figure 20 shows the circumferential arrangement of the probe of test section II of a certain core engine, the red line indicates the position not affected by the upstream air inlet struts and probes. Considering the relative angle between the circumferential positions of the probe of section II and the probe of section I, the circumferential positions with angles of 1.62°, 2.62°, 3.62°, 4.62° and 5.62° are selected for analysis.

Figure 21 shows the total pressure distribution curves at different circumferential positions of test section II, it can be seen that the section is not affected by the upstream air inlet struts and probes, and the total pressure distribution in the main flow area is more uniform than that in the affected area. Analyze the affected area, and the total pressure distribution at different relative angular position between the test section and the upstream probe is selected for analysis. The total pressure curve at the 5.62° angle is relatively uniform, while the flow field is greatly affected by the 1.62° angle because it is close to the probe a of test section I. therefore, the probe of compressor test section II should avoid the influence of the probe of section I and the wake area of inlet support plate in space.

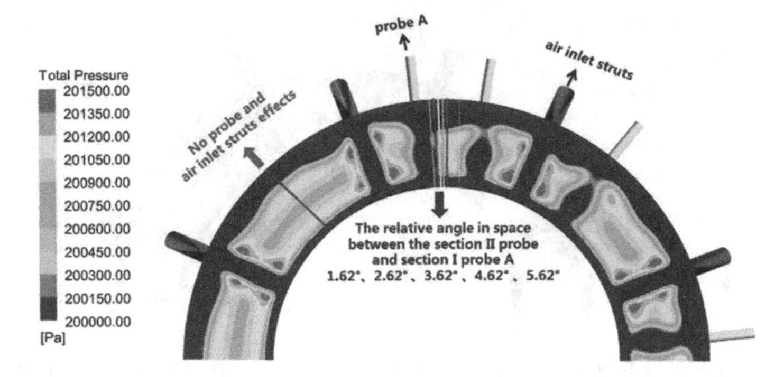

Fig. 20. Relative angle between section II probe and section I probe after air inlet support plate.

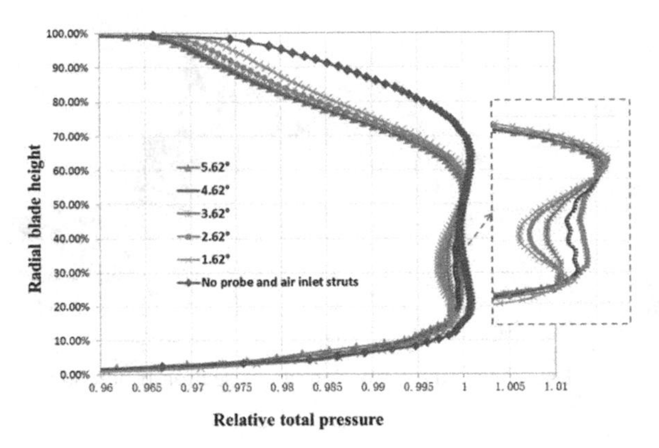

Fig. 21. Total pressure distribution curve at different circumferential positions of probe of test section II.

6 Conclusion

In this paper, the core engine test intake system, including intake device and compressor inlet, is studied by numerical simulation. Through detailed flow field analysis, the following conclusions can be obtained:

1) The air flow measurement method of core engine test is slightly different from that of conventional compressor test, there is some variation in the size of the inlet flow pipe and the selection of the test section. Since pipeline air inlet is adopted for core engine test, it is generally recommended that inlet flow pipe sizes are not too small. Consider test accuracy requirements, Mach number in the flow pipe should not be selected too low, it is recommended to choose a section with a relatively uniform static pressure.
2) The static pressure distribution, static pressure nonuniformity and flow coefficient of different test sections of the intake flow pipe are analyzed. The selection of different test sections has different static pressure distribution uniformity, static pressure

non-uniformity and flow coefficient. The test sections with uniform static pressure distribution, static pressure non-uniformity (less than 1%) and high flow coefficient should be selected. Therefore, based on the above factors, the core engine test air intake device selects 0.3D measurement section as the test section of the air intake device, which effectively ensuring the accurate measurement of the core engine inlet air flow.

3) The inlet of the compressor is simulated and analyzed. The inlet flow pipe, the inlet probe of test section I and the air inlet struts have a certain influence on the inlet flow field of the compressor. The connecting step of the inlet flow pipe has an impact on the flow field near the compressor casing. Therefore, during the core engine test, the inlet device shall be designed to ensure a smooth connection transition with the inlet of the compressor.
4) The inlet probe of test section I and the air inlet struts cause the wake low-pressure area of compressor test section II in space, and the Mach number is low. The test uniformity should be considered when the circumferential position of the probe of the main test section of the compressor is arranged. The total pressure curves at different circumferential positions should be analyzed through numerical simulation, the probe of compressor test section II should avoid the influence of the wake area of the probe of section I and the air inlet struts in space, so as to reduce the test error.

References

1. Huang, S., Hu, J., Jiang, H.: Investigation of core engine and derivative aero-engine development. J. Aero. Power. **21**(2), 242–247 (2006)
2. Wang, T., Zhao, X., Bao, Y.: Design Techniques of the Core Engine Testing Machine, pp. 290–292. Chian Academic Journal Electronic Publishing House (2015)
3. Shi, Y.: A preliminary study on measuring deviation of important parameters for a core engine. J. Shenyang Aero. Univ. **28**(4), 12–16, 24 (2006)
4. Brokopp, R.A., Gronski, R.S.: Small engine components test facility compressor testing cell at NASA lewis research center. AIAA-92-3980 (1992)
5. Xiang, H., et al.: Aerodynamic design and application validation of flow tube used in compressor experiment. Gas Turb. Technol. **28**(4), 28–34 (2015)
6. Xiao, M., Zhuang, H., Guo, X.: Numerical and experimental study of air flow field of a core engine with by-pass duct. J. Aero. Power. **24**(4), 836–842 (2009)
7. Wang, A., Zhang, Z., Chen, Y., Wu, H.: Numarical Simulation of Aerodynamic Performance of Inlet Flow Tube Bell Mouth Profile Design, pp. 590–597. Chian Academic Journal Electronic Publishing House (2020)
8. HB7115-1994, Compressor Aerodynamic Performance Test

Big Data Challenges and Requirements for Simulation and Knowledge Services of Big Data Ecosystem

A Review of Failure Prediction in Distributed Data Centers

Yuqing Ma, Xu Xie(✉), and Miao Zhang

College of Systems Engineering, National University of Defense Technology, Changsha, China
{mayuqing21,zhangmiao15}@nudt.edu.cn

Abstract. With the advent of the era of big data, distributed data centers with advantages in computing, storage, security, etc. have become the trend of future development. In the civil field, when distributed data centers are abnormal, problems such as service interruption, data leakage and capital loss will be caused. Therefore, it is necessary to predict failure for distributed data centers. In the military field, when the LVC system is applied to distributed data centers, the live assets will be lost if the simulation system fails, and the loss is very costly. Hence, failure prediction is more necessary for the simulation system. In summary, first, the failure prediction in distributed data centers is systematically reviewed from four aspects: overall architecture, data types, mainstream prediction methods, and existing difficulties and challenges, to analyze the current research status at home and abroad. Subsequently, from LVC system perspective, the importance and challenges of failure prediction for the simulation application layer in distributed data centers are analyzed.

Keywords: Distributed data centers · Failure prediction · Statistical learning · Machine learning · Deep learning · LVC · Simulation application layer

1 Introduction

At present, emerging technologies such as cloud computing and the Internet of Things (IoT) are being used on a large scale. These technologies make the data center (DC) become the focus of the construction in many industries, such as communication, medical care, transportation, smart grid and so on [1]. However, the traditional centralized data center (CDC) usually adopts the architecture of "active-standby/dual-active data centers in the same city" or "three centers in two places" [1]. The architecture of "active-standby data centers" is that only one data center is enabled, while the other is used for backup; the architecture of "dual-active data centers" is that two data centers deploy and run the same business; the architecture of "three centers in two places" also uses one of three data centers as a backup. Although these architectures achieved high availability and continuity of business, they resulted in a serious waste of resources due to an idle data center [1]. Accordingly, distributed data centers (DDC) have become the trend of future development, which have advantages in computing, storage, security, etc.

DDC refers to use cloud computing technology, network interconnection technology and data replication technology to form multiple DCs into a distributed cross-regional

W. Fan et al. (Eds.): AsiaSim 2022, CCIS 1712, pp. 497–509, 2022.
https://doi.org/10.1007/978-981-19-9198-1_37

"virtual resource pool". All services and data are allocated to different DCs on demand [1]. Therefore, compared with traditional CDC, DDC can save resources effectively and have excellent scalability.

Because of the excellent scalability, DDC can handle big data business well, but have led to a dramatic increase in the demand for data transmission and digital services [2]. DDC contains a large amount of computing infrastructures. These infrastructures are heterogeneous and interconnected, which makes interaction of components in distributed systems highly complicated [3]. Thus, failure in such tightly coupled system is no longer an accident, but an expected behavior with a certain trend to follow. In addition, in medical care, banking, autonomous driving and other civil industries, demand for the reliability of data communication is pretty high, and it is necessary to ensure that data communication is not interrupted and can respond quickly to business. If a system occurs failure, it will cause a large amount of data leakage, capital loss and other issues [4]. At present, operation and maintenance personnel repair the fault by monitoring the relevant operating parameters after the data center platform fails. However, it takes a certain amount of time from the data center platform fails to operation and maintenance personnel repair the failure. During this period, the failure may affect the normal and stable operation of the entire network. Accordingly, how to prevent failures before the data center platform fails has become a technical problem to be urgently solved. We need to predict the fault of DDC to achieve efficient data processing and emergency disaster mitigation, and greatly improve the availability of the system, which is also the current research focus of DDC [5].

In the military field, some systems (such as nuclear power plants, manned spacecraft, weapons and equipment, etc.) have a huge loss caused by the failure of direct experiments. Thus, simulation technology is widely used in the military field. Generally, we use simulation technology to build a simulation application layer in DDC. Simulation systems that implement different functions are built on this layer. The loss caused by the failure in simulation systems is pretty small. But with the emergence of LVC system, the live asset will be lost when the simulation system fails and the loss is quite costly. Accordingly, in the context of LVC, it is very necessary to predict failure of the simulation application layer in DDC.

In summary, this paper presents a systematic review of the current research status of failure prediction in DDC. The rest of this paper is organized as follows. Section 2 presents the overall architecture of failure prediction in DDC. Section 3 sorts out the data types. Section 4 generalizes the mainstream fault prediction methods. Section 5 concludes the existing difficulties and challenges. Section 6 summarizes the paper, and then extends to failure prediction of the simulation application layer in DDC.

2 Overall Architecture

DDC are composed of infrastructure, VM, middleware and applications. Failure prediction in DDC is divided into three parts, namely data collection, data processing, and failure prediction. The overall architecture of failure prediction in DDC is shown in Fig. 1.

In the three parts, the following three points need to be considered. First, the architecture of DDC includes the infrastructure layer and the application layer, and the two layers consist of various software and hardware resources. Components of the two layers are not completely consistent, so data types that need to be collected are not the same. Second, different data types require different feature extraction methods to extract effective information. Third, there are a lot of software and hardware in DDC. They cooperate with each other in order to accomplish the same task. This requirement is called interactivity. Due to the interactivity, using only one prediction method usually does not address all failures that occur in DDC. Hence, a combination of multiple methods is required.

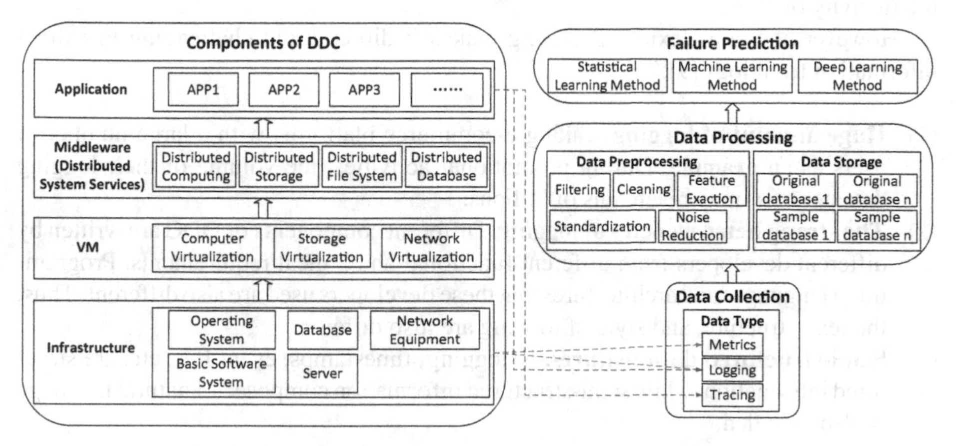

Fig. 1. The overall architecture of failure prediction in DDC.

3 Data Type

Different data types have different characteristics. According to the data characteristics, designing or selecting rational methods will achieve better results. Thus, it is necessary to analyze the characteristics of various data types involved in failure prediction. There are three data types in the field of failure prediction, namely metrics, logging, and tracing [6].

3.1 Metrics

Metrics refer to key performance indicators (KPI) measured in consecutive time periods. It can be used to observe the status and trend of the system, such as CPU usage, memory usage, average disk I/O time, network throughput, etc. Benefiting from their essential feature of atomicity and accumulation [7], metrics can compress data space and save a lot of resources, which is an important advantage for failure prediction.

It should be noted that metrics themself do not have correlation, and directly inputting them into the prediction model as raw data will lead to low generalization [8]. Hence, the complex data preprocessing is required before using metrics, to obtain the data containing the time period state.

3.2 Logging

Logging refers to the textual data generated by the output code embedded in the program, to help program developers to debug and record variable information and program execution status [9]. Logging can capture the running status and abnormal information of programs across components and services. It can reflect the dynamic characteristics of the system and effectively infer the unknown faults of the system through the logical relationship of the context. Hence, failure prediction based on logging can meet the interactivity of DDC.

However, the complexity of logging makes it difficult and challenging to extract information from logs [9]:

(1) **Huge amount of logging.** Taking e-commerce platforms with a large number of users as an example. During promotional activities, the volume of their logging data per hour may reach tens of billions.
(2) **The strong heterogeneity of logging.** Different components of DDC are written by different developers from different functional and logical requirements. Programming languages and architectures that these developers used are also different. Thus, the text, structure, and style of logging are also different.
(3) **Semi-structured data features.** In logging, timestamps, event IDs, etc. are structured information, while semi-structured information composed of natural language is also included.

Therefore, it is very important to process logging and extract useful information. Based on the above analysis, two logging processing methods are proposed, namely log template mining and log feature extraction [9]. Benefiting from the textual similarity, temporal proximity and parameter flow mining of logging, literatures [10–12] use a clustering approach for template mining to address the heterogeneity and semi-structure of logging. Mining log templates can greatly compress log data space and save resources.

The log template mining method is also used in literature [13], but it is based on static source code analysis. First, the source code is transformed into the abstract syntax tree (AST). The AST is used to record method call information including method call relationships, object names, class names, etc. Each method call information generates a log template fragment. Subsequently, according to the log template, the literature [13] uses log feature extraction method to count the occurrence frequency of a set of state variables within a time window for each log template, and calculates its ratio to get state ratio vectors. The literature also uses log feature extraction method to group messages with the same event ID to get message count vectors. According to these two feature vectors, principal component analysis (PCA) is used to predict anomaly.

3.3 Tracing

Tracing is actually the service call link data. Behind the realization of a user's request, multiple services may be invoked, such as Google search, Taobao shopping, etc. These services may be provided by different applications, implemented in different programming languages, and even processed in different data centers [14]. When a request fails during a service invocation process, the root cause of the failure may not be in this process. At this time, it is necessary to backtrack the invocation path of the service to find the failure. Accordingly, it makes sense to study service call link data, namely tracing.

When using tracing for fault prediction in DDC, the key is to obtain the entire service request execution path. The path is usually obtained by using information contained in spans or events [15]. The information can infer causal relationships between events [15]. Span refers to the specific operation process of a call from sending a request to receiving a reply, which is a component of the service call link. Literature [16] uses the causal relationship between service response time information and service call path information to construct service tracing vectors, so as to obtain the call path. Literature [17] adopts an encoder-decoder neural network called the mask span prediction (MSP) task solver. According to the causal relationship between context of the "hidden" span, MSP task solver deduces the probability distribution of possible events at the "hidden" position, to obtain the complete service call link path.

Although tracing shows great strength in positioning capability [15], its end-to-end processing makes the amount of data pretty large, the processing of noise pretty difficult, and the calling path span pretty long. As the path gets longer, the earlier spans of the calling link in the behavior model are forgotten, which reduces the accuracy of the model.

3.4 Relationship and Distinction

Although metrics, logging, and tracing differ in their characteristics, they are also related to each other. Metrics and tracing can be obtained from logging. In addition, metrics may also be gradually converted into tracing, and then perform in-request indicator statistics [7]. To sum up, the advantages and disadvantages of using different data types to predict faults in DDC are summarized in Table 1.

4 Failure Prediction Methods

At present, there are many failure prediction methods. Different scholars and research institutions have slightly different classifications, including model-based failure prediction methods, data-based failure prediction methods, artificial intelligence methods, etc. [18]. In view of the characteristics of massive data and complex data structure in DDC, it is particularly important to extract fault data from the massive data and obtain a reliable prediction model. Therefore, from different perspectives of the establishment of failure prediction models, this paper divides the failure prediction methods into statistical learning method and machine learning method.

Statistical learning method is also called the learning method based on probability and statistics. According to the statistical characteristics of historical fault data, the method explores the relationship between variables and establishes a failure prediction model. Machine learning method divides the acquired failure data set into a training set and a test set. A prediction model is obtained by training on the training set and testing its quality on the test set, resulting in a failure model that can predict repeatedly. This section will analyze the principles, advantages and disadvantages of the two methods.

Table 1. Comparison of three data types.

Data type	Advantages	Disadvantages	Related literature
Metrics	(1) Accumulativeness; (2) Compression of data space to save resources	(1) Strong heterogeneity. When extracting data features, a large amount of information may be lost; (2) Low generalization. Requiring complex data preprocessing; (3) The method of clustering metrics is complicated	[7, 8]
Logging	(1) Logging can capture the running status and abnormal information of programs across components and services; (2) Logging reflects the dynamic characteristics of the system, and effectively infers the unknown faults of the system through the logical relationship of the context	(1) Massive data makes it difficult to extract valid information; (2) Strong heterogeneity makes it difficult to mining common log template; (3) Semi-structured feature; (4) Due to the fast iteration speed of the system, the log data is also frequently updated, which makes the accuracy of the existing offline prediction model not high	[9–13]
Tracing	(1) Completely describe the request execution path; (2) Strong fault location capability and accurate analysis of the end-to-end processing process; (3) Meet the interactivity of DDC	(1) Massive data makes it difficult to deal with noise; (2) The call path span is long, which will reduce the accuracy of the model	[14–17]

4.1 Statistical Learning Method

Statistical learning method first analyzes the relationship changes between data features based on historical fault data, and then establishes a prediction model to analyze whether the system will fail in the future. The algorithm flow is shown in Fig. 2. Statistical learning method can be divided into **time series prediction method** and **regression prediction method**.

Time series prediction method. According to the time series data characteristics of known failures, time series prediction method uses the relevant algorithm to establish a prediction model, and then predicts the future change trend of the sequence to determine whether a failure occurs. For example, literature [19] uses autoregressive moving average (ARMA) model to analyze a set of regularly collected time series data of components in

DDC to obtain a prediction model. Literature [20] randomly selects a certain attribute of time series data to establish an isolation forest, and predicts faults according to the path length of normal data is longer than that of abnormal data. However, since time series prediction method is highly dependent on historical fault characteristics, short-term forecasting is better.

Regression prediction method. Regression prediction method analyzes the relationship between the independent variable (prediction model) and the dependent variable (target) to establish a regression model. According to the change of the independent variable, the regression model analyzes the change of the dependent variable. For example, literature [13] uses PCA to obtain state ratio vectors and message count vectors of logging. Then a regression model is established based on the relationship between the two vectors, and the state of the system is predicted by analyzing the input deviation of the predictive data on the model. When the deviation exceeds the threshold, the system occurs faults.

Statistical learning method can only make probabilistic predictions on small datasets for specific types of failures (such as disk failures, host failures, etc.). Hence, it has low generalization.

4.2 Machine Learning Method

Machine learning method usually convert prediction problems into classification problems. It uses various classification algorithms to learn a large amount of data and automatically create models. It can predict the future trend of the system state, so it has become the mainstream method in the field of fault prediction. The algorithm flow of machine learning method is shown in Fig. 3. Machine learning method is usually divided into two kinds of algorithms, namely **supervised learning** and **unsupervised learning**.

Supervised learning. The supervised learning algorithm inputs a labeled dataset into a classifier and trains the classifier to classify predicted data (normal or abnormal). Literature [21] uses the extreme gradient boosting algorithm (XGBoost) to train the host performance metrics collected in real time within 3 minutes to obtain a prediction model. Then the real-time data is inputted into the model to calculate the service invocation time. According to the time threshold, the host is classified as normal or abnormal. Literature [22] uses SMOTE sampling and cost-sensitive learning to improve the XGBoost algorithm to solve the problem of extremely imbalanced data. Literature [23] adopts the feature selection technique to obtain the feature matrix, and then input the matrix into the robust multi-task learning (RMTL) model. The RMTL model uses cross-validation method to obtain the weight matrix and regularization parameters to determine which tasks in the test set are relevant to the labelled tasks in the training set. According to the correlation of tasks between the test set and the training set, the termination status of the predicted task can be classified as completed or failed. Literature [24] uses random index (RI) to represent the operation sequence extracted from the log file, and then uses the vector representing the operation sequence to train the support vector machine (SVM). According to the trained SVM, the classified results of the predicted samples is obtained.

The supervised learning algorithm is based on the baseline model to compare with the new model to find faults. It assumes that the training data of the distributed system is consistent with the actual operating data. However, the update speed of the baseline model is slow and inefficient, and it cannot adapt to frequent updates of the system, so the accuracy of failure prediction is not high.

Unsupervised learning. Instead of providing labeled data in advance, unsupervised learning algorithms cluster different clusters based on the similarity or regularity of the data, and then classify the predicted samples into clusters with the same features, thereby achieving failure prediction. Literature [5] uses KNN and hierarchical clustering theory to cluster the collected metrics of physical hosts of each node in DDC, and divides them into low-risk clusters (LR) and high-risk clusters (HR). Based on the fault prediction model (LR, HR) and cluster similarity, the predicted data is classified as normal or abnormal.

It is worth mentioning that **deep learning method**, as a kind of unsupervised learning methods, has been widely used in various fields in recent years because of its pretty good robustness, versatility and scalability.

Deep learning method. Deep learning method can handle large datasets. The more data there is, the more stable the method is. Therefore, many scholars currently use deep learning method to predict faults in DDC. Slightly different from the other machine learning methods mentioned above, deep learning method has been improved. The deep learning method uses a multi-layer neural network. The neural network can meet the functions of feature extraction and classification at the same time [22].

The key to using deep learning method is to train the neural network to obtain the baseline behavior model. Subsequently, according to the deviation between the training sample and the actual sample, the model determines the dynamic threshold and then inputs the predicted sample to obtain predicted results. If the deviation of the sample exceeds the given threshold, the system fails. The algorithm flow is shown in Fig. 4.

Literature [26] obtains a normal behavior model by training an improved variational autoencoder, and calculates the deviation between the output result and the actual sample to form a Gaussian distribution. The Gaussian distribution is used as a dynamic threshold for judging whether the predicted sample fails. Literature [27] uses the feature vector containing static and dynamic features as the input data of the Bi-LSTM neural network model to obtain the baseline model of the task termination state. Then the model uses the logistic regression function to calculate the probability value of the failure feature. If the probability value of the failed task exceeds the given threshold, the termination status of the task is judged to be failed, and the system is faulty. Literature [28] collects the fault time series data of the system and inputs it into the LSTM neural network model for training. The training results are continuously re-input into the model in an iterative manner to obtain the fitting sequence for the next time period. Finally, based on the deviation between the fitting sequence and the actual sequence, the failure of the system is predicted by analyzing whether the deviation exceeds the given threshold.

The deep learning method can realize fast feature analysis and deep mining of massive data [29], but because of the large amount of data, the model convergence speed is pretty slow. At the same time, deep learning is based on existing data to learn, it does not understand data. If the data used to train the model do not truly reflect the state of the system, the final prediction will be inaccurate.

To sum up, compared with the statistical learning method, the machine learning method usually does not need to reflect the relationship between variables when there are many input factors and complex relationships. It only needs to be able to predict the outcome accurately, so the prediction model built using the machine learning method is more accurate. But a significant drawback of the machine learning method is that it has a large number of learning functions, so it does not have sufficient control over actual usage time and has slow prediction times [25]. In addition, machine learning method can only achieve classification of the predicted results, but not predict the type of fault accurately [22].

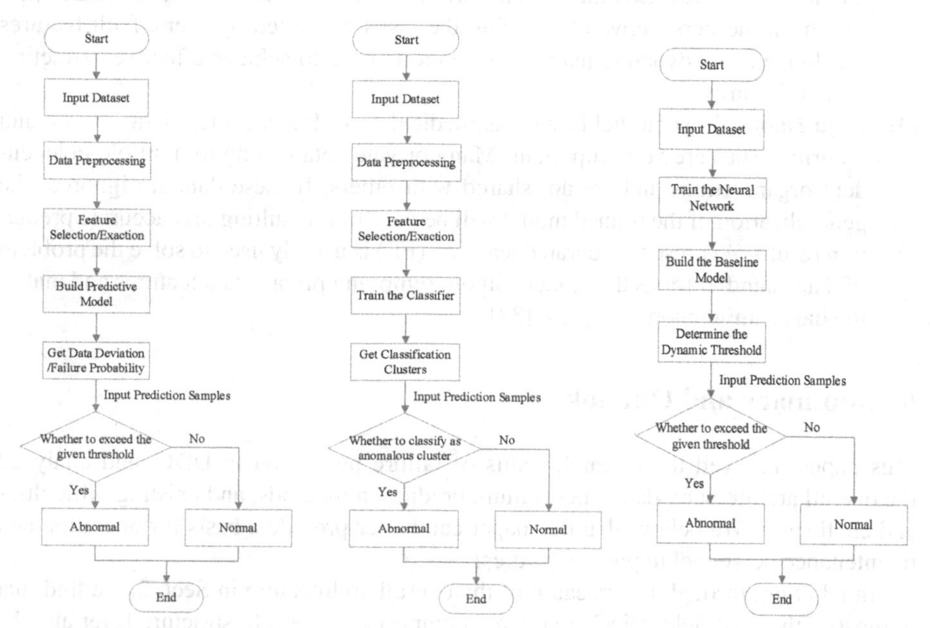

Fig. 2. Failure prediction algorithm flow based on statistical learning method.

Fig. 3. Failure prediction algorithm flow based on machine learning method.

Fig. 4. Failure prediction algorithm flow based on deep learning method.

5 Difficulties and Challenges

Since DDC involves complex interactions of various infrastructures and applications, there are many difficulties and challenges in failure prediction, such as multi-source data heterogeneity, data island, inaccurate fault location, highly unbalanced data distribution, complex fault causes, rough feature extraction, etc. [22], this paper mainly discusses the following three points:

(1) **Highly unbalanced data distribution.** Because the proportion of fault data is very small compared to normal data, for example, in Microsoft's large-scale cloud service system, only one thousandth of the nodes will fail every day [22], this situation will lead to prediction model instability and low generalization. Current studies usually adopt data rebalancing techniques (such as oversampling and undersampling) [22, 24, 30], virtual sample generation (VSG) techniques [8], or ranking models [31] to solve highly unbalanced data distribution.

(2) **Multi-source data heterogeneity and rough feature extraction.** In DDC, data may be stored in different computers, different formats, and even different geographic locations, so it has strong heterogeneity. In addition, rough feature extraction will lead to a large amount of information loss [31]. To solve these problems, different feature extraction methods need to be selected according to different data features. For example, literatures [23, 27, 32] divide data features into temporal feature and spatial feature from the perspective of node failure, and then adopt different feature extraction methods according to the two features. In literatures [21, 24], from the perspective of realizing the tasks requested by users, fault features are divided into dynamic feature and static feature to achieve effective extraction of data features.

(3) **Data island.** In some fields such as medical care, finance, etc., data privacy and security issues are very important. Many private data belong to multiple independent organizations and are not shared with others. If these data are ignored, the generalization of the trained model will be very low, resulting in inaccurate prediction results. At present, federated learning (FL) is usually used to solve the problem of data island. It trains the model without removing private data features and makes the data sample more complete [33].

6 Summary and Outlook

This paper surveyed the research status of failure prediction in DDC, and analyzed the overall architecture, data types, failure prediction methods, and existing difficulties and challenges. We believe that this paper can better provide a basis for operation and maintenance personnel to predict failures.

In addition, through the research of the overall architecture in Sect. 2, we find that in most of the civil field, DDC are mainly composed of the infrastructure layer and the application layer. At present, the research on the above two layers was quite mature. But in the military field, training directly on live assets will lead to the cost of failure being very high. Thus, on top of the application layer, some developers use distributed simulation technology to create a simulation application layer. At the simulation application layer, the cost of failure is not high, but the emergence of LVC system enables live, virtual and constructive assets to be combined into a seamless and consistent real-time operating environment [34]. The flow of time in the LVC system is consistent with the real world. In the LVC system, it is not allowed to decelerate, accelerate, pause, or even replay time as in the traditional distributed simulation system [35]. The needs of the real system dominate the LVC system. Once the simulation system fails, the virtual assets will not be generated, and the loss will be the live assets of the real world. Hence, how to ensure the smooth generation of virtual assets is the key to the successful operation of the LVC system.

The smooth generation of virtual assets depends on the correctness of developers to compile programs on the simulation platform. However, developers often generate syntax errors, runtime errors and logical errors in the process of compiling simulation programs. Syntax errors will be reported directly in the compiler. Runtime errors will also get corresponding exception prompts in the compiled program when the program is executed. But logical errors are only discovered when the program does not have the desired effect after execution. For example, in military training, a trainer wants to launch a real missile to attack a virtual target. The simulation program that creates the virtual target reports no errors when compiling and executing, but no virtual target is created after execution. Launching a real missile requires multi-party coordination and is costly. If the virtual target is not created successfully due to a logic error, the losses will be enormous. We need to predict and resolve failures caused by logical errors.

However, multiple simulation applications are created in large distributed simulation platforms. These programs are developed by different developers. The programming languages employed and the logical ideas followed may be quite different. The input data may also be diverse. The resulting logical errors are more complex and varied. Hence, how to collect valid data is crucial. It is also a problem to be urgently solved on how to integrate these logical errors effectively.

In view of the above situation, we need to solve the following two problems:

(1) **Multi-source data heterogeneity and feature extraction.** Different scenarios, or even different situations in the same scenario (such as tic-tac-toe intersections, roundabouts, multi-fork intersections and other traffic conditions in traffic simulation) will lead to different types and structures of input data. We need to judge which simulation data types will cause what kind of failures, and use appropriate methods to extract data features to avoid a large amount of data loss.
(2) **General failure template mining.** Although logical errors are complex and diverse, the types of faults are limited in specific scenarios (such as combat scenarios). We need to associate fault types with logical errors and dig out a general failure template. The general template needs to contain as many logical errors as possible.

On the basis of obtaining data and general failure templates, the failure prediction of the simulation application layer in DDC can be effectively solved by adopting appropriate failure prediction methods. At present, the research on failure prediction methods has been perfected, so how to obtain simulation data characteristics and general fault templates are the focus of our future work.

References

1. Liu, D.: Is Distributed Architecture the Future of Data Centers. https://www.talkwithtrend.com/Article/245739. Last Accessed 26 August 2019 (in Chinese)
2. Ahmed, K.M.U., Bollen, M.H.J., Alvarez, M.: A review of data centers energy consumption and reliability modeling. IEEE Access. **9**, 152536–152563 (2021)
3. Datskova, O., Grigoras, C., Shi, W.: Failure analysis for distributed computing environments. In: UCC Companion'17: Companion Proceedings of the10th International Conference on Utility and Cloud Computing (2017)
4. Inuzuka, F., et al.: Demonstration of a novel framework for proactive maintenance using failure prediction and bit lossless protection with autonomous network diagnosis system. J. Lightwave Technol. **38**(9), 2695–2702 (2020)
5. Zhao, J., et al.: Explore unlabeled big data learning to online failure prediction in safety-aware cloud environment. J. Parallel Distrib. Comput. **153**, 53–63 (2021)

6. Xue, L., et al.: Cloud-native intelligent operation and maintenance architecture and key technologies. Telecommun. Sci. **36**(12), 105–112 (2020). (in Chinese)
7. Peter Bourgon. Metrics, tracing, and logging. http://peter.bourgon.org/blog/2017/02/21/metrics-tracing-and-logging.html, last accessed 2017/02/21
8. Skydt, M.R., Bang, M., Shaker, H.R.: A probabilistic sequence classification approach for early fault prediction in distribution grids using long short-term memory neural networks. Measurement. **170**, 108691 (2021)
9. Jia, T., Li, Y., Wu, Z.H.: Survey of state-of-the-art log-based failure diagnosis. J. Softw. **31**(7), 1997–2018 (2020). (in Chinese)
10. Nandi, A., et al. Anomaly detection using program control flow graph mining from execution logs. In: KDD'16: Proceedings of the 22nd ACM SIGKDD International Conference on Knowledge Discovery and Data Mining (2016)
11. Tong, J., et al.: LogSed: anomaly diagnosis through mining time-weighted control flow graph in logs. In: IEEE International Conference on Cloud Computing (2017)
12. Mandal, A., et al.: Improved topology extraction using discriminative parameter mining of logs. Lecture Notes in Computer Science (including subseries Lecture Notes in Artificial Intelligence and Lecture Notes in Bioinformatics) **12712**, 333–345 (2021)
13. Xu, W., et al.: Detecting large-scale system problems by mining console logs. In: SOSP '09: Proceedings of the ACM SIGOPS 22nd Symposium on Operating Systems Principles (2009)
14. Sigelman, B.H., et al. Dapper, a large-scale distributed systems tracing infrastructure. In: Google Technical Report (2010)
15. Yang, Y., Li, Y., Wu, Z.H.: Survey of state-of-the-art distributed tracing technology. J. Softw. **31**(7), 2019–2039 (2020). (in Chinese)
16. Liu, P., et al.: Unsupervised detection of microservice trace anomalies through service-level deep Bayesian networks. In: 2020 IEEE 31st International Symposium on Software Reliability Engineering (ISSRE) (2020)
17. Bogatinovski, J., et al.: Self-supervised anomaly detection from distributed traces. In: 2020 IEEE/ACM 13th International Conference on Utility and Cloud Computing (UCC) (2020)
18. Li, L., et al.: Research on data-driven failure prediction method. Measure. Control Technol. **41**(5), 66–74 (2022). (in Chinese)
19. Chalermarrewong, T., Achalakul, T., See, S.: Failure prediction of data centers using time series and fault tree analysis. In: IEEE International Conference on Parallel and Distributed Systems (2012)
20. Zhang, T., Wang, E., Zhang, D.: Predicting failures in hard drivers based on isolation forest algorithm using sliding window. J. Phys. Conf. Ser. **1187**(4), 042084 (6pp) (2019)
21. Wang, X.F., et al.: Research and application of distributed service fault prediction model based on XGBoost algorithm. Telecommun. Technol. **10**, 13–16 (2019). (in Chinese)
22. Yang, Y., et al.: FP-STE: a novel node failure prediction method based on Spatio-temporal feature extraction in data centers. Comput. Model. Eng. Sci. **123**(3), 1015–1031 (2020)
23. Liu, C., Dai, L., Lai, Y., Lai, G., Mao, W.: Failure prediction of tasks in the cloud at an earlier stage: a solution based on domain information mining. Computing **102**(9), 2001–2023 (2020). https://doi.org/10.1007/s00607-020-00800-1
24. Fronza, I., et al.: Failure prediction based on log files using random indexing and support vector machines. J. Syst. Softw. **86**(1), 2–11 (2013)
25. Memon, M.A., et al.: Defects prediction and prevention approaches for quality software development. Int. J. Adv. Comput. Sci. Appl. **9**(8), 451–457 (2018)
26. Nedelkoski, S., Cardoso, J., Kao, O.: Anomaly detection and classification using distributed tracing and deep learning. In: 2019 19th IEEE/ACM International Symposium on Cluster, Cloud and Grid Computing (CCGRID) (2019)
27. Gao, J.C., Wang, H.Y., Shen, H.Y.: Task failure prediction in cloud data centers using deep learning. IEEE Trans. Serv. Comput. **15**(3), 1411–1422 (2020)

28. Wang, X., et al.: Exploring LSTM based recurrent neural network for failure time series prediction. J. Beijing Univ. Aeronaut. Astronaut. **44**(4), 772–784 (2018). (in Chinese)
29. Ping, Y.U., Jie, C.A.O.: Deep learning approach and its application in fault diagnosis and prognosis. Comput. Eng. Appl. **56**(3), 1–18 (2020). (in Chinese)
30. Yu, F.Y., et al.: DRAM failure prediction in large-scale data centers. In: 2021 IEEE International Conference on Joint Cloud Computing (JCC) (2021)
31. Lin, Q., et al. Predicting Node failure in cloud service systems. In: ESEC/FSE 2018: Proceedings of the 2018 26th ACM Joint Meeting on European Software Engineering Conference and Symposium on the Foundations of Software Engineering (2018)
32. Zheng, W., Wang, Z., Huang, H., Meng, L., Qiu, X.: SPSRG: a prediction approach for correlated failures in distributed computing systems. Clust. Comput. **19**(4), 1703–1721 (2016). https://doi.org/10.1007/s10586-016-0633-2
33. Ge, N., et al.: Failure prediction in production line based on federated learning: an empirical study. J. Intell. Manuf. **32**, 1–18 (2021)
34. Tolk, A.: Engineering Principles of Combat Modeling and Distributed Simulation. 1st edn. John Wiley & Sons, Inc. (2012)
35. Russell Noseworthy, J.: The test and training enabling architecture (TENA)-supporting the decentralized development of distributed applications and LVC simulations. In: 2008 12th IEEE/ACM International Symposium on Distributed Simulation and Real-Time Applications, pp. 259–268 (2008)

A Knowledge Graph Based Approach to Operational Coordination Recognition in Wargame

Chenye Song[1,3], Ludi Wang[2](✉), Yi Du[2], Xiao Xu[1], Shengming Guo[1], Xiaoyuan He[1], and Lin Wu[1]

[1] Joint War College, National Defense University, Beijing 100094, China
[2] Computer Network Information Center, Chinese Academy of Sciences, Beijing 100083, China
wld@cnic.cn
[3] Graduate School, National Defense University, Beijing 100094, China

Abstract. Recognizing military coordination relationships among adversarial operations is essential in task planning and decision making. Due to the complexity of modern informationized war, it is challenging to analyze the diverse relations between entities in the war situation. In this study, we propose a novel framework based on knowledge graph to predict operational coordinations. We first construct a novel large scale knowledge graph that consits of 29313 nodes and 191542 edges from Wargame Competition dataset. The embedding method jointly considers information from node attributes, local situations and global structure, and then combine the three parts with a self-attention mechanism. Experiments compared with baselines demonstrate that the proposed model is more accurate and robust than existing methods.

Keywords: Coordination recognition · Knowledge graph · Graph neural networks · Attributed heterogeneous network · Wargame

1 Introduction

Modern warfare is characterized by cross-domain operations and interactions. Military forces have to coordinate with others to achieve optimized effectiveness and efficiency. As a system of military units featuring complex associations in spatial, temporal and electromagnetic dimensions, the whole system can be seen as a network, indicating that they can be modeled with graph. Since structures determine functions, an acceptable hypothesis is that, if a large number of similar local structures can be found in many combat processes, these play similar functions that means the operations in these structures have to coordinate with others to achieve the same function, or military task. In the process of a war, no operation can be totally independent to others. Analyzing the relations among entities in battlefield and recognizing coordination relationships between

W. Fan et al. (Eds.): AsiaSim 2022, CCIS 1712, pp. 510–521, 2022.
https://doi.org/10.1007/978-981-19-9198-1_38

operations give military commanders a better understanding about the war situation. It can also support operation command decision making by task mode recommendation [1,11].

Conventional coordination recognition approaches are generally based on hard boundary conditions defined by military experts based on military manuals and knowledge, which are used as the conditions to determine whether there exists a coordination between two operations. These approaches are concise and can be understood easily by military commanders. However, at joint operation level, these approaches of fail due to the complex interaction modes in the system. In the hard boundary condition methods, too "strict" conditions often fail in covering all possible coordinations that might occur in the system, which can result in missed judgment of the coordination that has occured, while too "loose" conditions may cause wrong judgment of the coordination that has not occurred.

To overcome these shortages, this study provides a new framework to recognize coordinations in joint operation level wargames. First we build a large scale Situation Graph contains 29313 nodes and 191542 relations from wargame competition datasets. To extract features in the graph, an embedding method is designed to capture information in entitiy attributes, situation and structure. In coordination reasoning, random-walk is used to obtain node sequences, and a heterogeneous Skip-Gram model is used in training. Experiments compared with baselines demonstrate the applicability and effectiveness of the proposed model in coordination recognition. Experiments compared with baselines demonstrate the applicability and effectiveness in coordination recognition.

Rest part of this paper is as follows. Section 2 describes background and related work to this paper. Section 3 present the Situation Graph. Section 4 describes the collaborative recognition methods using Situation Graph. Section 5 shows the experiments and comparisons with baselines. Section 6 draws conclusions and discuss future trends in military use of knowledge graphs based approaches.

2 Backgrounds and Related Works

2.1 Wargame Dataset

Wargame is developed from manual board deduction which is used to simulate the battlefield environment by chessboard and the military units with different forces by chess pieces [6]. Computer Wargames are experimental tools in planned force testing, plan variation test, concept or force development and procurement [7]. Wargame competition datasets reflect the players knowledge and experience. It provides us a new insight to discover the process of military commanders' situation awareness during the dynamic confrontation in war.

2.2 Knowledge Graph

A Knowledge Graph represents real world facts with nodes and edges in a structural way. It has great potential in expressing military rules and knowledges in wargames. With the in-depth development of the knowledge graph algorithm, the graph em-bedding method based on graph representation learning model has attracted extensive attention. Graph representation learning model can be used to vectorize the operations in the operational situation graph, and the relationship recognition model can discover the coordination relationships between operations to analyze the operation system coordination network [2]. Zhu Tao constructed a decision-related knowledge network based on the information coordination relationships between the different command organizations, and analyzed the time evolution process of different coordination modes based on average path, clustering coefficient, degree distribution and other indicators synchronously [16]; Deller analyzed the relation-ship between the matrix eigenvalues and the deduction results under different node allocation decisions through IACM model and Agent model [17]; Dekker analyzed the relationship between the average path length and synchronization time in the coordination network under different commands and controls, based on the Kuramoto synchronization model in complex networks [5]. In civilian applications, the entity relationship recognition based on graph representation has made progresses [12–15]. By mining structure information and attribute information of multiple graph nodes, the low-dimensional characteristic vector is used to represent nodes, and the representation vector is input into the downstream task model for follow-up recognition task. The initial graph representation method is to map high-dimensional vectors into low-dimensional vectors through the matrix decomposition. However, such method needs to solve the eigenvalue of the eigenmatrix, which is hard to be calculated when the graph size is large; therefore, researchers proposed the graph node representation methods based on the random walk, represented by DeepWalk [3] and Node2Vec [4] methods. Such methods take the node to be represented as the starting point and obtain the walk path of the node through the random walk. The walk path is regarded as a sentence composed of nodes, and the classic algorithm Word2Vec in natural language processing is used for learning, and the final word vector is used as the vectorized representation of the nodes in the graph.

3 Situation Graph Construction

In order to get a structured representation of operations and their relations, we first constructured a Situation Graph (SG) from the dataset.

Situation Graph. Let $G = (V, E)$ denote a graph where V represents the node set and E represents the edge set. There are two subsets in V: an operational entity set and an situational entity set. At joint operation level, operations are performed by different military forces, which means nodes and edges in the graph are attributed and heterogeneous. Hence, let $v_i^{(n)} \in$ V represents an entity vi with type n and $e_{ij}^{(r)} \in$ E represents an edge $e_{ij} = \{v_i, v_j\}$ with type r. Edges between operational entities represents coordinations (see Fig. 1). Weight $w_{ij} \geq 0$ represents correlation of the two entities in semantic spaces.

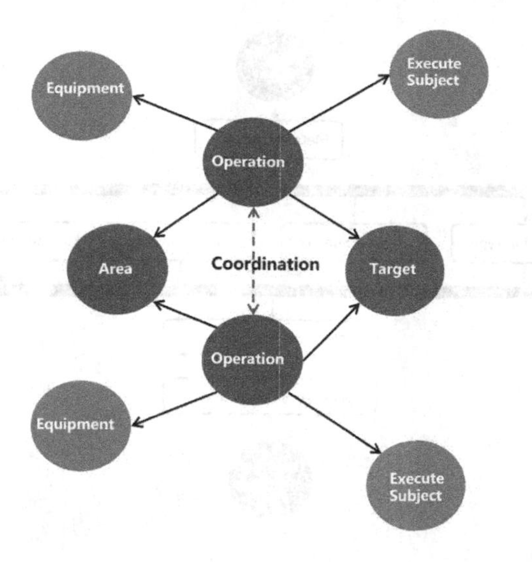

Fig. 1. Situation graph structure and coordination relationships.

Figure 2 describes the process of an SG construction. Analyze historical confrontation orders in the wargame system database that represent different military operations to obtain operations and their related attributes including execute subject, target, area and equipment. Operational entities are originally correlated through situational entities. Coordination relationships between operational entities are predefined by military experts (See in Table 1), including fire coordination, information coordination, electronic coordination, and cover coordination. In the SG, operational entities can be related through situational entities, and vice-versa, situational entities can jointly represent an operation concept. Thus we can get a structured representation of the whole wargame process. On a single temporal slice of the process, we can also describe the wargame situation structure. The task of coordination recognition is to predict the edge type between two operational entities in the SG by learning a representation of the graph. After constructing the situation graph, we next focus on embedding methods to represent the graph.

Table 1. Coordination type and number.

Coordination type	Number
Information	40411
Electronic confrontation	7472
Cover	53010
Fire	23674

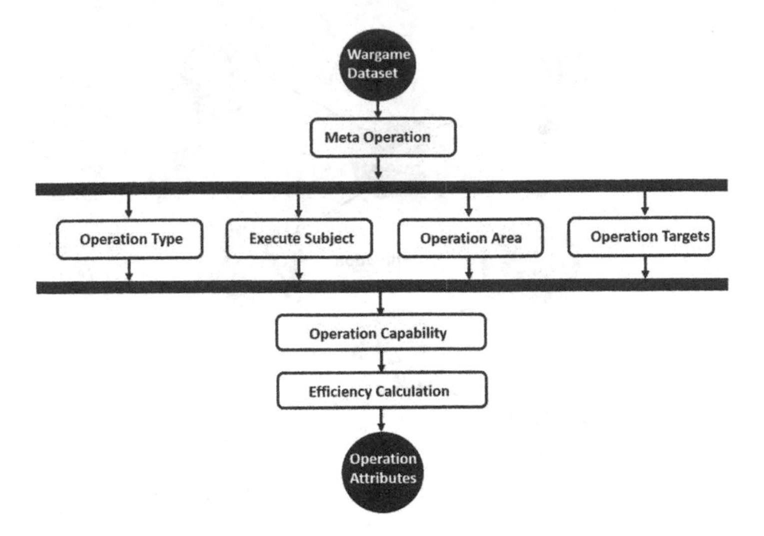

Fig. 2. Order data analysis process of the wargame system.

4 Situation Graph Embedding

Mapping the nodes in SG into an appropriate representation space is vital to the success of the follow-up prediction task. Figure 3 shows the overall structure of our model [9,10]. The embedding method is based on GATNE [10], including three parts: the attribute embedding, the situation embedding and the structure embedding. The outputs of the three parts are integrated into the final entity representation vector through a self-attention mechanism.

The attribute of each entity is vectorized through a multi-layer neural network, which represents the influence of the entity itself. The situation embedding considers the information of other entities without coordination relationship, which represents the influences of other entities in the battlefield. The structure embedding considers the influences of higher order correlation structure of the graph. In the structure embedding layer, the node representation is learned with a pre-training model. Informations from different embedding parts are combined with a self-attention mechanism. Furthermore, relationship weights are considered in random walk stage to increase the influences of the neighbor nodes with higher weights, thus achieving a more accurate vectorized representation of the graph.

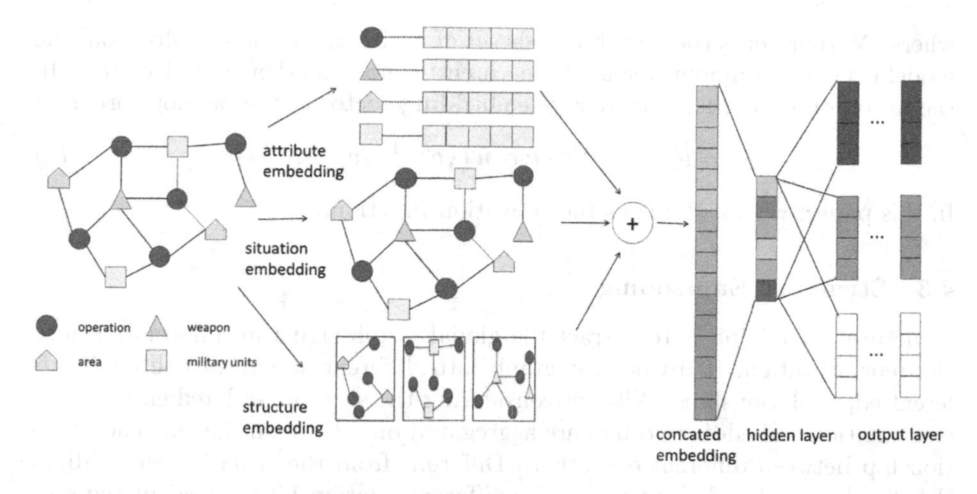

Fig. 3. Coordination relationship recognition framework.

4.1 Attribute Embedding

Attribute embedding aims at extracting the informations from the node attributes into the representation. Table 2 shows the main attributes of an operational entity. Each attribute is converted to an one-hot vector x_i and then get the attribute embedding v_i through a multi-layer neural network.

$$A_i = \sigma\left(W_a \cdot x_i\right) \tag{1}$$

Table 2. Operational entity attributes

Attribute name	Attribute interpretation
Operation type	Specific type of the operation, such as air-to-ship attack, air-to-ground attack, etc.
Order issuance time	Specific time for issuing operation orders
Execution duration	Duration for executing the operation, which is taken as the end time of the command minus order issuance time
Completion degree	Operation completion status

4.2 Situation Embedding

The main purpose of situation embedding is to extract the informations from local situation into the representation. The model calculates a k-order embedding vector by using the operational entity itself and its correlated situational entities. For each entity v_i, its situation embedding vector can be represented as:

$$b_i^k = agg\left(\left\{b_i^{k-1}, \forall v_j \in N_i\right\}\right) \tag{2}$$

where, N_i represents the neighbor node set of entity v_i. In the initialization, the model generates random weights for each entity node as their initial vectors. In the fusion of each order, the average embedding vector of the previous order is:

$$B_i = b_i^k = \sigma\left(W^k \cdot mean\left(\left\{b_i^{k-1}, \forall v_j \in N_i\right\}\right)\right) \tag{3}$$

In this paper, we use ReLU as the activation function.

4.3 Structure Embedding

Structure embedding is to extract the global graph structure information into the representation. It divides the graph into different sub-graphs through different edge relationships. Without considering the situation-related entities, the coordination embedding vectors are aggregated only through the existence relationship between different operations. Different from the situation embedding, the structure embedding vectors under different relationships are calculated separately, and finally aggregated by:

$$u_{i,r}^k = agg\left(\left\{u_{i,r}^{k-1}, \forall v_j \in N_{i,r}\right\}\right), u_{i,r}^k \in R^s \tag{4}$$

The structure embedding comprehensively considers the structure embedding vectors under all relationships, which is finally aggregated into:

$$U_i = \left(u_{i,1}, u_{i,2}, \ldots, u_{i,R}\right) \tag{5}$$

where, R represents the number of relationship types in the graph, which is 5 in this paper. In order to better calculate the structure embedding vectors under multiple relationships, a self-attention mechanism is used to capture the more influential relationship types. The calculation formula of its combination coefficient $att_{i,r}$ is:

$$att_i = softmax\left(w^T \cdot \tanh\left(W_i U_i\right)\right)^T \tag{6}$$

where, w_r and W_r are weights with training Above all, we get the structure embedding vector of the operational entity node v_i:

$$C_i = \alpha_r M_r^T U_i att_i \tag{7}$$

The embedding vectors obtained by the three embedding models are concated as a joint embedding vector V_i for the final representation of the operational entity v_i when the coordination relationship is r:

$$V_i = A_i + B_i + C_i \tag{8}$$

5 Model Training

During training, the path sequence of node v_i is derived through the coordination path by random walk. Then a heterogeneous Skip-gram model is used to train and optimize the embedding vectors in the node sequence.

In the path generation, random walk is a commonly used method. It starts from the generated node and jumps to another node according to a specific probability. For example, when describing the joint command decision-making graph from the perspective of fire coordination, a sub-graph $G_r = (V, E_r, A)$ generated under the fire coordination is first extracted; then in this sub-graph, the generated sequence ($P_{i,r} = (v_i, v_j^1, ..., v_j^{l-1})$ is extracted by random walk, where $v_j \in N_{i,r}$; the transition probability of the random walk at step t is:

$$p(v_j \mid v_i) = \begin{cases} \frac{w_{i,j}}{\left|N_{i,r} \sum_{v_j \in N_j} w_{j,i}\right|}, (v_i, v_j) \in E_r, v_j \in V_{t+1} \\ 0, (v_i, v_j) \in E_r, v_j \notin V_{t+1} \\ 0, \qquad (v_i, v_j) \notin E_r \end{cases} \tag{9}$$

Compared with the conventional random walk, we take the edge weight into consideration in "jumping". As a quantification of the influence of time, the weight is proportional to the absolute value of the time difference executed between the two operations. Moreover, for the operation nodes on the walk path, we use Skip-gram window filter method to generate the semantic context of the operation nodes, so that multiple triples can be constructed, and the ultimate objective of the model is to minimize the following loss function while introducing negative samples:

$$E = -\log \sigma\left(c_j^T \cdot v_{i,r}\right) - \sum_{l=1}^{L} E_{v_k \sim P_t(v)}\left[\log \sigma\left(-c_k^T \cdot v_{i,r}\right)\right] \tag{10}$$

The content in this part is the same as the training of the Skip-gram model in natural language processing [3].

6 Experiments

6.1 Dataset and Model Setting

A situation graph is constructed by analyzing and extracting the data of the historical confrontation cases in the "human in loop" wargame system. The graph includes operation represented by the orders issued by military commander, and operation-related entities and their attributes, such as combat units, operation phases, geographic areas, weaponry, etc. After the data from multiple simulated confrontations are analyzed and extracted, the joint operational situation graph, including 29,313 entities and 191,542 relationships was finally constructed.

In the setting of the model hyper parameters, the final embedding vector dimension is set to 200, and the coordination embedding dimension is set to 10. In random walk, the walk length of each node is set to 20, the window size of the semantic context of the Skip-gram model node is 5, the ratio of positive and negative sample sizes is set to 5, and the maximum number of iterations is 100. Moreover, the early stopping method based on the verification set is used to prevent over-fitting. The optimization function adopts adaptive moment estimator (Adam Optimizer), and the learning rate is set to 0.001.

6.2 Verification

For the prediction task, the coordination relationships of some entity are hidden; and the remaining entity set relationships are trained. We have constructed a training set, a verification set and a test set. The training set is used for model training. The verification set is used for early stopping and hyper-parameter tuning. The test set is used for verifying accuracy and robustness of the model. In terms of verification method, we selected the commonly-used evaluation criteria in the field of relationship recognition, namely, ROC-AUC, PR-AUC and F1 score. Figure 4 shows the model training process.

In terms of verification, we selected the commonly-used evaluation criteria in the field of relationship recognition, namely, ROC-AUC, PR-AUC and F1 score. In the comparison experiment, we use the non-graph embedding method as the comparison model, and used the multi-layer neural network for relationship recognition as the reference model of the graph embedding method; moreover, we selected the classic graph embedding method LINE [8] as a baseline model to prove that the joint embedding method adopted in this paper can better capture topology information and attribute information of military operation in the graph to acquire the higher recognition accuracy.

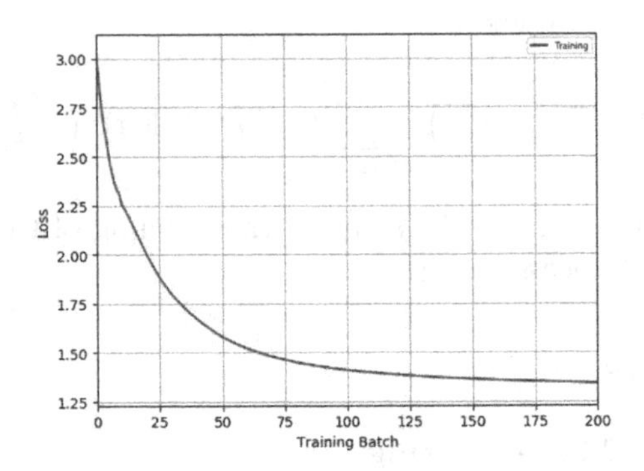

Fig. 4. Loss curve in training.

Multi-layer Neural Network. As the most studied neural network model with classic structure, the multi-layer neural network generally consists of input layer, hidden layer and output layer. The network model can be regarded as a mapping from the input space X to the output space O, and its mathematical description is as follows:

$$o_j = g_i\left(x_i, \theta\right) = \sigma_j\left(\sum_{k=1} w_k \sigma_k\left[\sum_{i=1} v_i x_i + b_1\right] + b2\right) \quad (11)$$

We generate the operation entity pairs through the operational situation graph. During model training, we only constructed the multi-layer neural network for the entity attributes to generate vector representation.

LINE. As one of the classic graph embedding models, LINE [8] has been widely applied in the field of relationship prediction. LINE constructs a loss function for training based on the first-order similarity and the second-order similarity between the nodes, obtains different embedding vectors respectively, and splices the two as the final node representation vector.

The first-order similarity mainly represents the direct correlation of the two nodes in the graph, and the weight of the edge is measured by the similarity of the two nodes. For example, in Fig. 5, there exists a correlation relationship between the nodes 5 and 6, then the first-order similarity between them is greater than zero, and the similarity is measured by weight. The second-order similarity mainly represents the similarity between the neighbors of the two nodes in the graph. For example, although there exists no direct correlation relationship between the nodes 4 and 5 in the graph, but the neighbors of the two are the same, then there exists the second-order similarity between them.

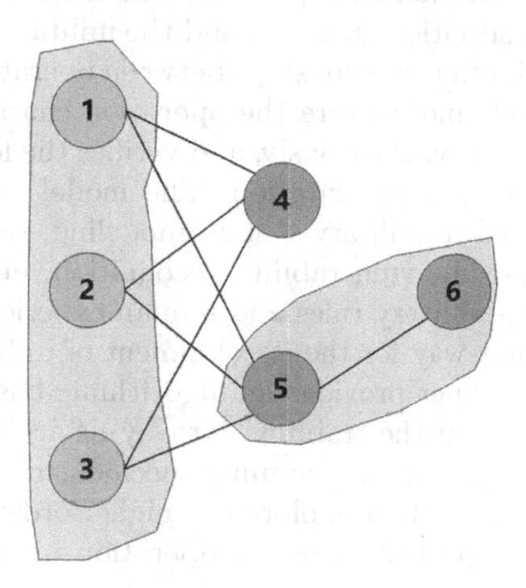

Fig. 5. Embedding of LINE.

In comparison model designing, we calculate the second-order similarity embedding vector of each operation entity through the operational entities and their related situational entities (execute subject, areas, etc.) as the representation vectors.

Results demonstrate that our model performs better in recognition task due to its ability in capturing topology information and attribute information of the entities in the graph (see in Table 3).

The proposed model can significantly improve the recognition results, which are 0.9601, 0.8041 and 0.8402 respectively. It jointly considers the topological structure of operational entity in the situation graph, fuses the three parts hierarchically, and further mines the implicit relations of the operational entity.

Table 3. Experimental results

Evaluation criteria	Multi-layer neural network	LINE	Proposed model
ROC-AUC	0.6163	0.7737	0.9601
PR-AUC	0.4970	0.7045	0.8041
F1	0.4883	0.7595	0.8402

7 Conclusions

Multi-forces and arms coordination operation is the main mode of cross-domain joint operations in the future. This paper proposes an effective recognition model of military operation coordination relationship, which makes full use of basic military rules, knowledge and human experience contained in wargame data, constructs the operational situation graph around the military operation, recognizes and predicts the coordination relationships between operations through the joint graph embedding model, and capture the operation characteristics and battlefield situation information synchronously, and verifies the feasibility of command decision-making knowledge representation. The model can be further used in the research fields related to military system modeling, such as task mode mining, system risk assessment, vulnerability recognition, etc. Making full use of existing data and fusing military rules and human experience in the model construction is an important way for the development of military intelligence. The model proposed in this paper provides an algorithmic basis to deeply mine the implicit modes contained in the training data by using intelligent means. It is an experimental exploration of the winning mechanism of the joint operation. In the future, we'd like to further explore the higher order structure in military systems to find better representations for operation mode research with more wargame data.

References

1. Hu, X., Si, G., Wu, L., et al.: Wargaming and Simulation Principle and System. National Defense University Press, Beijing (2009)
2. Wang, B., Wu, L., Hu, X., et al.: Knowledge representation method of joint operation situation based on knowledge graph. J. Syst. Simul. **31**(11), 2228 (2019)
3. Perozzi, B., Al-Rfou, R., Skiena, S.: DeepWalk: online learning of social representations. In: Proceedings of the 20th ACM SIGKDD International Conference on Knowledge Discovery and Data Mining, KDD 2014, pp. 701–710. Association for Computing Machinery, New York (2014). https://doi.org/10.1145/2623330.2623732
4. Grover, A., Leskovec, J.: Node2vec: scalable feature learning for networks. In: Proceedings of the 22nd ACM SIGKDD International Conference on Knowledge Discovery and Data Mining, KDD 2016, pp. 855–864. Association for Computing Machinery, New York (2016). https://doi.org/10.1145/2939672.2939754

5. Dekker, A.H.: Analyzing C2 structures and self-synchronization with simple computational models. Technical report, Defence Science and Technology Organisation (Australia) Joint Operations (2011)
6. Liu, X., Zhao, M., Dai, S., Yin, Q., Ni, W.: Tactical intention recognition in wargame. In: 2021 IEEE 6th International Conference on Computer and Communication Systems (ICCCS), pp. 429–434(2021)
7. Goodman, J., Risi, S., Lucas, S.: AI and wargaming. arXiv preprint arXiv:2009.08922 (2020)
8. Tang, J., Qu, M., Wang, M., Zhang, M., Yan, J., Mei, Q.: LINE: large-scale information network embedding. In: Proceedings of the 24th International Conference on World Wide Web, WWW 2015, pp. 1067–1077. International World Wide Web Conferences Steering Committee, Republic and Canton of Geneva, CHE (2015). https://doi.org/10.1145/2736277.2741093
9. Yang, C., Liu, Z., Zhao, D., Sun, M., Chang, E.Y.: Network representation learning with rich text information. In: Proceedings of the 24th International Conference on Artificial Intelligence, IJCAI 2015, pp. 2111–2117. AAAI Press (2015)
10. Cen, Y., Zou, X., Zhang, J., Yang, H., Zhou, J., Tang, J.: Representation learning for attributed multiplex heterogeneous network. In: Proceedings of the 25th ACM SIGKDD International Conference on Knowledge Discovery & Data Mining, pp. 1358–1368. Association for Computing Machinery (2019)
11. Wu, W., Hu, X., Guo, S., et al.: Analysis of combat SoS coordination based on multi-layered temporal networks. Complex Syst. Complex. Sci. **14**(2), 1–10 (2017)
12. Lin, Z., et al.: A structured self-attentive sentence embedding. arXiv preprint arXiv:1703.03130 (2017)
13. Zhang, H., Qiu, L., Yi, L., Song, Y.: Scalable multiplex network embedding. In: Proceedings of the 27th International Joint Conference on Artificial Intelligence, IJCAI 2018, pp. 3082–3088. AAAI Press (2018)
14. Shi, C., Hu, B., Zhao, W.X., Philip, S.Y.: Heterogeneous information network embedding for recommendation. IEEE Trans. Knowl. Data Eng. **31**(2), 357–370 (2018)
15. Tang, J., Qu, M., Mei, Q.: PTE: predictive text embedding through large-scale heterogeneous text networks. In: Proceedings of the 21th ACM SIGKDD International Conference on Knowledge Discovery and Data Mining, pp. 1165–1174 (2015)
16. Zhu, T., Chang, G., Zhang, S., et al.: Research on model of command and control information coorperation based on complex networks. J. Syst. Simul. **20**(22), 6058–6060 (2008)
17. Deller, S., Bowling, S.R., Rabadi, G.A., Tolk, A., Bell, M.I.: Applying the information age combat model: quantitative analysis of network centric operations. Int. C2 J. **3**(1), 1–25 (2009)

Artificial Intelligence for Simulation

Heavy-Duty Emission Prediction Model Using Wavelet Features and ResNet

Ruibin Wang[1,2], Xiushan Xia[3(✉)], and Zhenyi Xu[2(✉)]

[1] School of Computer Science and Technology, Anhui University, Hefei 230039, China
e20201061@stu.ahu.edu.cn

[2] Institute of Artificial Intelligence, Hefei Comprehensive National Science Center, Hefei 230088, China
xuzhenyi@mail.ustc.edu.cn

[3] Institute of Advanced Technology, University of Science and Technology of China, Hefei 230088, China
xiaxiushan@iat.ustc.edu.cn

Abstract. To solve the problem that the COPERT model cannot be applied to heavy-duty diesel vehicle OBD monitoring data to obtain accurate NOx emission factors, we propose a ResNet amendment model based on historical driving time-frequency wavelet features. First, for multiple consecutive driving segments of the actual collected diesel vehicle single-vehicle trip data, the attributes with high correlation with the actual OBD emission factors were obtained using Spearman rank correlation analysis based on the data volume. Then, the highly correlated attributes and the emission factors predicted by the COPERT model were combined by constructing the historical information matrix and using the continuous wavelet transform for the time-frequency representation; Finally, the time-frequency representation was used as the input to complete the amendment of the COPERT model on the ResNet50 network. The experimental results show that the proposed method can effectively amend the COPERT model so that the COPERT model can be applied to OBD data to achieve accurate prediction of NOx emission factors.

Keywords: COPERT · OBD · Time-frequency information · Amendment model

1 Introduction

In the past few years, the transportation industry has achieved significant emission reductions, but NOx emissions still account for a great proportion. In terms of NOx emissions, diesel vehicles account for more than 80% of total vehicle emissions and are the main contributor to NOx emissions. For the exhaust emission monitoring of diesel vehicles, the OBD (On Board Diagnostics) [1] vehicle

This work was supported in part by the National Natural Science Foundation of China (62103124, 62033012, 61725304), Major Special Science and Technology Project of Anhui, China (201903a07020012, 202003a07020009, 2022107020030), China Postdoctoral Science Foundation (2021M703119).

The original version of this chapter was revised: The reference [8] has been added to the text and source citation has been added in the bibliography. The correction to this chapter is available at https://doi.org/10.1007/978-981-19-9198-1_49

W. Fan et al. (Eds.): AsiaSim 2022, CCIS 1712, pp. 525–537, 2022.
https://doi.org/10.1007/978-981-19-9198-1_39

network online monitoring method, which has the advantages of representative data conditions, low testing cost and high accuracy, has become mainstream in China.

The COPERT emission model [2] was developed with funding from the European Environment Agency and is compatible with different national standards and parameter variables. Since the emission standards and vehicle control technologies adopted in China are close to those in Europe, COPERT has been widely used in domestic motor vehicle pollution emission studies [3–5], and in addition, the parameters of COPERT model are relatively easy to obtain compared with other models. While the parameters of COPERT model are obtained based on bench test, it is not practical to use COPERT emission model directly on OBD data, and we need to amend the COPERT emission model to make it applicable to OBD monitoring data.

On-road mobile source operation status is the main factor affecting exhaust emissions, and the modeling of on-road mobile source emissions usually establishes the relationship between mobile source operation status and pollutant emission level, and the validity of the model depends on the accurate representation of mobile source operation status [6]. However, the vehicle driving status is affected by driving behavior, external environment and other factors and is complex and variable, so only relying on a moment of monitoring data is not enough to accurately represent the vehicle driving status.

To address the above problems, this paper proposes a ResNet50 correction model (HTFResNet50) using historical time-frequency information to realize the amendment of COPERT emission model and make it successfully applied on OBD data. First, for the actual diesel vehicle single-vehicle collection data, we perform the segmentation of continuous driving segments and filter out the attributes with strong correlation with the actual NOx emission factor based on the data volume using Speaman rank correlation analysis; Then, the historical information matrix is constructed by the relevant attributes under the same moment of the actual NOx emission factor, and the time-frequency matrix of each time-series matrix is obtained by continuous wavelet transform; finally, the time-frequency matrix is extracted by using ResNet50 network for feature extraction, and the NOx emission factor obtained after the COPERT model is amended. The experimental results show that HTFResNet50 can make a reasonable amendment to COPERT.

2 Method Background

2.1 Continuous Wavelet Transform

Compared to the one-dimensional temporal signal, its two-dimensional form carries a greater amount of information [7]. When using CNNs for feature extraction, we want the more adequate information that can be obtained, the better. Therefore, it is necessary to transform the two-dimensional historical information matrix, which is combined from one-dimensional information, into a higher dimensional representation [8].

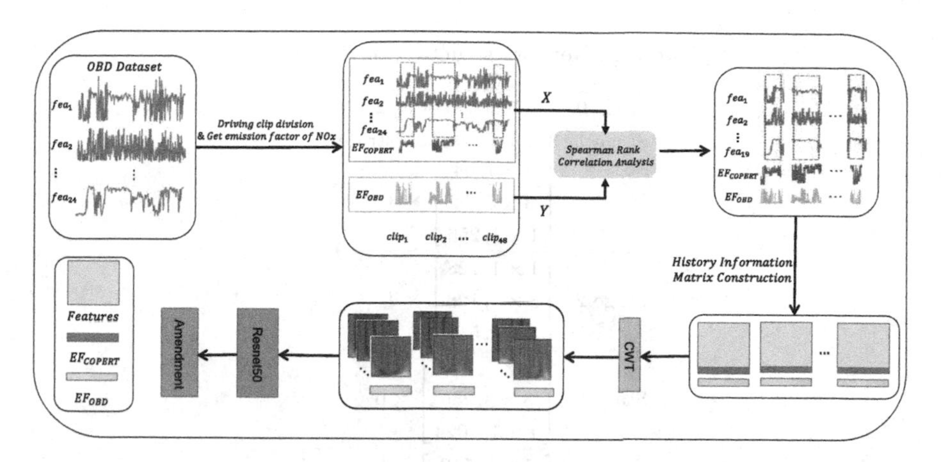

Fig. 1. The architecture of the HTFResNet50 network

The continuous wavelet transform (CWT) is an adaptive time-frequency analysis method that decomposes the original signal into a time-scale plane by scaling and translating the wavelet basis function. For any $L^2(R)$ space with function $f(t)$, the conversion details of CWT are shown in Eq. 1 [9].

$$CWT_f(a,\tau) = \langle f(t), \psi_{a,\tau}(t) \rangle = \frac{1}{\sqrt{a}} \int f(t)\psi(\frac{t-\tau}{a})dt \tag{1}$$

where $\psi_{a,\tau}(t) = \frac{1}{\sqrt{a}}\psi\left(\frac{t-\tau}{a}\right)$, a and τ are the stretching and translation factors, respectively, and $\psi(t)$ is a wavelet basis function that can be obtained by stretching and translating the mother wavelet.

2.2 ResNet

CNN is one of the most commonly used structures in deep learning, which solves the disadvantages faced by fully connected neural networks in dealing with processing two-dimensional data, such as many parameters, difficult training and easy overfitting. In recent years, CNN network structures have emerged, such as AlexNet [10], GooLeNet [11], VGGNet [12], ResNet [13], etc. have been proposed one after another. Among all the deep networks, ResNet uses the idea of residual learning to solve the degradation phenomenon that the number of layers stacked in a deep network rises while the performance decreases.

ResNet is almost the most widely used feature extraction network nowadays. The commonly used ResNet network models are ResNet18, ResNet34, ResNet50, ResNet101 and ResNet152. In this paper, we choose to use ResNet50 as the final backbone network, and its network structure is shown in Table 1.

Table 1. Network structure of ResNet50

Stage 0	7 × 7, 64, strider 2
Stage 1	3 × 3 max pool, stride 2 $\begin{bmatrix} 1 \times 1, 64 \\ 3 \times 3, 64 \\ 1 \times 1, 256 \end{bmatrix} \times 3$
Stage 2	$\begin{bmatrix} 1 \times 1, 128 \\ 3 \times 3, 128 \\ 1 \times 1, 512 \end{bmatrix} \times 4$
Stage 3	$\begin{bmatrix} 1 \times 1, 256 \\ 3 \times 3, 256 \\ 1 \times 1, 1024 \end{bmatrix} \times 6$
Stage 4	$\begin{bmatrix} 1 \times 1, 512 \\ 3 \times 3, 512 \\ 1 \times 1, 2048 \end{bmatrix} \times 3$

Table 2. Attribute comparison table

Name	Label
Engine speed	E_{speed}
Actual output torque percentage	AOTP
Engine water temperature	EWT
Engine fuel temperature	EFT
Post-treatment downstream NOx	NOx
Post-treatment downstream oxygen	O_2
Atmospheric pressure	AP
Environmental temperature	ET
Post-treatment exhaust gas mass flow rate	PT_{EGM}
Urea tank level	UTL
Urea tank temperature	UTT
Vehicle speed	V_{speed}
Gas peddal opening	GPO
Single Driving Miles	SDM
Engine fuel consumption rate (instantaneous)	EFCR
Average Engine fuel consumption rate	$EFCR_{avg}$
Engine fuel consumption for single driving	EFC_{SD}
Total engine fuel consumption	EFC_{total}
Battery voltage	BV
Fuel tank level	FTL
Cumulative engine runtime	CER
Longitude	LNG
Latitude	LAT

3 Methodology

As shown in Fig. 1, a heavy-duty diesel vehicle exhaust emission amendment model based on historical time-frequency information is proposed in this paper. First, for the OBD monitoring dataset, we reasonably partition the dataset to obtain multiple consecutive travel segments, calculate the corresponding COPERT model emission factors and OBD emission factors for each record of each travel segment, and filter out the attributes with high correlation with the OBD emission factors by Spearman correlation rank analysis based on the data volume. Then, the historical information matrix is constructed for the dataset consisting of highly correlated attributes and emission factors, and the CWT is used to convert the historical information matrix into the corresponding time-frequency matrix; Finally, the time-frequency matrix is used as the input of the ResNet50 model and the final amended model is trained. The comparison experiments show that our model has excellent performance for COPERT model to obtain NOx emission factors on the OBD dataset.

3.1 Data Processing

a) Driving Clip Division. The collected attributes include: license plate, terminal number, date, engine speed (rpm), actual output torque percentage (%), engine water temperature (°C), engine fuel temperature (°C), etc. After data preprocessing means such as removing irrelevant attributes and removing invalid records, each attribute with relevant abbreviations is shown in Table 2.

Since the actual data consisted of multiple trip records, each starting phase was not specifically labeled on the dataset, which was not ideal for our later use of the data. Therefore, we need to divide the whole data set into driving segments, because we delete some illegal records in the data pre-processing stage, so in the original continuous period of time will also cause the time interval is greater than the sampling interval, if the adoption interval is set to the maximum value of continuous time will split the original continuous driving segments again, which is what we do not want to see, so we set the maximum value of the time interval to 180s here, and the adjacent records with interval greater than 180 are considered to be divided into different driving segments. In addition, we hope that each selected driving segment can be as long as possible to meet the emission factor calculation based on the historical information of a certain length later, so we set the minimum number of records for driving segments to 180, i.e., 15 min. 48 driving segments were obtained after division, and the amount of data for each segment is shown in Table 3.

b) Get Emission Factor of NOx. Since the value of NOx emission factor is not an attribute directly monitored by OBD, it cannot be obtained directly from the OBD data set. Therefore, we refer to the COPERT model [14] and the existing method of estimating NOx emission factor using OBD data [15] to calculate the NOx emission factor EF_{COPERT} and OBD emission factor EF_{OBD} for COPERT. In the process of estimating the emission factor, we consider that

Table 3. Valid driving clip display

Clip number	Sample size	Clip number	Sample size
clip 1	580	clip 2	654
clip 3	314	clip 4	813
clip 5	342	clip 6	781
clip 7	258	clip 8	218
clip 9	261	clip 10	604
clip 11	303	clip 12	190
clip 13	371	clip 14	485
clip 15	362	clip 16	262
clip 17	242	clip 18	585
clip 19	301	clip 20	261
clip 21	226	clip 22	204
clip 23	287	clip 24	256
clip 25	236	clip 26	322
clip 27	281	clip 28	185
clip 29	435	clip 30	447
clip 31	453	clip 32	461
clip 33	206	clip 34	217
clip 35	398	clip 36	428
clip 37	378	clip 38	285
clip 39	203	clip 40	495
clip 41	470	clip 42	271
clip 43	258	clip 44	209
clip 45	522	clip 46	232
clip 47	198	clip 48	193

the concentration of the emission factor of the current record is jointly determined by the first to the kth record, so we calculate the emission factor of the kth record by averaging the data of the first to the kth record. The calculation results (k = 48) are shown in Fig. 2. It is easily known from the figure that the emission factors estimated by the COPERT model are poor on the OBD data set and fit poorly with the actual estimated emission factors of OBD.

3.2 Screening of Correlation Factors

The various attributes collected by OBD interfere with the accurate estimation of the concentrations of emission factors, and the weakly correlated attributes have a low contribution in the model training process, which not only makes it difficult to provide help to the model performance improvement, but also makes the amount of model parameters more. Therefore, it is necessary to select factors with strong correlation with EF_{OBD} before starting the model training. In this paper, we propose a Spearman rank correlation analysis method based on data

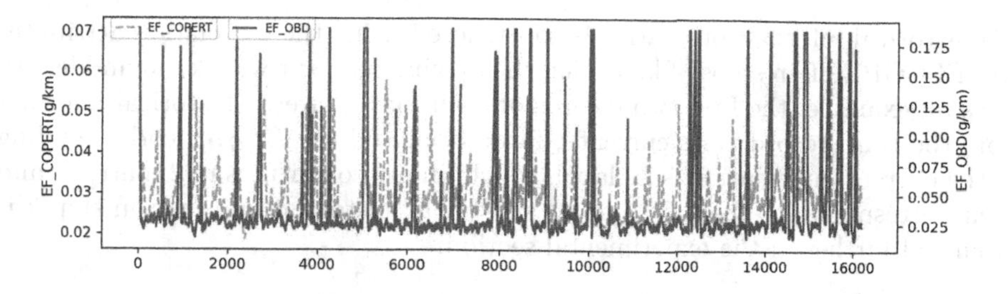

Fig. 2. Emission factor of NOx

volume for correlation factor selection for multiple consecutive driving segments with differences in data volume.

First, we performed Spearman correlation analysis for all driving segments and performed hypothesis testing at $t = 1.645$ to select attributes with high correlation with the actual emission factor of OBD, based on which we constructed a correlation matrix C. Each row of C corresponds to the correlation results of each attribute in a driving segment and EF_{COPERT} with EF_{OBD} (correlation record is 1 and vice versa is 0), and each column represents a class of attributes. For different driving segment data volumes, we build the data volume matrix D. The i-th row represents the i-th driving segment, and the values of each column in each row are equal to the data volume of the i-th driving segment. Matrix C and matrix D are homotypic matrices. The Hadamard product of matrices C and D is calculated, and the columns in its result represent the correlation between each attribute and NOx emission factor about the data volume, and each column of this matrix is summed to obtain the final correlation data volume matrix R_C. The mathematical description is shown below.

$$\begin{aligned} R_C &= \sum_{i=0}^{num(fea)} C \odot D \\ &= [nums_{f1}, nums_{f2}, \cdots, nums_{fn}, nums_{EF_{COPERT}}] \end{aligned} \tag{2}$$

where $nums_{fi}$ represents the amount of data associated with EF_{OBD} for the i-th attribute, $nums_{EF_{COPERT}}$ is the amount of data associated with EF_{OBD} for the emission factor calculated by the COPERT model, and $\odot$ denotes the Hadamard product. The amount of data for each attribute in R_C is compared with the set amount of relevant data P_C, which is considered as relevant when $P_C \geq nums_{f_i}$, and then the relevant attributes are combined with emission factors to obtain the final data set.

3.3 Historical Time-Frequency Feature Construction

a) History Information Matrix Construction. For the final dataset with the combination of correlation factors and emission factors, we construct the historical information matrix for each moment by considering the impact that historical information brings to the current moment. For each driving segment, the historical information matrix is constructed using the correlation attributes and EFCOPERT in steps of k. Taking the driving segment with k = 3 and length m as an example, the first two records of each matrix are its historical information, the third record is its current information, and the EF_{OBD} corresponding to the current information is its label, which finally constitutes m-2 matrices and their corresponding labels. The exact size of the historical information step k is discussed further in the experimental section.

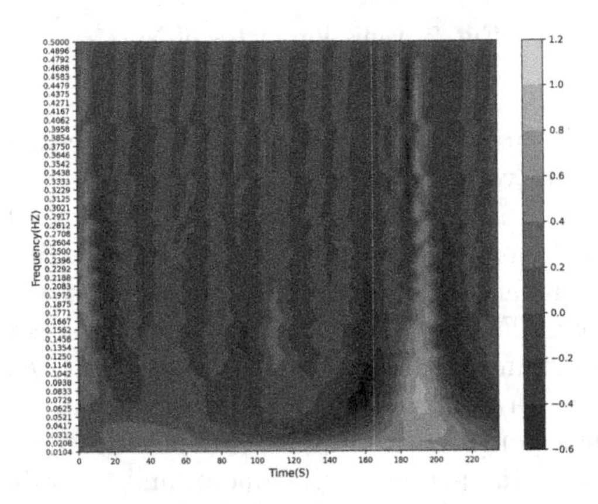

Fig. 3. Visualization of the time-frequency matrix

b) CWT. For each column of the history information matrix, CWT can be used to transform the one-dimensional time-series signal into a two-dimensional time-frequency matrix, and the transformed time-frequency matrix can effectively visualize the time-domain information and frequency-domain information of the mobile source emission attribute signal, and clearly describe the relationship between signal frequency and time transformation.

In this paper, we use the mexh function as the wavelet basis function, and visualize the matrix after transforming the column attributes of the historical information matrix by continuous wavelet transform as shown in Fig. 3, where the horizontal axis indicates time, the vertical axis indicates frequency, and the color shade indicates the size of the wavelet coefficients, and the larger the coefficients the darker the color. For each relevant attribute and the time-frequency matrix obtained by EF_{COPERT}, it is superimposed into a multi-channel matrix as the final input to the ResNet50.

3.4 Metrics

For the correction model in this paper, we use MAE (Mean Absolute Error), RMSE (Root Mean Squared Error), and MAPE (Mean Absolute Percentage

Error) to evaluate the validity of the calculation of the correction model on NOx, and its specific calculation method is described as follows.

$$MAE = \frac{1}{n}\sum_{i=1}^{n}|\hat{y}_i - y_i| \tag{3}$$

$$MAPE = \frac{100\%}{n}\sum_{i=1}^{n}\left|\frac{\hat{y}_i - y_i}{y_i}\right| \tag{4}$$

$$RMSE = \sqrt{\frac{1}{n}\sum_{i=1}^{n}(\hat{y}_i - y_i)^2} \tag{5}$$

where n is the number of samples, y_i is the true value of the label, and $\hat{y}_i$ is the predicted value of the label.

4 Experiments

In this paper, we normalize the dataset by Min-Max normalization and divide it into training, validation and test sets according to 7:2:1, with bathsize of 64, optimizer of Adam, learning rate of 0.0001, loss of MSE, maximum number of iterations of 200, and stop training when the loss stops decreasing for 15 consecutive times.

4.1 Setting of History Information Step

The historical information step size has a large impact on the performance of the model. As the step size of the historical information increases, the number of parameters input to the model increases, but the accuracy and stability do not strictly increase with it, and the existing knowledge and data do not support us to choose the appropriate step size directly. Therefore, we designed experiments before the relevant attribute filtering, setting 1 min length (12 data) of historical information as an interval and the set $K_L = 12, 24, 36, 48, 60, 72$ of historical information steps K. The model performance under different k is shown in Table 4.

From the data in the Table 4, it is easy to know that when the history step is too small, the model performance is poor due to insufficient history information. As K increases, the model performance starts to rise gradually, and the model performance reaches the optimum when K = 48. Increasing the history information step can no longer improve the model performance except for increasing the data volume. Therefore, we set the optimal value of the history information step K to 48.

Table 4. Model performance at different K values

K	Metrics		
	MAE	MAPE	RMSE
12	0.0326	34.45%	0.04574
24	0.0263	39.66%	0.0392
36	0.0152	17.36%	0.0266
48	**0.0130**	**16.4%**	**0.0202**
60	0.0143	25.33%	0.0248
72	0.0176	25.12%	0.0267

4.2 Setting of Correlation Data Volume

In the Spearman order correlation analysis based on data volume, the correlation data volume P_C needs to be preset. Too large setting of P_C will lose the attributes that have a positive effect on COPERT model amendment and degrade the performance in the model amendment process; too small setting of P_C will lead to invalid attributes participating in the model amendment increasing the computation, which further may also lead to the deterioration of model performance. Therefore, it is important to choose the appropriate P_C for model amendment, and we set different P_C according to the quartile of total data volume, and the corresponding correlation of each attribute under different P_C is shown in Table 5.

Table 5. The correlation of each feature under different P_C

Feature \ P_C	Q_1	Q_2	Q_3	Feature \ P_c	Q_1	Q_2	Q_3	Feature \ P_c	Q_1	Q_2	Q_3
E_{speed}	✓	✓		AOTP	✓	✓		EWT	✓	✓	
EFT	✓	✓	✓	NO_x	✓	✓		O_2	✓	✓	✓
AP				ET	✓	✓		PT_{EGM}	✓		
UTL	✓			UTT	✓			V_{speed}	✓	✓	✓
GPO	✓	✓		SDM	✓	✓	✓	EFCR	✓	✓	
$EFCR_{avg}$	✓	✓		EFC_{SD}	✓	✓	✓	EFC_{total}	✓	✓	✓
BV	✓	✓		FTL	✓	✓		CER	✓	✓	
LNG	✓	✓	✓	LAT	✓	✓	✓	EF_{COPERT}	✓	✓	✓

After determining the historical information step K, experiments were conducted for different P_C with K = 48, and the experimental results are shown in Table 5, where 0 means no attribute is removed, Q1 is to remove the attribute whose relevant data volume is below the lower quartile of the total data volume, Q2 is to remove the attribute whose relevant data volume is below the median of the total data volume, Q3 is to remove the attribute whose relevant data volume is below the upper quartile of the total data volume, and 1 means only EF_{COPERT} was used to fit the EF_{OBD}.

Table 6. Model performance at different P_C

P_C	Metrics		
	MAE	MAPE	RMSE
0	0.0130	16.4%	0.0202
Q_1	0.0127	16.79%	0.0202
Q_2	**0.0109**	**18.45%**	**0.0175**
Q_3	0.0153	25.8%	0.0253
1	0.1571	236.5%	0.2506

As can be seen from Table 6, the model performance of Q1 state is similar to that of the model without irrelevant factor deletion because only one attribute of AP is deleted; the best performance of the model is at Q2, where all indicators are the smallest, and the rest of the relevant data volume has a significant gap compared with it; while at Q3, the model performance becomes worse because the relevant data volume is too large and too many relevant attributes are deleted; and at $P_C = 1$ shows that using only EF_{COPERT} to fit EF_{OBD} is the least effective and the model basically does not learn anything in favor of COPERT amendment. Therefore, we finally choose the amount of relevant data P_C as Q2.

4.3 Comparative Experiment

In order to analyze the advantages and disadvantages of this model with other mainstream models, we have chosen the following model to compare with our proposed HTFResNet50 model.

a) Support Vector Regression (SVR).
b) Artificial Neural Network (ANN).
c) Convolutional Neural Network (CNN).
d) ResNet50.
e) TSFFResNet [16].

where models a) and b) do not consider the effect of historical information, and the inputs to the training process are a combination of external factors and EF_{COPERT} at the current moment; models c) and d) consider the influence of historical information on the current moment, and the inputs to the training process are the historical information matrix; model e) considers the impact of historical information on the current moment by piecing together each column of the historical information matrix into a 6*8 matrix [16] and superimposing the two-dimensional representations of all attributes into a multichannel matrix, and inputting the historical information matrix and the multichannel matrix into two Resnet50 parallel structures.

Table 7 shows the comparison between the above five models and our proposed model correction prediction results, from the data in the table, we can see that our proposed HTFResNet50 model has the best correction effect. Among

Table 7. Performance of Mainstream model vs. HTFResNet50 (Ours) for COPERT modification

Model	Metrics		
	MAE	MAPE	RMSE
SVR	0.0530	87.18%	0.1524
ANN	0.0229	138.43%	0.0275
CNN	0.01963	60.6%	0.0292
ResNet50	0.0144	19.33%	0.0236
TSFFResNet	0.0163	20.13%	0.0274
HTFResNet50 (Ours)	**0.0109**	**18.45%**	**0.0175**

them, the SVR and ANN without considering the influence of historical information have poorer results, and there is a significant gap with other models considering historical information, while the ANN model can obtain better amendment results than the SVR due to sufficient experimental data. The CNN model has less depth than ResNet50 and is relatively weak in feature extraction, so it has the worst performance under the experiments considering the influence of historical information on the current. The performance comparison of HTFResNet50 with ResNet50 and TSFFResNet50 shows that our proposed model trained with historical time-frequency information can get better amendment effect.

5 Conclusion

In this paper, a modification method based on historical time-frequency information is proposed to address the inability of the COPERT model to accurately predict the NOx emission factors on OBD data. By using the time-frequency representation of the historical information of the attributes monitored by OBD that have high correlation with the emission factors, the amendment of the COPERT model is accomplished on the ResNet50 network, so that the COPERT model can be applied to the OBD data. The excellent amendment performance of our proposed method is demonstrated by comparing it with other models.

References

1. Oluwaseyi, M.M., Sunday, A.M.: Specifications and analysis of digitized diagnostics of automobiles: a case study of on board diagnostic (OBD II). Int. J. Eng. Tech. Res. **9**(1) (2020)
2. Ntziachristos, L., Gkatzoflias, D., Kouridis, C., Samaras, Z.: COPERT: a European road transport emission inventory model. In: Athanasiadis, I.N., Rizzoli, A.E., Mitkas, P.A., Gómez, J.M. (eds.) Information Technologies in Environmental Engineering. ENVENG, pp. 491–504. Springer, Heidelberg (2009). https://doi.org/10.1007/978-3-540-88351-7_37

3. Liu, Y.H., Liao, W.Y., Li, L., et al.: Vehicle emission trends in China's Guangdong Province from 1994 to 2014. Sci. Total Environ. **586**, 512–521 (2017)
4. Sun, S., Jiang, W., Gao, W.: Vehicle emission trends and spatial distribution in Shandong province, China, from 2000 to 2014. Atmos. Environ. **147**, 190–199 (2016)
5. D'Angiola, A., Dawidowski, L.E., Gómez, D.R., et al.: On-road traffic emissions in a megacity. Atmos. Environ. **44**(4), 483–493 (2010)
6. Wang, Z., Guan, W., Jun, B.I., et al.: Microscopic emission model of motor vehicle based on short-time real driving cycle. Chin. J. Environ. Eng. (2016)
7. Liang, P., Deng, C., Wu, J., et al.: Intelligent fault diagnosis of rotating machinery via wavelet transform, generative adversarial nets and convolutional neural network. Measurement **159**, 107768 (2020)
8. Xu, Z., Wang, R., Cao, Y., et al.: Amending the COPERT model for heavy-duty vehicle emissions using a time frequency fusion network. Front. Inf. Technol. Electron. Eng. **1**(1) (1998). https://doi.org/10.1631/FITEE.2200218
9. Miao, R., Gao, Y., Ge, L., et al.: Online defect recognition of narrow overlap weld based on two-stage recognition model combining continuous wavelet transform and convolutional neural network. Comput. Ind. **112**, 103115 (2019)
10. Krizhevsky, A., Sutskever, I., Hinton, G.E.: ImageNet classification with deep convolutional neural networks. In: Advances in Neural Information Processing Systems, vol. 25 (2012)
11. Szegedy, C., Liu, W., Jia, Y., et al.: Going deeper with convolutions. In: Proceedings of the IEEE Conference on Computer Vision and Pattern Recognition, pp. 1–9 (2015)
12. Yue-Hei Ng, J., Hausknecht, M., Vijayanarasimhan, S., et al.: Beyond short snippets: deep networks for video classification. In: Proceedings of the IEEE Conference on Computer Vision and Pattern Recognition, pp. 4694–4702 (2015)
13. He, K., Zhang, X., Ren, S., et al.: Deep residual learning for image recognition. In: Proceedings of the IEEE Conference on Computer Vision and Pattern Recognition, pp. 770–778 (2016)
14. Li, R., Yang, F., Liu, Z., et al.: Effect of taxis on emissions and fuel consumption in a city based on license plate recognition data: a case study in Nanning, China. J. Clean. Prod. **215**, 913–925 (2019)
15. Zhang, S.J., He, L.Q., Wu, Y.: Patent application disclosure, CN 110823585A (2019). (in Chinese)
16. Zhang, W., Xue, F., Xue, F., et al.: A novel intelligent fault diagnosis method of rolling bearing based on two-stream feature fusion convolutional neural network. Measurement **176**, 109226 (2021)

Solder Paste Printing Quality Prediction Model Based on PSO Optimization

Wei Wang[1,2], Wangyou Gui[2,3], and Zhenyi Xu[2,3](✉)

[1] AHU-IAI AI Joint Laboratory, Anhui University, Hefei 230601, China
[2] Institute of Artificial Intelligence, Hefei Comprehensive National Science Center, Hefei 230088, China
[3] Institute of Advanced Technology, University of Science and Technology of China, Hefei 230088, China
xuzhenyi@mail.ustc.edu.cn

Abstract. Statistics show that about 70% of the surface mount quality problems are in the solder paste printing process, so it is particularly important to optimize the solder paste printing process parameters. The traditional parameter adjustment method still relies on the engineer's own experience to manually adjust, which is inefficient. In recent years, artificial intelligence technology has achieved excellent performance in various cities, providing a new idea for the optimization of key process parameters of SMT production lines. Therefore, this paper proposes a method for optimizing process parameters of solder paste printing based on machine learning. It mainly includes two parts, one is the construction of the solder paste prediction model, and the other is the optimization algorithm based on the model prediction parameters. We choose the SVR model as the regressor of parameter prediction, and use the value predicted by the model as the evaluation standard, and use the PSO algorithm to optimize the search in the search space of process parameters. Finally, after experimental tests, the process parameters optimized by this method can well meet the actual production requirements and effectively improve the production efficiency.

Keywords: Solder paste · Support vector regression · Optimization algorithm

1 Introduction

Surface Mount Technology (SMT) is a series of technological processes for processing on the basis of PCB. It is the most popular process and technology in the current electronic assembly industry. The process flow of SMT mainly includes solder paste printing, component placement, reflow soldering, automatic optical

This work was supported in part by the National Natural Science Foundation of China (62103124), Major Special Science and Technology Project of Anhui, China (202104a05020064), China Postdoctoral Science Foundation (2021M703119).

W. Fan et al. (Eds.): AsiaSim 2022, CCIS 1712, pp. 538–547, 2022.
https://doi.org/10.1007/978-981-19-9198-1_40

inspection and other links. The purpose is to use the viscosity of the solder paste to firmly solder surface mount devices (SMD) such as capacitors and resistors to the specified positions on the PCB, and to detect the defects of the soldered devices. Among them, solder paste printing, as the first process of SMT, is also the most critical process, which directly affects whether the entire SMT can proceed normally. The process mainly relies on the squeegee and stencil on the solder paste printer. The small holes on the stencil correspond to the pads at the corresponding positions of the PCB. The squeegee moves on the stencil to deposit the solder paste on the PCB pads through the small holes [9]. In this process, process parameters such as blade pressure, demolding speed, stencil cleaning frequency, and offset need to be set in advance for the solder paste printing machine to control the operation of the printing machine. Obviously, the setting of these process parameters plays a crucial role in the solder paste printing process. The current method adopted by the industry is to manually adjust the solder paste printing process parameters by engineers after the solder paste quality inspection fails. This method has great drawbacks. Different solder paste quality defects are usually associated with multiple process parameters. Engineers cannot accurately locate the corresponding process parameters at the first time. Even if they locate the corresponding process parameters, they cannot The time is adjusted accordingly, which usually requires multiple trial and error, which is inefficient.

In order to solve the above problems, this paper proposes an optimization method of solder paste printing process parameters based on SVR. SVR can efficiently solve various high-dimensional nonlinear problems in practical engineering using only a small amount of sample data [2]. With its powerful nonlinear fitting ability, it automatically learns the relationship between printing machine parameters and solder paste quality parameters. At the same time, the global search of PSO algorithm is used to conduct dynamic search in the parameter search space to obtain the most suitable parameters [4], and then establish the PSO-SVR solder paste quality prediction model. Figure 1 shows the structure of our model.

In Sect. 2, We propose a solder paste printing quality prediction model based on SVR. In Sect. 3, we propose an optimization algorithm called improved PSO algorithm, which has better performance than the traditional PSO algorithm, and use it to dynamically search for parameters. In Sect. 4, we conduct experiments with real data to verify the effectiveness of the model.

2 Related Work

2.1 Support Vector Regression

With the development of support-vector machine (SVM), two models for different tasks have been derived, namely support-vector classification (SVC) and support-vector regression (SVR). The purpose of SVR is to solve the regression problem, and its core idea is to estimate a relationship between the system input and output from the available samples or training data [3]. It is desirable that the relationship should be determined so that the system output matches the real value as closely as possible [6].

Given a set of training and samples $(x_i, y_i)|_{i=1}^{n}, i = 1, \ldots, n$, where x_i represents the input feature vector and y_i represents the output feature vector [6], the purpose of SVR is to map the linearly inseparable input data in the low-dimensional space to the high-level through the mapping function. dimensional space, and then fit a linear regression function. The description of the function is as follows

$$\mathrm{f(x)} = \omega \cdot \Phi(\mathrm{x}) + \mathrm{b} \tag{1}$$

The purpose of training the SVR model is essentially to find the optimal parameters ω and b, so that $f(x_i)$ is as close to y_i as possible, so the solution of Eq. (1) is transformed into minimizing the regression risk, so that for any $\varepsilon > 0$, there are $|y_i - f(x)| \le \varepsilon$, the following formula is introduced for this

$$\frac{1}{2}\|w\|^2 + C\sum_{i=1}^{n}\Gamma(f(x_i) - y_i) \tag{2}$$

where $\frac{1}{2}\|w\|^2$ is the regularization term, C is the penalty factor, and Γ is the cost function.However, since it is usually difficult to determine a suitable ε, a slack variable ξ is introduced to adapt to the unpredictable errors on the training set, the problem can be described as

$$|y_i - w \cdot \Phi(x_i) - b| \le \varepsilon + \xi_i, \quad i = 1, 2, \ldots, n \quad \xi_i, \xi_i^* \ge 0 \tag{3}$$

Using a kernel function, the required decision function can be expressed as

$$f(x) = \sum_{i=1}^{n}(a_i^* - a_i)K(x, x_i) + b \tag{4}$$

The radial-basis function (RBF) has been used in this paper as a kernel []

$$K(\mathbf{x}_1, \mathbf{x}_2) = \exp\left(-\gamma\|\mathbf{x}_1 - \mathbf{x}_2\|_2^2\right) \tag{5}$$

where the σ is called as kernel parameter.

2.2 Particle Swarm Optimization

The particle swarm optimization algorithm (PSO) belongs to the meta-heuristic algorithm, which is derived from the research on the behavior of biological groups in the idea of bionics [1]. The PSO algorithm has the advantages of simple iterative format, few parameters to be adjusted, and can quickly converge to the region where the optimal solution is located [11].

First we need to determine two core properties of the algorithm, namely the velocity and position of the particle. Velocity represents the direction and distance the particle will move in the next iteration, and position is a solution to the problem being solved. The dimensions of the velocity vector and the position vector are determined by the dimension of the search space. For the ith particle, its position is $X_{id} = (x_{i1}, x_{i2}, \ldots, x_{iD})$, velocity is $V_{id} = (v_{i1}, v_{i2}, \ldots, v_{iD})$. For each particle, during the search process, it will continuously adjust its position and velocity according to the fitness value until an optimal position is found. The velocity of the particle at the next moment is determined by three factors, including the inertial direction, the individual optimal direction and the group optimal direction. The following is the optimization formula of the PSO algorithm [10].

$$\mathbf{v}_i(t+1) = \omega(t+1)\mathbf{v}_i(t) + c_1(\mathbf{p}_i - \mathbf{x}_i(t))\mathbf{R}_1 + c_2(\mathbf{g} - \mathbf{x}_i(t))\mathbf{R}_2 \quad (6)$$

where ω is the inertia weight, c_1 is the individual learning factor, c_2 is the group learning factor, t represents the current moment, t+1 represents the next moment, R_1 and R_2 are random constants, and p_i is the historical best position of particle i at the current moment, g is the optimal position of the entire particle swarm at the current moment, v is the velocity vector, and x is the position vector.

3 Solder Paste Printing Process Parameter Optimization Method

3.1 Solder Paste Volume Prediction Model

SVR is a structural risk minimization model, which can well adapt to the situation with a small sample size of process parameters, and effectively improve the generalization ability of the model. When choosing SVR as our regressor, in addition to preparing the training samples and output of the model, two core parameters of SVR must be determined, namely the penalty factor C and the kernel function parameter γ. However, we don't know how to choose the value. It is too time-consuming and laborious to rely solely on manual debugging. To this end, we refer to the GridSearchCV method to automatically adjust the model parameters.

The name of GridSearchCV can actually be split into two parts, namely Grid Search and Cross-validation. GridSearchCV is a parameter tuning method based on the exhaustive method. It requires us to specify the parameter space, adjust the parameters according to the step size, and use the adjusted parameters to train the learner, and find the parameter with the highest accuracy on the validation set from all the parameters. However, because it is too time-consuming, it is only suitable for cases with no more than four hyperparameters. But the parameters of the general SVR model are selected from several parameters, which just fits the GridSearchCV method.

In this paper, we take some parameters of the printing machine, including the three features of X-Offset, Y-Offset and Theta-Offset as the input of the model, and take the volume of the SPI detection result and the percentage of the standard value (Volume) as the output of the model to get a A single-output regressor. Use the regressor to fit an abstract function

$$y = g(x_1, x_2, x_3) \tag{7}$$

where y is the model output, representing the percentage of solder paste volume and standard value; x1, x2, x3 are model inputs, representing X-Offset, Y-Offset, Theta-Offset respectively. We use the SVR parameters determined by GridSearchCV to train the model on the training set to obtain an optimal solder paste volume prediction model.

3.2 Solder Paste Printing Process Parameter Optimization Method

This chapter builds the overall structure of the solder paste printing process parameter optimization method, as shown in Fig. 1.

In the standard particle swarm optimization algorithm, the inertia weight w has a great influence on the search ability of the algorithm. If it is not set properly, the search ability of the local search in the early stage of the iteration and the global search in the later stage will be reduced to varying degrees. Therefore, this paper adopts the method of randomly adjusting the inertia weight, sets its upper and lower bounds, and changes randomly within the range,

$$\omega = \omega_{\min} + (\omega_{\max} - \omega_{\min}) \times \mathrm{rand}() + \sigma \times \mathrm{randn}() \tag{8}$$

where $\omega_{\max}$ is the maximum inertia weight, $\omega_{\min}$ is the minimum inertia weight, and σ represents the degree of deviation between the random inertia weight and its mathematical expectation []. Experiments show that the convergence speed of the algorithm is significantly improved after citing random inertia weights.

The premise of PSO algorithm optimization is to construct a fitness function. In this section, based on the predicted value of the SVR model, the fitness function will be constructed through function transformation, and then the optimal process parameters will be reversely searched through the improved PSO algorithm. Since the predicted value of SVR is the percentage of solder paste volume and standard value, we proceed from reality. When the predicted value is closer to 1, it indicates that the quality of solder paste printing is better. Therefore, the fitness function we construct is

$$Fitness = |y_{optimize}/100 - 1\,| \tag{9}$$

where Fitness represents the fitness value, and $y_{optimize}$ is the percentage of the solder paste volume predicted by the model under the current process parameters to the standard value. When the value of Fitness is closer to 0, it means that the effect of the current particle is better.

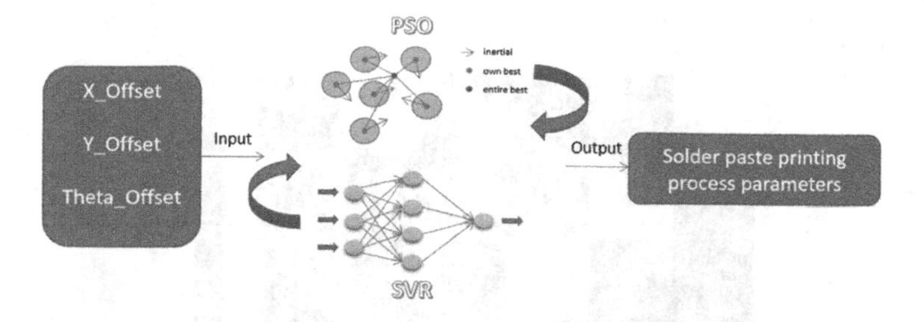

Fig. 1. Solder paste printing quality prediction model based on PSO optimization

4 Experiments

In this section, We use the data on the actual production line to verify the model.

4.1 Experimental Data

We use the solder paste data collected in a certain week on the SMT production line of a technology company to conduct experiments on our model. There are a total of 8949 raw data and 22 characteristic parameters. Among them, there are 8 features of material attribute data, 10 features of printing machine parameters, and 4 features of SPI detection results. After data cleaning, only 405 pieces of data were retained. We choose to use some parameters of the printing press, including the three features of X-Offset, Y-Offset, and Theta-Offset as the input of the model. Due to the small sample size, we cannot make predictions for all parameters of SPI. Therefore, the correlation analysis of the sample data is carried out first, and the pearson correlation coefficient [5] between the input parameters and the pre-output parameters is obtained, and the result is shown in Fig. 2. The pearson correlation coefficient can be used to measure the degree of correlation between two random variables. The closer the value is to 0, the smaller the correlation. After comparison, it is found that the volume and standard value percentage have the strongest correlation with the input, and their fluctuations are more stable. Therefore, the volume of the SPI detection result and the standard value percentage (Volume) are used as the output of the model.

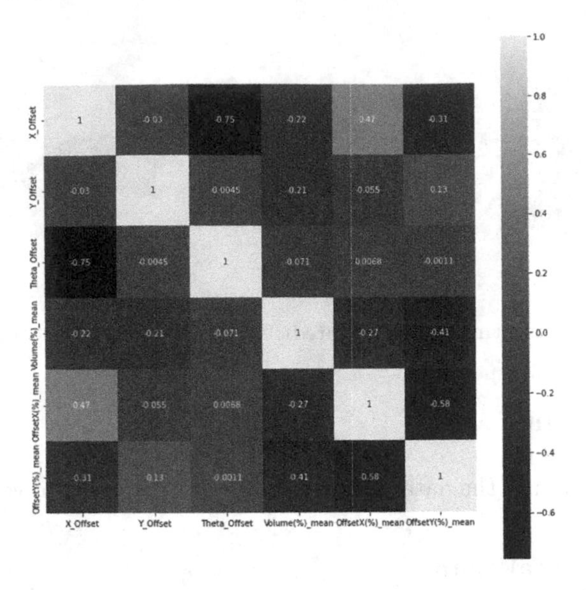

Fig. 2. Correlation analysis of SPI parameter

4.2 Results

In this paper, the SVR model is used to predict the volume and standard value percentage of the solder paste. First, the GridSearchCV method is used to determine the penalty factor C and the kernel function parameter gamma of the SVR. After the parameters of SVR are determined, we need to evaluate the effectiveness of the model. The evaluation indicators of common regression models are RMSE [7], MAPE [8], explained-variance-score and r2-score

$$RMSE = \sqrt{\frac{1}{n}\sum_{i=1}^{n}\left(\hat{y}_i - y_i\right)^2} \tag{10}$$

$$MAPE = \frac{100\%}{n}\sum_{i=1}^{n}\left|\frac{\hat{y}_i - y_i}{y_i}\right| \tag{11}$$

$$\text{explained-variance-score }(y, \hat{y}) = 1 - \frac{\text{Var}\{y - \hat{y}\}}{\text{Var}\{y\}} \tag{12}$$

$$R^2(y, \hat{y}) = 1 - \frac{\sum_{i=1}^{n}\left(y_i - \hat{y}_i\right)^2}{\sum_{i=1}^{n}\left(y_i - \bar{y}_i\right)^2} \tag{13}$$

The value range of RMSE is $[0, +\infty)$, and it is equal to 0 when the predicted value is completely consistent with the actual value, that is, a perfect model; the larger the error, the larger the value. The MAPE range is $[0, +\infty)$. A MAPE of 0 indicates a perfect model, and a MAPE greater than 1 indicates an inferior model. Explained-variance-score is used to measure how well our model explains the fluctuation of the dataset. If the value is 1, the model is perfect, and the smaller the value, the worse the effect. r2-score is the ratio of the regression sum of squares to the total sum of squares in multiple regression. The closer the value is to 1, the greater the proportion of the regression sum of squares to the total sum of squares, the closer the regression line is to each observation point, and the change in x to explain y The more the values vary, the better the regression fit. The parameter combination (C, γ) determined by GridSearchCV is (10.0, 10.0). In order to prove the effectiveness of this method, we compare this combination with other combinations, and the results are shown in Table 1. According to the above description, the effect of (10.0, 10.0) is the best, proving the effectiveness of GridSearchCV.

Table 1. Comparison of multiple sets of SVR parameters

	RMSE	MAPE	explained_variance_score	r2_score
(10.0, 10.0)	3.71	3.49	0.89	0.89
(1.0, 10.0)	4.61	4.47	0.83	0.83
(10.0, 1.0)	15.2	16.1	0.08	−0.9
(100.0, 10.0)	3.89	3.95	0.88	0.88

We import the cleaned data into the data set, and divide the training set and test set with a ratio of 8:2, and then use the parameter combination of C = 10.0, γ = 10.0 to train the SVR model, and verify it on the test set, the results are shown in the Fig. 2 shown.

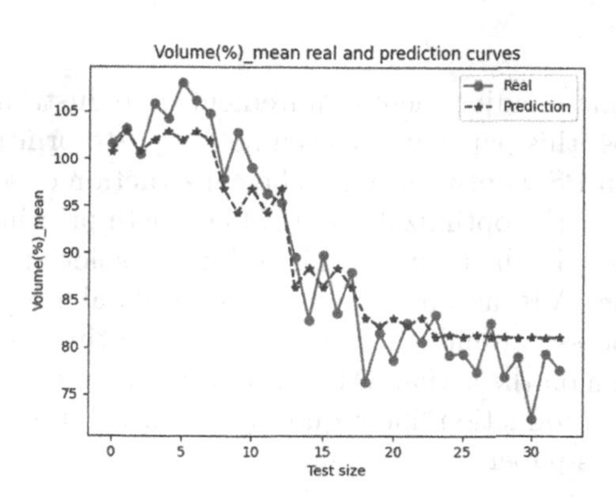

Fig. 3. Comparison of predicted and actual SVR values

Next, we need to optimize the parameters of the printing machine. In actual industrial scenarios, each printing machine parameter has its specified upper and lower bounds, and when it exceeds this range, it is considered unqualified. Therefore, we set the search space of PSO as the upper and lower bounds of the parameters, and in the search space, according to the predicted value of SVR, we constantly search for the optimal combination of printing press parameters. Since PSO has the advantage of fast convergence, we set the initial number of particles to 100 and the number of iterations to 100. After many experiments, it was found that the PSO was close to convergence at the 20th time, and the results are shown in Fig. 3.

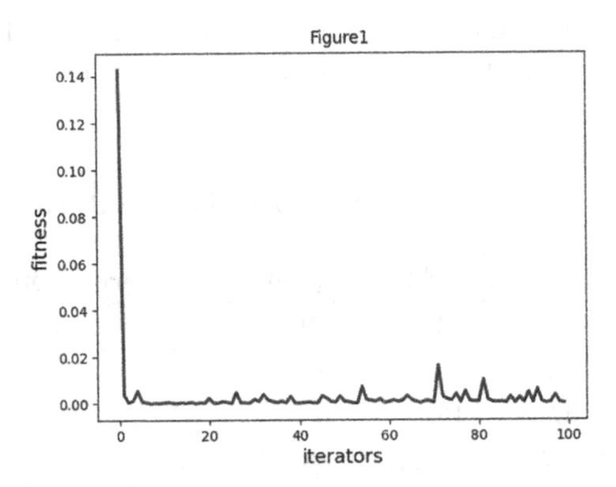

Fig. 4. At the 20th iteration, the PSO algorithm is close to convergence

We test the optimized parameters on the actual SMT production line, the results show that it can meet the actual production line requirements, which proves the effectiveness of the method in this paper.

5 Conclusion

Aiming at the parameter adjustment requirements in industrial SMT solder paste inspection scenarios, this paper proposes a solder paste printing quality prediction model based on PSO optimization. The construction of solder paste volume prediction model and the optimization of solder paste printing process parameters are mainly studied. The main work is to build a solder paste quality prediction model based on SVR, and use the improved PSO algorithm to optimize the search space of process parameters, and obtain the optimal process parameters. After the verification on the actual SMT production line, the results show that it can meet the actual production line requirements, which proves the effectiveness of the method in this paper.

References

1. Mousakazemi, S.M.H., Ayoobian, N.: Robust tuned PID controller with PSO based on two-point kinetic model and adaptive disturbance rejection for a PWR-type reactor. Prog. Nucl. Energy **111**, 183–194 (2019)

2. Cheng, R., Yu, J., Zhang, M., Feng, C., Zhang, W.: Short-term hybrid forecasting model of ice storage air-conditioning based on improved SVR. J. Build. Eng. **50**, 104194 (2022)
3. Abo-Khalil, A.G., Lee, D.C.: MPPT control of wind generation systems based on estimated wind speed using SVR. IEEE Trans. Ind. Electron. **55**(3), 1489–1490 (2008)
4. Anbo, M., Yongfeng, C., Jiajing, F., De, C., Hao, Y., Zihui, C.: Harmonic loss evaluation of low voltage overhead lines based on CSO-SVR model. Power Eng. Technol. **41**(3), 202 (2022)
5. Benesty, J., Chen, J., Huang, Y., Cohen, I.: Pearson correlation coefficient. In: Noise Reduction in Speech Processing, pp. 1–4. Springer, Heidelberg (2009). https://doi.org/10.1007/978-3-642-00296-0_5
6. Bi, D., Li, Y.F., Tso, S.K., Wang, G.L.: Friction modeling and compensation for haptic display based on support vector machine. IEEE Trans. Ind. Electron. **51**(2), 491–500 (2004)
7. Chai, T., Draxler, R.R.: Root mean square error (RMSE) or mean absolute error (MAE)?–Arguments against avoiding RMSE in the literature. Geosci. Model Dev. **7**(3), 1247–1250 (2014)
8. Goodwin, P., Lawton, R.: On the asymmetry of the symmetric MAPE. Int. J. Forecast. **15**(4), 405–408 (1999)
9. Liu, C.: Research on optimization method of solder paste printing process parameters for SMT. Ph.D. thesis, Xidian University
10. Marini, F., Walczak, B.: Particle swarm optimization (PSO). A tutorial. Chemom. Intell. Lab. Syst. **149**, 153–165 (2015)
11. Ülker, E.D.: A PSO/HS based algorithm for optimization tasks. In: 2017 Computing Conference, pp. 117–120. IEEE (2017)

Hierarchy SeparateEMD for Few-Shot Learning

Yaqiang Sun[1,2], Jie Hao[1,2](✉), Zhuojun Zou[2,3], Lin Shu[1,2], and Shengjie Hu[1,2]

[1] Guangdong Institute of Articial Intelligence and Advanced Computing, Guangzhou, China

[2] Institute of Automation, Chinese Academy of Sciences, Beijing, China
jie.hao@ia.ac.cn

[3] School of Articial Intelligence, University of Chinese Academy of Sciences, Beijing, China

Abstract. Few-shot learning methods are studied for the problem of insufficient samples in neural network tasks. Taking advantage of the excellent feature extraction capabilities of neural networks, meta-learning was proposed and became the mainstream for few-shot learning. Among the few shot learning methods, the metric-based method has the characteristics of simple mode and no need for iterative training in the inference process, which is more in line with the original intention of few-shot learning task and has been widely studied. The metric-based method with Euclidean distance has established a relatively complete training and inference system, but the accuracy has gradually stagnated. The Earth Mover's Distance has gradually emerged in the field of few-shot learning, and greatly surpassed Euclidean distance methods in terms of accuracy. Aiming at the problem that the current EMD method can only be trained with 1-shot learning and cannot be trained with multi-shot, we proposed a SeparateEMD that can train in multi-shot mode. In order to make up for the lack of global information of the SeparateEMD, Hierarchy Attention Module for proto is constructed to increase the intra-category correlation and intra-category global information of the support samples, and also improve the sample discrimination. We build the Half Pyramid Merge Module for query to increase the correlation between local information and global information of a single query sample. The Global Sampling is constructed on the basis of the Random Sampling of the EMD, which can not only increase the randomness of the samples, but also ensure the stability of the information. Experiments show that our proposed method can achieve 5-shot training of EMD-like method and outperform the current state-of-the-art few-shot learning methods on popular benchmark datasets.

Keywords: Few shot learning · SeparateEMD · Proto

1 Introduction

In the past few years, with the development of parallel computing hardware, deep neural networks have achieved widespread breakthroughs in visual

W. Fan et al. (Eds.): AsiaSim 2022, CCIS 1712, pp. 548–560, 2022.
https://doi.org/10.1007/978-981-19-9198-1_41

understanding tasks [1–3]. Especially in tasks with massive amounts of data, neural networks have achieved superhuman accuracy in multiple domains. However, neural networks often fail to perform well on tasks without enough labeled training data. In fact, it is difficult to obtain enough data or annotations in many practical tasks, such as image recognition of a special disease, recognition of a special type of vehicle, and so on.

As the shortage of training data grows, meta-learning is proposed, which can achieve good results with few data, enabling it to be quickly generalized to new tasks. Few-shot image classification is one of the fundamental and important tasks in meta-learning, where the goal is to classify new classes of images using only a small amount of labeled data.

In recent work, the representative work, DeepEMD [4], has achieved good results, and it has been used in some latest methods and achieved the best results. DeepEMD adopts Earth Mover's Distance (EMD) to calculate their structural similarity of two images. EMD is a metric that computes the distance between representations of structures and was originally proposed for image retrieval. Given the cell pair spacing, EMD can obtain the optimal matching flow between the two structures with the least cost. The global minimum EMD can be achieved by solving a linear programming problem.

However, this method currently has some obvious shortcomings. This method belongs to the metric-based method in the 1-shot task, while in the multi-shot task (more than 1 shot), the method uses the optimization-based method locally. In the multi-shot task, this method builds a Structured Fully Connected Layer (SFC) to generate one feature from multiple features. The SFC generates one feature with the closest EMD distance from multiple features through iterative training, and uses this feature as the representative feature of the class. Since the feature vector in the SFC method needs to be iteratively trained during forward propagation in multi-shot task, the SFC iterative training is nested in the network iterative training, which makes it difficult to train. In the DeepEMD method, this is also considered and the training of multi-shot task is cancelled. Only after one shot task, the model trained in 1-shot mode is used for multi-shot task. Although this method can obtain good accuracy, it needs to be retrained using the SFC for new tasks.

To address the above problems, we construct the Dynamic Hierarchy Integration Network (DHI). Our method avoids the problem that the mulit-shot task in DeepEMD requires SFC iteration by building the method of SeparateEMD, so that the model can accomplish the training of the multi-shot task; on the basis of SeparateEMD, Hierarchy Integration Module is proposed to integrate multi-sample hierarchical features of proto in multi-shot tasks; in order to further mine local features, a Half Pyramid Merge Module is constructed, which can include both local and global features when just smaller features are added. Our innovations mainly include the following aspects:

- Construct the SeparateEMD method to solve the problem that the mulit-shot task cannot be trained in the SFC method;

- The Hierarchy Integration Module is proposed to increase the multi-sample hierarchical features of proto and improve the sample discrimination of the SeparateEMD method;
- Build the Half Pyramid Merge Module, so that the predicted features can contain both local features and global features with the small features increase;

2 Related Work

The research on few-shot image classification can mainly be divided into metric-based methods and optimization-based methods. Optimization-based methods have been employed in previous literature [5–8]. It learns to represent image data in appropriate feature space and treats the task as an optimization problem, focusing on extracting the meta-knowledge needed to improve optimization performance. This kind of methods has fine-tuning for the test samples, so the robustness and generalization ability of such methods are strong. Although the optimization-based method has better robustness and generalization ability, the method needs to be fine-tuned on the test samples, which make the computational load significantly increase compared with the metric-based method. The main research on few shot image classification is currently on metric-based approach.

Metric-based methods [9,10] map image data into a specific feature space, and use a distance function to calculate the distance between features to obtain category similarity. After standard image classification networks [11,12], metric-based methods usually build on the basic classification network by removing the fully connected layer of the network and using the output of the convolutional layer as a representation feature. After that, it use distance functions, such as cosine distance and euclidean distance, to measure the distance between features. This method utilizes the training sample class for training in order to gain generalization ability to unknown classes. This method does not require retraining on unknown classes, and can use the model trained on the training set to make predictions on new classes. Obviously, from a pragmatic point of view, this method without retraining can greatly improve the application efficiency in new task, and it is also popular in the current research. Our method is also a metric-based method.

To achieve this, most previous methods represent the entire image as one data point in the feature space. There are also methods that make predictions based on local features. For example, Lifchitz et al. [8] directly exploit each local feature for prediction and fuse their results. Li [9] adopted k-NN to fuse local distances.

Apart from these two popular branches, many other promising approaches have been proposed to deal with few-shot classification problems, such as those based on graph theory [13], differentiable support vector machines [14], temporal convolution [15], etc. [16,17].

3 Method

We describe the problem setting for few-shot learning along with the training description, and then we present our method.

3.1 Problem Definition

Few-shot learning involves three things: dataset division, task description, and training methods.

Dataset Division. The few shot classification data set D_{base} is divided into three parts: D_{train}, D_{val}, and D_{test}. For each part of the division, it will be divided into support set and query set. For example, D_{train} will be divided into S_{train} and Q_{train}, and the category of S_{train} is the same as that of Q_{train}.

The Task Description. Taking 5way-1shot as an example, during the training process, one sample is extracted from each of the 5 categories of S_{train}, and the support sample set is $[s_1, s_2, s_3, s_4, s_5]$, which is obtained from Q_{train} and S_{train}. Multiple samples are extracted from the 5 categories to form a query sample set $[q_{11}, q_{12}, q_{13}, q_{21}, \ldots\ldots, q_{53}]$, where i in q_{ij} represents the number of classes in the data, and j represents the number of samples in this class.

Assuming that the model is $f(\theta)$, the support sample set generates corresponding features respectively $[f_{s1}, f_{s2}, f_{s3}, f_{s4}, f_{s5}]$ and the feature f_{qij} generated by the query sample set are generated respectively. According to the distance function D, the similarity between the query sample and the support sample can be obtained, so as to obtain the classification result of the query sample.

$$s_{ij} = D(f_{si}, f_{qij}) \tag{1}$$

For sample q_{ij},

$$L_{ij} = softmax(s_{0j}, s_{1j}, s_{2j}, s_{3j}, s_{4j}, s_{5j}) \tag{2}$$

$$L = sum(L_{ij}) \tag{3}$$

By optimizing L, the network parameters are updated. Similarly, after obtaining s_{ij} during prediction, the corresponding category can be obtained.

$$cls = argmax([s_{0j}, s_{1j}, s_{2j}, s_{3j}, s_{4j}, s_{5j}]) \tag{4}$$

3.2 SeparateEMD

In the 1-shot task, it is easy to calculate the distance between the sample and each category, whether by EMD or Euclidean distance. Taking the 3way-1shot task as an example, there are 3*1 samples in the 3way-1shot task. The support samples in each task contain 3 categories, and each category contains 1 sample. For a query sample, compare it with the 3*1 support sample, and get 3*1 corresponding results, which is the similarity between the query and each category.

In the multi-shot task, such as the 3way-4shot task, the sample size is 3*4. The support samples in each task contain 3 categories, and each category contains 4 samples. The query sample still needs to be compared with the support sample, and finally a result of 3*1 is obtained, which is the similarity between the query sample and each category.

In order to compare the query sample with the 3*4 support sample to obtain a 3*1 similarity result, a common method is to average the support samples. The 3*4 support samples are averaged at 4way to obtain 3*1 supports. The samples are converted into the same situation as the one shot, and the similarity between the query and each category can be obtained.

However, in the EMD, the averaging method does not achieve a good accuracy. The average method can obtain the center of multiple features under Euclidean distance, which can ensure the shortest distance from the center point to each feature; while for the EMD metric, the average method cannot get the position with the shortest distance.

In fact, it is not easy to obtain the center under the EMD metric. In the DeepEMD method, in order to solve this problem, the SFC method is constructed, and this method uses gradient descent to iteratively obtain the optimal center under the EMD metric. With SFC, each inference needs to iteratively find the center under the EMD metric. And the training of the neural network is also iterative process, which makes the training process extremely slow and difficult to complete (Fig. 1).

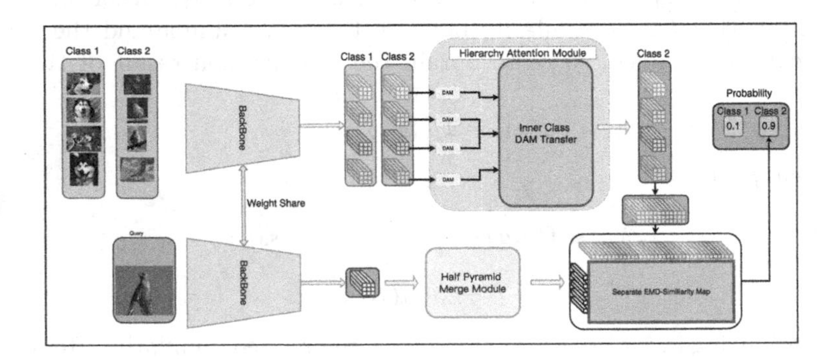

Fig. 1. SeparateEMD

To solve the inference iteration problem caused by EMD distance, we construct the method SeparateEMD. SeparateEMD uses feature maps with local information as representation vectors for each image. The local feature is used as an independent module, and the optimal matching distance is calculated with the query vector.

The optimal matching distance calculation includes two parts, Weight calculation and Similarity calculation.

Each image representation contains a set of local feature vectors $[u_1; u_2; \ldots u_{HW}]$, and each vector u_i can be seen as a node in the set. Therefore, the optimal matching distance between image groups can be expressed as the optimal matching cost between the two sets of vectors.

$$c_{ij} = 1 - \frac{{u_i}^T v_j}{||u_i||\,||v_j||} \tag{5}$$

$$s(U,V) = \sum_{i=1}^{HW} \sum_{i=1}^{HW} (1 - c_{ij}) \widetilde{x}_{ij} \tag{6}$$

Among them, Weight measures the inner product of the Euclidean space between vectors, and Similarity measures the cosine distance between vectors. Taking 3way-4shot as an example, for 4-shot samples of a certain class, there are vectors $([u_{11}; u_{12};...;u_{1HW}], [u_{21}; u_{22};...;u_{2HW}], [u_{31}; u_{32};...;u_{3HW}], [u_{41};u_{42}; ...;u_{4HW}])$. The effects of Weight and Similarity are measured separately.

For Weight, the inner product of Euclidean space is mean invariant.

For similarity, the separate calculation of the cosine distance after the average method can increase the intra-class constraints.

SeparatED uses the feature expansion method to move the center acquisition step backward, avoiding the need to find the center of multiple features, thus avoiding the iterative problem.

The accuracy of the basic SeparateEMD is not comparable to the iterative center finding of SFC. The main problem is that the calculation of SeparateEMD has no difference description between 4-shot samples. For $([u_{11}; u_{12};...;u_{1HW}], [u_{21}; u_{22};...;u_{2HW}], [u_{31}; u_{32};...;u_{3HW}], [u_{41};u_{42}; ...;u_{4HW}])$ vector, there is no distinction between u_{11} and u_{21}, u_{32} and other non-first images. u_{11}, u_{12},...,u_{1HW} come from the same feature and have high-intensity correlation, but weak correlation with features from different images, this method cannot distinguish the features of different samples. In the optimization process, the SeparateEM method contains more local information, but lacks global information. Therefore, compared with the method of finding the center, in the classification task with a large background area, this method has weak suppression of background noise.

However, this leaves a lot of room for subsequent accuracy improvement. In order to increase the correlation of the features from one image, avoid the lack of global information, and reduce the impact of noise, we added Hierarchy Integration Module on the basis of SeparateEMD.

3.3 Hierarchy Attention Module for Proto

In the SeparateEMD method, the Hierarchy Attention Module for proto contains Hierarchy Integration Module (HIM) and Dynamic Attention Module (DAM).

Hierarchy Integration Module. The SeparateEMD method is weak in distinguishing homologous features and non-homologous features, and only local features are included, while in few-shot classification tasks. Images will contain large-area environmental information in classification task, which will affect the classification results. In order to increase the content of support homologous features, we propose Hierarchy Integration Module (HIM) for support features (i.e. proto) (Fig. 2).

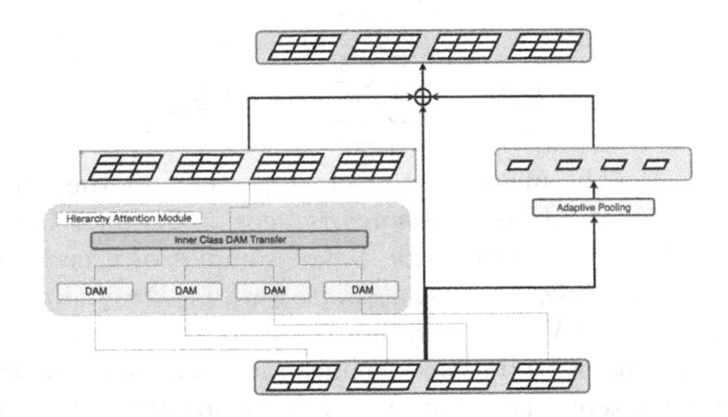

Fig. 2. Hierarchy attention module

The feature extraction network maps the image into a feature map. For the features without global pooling, DAM uses the attention mechanism to increase the correlation of its intra-class features, and converts the local features into image-level global features through Adaptive Pooling. Finally, the original features, DAR features and Adaptive Pooling feature are fused, so that the new features have intra-class feature correlation and image-level global features.

Dynamic Attention Module. For classification tasks, there are often cases where the target object occupies only a small area of the image. After sampling and segmentation of images or features, there will be more background noise in the obtained data (Fig. 3).

Fig. 3. Background noise

Obviously, there is still semantic correlation between partial images after segmentation or sampling. We use self-attention to improve the semantic correlation between local information of images. It also can achieve the attention of target features and the suppression of noise.

In this process, in order to deal with the fusion of different features appropriately, we built a DynamicAdd module. The network can change the weight of each branch, and with self-attention module, we constructed the Dynamic Attention Module (DAM) (Fig. 4).

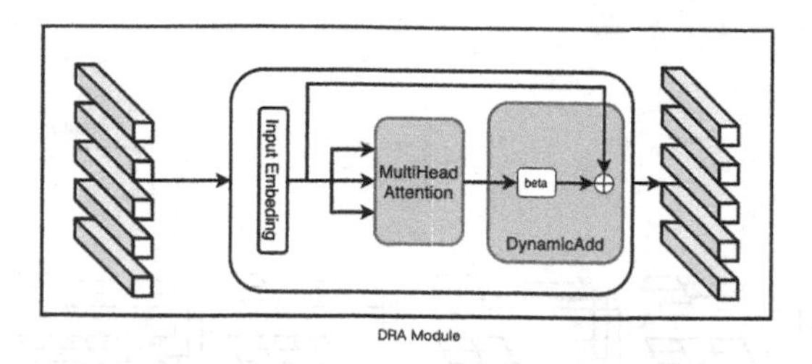

Fig. 4. Dynamic attention module

DRA fuses the attention branch and the short cut branch, and adjusts the weight of the attention branch dynamically. So the DRA can adjust the multi-branch fusion parameters.

Based on the combination of features in SeparateEMD, DRA improves the feature correlation between 5-shot samples, which can improve the feature similarity between 5-shot samples in the 5-shot task, and also enhance intra-class attention and inter-samples attention.

3.4 Half Pyramid Merge Module for Query

In order to increase local information for query, we propose the Half Pyramid Merge Module to add information from different scales. For previous pyramid networks, such as FPN, a full-scale pyramid structure is constructed. However, this pyramid structure greatly increases the number of features (Fig. 5).

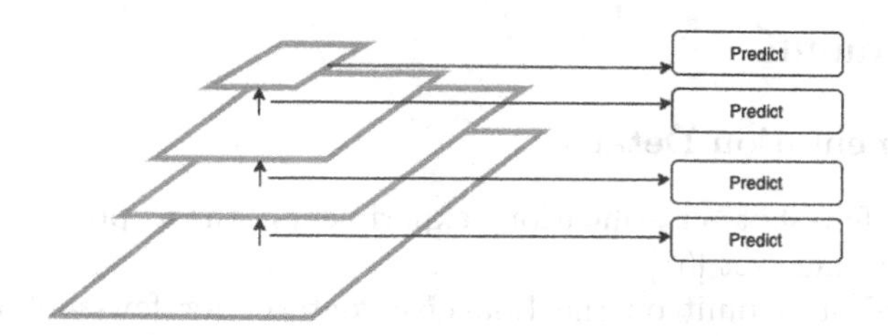

Fig. 5. Feature pyramid networks

The HPM Module is constructed on one side to form a semi-pyramid structure. For the constructed half-pyramid structure, the fusion of local information and global information can be obtained by only taking the leftmost structure of each layer of the pyramid. In the HPM Module in this task, a 4-layer structure is constructed, and the top-level features and the left-side features of the second layer are obtained as fusion features. The features obtained by HPM include both global features and local features (Fig. 6).

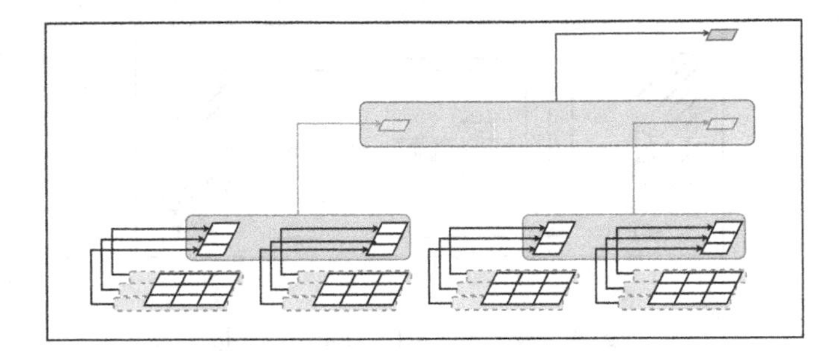

Fig. 6. Half pyramid merge module

3.5 Global Sampling Mode

In the DeepEMD, a single image is sampled using Random Sampling mode to increase the accuracy of EMD matching. However, the Random Sampling mode has serious problems. The size of random sampling is between (0.1, 0.9). This random sampling mode has a certain probability, which leads to the situation that the target in the image may not be sampled or incomplete.

In order to ensure that the target class information can be included as much as possible in the sampling results, this paper changes the Random Sampling mode to the Global Sampling mode. In the training task, the method ensures that among the images sampled multiple times, there must be an image whose size is more than 0.8 of the original size. Based on the random sampling, the semantic stability of sampling image is improved.

4 Experiments

4.1 Implementation Details

We conduct few-shot classification experiments on a popular benchmark datasets: miniImageNet [67].

MiniImageNet is built on the ImageNet dataset for few-shot tasks. It was proposed by [67] and contains a total of 100 classes with 600 images in each class. Among the 100 classes, 64 classes are used as the training set, 16 as the validation set, and 20 as the test set. Currently, it is the most popular benchmark for few-shot classification tasks.

4.2 Analysis

In the ablation experiments, we conduct various experiments to evaluate the effectiveness of our algorithm. All experiments are performed on the miniImageNet dataset (Table 1).

Table 1. Ablation experiments

Baseline	SeparateEMD	DAM	HIM	HPM	10s	9s+GS	MiniImageNet5w-5s
✓							82.76 ± 0.20
	✓						83.55 ± 0.22
	✓		✓				84.65 ± 0.17
	✓	✓	✓				84.99 ± 0.17
	✓	✓	✓	✓			85.13 ± 0.18
	✓	✓	✓	✓	✓		85.16 ± 0.18
	✓	✓	✓	✓		✓	85.27 ± 0.18

Our baseline is based on DeepEMDv2, modified by averaging, and modified into a 5-shot training mode. The accuracy of this baseline is 82.76, which is much lower than the accuracy of the original version of DeepEMDv2 84.13. This is also the reason why the DeepEMD proposes SFC to generate new proto.

After replacing the mean method of Baseline with the SeparateEMD method, the test accuracy of 5-shot is 83.55, which is a certain improvement compared to the mean method, but it is still lower than the accuracy of the original version of DeepEMDv2. However, DeepEMDv2 cannot be trained in 5-shot and the 5-shot is iteratively transformed into a 1-shot problem, and the SeparateEMD method can accomplish the 5-shot training and avoid the iteration in the inference process.

The HIM is added to improve the global information, and the accuracy is improved to 84.65. It can be proved that on the basis of SeparateEMD, the global information introduced by HIM can greatly improve the accuracy. In order to enhance global information, a DAM module based on the self attention mechanism is introduced, which can improve the accuracy to 84.99.

The HPM module add global feature for the query feature. It can increase the global feature expression with slightly increasing the number of features, thereby improving the accuracy to 85.16, which has surpassed the current best results.

Increasing the number of random sampling can improve the accuracy slightly. After increasing the number of sampling from 9 to 10, the accuracy has been slightly improved, from 85.13 to 86.16. After using global sampling, the amount of calculation does not increase and the accuracy can be improved to 85.27 (Table 2).

Table 2. Experiments in MiniImageNet

Method	Type	Backbone	MiniImageNet	
			5-way 1-shot	5-way 5-shot
MetaOpt (Lee et al. 2019) [14]	O	ResNet-12	62.64±0.62	78.63±0.46
IFSL (Yue et al. 2020) [18]	O	WRN-28	64.40	80.20
FEAT (Ye et al. 2020) [19]	M	ResNet-12	66.78 ± 0.20	82.05 ± 0.14
EMD-9s (Zhang et al. 2020a) [4]	M	ResNet-12	67.82 ± 0.27	81.93 ± 0.36
EMDv2-9s (Zhang et al. 2020a) [20]	M	ResNet-12	68.77 ± 0.29	84.13 ± 0.53
ECSIER (Mamshad et al. 2021) [21]	M	ResNet-12	66.82 ± 0.80	84.35 ± 0.51
ECSIER+Distill [21]	M	ResNet-12	67.28 ± 0.80	84.78 ± 0.52
CSEI+CAN (Li et al. 2021) [22]	M	ResNet-12	67.59 ± 0.83	81.93 ± 0.36
CSEI+EMD-9s (Li et al. 2021) [22]	M	ResNet-12	68.94 ± 0.28	85.07 ± 0.50
Ours	M	ResNet-12	**69.03± 0.33**	**85.27 ± 0.18**

4.3 Comparison with the State-of-the-Art

We compare our method with existing methods. Compared with the base EMDv2-9s, our method achieves training in 5-shot mode, thereby avoiding the iterative generation of the 5-shot task, and at the same time, the accuracy is greatly improved. Compared with the CSEI, the accuracy of the CSEI method is strongly rely on the basic method. The highest accuracy obtained by the CSEI takes the EMD-9s for base model. Therefore, this method is similar to the EMD-9s and cannot achieve 5-shot training. At the same time, this method adds self-supervised learning and additional training tasks. Our method does not introduce too many additional training tasks, such as self-supervised training, knowledge distillation, and still achieves the state-of-the-art result.

5 Conclusion

We propose a method SeparateEMD to solve the problem that EMD methods cannot be trained in multi-shot task mode. On the basis of the SeparateEMD structure, in order to improve the potential of this method, we build Hierarchy Integration Module, Dynamic Attention Module for proto features, and Half Pyramid Merge Module for query features, respectively, to solve the global information missing of SeparateEMD structure. The results of ablation experiments show that our proposed SeparateEMD structure can achieve multi-shot training and have large room for improvement in accuracy. With the help of global feature integration modules such as the Hierarchy Integration Module, our method achieves state-of-the-art results in few-shot image classification tasks.

Acknowledgements. This work was supported by the Guangdong Provincial Key Research and Development Plan (Grant No. 2019B090917009), the National Science and Technology Major Project from Minister of Science and Technology, China (Grant No. 2018AAA0103100), the Strategic Priority Research Program of Chinese Academy of Science (Grant No. XDB32070203).

References

1. Ge, Z., Liu, S., Wang, F., Li, Z., Sun, J.: Yolox: exceeding yolo series in 2021. arXiv preprint arXiv:2107.08430 (2021)
2. Zhang, H., et al.: Resnest: split-attention networks. arXiv preprint arXiv:2004.08955 (2020)
3. Shen, X., et al.: DCT-Mask: discrete cosine transform mask representation for instance segmentation. In: Proceedings of the IEEE/CVF Conference on Computer Vision and Pattern Recognition, pp. 8720–8729 (2021)
4. Zhang, C., Cai, Y., Lin, G., Shen, C.: Deepemd: few-shot image classification with differentiable earth mover's distance and structured classifiers. In: IEEE/CVF Conference on Computer Vision and Pattern Recognition (CVPR) (2020)
5. Antoniou, A., Edwards, H., Storkey, A.: How to train your maml. arXiv preprint arXiv:1810.09502 (2018)
6. Chen, W.Y., Liu, Y.C., Kira, Z., Wang, Y.C.F., Huang, J.B.: A closer look at few-shot classification. arXiv preprint arXiv:1904.04232 (2019)
7. Jamal, M.A., Qi, G.J.: Task agnostic meta-learning for few-shot learning. In: Proceedings of the IEEE/CVF Conference on Computer Vision and Pattern Recognition, pp. 11719–11727 (2019)
8. Lifchitz, Y., Avrithis, Y., Picard, S., Bursuc, A.: Dense classification and implanting for few-shot learning. In: Proceedings of the IEEE/CVF Conference on Computer Vision and Pattern Recognition, pp. 9258–9267 (2019)
9. Li, W., Wang, L., Xu, J., Huo, J., Gao, Y., Luo, J.: Revisiting local descriptor based image-to-class measure for few-shot learning. In: Proceedings of the IEEE/CVF Conference on Computer Vision and Pattern Recognition, pp. 7260–7268 (2019)
10. Sung, F., Yang, Y., Zhang, L., Xiang, T., Torr, P.H., Hospedales, T.M.: Learning to compare: Relation network for few-shot learning. In: Proceedings of the IEEE Conference on Computer Vision and Pattern Recognition, pp. 1199–1208 (2018)
11. He, K., Zhang, X., Ren, S., Sun, J.: Deep residual learning for image recognition. In: Proceedings of the IEEE Conference on Computer Vision and Pattern Recognition, pp. 770–778 (2016)
12. Simonyan, K., Zisserman, A.: Very deep convolutional networks for large-scale image recognition. arXiv preprint arXiv:1409.1556 (2014)
13. Kim, J., Kim, T., Kim, S., Yoo, C.D.: Edge-labeling graph neural network for few-shot learning. In: Proceedings of the IEEE/CVF Conference on Computer Vision and Pattern Recognition, pp. 11–20 (2019)
14. Lee, K., Maji, S., Ravichandran, A., Soatto, S.: Meta-learning with differentiable convex optimization. In: CVPR (2019)
15. Mishra, N., Rohaninejad, M., Chen, X., Abbeel, P.: A simple neural attentive meta-learner. arXiv preprint arXiv:1707.03141 (2017)
16. Shen, W., Shi, Z., Sun, J.: Learning from adversarial features for few-shot classification. arXiv preprint arXiv:1903.10225 (2019)
17. Sun, X., Yang, Z., Zhang, C., Ling, K.V., Peng, G.: Conditional gaussian distribution learning for open set recognition. In: Proceedings of the IEEE/CVF Conference on Computer Vision and Pattern Recognition, pp. 13480–13489 (2020)
18. Yue, Z., Zhang, H., Sun, Q., Hua, X.S.: Interventional few-shot learning. In: NeurIPS (2020)
19. Ye, H.J., Hu, H., Zhan, D.C., Sha, F.: Few-shot learning via embedding adaptation with set-to-set functions. In: IEEE/CVF Conference on Computer Vision and Pattern Recognition (CVPR), pp. 8808–8817 (2020)

20. Zhang, C., Cai, Y., Lin, G., Shen, C.: Deepemd: differentiable earth mover's distance for few-shot learning (2020)
21. Rizve, M.N., Khan, S., Khan, F.S., Shah, M.: Exploring complementary strengths of invariant and equivariant representations for few-shot learning. In: Proceedings of the IEEE/CVF Conference on Computer Vision and Pattern Recognition (CVPR), pp. 10836–10846 (2021)
22. Li, J., Wang, Z., Hu, X.: Learning intact features by erasing-inpainting for few-shot classification (2021)

Improving the Accuracy of Homography Matrix Estimation for Disturbance Images Using Wavelet Integrated CNN

Mikichika Yokono[1] and Hiroyuki Kamata[2](✉)

[1] Graduate School of Science and Technology, Meiji University, Tokyo, Japan
ce211069@meiji.ac.jp
[2] School of Science and Technology, Meiji University, Tokyo, Japan
kamata@meiji.ac.jp

Abstract. The objective of this study is to improve the robustness of homography estimation using deep learning for images superimposed with various types of disturbances. In conventional deep learning homography estimation, the original image and the image with perturbations and disturbances are simultaneously input to the model for estimation. The disadvantages of this approach are that the original image is affected by noise and the model itself is unclear. In this study, features are extracted separately for each of the two images, and a model is constructed based on ResNet using these features as input. In addition, when extracting the features of the perturbed and disturbed images, WaveCNet, which integrates the discrete wavelet transform into the CNN, is used to add pinpoint tolerance to the disturbance. The estimation accuracy of the homography matrix in the method proposed in this study shows improved accuracy in various noises. These results suggest that the proposed method is effective in reducing the effect of disturbance by extracting features robust to disturbance for each image.

Keywords: Homography matrix estimation · Deep learning · Discrete wavelet transform

1 Introduction

In this study, we aim to improve the accuracy of homography estimation using deep learning for various types of disturbances. Homography is a technique for mapping two images on a plane from different perspectives [1] and plays an important role in computer vision [2]. Conventional deep learning methods [3] show a decrease in matching accuracy when the images are subjected to disturbances compared to when high-definition images are used. This is thought to be due to the fact that the original image and the image with perturbation and disturbance are input simultaneously at the time of input to the model, and the original image is also affected by the noise. Another problem is that the behavior near the input layer is unclear [4].

In order to create robust deep learning under various types of disturbances, this study creates a model that extracts features separately from the original image and

W. Fan et al. (Eds.): AsiaSim 2022, CCIS 1712, pp. 561–571, 2022.
https://doi.org/10.1007/978-981-19-9198-1_42

the image with disturbances and perturbations, and uses them as input to estimate the homography matrix. When extracting features from perturbed and disturbed images, WaveCNet, which integrates discrete wavelet transforms and CNNs, is used to add robust characteristics to the noise with pinpoint accuracy so that low-frequency components containing key information, including the basic structure of the object, are extracted. This has three possible advantages: (1) The effect of noise on the original image is reduced by extracting features separately from the original image and the image with disturbances and perturbations. (2) The model can be tuned to be more robust to noise by forming a noise-robust model during the feature extraction phase. (3) The role of each phase of the model is clarified to some extent.

2 Methods

2.1 Deep Homography Estimation

Deep homography estimation is a model that estimates the homography matrix between two images (Fig. 1). In this case, the conventional 3×3 parameters have a large variance, so in the deep learning model, a 4-point homography parameterization parameterized by taking the difference of the corresponding 4 points is used [5].

Assuming that $u_k^A = \left(u_k^A, v_k^A, 1\right)$ and $u_k^B = \left(u_k^B, v_k^B, 1\right)$ are four fixed points in images I_A and I_B, respectively, the four-point homography parameterization is as follows:

$$H_{4pt} = \begin{pmatrix} u_k^B - u_k^A \\ v_k^B - v_k^A \end{pmatrix} = \begin{pmatrix} \Delta u_k \\ \Delta v_k \end{pmatrix} \quad (k = 1, 2, 3, 4) \tag{1}$$

This reduces the variance of the homography matrix and the number of parameters to 2×4. Also, the original homography matrix and the 4-point parameterized homography have a one-to-one correspondence.

In the estimation flow, the original image I_A and the image I_B perturbed to the original image are simultaneously input to the model, and the four-point parameterized homography matrix H_{4pt_pred} is estimated. The model then takes the H_{4pt_truth} and L2 norm prepared as the correct labels and uses the following loss function:

$$Loss_{L2} = \frac{1}{2}\left\| H_{4pt_{pred}} - H_{4pt_{truth}} \right\|^2 \tag{2}$$

2.2 Filter Characteristics Using the Discrete Wavelet Transform

Wavelet analysis has a wide range of applications in signal processing, pattern recognition, and other fields because of its superiority in time-frequency analysis [6]. The wavelet transform uses the scaling function $\varphi(x)$, and the wavelet function $\psi(x)$ to construct a basis by scaling and shifting, and this basis allows the signal to be decomposed and reconstructed by a multiple resolution representation. From the scaling function $\varphi(x)$ and the wavelet function $\psi(x)$, low-pass filter l_k and high-pass filter h_k are deduced, and extending these filters to two dimensions, they are expressed as the following matrix equation.

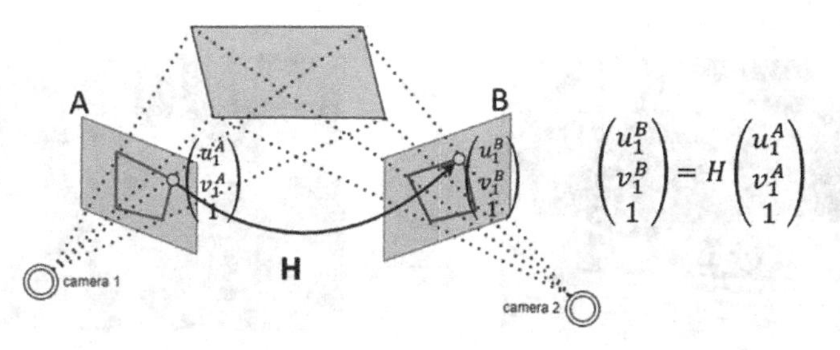

Fig. 1. Homography matrix

$$L = \begin{pmatrix} \cdots\cdots\cdots & & \\ \cdots l_{-1}\ l_0\ l_1 \cdots & & \\ & \cdots l_{-1}\ l_{-1}\ l_{-1} \cdots & \\ & & \cdots\cdots \end{pmatrix} \tag{3}$$

$$H = \begin{pmatrix} \cdots\cdots\cdots & & \\ \cdots h_{-1}\ h_0\ h_1 \cdots & & \\ & \cdots h_{-1}\ h_0\ h_1 \cdots & \\ & & \cdots\cdots \end{pmatrix} \tag{4}$$

Given two-dimensional data X, the 2D discrete wavelet transform (DWT) is used with the filters in Eqs. (3) and (4) to decompose the data into low-frequency components X_{ll} and high-frequency components X_{lh}, X_{hl}, X_{hh} as follows [7].

$$X_{ll} = LXL^{\mathrm{T}} \tag{5}$$

$$X_{lh} = HXL^{T} \tag{6}$$

$$X_{hl} = LXH^{T} \tag{7}$$

$$X_{hh} = HXH^{T} \tag{8}$$

The corresponding 2D discrete inverse wavelet transform (IDWT) is expressed as follows.

$$X = L^T X_{ll} L + H^T X_{lh} L + L^T X_{hl} H + H^T X_{hh} H \tag{9}$$

The original data is reconstructed from the decomposed elements of Eqs. (5)–(7). When a two-dimensional discrete wavelet transform is used for the image, it can be decomposed and reconstructed into each frequency component as shown in Fig. 2 This decomposition performance is also used for noise reduction, which drops the high-frequency component of the signal, and for super-resolution by using the main (low-frequency) and detailed (high-frequency) components of the separated image [8].

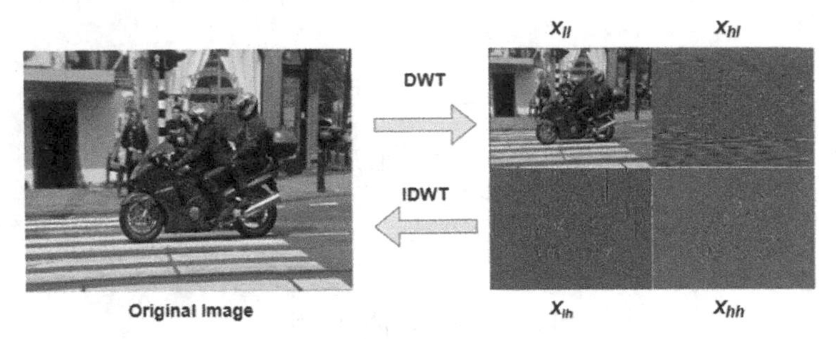

Fig. 2. 2D discrete wavelet transform

2.3 WaveCNet

Downsampling operations in ordinary CNN models ignore the sampling theorem [9] and may have adverse effects due to aliasing between data components, destruction of basic object structure, and accumulation of effects due to noise in the CNN model.

WaveCNet [10] replaces the max pooling, average pooling, and slide 2 convolution with downsampling of the CNN model with $\mathrm{DWT_{ll}}$ such that the low frequency components are extracted.

$$\mathrm{Maxpooling_{s=2}} \rightarrow DWT_{ll} \tag{10}$$

$$\mathrm{Conv_{s=2}} \rightarrow DWT_{ll} \circ \mathrm{Conv_{s=1}} \tag{11}$$

$$\mathrm{AvgPool_{s=2}} \rightarrow DWT_{ll} \tag{12}$$

Extracting features with high-frequency components removed and propagating the features allows us to maintain the basic object structure of the feature map and construct a CNN that is robust to noise. Adapting 2D DWT/IDWT to deep neural networks, the back propagation of DWT/IDWT is as follows becomes.

$$DWT : \frac{\partial X_{ll}}{\partial \mathrm{X}}(\mathrm{G}) = L^{\mathrm{T}}GL, \qquad IDWT : \frac{\partial \mathrm{X}}{\partial X_{ll}}(\mathrm{G}) = \mathrm{LG}L^{\mathrm{T}} \tag{13}$$

$$DWT : \frac{\partial X_{lh}}{\partial \mathrm{X}}(\mathrm{G}) = H^{\mathrm{T}}GL, \qquad IDWT : \frac{\partial \mathrm{X}}{\partial X_{lh}}(\mathrm{G}) = \mathrm{HG}L^{\mathrm{T}} \tag{14}$$

$$DWT : \frac{\partial X_{hl}}{\partial \mathrm{X}}(\mathrm{G}) = L^{\mathrm{T}}GH, \qquad IDWT : \frac{\partial \mathrm{X}}{\partial X_{hl}}(\mathrm{G}) = \mathrm{LG}H^{\mathrm{T}} \tag{15}$$

$$DWT : \frac{\partial X_{hh}}{\partial \mathrm{X}}(\mathrm{G}) = H^{\mathrm{T}}GH, \qquad IDWT : \frac{\partial \mathrm{X}}{\partial X_{hh}}(\mathrm{G}) = \mathrm{HG}H^{\mathrm{T}} \tag{16}$$

(13)–(16) make it possible to incorporate the discrete wavelet transform into the training of deep neural networks, such that the main structure of objects can be extracted.

3 Experiments

In the experiments, we will compare the proposed method with the conventional method and also consider multiple mother wavelets in the proposed method to evaluate their estimation accuracy and noise tolerance. The model is also constructed based on ResNet34 [11].

The models of the conventional and proposed methods are shown in Fig. 3(a) and (b), respectively. In the conventional method (a), the original image and the perturbed and disturbed images are input simultaneously and directly to the model to estimate the homography matrix. Therefore, it is considered that the original image is also affected by the noise and that the operation in the area close to the input layer is unclear.

On the other hand, in the proposed method in (b), the phases to extract the features of each image are sandwiched as "Original Layer" and "Noise Layer" at the beginning, and then the model is configured to perform homography estimation using the extracted features as input. Furthermore, in the "Noise Layer," WaveCNet is introduced to perform noise-robust feature extraction, and pinpoint noise tolerance is added by performing noise-robust feature extraction. Furthermore, in the "Noise Layer," WaveCNet is introduced to perform noise-robust feature extraction and add pinpoint noise tolerance by performing noise-robust feature extraction. In this study, three types of mother wavelets, "Haar," "Daubechies," and "Cohen," are introduced and compared when the order of each wavelet is varied. For training, the dataset will be Microsoft MS-COCO2014 provided by Microsoft. The image data will be subjected to perturbations and disturbances as shown in Table 1, and robustness to each noise will also be considered. The training conditions will be 150 epochs using Adam with a batch size of 64.

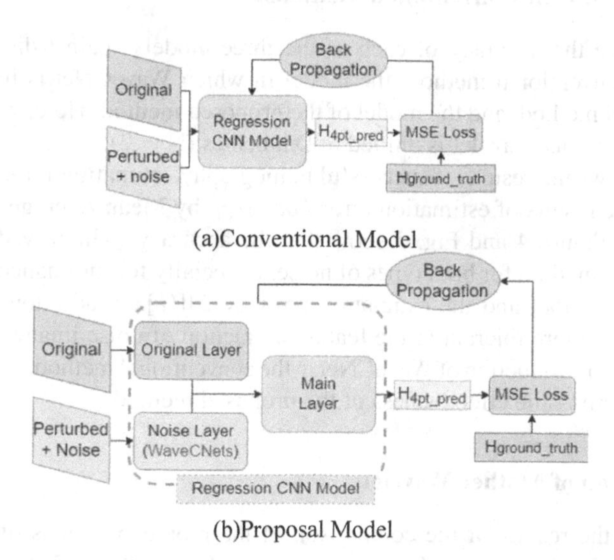

Fig. 3. Deep Homography Model ((a) Conventional method: source image and noise perturbed image are input simultaneously, (b) Proposal method: features of source image and noise perturbed image are extracted separately and their features are input)

Table 1. Additional noise

Noise	Summary	Degree
Brightness	$J = I + n_{brightness}$	$n_{brightness}(-0.5\sim0.5)$
Contrast	$J = ((I - 0.5)) \times 10^{n_{contrast}}) + 0.5$	$n_{contrast}(-1\sim1)$
Gaussian blur	Gaussian filter with standard deviation n_{blur}	$n_{blur}(0\sim5)$
Gaussian noise	Variation of pixel luminance with variance n_{noise}	$n_{noise}(0\sim0.01)$
Radiation noise	Add sesame salt noise to pixels with $n_{radiation} \times 100$	$n_{radiation}(0\sim0.1)$

During homography estimation, the Mean Average Corners Error, which is a loss of four points between the corresponding two images, is considered successful if the following conditions are met:

$$MACE = \frac{1}{4}\sum_{i=4} \left\| (x_i, y_i)_{pred} - (x_i, y_i)_{truth} \right\|_2^2 \leq 3 \tag{17}$$

Based on Eq. (17), we calculate the estimation success accuracy and estimation error and compare the performance of each model.

4 Results

4.1 Comparison with Conventional Methods

We will examine the accuracy of each of the three models against disturbances: the model of the conventional method, the model in which WaveCNet is introduced into the conventional method, and the model of the proposed method. Here, we compare the case where the mother wavelet is unified to Daubchies (N = 2).

Figure 4 shows the results of successful homography estimation for each noise, and Fig. 5 shows the results of estimation error $Loss_{MACE}$ by Mean Average Corners Error for each noise. Figure 4 and Fig. 5 show that the accuracy is improved compared to the conventional method for both types of noise. Especially for luminance, the accuracy is improved by 0.8[%] and the estimation error by 24[%]. In addition, the proposed method, which is more tolerant to the feature extraction of noise images with pinpoint accuracy than the introduction of WaveCNet in the conventional method, shows improved accuracy, confirming the effectiveness of the proposed method.

4.2 Comparison of Mother Wavelets

Table 2 shows the results of the comparison of the mother wavelets of the proposed method model with three types of mother wavelets, "Haar," "Daubchies," and "Cohen," when the order of each wavelet is varied. Table 2 shows that the optimal mother wavelet and its order differ depending on the noise, but Cohen's "ch2.2" shows high accuracy for most noises.

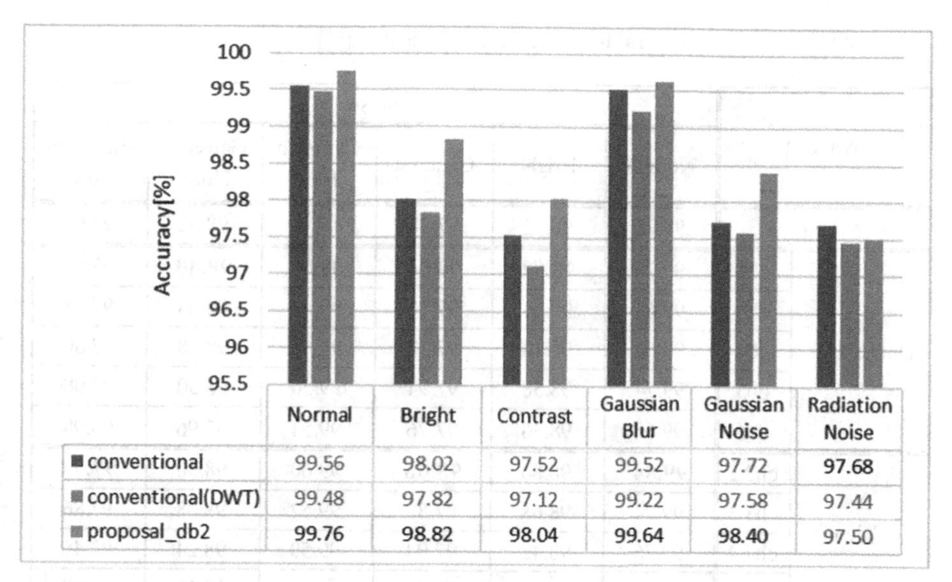

	Normal	Bright	Contrast	Gaussian Blur	Gaussian Noise	Radiation Noise
conventional	99.56	98.02	97.52	99.52	97.72	**97.68**
conventional(DWT)	99.48	97.82	97.12	99.22	97.58	97.44
proposal_db2	**99.76**	**98.82**	**98.04**	**99.64**	**98.40**	97.50

Fig. 4. Comparison of the Accuracy of conventional and proposed methods

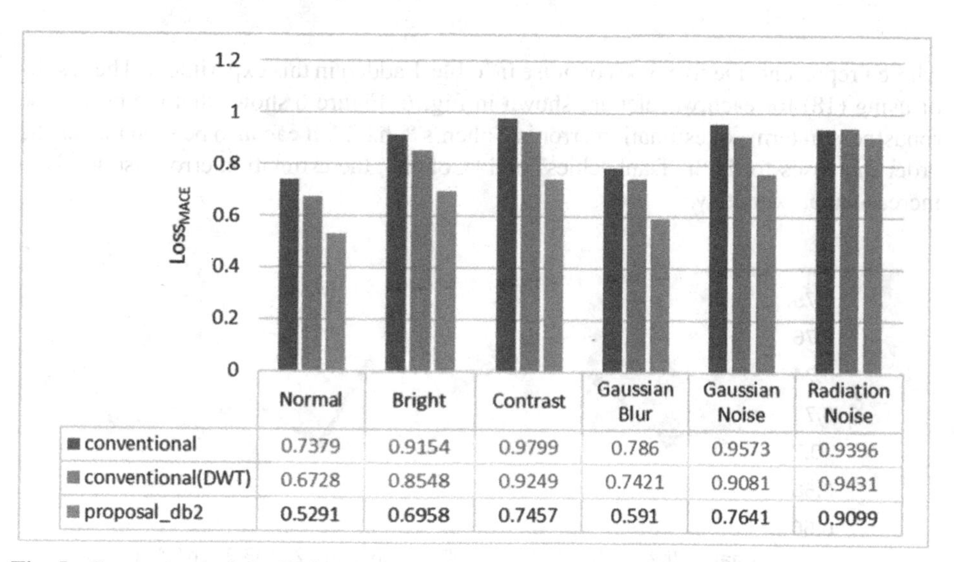

	Normal	Bright	Contrast	Gaussian Blur	Gaussian Noise	Radiation Noise
conventional	0.7379	0.9154	0.9799	0.786	0.9573	0.9396
conventional(DWT)	0.6728	0.8548	0.9249	0.7421	0.9081	0.9431
proposal_db2	**0.5291**	**0.6958**	**0.7457**	**0.591**	**0.7641**	**0.9099**

Fig. 5. Comparison of the Estimation Error of conventional and proposed methods (Lower is better model)

To compare the robustness of each noise, we define the score as the average of the estimation error by Mean Average Corners Error calculated for each noise as follows:

$$Loss = \frac{1}{5}\left(\sum Loss^{f}_{MCSE}\right) \tag{18}$$

Table 2. Accuracy of wavelet [%]

Wavelet		Noise					
		Normal	Bright	Contrast	Gaussian Blur	Gaussian Noise	Radiation Noise
Haar		99.78	98.72	97.98	99.6	98.32	97.66
Daubechies	db2	99.76	**98.82**	98.04	99.64	**98.40**	97.50
	db3	99.68	98.44	97.86	99.42	98.08	97.08
	db4	99.64	98.18	97.38	99.48	98.18	97.60
	db5	99.74	98.58	97.94	99.50	98.30	97.08
	db6	99.60	98.36	97.76	99.54	97.96	96.98
Cohen	ch2.2	**99.84**	98.80	**98.06**	99.60	98.30	**97.94**
	ch3.3	99.74	98.68	97.82	99.52	98.08	97.86
	ch4.4	99.76	98.42	97.94	99.50	98.24	97.46
	ch5.5	99.80	98.56	98.02	**99.66**	98.30	97.08

where f represents the five types of noise in Table 1 added in this experiment. The results of using (18) for each wavelet are shown in Fig. 6. Figure 6 shows that the best noise robustness in terms of estimation error is Cohen's "ch2.2". It can also be seen that as the order increases for both "Daubechies" and "Cohen", the estimation error also tends to increase approximately.

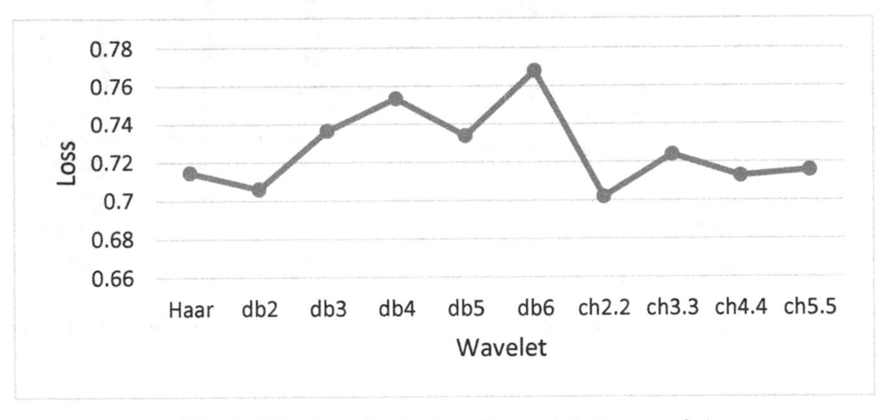

Fig. 6. Wavelets of noise loss (Lower is better wavelet)

4.3 Visualization of Homography Matrix Estimation

We visualize and compare the homography matrix estimation using actual images with the model of the conventional method and the model proposed in this experiment. The

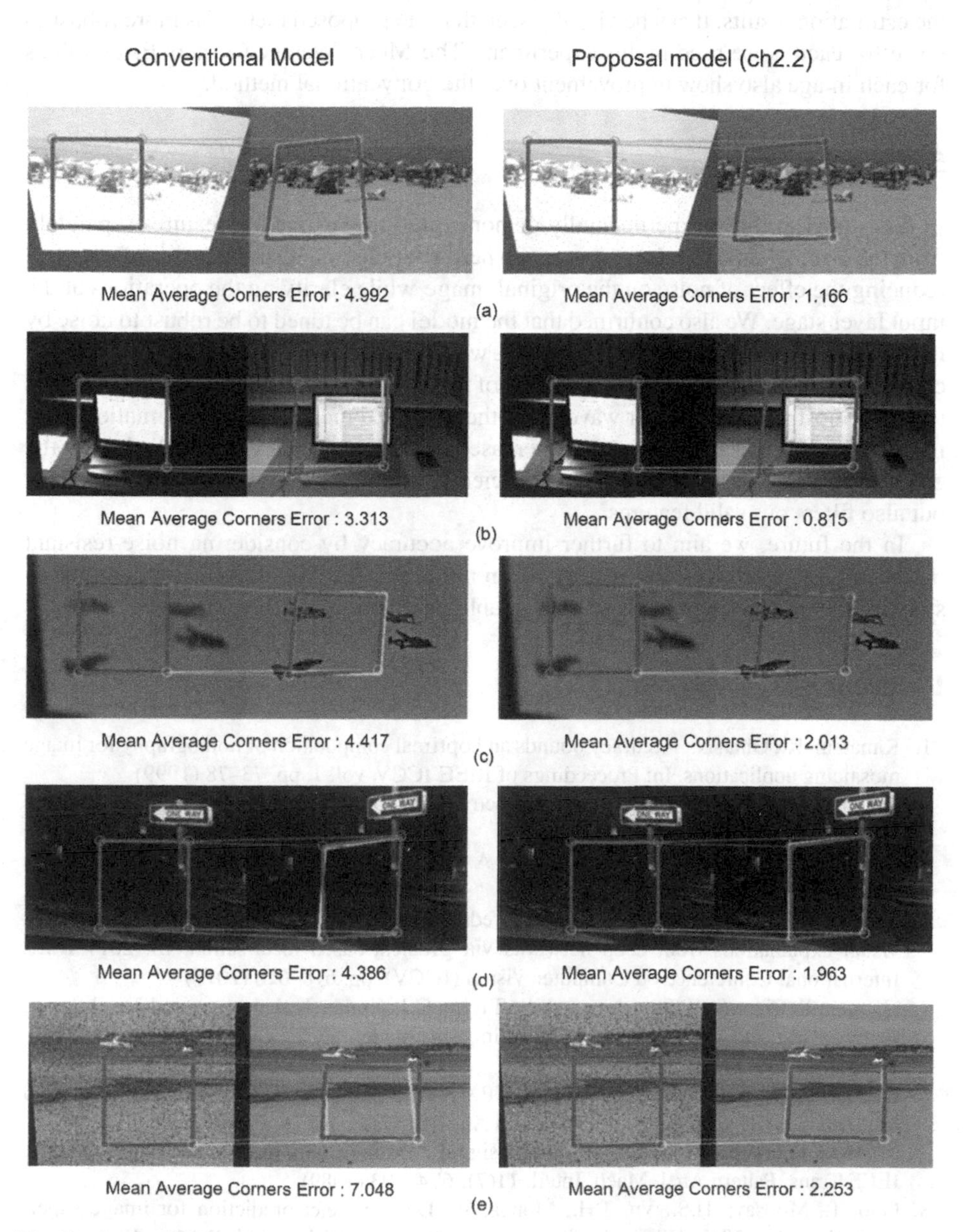

Fig. 7. Conventional model vs Proposal model (ch2.2) ((a): Brightness, (b) Contrast, (c) Gaussian Noise, (d) Gaussian blur, (e) Radiation noise)

mother wavelet of the proposed model is Cohen 2.2, which has the highest noise robustness in this experiment. The left side of Fig. 7 shows the estimation results of the conventional method and the right side shows the results of the proposed method. The red squares in the image represent the correct answer labels, and the yellow ones represent the estimation results. It can be visually seen that the proposed method is more robust to noise for each noise used in this experiment. The Mean Average Corners Error values for each image also show improvement over the conventional method.

5 Conclusion

In this study, we have experimentally demonstrated that extracting features separately from the original image and the image with perturbations and disturbances is effective in reducing the effect of noise on the original image while clarifying the operations at the input layer stage. We also confirmed that the model can be tuned to be robust to noise by introducing WaveCNet, which uses discrete wavelet transforms to extract low-frequency components in the feature extraction stage of the input image. Cohen's ch2.2 was found to be the most accurate mother wavelet for the proposed model, and the estimation error increased as the order of the wavelet increased. This is expected because the higher the order, the greater the number of filter coefficients, which improves filtering performance, but also filters out valid features.

In the future, we aim to further improve accuracy by considering noise-resistant models for layers that extract features from noisy images. We also plan to continue to study more appropriate models of homography estimation and their structure.

References

1. Kanatani, K., Ohta, N.: Accuracy bounds and optimal computation of homography for image mosaicing applications. In: Proceedings of IEEE ICCV, vol. 1, pp. 73–78 (1999)
2. Chum, O., Pajdla, T., Sturmb, P.: The geometric error for homographies. Comput. Vis. Image Underst. **97**, 86–102 (2005)
3. DeTone, D., Malisiewicz, T., Rabinovich, A.: Deep image homography estimation. arXiv: 1606.03798 (2016)
4. Selvaraju, R.R., Cogswell, M., Das, A., Vedantam, R., Parikh, D., Batra, D.: Grad-CAM: visual explanations from deep networks via gradient-based localization. In: 2017 IEEE International Conference on Computer Vision (ICCV), pp. 618–626 (2017)
5. Nguyen, T., Chen, S.W., Shivakumar, S.S., Taylor, C.J., Kumar, V.: Unsupervised deep homography: a fast and robust homography estimation model. IEEE Robot. Autom. Lett. **3**(3), 2346–2353 (2018)
6. Daubechies, I.: Ten Lectures on Wavelets, CBMS-NSF Regional Conference Series in Applied Mathematics 61. Society for Industrial and Applied Mathematics, Philadelphia (1992)
7. Mallat, S.G.: A theory for multiresolution signal decomposition: the wavelet representation. IEEE Trans. Pattern Anal. Mach. Intell. **11**(7), 674–693 (1989)
8. Guo, T., Mousavi, H.S., Vu, T.H., Monga, V.: Deep wavelet prediction for image super-resolution. In: 2017 IEEE Conference on Computer Vision and Pattern Recognition Workshops (CVPRW), pp. 1100–1109 (2017)
9. Azulay, A., Weiss, Y.: Why do deep convolutional networks generalize so poorly to small image transformations? arXiv preprint arXiv:1805.12177 (2018)

10. Li, Q., Shen, L., Guo, S., Lai, Z.: Wavelet integrated CNNs for noise-robust image classification. In: Proceedings of the IEEE/CVF Conference on Computer Vision Pattern Recognition (CVPR), pp. 7245–7254 (2020)
11. He, K., Zhang, X., Ren, S., Sun, J.: Deep residual learning for image recognition. arXiv:1512.03385 (2015)

Defect Detection of Tire Shoulder Belt Cord Joint Based on Periodic Texture

Zhen Zhang[1], Chen Peng[2(✉)], Miao Rong[2], and Liang Xiao[1]

[1] Shanghai Key Laboratory of Power Station Automation Technology, School of Mechatronic Engineering and Automation, Shanghai University, Shanghai 200444, China

[2] School of Mechatronic Engineering and Automation, Shanghai University, Shanghai 200444, China

c.peng@i.shu.edu.cn

Abstract. Tire quality plays an important role in traffic safety. Among the categories of defects that often occur in the actual production process of tires, the tire shoulder belt layer cords joint opening defect is the most common and serious defect. In this paper, a tire shoulder belt layer cords joint opening defect detection algorithm based on grayscale feature statistics and threading method is proposed based on the machine vision nondestructive testing technology. It first combines the periodic texture grayscale feature to accurately pre-locate the defect position, followed by performing a series of pre-processing operations on the target area to accurately determine the tire X-ray image. Through comparative experimental analysis, the detection algorithm has high recognition and accuracy. The detection speed of the algorithm has also reached a satisfactory level.

Keywords: Tire X-ray image · Defect detection · Tire belt cords · Periodic textures · Accurate grading

1 Introduction

The production process of radial tires is complex, and the quality of tire molding is highly related to the molding quality of radial cords. Due to good visual performance and high detection efficiency, X-ray imaging technology in non-destructive testing technology has been widely used in major tire manufacturers.

According to statistics, most tire manufacturers still use visual inspection methods to detect tire X-ray image defects due to cost control and other reasons. This inspection method has many shortcomings: high cost, low efficiency, and inability to meet the standardization of product quality control.

Inspired by target detection technology, this paper analyzes the grayscale features of tire X-ray images from a visual perspective. Based on this, we design a belt joint opening defect detection algorithm based on grayscale feature statistics and a designed threading method. The algorithm is not only robust to background noise, but also extremely sensitive to defect features, and its running speed and grading accuracy exceed the existing level of the factory. The main contributions of this paper are as follows:

W. Fan et al. (Eds.): AsiaSim 2022, CCIS 1712, pp. 572–582, 2022.
https://doi.org/10.1007/978-981-19-9198-1_43

1. A two-step method structure of defect detection is proposed, that is, the position of the defect is accurately pre-positioned, and then the target area is morphologically processed for pixel-level judgment.
2. In the pre-positioning stage, a nonlinear template is constructed to enhance the useful features of defects and weaken irrelevant features. The region of interest is accurately located based on grayscale statistical features, laying the foundation for grading;
3. In the grading stage, the Sobel operator is proposed to remove the transverse cord, and the shortcomings of the maximum filtering are compared horizontally; then, in order to solve the new problem of broken segments of the oblique cords, the improved morphological closing operation is used to process the target image;
4. Refinement operation is performed on the morphological processed target image containing defects, based on the threading method and the box method to count the standard spacing of the cords and pixel-level grading is performed to meet the production standards;
5. The belt joint opening detection method proposed in this paper is based on the previous work, which has successfully achieved the accurate segmentation of the tire area. This algorithm has been integrated into the entire inspection system and has been applied in Shandong BaYi Rubber Co. Ltd.

2 Related Work

Tire defect detection algorithms can be divided into two categories: tire defect detection algorithms based on machine vision image processing and tire defect detection algorithms based on deep learning.

2.1 Tire Defect Detection Algorithm Based on Image Processing

In the field of tire X-ray inspection, traditional machine vision technology is playing an increasingly important role in non-destructive testing (NDT) and automatic vision inspection applications due to its efficiency, accuracy and real-time performance. In recent years, researchers have designed corresponding algorithms for different types of tire defects. Q Guo [1] of North Central University identified the defects in the sidewall region by pre-processing threshold transformation, thinning and other operations on the original image with the extracted feature parameters, but these methods can only identify and classify sidewall cord defects. For the complex background texture of tire bubble images, Z Yang [2] proposed a method based on gray image morphological processing and region growing. By using the eroded grayscale image and the original image to be superimposed, the contrast with the bubbles and the background is enhanced, the texture of the points of the image is effectively removed, the seed points are selected on this basis, and the bubbles are effectively segmented by the region growing method. However, due to the various occurrences of tire bubbles in the actual production process, the gray value is always different. This method is not universal and cannot often achieve the expected results. The seed selection of the regional growth method still requires manual intervention, and manpower cannot be liberated. Y Zhang [3] proposed a method based on total variation image decomposition and edge detection to detect foreign objects and bubble defects in tire X-ray images. Considering the problem of tire defect characterization from the perspective of local regularity analysis and scale features, Y Zhang [4] uses the defect edge measurement model to select the optimal scale and threshold parameters for frame defect edge detection. Finally, the wavelet multi-scale analysis method is used to detect defects. Detection. Q Guo [5] proposed a method combining curve transformation and Canny edge detection operator to detect defects in tire laser ultrasonic images. Based on the edge feature, which can be represented by a larger coefficient in the sub-high frequency band, modify the curve coefficient to enhance the curve feature of the reconstructed image. This method has obvious effects on the detection of impurities and bubble defects in complex backgrounds. However, non-destructive testing methods based on laser ultrasound have serious limitations: the vacuum environment required for testing is a big challenge, and the overall testing accuracy is insensitive.

2.2 Tire Defect Detection Algorithm Based on Deep Learning

In recent years, deep learning technology has also been widely used in the field of tire defect detection. The target detection algorithm was first applied in the classification of tire X-ray image defects. After X Cui [6, 7] first introduced the convolutional neural network technology to verify the effectiveness of the deep learning technology, G Yao [8] adopted the YOLOv3 model for tire defect detection, the network structure is modified from the original three-scale detection to four-scale detection. The YOLOv3 model effectively integrates high-resolution shallow features and high-level information with high semantic information, effectively solving the problem of small target detection. At the same time, the Faster R-CNN model is also widely used in the field of tire defect detection. Q Zhu [9] improved the region proposal network in Faster R-CNN, using the image features extracted after image convolution to select the proposal frame, reducing the network training time; Y Li [10] proposed a method for An end-to-end method for automatic tire defect detection from X-ray images (TireNet), introducing an object detection model as a baseline. Inspired by the periodic features of tire X-ray images, a Siamese network is used in the new model as part of a downstream classifier to capture Defect features.

Compared with traditional detection methods based on image processing, the neural network structure used by deep learning needs to be trained. The training process often requires a large data set [11]. In the actual production process, the probability of occurrence of tire images containing defects is extremely low. Each type of defect is often random, so tire manufacturers cannot provide a large number of original tire X-ray images of each defect type. There are few defect samples in the provided images. If the sample set is too small, the network will be over-fitted, which will increase the learning burden of the neural network.

Inspired by target detection technology, this paper analyzes the grayscale features of tire X-ray images from a visual perspective. A belt joint opening defect detection algorithm is proposed based on grayscale feature statistics and a designed threading method. The algorithm is not only robust to background noise, but also extremely sensitive to defect features. Its running speed and grading accuracy exceed the existing level of the factory.

3 Proposed Algorithm

In this section, a tire X-ray image 2# belt cord defect detection algorithm is introduced. The preparatory work of this paper has successfully divided the entire tire X-ray tread image into the turn-up area, the sidewall area and the crown area (Fig. 1). The research object of this paper, namely the 2# belt layer (Fig. 2), is in the segmented crown area. The texture of the crown area is complex, and the cords are overlapped and crossed. The result is that the X-ray transmission effect is poor, the brightness of the entire area is dark. Through the structural characteristics of tire production, we know that the 2# belt layer has the largest width value among all belt layers, and only carcass cords and 2# belt layer cords are contained at the tire shoulder area where the cord texture has a high overall gray value. In view of this feature, we propose a detection algorithm for shoulder belt cord joint opening defects based on the gray feature statistics and threading method. The overall algorithm structure is divided into two parts: (Stage1) pre-positioning the defect position accurately; (Stage2) judging the defect existing area at the pixel level.

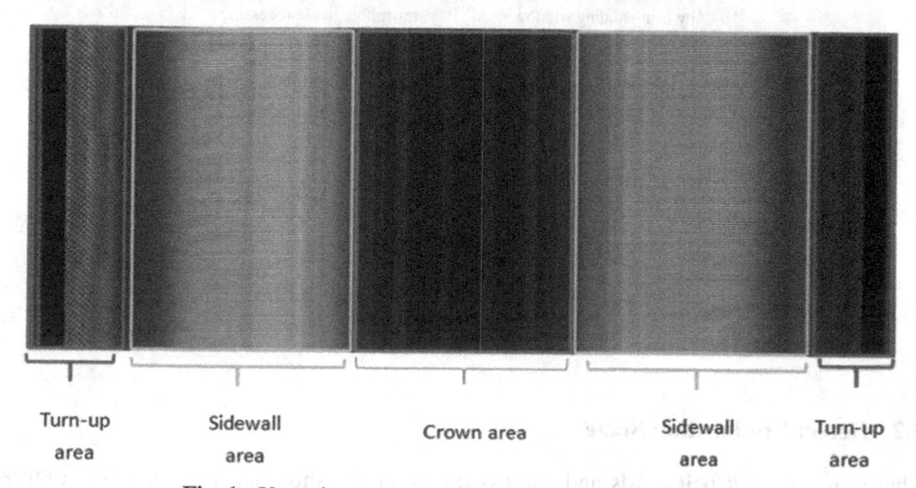

Fig. 1. X-ray image area segmentation map of radial tire.

Fig. 2. Tire 2# belt layer edge defects.

3.1 Algorithm Structure Design

As shown in Fig. 3, the algorithm is divided into two stages: defect pre-location stage based on gray feature statistics and accurate grading stage based on digital image processing. The first stage of the algorithm aims at locating the approximate location of the 2# belt defect and selecting the region of interest accordingly. The second stage is designed to accurately determine the defect level based on the cord spacing. The the two-stage technique has the advantage that the time complexity of the algorithm is greatly reduced, while the detection accuracy is improved.

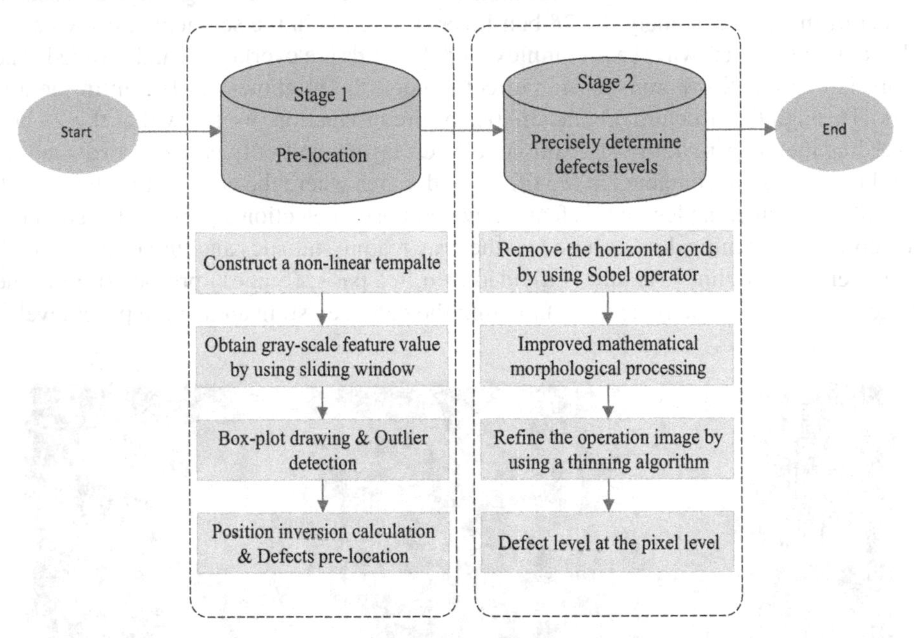

Fig. 3. The proposed tire 2# belt defects detection method architecture.

3.2 Defect Pre-location Stage

There are only 2# belt cords and carcass cords in the shoulder area, and the texture of the foreground cords is clearly discernible. Based on the research and statistical evaluation of pixel grayscale characteristics, a pre-positioning method for 2# belt joint opening defects is proposed. This method can pinpoint the location of defects and select regions of interest as input to algorithms in subsequent stages. The goal of this method is to shorten the morphological processing time of the second-stage digital images to improve the overall efficiency of the algorithm.

In the pre-positioning stage, in order to fully consider the influence of the grayscale characteristics brought by the direction of the cords, a window is firstly constructed that is almost the same as the direction of the cords of the 2# belt layer (65°–75°), Then the sliding window method is used to calculate grayscale characteristics to obtain a grayscale statistical scatterplot. An abnormal value detection algorithm based on boxplot is used to detect and locate the abnormal window of gray eigenvalues. Finally, the position of the pre-positioned 2# belt layer defect is calculated by position inversion, and the region of interest is selected as the input for the next stage.

As an outlier detection method, the boxplot method uses five statistics in the data: the minimum value, the first quartile Q1, the median, the third quartile Q3 and the maximum value to describe the data information such as the symmetry and the degree of dispersion. The detection principle is as follows:

(1) Cut out the 2# belt edge area from the obtained tire X-ray image;
(2) Construct a nonlinear kernel with almost the same direction as the 2# belt cord (65°–75°);
(3) The gray value of the image is calculated by the sliding window method, and the two-dimensional defect information is reduced to one-dimensional gray value information to meet the requirements of rapidity and real-time performance in the engineering field. The process is shown in Fig. 4;
(4) Apply an outlier detection method based on boxplots to detect and locate abnormal windows of grayscale eigenvalues;
(5) Perform position inversion calculation to determine the region of interest.

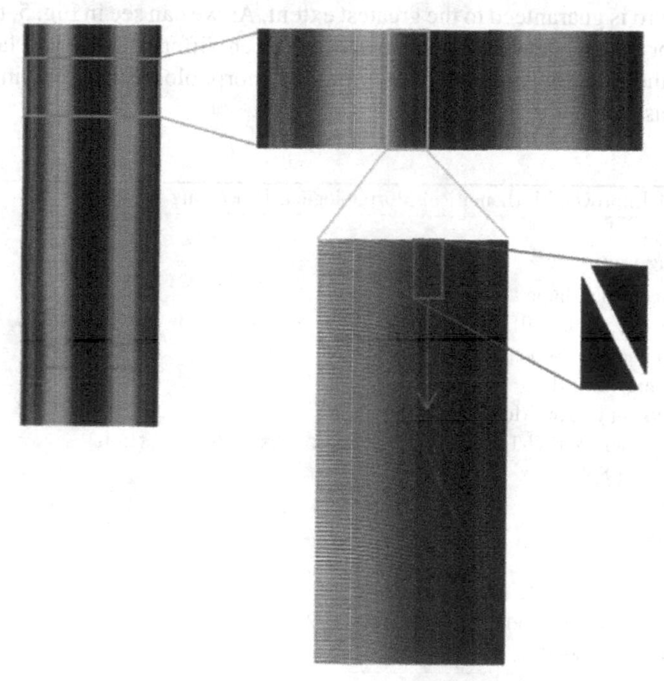

Fig. 4. Schematic diagram of pre-positioning 2# belt defects by sliding window method.

3.3 Defect Accurate Grading Stage

Step1: Removal of Horizontal Cords Using Sobel Operator. The shoulder area of the tire X-ray image only contains transverse cords and 2# belt slanted cords. The proposed algorithm uses the Sobel operator to differentiate the horizontal direction, and only the edge details in the vertical direction are obtained, that is, 2# belt cord edge, achieving a better cord layering effect.

Step2: Improved Mathematical Morphological Processing. The image after the cord layering only contains the oblique cord of the 2# belt layer, but a typical situation that may affect the subsequent cord search algorithm is inevitably cord breakage. This is indicated by the red circle in Fig. 5. In order not to have a shattering effect on subsequent threading methods, an improved mathematical morphological processing technique to achieve consistent adhesion of diagonal cords and maintain cord integrity was addressed.

Morphological closing operations are often used to fill small voids in objects, connect adjacent objects, and smooth object edges. Inspired by the closing operation principle, we propose an improved morphological processing algorithm to effectively bond the 2# belt obliquely broken cords. In the expansion stage, we construct a 5 * 5 square structuring element that is consistent with the direction of the oblique cords and use a 3 * 3 cross-shaped structuring element in the subsequent erosion stage. The binarized image is continuously operated twice in this way. So that the oblique adhesion effect of the oblique cord is guaranteed to the greatest extent. As we can see in Fig. 5, the different cords are kept independent and the distances between different cords are large enough to facilitate the thinning process. The improved morphological processing operation algorithm is as follows:

Algorithm 1 Improved Mathematical Morphological Processing

```
Input: img-Sobel
Output: img-processed
1: h, w = img-Sobel.shape
2: kernel-1 = ((0, 1, 0, 0, 0), (0, 1, 1, 0, 0), (0, 1, 1, 1, 0), (0, 0, 1, 1, 0), (0, 0, 0, 1, 0))
3: tmp = np.pad(img-Sobel, (2,2),'edge')
4: for y in range(2,h) do
5:     for x in range(2,w) do
6:         if np.sum(kernel-1 * tmp[y - 2:y + 3, x - 2:x + 3]) != 0 then
7:             img-Sobel[y, x] = 255
8:         end if
9:     end for
10: end for
11: kernel-2 = ((0, 1, 0), (1, 1, 1), (0, 1, 0))
12: img-processed = cv2.erode(img-Sobel, kernel-2)
13: return img-processed
```

Step3: The Refinement Operation Extracts the Cord Frame. The morphologically processed cords achieve a better oblique consistency adhesion effect while ensuring the independence between the cords, but due to the large width of the foreground cords, the background cord spacing is relatively small. It is not conducive to the subsequent accurate classification, so it is necessary to perform thinning processing to extract the single-pixel skeleton of the cord. The skeletonized cord perfectly retains the structural characteristics of the cord and has good connectivity which lays the foundation for the next pixel-level cord traversal algorithm.

Step4: Determination of Defect Level at Pixel Level by Horizontal Threading Method. The threading method proposed scans the thinning map by row by pixel, and counts the number of encounters of the cords, which is the number of intersection points in each row. The innovation is to determine the starting line and end of each row threading before scanning. The advantage of doing this is to ensure the accuracy of the number of intersections.

The threading method proposed in this paper is to first determine the start line and end line of the horizontal scan. We select the position of the left cord edge of the refined image to shift inward by 10 pixels as the scan start line. The scan termination line is also shifted inward by 10 pixels to ensure that the distance between the scanned cords is not affected by the cluttered boundaries on both sides. All spacing are calculated by pixel-by-pixel scanning, and the normal cord spacing is calculated by a statistical-based method, and the standard spacing is used as the detection standard, the abnormal point location and level judgment are carried out line by line, and the final test result (qualified product/defective product) is returned. Different with the traditional direct calculation of the average distance as the standard distance, this algorithm uses the statistical principle to calculate the standard distance between the cords. This approach not only reduces the computational complexity, but also fundamentally avoids the adverse effect of the slight skeleton discontinuity caused by the refinement operation on the calculation of the standard distance.

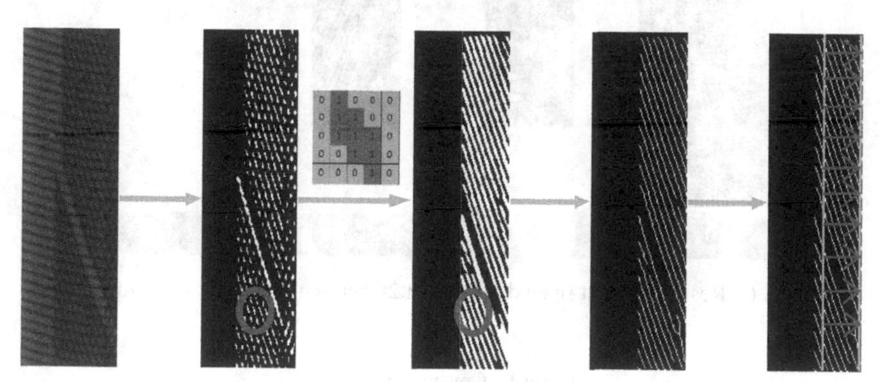

Fig. 5. Schematic diagram of the preprocessing process and threading method in the classification stage of the algorithm.

4 Experiments

4.1 Experimental Environment, Datasets, and Experimental Metrics

The hardware configuration of the experiment is as follows: AMD Ryzen 5 2600 Six-Core Processor clocked at 3.40 GHz and 16 GB of RAM. Using the Python 3.7 integrated development environment. The tire X-ray image self-built data set used in this paper has a total of 200 images, including 40 2# belt cord joint opening defect images and 160 normal images with different structural texture backgrounds. The experimental steps are as follows: firstly, the image that has been segmented is obtained as input. Then the image is pre-located in the first stage. In the second stage, the pixel-level detection is performed. Finally, the false acceptance rate (FAR) and recall rate are counted according to the final return results.

4.2 Experimental Results and Analysis

Figure 6 shows the detection process and results of 2# belt joint opening defects. The detection results of the entire self-built data set are shown in Tables 1 and 2 is the comparison results of the image processing method using the maximum value filter in the judging stages and the proposed method respectively. Figure 7 is a schematic diagram of the comparison between the maximum filtering and Sobel operator method adopted in the precise judging stage.

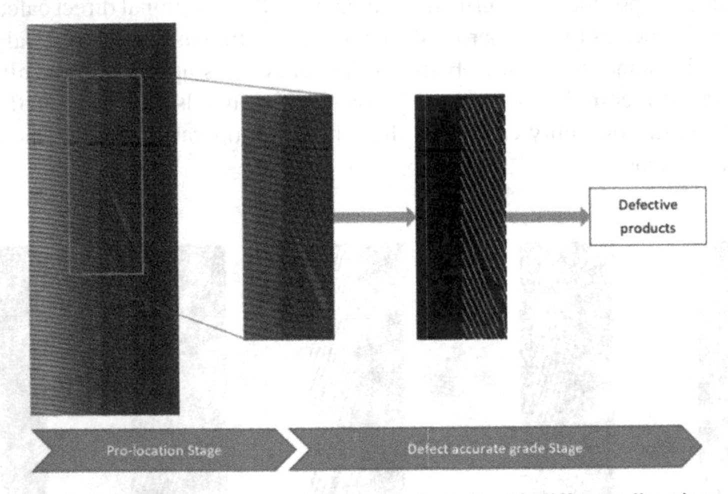

Fig. 6. Results of joint-open defection in 2# belt with different directions.

Table 1. Confusion matrix.

Actual class	Detected class	
	Normal	Defective
Normal	157	3
Defective	1	39

It can be seen from Tables 1 and 2 that dividing the algorithm into two independent stages, namely, the pre-positioning stage and the precise grading stage, can greatly

Table 2. Comparison results with various methods on the same datasheet.

Method	Time consuming (ms)	Recall rate	FAR
Without pre-location stage	13125	–	–
Maximum filtering method	861	97.5%	62.5%
Proposed method	638	98.12%	2.5%

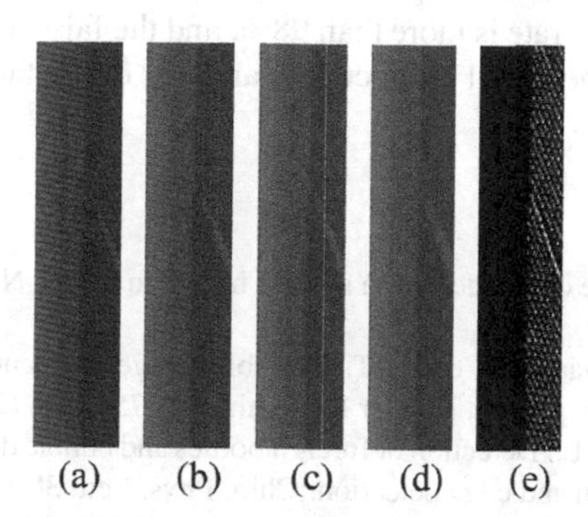

Fig. 7. Comparison results of joint-open defection in 2# belt with different methods: (a) original image (b) maximum filtering(kernel size:3 * 3) (c) 5 * 5 (d) 7 * 7 (e) Sobel operation.

reduce the time consumption. In the image processing process of the second stage, the maximum filtering method and the Sobel operation method are compared. It can be seen from Table 2 that the false detection rate of the maximum filtering algorithm is as high as 62.5%, and a large number of actual defect maps are incorrectly detected. The result is far lower than the actual production requirements of the factory. The reason can be explored through Fig. 7: the maximum value filtering improves the overall brightness of the target area, which is not conducive to the implementation of binarization. The maximum value filtering increases with the increase of the filter kernel size. The oblique cord is blurred while removing the horizontal cord, which is not conducive to the subsequent refinement operation. The phenomenon results in a high false detection rate.

The proposed method can not only locate the defect position accurately, but also realize the pixel-level layering of the cord in the complex background. The entire algorithm takes less than 1 s on average. The detection recovery rate on the self-built data set reaches 98.12%, and the false detection rate is 2.5%, which meets the actual production requirements of the factory.

5 Conclusions

In this paper, a 2# belt joint opening defect detection algorithm is proposed based on the tire X-ray image with periodic texture. The algorithm is divided into a pre-positioning stage and an accurate grading stage. In the pre-positioning stage, a nonlinear template is used to enhance the useful features of defects and weaken irrelevant features. The region of interest is accurately located based on gray statistical features. In the stage of accurate grading, Sobel operation is used to solve the problem of cord delamination. Then, the improved morphological closing operation is proposed for the new problem of oblique cord breakage and segmentation. Finally, the target image including defects after morphological processing is refined and graded accurately based on the proposed threading method.

The proposed algorithm is sensitive to defect features and has high robustness to background noise. It can obtain a high accuracy rate with the designed framework. The overall detection recovery rate is more than 98%, and the false detection rate is less than 3%, which fully meets the actual production standards of the factory.

References

1. Guo, Q.: Design of tire defect detection system based on X-ray. North Central University of China, China (2015)
2. Yang, Z., Zhou, S., Wang, G., et al.: Tire bubble image segmentation based on grayscale morphology and region growth. Rubber Ind. China **59**, 754–756 (2012)
3. Zhang, Y., Li, T., Li, Q.L.: Detection of foreign bodies and bubble defects in tire X-ray images based on total variation and edge detection. Chin. Phys. Lett. **30**(8), 084205 (2013)
4. Zhang, Y., Lefebvre, D., Li, Q.: Automatic detection of defects in tire radiographic images. IEEE Trans. Autom. Sci. Eng. **14**(3), 1378–1386 (2017)
5. Guo, Q., Zhang, C., Liu, H., et al.: Defect detection in tire X-ray images using weighted texture dissimilarity. J. Sens., Article ID 4140175, 12 p. (2016)
6. Cui, X., Liu, Y., Whang, C., et al.: Defect classification for tire X-ray image using convolutional neural networks. Electron. Meas. Technol. **40**(5), 168–173 (2017)
7. Cui, X., Liu, Y., Zhang, Y., et al.: Defect classification with multi-contrast convolutional neural network. Int. J. Pattern Recognit. Artif. Intell. **32**(4), 84–95 (2017)
8. Yao, G.: Research on tire defect detection technology of small sample set based on deep learning. Hangzhou University of Electronic Science and Technology (2020)
9. Zhu, Q., A,i X.: The defect detection algorithm for tire X-Ray images based on deep learning. In: 2018 3rd IEEE International Conference on Image Vision and Computing (2018)
10. Li, Y., Fan, B., Zhang, W., et al.: TireNet: a high recall rate method for practical application of tire defect type classification. Future Gener. Comput. Syst. Int. J. Sci. **125**, 1–9 (2021)
11. Sun, H.: Research on tire X-ray image defect detection algorithm. University of Science and Technology of China (2021)

A 3D Reconstruction Network Based on Multi-sensor

Yuwen Zhou[1,2], Jianhao Lv[1,2], Yaofei Ma[1,2](✉), and Xiaole Ma[1,2]

[1] BeiHang University, Beijing 100191, China
18373795@buaa.edu.cn, mayaofeibuaa@163.com
[2] CASIC Research Institute of Intelligent Decision Engineering, Beijing 100074, China

Abstract. To reconstruct the 3D model of specific targets in real time, a multi-sensor data fusion-based 3D reconstruction algorithm is proposed in this paper. This network-based algorithm takes the camera image and lidar point-cloud data as inputs, employing RGB channel and lidar channel to process each type of data separately, and finally obtains the targets' dense depth map by fusion. In RGB channel, the transformer network rather than CNN (convolutional neural network) is used to obtain multi-scale image features with global receptive field and high resolution, and generate monocular depth, guidance map and semantic segmentation. In the lidar channel, the sparse lidar data is fused with the guidance map to generate the final prediction of dense depth. In the test, our algorithm achieved a high ranking on the leaderboard. In application, under the condition of equal reconstruction quality, a five times faster speed is obtained in 3D reconstruction comparing to the traditional image-based method.

Keywords: 3D reconstruction · Multi-sensor · Real-time · Deep learning

1 Introduction

3D reconstruction is an important research topic in the field of computer vision, which has attracted more and more attention in many application fields, such as face recognition, building mapping and state evaluation, scenario restoration of cultural relics or archaeological scenes, etc. In our application, the 3D reconstruction technology is used to evaluate the damage of combat vehicles, that is, to evaluate the damage degree of specific vehicles through 3D dense reconstruction of them, so as to provide evaluation information for the subsequent operations, such as target-fire reallocation.

In our application, the time performance is crucial in 3D reconstruction, while accuracy is also required. This is because: 1) The battlefield situation is dynamic and time sensitive, and the shorter the reconstruction process, the better the effectiveness of the generated target data. 2) The battlefield is highly confronted, and the detecting equipment should complete its reconstruction task as soon as possible to ensure its own safety.

Y. Ma(1981)—-Associated professor of BeiHang University, researching in fields of M&S theory and practice, and intelligent behavior modeling.

W. Fan et al. (Eds.): AsiaSim 2022, CCIS 1712, pp. 583–594, 2022.
https://doi.org/10.1007/978-981-19-9198-1_44

The 3D reconstruction based on pure vision methods belongs to inactive methods, which uses feature points and geometric transformation to establish the 3D model of the target. Although vision-based methods have the advantages of extracting features from rich image information and generating dense depth map, it generally has a problem of low efficiency. We have tested the binocular vision method, where the 1280 * 720 resolution binocular images are processed in pipeline of SGBM (semi-global block matching), bilateral filtering and dense point cloud construction. Although several acceleration measures such as down-sampling, image segmentation for reducing reconstruction area, GPU parallel computing and so on are adopted in the pipeline, the reconstruction process still takes about 5 s. It is far from meeting the real-time requirements. In addition, pure vision methods have the problems of low accuracy and even encounter failure in some extreme scenes, such as highlight reflection, sudden change in depth. Active 3D reconstruction methods, such as lidar scanning based on TOF (time of flight) mechanism, can obtain more accurate 3D data, but generally generate sparse data, which is insufficient for accurate evaluation of target state.

We propose a high-speed, high-quality 3D reconstruction model using deep network, which fuse image data and lidar data to generate the depth prediction map of specific targets. Deep neural network works quite fast and could save several folds of time. Lidar has the advantages of high precision, large scanning range and less influence by environmental factors. As the result, lidar data can be used for image depth correction and completion. On the other hand, lidar data is sparse and unevenly distributed, which makes it almost impossible to produce dense depth map directly. How to effectively fuse image data and lidar data is a major challenge.

Many related studies have proposed a variety of deep completion architectures, which usually employed independent encoder and decoder networks to perform data fusion. The encoder, which is often a CNN component, adopts pre-trained large-scale image classification network as the backbone to extract image features. The decoder aggregates the features from encoder to predict the final dense depth map. However, CNN encoder has its disadvantages in 3D reconstruction task, i.e., with the deepening of network level, the feature resolution and granularity will be lost and cannot be restored. Although the loss of feature resolution and granularity may not matter for tasks such as image classification, it will be fatal for depth prediction.

Transformer [1] network has a natural advantage in depth prediction task, i.e., its feature of each layer has a global receptive field and can maintain a representation with constant dimensionality. Therefore, the transformer network is employed as the encoder in our method. The method is a two-channel depth prediction network, which includes RGB processing channel and lidar processing channel. The RGB channel aims to extract color correlation information to generate monocular depth prediction and depth guidance information. The lidar channel fuses the guidance information and lidar sparse depth information to generate the final prediction of dense depth map. In the RGB channel, we also employed a pretrain-downstream style like BeRT [2] to perform image semantic segmentation in RGB channel. Combining depth map predication and semantic segmentation, the 3D reconstruction of specific targets can be achieved. The pre-trained backbone network guarantees the faster migration to our datasets and scenarios.

2 Related Works

2.1 Dense Prediction Model

Different from classification or regression tasks, dense prediction tasks predict by pixels and focus on the global relationship of the whole image. While CNN is a mainstream choice to deal with such tasks, transformer, proposed for NLP (natural language processing) tasks in recent years, has attracted increased attention in the field of dense prediction of computer vision in recent years because of its ability to extract features with global receptive field and maintain a representation with constant dimensionality.

CNN-Based Models
In dense prediction tasks, a classical method is based on FCN (full convolution network) [3]. Many similar frameworks adopt a pixel-wise automatic encoder-decoder structure, where the encoder extract features, and the decoder recover map size. The U-Net [4] model, which owns such encoder-decoder structure diagram and is often used in medical images processing, has been proved to have good dense prediction ability.

Both depth completion and semantic segmentation are dense prediction tasks. Depth completion is to generate a dense depth map from a sparse depth map with or without the guidance of reference images. Semantic segmentation is to segment the image at the pixel or sub-pixel level into regions that conform to different semantics. In recent years, the representative models for depth completion include the CSPN (convolutional spatial propagation network) [5] proposed by Cheng et al. and the Sparse2Dense [6] network proposed by Tang et al. The representative models in the field of semantic segmentation include Segnet [7] and Deeplabv3 [8] et al. These methods are all CNN-based encoders-decoder networks. Despite the introduction of transformer, our network also follows the paradigm of the encoder-decoder structure.

Vision Transformer
In recent years, network models based on attention mechanism, especially transformer, has become the preferred architecture in the NLP field, which has performed well in large datasets. Several tasks have applied the attention mechanism to images and the most representative model is ViT (Vision Transformer) [9]. In many computer visual tasks ViT has been proved to performing better than CNN. To deal with dense prediction tasks, Intel Co. Proposes the DPT (dense prediction transformer) architecture [10], which combines the ViT and CNN methods. This network contains an encoder composed of multiple transformers and a decoder composed of convolutional layers. The feature fusion method of DPT is similar to U-Net. In our work, the RGB channel also refers to the DPT architecture to obtain better feature extraction performance. At the same time, a multiple-head design is applied in our model to obtain monocular depth prediction, guidance information and semantic segmentation simultaneously.

2.2 Depth Completion

How to fuse the RGB image with the sparse depth data efficiently is the most important topic in depth completion tasks. To be more specific, it should be decided that in which way RGB information could better guide the sparse depth and at which stage two types of inputs get fused.

Guidance
There are a lot of research on leveraging RGB information to expand sparse depth to dense depth. Schneider et al. use RGB information to generate accurate object boundary information for depth prediction. The specific method is to use semantic annotations of pixels to distinguish multiple objects. Ma et al. [11] use a neural network based on Resnet [12], combining lidar data and RGB images in the same feature space to get better prediction results. Zhang et al. [13] use RGB data to predict the surface normal to provide better prior information for depth prediction. From the continuous trials of various tasks, it can be found that it is a challenging task to recover dense depth from the independent sparse depth sample, which proves the importance of RGB guidance. Inspired by these tasks, we also use RGB information to predict a guidance information and integrate with sparse depth information.

Fusion
The fusion of multi-sensor data is always not straight forward. Previous works have shown different fusion strategies: early fusion (fusion of RGB features and depth features before final prediction), late fusion (the fusion of the monocular depth prediction and the sparse depth prediction results) or multi-scale fusion. Valada et al. [14] extract and combine the features in various stages of the encoder to achieve a multi-scale fusion. On a single dataset, late fusion leads to slightly better performance, because the confidence of later integration provides robustness. However, due to the problem of scale, the depth prediction of RGB is not reliable for different datasets due to the fact that differences in scale will make the weight of network change violently while training on a new dataset. For the generalization and adaptability of the network, we only use the early fusion method to integrate the guidance information generated by the RGB channel with the sparse depth. There are many ways of fusion in a mathematical sense. (connect, add or multiply, etc.).

3 Methods

3.1 Architecture

As Fig. 1 shows, the structure of the entire network is a dual channel network composed of a RGB channel and a depth channel. The RGB channel obtains monocular prediction, sparse depth guidance information and semantic segmentation map through RGB, of which monocular predictions are only used to calculate loss. The depth channel is a simple CNN based encoder-decoder network. It translates the fusion result of depth guidance information and sparse point cloud into the final depth prediction.

RGB Channel
As shown in Fig. 2, The structure of the RGB channel refers to DPT. In our method, the input image is divided into 16x16 patches, and the flattened representation of each patch is passed into the ViT encoder. We extract the feature presentation of each patch

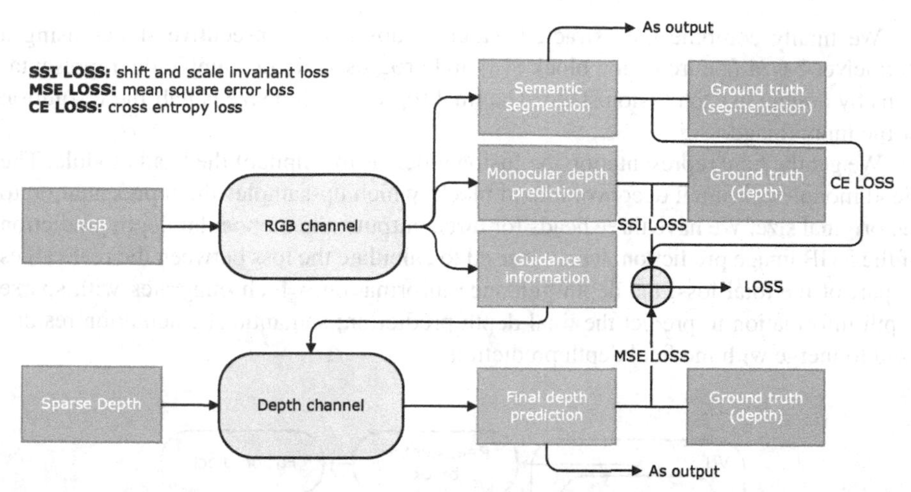

Fig. 1. The architecture of the whole network. RGB input and the sparse depth input are fed into RGB channel (yellow), and depth channel(green) separately firstly get semantic segmentation, moo depth prediction and guidance information (orange) by RGB channel. Then we get guidance information and sparse depth fused and obtain final depth prediction by depth channel. Monocular depth prediction and final depth prediction are all used to calculate loss, but we only take the final depth prediction as output. (Color figure online)

at different layers of ViT. These feature presentations are resampled to different scales in the reassemble block, get fused in the fusion blocked, and eventually up-sampled to the original image size. The reassemble block consists of three sub-blocks:

Read Block: Map the sequence with a length of $N_p + 1$ to a sequence with a length of N_p to prepare for the deformation of the next step. The dimension of feature map is transformed as below, where D is the feature dimension of transformer:

$$\mathbb{R}^{(N_p+1)\times D} \rightarrow \mathbb{R}^{N_p\times D}$$

Concatenate Block: The sequence obtained by read block is reshaped into a representation in the form of an image (the same ratio as the original image), we align each token according to the initial position of its picture. The dimension of feature map is transformed as below, where H and W are the height and weight of the original image and p is the number of patches:

$$\mathbb{R}^{N_p\times D} \rightarrow \mathbb{R}^{\frac{H}{p}\times\frac{W}{p}\times D}$$

Resample Block: Resample feature representations obtained by the concatenate block at different scales. The dimension of feature maps is transformed as below, where s is the expected scale factor:

$$\mathbb{R}^{\frac{H}{p}\times\frac{W}{p}\times D} \rightarrow \mathbb{R}^{\frac{H}{s}\times\frac{W}{s}\times D}$$

We finally combine the extracted feature maps from consecutive stages using a RefineNet-based feature fusion block [15] and progressively up-sample the representations by double in each fusion stage. The final representation size is half the resolution of the input image.

We get the final representation the fusion block as the input of the head module. The Head module is a small deconvolutional block, which up-samples the representation to the original size. We have three heads for three outputs: the monocular depth prediction of the RGB image prediction, which is used to calculate the loss between the real values as part of the total loss; the depth guidance information, which integrates with sparse depth information to predict the final depth prediction; semantic segmentation results, used to merge with the final depth prediction.

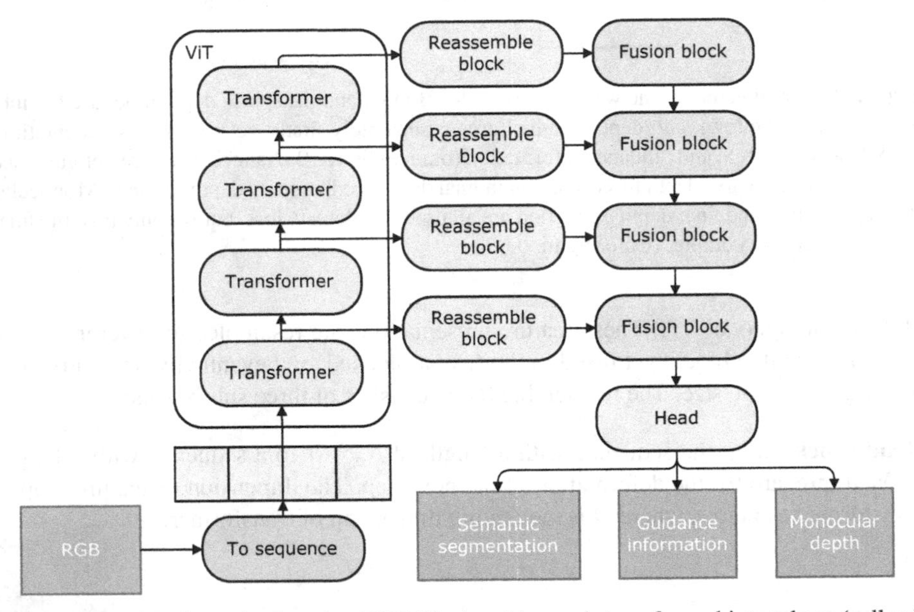

Fig. 2. The RGB channel referred to DPT. The input image is transformed into tokens (yellow) and then the tokens are passed through multiple transformer stages (green). We reassemble tokens from various stages into an image-like representation at multiple resolutions (red). Fusion modules (blue) progressively fuse and up-sample the representations to generate a fine-grained prediction. We implement different tasks through heads (grey). (Color figure online)

Depth Channel

The depth channel is a relatively simple encoder-decoder structure. The encoder and the decoder down-samples and up-samples feature maps, which are performed through convolution and deconvolution. Each layer of the network adopts a skip-connection structure like ResNet. At the same time, like U-Net, the features extracted from the encoder is fused with feature maps of the decoder, which improves the robustness of the network. The output of the depth channel is directly used as the final result, instead of fusing with the monocular depth prediction of the RGB channel (Fig. 3).

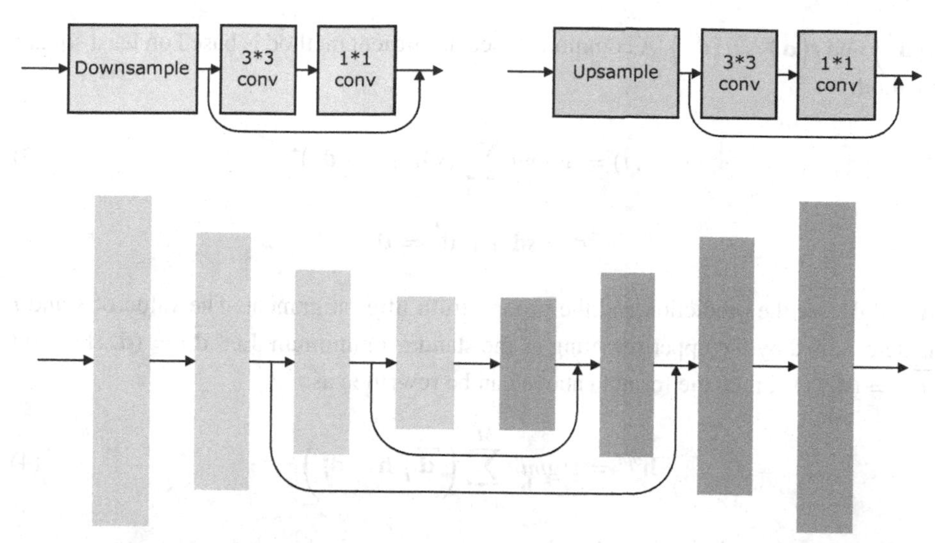

Fig. 3. The depth channel. A yellow refers to a down-sample block and a blue block refers to an up-sample block. The specific structure of each block is shown at the top of figure. (Color figure online)

3.2 Loss

In addition to the errors between predictions and ground truths of depth prediction and semantic segmentation, our work also uses the monocular depth loss as a part of the final loss. We take scaled and shift invariant loss [16] as the monocular depth loss, which is proposed recently. This loss focuses on the accuracy of the distribution of reconstruction results without calculating the difference in value, and its adaptability to different data sets is much better. We use MSE (mean square error) and cross-entropy as loss for the deep prediction and the semantic segmentation prediction. The final loss can be calculated as:

$$\mathcal{L} = \mathcal{L}_{mse} + \lambda\mathcal{L}_{ce} + \mu\mathcal{L}_{ssi} \tag{1}$$

The best values of the λ and μ are obtained by experiments.

Shift and Scale Invariant Loss

Shift and scale invariant loss is defined as the following form:

$$\mathcal{L}_{ssi}\left(\hat{\mathbf{d}}, \hat{\mathbf{d}}^{*}\right) = \frac{1}{2M}\sum_{i=1}^{M}\rho\left(\hat{\mathbf{d}}_i - \hat{\mathbf{d}}_i^{*}\right) \tag{2}$$

In the formula, $\hat{\mathbf{d}}$ and $\hat{\mathbf{d}}^{*}$ are the predictions and ground truths that have been scaled and shifted, and ρ define a specific loss function. To make the loss sensable, a reasonable requirement is that the predictions and the ground truths should be aligned at the same scale when calculating the shift and the scale. That is, we need to ensure that $s\left(\hat{\mathbf{d}}\right) =$

$s\left(\widehat{\mathbf{d}}^*\right)$ and $t\left(\widehat{\mathbf{d}}\right) = t\left(\widehat{\mathbf{d}}^*\right)$. A commonly used alignment method is based on least square method:

$$(s, t) = \arg\min_{s,t} \sum_{i=1}^{M} \left(s\mathbf{d}_i + t - \mathbf{d}_i^*\right)^2 \tag{3}$$

$$\widehat{\mathbf{d}} = s\mathbf{d} + t, \widehat{\mathbf{d}}^* = \mathbf{d}^*$$

$\widehat{\mathbf{d}}$ and $\mathbf{d}^*$ are the prediction and the ground truth after alignment. The value of s and t is determined by the upper rewriting as the standard minimum. Let $\vec{\mathbf{d}}_i = (\mathbf{d}_i, 1)^\top$ and $\vec{\mathbf{d}}_i = (\mathbf{d}_i, 1)^\top$, then the formula above can be rewritten as:

$$\mathbf{h}^{opt} = \arg\min_{\mathbf{h}} \sum_{i=1}^{M} \left(\vec{\mathbf{d}}_i^\top \mathbf{h} - \mathbf{d}_i^*\right)^2 \tag{4}$$

We can get a solution in a closed form:

$$\mathbf{h}^{opt} = \left(\sum_{i=1}^{M} \vec{\mathbf{d}}_i \vec{\mathbf{d}}_i^\top\right)^{-1} \left(\sum_{i=1}^{M} \vec{\mathbf{d}}_i \mathbf{d}_i^*\right) \tag{5}$$

In our work we choose $\rho(x) = \rho_{mse}(x) = x^2$ as the loss function of $\widehat{\mathbf{d}}$ and $\mathbf{d}^*$, and the shifted and scale invariant loss is confirmed to be:

$$\mathcal{L}_{ssi} = \frac{1}{2M} \sum_{i=1}^{M} \rho_{mse}\left(\widehat{\mathbf{d}}_i - \widehat{\mathbf{d}}_i^*\right) \tag{6}$$

4 Experiments

4.1 Dataset

In order to complete the training process, our dataset should include: RGB images, lidar point clouds, depth ground truth and semantic ground truth. The existing public dataset generally cannot meet this requirement. In order to get a multi-mode dataset, we use the AirSim plugin on Unreal Engine 4 to obtain the dataset in the virtual scenario. With tank and airplane as our targets, we have built two datasets, respectively with 3813 and 2166 sets of data.

4.2 Training

We adopt a fine-tuning strategy during the training stage. To take the advantages of the features learned from big datasets, The transformer part of the RGB channel is initialized by the weight pretrained on ImageNet dataset. The decoder of the RGB channel and the whole depth channel are trained from scratch. To achieve this, we divide the parameters of the network into two parts: parameters of transformer and other. Two ADAM optimizers are used to update the two parts of parameters with different learning rates. The initial learning rate of the transformer part is set to 10^{-8}(weight freeze), and the learning rate of rest part is set to $7 * 10^{-4}$. At the same time, we use a learning rate scheduler, which makes the learning rate reduce by half in the round of 10, 20, 30, and 60.

We perform the experiment on a single NVIDIA GTX 1080 GPU. The training batch size and epoch was set to 4 and 120. Considering the flip and crop invariance of the depth map and semantic segmentation, we adopt a strategy of random crop (384 * 384) and random flip (30%) to the whole dataset. At the same time, we apply a data augmentation including normalization and standardization to all RGB images. We have kept 70% of the dataset for the training set, 20% for the evaluation set and 10% for the testing set. At the same time, we use wandb to visualize the decrease of loss and metrics and the depth map predictions during validation (Fig. 4).

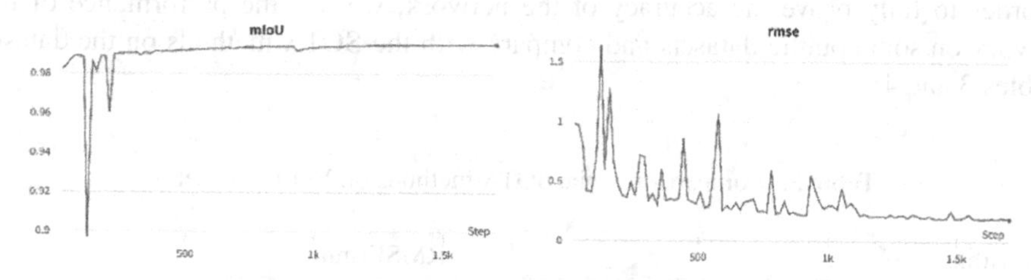

Fig. 4. The decrease curve of mIoU and RMSE visualized by wandb

4.3 Results

Inference Speed Performance

We expect to get a boosting improvement on speed and an acceptable accuracy when testing with our algorithm compared to the traditional methods. For each set of data, we use a stereo matching 3D reconstruction method (OpenMVS) to perform the 3D reconstruction for the binocular RGB images and compare the result with the depth prediction obtained by our method in terms of speed and accuracy. The comparison metrics are the time (seconds) and RMSE (meters), and the results are shown below (Tables 1 and 2):

Table 1. The speed and precision performance comparison of to the traditional method on AirSim tank dataset.

Algorithm	RMSE(m)	Time(s)
Ours	0.19	0.16
Stereo vision method	0.182	0.90

Table 2. Comparison on AirSim airplane dataset

Algorithm	RMSE(m)	Time(s)
Ours	0.26	0.163
Stereo vision method	0.23	0.88

During the process of comparison, we artificially eliminate the impact of the noise points caused by traditional methods. By comparing the results, we can see that the accuracy of our framework is quietly close to the result of the stereo vision method, while the time cost is five times less, which means a huge improvement in speed.

Accuracy Performance

In order to fully prove the accuracy of the network, we test the performance of the network on some public datasets and compare with the SOTA methods on the dataset (Tables 3 and 4):

Table 3. Comparison to the SOTA methods on VOID dataset

Algorithm	RMSE(mm)
NLSPN(1st)	79.121
KBNet(2nd)	95.86
Ours	**116.24**
ScaffNet-FusionNet(3rd)	119.14

Table 4. Comparison to the SOTA methods on KITTI dataset

Algorithm	RMSE(mm)
SemAttNet(1st)	709.41
HMSNet(9th)	937
Ours	**988**
Spade-Sd (10th)	1035

Restricted to the hardware condition, we just take a part of dataset and could not train for many iterations. Even though it can be seen that our work has got a high ranking on these leaderboards, and once again verifies the accuracy of the framework (Fig. 5).

Figure 6 shows the target depth map obtained by the fusion of depth prediction and semantic segmentation. And then we back-project the depth map to 3D space to obtain the point-clouds and fuse the point-clouds generated from each frame together to get a whole 3D structure of the target.

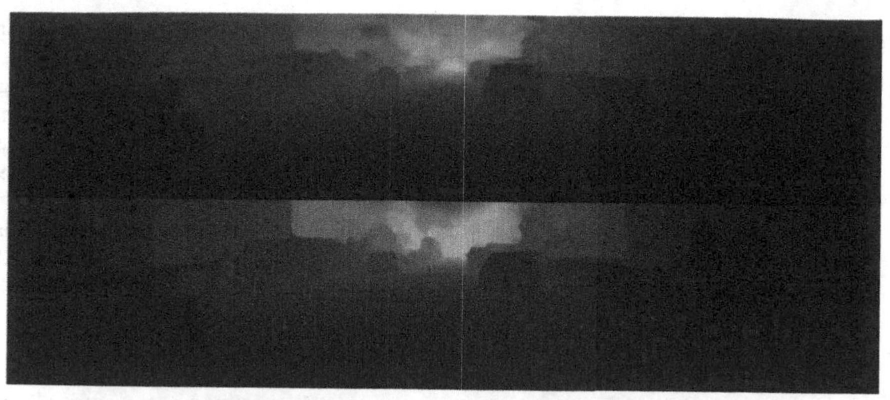

Fig. 5. Depth map prediction comparison. *Top*: our work. *Bottom*: PENet(3rd on leaderboard)

Fig. 6. The 3D reconstruction result for an airplane. *Left*: The depth map of a specific target. *Right*: its point-clouds generated from the depth map.

5 Conclusions

In this project, we have built a deep-learning framework based on multi-sensor to get the depth map with high speed and high quality. It has been proved that the framework can be used for real-time 3D reconstruction, which cannot be achieved by the traditional methods like stereo matching.

Apart from the efficiency, our work is also meaningful for taking advantage of transformer to obtain depth prediction and semantic segmentation simultaneously. Thus, we could get the depth map of a specific target directly, without any other effort to deal with the issue of segmentation.

References

1. Vaswani, A., et al.: Attention is all you need. In: Advances in Neural Information Processing Systems, vol. 30 (2017)
2. Devlin, J., et al.: Bert: pre-training of deep bidirectional transformers for language understanding. arXiv preprint arXiv:1810.04805 (2018)

3. Long, J., Evan, S., Trevor, D.: Fully convolutional networks for semantic segmentation. In: Proceedings of the IEEE Conference on Computer Vision and Pattern Recognition (2015)
4. Ronneberger, O., Philipp, F., Thomas, B.: U-net: convolutional networks for biomedical image segmentation. In: Navab, N., Hornegger, J., Wells, W., Frangi, A. (eds.) Medical Image Computing and Computer-Assisted Intervention – MICCAI 2015. MICCAI 2015. Lecture Notes in Computer Science, vol. 9351. Springer, Cham (2015). https://doi.org/10.1007/978-3-319-24574-4_28
5. Cheng, X., Wang, P., Yang, R.: Depth estimation via affinity learned with convolutional spatial propagation network. In: Ferrari, V., Hebert, M., Sminchisescu, C., Weiss, Y. (eds.) ECCV 2018. LNCS, vol. 11220, pp. 108–125. Springer, Cham (2018). https://doi.org/10.1007/978-3-030-01270-0_7
6. Tang, J., Folkesson, J., Jensfelt, P.: Sparse2dense: From direct sparse odometry to dense 3-d reconstruction. IEEE Robot. Autom. Letters **4**(2), 530–537 (2019)
7. Badrinarayanan, V., Kendall, A., Cipolla, R.: Segnet: A deep convolutional encoder-decoder architecture for image segmentation. IEEE Trans. Pattern Anal. Mach. Intell. **39**(12), 2481–2495 (2017)
8. Chen, L.-C., Zhu, Y., Papandreou, G., Schroff, F., Adam, H.: Encoder-decoder with atrous separable convolution for semantic image segmentation. In: Ferrari, V., Hebert, M., Sminchisescu, C., Weiss, Y. (eds.) ECCV 2018. LNCS, vol. 11211, pp. 833–851. Springer, Cham (2018). https://doi.org/10.1007/978-3-030-01234-2_49
9. Dosovitskiy, A., et al.: An image is worth 16x16 words: transformers for image recognition at scale. arXiv preprint arXiv:2010.11929 (2020)
10. Ranftl, R., Alexey, B., Vladlen, K.: Vision transformers for dense prediction. In: Proceedings of the IEEE/CVF International Conference on Computer Vision (2021)
11. Ma, F., Guilherme, V.C., Sertac, K.: Self-supervised sparse-to-dense: Self-supervised depth completion from Lidar and monocular camera. In: 2019 International Conference on Robotics and Automation (ICRA). IEEE (2019)
12. He, K., et al.: Deep residual learning for image recognition. In: Proceedings of the IEEE Conference on Computer Vision and Pattern Recognition (2016)
13. Zhang, Y., Thomas, F.: Deep depth completion of a single RGB-D image. In: Proceedings of the IEEE Conference on Computer Vision and Pattern Recognition (2018)
14. Valada, A., Mohan, R., Burgard, W.: Self-supervised model adaptation for multimodal semantic segmentation. Int. J. Comput. Vision **128**(5), 1239–1285 (2020)
15. Lin, G., et al.: RefineNet: multi-path refinement networks for high-resolution semantic segmentation. In: Proceedings of the IEEE Conference on Computer Vision and Pattern Recognition (2017)
16. de Brébisson, A., Pascal, V.: The z-loss: a shift and scale invariant classification loss belonging to the spherical family. arXiv preprint arXiv:1604.08859 (2016)

3D Point Cloud Registration Method Based on Structural Matching of Feature Points

Kezhi Wang, Jianyu Li, Zonghai Chen, and Jikai Wang(✉)

University of Science and Technology of China, Hefei 230026, Anhui, China
{wangkz,jyli1998}@mail.ustc.edu.cn, {chenzh,wangjk}@ustc.edu.cn

Abstract. 3D point cloud is an important expression of three-dimensional environment information. The registration of point cloud is the basis of realizing the functions such as localization, map construction and target detection based on 3D point cloud. For better registration, a method based on feature points and their spatial structure properties is proposed, which is divided into two stages: coarse registration and fine registration. Firstly, a feature point extraction network based on PointNet++ and Probabilistic Chamfer Loss is designed to extract robust and highly repetitive feature points. On the extracted feature points, the Super 4PCS method based on the spatial structure features of point cloud is used for global coarse registration. Then, taking the result obtained by coarse registration as the initial solution, the fine registration method based on NDT is used to register the point cloud more accurately. Finally, accurate registration results are obtained. This method combines the advantages of deep learning method and traditional method based on structure of point cloud, and makes full use of the spatial distribution properties of feature points. Experiments show that this method can achieve good registration results in a variety of application scenarios. And the method also shows the application potential of global location for initialization purpose in small-scale environments.

Keywords: 3D point cloud · Point cloud registration · Neural network · Feature points · Global localization

1 Introduction

With the development and maturity of 3D laser scanning [1] and image depth estimation [2] technologies, point cloud has become a common and effective way to express 3D environmental information. 3D point cloud registration is the process of finding the correspondence between different point clouds, which is an important part of 3D reconstruction [3], 3D localization and pose estimation [4], and is widely used in robot navigation [5], unmanned driving [6] and virtual reality [7]. How to achieve point cloud registration efficiently and accurately is a key problem that researchers have been working on. The mainstream point cloud registration methods can be generally divided into two stages: coarse registration and fine registration [8].

The fine registration methods for point clouds are ICP (Iterative Closest Point) [9], NDT (Normal Distribution Transform) [10] and their related extensions [11] mostly used.

W. Fan et al. (Eds.): AsiaSim 2022, CCIS 1712, pp. 595–608, 2022.
https://doi.org/10.1007/978-981-19-9198-1_45

Point cloud coarse registration algorithms mainly include methods based on search and methods based on geometric feature description. The search-based registration methods mainly select a number of points from the source point cloud data, and find the corresponding points according to the exhaustive search of the target point cloud or other specific search methods. Then calculate the possible variation matrix and possibility and finally determine the optimal correspondence, such as RANSAC [12] and 4PCS [13] algorithms. The registration methods based on geometric feature descriptions form feature descriptors based on the distribution characteristics of the point cloud around the feature points and the intensity or color variation characteristics, such as SHOT [14] and FPFH [15] algorithms. And the correspondence is constructed based on the descriptors, but there are problems such as sensitivity to noise, large computational effort, and possible lack of discrimination. In particular, most of such methods focus on local features almost [16], while the overall distribution of feature point clouds and their spatial structure features are not sufficiently characterized. With the development and wide range of applications of deep learning, it has played a significant role in 3D point cloud registration due to its good robustness and generalization [17]. However, there are difficulties such as the lack of common and well-labeled datasets, unclear feature definitions, and the influence of scale and rotation factors, which adversely affect the robustness and repeatability of feature point extraction in point clouds and the corresponding registration effect.

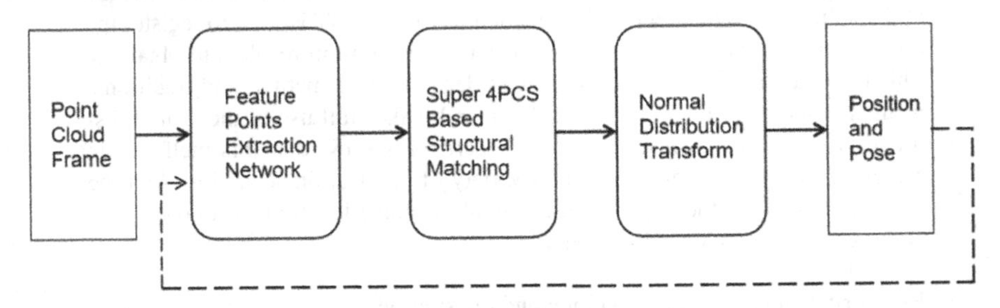

Fig. 1. Algorithm process

As a global registration algorithm, Super 4PCS [18] is insensitive to the initial poses. Super 4PCS algorithm can align scans of arbitrary poses, achieve fast extraction of point pairs by rasterized meshing in the target point cloud, and avoid global search of point cloud data. It also uses angular constraints to reduce the generation of invalid point pairs and improve the registration efficiency and accuracy of the overall point cloud. On the contrary, in the case of low overlap and high outlier point clouds, local registration algorithms such as ICP, NDT and their derivative algorithms may fall into local minima. However, Super 4PCS can still take some time when applied to complex or large-scale point clouds.

In this paper, a 3D point cloud registration method based on structural matching of feature points is proposed to address the above problems. Firstly, a feature point extraction network is constructed based on PointNet++ [19]. The network can perform unsupervised learning on the dataset without explicit feature point annotation, and it can robustly extract repeatable feature points in 3D point clouds. Then a point cloud coarse registration method based on the Super 4PCS algorithm for feature points is designed to make full use of the spatial distribution characteristics of feature points. Under the condition that a good initial solution of the pose is obtained from the coarse registration, the NDT method is applied to the fine registration of the point cloud to finally obtain accurate position and pose information. In addition, if the point cloud registration fails, return to the feature points extraction step to re-extract new feature points and perform the registration algorithm (as the dotted line in Fig. 1) until the set matching times or time is exceeded. Finally, we design experiments to verify the performance of feature point extraction algorithm, point cloud registration algorithm and small-scale scenarios-orientated localization used for initialization. The overall architecture of the algorithm is shown in Fig. 1.

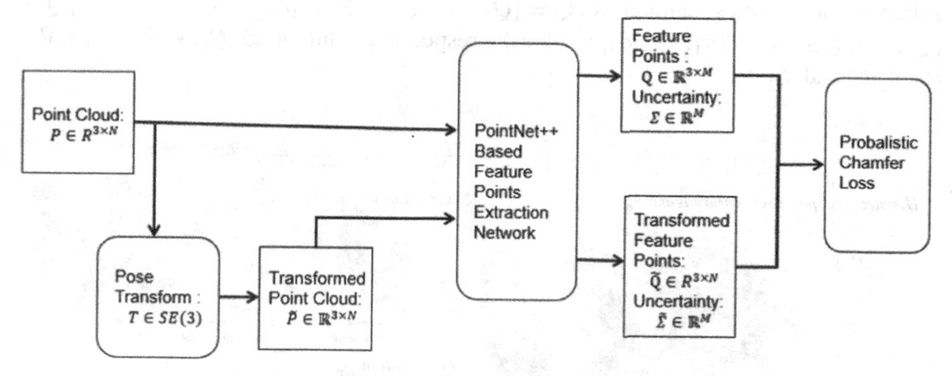

Fig. 2. Pipeline of feature points extraction network

2 PointNet++ Based Feature Extraction Network

2.1 Network Architecture

The general pipeline architecture of the feature point extraction network in this paper is shown in Fig. 2. $P \in \mathbb{R}^{N\times 3}$ is the original point cloud data, which is used as the input. The robustness of feature points to rotation has been one of the challenges for feature points extraction in 3D point clouds [20]. It is necessary to ensure the robustness of feature points under different viewpoints, i.e., in different viewpoints, try to ensure that the feature points are almost the same points in the world coordinate system. The original point cloud is transformed using a randomly generated transformation matrix $T \in SE(3)$ to obtain $\tilde{P} \in \mathbb{R}^{N\times 3}$ for the purpose of data enhancement and improving the robustness of the feature point extraction network to rotational transformations. Experiments demonstrate that the method effectively improves the repeatability of the extracted feature points under different viewpoints.

The main network for initial feature point extraction and uncertainty calculation is based on PointNet++, which is divided into encoder and decoder modules. The network frame diagram of PointNet++ is shown in Fig. 3 [19]. The Encoder module mainly consists of two or more steps of set abstraction, and each set abstraction consists of three steps: sampling, grouping and PointNet. In the Sampling step, FPS [21] is used to downsample the point cloud. Grouping step takes the downsampled points as the centroids and performs KNN [22] or query ball point grouping to form local neighbor patches. Then PointNet [23] is used to extract features for each patch. In the decoder module, the backward interpolation method is used to achieve feature adoption back to obtain the global features of all points in the original point cloud, and then combined with the local features obtained in the set abstraction to form a complete feature representation of each point and classify out the feature points. Combining the method in SO-Net [24] to learn the uncertainty vector Σ, $\tilde{\Sigma} \in \mathbb{R}^{M}$ corresponding to the feature points Q, $\tilde{Q} \in \mathbb{R}^{M\times 3}$, their elements represent the uncertainty of each feature point. Outputs of PointNet++ based feature points extraction network, M proposal keypoints and its saliency uncertainties denoted as $\{Q = [Q_1, \cdots, Q_M], \Sigma = [\sigma_1, \cdots, \sigma_M]^T\}$ and$\{\tilde{Q} = [\tilde{Q}_1, \cdots, \tilde{Q}_M], \tilde{\Sigma} = [\tilde{\sigma}_1, \cdots, \tilde{\sigma}_M]^T\}$ for the respective point cloud. $Q_m \in R^3$, $\tilde{Q}_m \in R^3$, $\sigma_m \in R^+$ and $\tilde{\sigma}_m \in R^+$.

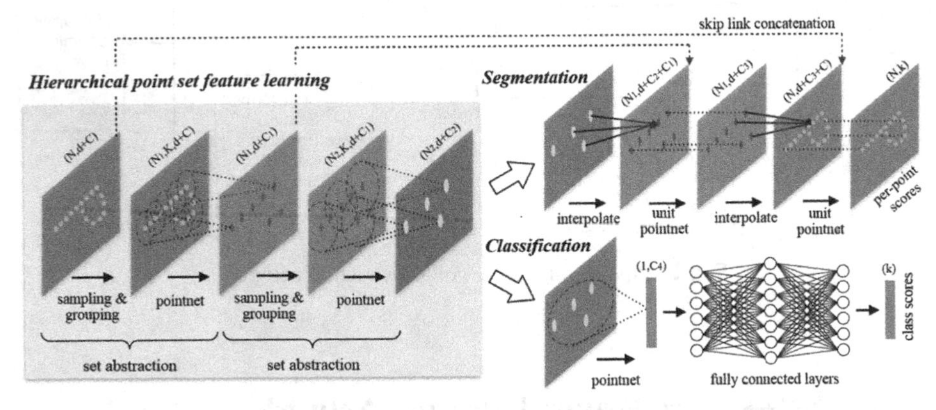

Fig. 3. The network frame of PointNet++ [19]

2.2 Loss Function

The uncertainty of feature points varies. For example, on a table, the uncertainty of the feature points at the corners of the table is low, which can be called good feature points, while the uncertainty of the feature points on the table is high, which can be called bad feature points. Therefore, in the process of duplicating the feature points in different views as much as possible, if each feature point is treated equally, for example, using the standard Chamfer Loss [25] as the loss function, the feature points with low uncertainty will have a bad effect on the training of the network. Let T^{-1} be the inverse transformation of the bitwise transformation corresponding to T. The resultant feature point cloud obtained by the pose transformation corresponding to T^{-1} acting on $\tilde{Q}$ is Q'. And assume that the corresponding uncertainty remains constant before and after the transformation. Then the standard Chamfer Loss function takes the following form, where Q'_j is the nearest neighbor of Q_i in Q′.

$$\sum_{i=1}^{M} \min_{Q'_j \in Q'} \left\| Q_i - Q'_j \right\|_2^2 + \sum_{j=1}^{M} \min_{Q_i \in Q} \left\| Q_i - Q'_j \right\|_2^2 \tag{1}$$

To solve this problem, this paper draws on Probabilistic Chamfer Loss [26] as the loss function in network training. Firstly, the probability distribution between Q_i and Q'_j for $i = 1, \cdots, M$ is given by:

$$p(d_{ij}|\sigma_{ij}) = \frac{1}{\sigma_{ij}} \exp\left(-\frac{d_{ij}}{\sigma_{ij}}\right), \text{ where } \sigma_{ij} = \frac{\sigma_i + \sigma'_j}{2} > 0, d_{ij} = \min_{Q'_j \in Q'} \left\| Q_i - Q'_j \right\|_2 \tag{2}$$

$p(d_{ij}|\sigma_{ij})$ is a valid probability distribution since it integrates to 1. When the order of the point clouds is exchanged, i.e., Q′ and Q, the probability distributions are asymmetric because the sets of nearest neighbor points are already different sets, i.e., $d_{ij} \neq d_{ji}$ and $\sigma_{ij} \neq \sigma_{ji}$. Then the specific form of Probabilistic Chamfer Loss Function L is as follows.

$$\begin{aligned} L &= \sum_{i=1}^{M} - \ln p(d_{ij}|\sigma_{ij}) + \sum_{i=1}^{M} - \ln p(d_{ji}|\sigma_{ji}) \\ &= \sum_{i=1}^{M} \left(\ln \sigma_{ij} + \frac{d_{ij}}{\sigma_{ij}} \right) + \sum_{i=1}^{M} \left(\ln \sigma_{ji} + \frac{d_{ji}}{\sigma_{ji}} \right) \end{aligned} \tag{3}$$

Probabilistic Chamfer Loss for assigning different weights based on the distance and confidence of the feature points to guide the network training in the right direction.

3 Point Cloud Registration Based on Structural Property of Feature Points

3.1 Feature Points-Oriented Super 4PCS Coarse Registration Algorithm

In the case of low overlap and high outlier point clouds, local registration algorithms such as ICP, NDT and their derivatives are prone to fall into local minima and thus require good initial solutions for the positions and poses. Usually, researchers combine feature points with specific point descriptors to perform initial registration of point clouds. However, the feature point descriptors do not provide sufficient representation of the relationships between feature points and the overall structural properties of the feature point cloud distribution in space. To address this problem, this paper chooses to use the Super 4PCS method for coarse registration based on the feature point cloud in the case of the feature extraction network extracts robust feature points. In order to make full use of the overall spatial structure features of the feature point cloud and solve a better initial solution of the pose.

In this section, the source point cloud is denoted as P and the target point cloud is denoted as P′. The source point cloud is generally a 3D point cloud frame or a reconstructed point cloud model obtained from lidar scan or other methods. The target cloud can be a lidar scan frame with overlapping parts, a model with different position and pose, or a priori point cloud map of small scale, etc. The feature point cloud obtained by feature extraction of P through the feature point extraction algorithm in this paper is denoted as Q, and the corresponding feature point cloud of P′ is denoted as Q′. The flow of Super 4PCS coarse registration algorithm based on feature points is as follows:

1. When selecting the coplanar four-point base in the feature point cloud Q corresponding to the source point cloud P, three points are randomly selected first, and then the fourth point is selected so that the four points are coplanar and not co-linear. Note that we give preference to good feature points with low uncertainty to be used as the point base. The distance between the four selected points should be as large as possible, and they all lie within the overlap region. This results in a coplanar four-point base B = $\{P_a, P_b, P_c, P_d\}$. According to Eqs. (4) and (5), the distances d_1, d_2 and intersection ratios r_1, r_2 of the two lines in the coplanar four-point base B can be calculated.

$$d_1 = \|P_a - P_b\|,\ d_2 = \|P_c - P_d\| \tag{4}$$

$$r_1 = \frac{\|P_a - e\|}{\|P_a - P_b\|},\ r_2 = \frac{\|P_c - e\|}{\|P_c - P_d\|} \tag{5}$$

2. The sets S_1 and S_2 of point pairs are determined in the feature point cloud Q′ corresponding to the target point cloud P′. The sphere is drawn with each point $q_i' \in$ Q′ as the center of the sphere and with $R_1 = d_1 \pm \varepsilon$ and $R_2 = d_2 \pm \varepsilon$ as the radius, respectively. The set S_1 is q_i' with points distributed in the range $[d_1 - \varepsilon, d_1 + \varepsilon]$, and q_i' with points distributed in the range $[d_2 - \varepsilon, d_2 + \varepsilon]$ is S_2. The point cloud surface is rasterized to create a cell size ε three-dimensional grid G.
3. Extract the set of four points corresponding to the base B in the target feature point cloud Q′. Iterate through all candidate point pairs in point pair sets S_1 and S_2, calculate all intersection nodes e according to the intersection ratio consistency, and store them in the grid G. Based on the angle of two corresponding point pairs in base B as θ, the four-point base with nodes approximately equal, while the angle of the line between two point pairs is approximately equal to θ, and whose error range is within ξ is extracted in grid G for matching.
4. The set U = $\{U_1, U_2, \cdots, U_n\}$ of all four points in the target feature point cloud Q′ that satisfy the conditions corresponding to the base B. The transformation matrix between the base B and each U_i is found. By comparing the Largest Common Pointset (LCP), the transformation matrix with the highest registration accuracy is selected for global transformation.

Super 4PCS is based on the improved 4PCS algorithm. 4PCS also utilizes a specific set of congruent points for global registration. However, the complexity of the 4PCS algorithm is $O(n^2 + k)$, where n denotes the size of the point cloud and k denotes the candidate congruent 4-point set. For larger point clouds with less overlap, the quadratic complexity of the 4PCS algorithm can easily become its bottleneck. Super 4PCS outperforms 4PCS in terms of both robustness and speed. It lowers the quadratic complexity to optimal linear complexity; and lowers the number of candidate conjugate pairs generated. The time complexity of Super 4PCS is $O(n + k_1 + k_2)$, where k_1 is the number of pairs in the target point cloud at a given distance and k_2 is the number of allometric sets.

3.2 NDT-Based Point Cloud Fine Registration Algorithm

After the coarse registration algorithm of feature points-oriented Super 4PCS to obtain the rough correspondence of point clouds and an expected initial solution of the poses, the NDT-based algorithm is chosen to accurately align the point clouds. The core idea of Normal Distribution Transformation (NDT) is to deposit the point cloud data into a grid composed of small voxels, and the point clouds in each voxel are converted into a probability density function, and then the matching relationship between the point clouds is found mathematically.

Given the source point cloud $\mathrm{P} = \{P_1, P_2, \cdots, P_N\} \subset \mathbb{R}^3$ and the target point cloud $\mathrm{P}' = \{P'_1, P'_2, \cdots, P'_L\} \subset \mathbb{R}^3$, the classical NDT algorithm voxelizes the target point cloud and then calculates the mean μ and variance Σ of the points in the voxel, as:

$$\mu = \frac{1}{N}\sum_{k=1}^{N} P'_k \tag{6}$$

$$\Sigma = \frac{1}{N-1}\sum_{k=1}^{N} (P'_k - \mu)(P'_k - \mu)^T \tag{7}$$

This gives a probability density function that describes the distribution of points within the voxel and does not need to be recalculated at each iteration. Specifically as:

$$\rho(x) = \frac{1}{\sqrt{2\pi}\sqrt{|\Sigma|}} \times \exp\left(-\frac{1}{2}(x-\mu)^T \Sigma^{-1}(x-\mu)\right) \tag{8}$$

The optimal transformation matrix maximizes the overall likelihood of the source point cloud in its corresponding probability density function, as:

$$T^* = \underset{T}{\operatorname{argmin}} \prod_{k=1}^{L} \rho(TP_i) \tag{9}$$

The optimal transformation is applied to the source point cloud, and the new optimal transformation is calculated by iterating Eq. (9).

4 Experiments and Analysis

4.1 Experiment of Feature Points Extraction

The repeatability of feature points is an important metric for the performance of the feature point extraction algorithm when the point cloud data contains viewpoint variations, and it does not depend on the feature point descriptors. Therefore, in the experiments, given two point clouds P and P′ with different viewpoints. The two point clouds are associated by a rotation matrix $R \in SO(3)$ and a translation vector $t \in \mathbb{R}^3$.

A feature point detector detects a set of feature points Q = $[Q_1, \ldots, Q_M]$ and Q′ = $[Q'_1, \ldots, Q'_M]$ from P, P′ respectively. A feature point $Q_i \in$ Q is repeatable if the distance between $RQ_i + t$ and its nearest neighbor $Q'_j \in$ Q′ is less than a threshold τ, i.e.:

$$\left\| RQ_i + t - Q'_j \right\|_2 < \tau \tag{10}$$

In this paper, to verify the effect of the algorithm on the alignment of the object model, we use 70% of the ModelNet40 dataset as the training set and the other 30% as the test set. In addition, for outdoor scenarios, 70% frames of sequences 00 to 10 of the KITTI dataset are used as the training dataset, and the other 30% of the KITTI dataset are used as the test dataset. The obtained feature point extraction results are shown in Fig. 4 (the point cloud in the first column on the left is from the piano category of ModelNet40, and the two columns on the right are from the KITTI dataset, where red points in the figure are extracted feature points, and please note that we have adjusted perspective of point cloud in the figure for clarity of presentation, similar to the following). Considering the scale of the scene and the performance of the subsequent point cloud registration, 128 feature points are extracted from the point cloud for each frame in the experiment as a comparison. The comparison results with feature point extraction methods such as 3DFeat-Net, ISS [27] and Harris-3D are shown in Table 1.

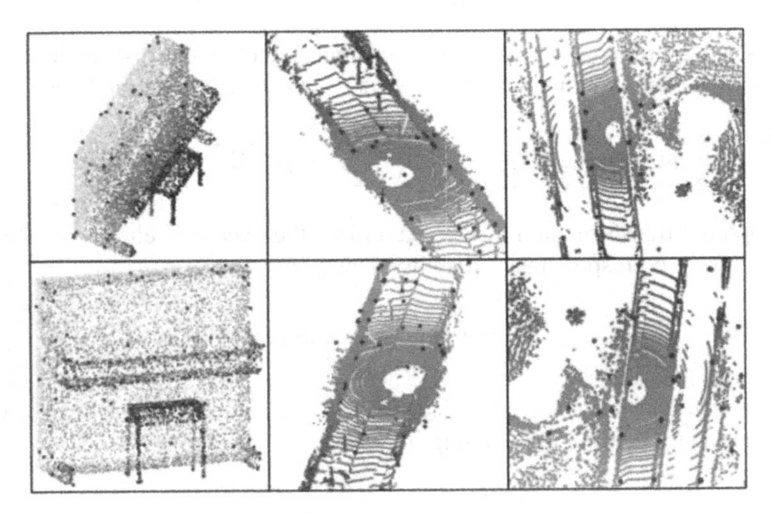

Fig. 4. Results of feature points extraction

From the experimental results, it can be seen that the method in this paper achieves the best results among the methods involved in the comparison. And its relative repeatability of feature points is much higher than all other methods. In addition, we ran the algorithm on a computer with an RTX1080Ti GPU and the average time taken to extract 128 feature points per frame on the KITTI dataset was 0.017s, 64 on ModelNet40 was 0.0089s.

Table 1. Feature points repeatability experiment results.

Detector	Feature points relative repeatability	
	ModelNet40	KITTI
Ours	88%	60%
3DFeat-Net	19%	20%
ISS	34%	13%
Harris-3D	37%	14%

Fig. 5. Sample of point cloud coarse registration results

4.2 Experiment of Point Cloud Registration and Localization

In the experiments for point cloud registration, we apply arbitrary translational and rotational transformations to frames of ModelNet40 and KITTI dataset, and then perform point cloud registration algorithm to align them. Samples of point cloud coarse registration results are shown in Fig. 5, and the corresponding fine registration results are shown in Fig. 6 (the two point clouds in the first row are from ModelNet40, and the two point clouds in the second row are from KITTI dataset).

There are only a few methods based on structural matching of feature points, and we chose to compare with the method of paper [28] of this type denoted as ISS_4PCS, which uses ISS as a feature point extractor and uses 4PCS as a coarse matching algorithm. To be fair, we did not use the DNT fine matching algorithm in the comparison. Table 2 shows the success rate and average registration time of 200 experiments for sofa0770 to sofa0773 of the sofa category in ModelNet40 (50 times of experiments for each and with different rotations and translations) dataset conducted by the proposed algorithm and ISS_4PCS, where RT denotes the registration time and SR denotes the success rate.

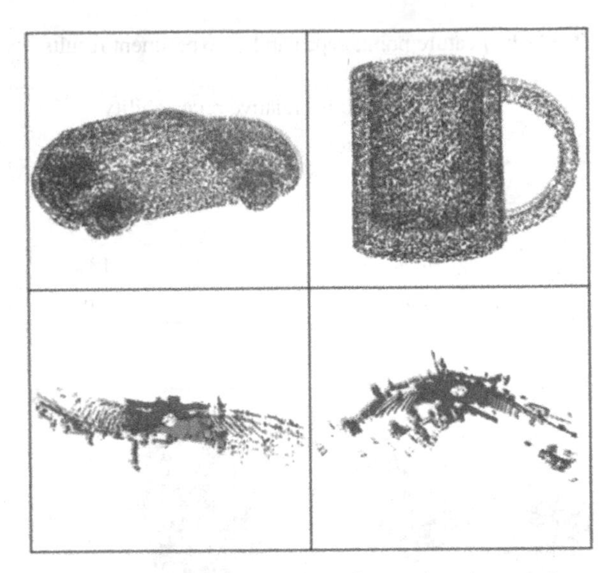

Fig. 6. Sample of point cloud fine registration results

Please note that through the comparison of several groups of experiments, a good initial registration position can be obtained when the root mean square error (RMSE) is less than 0.15 m, and we use it as the criterion for evaluating SR. The point cloud RMSE plots for 50 experiments conducted for sofa0770 are shown in Fig. 7.

Table 2. Experiment results of comparison with ISS_4PCS method.

Registration method	sofa0770		sofa0771		sofa0772		sofa0773	
	RT	SR	RT	SR	RT	SR	RT	SR
Ours	4.9s	90%	5.6s	86%	4.7s	90%	3.8s	88%
ISS_4PCS	9.9s	78%	10.1s	72%	12.4s	62%	10.5s	68%

In order to demonstrate the positive effect of Probabilistic Chamfer Loss, we designed the following ablation experiment. In the experiment, the first group uses the coarse registration method proposed in this paper. While in the second group we change the loss function in the feature extraction network to the standard Chamfer Loss as shown in formula (1), and other parts are the same as those in the first group. We selected piano0100 to piano0102 of the piano category of the ModelNet40 dataset as the test data samples in the experiments. And we conducted 50 experiments for piano0100 to piano0102 respectively (extract 256 feature points each for registration), and the success rate results are shown in Table 3 (the results with RMSE less than 0.12 m are considered successful). For simplicity of representation, we denote the second method as SCLG in Table 3. The experimental results show that the registration method using Probabilistic Chamfer Loss has a higher success rate. It can be inferred that, compared with standard

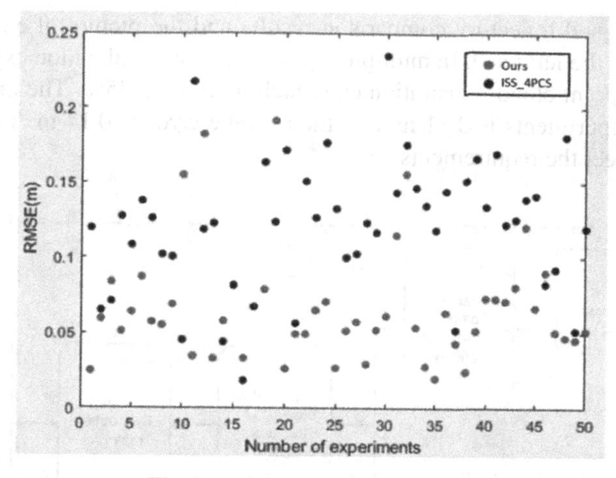

Fig. 7. RMSE plots for sofa0770

Chamfer Loss, the Probabilistic Chamfer Loss can better guide the network to extract high-quality feature points that are conducive to registration.

Table 3. Results of ablation experiment.

Registration method	Success rate (SR)		
	piano0100	piano0101	piano0102
Ours	84%	88%	82%
SCLG	72%	70%	74%

Initial localization refers to the process of obtaining its own initial position and pose before the mobile robot starts running. In the point cloud registration and localization experiment, we use an unmanned mobile platform equipped with RTK (Real-time kinematic) and 32-line velodyne lidar to collect laser point cloud data and perform point cloud registration and localization in an area of campus scenario.

The process of localization based on our registration algorithm is shown in Fig. 8. First, an offline map is constructed for the specific area about 100 m^2, and the feature point cloud is obtained using the feature point extraction algorithm proposed in this paper every 2 m apart (512 feature points each time). In order to avoid excessive repetition of feature points, non-maximal value suppression method is used to filter the regions with too dense feature points and retain a number of feature points with the lowest uncertainty. We use the point cloud data collected by the unmanned mobile platform at different locations in map to calculate the localization results based on our registration algorithm (red points in right part of Fig. 9), and compare it with the solution obtained by RTK (black points in right part of Fig. 9). Since we finally use the NDT-based point cloud fine registration method, if the coarse registration stage can be successfully aligned, the final registration accuracy of the algorithm can basically meet the localization requirements. One of the successfully aligned trajectory comparison results and the picture of environment are shown in Fig. 9 (the left part). In multiple registration and localization experiments, the success rate of point cloud registration can reach more than 85%. The maximum error in successful experiments is 0.31 m, and the average error is 0.14 m. The positioning accuracy can meet the requirements.

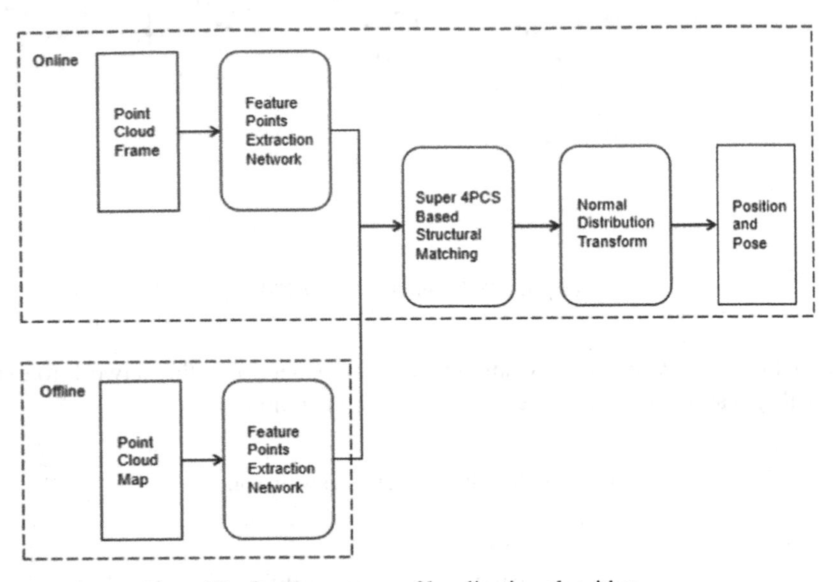

Fig. 8. The process of localization algorithm

Fig. 9. Result of localization experiment

5 Conclusion

In this paper, a 3D point cloud matching algorithm based on structural matching of feature points is proposed. A robust and efficient feature point extraction network is designed based on PointNet++ and Probabilistic Chamfer Loss Function for extracting highly repetitive feature points in 3D point clouds. Then, the feature points-oriented

Super 4PCS coarse registration algorithm and NDT fine registration algorithm are used to perform point cloud registration and pose solution. And the experiments are designed to verify the effectiveness of the algorithms. The algorithm designed in this paper makes full use of the structural properties of feature points and shows the application potential of global localization for initialization purposes in small-scale scenarios. This paper shows that the distribution of feature points can be used as the characteristic of point cloud to a certain extent, and it can be applied to point cloud registration.

Acknowledgments. This work was supported by the National Natural Science Found of China (Grant No. 62103393).

References

1. Dong, Z., Liang, F., Yang, B., Xu, Y., Stilla, U.: Registration of large-scale terrestrial laser scanner point clouds: a review and benchmark. ISPRS J. Photogramm. Remote Sens. **163**, 327–342 (2020)
2. Cao, Y., Wu, Z., Shen, C.: Estimating depth from monocular images as classification using deep fully convolutional residual networks. IEEE Trans. Circuits Syst. Video Technol. **28**(11), 3174–3182 (2017)
3. Berger, M., et al.: A survey of surface reconstruction from point clouds. Comput. Graph. Forum **36**(1), 301–329 (2016)
4. Behley, J., Stachniss, C.: Efficient surfel-based SLAM using 3D laser range data in urban environments. Rob. Sci. Syst. **2018**, 59 (2018)
5. Teng, X., Guo, D., Guo, Y., Zhou, X., Liu, Z.: CloudNavi: toward ubiquitous indoor navigation service with 3D point clouds. ACM Trans. Sens. Netw. **15**, 1–28 (2019)
6. Wang, X., Wang, J., Zhang, Y., Li, C., Wang, L.: 3D LIDAR-based intersection recognition and road boundary detection method for unmanned ground vehicle. In: 2015 IEEE 18th International Conference on Intelligent Transportation Systems (ITSC), pp. 499–504 (2015)
7. Bolkas, D., Chiampi, J., Chapman, J., Pavill, V.F.: Creating a virtual reality environment with a fusion of sUAS and TLS point-clouds. Int. J. Image Data Fusion **11**(2), 136–161 (2020)
8. Zhao, F.Q.: Point cloud registration algorithm from coarse to fine. Transducer and Microsystem Technologies (2018)
9. Besl, P.J., Mckay, H.D.: A method for registration of 3-D shapes. IEEE Trans. Pattern Anal. Mach. Intell. **14**, 239–256 (1992)
10. Biber, P.: The normal distributions transform: a new approach to laser scan matching (2003)
11. Pang, S., Kent, D., Cai, X., Al-Qassab, H., Radha, H.: 3D scan registration based localization for autonomous vehicles - a comparison of NDT and ICP under realistic conditions. In: 2018 IEEE 88th Vehicular Technology Conference (VTC-Fall), pp. 1–5 (2018)
12. Fischler, M.A., Bolles, R.C.: Random sample consensus: a paradigm for model fitting with applications to image analysis and automated cartography. Commun. ACM **24**, 381–395 (1981)
13. Aiger, D., Mitra, N.J., Cohen-Or, D.: 4-points congruent sets for robust pairwise surface registration. ACM Trans. Graph. **27**, 1–10 (2008)
14. Salti, S., Tombari, F., Stefano, L.D.: SHOT: unique signatures of histograms for surface and texture description. Comput. Vis. Image Underst. **125**, 251–264 (2014)
15. Rusu, R.B., Blodow, N., Beetz, M.: Fast point feature histograms (FPFH) for 3D registration. In: IEEE International Conference on Robotics and Automation, pp. 3212–3217 (2009)

16. Sipiran, I., Bustos, B.: Harris 3D: a robust extension of the Harris operator for interest point detection on 3D meshes. Vis. Comput. **27**, 963 (2011)
17. Yew, Z.J., Lee, G.H.: 3DFeat-Net: weakly supervised local 3D features for point cloud registration. In: Ferrari, V., Hebert, M., Sminchisescu, C., Weiss, Y. (eds.) ECCV 2018. LNCS, vol. 11219, pp. 630–646. Springer, Cham (2018). https://doi.org/10.1007/978-3-030-01267-0_37
18. Mellado, N., Aiger, D., Mitra, N.J.: Super4PCS: fast global pointcloud registration via smart indexing. Comput. Graph. Forum **33**, 205–215 (2015)
19. Qi, C.R., Li, Y., Hao, S., Guibas, L.J.: PointNet++: deep hierarchical feature learning on point sets in a metric space. Adv. Neural Inf. Process. Syst. **30** (2017)
20. Chen, Z., Yang, F., Tao, W.: DetarNet: decoupling translation and rotation by Siamese network for point cloud registration (2021)
21. Fd, C.C., Dodgson, N.A., Moenning, C.: Fast marching farthest point sampling. In: Proc. Eurographics (2003)
22. Abeywickrama, T., Cheema, M.A., Taniar, D.: K-nearest neighbors on road networks: a journey in experimentation and in-memory implementation. In: Proceedings of the VLDB Endowment 9 (2016)
23. Qi, C.R., Su, H., Mo, K., Guibas, L.J.: PointNet: deep learning on point sets for 3D classification and segmentation. In: 2017 IEEE Conference on Computer Vision and Pattern Recognition (CVPR), pp. 652–660 (2017)
24. Li, J., Chen, B.M., Lee, G.H.: SO-Net: self-organizing network for point cloud analysis. In: Computer Vision and Pattern Recognition (CVPR), pp. 9397–9406 (2018)
25. Qi, C.R., Litany, O., He, K., Guibas, L.J.: Deep hough voting for 3D object detection in point clouds. In: 2019 IEEE/CVF International Conference on Computer Vision (ICCV), pp. 9277–9286 (2019)
26. Li, J., Lee, G.H.: USIP: unsupervised stable interest point detection from 3D point clouds. In: Proceedings of the IEEE/CVF International Conference on Computer Vision, pp. 361–370 (2019)
27. Yu, Z.: Intrinsic shape signatures: a shape descriptor for 3D object recognition. In: IEEE International Conference on Computer Vision Workshops, pp. 689–696 (2010)
28. Yang, Z., Wang, X., Hou, J.: A 4PCS coarse registration algorithm based on ISS feature points. In: 40th Chinese Control Conference, pp. 285–289 (2021)

Research on Navigation Algorithm of Unmanned Ground Vehicle Based on Imitation Learning and Curiosity Driven

Shiqi Liu, Jiawei Chen, Bowen Zu, Xuehua Zhou, and Zhiguo Zhou(✉)

School of Integrated Circuits and Electronics, Beijing Institute of Technology, Beijing 100081, China
zhiguozhou@bit.edu.cn

Abstract. The application of deep reinforcement learning (DRL) for autonomous navigation of unmanned ground vehicle (UGV) has the problem of sparse rewards, which makes the trained algorithm model difficult to converge and cannot be transferred to real vehicles. In this regard, this paper proposes an effective exploratory learning autonomous navigation algorithm Double I-PPO, which designs pre-training behaviors based on imitation learning (IL) to guide UGV to try positive states, and introduces the intrinsic curiosity module (ICM) to generate intrinsic reward signals to encourage exploratory learning strategies. Build the training scene in Unity to evaluate the performance of the algorithm, and integrate the algorithm strategy into the motion planning stack of the ROS vehicle, so as to extend to the actual scene for testing. Experiments show that in the environment of random obstacles, the method does not need to rely on prior map information. Compared with similar DRL algorithms, the convergence speed is faster and the navigation success rate can reach more than 85%.

Keywords: Deep reinforcement learning · Unmanned ground vehicle · Navigation · Unity · Spare reward · ROS

1 Introduction

The autonomous navigation algorithm can make UGV reach the target position with a reasonable planned path without collision with obstacles. At present, the widely used autonomous navigation and planning algorithms include A* algorithm, graphical method, dynamic window approach, etc. [1–4]. In recent years, some intelligent bionic planning algorithms have also been proposed and effectively promoted, such as genetic algorithm and ant colony optimization [5,6]. The current mainstream solutions for autonomous navigation of UGV usually combine a variety of algorithms, including simultaneous localization and mapping (SLAM), global and local path planning algorithms, and motion control

Supported by Equipment Pre-research Field Foundation (61403120109).

W. Fan et al. (Eds.): AsiaSim 2022, CCIS 1712, pp. 609–621, 2022.
https://doi.org/10.1007/978-981-19-9198-1_46

algorithms. Most of these methods need to rely on global map information, and there are problems such as large program size and poor generality. The robustness and generalization ability also need to be improved. With the development of deep learning and reinforcement learning technology, DRL is applied to the autonomous navigation of UGV. DRL optimizes the state-to-action mapping through the interaction between the agent and the environment, and uses neural networks to extract environmental information features as the input of the state to maximize cumulative reward or achieve specific goals [7]. In the autonomous navigation task of UGV, the DRL method can reach the target point in an autonomous exploration way by learning the control strategy of the agent without prior information such as the global map. For example, Tai et al. [8] used 10-line laser ranging data as input based on the ADDPG algorithm to realize autonomous navigation and obstacle avoidance for UGV in the actual environment; Josef et al. [9] adopted the structure of the DQN algorithm. The attention module is introduced and the reward is shaped to realize the autonomous navigation of UGV in the unknown rough terrain.

However, DRL has the problem of low sampling efficiency and requires a large amount of data to obtain good results. At the same time, UGV has the problem of sparse rewards when using the DRL algorithm for mapless navigation tasks in complex environments. The reward targets are scattered on the map, and the probability of reaching the target using random methods such as greedy strategies is low, which leads to less effective experience. Under the influence of the two, the algorithm iterates slowly and even fails to converge. It is easy to fall into a local optimum, and UGV cannot successfully reach the target location during deployment.

In order to solve the problem that the agent is difficult to obtain rewards during initial training, resulting in slow training or even failure to converge, we introduce IL [10,11] to provide effective experience for UGV. IL is a way of learning by observing the behavior of other agents and then learning to reproduce. Behavioral cloning and inverse reinforcement learning are the two main categories of IL, where behavioral cloning uses supervised learning to directly learn policies from taught sequences, while inverse reinforcement learning evaluates reward functions from expert data to guide agents to learn skills. Wei et al. [12] developed an IL framework and used it to train UAV navigation strategies in complex and GPS-free river environments, enabling vision-based UAV navigation. Wu et al. [13] proposed an IL based object-driven navigation system to improve mapless visual navigation in indoor scenes, enabling the robot to move to the target location without the need for an odometer or GPS. In this paper, we adopt IL based on generative adversarial networks, allowing the agent to continuously learn from the behaviors demonstrated by pre-trained experts, improve the strategy to generate behavioral states similar to the experts, and avoid learning a lot of wrong or meaningless experiences from scratch.

In an environment with sparse rewards, most actions cannot be rewarded, and if the agent cannot obtain rewards for a long time, it will not be able to learn a good policy and achieve the desired effect. Therefore, an additional reward function needs to be defined to make the reward denser. Pathak et al. [14] proposed a new prediction error-based ICM, which leverages the inverse dynamics model and the forward dynamics model on the original problem space to learn a new feature space, so that the learned feature space only encodes the part that affects the decision-making of the agent, and ignores irrelevant interference such as noise in the environment. When the next state is inconsistent with the agent's prediction, a certain reward will be given to satisfy the "curiosity" of the agent. In the process of autonomous navigation without a map, using the curiosity mechanism can make it easier for the UGV to try new actions and reach new locations, thereby speeding up the exploration process and reducing local convergence.

This paper describes our work in the following structure. In the first section, we will introduce the basic principles of the algorithm; in the second section, we will evaluate and verify the algorithm in the constructed virtual simulation environment to realize the training of the navigation algorithm model based on DRL; in the third section, based on the model trained in the virtual simulation environment, according to our proposed training and testing architecture, the model will be transplanted to the real UGV for testing; finally, we will summarize in Sect. 5 and discussions to propose future directions for improvement.

2 Algorithm Principle

2.1 Deep Reinforcement Learning

DRL is a combination of deep learning and reinforcement learning, and has the advantages of both. On the basis of reinforcement learning, deep reinforcement learning introduces neural network to extract features of environmental information, which greatly improves the perception ability of the agent. As shown in Fig. 1, DRL is an end-to-end training process. Among them, deep learning represented by neural network is responsible for perception, and reinforcement learning is responsible for decision-making.

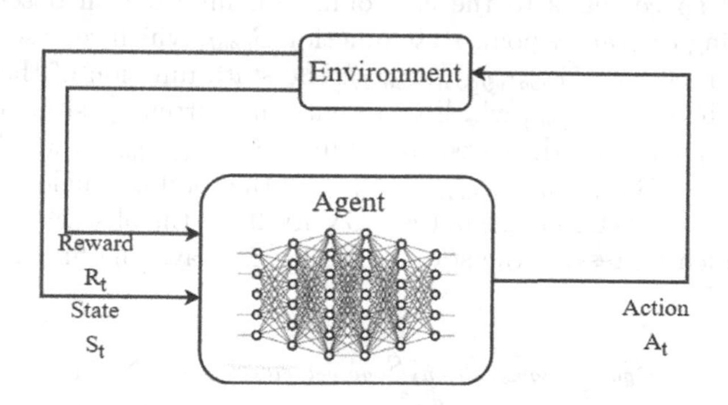

Fig. 1. Schematic of DRL.

Reinforcement learning tasks are often described as a Markov Decision Process (MDP). The MDP is evolved from a Markov chain and includes five key elements, namely the state space S, the state transition function P, the action set A, the reward function R and the discount factor γ, which can be represented as a quintuple $<S, P, A, R, \gamma>$. In the MDP, the next state of the system is only related to the current state and the current action.

$$P_{ss'}^{a} = P[S_{t+1} = s' | S_t = s, A_t = a] \tag{1}$$

The agent chooses an action in a certain state based on the policy function, denoted by the π function:

$$\pi(a|s) = P[A_t = a | S_t = s] \tag{2}$$

After the agent performs the action, it will obtain the delayed reward. The reward function is shown in formula 3.

$$R_s^a = E[R_{t+1} | S_t = s, A_t = a] \tag{3}$$

In order to reflect the effect of delay on cumulative reward, a discount factor γ is introduced. The impact of future rewards on the current moment is discounted and needs to be multiplied by the discount factor γ.

$$G_t = \sum_{k=0}^{\infty} \gamma^k R_{t+k+1} \tag{4}$$

The agent's only goal is to improve the policy to maximize the delayed cumulative reward. The quality of a strategy is evaluated by a value function. In fact, finding the optimal policy is equivalent to finding the optimal value function. There are two commonly used value functions, namely the state value function v and the action value function q.

This paper abstracts the navigation problem of UGV based on the MDP, and designs the state space, action set and reward function.

The state space refers to the environmental information observed by the UGV, including the target point state function S_{goal}, which represents the position of the target point$(x_{goal}, y_{goal}, z_{goal})$; the state function of the position of the UGV itself $S_{ugv\_position}$, which represents the current position of the UGV $(x_{ugv-p}, y_{ugv-p}, z_{ugv-p})$; the speed state function $S_{ugv\_velocity}$, which represents the speed of the UGV $(x_{ugv-u}, y_{ugv-v}, z_{ugv-r})$; the rotation angle of the steering wheel S_{ϕ} and the rotation angle of the UGV itself S_{δ}; the obstacle state function $S_{rayarray}$, which represents the state of the lidar array. The final state space is shown in Eq. 5.

$$S = \{S_{goal}, S_{ugv\_position}, S_{ugv\_velocity}, S_{\phi}, S_{\delta}, S_{rayarray}\} \tag{5}$$

Action space refers to the set of actions taken by the UGV, including the standardized rotation angle of the steering wheel $\theta(-1 \leq \theta \leq 1)$ and the standardized acceleration of the UGV $a(-1 \leq a \leq 1)$.

The reward function is the key element of training, which is directly related to the training effect of the agent. In order to make the UGV drive to the destination, the core idea of designing the reward function in this paper is that the agent will be rewarded when it reaches the target point, and punished when it collides with an obstacle. Furthermore, to avoid stalling, set the maximum time step 'Maxstep' and give the UGV a fixed time penalty within each time step. For different indoor and outdoor navigation training environments, this paper makes targeted adjustments in the design of the reward function, and introduces it in detail in the second section of the environment and parameter settings.

2.2 Navigation Algorithm Framework

Aiming at the problem of sparse rewards in autonomous navigation training of UGV, this paper adds ICM and IL to the proximal policy optimization algorithm, and designs a DRL algorithm Double I-PPO. The principle block diagram of the whole algorithm is shown in Fig. 2.

Proximal Policy Optimization. Proximal policy optimization (PPO) is one of the most widely used DRL algorithms. It borrows the idea of different-policy algorithms and introduces importance sampling to improve the policy gradient algorithm, which overcomes the defect of low sampling efficiency. At the same time, the PPO limits the differences between the two strategies. The two restriction methods correspond to PPO penalty and PPO clipping respectively. In practice, PPO clipping shows better results, so this paper selects PPO clipping as the main body of the UGV navigation algorithm.

PPO contains two policy networks and one review network. The policy network A is responsible for interacting with the environment to record the trajectory information, and the comment network uses the information to optimize the value function by minimizing the mean square error, as shown in Eq. 6. The policy network B is optimized by maximizing the objective function, as shown in Eq. 7, where A^{π_θ} is the advantage function. Before the next iteration starts, strategy B passes the network parameters to strategy A for network update.

$$\arg\min_{\theta} \frac{1}{|D_k|T} \sum_{\tau \in D_k} \sum_{t=0}^{T} (V_\phi(S_t) - G_t^{'})^2 \tag{6}$$

$$E_{\pi_\theta}[min(\ell_t(\theta^{'})A^{\pi_\theta}(S_t, A_t), clip(\ell_t(\theta^{'}), 1-\varepsilon, 1+\varepsilon)A^{\pi_\theta}(S_t, A_t))] \tag{7}$$

Curiosity Drive Module. The curiosity module essentially provides an additional reward to the self-driving car to overcome the sparse reward problem. The ICM module contains two networks and two feature extractors. The basis of the curiosity reward is the similarity of the state characteristics of the UGV at the two moments before and after. The smaller the similarity, the bigger the reward. Among them, the function of the forward network is to predict features from actions; the function of the reverse network is to predict actions based on features to prevent UGV from being curious about irrelevant changes.

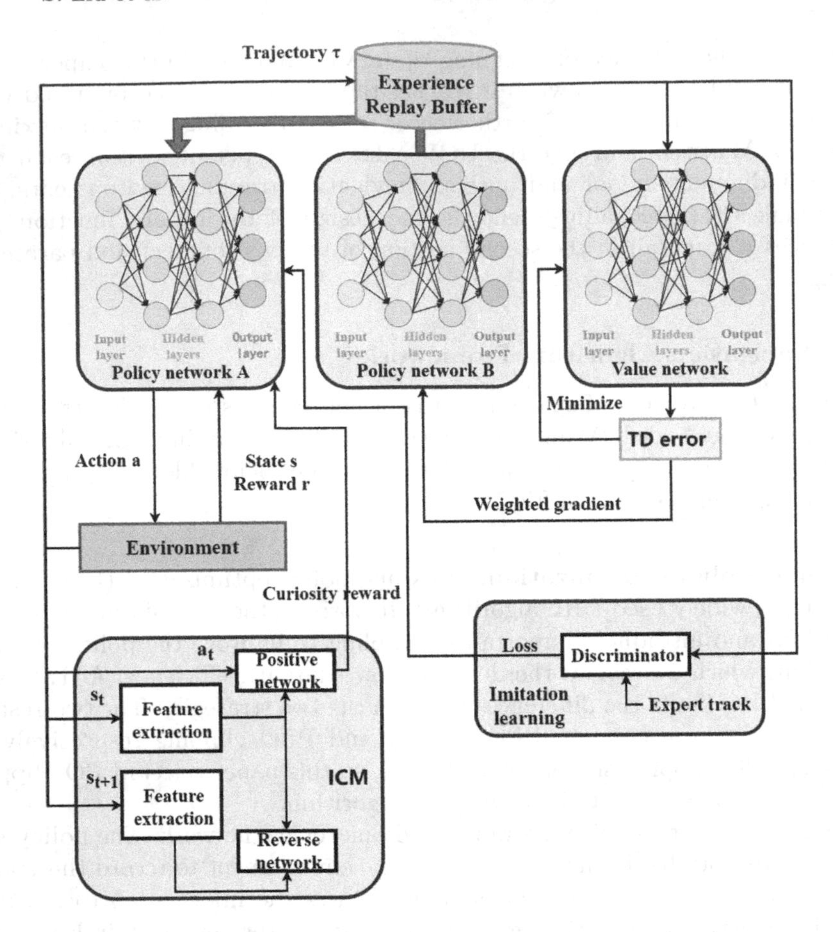

Fig. 2. Schematic of Double I-PPO.

Imitation Learning. In order to accelerate the learning speed of UGV, this paper adopts Generative Adversarial Imitation Learning (GAIL) to introduce expert demonstrations as the imitation objects of UGV. The core of generative adversarial imitation learning is the generative adversarial network (GAN). The GAN consists of a generator and a discriminator, and the performance of the generator is improved by training the discriminator and the generator successively, so as to achieve the purpose of confusing the real. In GAIL, the agent acts as a generator, and the input of the discriminator is the actual trajectory of the agent and the expert demonstration trajectory. Applied to the scene of this paper, the purpose of GAIL training is to make the trajectory of the UGV closer to the expert demonstration trajectory, so as to achieve the purpose of learning.

3 Unity Simulation Experiments

This paper uses Unity to build a training scene and completes the training based on its built-in ML-Agents Toolkit. Unity is a professional virtual reality software and game engine with powerful virtual reality capabilities and compatibility. ML-Agents Toolkit is an open-source machine learning toolkit included with Unity [15]. As shown in Fig. 3, the Python API acts as a bridge to link the DRL algorithm with the agents in the Unity scene. The environment refers to the training scene built in Unity, which includes agents and other objects. The agent communicates with the Python side through an external communicator, so as to realize the algorithm control of the agent. The agent has three behavior modes, namely, the learning mode during the training process, the model-driven inference mode after training, and the keyboard-controlled heuristic mode.

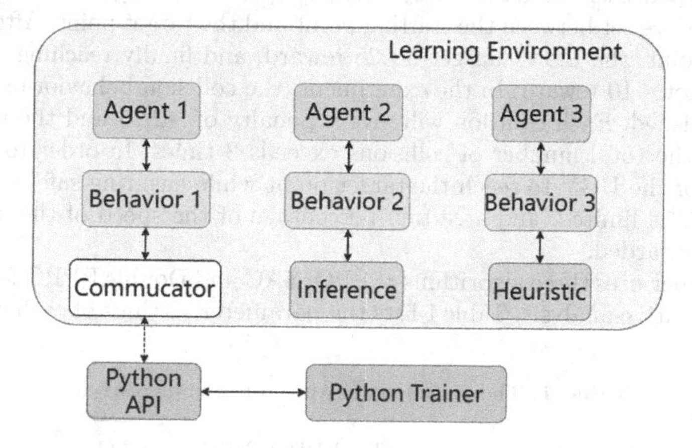

Fig. 3. Algorithm training implementation architecture based on ML-Agents.

3.1 Environment and Parameter Settings

In order to verify the effectiveness of the Double I-PPO algorithm, this paper builds two different indoor and outdoor scenes in Unity for navigation training and test evaluation, as shown in Fig. 4.

The indoor training scene is shown in Fig. 4(a). The size of the scene is 20 m × 40 m, which includes 5 randomly distributed obstacles, 2 cylinders and 3 baffles. The red car model represents the UGV, and its task is to avoid obstacles and successfully reach the destination, which is represented by a blue square. When the UGV reaches the target point, it will give a +5 reward, and when it encounters an obstacle, it will be given a -2 penalty, reset and start the next training round.

The outdoor training scene is shown in Fig. 4(b). On a road of about 250m in length, 5 to 7 obstacles such as warning signs and cones are randomly placed, and

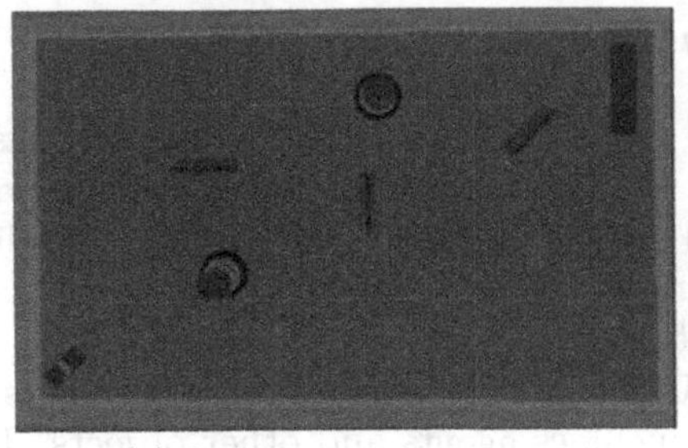

(a) Indoor training scene.

(b) Outdoor training scene.

Fig. 4. Algorithm training and evaluation scenes built in Unity.

the UGV reaches the target point by navigating and avoiding obstacles. Multiple checkpoints are set between the starting point and the target point. After passing the checkpoint, the UGV can get +1.25 reward, and finally reaching the target point can get +10 reward. In the experiment, the collision behavior of the UGV is also punished. Each collision will give a penalty of −2.75 and the round will end when the total number of collisions exceeds 3 times. In order to speed up the speed of the UGV to reach the target point while ensuring safety, the speed of the UGV is limited, and a certain percentage of the speed of the unmanned vehicle is rewarded.

This paper uses three algorithms of PPO, SAC and Double I-PPO for training and comparative analysis. Table 1 lists the parameter settings of each algorithm.

Table 1. The parameter settings of each algorithm.

Parameter	Double I-PPO	PPO	SAC
batch_size	128	128	128
learning_rate	0.0003	0.0003	0.0003
max_steps	3000000	3000000	3000000
hidden_units	512	512	512
num_layers	2	2	2
Curiosity_strength	0.01	/	/
Curiosity_gamma	0.99	/	/
Gail_strength	0.01	/	/
Gail_gamma	0.99	/	/

3.2 Algorithm Training Experiments

The DRL model proposed in this paper designs pre-training behaviors based on IL, and applies the curiosity mechanism for optimization. In the IL process, we

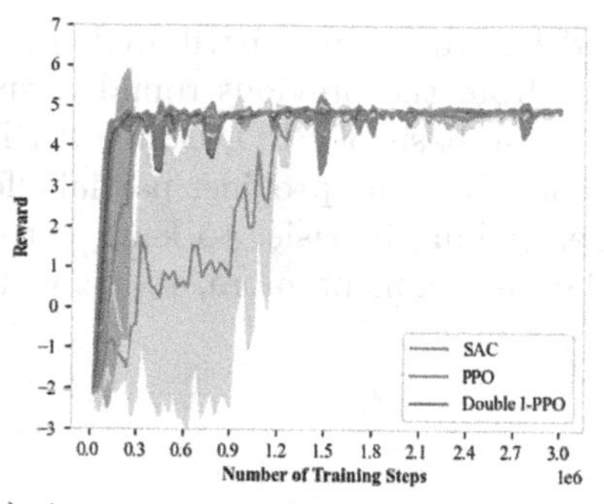

(a) Average reward curve of each algorithm in indoor scene.

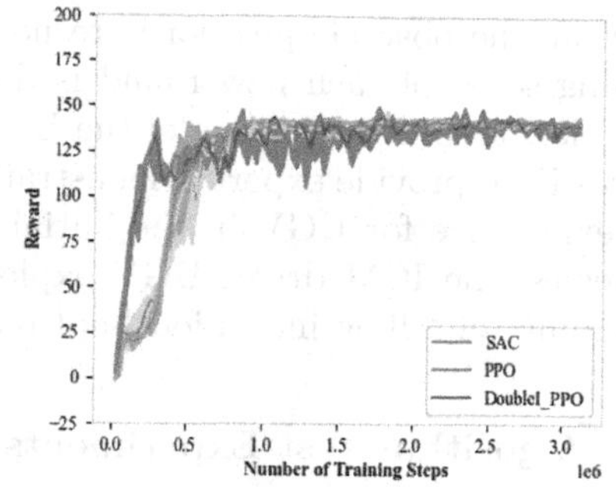

(b) Average reward curve of each algorithm in outdoor scene.

Fig. 5. Average reward curve of each algorithm.

added 5000 steps of data for IL of the agent, and compared the training conditions of the PPO model, the SAC model and the Double I-PPO model. In order to eliminate chance, we use the method of random sampling hyperparameters in both indoor and outdoor training scene, and calculate the average and standard deviation of cumulative rewards respectively, and the length of each training is 3 million. The average reward curve is shown in Fig. 5, the horizontal axis shows the number of training steps, and the vertical axis shows the reward accumulated by the agent in each round.

Indoor Training Situation. It can be seen from Fig. 5(a) that the three algorithms all completed the convergence in the 3 million training process. Among them, the SAC algorithm has the worst performance and the longest convergence time, and the training process is unstable and accompanied by large fluctuations; the performance of the PPO algorithm is in the middle, and its convergence speed and stability are between Double I-PPO and SAC; Double I-PPO has the best performance, and the convergence speed has been improved to a certain extent compared with traditional PPO and SAC.

Outdoor Training Situation. It can be seen from Fig. 5(b) that the Double I-PPO algorithm performs the best and can converge at a faster speed. In the training process of the outdoor scene, the convergence of PPO is not stable, because it is difficult to obtain effective experience by randomly selecting actions initially. When the training exceeds a certain number of steps, the UGV is easy to be in a local optimal state.

Summary. PPO and SAC are the current mainstream deep reinforcement learning algorithms, but in the case of sparse rewards, the convergence time is long. Compared with the outdoor training scene, it is difficult to add secondary rewards to guide the agent in the indoor scene during the training process. In addition, the obstacle positions are not fixed but randomly distributed, so the training scene of each new round is different from the previous round, resulting in low learning efficiency of the UGV. On the basis of PPO, Double I-PPO applies IL to provide expert demonstrations for UGV, and provides partial effective experience for UGV in the initial stage, making it easier to learn correct strategies; the ICM drives UGV explore the environment more, reducing the probability of falling into a local optimum.

3.3 Algorithm Test Experiments

In order to further test the performance of the algorithm, this paper designs navigation obstacle avoidance experiments for testing: the models trained by each algorithm are mounted on the indoor and outdoor UGVs, and the number of obstacles is increased on the basis of the original scene. Among them, the indoor test scene completed 500 navigation rounds. The number of rounds that successfully reached the destination, lost and crashed was counted, as shown in Fig. 6; the outdoor test scene completed 1000 navigation rounds. The arrival rate, the total number of collisions and collision rate were counted, as shown in Table 2.

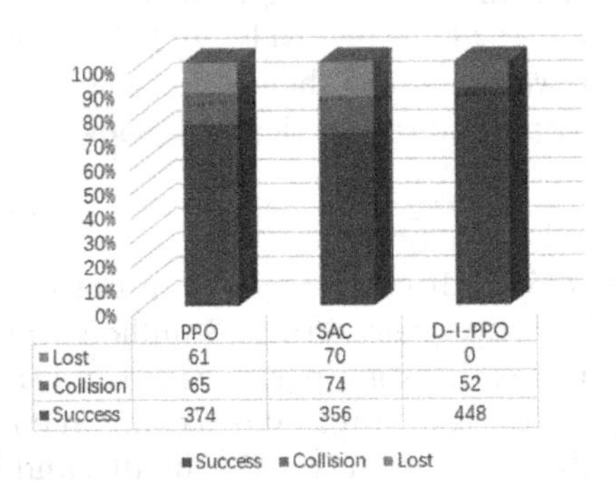

Fig. 6. Statistics of algorithm test results in the indoor scene.

Table 2. Statistics of algorithm test results in the outdoor scene.

Type	Arrival rate	Total collisions	Collision rate
SAC	100%	1996	83%
PPO	100%	495	49.5%
DoubleI-PPO	100%	200	11.3%

From the results in Fig. 6, the model trained by SAC and PPO has a success rate of more than 70%, and the Double I-PPO model has a success rate of more than 85%, showing higher environmental adaptability and generalization. From the data in Table 2, it can be seen that due to the checkpoint setting, all three

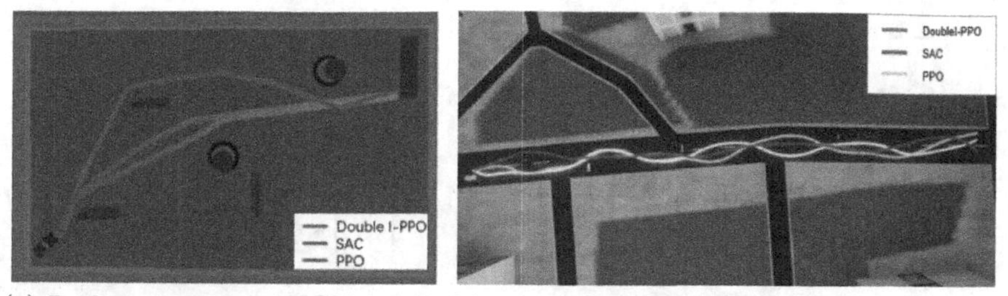

(a) Path trajectory of UGV in the indoor test scene.

(b) Path trajectory of UGV in the outdoor test scene.

Fig. 7. Path trajectory of UGV in indoor and outdoor test scenes.

types of algorithms show a 100% arrival rate during the test, but the total number of collisions and the collision rate of Double I-PPO are significantly better than the other two. The test trajectory path of the UGV is shown in Fig. 7.

4 ROS Vehicle Experiments

4.1 Communication Process

In order to test the navigation effect of the algorithm in actual scenes and confirm whether it has practical application value, this paper designs a test architecture of the UGV navigation algorithm based on ROS-TCP-Connector, as shown in Fig. 8. The actual Ackerman UGV is used as the server, and the Unity algorithm training platform is used as the client. Among them, the Unity algorithm training platform subscribes to and reads the laser data. After processing, the Instantiate function of Unity is used to display the obstacles detected by the real vehicle. The Double I-PPO algorithm decision model gives the advised speed and heading of the uUGV according to the Unity scene, and publish it to the real vehicle, which is executed by the control node of the real vehicle.

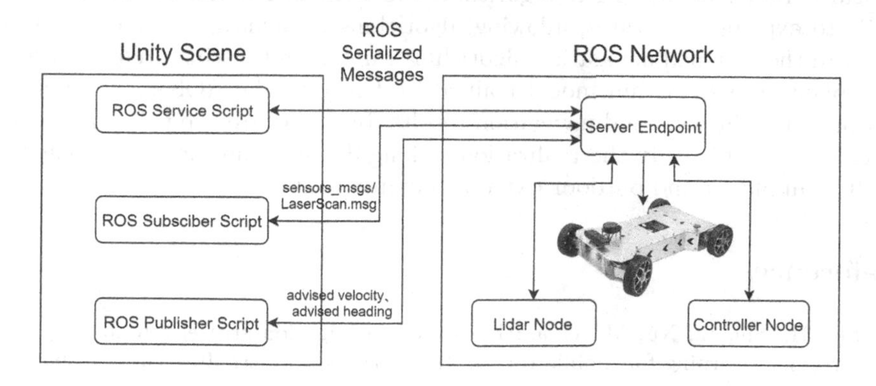

Fig. 8. Test flow of the UGV navigation algorithm.

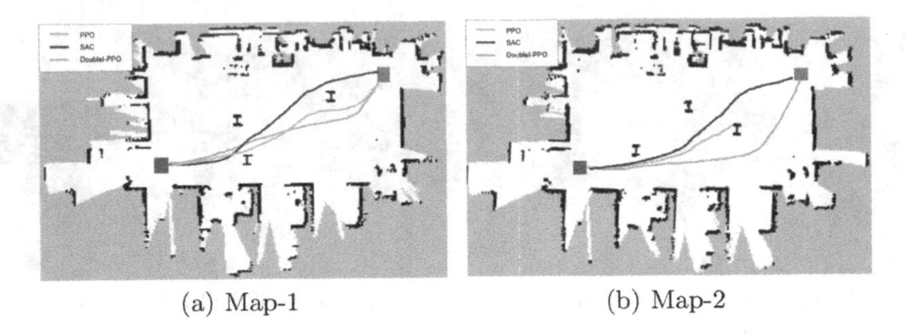

(a) Map-1 (b) Map-2

Fig. 9. The trajectory paths of the three types of algorithms in different actual scenes.

4.2 Real Environment Test

In the actual environment, we built an experimental site with a similar size to the Unity scene, including walls and chairs. Among them, 3 chairs were selected as random obstacles, and the navigation effects of the 3 algorithms were tested. The execution trajectory of each algorithm model is shown in Fig. 9. The red position is the starting point and the blue is the target position.

It can be seen from Fig. 9 that the Double I-PPO algorithm can perform navigation tasks more stably, and its trajectory is smoother and the probability of collision is lower than that of the SAC algorithm, and the PPO algorithm still did not learn enough navigation strategies under the same training conditions, and did not turn in time when approaching obstacles in some test scenes, resulting in collisions.

5 Conclusion

In this paper, we propose a DRL navigation algorithm for UGV without map prior information, and design pre-training behaviors based on IL to guide UGV to learn. Based on the PPO algorithm, the ICM is introduced to encourage UGV to explore new states, allowing algorithms to achieve faster convergence rates. In the designed navigation algorithm training and testing architecture, we transplante the algorithm model trained by Unity to the ROS real vehicle for testing, and achieved good navigation results. In the future, we will combine the global planner to study the realization of long-distance autonomous navigation of UGV in indoor and outdoor dynamic environments.

References

1. Shi, H., Shi, L., Xu, M., et al.: End-to-end navigation strategy with deep reinforcement learning for mobile robots. IEEE Trans. Industr. Inf. **16**(4), 2393–2402 (2019)

2. Weiwei, Z., Wei, W., Nengcheng, C., et al.: Path planning strategies for UAV based on improved A* algorithm. Geomat. Inf. Sci. Wuhan Univ. **40**(3), 315–320 (2015)
3. Song, P., Huang, J., Mansaray, L.R., et al.: An improved soil moisture retrieval algorithm based on the land parameter retrieval model for water-land mixed pixels using AMSR-E data. IEEE Trans. Geosci. Remote Sens. **57**(10), 7643–7657 (2019)
4. Xinyi, Y., Yichen, Z., Liang, L., et al.: Dynamic window with virtual goal (DW-VG): a new reactive obstacle avoidance approach based on motion prediction. Robotica **37**(8), 1438–1456 (2019)
5. Xu, Y., Liu, X., Hu, X., et al.: A genetic-algorithm-aided fuzzy chance-constrained programming model for municipal solid waste management. Eng. Optim. **24**(5), 652–668 (2019)
6. Luo, Q., Wang, H., Zheng, Y., et al.: Research on path planning of mobile robot based on improved ant colony algorithm. Neural Comput. Appl. **32**(6), 1555–1566 (2020)
7. Liu, J., Gao, F., Luo, X.: Survey of deep reinforcement learning based on value function and policy gradient. Chin. J. Comput. **42**(6), 1406–1438 (2019)
8. Tai, L., Paolo, G., Liu, M.: Virtual-to-real deep reinforcement learning: continuous control of mobile robots for mapless navigation. In: 2017 IEEE/RSJ International Conference on Intelligent Robots and Systems (IROS), pp. 31–36. IEEE (2017)
9. Josef, S., Degani, A.: Deep reinforcement learning for safe local planning of a ground vehicle in unknown rough terrain. IEEE Robot. Autom. Lett. **5**(4), 6748–6755 (2020)
10. Cèsar-Tondreau, B., Warnell, G., Stump, E., et al.: Improving autonomous robotic navigation using imitation learning. Front. Robot. AI **146** (2021)
11. Tai, L., Zhang, J., Liu, M., et al.: Socially compliant navigation through raw depth inputs with generative adversarial imitation learning. In: 2018 IEEE International Conference on Robotics and Automation (ICRA), pp. 1111–1117. IEEE (2018)
12. Wei, P., Liang, R., Michelmore, A., et al.: Vision-based 2D navigation of unmanned aerial vehicles in riverine environments with imitation learning. J. Intell. Robot. Syst. **104**(3), 1–19 (2022)
13. Wu, Q., Gong, X., Xu, K., et al.: Towards target-driven visual navigation in indoor scenes via generative imitation learning. IEEE Robot. Autom. Lett. **6**(1), 175–182 (2020)
14. Pathak, D., Agrawal, P., Efros, A.A., et al.: Curiosity-driven exploration by self-supervised prediction. In: International Conference on Machine Learning, pp. 2778–2787. PMLR (2017)
15. Juliani, A., Berges, V.P., Teng, E., et al.: Unity: a general platform for intelligent agents. arXiv preprint arXiv:1809.02627 (2018)

Improving Depth Perception Using Edge Highlighting in Transparent Stereoscopic Visualizations of Laser-Scanned 3D Point Clouds

Daimon Aoi[1](✉), Kyoko Hasegawa[2], Liang Li[2], Yuichi Sakano[3], Naohisa Sakamoto[4], and Satoshi Tanaka[2]

[1] Graduate School of Information Science and Engineering, Ritsumeikan University, 1-1-1 Noji-higashi, Kusatsu, Shiga, Japan
is0291fe@ed.ritsumei.ac.jp

[2] College of Information Science and Engineering, Ritsumeikan University, 1-1-1 Noji-higashi, Kusatsu, Shiga, Japan

[3] Department of Psychology, Aichi Gakuin University, 12 Araike, Iwasaki-cho, Nisshin, Aichi, Japan

[4] Graduate School of System Informatics, Kobe University, 1-1 Rokkoudai, Nada, Kobe, Hyogo, Japan

Abstract. Digital archiving is the activity of digitizing cultural heritage to preserve and utilize it. Visualizing these data helps us to understand the complicated internal structure of the cultural heritage. Transparent stereoscopic visualization based on 3D point clouds is effective for understanding the complex structures of cultural heritage. However, the position and depth information often become unclear when three-dimensional data are rendered transparently. We examined whether perceived depth of transparently and stereoscopically visualized objects can be improved by highlighting 3D edges of the structure. We also investigated the effect of edge opacity on perceived depth. We tested two types of figures: vertically and horizontally combined shapes and diagonally combined shapes, which is extracted from the point cloud data of cultural heritage. We conducted psychophysical experiments, and the results suggest that edge highlighting improves the accuracy of perceived depth. Moreover, the effect of edge highlighting was more significant when binocular disparity and motion parallax were available.

Keywords: Transparent stereoscopic visualization · Depth perception · Visual guide · Edge highlighting

1 Introduction

The activities to scan cultural heritage properties and preserve them as digital data for inheritance and analysis are called digital archiving [1,2]. Visualizing the digital data obtained by scanning allows the observer to understand the structure easily. Transparent visualization can visualize both internal and external structures simultaneously, which traditional visualization cannot achieve.

W. Fan et al. (Eds.): AsiaSim 2022, CCIS 1712, pp. 622–631, 2022.
https://doi.org/10.1007/978-981-19-9198-1_47

Stochastic point-based rendering (SPBR) method has been developed as a transparent visualization method for high-definition and high-speed visualization [3,4]. The method is good at visualizing large-scale three-dimensional (3D) point clouds because of its advantage of low calculation cost. SPBR enables transparent visualization with correct depth feel [5]. In addition, we propose transparent stereoscopic visualization, combining transparent and stereoscopic visualizations to make it easier to recognize shapes in transparent visualization [6]. In our recent papers [7,8], we have reported that transparent stereoscopic visualization is effective in reducing the complexity of visualized images.

It is well-known that depth perception is produced via many depth cues. For instance, binocular disparity, which is the differences in the two eyes' retinal images, and motion parallax, which causes temporal changes in the retinal images due to the observer's motion, are well-known cues for depth [9]. The cue produced when the front object hides the behind object is called occlusion, which is an important depth cue [10].

Transparent visualization enables us to understand the structure of an object easily because the internal and external structures can be visualized simultaneously. On the other hand, objects that overlap each other are difficult to distinguish between their front and rear positions. It is considered that the effect of occlusion and motion parallax, which are depth cues by stereoscopic vision, is lost, and the accuracy of depth perception is reduced. In previous studies, it has been reported that depth is underestimated in see-through stereoscopic vision using SPBR [7]. In this paper, we focused on providing a visual guide to improve depth perception. Specifically, the idea is to improve depth perception by highlighting edges as visual guides. This method is confirmed to be effective in suppressing the decrease in the visibility of shapes that occur as the opacity decreases in 2D images [11].

2 Visualization Method

2.1 Stochastic Point-Based Rendering

SPBR is a visualization method based on a stochastic algorithm using opaque luminescent particles. In SPBR, it is necessary to project the opaque particle group onto the image plane and perform ensemble averaging on the transparent image. Therefore, transparent visualization can provide correct depth information because it is necessary to sort the particles. In SPBR, the opacity is controlled by the point density, and the image quality is controlled by the number of averaged images. The four major steps of SPBR are as follows [12].

Step 1: Generation of points
Different methods for generating points depend on the data type to be visualized. The size of a point is the same as that of one pixel of the projected image.

Step 2: Division of point clouds
The point cloud generated in Step 1 is randomly divided into multiple point groups. We call the number of point clouds the "repeat level (L_R)".

Step 3: Projection of the points with hidden point processing
Hidden points are removed from each point group and create an intermediate image projected onto the image plane.

Step 4: Averaging the image
The luminance value of each intermediate image created in Step 3 is averaged for each pixel to generate an average image. Here, each pixel value is the expected value of each color and background color of many points.

We denote the opacity by α, the number of points that were projected is denoted by n, the cross-sectional area of the point is denoted by s_p, and the surface area of the iso-surface is denoted by s_A. In addition, if all the generated points are uniformly divided into L_R groups, the opacity is defined by

$$\alpha = 1 - \left(1 - \frac{s_\mathrm{A}}{s_\mathrm{p}}\right)^{\frac{n}{L_\mathrm{R}}}. \quad (1)$$

2.2 Point Feature Extraction

We describe the point feature extraction method (PFE) used in this research. Feature regions are areas of high curvatures, such as vertices or boundaries between surfaces. In this method, only areas of high curvatures are extracted. Consider a sphere with each point in the point cloud as its center. First, principal component analysis is performed on the coordinate values of the point cloud inside the sphere, and the feature value is defined by the combination of the obtained covariance matrix eigenvalues. Next, feature values are calculated for all points, and only points with feature values larger than an arbitrarily set threshold are extracted. Several types of 3D features based on eigenvalues have been proposed, and edge linearity is used in this study.

$$L_\lambda = \frac{\lambda_2 - \lambda_1}{\lambda_1}. \quad (2)$$

The eigenvalues of the first and second principal components obtained by principal component analysis are λ_1 and λ_2. The normalized value of L_λ is used as the feature f ($0 \leq f \leq 1$). Regions where this feature is larger than a threshold value are considered as edges.

3 Experiment

3.1 Experimental Conditions

To present stimulus images, we used a 42-in. autostereoscopic display utilizing a parallax barrier (TRIDELITY Display Solutions LLC, New Jersey, United States). This 3D display presented five views that provided binocular disparity and motion parallax so that the observers could perceive the presented images in 3D without glasses [13–18]. The image resolution for each view was 1920×1080. The viewing distance for the best 3D image quality of the display was 350 cm. In the experiment, the participants viewed the display at this distance.

The experimental conditions were the following four types: (1) monocular without motion parallax, (2) binocular without motion parallax, (3) monocular with motion parallax, and (4) binocular with motion parallax. In the case of without motion parallax, the participant placed his or her chin on a chin rest and observed the stimuli without moving his or her head. In the case of without binocular disparity, the non-dominant eye was covered by a blinder and the stimuli were observed with a single eye.

The participants included 15 male and 4 female in their 20 s to 40 s. The participants could use their naked eyes or glasses and contacts. All of the participants had normal or corrected-to-normal visual acuity and normal stereo vision [19]. In addition, the participant sat at a distance of 350 cm from the display. It was explained that an autostereoscopic display could be viewed with multiple viewpoints in advance. We also prepared a test stimulus image and confirmed that the participants could perceive the image in 3D correctly with multiple viewpoints when they move their heads horizontally. The stimulus images were presented in a random order for each participant. For each experimental image, the exposure time was 15 s.

3.2 Experimental Data

The 3D point data used in this study was "Fune-hoko", which is one of the floats participant in the Gion Festival held in Kyoto as shown in Fig. 1. Two types of data were used to combine the data extracted from the decomposed data. The two data were designated as data A and data B, respectively, and are shown in Fig. 2.

Fig. 1. Transparent Visualizations of Fune-hoko

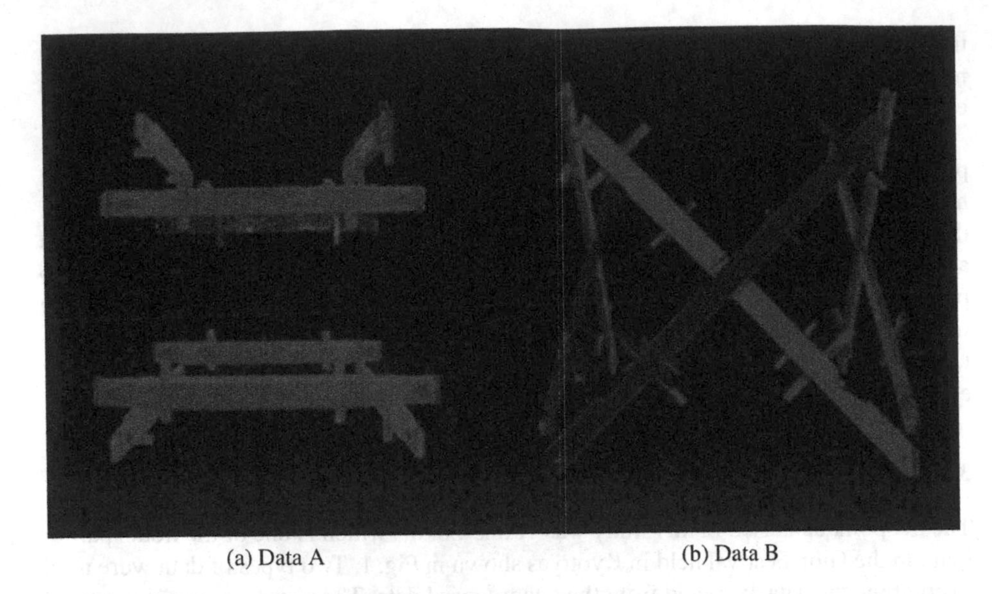

Fig. 2. Two types of component data extracted from Fune-hoko

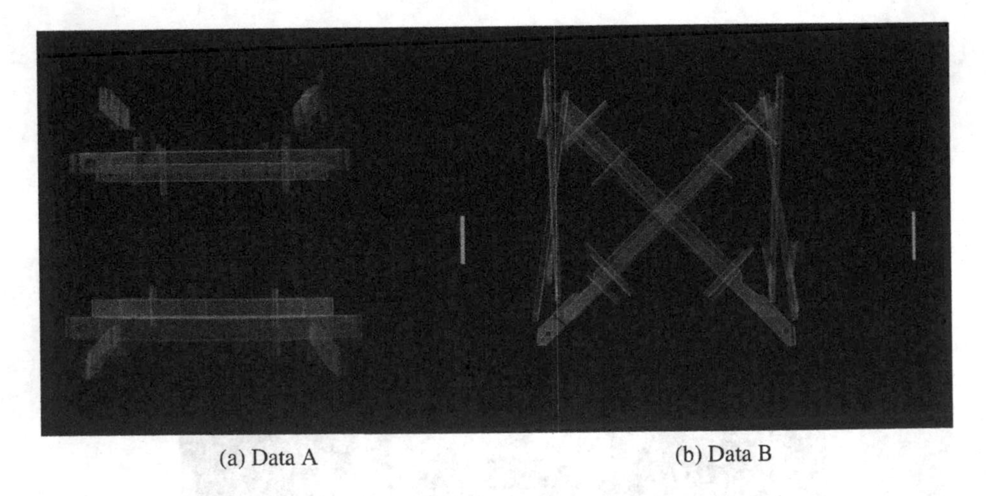

Fig. 3. Transparent visualization

The data of A and B in the case of transparent visualization are shown in Fig. 3a and 3b. Data A was processed in the depth direction, and three magnitudes of depth (1.0, 1.25, and 1.5 times the depth of the original data) are provided (see Fig. 4). All images Fig. 4 have been created with the same viewpoint. The parts arranged along the viewing direction are stretched to the back to three different extents, resulting in the three data cases. The parts processed from the original data are shown in yellow. Hereafter, we call them data A (case a), data A (case b), and data A (case c). In addition, three levels of different opacity (α) were prepared for each edge enhancement. (α = 0.050, 0.125, 0.300) The edges of data A (case a) with three different levels of opacity are shown in Fig. 5. The edges of data B with three different levels of opacity are shown in Fig. 6.

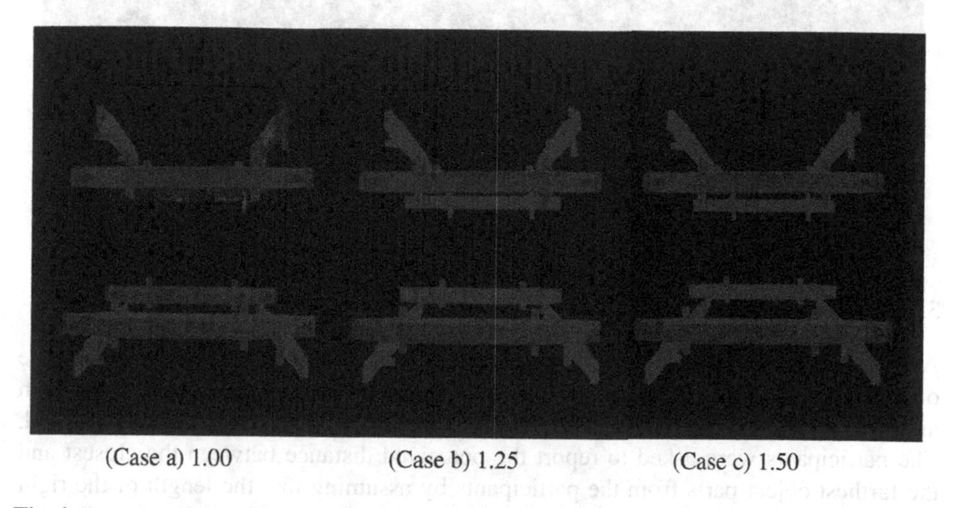

Fig. 4. Data A processed in 3 levels (the part where yellow is processed in the depth direction). (Color figure online)

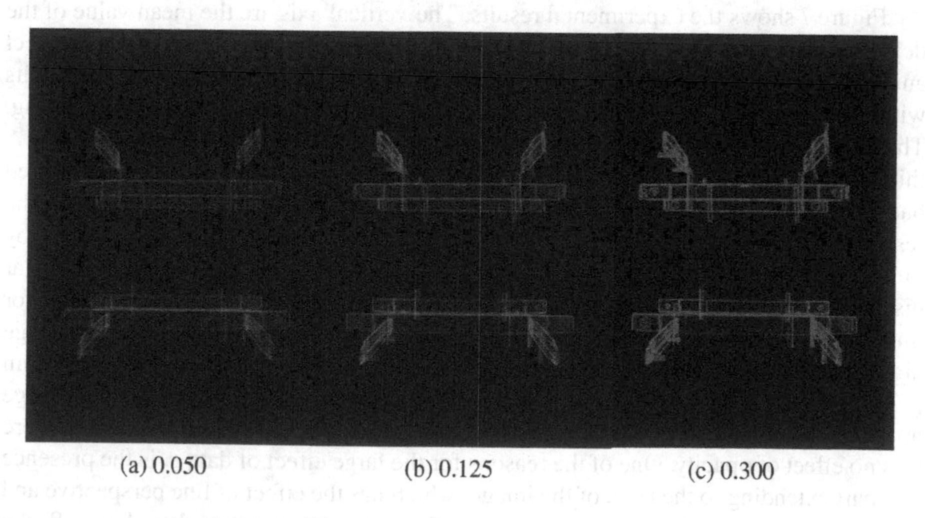

Fig. 5. Edge of data A (case a) with 3 different opacity levels

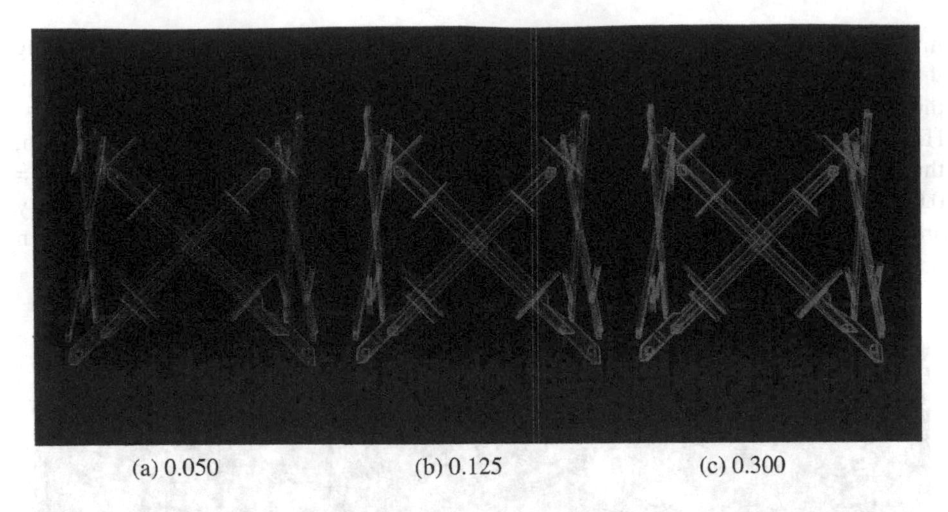

(a) 0.050 (b) 0.125 (c) 0.300

Fig. 6. Edge of data B with 3 different opacity levels

3.3 Experimental Result

A total of 64 experimental cases (three magnitude of depth of data A and one magnitude of depth of data B, three levels of opacity, with and without edge highlighting, with and without binocular disparity, and with and without motion parallax) were examined. The participants were asked to report the perceived distance between the closest and the farthest object parts from the participants by assuming that the length of the right vertical reference line was one (Fig. 3). The correct values for (case a), (case b), and (case c) of data A were 4.0, 5.0, and 6.0, respectively. The correct value for data B was 4.46.

Figure 7 shows the experimental results. The vertical axis are the mean value of the depth perceived by the participants, and the orange dashed lines indicate the correct answers. The error bars indicate the standard errors. The gray-only bars are the results without edge highlighting. The red bordered bars are the results of edge highlighting. There are three types of red bordered bars: thin, neutral, and thick bordered bars. A thin bordered bar shows the result when the edge opacity was 0.050, a neutral bordered bar showed the result when the edge opacity was 0.125, a thick bordered bar shows the result when the edge opacity was 0.300. This depth underestimation was alleviated by introducing binocular disparity and motion parallax. This result suggests that binocular disparity and motion parallax provided by the stereoscopic display were effective for improving the accuracy of the perceived depth of a transparently visualized object. Edge highlighting improved the perception of depth. In addition, the p-values were within $p < 0.05$. It is considered that the effect of binocular disparity increased when edge highlighting was applied, and the perceived value was closer to the correct value. There was no effect of opacity. One of the reasons for the large effect of data A is the presence of a part extending to the back of the image, which has the effect of line perspective and is considered to have improved the depth of the image. Comparing data A and B, the

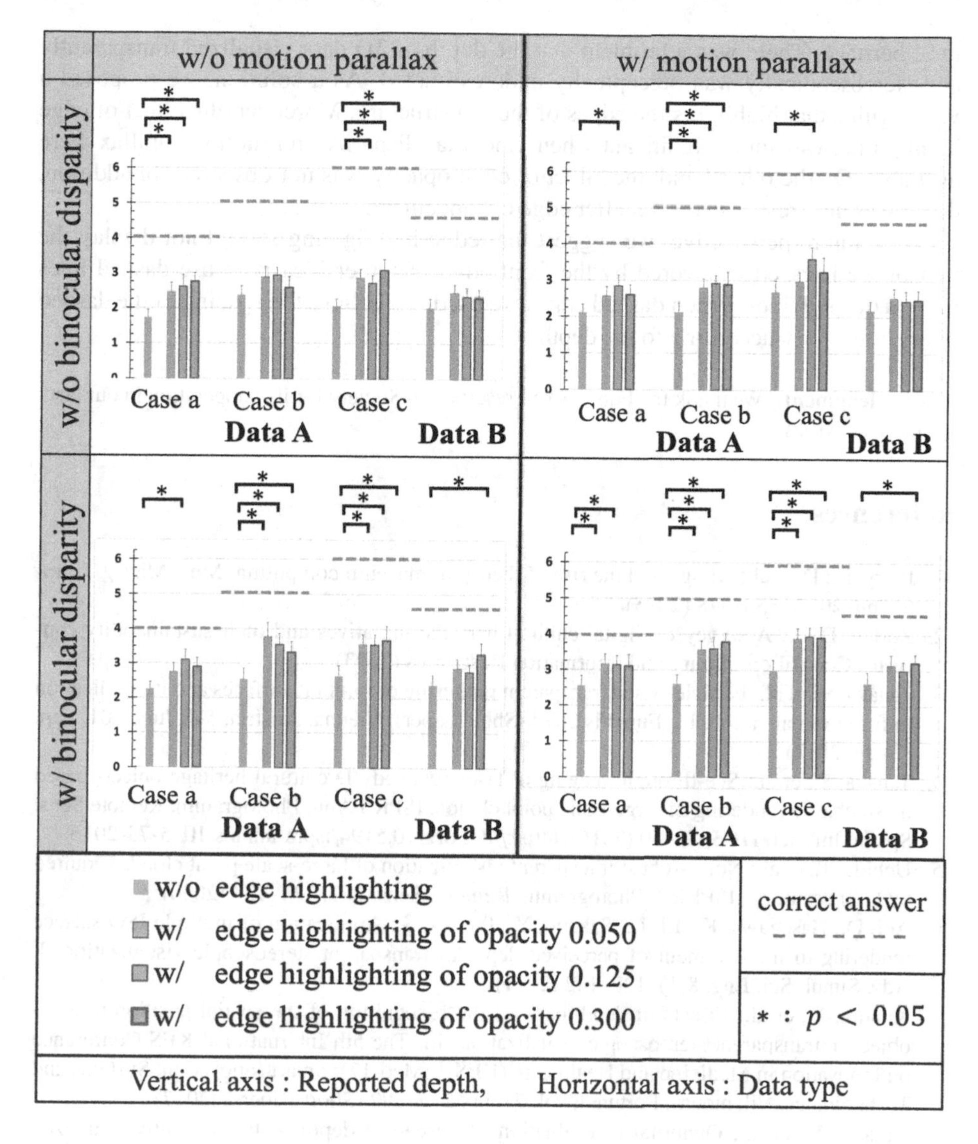

Fig. 7. Experimental results. The error bars indicate the standard error of the mean (SEM).

effect of edge highlighting on data B is small. This is considered to be due to the fact that the shape of data B itself was difficult to distinguish the spatial relationship of its parts.

4 Conclusion

In this study, we proposed transparent stereoscopic visualizations as a visualization method to accurately understand the structure of data used in digital archives of cultural heritage. There was a problem that the depth of 3D data visualized transparently and stereoscopically was perceptually underestimated. As a solution, we proposed a visual guide that highlights the edges of the 3D structure. Moreover, the effect of edge highlighting was more significant when binocular disparity and motion parallax were available. On the other hand, the effect of edge opacity was not observed. In addition, depth was underestimated even after edge enhancement.

As a future perspective, we suggest that edge highlighting should not display the part of the back edge covered by the front edge. Another idea is to use dashed lines instead of solid lines. When dashed edges are used, we believe the spacing of the dashed lines should vary according to the depth.

Acknowledgment. We thank the Fune-hoko Preservation Society for its cooperation in our laser scanning activities.

References

1. Parry, R.: Digital heritage and the rise of theory in museum computing. Mus. Manag. Curatorship **20**(4), 333–348 (2005)
2. Zorich, D.M.: A survey of digital cultural heritage initiatives and their sustainability concerns. Council on Library and information Resources (2003)
3. Tanaka, S., et al.: Particle-based transparent rendering of implicit surfaces and its application to fused visualization, In: EuroVis 2012 (Short Paper), Vienna, Austria, 5–8 June 2012, pp. 35–39 (2012)
4. Tanaka, S., et al.: See-through imaging of laser-scanned 3D cultural heritage objects based on stochastic rendering of large-scale point clouds. ISPRS Ann. Photogramm. Remote Sens. Spatial Inf. Sci. **III-5**, 73–80 (2016). https://doi.org/10.5194/isprs-annals-III-5-73-2016
5. Uchida, T., et al.: Noise-robust transparent visualization of large-scale point clouds acquired by laser scanning. ISPRS J. Photogramm. Remote. Sens. **161**, 124–134 (2020)
6. Aoi, D., Hasegawa, K., Li, L., Sakano, Y., Tanaka, S.: Application of multiple ISO-surface rendering to improvement of perceived depth in transparent stereoscopic visualization. J. Adv. Simul. Sci. Eng. **8**(1), 128–142 (2021)
7. Kitaura, Y., et al.: Effects of depth cues on the recognition of the spatial position of a 3D object in transparent stereoscopic visualization. In: The 5th International KES Conference on Innovation in Medicine and Healthcare (KES-InMed-17), Smart Innovation, Systems and Technologies, Vilamoura, Portugal, vol. 71, pp. 277–282 (Short Papers) (2017)
8. Sakano, Y., et al.: Quantitative evaluation of perceived depth of transparently-visualized medical 3D data presented with a multi-view 3D display. Int. J. Model. Simul. Sci. Comput **9**(3), 1840009, 16 p. (2018)
9. Howard, I.P., Rogers, B.J.: Seeing in Depth, I, Porteous, Thornhill, Ontario (2002)
10. Heine, L.: Uber Wahrnehmung und Vorstellung von Entfernungsunterschieden. Experimentelle Ophthalmologie **61**, 484–498 (1905)
11. Okamoto, N., Hasegawa, K., Li, L., Okamoto, A., Tanaka, S.: Highlighting feature regions combined with see-through visualization of laser-scanned cultural heritage. In: Proceedings of 2017 International Conference on Culture and Computing, 10 September 2017, pp. 7–12 (2017)
12. Hasegawa, K., Ojima, O., Shimokubo, Y., Nakata, S., Hachimura, K., Tanaka, S.: Particle-based transparent fused visualization applied to medical volume data. Int. J. Model. Simul. Sci. Comput. **4**, 1341003, 11 p. (2013)

13. Dodgson, N.A., Moore, J.R., Lang, S.R.: Multi-view autostereoscopic 3D display. Int. Broadcast. Convention **99**, 497–502 (1999)
14. Dodgson, N.A.: Autostereoscopic 3-D display. IEEE Comput. **38**(8), 31–36 (2005)
15. Son, J.Y., Javidi, B.: Three-dimensional imaging methods based on multiview images. J. Disp. Technol. **1**(1), 125–140 (2005)
16. Hill, L., Jacobs, A.: 3-D liquid crystal displays and their applications. Proc. IEEE **94**(3), 575–590 (2006)
17. Konrad, J., Halle, M.: 3-D displays and signal processing. IEEE Signal Process Mag. **24**(6), 97–111 (2007)
18. Jain, A., Konrad, J.: Crosstalk in automultiscopic 3-D displays: blessing in disguise? In: Proceedings of the SPIE, pp. 6490–649012 (2007)
19. Vancleef, K., Read, J.C.A.: Which stereotest do you use? A survey research study in the British isles. Br. Irish Orthoptic J. **15**(1), 15–24 (2019)

Modeling of Stepper Motor Fault Diagnosis Based on GRU Time Series Analysis

Zilong Liu, Gang Chen, Baoran An(✉), and Yunfei Liu

Institute of Computer Application, Chinese Academy of Engineering Physics, Mianyang, China
anbaoran@qq.com

Abstract. There are few data sources for fault diagnosis of stepper motor motion control systems, conventional voltage, current, vibration signals cannot well characterize stepper motor faults, and the fault diagnosis capabilities of stepper motor controllers on the market are limited. In view of the problem that the stepper motor data source is limited, this paper uses the stepper motor load value to provide an effective data source for fault diagnosis and condition monitoring. The time series decomposition of motor load value data is carried out, the time series data is analyzed by GRU recurrent neural network algorithm, fault classification is performed, and a GRU stepper motor fault diagnosis model is constructed to conduct online self-diagnosis of intermittent vibration, instantaneous overload and connection loosening fault, and fault prediction is carried out on motor overload stall fault.

Keywords: GRU stepper motor fault diagnosis model · Time series decomposition · Online self-diagnosis

1 Introduction

Common data used for motor fault diagnosis include vibration data, current and voltage data. Transfer principal component analysis (TPCA) under multiple-single working condition analysis vibration data to realize motor fault diagnosis, and the classification accuracy is 97.17%, but the average time is 1.906 s [1] which cannot meet the real-time requirements of motor control system. Artificial neural network diagnose unbalanced voltage in motor by current data, but it must use an external sensor called two-axis PASPORT sensor, and the classification accuracy is 82.75% [2].

Stepper motor is an open-loop control element stepper motor that converts electrical pulse signal into angular displacement or linear displacement. Its current pulse frequency is very high. General current transformers cannot accurately collect the original current waveform, but only the equivalent current and voltage waveform. Voltage and current data can only characterize electrical faults such as short circuit and open circuit.

The load value signal can accurately represent the load related parameter indicators and load related changes, and the load is an important link that causes the stepping motor

This work is supported by National Nature Science Foundation under Grant No. 61903348.

W. Fan et al. (Eds.): AsiaSim 2022, CCIS 1712, pp. 632–639, 2022.
https://doi.org/10.1007/978-981-19-9198-1_48

fault. Processing the load value of the stepper motor can analyze the load related fault and effectively predict the fault, provide effective data support for predictive maintenance, and improve the service life and online operation time of the stepper motor in major devices, improve the effective operation efficiency of equipment. Taking overload as an example, the current and vibration load data of stepper motor under transient overload fault are collected respectively. The data waveform is shown as follows.

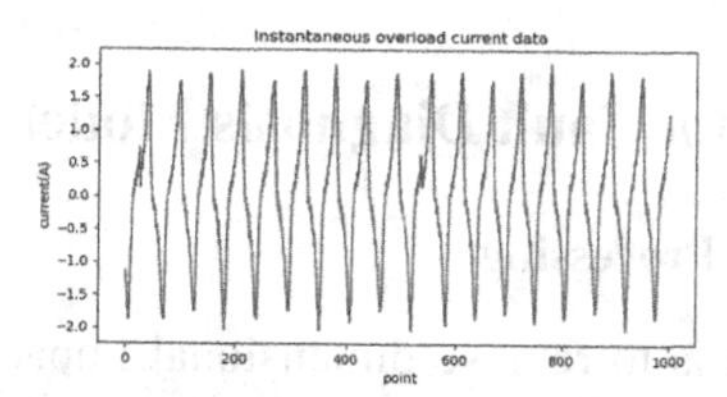

Fig. 1. Instantaneous overload current data waveform

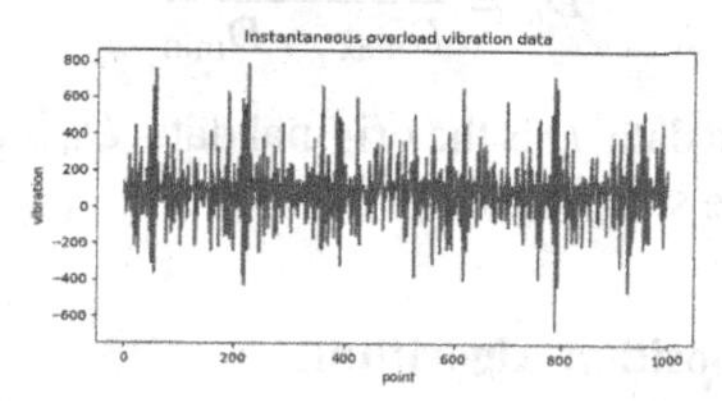

Fig. 2. Instantaneous overload vibration data waveform

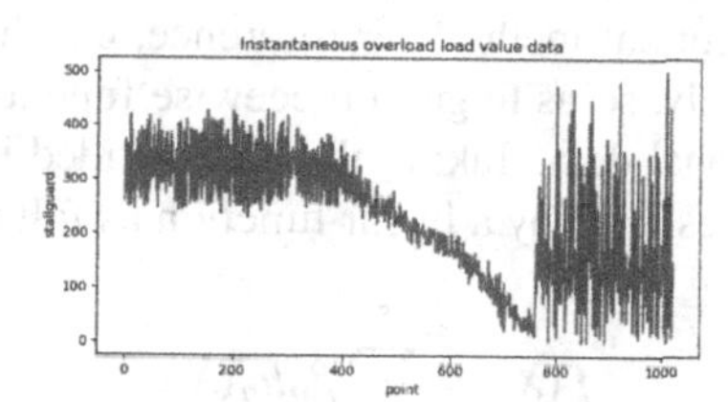

Fig. 3. Instantaneous overload load value data waveform

It can be seen from Figs. 1, 2 and 3 that the load value can more intuitively show the common faults of stepper motor compared with the vibration and current signals. At the same time, the stepper motor load value data sampling rate is as low as 20 Hz, which has the characteristics of simple and efficient analysis, and provides data support for distributed fault diagnosis. Low data volume and low sampling rate ensure the timeliness of control communication.

This paper uses the stepper motor load value to provide an effective data source for fault diagnosis and condition monitoring. In this paper, we study the Gate Recurrent Unit Network Algorithm (GRU) algorithm based on time series analysis, which has a simple structure and efficient algorithm. The time series decomposition method is used to decompose the original sample data into three parts: trend data, seasonal data and residual data, and the GRU stepper motor fault diagnosis model is constructed, and the above data is analyzed to achieve the diagnosis of three types of faults: intermittent vibration, instantaneous overload, and loose connection, and the overload fault prediction is carried out [3], and the diagnostic success rate reaches 97.68%.

By transplanting the GRU fault diagnosis model based on time series analysis to the embedded CPU, the fault self-diagnosis technology can be implemented in the embedded platform, and the adaptability of the fault diagnosis algorithm to different load conditions and motors can be improved by adjusting parameters and online training.

2 GRU Stepper Motor Fault Diagnosis Model

2.1 Data Normalization Processing

The role of normalization is to remove dimensional impact between indicators. The mathematical expression is as follows.

$$D_n = \frac{D - D_{\min}}{D_{\max} - D_{\min}} \tag{1}$$

where, D_n is the normalized data, D is the original data, $D_{\min}$ and $D_{\max}$ are the minimum and maximum values in the sample data [4].

2.2 Time Series Decomposition Algorithm

Time series decomposition and downsampling are performed on the collected stepper motor load data to obtain trend data, seasonal data and residual data.

Using piecewise polynomial method for reference, cut the definition domain, and then fit each block separately, so as to get a piecewise function. In this paper, the load value data is one-dimensional data. Taking the data divided into three segments as an example, the model is represented by a linear function as follows.

$$f(X) = \sum_{n=1}^{5} \beta_n h_n(X) \tag{2}$$

$$h_1(x) = I(X < e_1) \tag{3}$$

$$h_2(x) = I(e_1) \tag{4}$$

$$h_3(x) = I(e_1 \leq X < e_2) \tag{5}$$

$$h_4(x) = I(X = e_2) \tag{6}$$

$$h_5(x) = XI(X > e_2) \tag{7}$$

Because the data may be discontinuous at the cutting point, this paper adds conditional restrictions to ensure that the left and right limit data at the cutting point are equal. The least square method is used to fit the third-order polynomial curve, and the residual data can be obtained by subtracting the fitted trend data and seasonal data from the original data. Using the sliding window mechanism, the trend data is downsampled by using the median in the current window, and the fixed length fitting trend data for fault prediction processing is obtained.

2.3 GRU Gate Recurrent Unit Network Algorithm

The Gate Recurrent Unit neural network (GRU) was improved by CHO et al. On the basis of LSTM network in 2014. The Gate Recurrent Unit neural network (GRU) is an improved model of the long-term and short-term memory network (LSTM). The update gate of GRU consists of forgetting gate and input gate, including neuron state and hidden state [5, 6]. It can effectively alleviate the problem of gradient disappearance in the cyclic neural network, and can reduce the training parameters while maintaining the training effect [7]. At the same time, as a kind of RNN, It inherits the advantages that RNN model can automatically learn features and effectively model long-distance dependent information [8]. The meanings of update door and reset door are as follows (Fig. 4).

The logic diagram of GRU is shown in the figure.

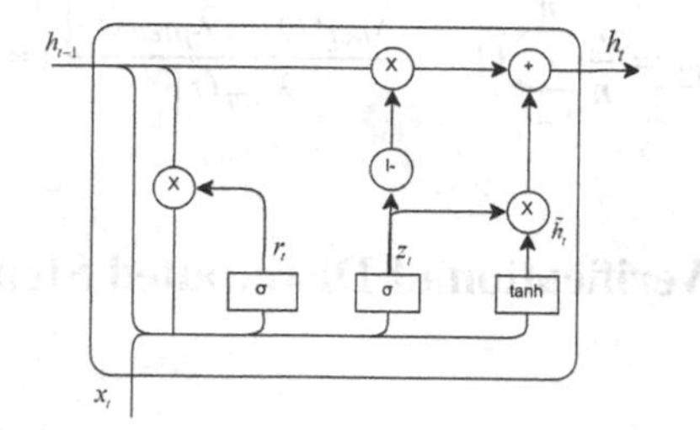

Fig. 4. GRU neural network logic diagram

The forward propagation formula of GRU is as follows.

$$z_t = \sigma\left(w_z.\left[h_{t-1}, x_t\right]\right) \tag{8}$$

$$r_t = \sigma\left(w_r.\left[h_{t-1}, x_t\right]\right) \tag{9}$$

$$\widetilde{h}_t = \tanh\left(w * \left[r_t * h_{t-1}, x_t\right]\right) \tag{10}$$

$$h_t = (1 - z_t) * h_{t-1} + z_t * \widetilde{h}_t \tag{11}$$

where [] indicates that two vectors are connected, and * indicates that matrix elements are multiplied

It can be seen from the above formula that the parameters to be learned in GRU training are W_r, W_z, W_h and W_o. These weight parameters are spliced, so they need to be separated during learning, that is:

$$w_r = w_{rx} + w_{rh} \tag{12}$$

$$w_z = w_{zx} + w_{zh} \tag{13}$$

$$w_{\tilde{h}} = w_{\tilde{h}x} + w_{\tilde{h}h} \tag{14}$$

The input of the output layer is $y_t^i = w_0 h$ and the output is $y_t^0 = \sigma(y_t^i)$. Let the loss function at a certain time be $E_t = \frac{1}{2}(y_t - y_t^0)^2$, then the loss is [9, 10] $E = \sum_{t=1}^{T} E_t$.

2.4 Performance Index and Evaluation Standard

To evaluate the accuracy of GRU stepper motor fault diagnosis model, this paper selects the maximum percentage of prediction error $\sigma_{E\max}$ and the average value of prediction accuracy $\sigma_{FA\arg}$ as the evaluation criteria, and Its mathematical expression is as follows [11].

$$\sigma_{E\max} = \max\left(\frac{|\lambda_{act}(t) - \lambda_{pred}(t)|}{\lambda_{act}(t)}\right) * 100\% \tag{15}$$

$$\sigma_{FA\arg} = \frac{1}{n}\sum_{t=1}^{n}\left(1 - \frac{|\lambda_{act}(t) - \lambda_{pred}(t)|}{\lambda_{act}(t)}\right) * 100\% \tag{16}$$

3 Modeling and Verification of Distributed Stepper Motor Fault Diagnosis

Run the Python programming environment in the embedded platform Raspberry Pi, realize time series decomposition and GRU fault classification algorithms, realize SPI communication by controlling the GPIO pin of the Raspberry Pi, control the TMC5160 chip drive stepper motor, call the TMC5160 toolkit to obtain load feedback value CALL-GUARD, speed and other data, build and transplant the GRU stepper motor fault diagnosis model, and control and fault diagnosis of stepper motor in the Raspberry Pi. To implement the function of distributed fault diagnosis, the model fault classification process structure is shown in the following figure (Fig. 5).

3.1 GRU Stepper Motor Fault Diagnosis Model Training and Testing

The motor load value and position residual are selected as the training sample set and the test sample set, the GRU neural network fault classification model is trained, the online fault diagnosis model is constructed, and the GRU stepper motor fault diagnosis model is transplanted to the embedded controller. The embedded controller communicates with the TMC5160 stepper motor drive chip through SPI communication, controls the drive chip to complete the functions of parameter setting, data acquisition, and drive output control [12], and runs the fault classification algorithm model and fault prediction algorithm in the embedded controller to achieve the functions of fault diagnosis and fault

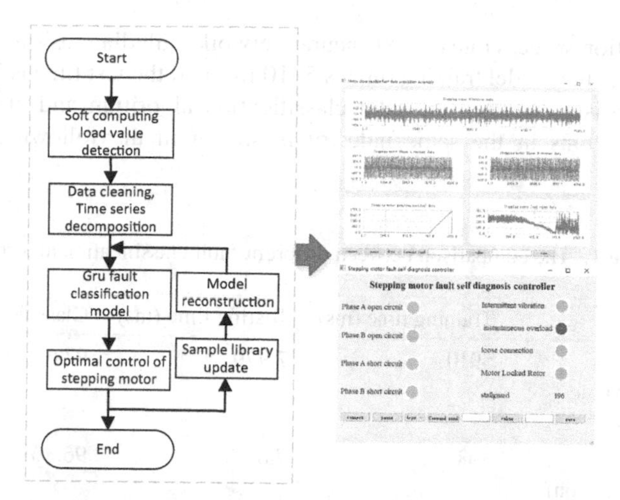

Fig. 5. The process structure of model fault classification

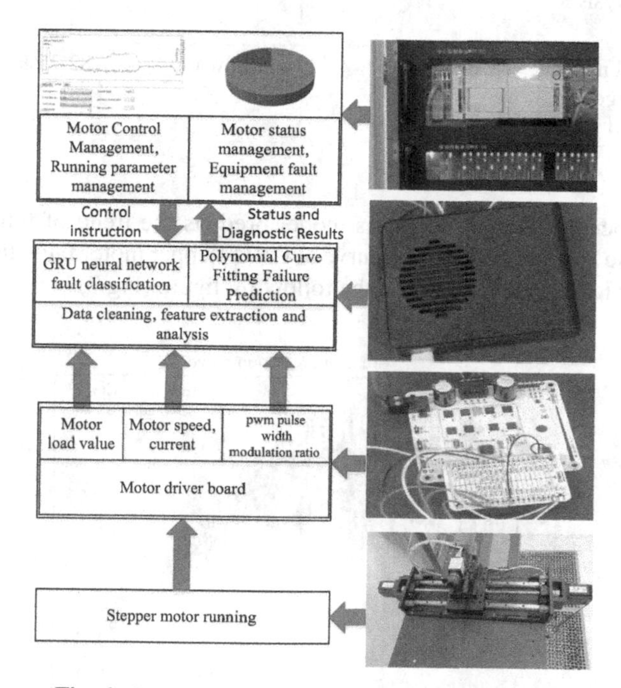

Fig. 6. Structure diagram of experimental prototype

prediction, and the experimental prototype structure is shown in the following figure (Fig. 6).

The fault classification model of GRU neural network is trained, and the test samples are input into the trained classification model to realize the classification and identification of instantaneous overload, intermittent vibration and loose connection. The fault classification success rate of GRU neural network fault diagnosis model algorithm reaches 97.68%, the model training time is 5910 ms, and the test time is 7.979 ms.

The comparison between the fault classification algorithm and other fault classification algorithms in the same industry is shown in the following table (Table 1).

Table 1. The comparison between different fault classification algorithm.

Classifier	Training time (ms)	Testing time (ms)	Classification accuracy
GRU neural network(Load value detection)	5910	7.979	97.68%
PSO-LSSVM (Load value detection)	308	7.8	96.3%
The Transfer principal component analysis (TPCA)	Unknown	1906	97.17%
Artificial neural network (Balanced voltage and 15% unbalanced voltage)	Unknown	Unknown	82.75%

GRU stepper motor fault diagnosis model predicts the trend of future load value according to the historical curve. The curve of GRU stepper motor fault diagnosis model predicts motor load value is shown in the following figure (Fig 7).

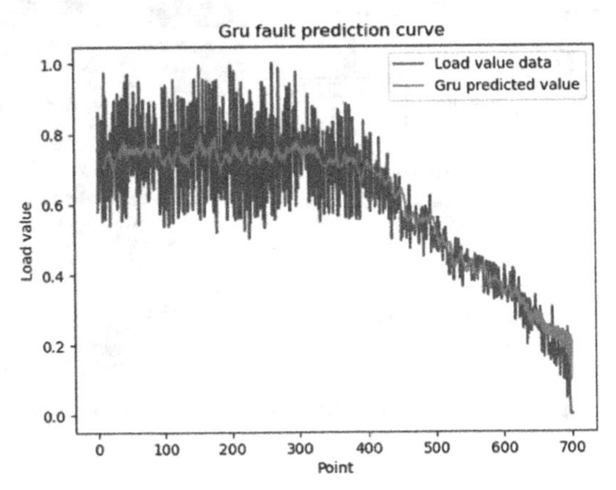

Fig. 7. The curve of GRU stepper motor fault diagnosis model predicts motor load

4 Conclusion

Aiming at the problem that the fault diagnosis ability of the stepper motor driver on sale is limited and it is impossible to predict the locked rotor fault, this paper reads the motor load value data with a low sampling rate, monitors the change of the motor load state, and proposes to use the time series solution to process and analyze the load value data. Through the GRU steppe motor fault diagnosis model, it identifies three types of faults: instantaneous overload, intermittent vibration and loose connection, According to the motor running curve, the occurrence time of motor locked rotor is predicted, and the on-line fault diagnosis model of stepping motor is constructed. After testing, the fault classification time of the model is 7.979 ms, and the accuracy of fault classification is 97.68%.

References

1. Ruqiang, Y., Fei, S., Mengjie, Z.: Induction motor fault diagnosis based on transfer principal component analysis. Chin. J. Electron. **30**(1), 18–25 (2021)
2. Sheikh, M.A., Nor, N.M., Ibrahim, T., et al.: Unsupervised on-line method to diagnose unbalanced voltage in three-phase induction motor. Neural Comput. Appl. **30**(12), 3877–3892 (2018). https://doi.org/10.1007/s00521-017-2973-0
3. Liu, Z., Wang, Y., Zhou, X., et al.: Load value analyse of stepter motor fault diagnose based on PSO-LSSVM. In: Chinese Control and Decision Conference (CCDC), pp. 689–692. IEEE (2020)
4. Ke, K., Hongbin, S., Chengkang, Z., et al.: Short-term electrical load forecasting method based on stacked auto-encoding and GRU neural network. Evol. Intel. **12**(3), 385 (2019). https://doi.org/10.1007/s12065-018-00196-0
5. Jin, C., Zhao, W., Wang, H.: Research on objective evaluation of recording audio restoration based on deep learning network. Adv. Multimedia **2018** (2018)
6. Cui, J., Long, J., Min, E., et al.: Comparative study of CNN and RNN for deep learning based intrusion detection system. In: International Conference on Cloud Computing and Security, pp. 159–170. Springer, Cham (2018) https://doi.org/10.1007/978-3-030-00018-9_15
7. Deng, W., Xu, J., Song, Y., et al.: Differential evolution algorithm with wavelet basis function and optimal mutation strategy for complex optimization problem. Appl. Soft Comput. **100**, 106724 (2021)
8. Liu, H., Xu, J., Wu, Y., et al.: Learning deconvolutional deep neural network for high resolution medical image reconstruction. Inf. Sci. **468**, 142–154 (2018)
9. Zeng, C., Ma, C., Wang, K., et al.: Parking Occupancy Prediction Method Based on Multi Factors and Stacked GRU-LSTM. IEEE Access **10**, 47361–47370 (2022)
10. Nguyen, V.H., Nguyen, M.T., Choi, J., et al.: NLOS identification in WLANs using deep LSTM with CNN features. Sensors **18**(11), 4057 (2018)
11. Zhang, W., Li, H., Tang, L., et al.: Displacement prediction of Jiuxianping landslide using gated recurrent unit (GRU) networks. Acta Geotech. **17**(4), 1367–1382 (2022)
12. Zhang, X., Pan, H.: Design of a two-phase stepper motor drive controller for 6000m deep-sea electric pan&tilt. In: International Conference on Artificial Intelligence and Electromechanical Automation (AIEA), pp. 148–151. IEEE (2021)

Correction to: Heavy-Duty Emission Prediction Model Using Wavelet Features and ResNet

Ruibin Wang, Xiushan Xia, and Zhenyi Xu

Correction to: Chapter "Heavy-Duty Emission Prediction Model Using Wavelet Features and ResNet" in: W. Fan et al. (Eds.): *Methods and Applications for Modeling and Simulation of Complex Systems*, CCIS 1712, https://doi.org/10.1007/978-981-19-9198-1_39

In the originally published version of the chapter 39 the reference to the source file was missing. The reference [8] has been added to the text and source citation has been added in the bibliography.

The updated original version of this chapter can be found at
https://doi.org/10.1007/978-981-19-9198-1_39

W. Fan et al. (Eds.): AsiaSim 2022, CCIS 1712, p. C1, 2023.
https://doi.org/10.1007/978-981-19-9198-1_49

Author Index

Printed in the United States
by Baker & Taylor Publisher Services